Merrill
Algebra One

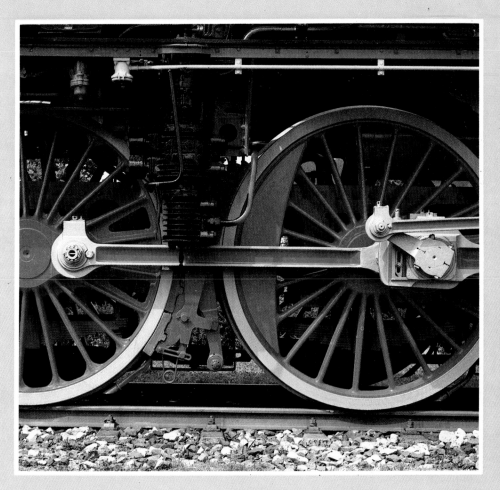

Foster • **Rath** • **Winters**

Merrill Publishing Co.
Columbus, Ohio

ISBN 0-675-05596-2

Published by

Merrill Publishing Company
Columbus, Ohio 43216

Printed in the United States of America

4 5 6 7 8 9 10 11 12 13 14 15 VH 00 99 98 97 96 95 94 93 92

Authors

Alan G. Foster is a chairman of the Mathematics Department at Addison Trail High School, Addison, Illinois. He has taught mathematics courses at every level of the high school curriculum. Mr. Foster obtained his B.S. from Illinois State University and his M.A. in mathematics from the University of Illinois. Mr. Foster is active in professional organizations at local, state, and national levels. He is past-president of the Illinois Council of Teachers of Mathematics. Mr. Foster is a recipient of the Illinois Council of Teachers of Mathematics T. E. Rine Award for excellence in the teaching of mathematics. He received the Illinois State Presidential Award for Excellence in Teaching of Mathematics in 1987. He is co-author of *Merrill Geometry, Merrill Algebra Essentials,* and *Merrill Algebra Two.*

James N. Rath has 30 years of classroom experience in teaching mathematics at every level of the high school curriculum. Mr. Rath is a former head of the mathematics department at Darien High School, Darien, Connecticut. He earned his B.A. in philosophy from the Catholic University of America and his M.Ed. and M.A. in mathematics from Boston College. Mr. Rath is active in professional organizations at local, state, and national levels. He is a co-author of *Merrill Pre-Algebra, Merrill Algebra Essentials,* and *Merrill Algebra Two.*

Leslie J. Winters is the Secondary Mathematics Specialist for the Los Angeles Unified School District. He has thirty years of classroom experience in teaching mathematics at every level from junior high school to college. He received his B.A. in mathematics from Pepperdine University, and advanced degrees from the University of Dayton, University of Southern California, and Boston College. He is a past president of the California Mathematics Council–Southern Section. He was a recipient of the 1983 Presidential Award for Excellence in Mathematics Teaching in the state of California. He is a co-author of *Merrill Algebra Essentials* and *Merrill Algebra Two.*

Consultants

Gail F. Burrill
Mathematics Department Chairperson
Whitnall High School
Greenfield, Wisconsin

Daniel T. Dolan
State Mathematics and Educational
 Technology Specialist
Montana Office of Public Instruction
Helena, Montana

Reviewers

Bert Anderson
Mathematics Department Chairperson
Foxboro High School
Foxboro, Massachusetts

Linda Coleman
Teacher
Hill Junior High School
Tampa, Florida

Betty Morgan
Teacher
Aiken High School
Aiken, South Carolina

Will Thompson
Teacher
Apollo High School
St. Cloud, Minnesota

Robert Wells
Teacher
San Juan High School
Citrus Heights, California

D'Ann Clay
Teacher
Tates Creek Senior High School
Lexington, Kentucky

Abner Korsness
Teacher
Rex Putnam High School
Milwaukie, Oregon

Frederick Sanders Jr.
Teacher
Marshall High School
San Antonio, Texas

Dr. Chuck Vonder Embse
Department of Mathematics
Central Michigan University
Mt. Pleasant, Michigan

Herbert Wierzbach
Mathematics Department Chairperson
Crystal Lake South High School
Crystal Lake, Illinois

Classroom Learner Verification Teachers

Peggy Collins, Westbury High School, Houston, TX;
Judy Crowell, Jeannette Junior High School, Sterling Heights, MI;
Nestor Diaz, Coral Gables High School, Miami, FL;
Sharyn Gibson, King High School, Tampa, FL;
Loraine Hunter, Central High School, Detroit, MI;
Paul Jargstorf, Carl Sandburg Junior High School, Rolling Meadows, IL;
David H. Kaplan, Wells Community Academy, Chicago, IL;
Mark Lazur, Schenley High School, Pittsburgh, PA;
Judith O'Neil, North Miami Beach High School, Miami, FL;
Jean Spiwak, Stanton College Prep School, Jacksonville, FL;
Daisy Upshaw, Longfellow Academy, Dallas, TX

Staff

Editorial

Project Editors: Nancy E. Dawson, Martin Sopher; *Editors:* Frederick D. Suter, Susan E. Bailey; *Photo Editor:* Barbara Buchholz; *Production Editors:* Carole R. Hill, Jean A. Mulligan

Art

Project Artist: Mark D. Clingan; *Designer:* Larry W. Collins

Photo Credits

Merrill Algebra One is designed for use by students in first-year high school algebra courses. The text, which was developed in the classroom by experienced high school teachers, is based on the successful prior editions of *Merrill Algebra One*. The goals of the text are to develop proficiency with mathematical skills, to expand understanding of mathematical concepts, to improve logical thinking, and to promote success. To achieve these goals, the following strategies are used.

Build upon a Solid Foundation. This program develops and utilizes the learning spiral. Students first review those concepts basic to the understanding of algebra. Students' understanding is thus solidified before the introduction of more difficult concepts.

Utilize Sound Pedagogy. *Merrill Algebra One* covers, in logical sequence, all topics generally presented at this level. Concepts are introduced when they are needed. Each concept presented is then used within that lesson and in later lessons.

Gear Presentation for Learning. An appropriate reading level has been maintained throughout the text. Furthermore, many photographs, illustrations, charts, graphs, and tables provide visual aids for the concepts and skills presented. Hence, students are able to read and learn with increased understanding.

Use Relevant Real-Life Applications. Applications are provided not only for practice, but also to aid understanding of how concepts are used.

Give Ample Opportunity for Review. A one-page Cumulative Review in each chapter provides students with a review of the skills and concepts presented up to that point. This research-proven method is most effective in increasing student retention of mathematical skills. Each chapter also includes two to four Mini-Reviews. Each of these provides students with a quick review of previously-presented concepts.

Use Relevant Technology. Using Calculator examples instruct students in using a calculator to solve problems that relate to the concepts taught within the lesson. Graphing Calculator Applications instruct students in using a graphing calculator by supplying step-by-step directions. Students can experience and apply modern technology to algebra concepts. A Using Computer feature in each chapter provides a computer program and exercises related to the objectives of the chapter. An Appendix on BASIC provides instruction in writing programs using the BASIC computer language.

The **Algebraic Skills Review** at the back of the text provides review exercises to help students maintain algebraic skills. These exercises can be used for remediation, to supplement exercises in the lessons, and as review material at the end of the school year.

Teacher and students familiar with the prior editions of this text will be pleased to see that the clarity of explanations and careful sequencing of topics have been retained in this new edition of *Merrill Algebra One*.

Table of Contents

1 The Language of Algebra———————2

Lessons

Expressions and Equations
- 1-1 Variables and Expressions. 3
- 1-2 Evaluating Expressions 6
- 1-3 Open Sentences and Equations. 9

Properties
- 1-4 Identity and Equality
       Properties 12
- 1-5 The Distributive Property. 15
- 1-6 Commutative and Associative
       Properties 19

Formulas and Verbal Problems
- 1-7 Using Formulas. 23
- 1-8 Exploring Verbal Problems 26
- 1-9 Problem Solving:
       Writing Equations 30

Review and Testing

Mini-Reviews   13, 21, 34
Cumulative Review 18
Vocabulary 36
Chapter Summary. 36
Chapter Review. 38
Chapter Test. 39

Applications and Extensions

Excursions in Algebra:
    Sets. 11
    Length in the Metric System 14
    History 34
Reading Algebra: Parentheses 22
Applications in Computer Science:
    Hexadecimal Numbers. 29

Technology

Using Calculators:
    Evaluating Expressions 7
    The Distributive Property 17
Using Computers: Formulas. 35

2 Adding and Subtracting Rational Numbers 40

Lessons

Integers and the Number Line
- 2-1 Integers on the Number Line 41
- 2-2 Addition on the Number Line 44

Addition and Subtraction
- 2-3 Adding Integers 46
- 2-4 Adding Rational Numbers. 49
- 2-5 More About Addition. 53
- 2-6 Subtraction 55

Equations and Problem Solving
- 2-7 Solving Equations Involving
       Addition 60
- 2-8 Solving Equations Involving
       Subtraction 63
- 2-9 Problem Solving:
       Rational Numbers 67

Review and Testing

Mini-Reviews   48, 58, 71
Cumulative Review 59
Vocabulary 73
Chapter Summary. 73
Chapter Review. 74
Chapter Test. 76
Standardized Test Practice Questions 77

Applications and Extensions

Problem Solving: Guess and Check 52
Applications in Meteorology:
    Wind Chill Factor 66
Excursions in Algebra: Magic Squares. 71

Technology

Using Calculators:
    Solving Equations Involving Subtraction . . . 64
Using Computers: Temperature. 72

3 Multiplying and Dividing Rational Numbers _____78

Lessons

Multiplication and Division
3-1   Multiplying Rational Numbers . .   **79**
3-2   Dividing Rational Numbers **83**

Solving Equations
3-3   Solving Equations Using
         Multiplication and Division . . .   **86**
3-4   Solving Equations Using More
         Than One Operation **90**
3-5   More Equations **93**
3-6   Still More Equations **97**

Problem Solving and Percents
3-7   Problem Solving:
         Using Equations **100**
3-8   Ratio and Proportion **105**
3-9   Percent **110**
3-10 Problem Solving:
         Percents **114**

Review and Testing

Mini-Reviews   **82, 89, 99, 108**
Cumulative Review **96**
Vocabulary **119**
Chapter Summary **119**
Chapter Review **120**
Chapter Test **121**

Applications and Extensions

Excursions in Algebra:
   Mental Computation **99**
   Capacity in the Metric System **109**
Problem Solving: Diagrams **104**

Technology

Excursions in Algebra:
   Reciprocals **85**
Using Calculators:
   Ratio and Proportion **106**
   Percent **110, 112**
Using Computers:
   Successive Discounts **118**

4 Inequalities _____122

Lessons

Solving Inequalities
4-1   Inequalities and Their Graphs **123**
4-2   Solving Inequalities Using
         Addition and Subtraction **128**
4-3   Solving Inequalities Using
         Multiplication and Division **131**
4-4   Inequalities with More Than
         One Operation **134**

Compound Sentences and Problem Solving
4-5   Comparing Rational Numbers **136**
4-6   Compound Sentences **140**
4-7   Open Sentences Containing
         Absolute Value **145**
4-8   Problem Solving: Inequalities . . .   **149**

Review and Testing

Mini-Reviews   **130, 139**
Cumulative Review **144**
Vocabulary **151**
Chapter Summary **152**
Chapter Review **153**
Chapter Test **154**
Standardized Test Practice Questions **155**

Applications and Extensions

Reading Algebra: Compound
   Statements **127**
Applications in Industry:
   Tolerance **148**

Technology

Excursions in Algebra:
   Comparing Numbers **139**
Using Computers:
   Comparisons **151**

5 Powers ———————————————————— 156

Lessons

Operations with Monomials

5-1   Multiplying Monomials 157
5-2   Powers of Monomials 160
5-3   Dividing Monomials 162
5-4   Scientific Notation 165

Problem Solving

5-5   Problem Solving:
          Age Problems.............. 169
5-6   Problem Solving: Mixtures 171
5-7   Problem Solving:
          Mixtures Using Percent 174
5-8   Direct Variation 178
5-9   Inverse Variation............. 181
5-10 Lever Problems 184

Review and Testing

Mini-Reviews   167, 183
Cumulative Review 177
Vocabulary 187
Chapter Summary..................... 187
Chapter Review...................... 188
Chapter Test........................ 189

Applications and Extensions

Problem Solving: Look for a Pattern 159
Applications in Business:
    Consumer Price Index 168

Technology

Using Calculators:
    Scientific Notation 166
Using Computers:
    Powers and E Notation 186

6 Polynomials ———————————————————— 190

Lessons

Addition and Subtraction

6-1   Polynomials 191
6-2   Adding and Subtracting
          Polynomials 194

Multiplication and Division

6-3   Multiplying a Polynomial by
          a Monomial 199
6-4   Multiplying Polynomials........ 201
6-5   Some Special Products 205
6-6   Solving Equations 210

Problem Solving

6-7   Problem Solving:
          Perimeter and Area 212
6-8   Problem Solving:
          Simple Interest.............. 216
6-9   Problem Solving:
          Uniform Motion.............. 220

Review and Testing

Mini-Reviews   193, 204, 215
Cumulative Review 209
Vocabulary 224
Chapter Summary..................... 224
Chapter Review...................... 225
Chapter Test........................ 226
Standardized Test Practice Questions 227

Applications and Extensions

Reading Algebra: Powers.............. 198
Excursions in Algebra:
    History 200
    Squaring Special Numbers............. 218
Applications in Banking:
    Checking Accounts 219

Technology

Excursions in Algebra:
    The Store and Recall Keys 208
Using Computers:
    Compound Interest 223

7 Factoring —————————————— 228

Lessons

Factors and the Distributive Property
7-1   Factors and Greatest
      Common Factor **229**
7-2   Factoring Using the
      Distributive Property **232**

Factoring Trinomials
7-3   Factoring Differences
      of Squares **235**
7-4   Perfect Squares and Factoring **239**
7-5   Factoring $x^2 + bx + c$ **243**

More Factoring Trinomials
7-6   Factoring by Grouping **247**
7-7   Factoring $ax^2 + bx + c$ **251**
7-8   Summary of Factoring **256**

Review and Testing

Mini-Reviews   **237, 254**
Cumulative Review **242**
Vocabulary **259**
Chapter Summary **259**
Chapter Review **260**
Chapter Test **261**

Applications and Extensions

Excursions in Algebra:
     Sieve of Eratosthenes **231**
     Divisibility Rules **233**
     Mass in the Metric System **238**
     Factoring Cubes **246**
     Grouping Three Terms **249**
Applications in Geometry: Area **234**
Reading Algebra: Reading Speed **250**

Technology

Excursions in Algebra:
     Checking Solutions **255**
     Using Computers: Bits and Bytes **258**

8 Applications of Factoring —————— 262

Lessons

Solving Equations
8-1   Solving Equations Already in
      Factored Form **263**
8-2   Factoring Review **266**
8-3   Solving Some Equations
      by Factoring **270**
8-4   Solving More Equations
      by Factoring **274**

Problem Solving and Rational Expressions
8-5   Problem Solving:
      Integer Problems **276**
8-6   Problem Solving:
      Velocity and Area **278**
8-7   Simplifying Rational
      Expressions **282**

Review and Testing

Mini-Reviews   **268, 280**
Cumulative Review **273**
Vocabulary **284**
Chapter Summary **284**
Chapter Review **285**
Chapter Test **286**
Standardized Test Practice Questions **287**

Applications and Extensions

Problem Solving: Using Tables **265**
Applications in Banking:
     Factoring and Compound Interest **269**
Excursions in Algebra:
     Inequalities **281**

Technology

Excursions in Algebra: Factoring **268**
Using Computers: Sums of Integers **284**

9 Functions and Graphs ———————— 288

Lessons

Relations
9-1   Ordered Pairs **289**
9-2   Relations **292**
9-3   Equations as Relations **296**
9-4   Graphing Relations **301**

Functions and Inequalities
9-5   Functions **305**
9-6   Graphing Inequalities in Two
       Variables **312**
9-7   Finding Equations from
       Relations **316**

Review and Testing

Mini-Reviews   **299, 315**
Cumulative Review **304**
Vocabulary **320**
Chapter Summary **320**
Chapter Review **321**
Chapter Test **323**

Applications and Extensions

Reading Algebra: Graphs **300**
Applications in Surveying: Area **309**
Excursions in Algebra:
       Misleading Graphs **315**

Technology

Graphing Calculator Applications:
       Absolute Value Function **310**
Using Computers:
       FOR/NEXT Statements **319**

10 Lines and Slopes ———————— 324

Lessons

Forms of Linear Equations
10-1   Slope **325**
10-2   Equations of Lines in
       Point-Slope and Standard
       Form **329**
10-3   Slope-Intercept Form **333**

Graphing and Writing Equations
10-4   Graphing Linear Equations **338**
10-5   Writing Slope-Intercept
       Equations of Lines **342**

Parallel and Perpendicular Lines, Midpoints
10-6   Parallel and Perpendicular
       Lines **346**
10-7   Midpoint of a Line Segment **350**

Review and Testing

Mini-Reviews   **337, 349**
Cumulative Review **341**
Vocabulary **354**
Chapter Summary **354**
Chapter Review **355**
Chapter Test **356**
Standardized Test Practice Questions **357**

Applications and Extensions

Problem Solving:
       Graphs and Estimation **332**
Excursions in Algebra:
       Rise Over Run **337**
       Slopes and Grades **349**
Applications in Aeronautics:
       Airplane Flying Time **345**

Technology

Using Calculators:
       Writing Equations **343**
Using Computers:
       Testing Points on a Line **353**

11 Systems of Open Sentences _____ 358

Lessons

Solving Systems by Graphing or Substitution

11-1   Graphing Systems of Equations. . **359**
11-2   Systems of Equations **362**
11-3   Substitution **365**

Solving Systems by Elimination

11-4   Elimination Using Addition **369**
11-5   Elimination Using
         Multiplication **372**

Problem Solving

11-6   Graphing Systems of
         Inequalities. **377**
11-7   Problem Solving:
         Digit Problems. **382**
11-8   Problem Solving:
         Using Systems of Equations. . . **385**

Review and Testing

Mini-Reviews   **364, 380**
Cumulative Review **371**
Vocabulary **391**
Chapter Summary. **391**
Chapter Review. **391**
Chapter Test. **393**

Applications and Extensions

Excursions in Algebra:
   Systems with Three Variables **376**
   Cramer's Rule **388**
Problem Solving: Using a Chart **381**

Technology

Applications in Business:
   Break-Even Analysis **361**
Excursions in Algebra: Solutions. **368**
Using Computers:
   Solving Systems of Equations **390**

12 Radical Expressions _____ 394

Lessons

Sets of Numbers, Approximating Square Roots

12-1   Square Roots **395**
12-2   Irrational Numbers and
         Real Numbers **398**
12-3   Approximating Square Roots. . . . **402**

Simplifying Radical Expressions

12-4   Simplifying Square Roots **404**
12-5   Simplifying Radical Expressions
         Involving Division **407**
12-6   Adding and Subtracting
         Radical Expressions **411**

Solving Equations and Finding Distances

12-7   Radical Equations. **415**
12-8   The Pythagorean Theorem. **418**
12-9   The Distance Formula **421**

Review and Testing

Mini-Reviews   **401, 417**
Cumulative Review **410**
Vocabulary **425**
Chapter Summary. **425**
Chapter Review. **426**
Chapter Test. **428**
Standardized Test Practice Questions **429**

Applications and Extensions

Excursions in Algebra:
   Radical Sign **397**
   Divide-and-Average Method. **403**
   Equations of Circles. **423**
Reading Algebra:
   Equivalent Expressions **406**
Applications in Physics:
   Escape Velocity **414**

Technology

Using Calculators:
   Approximating Square Roots. **402**
   Adding and Subtracting Radicals. **412**
Using Computers:
   Computing Square Roots. **424**

13 Quadratics _____ 430

Lessons

**Graphing, Solving Quadratics and the
Quadratic Formula**

13-1 Graphing Quadratic Functions... **431**
13-2 Solving Quadratic Equations
by Graphing **438**
13-3 Completing the Square **442**
13-4 The Quadratic Formula **445**

Discriminant and Problem Solving

13-5 The Discriminant **450**
13-6 Solving Quadratic Equations **454**
13-7 Problem Solving:
Quadratic Equations **456**
13-8 The Sum and Product of Roots.. **459**

Review and Testing

Mini-Reviews   **441, 458**
Cumulative Review **448**
Vocabulary **463**
Chapter Summary.................... **463**
Chapter Review..................... **464**
Chapter Test....................... **465**

Applications and Extensions

Applications in Business: Maximum Profit .. **435**
Problem Solving: List Possibilities **449**
Excursions in Algebra: Imaginary Roots **453**

Technology

Graphing Calculator Applications:
Graphing Quadratic Functions.......... **436**
Using Calculators:
Completing the Square **444**
Quadratic Formula................... **446**
Sum and Product of Roots............. **461**
Using Computers:
Graphing Quadratic Functions.......... **462**

14 Rational Expressions _____ 466

Lessons

Multiplication and Division

14-1 Simplifying Rational
Expressions **467**
14-2 Multiplying Fractions......... **471**
14-3 Dividing Fractions **475**
14-4 Dividing Polynomials **479**

Addition and Subtraction

14-5 Adding and Subtracting Fractions
with Like Denominators..... **482**
14-6 Adding and Subtracting Fractions
with Unlike Denominators... **486**
14-7 Mixed Expressions and Complex
Fractions **490**

Applying Rational Expressions

14-8 Solving Rational Equations.... **494**
14-9 Problem Solving:
Work and Uniform Motion .. **497**
14-10 Using Formulas.............. **501**
14-11 Rational Expressions
in Science **504**

Review and Testing

Mini-Reviews   **478, 493, 503**
Cumulative Review **485**
Vocabulary **508**
Chapter Summary....................... **508**
Chapter Review....................... **508**
Chapter Test.......................... **510**
Standardized Test Practice Questions **511**

Applications and Extensions

Problem Solving: Identify Subgoals....... **470**
Excursions in Algebra:
Does 2 = 1? **478**
Formulas **503**
Applications in Photography:
F-Stops........................... **500**

Technology

Excursions in Algebra:
Finding Factors **481**
Evaluating Sums and Differences........ **489**
Using Computers:
Adding Fractions..................... **507**

15 Statistics and Probability _____512

Lessons

Statistics and Organizing Data

15-1   What Is Statistics?.......... **513**
15-2   Line Plots **515**
15-3   Stem and Leaf Plots......... **517**

Central Tendency, Variation, and Prediction

15-4   Measures of Central Tendency:
        Mean, Median, and Mode ... **520**
15-5   Measures of Variation: Range,
        Quartiles, and Interquartile
        Ranges **524**
15-6   Box and Whisker Plots **528**
15-7   Scatter Plots **530**

Probability

15-8   Probability and Odds......... **534**
15-9   Drawing Conclusions
        from Experiments.......... **537**
15-10   Compound Events........... **539**

Review and Testing

Mini-Reviews   **519, 532, 541**
Cumulative Review **527**
Vocabulary **543**
Chapter Summary...................... **543**
Chapter Review....................... **544**
Chapter Test......................... **545**

Applications and Extensions

Applications in Statistics:
    Weighted Mean **523**

Technology

Graphing Calculator Applications:
    Regression Lines **533**
Using Computers:
    Bar Graphs **542**

16 Trigonometry _____546

Lessons

Angles and Triangles

16-1   Angles....................... **547**
16-2   30°–60° Right Triangles **550**
16-3   Similar Triangles.............. **554**

Trigonometric Ratios and Problem Solving

16-4   Trigonometric Ratios **557**
16-5   Using Tables and Calculators.... **561**
16-6   Solving Right Triangles **564**
16-7   Problem Solving:
        Using Trigonometry **567**

Review and Testing

Mini-Reviews   **553, 566**
Cumulative Review **560**
Vocabulary **572**
Chapter Summary...................... **572**
Chapter Review....................... **573**
Chapter Test......................... **574**

Applications and Extensions

Reading Algebra: Greek Letters.......... **563**
Applications in Surveying:
    Angle of Elevation **570**

Technology

Using Calculators:
    30°–60° Right Triangles **551**
    Trigonometric Ratios................. **558**
    Using Tables and Calculators.......... **562**
Using Computers:
    Trigonometric Functions **571**

Appendix: BASIC _____576

The BASIC Language 576
Assignment Statements 579
IF-THEN Statements 582
FOR-NEXT Statements 585
Special Features of BASIC 588

Symbols. 591
Squares and Approximate Square
    Roots 592
Trigonometric Ratios 593
Algebraic Skills Review. 594
Glossary 611
Selected Answers. 617
Index 634

CHAPTER 1

The Language of Algebra

Many people who are deaf use sign language to express thoughts and ideas. Each hand signal that they use has a meaning. In mathematics, numbers and math symbols have meanings. In this chapter, you will learn about symbols used in algebra.

1-1 Variables and Expressions

In algebra, verbal expressions are often translated into mathematical expressions. Consider the following sentence.

Kathy has three more dollars than Nick.

From the information given, you do not know exactly how much money Nick has or how much Kathy has. However, you can find the amount Kathy has if you know the amount Nick has.

If Nick has 6 dollars, then Kathy has $6 + 3$ dollars.
If Nick has 10 dollars, then Kathy has $10 + 3$ dollars.
If Nick has n dollars, then Kathy has $n + 3$ dollars.

The letter n is used as a **variable**, and $n + 3$ is an **algebraic expression**. In algebra, variables are symbols that are used to represent unspecified numbers. Any letter may be used as a variable.

$6 + 3$ and $10 + 3$ are numerical expressions.

An algebraic expression consists of one or more numbers and variables along with one or more arithmetic operations. Some algebraic expressions are listed below.

$$ab + 4 \qquad \frac{s}{t} - 1 \qquad 6a \times 4n \qquad 8rs \div 3k$$

In a multiplication expression, the quantities being multiplied are called **factors** and the result is called the **product**. In a division expression, the **dividend** is divided by the **divisor**. The result is called the **quotient**.

$$4 \times 5 \times 8 = 160 \qquad\qquad 12 \div 3 = 4$$

factors   product   dividend   divisor   quotient

In algebraic expressions, a raised dot or parentheses are often used instead of $\times$ to indicate multiplication. When variables are used to represent factors, the multiplication sign is usually omitted. Each expression below represents the product of a and b.

To indicate the product for 3 and a, write 3a.

$$ab \qquad a \cdot b \qquad a(b)$$

A fraction bar is often used to indicate division.

$$\frac{x}{3} \quad \text{means} \quad x \div 3$$

To solve verbal problems in mathematics, words must be translated into mathematical symbols. Thus, "the product of 3 and 4" becomes "3 · 4." The following chart shows some of the words that are used to indicate mathematical operations.

ADDITION	SUBTRACTION	MULTIPLICATION	DIVISION
the sum of	the difference of	the product of	the quotient of
increased by	decreased by	multiplied by	divided by
plus	minus	times	the ratio of
more than	less than		
the total of	subtracted from		
added to			

Examples

Write an algebraic expression for each verbal expression.

1 **the sum of *a* and *b***

The algebraic expression is $a + b$.

2 **a number *k* divided by a number *n***

The algebraic expression is $\frac{k}{n}$.

3 **the product of 4 and *x***

The algebraic expression is $4 \cdot x$ or simply $4x$.

4 **z less than 6**

The algebraic expression is $6 - z$.

In an expression like 10^3, the 10 names the **base** and the 3 names the **exponent**. An exponent indicates the number of times the base is used as a factor.

3^4　means　$3 \cdot 3 \cdot 3 \cdot 3$　　　　a^2　means　$a \cdot a$　　　*An expression in the form x^n is called a power.*

Study the following chart.

Symbols	Words	Meaning
8^1	8 to the first power	8
8^2	8 to the second power or 8 squared	$8 \cdot 8$
8^3	8 to the third power or 8 cubed	$8 \cdot 8 \cdot 8$
8^4	8 to the fourth power	$8 \cdot 8 \cdot 8 \cdot 8$
$6n^5$	6 times *n* to the fifth power	$6 \cdot n \cdot n \cdot n \cdot n \cdot n$

5 **Evaluate 2^5.**

$2^5 = 2 \cdot 2 \cdot 2 \cdot 2 \cdot 2$
$\quad = 32$

Using Calculators

A scientific calculator can be used to evaluate 2^5.

This key tells the calculator to raise the number 2 to the fifth power.

ENTER: 2 $\boxed{y^x}$ 5 $\boxed{=}$

Your calculator display should read *32*.

6 Write an algebraic expression for the verbal expression *the cube of y decreased by nine.*

The algebraic expression is $y^3 - 9$.

Exploratory Exercises

Write an algebraic expression for each verbal expression.

1. the product of x and 7
2. the quotient of r and s
3. the total of a and 19
4. a number n decreased by 4
5. a number b to the third power
6. 7 less than w
7. 25 squared
8. a number h cubed

State a verbal expression for each of the following.

9. x^2 **10.** n^4 **11.** 5^3 **12.** y^5 **13.** n^1 **14.** z^7

Written Exercises

Write each of the following as an expression using exponents.

15. $5 \cdot 5 \cdot 5$ **16.** $4 \cdot 4 \cdot 4 \cdot 4$ **17.** $7 \cdot a \cdot a \cdot a \cdot a$ **18.** $2(m)(m)(m)$
19. $\frac{1}{2} \cdot a \cdot a \cdot b \cdot b \cdot b$ **20.** $\frac{3}{4} \cdot x \cdot y \cdot y \cdot y \cdot y \cdot y$ **21.** $5 \cdot 5 \cdot 5 \cdot x \cdot x \cdot y$ **22.** $3 \cdot 3 \cdot a \cdot a \cdot a$

Write an algebraic expression for each verbal expression. Use x as the variable.

23. a number increased by 17
24. seven times a number
25. the cube of a number
26. a number to the sixth power
27. twice the square of a number
28. one-half the square of a number
29. six times a number decreased by 17
30. twice a number decreased by 25
31. 94 increased by twice a number
32. twice the cube of a number
33. three-fourths of the square of a number
34. the cube of a number increased by seven

Evaluate each expression.

35. 2^4 **36.** 6.2^6 **37.** 5^3 **38.** 10^4 **39.** 4.8^5 **40.** 3^6

Challenge Exercises

Write an algebraic expression for each verbal expression.

41. the sum of a and b, decreased by the product of a and b
42. the difference of a and b, increased by the quotient of a and b

1-2 Evaluating Expressions

The numerical expression $5 \cdot 6 + 3$ might be computed two ways. Only one is correct.

Multiply 5 and 6. Add 3.
$$5 \cdot 6 + 3 = 30 + 3$$
$$= 33$$

Add 6 and 3. Multiply by 5.
$$5 \cdot 6 + 3 = 5(9)$$
$$= 45$$

Each numerical expression should have a *unique* value. Therefore, the following order of operations has been agreed upon.

1. **Evaluate all powers.**
2. **Do all multiplications and divisions from left to right.**
3. **Do all additions and subtractions from left to right.**

Order of Operations

Thus, using the order of operations, the value of $5 \cdot 6 + 3$ is 33.

Example

1 Evaluate $12 \div 3 \cdot 5 - 4^2$.

$$12 \div 3 \cdot 5 - 4^2 = 12 \div 3 \cdot 5 - 16 \qquad \textit{Evaluate } 4^2.$$
$$= 4 \cdot 5 - 16 \qquad \textit{Divide 12 by 3.}$$
$$= 20 - 16 \qquad \textit{Multiply 5 by 4.}$$
$$= 4 \qquad \textit{Subtract 16 from 20.}$$

An algebraic expression can be evaluated when the value of each variable is known. The variables are simply replaced by their values. Then, the value of the numerical expression is calculated.

Example

2 Evaluate $2x^2 + 5xy$ if $x = 3$ and $y = \frac{2}{3}$.

$$2x^2 + 5xy = 2 \cdot 3^2 + 5(3)\left(\frac{2}{3}\right) \qquad \textit{Substitute 3 for each x and } \tfrac{2}{3} \textit{ for y.}$$
$$= 2 \cdot 9 + 5(3)\left(\frac{2}{3}\right) \qquad \textit{Evaluate } 3^2.$$
$$= 18 + 10 \qquad \textit{Multiply.}$$
$$= 28 \qquad \textit{Add.}$$

In mathematics, *grouping symbols* such as parentheses are used to clarify or change the order of operations.

$5 \cdot 6 + 3$ means   30 + 3 or 33

$5(6 + 3)$ means $5 \cdot 9$ or 45

The basic grouping symbols used are parentheses, ( ), brackets, [ ], and the fraction bar. Parentheses and brackets indicate that the expression within is to be treated as a single value and should be evaluated first. A fraction bar indicates that the numerator and denominator should each be treated as a single value.

When more than one grouping symbol is used, start evaluating within the innermost grouping symbols.

Examples

3 **Evaluate $3(2 + 5)^2 \div 7$.**

$$
\begin{aligned}
3(2 + 5)^2 \div 7 &= 3(7)^2 \div 7 & &\textit{Add 2 and 5.} \\
&= 3(49) \div 7 & &\textit{Evaluate } 7^2. \\
&= 147 \div 7 & &\textit{Multiply 49 by 3.} \\
&= 21 & &\textit{Divide 147 by 7.}
\end{aligned}
$$

4 **Evaluate $\frac{2}{3}[8(a - b)^2 + 3b]$ if $a = 5$ and $b = 2$.**

$$
\begin{aligned}
\frac{2}{3}[8(a - b)^2 + 3b] &= \frac{2}{3}[8(5 - 2)^2 + 3 \cdot 2] & &\textit{Substitute 5 for a and 2 for each b.} \\
&= \frac{2}{3}[8(3)^2 + 3 \cdot 2] & &\textit{Subtract 2 from 5.} \\
&= \frac{2}{3}[8 \cdot 9 + 3 \cdot 2] & &\textit{Evaluate } 3^2. \\
&= \frac{2}{3}[72 + 6] & &\textit{Multiply 9 by 8 and 2 by 3.} \\
&= \frac{2}{3}[78] & &\textit{Add 72 and 6.} \\
&= 52 & &\textit{Multiply 78 by } \frac{2}{3} \textit{ and simplify.}
\end{aligned}
$$

Scientific calculators follow the order of operations. It is necessary to analyze how a calculator would read entries made. It is sometimes necessary to add grouping symbols when using a calculator. This is shown in the next example.

Example

5 **Using Calculators**

Evaluate $\frac{b^7 - 5}{b^2 + x^4}$ if $b = 6$ and $x = 8$.

This key tells the calculator to square the number 6.

ENTER: (6 y^x 7 − 5) ÷ (6 x^2 + 8 y^x 4) =

Why is it necessary to have parentheses around the numerator and denominator?

0   6   7   279936   5   279931   6   36   8   4   4132   67.747096

Exploratory Exercises

State how to evaluate each expression. Do *not* evaluate.

1. $3 + 2 \cdot 4$ **2.** $2 \cdot 4 + 3$ **3.** $8 \div 4 \cdot 2$ **4.** $8 \cdot 2 \div 4$

5. $12 - 6 \cdot 2$ **6.** $(12 - 6) \cdot 2$ **7.** $(9 - 3)^2$ **8.** $9 - 3^2$

9. $9^2 - 3^2$ **10.** $3(3^2 - 3)$ **11.** $7 \cdot 3^2$ **12.** $7^2 \cdot 2$

13. $4(5 - 3)^2$ **14.** $8 \div 2 + 6 \cdot 2$ **15.** $(8 + 6) \div 2 + 2$ **16.** $8 + 6 \div (2 + 1)$

Written Exercises

Evaluate each expression.

17. $3 + 8 \div 2 - 5$ **18.** $4 + 7 \cdot 2 + 8$ **19.** $12 \div 4 + 15 \cdot 3$

20. $3 \cdot 6 - 12 \div 4$ **21.** $5(9 + 3) - 3 \cdot 4$ **22.** $7(8 - 4) + 11$

23. $29 - 3(9 - 4)$ **24.** $4(11 + 7) - 9 \cdot 8$ **25.** $12 \cdot 6 \div 3 \cdot 2 \div 8$

26. $16 \div 2 \cdot 5 \cdot 3 \div 6$ **27.** $288 \div [3(9 + 3)]$ **28.** $196 \div [4(11 - 4)]$

29. $5^3 + 6^3 - 5^2$ **30.** $6(4^3 + 2^2)$ **31.** $\frac{38 - 12}{2 \cdot 13}$

32. $\frac{9 \cdot 4 + 2 \cdot 6}{7 \cdot 7}$ **33.** $\frac{9 \cdot 3 - 4^2}{3^2 + 2^2}$ **34.** $\frac{2 \cdot 8^2 - 2^2 \cdot 8}{2 \cdot 8}$

35. $\frac{2}{3}(16) - \frac{1}{3}(6)$ **36.** $\frac{3}{4}(6) + \frac{1}{3}(12)$ **37.** $\frac{1}{2}(8 + 30) - 4$

38. $25 - \frac{1}{3}(18 + 9)$ **39.** $0.2(0.6) + 3(0.4)$ **40.** $7(0.2 + 0.5) - 0.6$

Evaluate each expression if $a = 6$, $b = 4$, $c = 3$, $d = \frac{1}{2}$, $e = \frac{2}{3}$, $x = 0.2$, and $y = 1.3$.

41. $a + b^2 + c^2$ **42.** $3ab - c^2$ **43.** $8(a - c)^2 + 3$

44. $\frac{a^2 - b^2}{2 + d^3}$ **45.** $\frac{2ab - c^3}{7}$ **46.** $12d + bc$

47. $d(a + b) - c$ **48.** $a(8 - 3e) + 4d$ **49.** $ax + bc$

50. $5y + 3$ **51.** $x^2 + 6d$ **52.** $c^2 + y^2$

53. $(100x)^2 + 10y$ **54.** $(15x)^3 - y$ **55.** $\frac{a + b^2}{3bc}$

56. $\frac{4d^2}{4ce}$ **57.** $\frac{8b^2c}{x}$ **58.** $\frac{6ab^2}{x^3 + y^2}$

Write an algebraic expression for each verbal expression. Then, evaluate the expression if $a = 3$, $b = \frac{1}{2}$, and $c = 0$.

59. twice the sum of a and b **60.** twice the product of a and b

61. the square of b increased by c **62.** the cube of a decreased by b

Challenge Exercises

63. Evaluate $3^2 + 4(8 - 2) - 6(3) + 4 \div 2 + (8 - 2)^2$.

64. Evaluate $a^3 + b^2 - (c^2 + abcde) - a \div b + d \div e + c$
    if $a = 2$, $b = \frac{1}{3}$, $c = 2.2$, $d = 4$, and $e = \frac{2}{5}$.

1-3 Open Sentences and Equations

Which of the following sentences are true?
Which are false?

A dog has three ears.

California is larger than Rhode Island.

$3 \times 7 = 21$

Twenty-seven divided by 3 is less than 7.

$\frac{3}{4} + \frac{1}{2} = 1\frac{1}{4}$

Note that each of the previous sentences are either true or false. However, it is not possible to say whether the following sentences are true or false.

Words	**Symbols**
A number x plus six is equal to eight.	$x + 6 = 8$
A number n is less than ten.	$n < 10$
Seven is greater than twice a number c.	$7 > 2c$

Before you can determine whether these sentences are true or false, you must know what numbers will replace the variables, x, n, and c. Mathematical sentences such as $x + 6 = 8$ are called **open sentences**. In an open sentence such as $x + 6 = 8$, the variable must be replaced in order to determine if the sentence is true or false.

Open sentences are neither true nor false.

> **Finding the replacements for the variable that make a sentence true is called solving the open sentence. Each replacement is called a solution of the open sentence.**

Solving an Open Sentence

Is 3 a solution of the open sentence $x + 6 = 8$?
Is 11 a solution of the open sentence $n < 10$?
Is 4 a solution of the open sentence $7 > 2c$?
Try naming some solutions of the three open sentences listed above.

An open sentence that contains an **equals sign**, =, is called an **equation**. Sometimes you can solve an equation by simply performing the indicated operations.

Examples

1 Solve $a = 9.4 - 3.06$.

$a = 9.4 - 3.06$

$\quad = 9.40 - 3.06$

$\quad = 6.34$

The solution is 6.34.

2 Solve $\dfrac{5 \cdot 3 + 3}{4 \cdot 2 - 2} = t$.

$\dfrac{5 \cdot 3 + 3}{4 \cdot 2 - 2} = t$

$\dfrac{15 + 3}{8 - 2} = t$

$\dfrac{18}{6} = t$

$3 = t$

The solution is 3.

Exploratory Exercises

State whether each sentence is *true* or *false*.

1. The capital of the U.S. is Houston.

2. Birds have wings.

3. $3(11 - 5) > 18$

4. $15 \div 3 + 7 < 13$

5. $0.01 + 0.01 = 0.0002$

6. $\frac{2}{3}(6) = 9 - 5$

7. $\frac{3 + 15}{6} = \frac{1}{2}(6)$

8. $\frac{1}{2} + \frac{3}{4} = \frac{3}{2} + \frac{1}{4}$

9. $3 - 1\frac{1}{2} = \frac{1}{2}$

10. $21 - 12 > 9$

11. $17 + 3 < 100 \div 4$

12. $0.11 + 1.1 > 1.2$

13. $10\left(\frac{2}{5}\right) = \frac{1}{5} \cdot 2 \cdot 10$

14. $0.101 > 0.110$

15. $7 = 7(0)$

16. $0 + 7 = 7$

Replace the variable to make each open sentence true.

17. x is President of the U.S.

18. There are y states in the U.S.

19. $18 + y = 20$

20. $3 \cdot a = 24$

21. t is an author of this book.

22. $\frac{1}{2}(m) = 10$

Written Exercises

Solve each equation.

23. $x = 8 + 3$

24. $y = 7 - 4$

25. $x = 12 - 0.03$

26. $x = 6 + 0.28$

27. $a = \frac{3}{4} \cdot 12$

28. $\frac{5}{6} \cdot 18 = m$

29. $8.2 - 6.75 = a$

30. $9.6 + 4.53 = b$

31. $a = \frac{12 + 8}{4}$

32. $m = \frac{64 + 4}{17}$

33. $\frac{21 - 3}{12 - 3} = x$

34. $\frac{14 + 28}{4 + 3} = y$

35. $14.8 - 3.75 = a$

36. $29.7 - 5.86 = t$

37. $x = \frac{84 \div 7}{18 \div 9}$

38. $x = \frac{96 \div 6}{8 \div 2}$

39. $\frac{2}{13} + \frac{5}{13} = x$

40. $\frac{3}{7} + \frac{2}{7} = y$

41. $\frac{5}{8} + \frac{1}{4} = x$

42. $\frac{7}{12} + \frac{1}{3} = x$

43. $a = 3\frac{1}{2} \div 2$

44. $m = 3\frac{1}{3} \div 3$

45. $r = 5\frac{1}{2} + \frac{1}{3}$

46. $s = 6\frac{3}{4} + \frac{1}{6}$

The set of numbers that a variable may represent is called a *replacement set*. Find all numbers from the replacement set {4, 5, 6, 7, 8} that are solutions to each open sentence.

47. $x + 2 > 7$

48. $10 - x < 7$

49. $\frac{x + 3}{2} < 5$

50. $x - 3 > \frac{x + 1}{2}$

51. $\frac{2(x - 2)}{3} = \frac{4}{7 - 5}$

52. $9x - 20 = x^2$

53. $0.3(x + 4) \le 0.4(2x + 3)$

54. $1.3x - 12 < 0.9x + 4$

55. $\frac{2x + 1}{7} \ge \frac{x + 4}{5}$

Excursions in Algebra   Sets

A collection of objects or numbers is often called a **set**. A set can be shown by using braces.

$$\{1, 2, 3\} \text{ is } \textit{the set of numbers 1, 2, 3.}$$

Each object or number in a set is called an **element**.

$$1 \text{ } \textit{is an element of } \{1, 2, 3\}.$$
$$1 \qquad \in \qquad \{1, 2, 3\}$$

Sets are often named by capital letters.

$$\text{Set A is } \{1, 2, 3\}.$$
$$A = \{1, 2, 3\}$$

If every element of set A is also an element of set B, then set A is a **subset** of set B.

$$\{1, 2\} \textit{ is a subset of } \{1, 2, 3\}. \qquad \{1\} \textit{ is a subset of } \{1, 2, 3\}.$$
$$\{1, 2\} \quad \subset \quad \{1, 2, 3\} \qquad\qquad \{1\} \quad \subset \quad \{1, 2, 3\}$$

The set with no elements is the **null** or **empty set**. It is shown by { } or ∅. The empty set is a subset of every set.

Exercises

Write in symbols.

1. the set of numbers 3, 4, 5, 6

2. the set of numbers 6, 7, 8, 9, 10, 11

3. Three is an element of the set of numbers 3, 4, 5, 6.

4. Six is an element of the set of counting numbers between and including 5 and 10.

List all subsets of each set to show that each statement is *true*.

5. {1, 2, 4} has 8 subsets.

6. {5, 6, 7, 8} has 16 subsets.

1-4 Identity and Equality Properties

In algebra, there are certain statements or **properties** that are true for any number. Some of the properties that you have already used will be examined in this lesson.

Solve each equation. $b + 15 = 15$ $4780 + c = 4780$

What value makes each statement true?

The solution of each equation is 0. The sum of any given number and 0 is the given number. The **additive identity** of any number is 0.

> **For any number a, $a + 0 = 0 + a = a$.**

Additive Identity Property

Solve each equation. $5x = 5$ $z \cdot 655 = 655$

What value makes each statement true?

The solution of each equation is 1. The product of any given number and 1 is the given number. The **multiplicative identity** of any number is 1.

> **For any number a, $a \cdot 1 = 1 \cdot a = a$.**

Multiplicative Identity Property

What role does 0 play in multiplication?

$0(11) = 0$ $5 \cdot 0 = 0$ $8 \cdot 3 \cdot 0 = 0$ $4 \cdot 0 \cdot 27 \cdot 8 = 0$

When a factor is 0, the product is 0.

> **For any number a, $a \cdot 0 = 0 \cdot a = 0$.**

Multiplicative Property of Zero

Think about the following statements.

1. $5 = 5$
2. If $7 + 3 = 10$, then $10 = 7 + 3$.
3. If $10 - 2 = 8$ and $8 = 5 + 3$, then $10 - 2 = 5 + 3$.

These are examples of the properties of equality.

> **The following properties are true for any numbers a, b, and c.**
> **Reflexive Property:** $a = a$
> **Symmetric Property:** If $a = b$, then $b = a$.
> **Transitive Property:** If $a = b$ and $b = c$, then $a = c$.

Properties of Equality

Which property of equality corresponds to each of the three numbered statements given above?

Another property of equality is the **Substitution Property**. This property permits substitution of a quantity for its equal. You use this property regularly. For example, you would simplify the expression $8 + (3 + 9)$ as follows.

$$8 + (3 + 9) = 8 + 12 \qquad \textit{Substitute 12 for (3 + 9).}$$
$$= 20 \qquad \textit{Substitute 20 for 8 + 12.}$$

For any numbers a and b, if $a = b$ then a may be replaced by b.	*Substitution Property of Equality*

Exploratory Exercises

State the property shown.

1. $7 + 6 = 7 + 6$

2. If $6 = 3 + 3$, then $3 + 3 = 6$.

3. $3 + (2 + 1) = 3 + 3$

4. $9 + 5 = (6 + 3) + 5$

5. If $8 = 6 + 2$ and $6 + 2 = 5 + 3$, then $8 = 5 + 3$.

Solve each equation.

6. $0 + x = 7$

7. $a \cdot 1 = 5$

8. $7b = 7$

9. $0(18) = n$

Written Exercises

State the property shown.

10. If $8 + 1 = 9$, then $9 = 8 + 1$.

11. $0 \cdot 36 = 0$

12. $1(87) = 87$

13. $9 + (2 + 10) = 9 + 12$

14. $14 + 16 = 14 + 16$

15. $0 + 17 = 17$

16. $(9 - 7)(5) = 2(5)$

17. $abc = 1abc$

18. $7(0) = 0$

19. If $3 = 4 - 1$, then $4 - 1 = 3$.

20. If $9 + 1 = 10$ and $10 = 5(2)$, then $9 + 1 = 5(2)$.

mini-review

Perform the indicated operation.

1. $345 + 567$

2. $238 + 789 + 2345$

3. $2345 - 567$

4. $789 - 234 + 345$

5. 78×89

6. $345 \div 15$

Solve each equation.

7. $a = 34 - 12$

8. $k = 5 + 6 + 12$

9. $34 - 18 = y$

10. $\frac{1}{2} + \frac{1}{6} = m$

11. $x = 13.7 - 2.3$

12. $t = \frac{34 + 12}{16 + 7}$

Evaluate each expression.

13. 4^3

14. $5(7 + 4) - 3$

15. $\frac{16 + 8}{4 - 2}$

Evaluate each expression if $a = 6$, $b = 5$, $c = 9$, and $d = 3$.

16. $a + b - c$

17. c^2

18. bcd

19. $\frac{b}{30}$

20. $\frac{ad}{2}$

21. $\frac{a^2 + b^2}{2}$

During the 1790's, a group of French scientists developed a system of measurement called the metric system. The basic unit of length is the meter.

The metric system is a *decimal* system. That is, units increase or decrease in size by multiples of ten. A prefix is attached to the word *meter* to define larger and smaller units. The table shows some metric prefixes and their meanings.

Prefix	Symbol	Increase or Decrease in Unit	
mega	M	1,000,000	(one million)
kilo	k	1000	(one thousand)
hecto	h	100	(one hundred)
deka	da	10	(ten)
deci	d	0.1	(one-tenth)
centi	c	0.01	(one-hundredth)
milli	m	0.001	(one-thousandth)
micro	μ	0.000001	(one-millionth)

When the prefix *centi* is added to *meter*, the result is *centimeter*. The symbol for centimeter is cm. A centimeter is *one-hundredth* of a meter.

$$1 \text{ cm} = 0.01 \text{ m} \qquad 1 \text{ m} = 100 \text{ cm} \qquad \textit{The symbol for meter is m.}$$
$$25 \text{ cm} = 0.25 \text{ m} \qquad 3.5 \text{ m} = 350 \text{ cm}$$

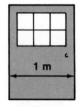

A door is about 1 m wide.

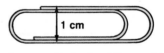

A paper clip is about 1 cm wide.

Exercises

1. Name 3 things that are best measured in meters.

2. Name 3 things that are best measured in centimeters.

Complete. *The symbol for kilometer is km and the symbol for millimeter is mm.*

3. 35 mm = ____ m   **4.** 57 cm = ____ m   **5.** 136 cm = ____ m

6. 395 cm = ____ m   **7.** 3 m = ____ cm   **8.** 4.7 m = ____ cm

9. 1 km = ____ m   **10.** 4 km = ____ m   **11.** 5000 m = ____ km

12. 1 cm = ____ mm   **13.** 14 cm = ____ mm   **14.** 55 mm = ____ cm

1-5 The Distributive Property

Angie and Maria are clerks in a local department store. Each one earns $4.95 per hour. Angie works 24 hours per week while Maria works 32 hours per week. What are their total earnings?

This problem can be solved in two ways. You could find Angie's earnings and Maria's earnings, then add to find the total earnings. Or, you could find the total number of hours worked, then multiply by the hourly wage.

Angie's Earnings	+	Maria's Earnings	=	Total Earnings
($4.95)(24)	+	($4.95)(32)	=	($4.95)(24 + 4)
$118.80	+	$158.40	=	($4.95)(56)
		$277.20	=	$277.20

The total earnings of $277.20 are distributed between Angie and Maria, each being paid $4.95 for each hour worked. This is an example of the **Distributive Property**.

For any numbers a, b, and c, $a(b + c) = ab + ac$ and $(b + c)a = ba + ca$.	*Distributive Property*

The Distributive Property can also be applied with the operation of subtraction.

$$a(b - c) = ab - ac \text{ and } (b - c)a = ba - ca$$

Example

1 **Evaluate 8(5 + 4) two different ways.**

$$8(5 + 4) = 8(9) \qquad\qquad 8(5 + 4) = 8(5) + 8(4)$$
$$= 72 \qquad\qquad\qquad\qquad = 40 + 32$$
$$= 72$$

The Symmetric Property of Equality allows the Distributive Property to be rewritten as follows.

If $a(b + c) = ab + ac$, then $ab + ac = a(b + c)$.

The Distributive Property can now be used to simplify an algebraic expression containing **like terms**.

A **term** of an expression is a number, a variable, or a

product or quotient of numbers and variables. Some examples of terms are $5x^2$, $\frac{ab}{4}$, $\frac{x}{y}$, and $7k$. Some pairs of *like terms* are $5x$ and $3x$, $2xy$ and $7xy$, and $7ax^2$ and $11ax^2$.

Like terms are terms that contain the same variables, with corresponding variables raised to the same power.	*Definitions of Like Terms*

An expression in **simplest form** has no like terms and no parentheses.

Examples

2 **Simplify $5a + 7a$.**

$$5a + 7a = (5 + 7)a \qquad \text{\textit{Use the Distributive Property.}}$$
$$= 12a \qquad \text{\textit{Use the Substitution Property of Equality.}}$$

3 **Simplify $\frac{6}{7}x^2y - \frac{2}{7}x^2y$.**

$$\frac{6}{7}x^2y - \frac{2}{7}x^2y = \left(\frac{6}{7} - \frac{2}{7}\right)x^2y \qquad \text{\textit{Distributive Property}}$$
$$= \frac{4}{7}x^2y \qquad \text{\textit{Substitution Property of Equality}}$$

The **coefficient**, or **numerical coefficient**, is the numerical part of a term. For example, in $6ab$, the coefficient is 6. In rs, the coefficient is 1 since, by the Multiplicative Identity Property, $1 \cdot rs = rs$. Like terms may also be defined as terms that are the same or that differ only in their coefficients.

Examples

4 **Name the coefficient in each term.**

a. $19g^2h$ The coefficient is 19.

b. xy^2z^2 The coefficient is 1 since $xy^2z^2 = 1xy^2z^2$.

c. $\frac{2x^3}{5}$ The coefficient is $\frac{2}{5}$ since $\frac{2x^3}{5}$ can be written as $\frac{2}{5}(x^3)$.

5 **Simplify $8n^2 + n^2 + 7n + 3n$.**

$$8n^2 + n^2 + 7n + 3n = 8n^2 + 1n^2 + 7n + 3n \qquad \text{\textit{Multiplicative Identity Property}}$$
$$= (8 + 1)n^2 + (7 + 3)n \qquad \text{\textit{Distributive Property}}$$
$$= 9n^2 + 10n \qquad \text{\textit{Substitution Property of Equality}}$$

14. Simplify $8a^2 + (8a + a^2) + 7a$.

a. $8a^2 + (8a + a^2) + 7a = 8a^2 + (a^2 + 8a) + 7a$
b. $\qquad\qquad\qquad = (8a^2 + a^2) + (8a + 7a)$
c. $\qquad\qquad\qquad = (8a^2 + 1a^2) + (8a + 7a)$
d. $\qquad\qquad\qquad = (8 + 1)a^2 + (8 + 7)a$
e. $\qquad\qquad\qquad = 9a^2 + 15a$

Written Exercises

State the property shown.

15. $5a + 2b = 2b + 5a$

16. $1 \cdot a^2 = a^2$

17. $(a + 3b) + 2c = a + (3b + 2c)$

18. $x^2 + (y + z) = x^2 + (z + y)$

19. $ax + 2b = xa + 2b$

20. $(3 \cdot x) \cdot y = 3 \cdot (x \cdot y)$

21. $29 + 0 = 29$

22. $5(a + 3b) = 5a + 15b$

23. $5a + 3b = 3b + 5a$

24. $5a + \left(\frac{1}{2}b + c\right) = \left(5a + \frac{1}{2}b\right) + c$

25. $(m + n)a = ma + na$

26. $0 + 7 = 7$

27. $1(a + b) = a + b$

28. $3m + nq = 3m + qn$

Simplify.

29. $5a + 6b + 7a$

30. $8x + 2y + x$

31. $3x + 2y + 2x + 8y$

32. $x^2 + 3x + 2x + 5x^2$

33. $\frac{2}{3}x^2 + 5x + x^2$

34. $\frac{3}{4}a^2 + 5ab + \frac{1}{2}a^2$

35. $3a + 5b + 2c + 8b$

36. $2x^2 + 3y + z^2 + 8x^2$

37. $5 + 7(ac + 2b) + 2ac$

38. $3(4x + y) + 2x$

39. $3(x + 2y) + 4(3x + y)$

40. $6(2x + y) + 2(x + 4y)$

41. $\frac{3}{4} + \frac{2}{3}(x + 2y) + x$

42. $0.2(3x + 0.2) + 0.5(5x + 3)$

43. $\frac{3}{5}\left(\frac{1}{2}x + 2y\right) + 2x$

44. $0.3(0.2 + 3y) + 0.21y$

45. $3[4 + 5(2x + 3y)]$

46. $4[1 + 4(5x + 2y)]$

mini-review

Simplify.

1. $6xxyyyy$

2. $\frac{3}{4}[17 - (11 - 6)]$

3. $4(12 - 6) - \frac{1}{2}(20 \div 5)$

4. $9a + 16b + 14a$

5. $x^2 - \frac{1}{4}x^2 + 3y^2$

6. $6(x^2 + 3x)$

Solve.

7. $t = 0.86(100)$

8. $a = \frac{17.76}{0.8}$

9. $(9.2)(4.5) = y$

State the property shown.

10. $3 + 4 = 4 + 3$

11. $8(2 \cdot 6) = (8 \cdot 2)6$

12. $23 \cdot 56 = 56 \cdot 23$

13. $A(1) = A$

14. $5(3 - 3) = 0$

15. $4(6) - 4(3) = 4(6 - 3)$

16. $3 + (b + c) = 3 + (c + b)$

17. If $a + b = c$ and $a = b$, then $a + a = c$.

18. $8 + (12 + 34) = (8 + 12) + 34$

19. $3(4 + 12) = 3(4) + 3(12)$

Suppose you are to write each of the following in symbols.

Words	Symbols
three times x plus y	$3x + y$
three times the sum of x and y	$3(x + y)$

In the second expression, parentheses are used to show that the *sum*, x plus y, is multiplied by three. In algebraic expressions, terms enclosed by parentheses are treated as one quantity. The expression $3(x + y)$ can be read *three times the quantity x plus y*. Suppose you are to read each expression below.

Symbols	Words
$a + 5^2$	a plus five squared
$(a + 5)^2$	the quantity a plus five squared

The phrase *the quantity* tells you that the sum, $a + 5$, is squared. To avoid confusion, $(a + 5)^2$ may also be read as *a plus five, the quantity squared*. Read $a + 5^2$ as *a plus five squared*.

Try an experiment. Have a classmate close his or her book. Read the following expressions to your classmate. Then have him or her write the symbols for each expression.

a. seven times the quantity b minus 4
b. the quantity 6 plus x cubed
c. two times the quantity a plus 3 times the quantity b minus 9

The correct answers are $7(b - 4)$, $(6 + x)^3$, and $2(a + 3)(b - 9)$, respectively. Compare your classmate's answers with the correct answers. How do they compare?

In verbal problems, look for key words that indicate that parentheses are to be used. Sometimes the words *sum*, *difference*, *quantity*, and *total* signal the use of parentheses. Study the following examples.

Words	Symbols
four divided by the difference of a number and 6	$4 \div (n - 6)$
the quantity a plus b divided by x	$(a + b) \div x$

Exercises

Write each expression in symbols.

1. eight times four plus x squared

2. eight times the quantity four plus x squared

3. Add n to your age and double it.

Write each algebraic expression in words.

4. $3a + 2$ **5.** $3(a + 2)$ **6.** $\frac{5}{9}(F - 32)$

7. $(r + s) - (r - s)$ **8.** $8 \div (4 - c)^2$ **9.** $8 \div 4 - c^2$

1-7 Using Formulas

Marie is making a patchwork quilt like the one shown at the right. The quilt is made up of squares of different fabrics. The squares will be arranged in a rectangle that is 16 squares in length and 12 squares in width. To find how many squares she will need to cut, Marie must find the area of the quilt. Marie knows that the area of a rectangle is equal to the product of the length and width. She uses the formula $A = \ell w$ to express this relationship. The variables A, ℓ, and w represent the measures of the area, length, and width, respectively.

$$A = \ell w$$
$$A = 16 \cdot 12$$
$$= 192$$

Marie finds that 192 squares of fabric are needed.

A **formula** is an equation that states a rule for the relationship between certain quantities.

Examples

1 **Find the area of a rectangle of length 17 cm and width 13 cm.**

$$A = \ell w$$
$$A = 17 \cdot 13$$
$$= 221$$

The area is 221 cm².

13 cm

17 cm

2 **The formula for the area of a triangle is $A = \frac{1}{2}bh$. Find the area of a triangle with a base (b) of 6 feet and a height (h) of 9 feet.**

$$A = \frac{1}{2}bh$$
$$A = \frac{1}{2}(6)(9)$$
$$= 3 \cdot 9$$
$$= 27$$

The area of the triangle is 27 ft².

9 ft

6 ft

Many sentences can be written as equations or formulas. Use variables to represent the unspecified numbers or measures referred to in the sentence. Then write the verbal expressions as algebraic expressions. Some verbal expressions that suggest the *equals sign* are listed below.

| is | is equal to | is as much as |
| equals | is the same as | is identical to |

Translate each sentence into an equation or formula.

3 The number z is equal to twice the sum of x and y.

$$\underbrace{z}\quad\underbrace{=}\quad\underbrace{2 \text{ times}}\quad\underbrace{\text{add } x \text{ and } y}$$

The equation is $z = 2(x + y)$.

4 The area of a circle equals the product of π and the square of the radius (r).

$$\underbrace{A}\quad\underbrace{=}\quad\underbrace{\text{multiply}}\quad\underbrace{\pi \text{ and}}\quad\underbrace{r^2}$$

The formula is $A = \pi r^2$.

Exploratory Exercises

Write a formula for each sentence. Use the variables indicated.

1. The area (A) of a square is the square of the length of one of its sides (s).

2. The perimeter (P) of a parallelogram is twice the sum of the lengths of two adjacent sides (a and b).

3. The perimeter (P) of a square is the product of 4 and the length of a side (s).

4. The circumference (C) of a circle is the product of 2, π, and the radius (r).

Written Exercises

The formula for the surface area (S) of a rectangular solid as shown at the right is $S = 2(\ell h + wh + \ell w)$. Copy and complete the charts.

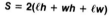

	ℓ	w	h	S
5.	5	8	6	
6.	18	10	4	
7.	$20\frac{1}{2}$	3	8	
8.	$5\frac{1}{2}$	12	$3\frac{1}{2}$	

	ℓ	w	h	S
9.	12.9	11	4.6	
10.	21.8	6.5	9.7	

The formula for the area (A) of a trapezoid as shown at the right is $A = \frac{1}{2}h(a + b)$. Copy and complete the charts.

$A = \frac{1}{2}h(a + b)$

	h	a	b	A
11.	6	24	19	
12.	11	37	23	
13.	12	24	40	
14.	10	19	54	
15.	$\frac{5}{8}$	$\frac{3}{4}$	$\frac{1}{2}$	

	h	a	b	A
16.	$3\frac{1}{3}$	12	$8\frac{1}{4}$	
17.	4	18.9	12.7	
18.	2.4	8.25	3.15	

The formula for the area (A) of the shaded region of the figure below is $A = \frac{1}{2}\pi a^2 - a^2$. Find the area of the region for each value of a. Use 3.14 for π. Round each answer to the nearest whole number.

19. 6   **20.** 4   **21.** 81   **22.** 64

23. 3.8   **24.** 5.6   **25.** 18.3   **26.** 27.4

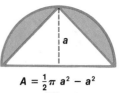

$A = \frac{1}{2}\pi a^2 - a^2$

Translate each sentence into an equation.

27. Twice x increased by the square of y is equal to z.

28. The square of a decreased by the cube of b is equal to c.

29. The sum of x and the square of a is equal to n.

30. The square of the sum of x and a is equal to m.

31. The number r equals the cube of the difference of a and b.

32. The number b equals x decreased by the cube of m.

33. The square of the product of a, b, and c is equal to k.

34. The product of a, b, and the square of c is f.

35. The sum of twice m and the square of n is y.

36. A is equal to the sum of m and the square of n.

37. The number 29 decreased by the product of x and y is equal to z.

38. R is the product of a and m decreased by z.

The distance (d) traveled is equal to the rate (r) times the time (t). Write the formula for this and use it to solve each problem.

39. Find the distance from Danville to the beach if it takes 3 hours to drive there at an average rate of 50 miles per hour.

40. Takeo runs for 30 minutes every day. Find the distance he runs if he averages 660 feet per minute.

41. The speed of sound through air is about 330 meters per second. Find the distance between Carlos and an explosion if it takes 10 seconds for the sound to reach him.

42. The speed of light is about 300,000 kilometers per second. Find the distance between Karen and a flash of light if it takes 6 seconds for Karen to see the light.

Challenge Exercises

Write a formula for the area of each figure.

43.

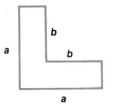

44.

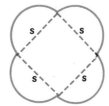

45.

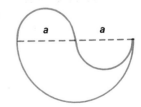

1-8 Exploring Verbal Problems

Many problems in algebra are solved by translating the verbal problem into symbols. To do this accurately, you must explore the problem until you completely understand the relationships among the given information. You can explore a problem situation by asking and answering questions. Study the following examples.

Examples

1 Peggy Richards has $3.25 in nickels and dimes. She has 7 more dimes than nickels.

Questions	*Answers*
a. Does she have more nickels or dimes?	**a.** more dimes
b. How much money does she have in all?	**b.** $3.25
c. Do you know how many nickels and dimes she has?	**c.** no
d. If she has n nickels, how many dimes does she have?	**d.** $7 + n$ or $n + 7$
e. If she has x dimes, how much money does she have in dimes?	**e.** $10x$ cents

2 One day John and Inger picked peaches for 4 hours. In all they picked 30 baskets of peaches. Inger picked 6 baskets less than John.

Questions	*Answers*
a. How many baskets were picked in all?	**a.** 30
b. How long did John and Inger work?	**b.** 4 hours
c. Who worked longer?	**c.** They worked the same.
d. Who picked more?	**d.** John
e. If Inger picked n baskets, how many did John pick?	**e.** $n + 6$
f. If John picked r baskets, how many did Inger pick?	**f.** $r - 6$

Written Exercises

For each problem, answer the related questions.

1. Mr. Limotta checked his cash register at the end of the day. He found that he had 7 fewer $5 bills than $1 bills. He had eleven $10 bills and no larger bills. In all he had $267.
   a. Did he have more $5 bills or $1 bills?
   b. How many more?
   c. How much money did he have in all?
   d. How many $20 bills did he have?

e. How much money did he have in $1 and $5 bills together?

f. When did he check his cash register?

g. If he had n $5 bills, how much money did he have in $5 bills?

2. Luisa has 20 books on crafts and cooking. She also has 21 novels. She has 6 more cookbooks than craft books.

   a. Does she have more cookbooks or craft books?

   b. How many more?

   c. Does she have more novels than craft books?

   d. How many books does she have in all?

   e. If she has n cookbooks, how many craft books has she?

   f. Of what kind of book does she have the least?

3. Two breakfast cereals, Kornies and Krispies, together cost $3.59. One of them costs 7¢ more than the other.

   a. Which one costs more?

   b. What is the difference in their prices?

   c. How much would two boxes of each cost?

   d. Would two boxes of the more expensive cereal cost more or less than $3.59?

   e. If the more expensive cereal costs n cents, what is the cost of the other cereal?

4. Miguel is 6 inches taller than Ira and Ira is 4 inches shorter than Bruce.

   a. Who is the tallest?

   b. Who is the shortest?

   c. Is Miguel taller or shorter than Bruce?

   d. How much taller than Ira is Miguel?

   e. How much taller than Bruce is Miguel?

5. Brenda Summer has 3 more nickels than dimes and 5 fewer pennies than nickels. She has 70 coins in all.

   a. Of which coin does she have the most?

   b. Of which coin does she have the fewest?

   c. How many fewer dimes than nickels does she have?

   d. How many fewer pennies than dimes does she have?

   e. How much money does she have in all?

   f. Does she have more money in nickels or dimes?

6. The Vegetable Mart offers corn at 18¢ per ear, cucumbers at 12¢ each, and tomatoes at 59¢ per basket. Mirna has only $3 to spend and wants one basket of tomatoes and 5 fewer ears of corn than cucumbers.

   a. How many baskets of tomatoes does she want?

   b. How much more is a basket of tomatoes than an ear of corn?

   c. After buying the tomatoes, how much money does she have remaining?

   d. Will she buy more cucumbers or ears of corn?

   e. If she buys n cucumbers, how many ears of corn will she buy?

7. Chris can mow the lawn in 4 hours and Mark can do it in 3 hours.

   a. Alone, how much of the lawn will Chris mow in 3 hours?

   b. Alone, how much of the lawn will Mark mow in an hour?

   c. In n hours, how much of the lawn will Chris mow?

   d. If they both start mowing at the same time, using two mowers, can they finish the lawn in 2 hours?

8. Phoebe goes to Pluto's Platters to buy her stereo records. She bought 3 more rock records than classical and 2 less western records than rock. Including 5 jazz records, she bought 18.

   a. Did she buy more classical than rock?

   b. Did she buy more rock than western?

   c. Which did she buy the most of?

   d. How many rock, classical, and western records did she buy?

   e. If she bought n classical records, how many rock records did she buy?

9. Ninety-six students signed up for football at East High School. Eight are not eligible because of grades. There are eight fewer sophomores than juniors. There are 26 seniors and no freshmen.

   a. Are there more sophomores than juniors?

   b. If there are x juniors, how many sophomores are there?

   c. Are there more juniors or seniors?

   d. Which class has the most players?

10. Craig said, "I am 24 years younger than my mom and the sum of our ages is 68 years."

   a. How old was Craig's mother when Craig was born?

   b. How much older than Craig is his mother now?

   c. What will be the sum of their ages in 5 years?

   d. How old was Craig's mother when Craig was 10 years old?

   e. In ten years, how much younger than his mother will Craig be?

11. Selam was working some math problems. She thought that she could finish 3 pages of 24 problems per page in 2 hours. After $1\frac{1}{2}$ hours, she had finished 2 pages.

   a. How many problems had Selam completed in $1\frac{1}{2}$ hours?

   b. How many problems are there in all?

   c. At the rate she is working, will she finish the three pages in 2 hours?

Challenge Exercises

12. Lorena can paint the house in 25 hours. Mia can paint the same house in 30 hours.

   a. Who paints faster, Mia or Lorena?

   b. Working alone, how much of the house can Lorena paint in 20 hours?

   c. Working alone, how much of the house can Mia paint in x hours?

   d. If Lorena and Mia work together, how many hours will it take to paint the house?

Laura Sheppard is a computer programmer for a large bank. She uses data from the computer to find any errors in her programs. The data is in the **hexadecimal numeration system**.

A *numeration system* is a method of expressing numbers. For example, the decimal system, which is a base 10 numeration system, uses ten number symbols 0, 1, 2, 3, 4, 5, 6, 7, 8, and 9 to represent all numbers. In this system, powers of ten determine the place values. Similarly, in the binary system, or base 2 system, the only number symbols are 0 and 1. Powers of two determine the place values.

The hexadecimal numeration system is a base 16 system. It uses the sixteen symbols 0, 1, 2, 3, 4, 5, 6, 7, 8, 9, A, B, C, D, E, and F to represent all numbers. Notice that $B_{sixteen} = 11_{ten}$.

The position of each symbol gives its place value as a power of sixteen. Study how the hexadecimal numeral 9F3 is changed to a decimal numeral.

$$9F3_{sixteen} = 9 \cdot 16^2 + F \cdot 16^1 + 3$$
$$= 9 \cdot 256 + 15 \cdot 16 + 3$$
$$= 2304 + 240 + 3$$
$$= 2547_{ten}$$

The hexadecimal numeration system provides a shorter way of representing numbers than the binary system. Suppose you want to change a binary numeral to a hexadecimal. Separate the numeral into groups of 4 digits starting at the right. Each group can be changed to one hexadecimal digit.

$$11\,0100\,1101\,0010 = \underline{11} \quad \underline{0100} \quad \underline{1101} \quad \underline{0010}$$
$$3 \quad\quad 4 \quad\ \nearrow D \quad\quad 2$$
$$1101_{two} = 13_{ten} = D_{sixteen}$$

Exercises

Change each hexadecimal numeral to a decimal numeral.

1. $53_{sixteen}$ **2.** $C6_{sixteen}$ **3.** $2E5_{sixteen}$ **4.** $10A3_{sixteen}$

Change each binary numeral to a hexadecimal numeral.

5. 1001_{two} **6.** 10101100_{two} **7.** 100101010_{two} **8.** 10111010111_{two}

Challenge

9. Subtract the hexadecimals $6E7 - 542$.   **10.** Add the hexadecimals $C41 + 25F$.

1-9 Problem Solving: Writing Equations

Four steps that can be used to solve verbal problems are listed below.

1. **Explore the problem.** 2. **Plan the solution.** 3. **Solve the problem.** 4. **Examine the solution.**	*Problem-Solving Plan*

Let us examine the first two steps of this plan.

explore   To solve a verbal problem, first read the problem carefully and *explore* what the problem is about.

- Identify what information is given.

- Identify what you are asked to find.

- Choose a variable to represent one of the unspecified numbers in the problem. This is called *defining the variable.*

- Use the variable in writing expressions for other unspecified numbers in the problem.

plan   Next, *plan* how to solve the problem. One way to solve a problem is to use an equation.

- Read the problem again. Decide how the unspecified numbers relate to other given information.

- Write an equation to represent the relationship.

Example

1   **Write an equation for the following problem.**

One number is 12 greater than a second number. The sum of the two numbers is 86. Find the numbers.

explore   Read the problem carefully.
Let n = the lesser number.
Then, $n + 12$ = the greater number.

plan   The problem states the sum of the two numbers is 86.
So, the equation is $n + (n + 12) = 86$.

Example

2 **Write an equation for the following problem.**

Jennifer Winters is in the ninth grade. Six years ago, twice her age was 16 years. How old is she now?

explore — Let j = Jennifer's age now.
Then, $j - 6$ = Jennifer's age 6 years ago.

plan — The problem states that twice her age six years ago was 16 years. So, the equation is $2(j - 6) = 16$.

Verbal problems can be developed from equations. First, you must know what the variable in the equation represents. Then you can use the equation to establish the conditions of the problem. Study the following examples.

Examples

3 **Write a problem based on the given information.**

Let x = Mary's age now.
$x - 5 = 17$

Since Mary's age is represented by x, then $x - 5$ must be her age 5 years ago, which was 17.
Thus, the following problem can be written.
If Mary's age 5 years ago was 17, how old is she now?

4 **Write a problem based on the given information.**

Let w = Steve's weight in kilograms.
Then, $w - 12$ = Perry's weight in kilograms.
$w + (w - 12) = 118$

The following problem can be written.
Perry weighs 12 kg less than Steve. The sum of their weights is 118 kg. How much does Steve weigh? How much does Perry weigh?

Exploratory Exercises

Write an expression for each problem.

1. Mr. Jackson is now 49 years old. How old was he n years ago?
2. The sum of two numbers is 18. The lesser number is t. What is the greater number?
3. The length of a rectangle is 4 more than the width. The width is w. Find the length.
4. Sheldon is 8 years older than Atepa. If Sheldon is n years old, how old is Atepa?
5. Alonso has 8 more than twice as many red marbles as blue marbles. He has t blue marbles. How many red marbles does he have?

6. This year's senior class has 117 fewer students than last year's class. This year's class has 947 students. How many were in last year's class?

7. The cost of gasoline has tripled in the past 8 years. Gasoline now costs $1.11 per gallon. What was the cost 8 years ago?

8. Tess types 42 words per minute. How many minutes would it take for her to type a 3000-word paper?

9. Anthony is paid $5.65 per hour. How much is he paid for working n hours?

10. Ricardo graduated from high school y years ago at the age of 17. How old is he now?

Written Exercises

Define the variable, then write an equation for each problem.

11. A number increased by 24 is 89. Find the number.

12. Twice a number decreased by 84 is 40. Find the number.

13. A number decreased by 19 is 83. Find the number.

14. 67 decreased by twice a number is 39. Find the number.

15. How old is Tyrone if twice his age increased by 17 is 53?

16. Twenty-seven years ago Kimiko was 21. How old is she now?

17. Sonia is 3 years older than Melissa. The sum of their ages in 4 years will be 59 years. How old is Sonia now?

18. Three times Cecile's age 4 years ago is 42. How old is Cecile now?

19. Bill is 5 inches taller than Bob and the sum of their heights is 137 inches. How tall is Bob?

20. Twice Mary Lou's height increased by 17 inches is 141 inches. How tall is Mary Lou?

21. Ling's dad is 27 years older than Ling. The sum of their ages 5 years ago was 45 years. How old is Ling now?

22. The sum of Mrs. Black's age and her daughter's age is 54 years. In 8 years, Mrs. Black will be twice as old as her daughter. How old is her daughter now?

23. In a football game, Michael gained 134 yards running. This was 17 yards more than the previous game. How many yards did he gain in both games?

24. Megan scored 12 points in a basketball game. This was 4 points less than the previous game. How many points did Megan score in both games?

25. Ponderosa pines grow about $1\frac{1}{2}$ feet each year. If a pine is now 17 feet tall, about how long will it take the tree to become $33\frac{1}{2}$ feet tall?

26. Kerri has 33 records and tapes altogether. If she has 4 more than half as many records as tapes, how many tapes does Kerri have?

27. Shannon has 4 more dimes than quarters and 7 fewer nickels than dimes. She has 28 coins in all. How many quarters does Shannon have?

28. Juan has 15 pennies and 8 more nickels than dimes. He has 51 coins in all. How many dimes does he have?

29. For 6 consecutive weeks, Connie lost the same amount of weight. Six weeks ago she weighed 145 pounds. She now weighs 125 pounds. How many pounds did Connie lose each week?

30. Each week for 7 weeks Sav-A-Buk stores reduced the price of a sofa by $18.25. The final reduced price was $252.50. What was the original price?

Write a problem based on the given information.

31. Let x = Olivia's age now.
$x + 7 = 29$

32. Let a = Quincy's age now.
$2(a - 7) = 58$

33. Let x = greater of two numbers.
$x - 7$ = lesser of two numbers.
$x + (x - 7) = 33$

34. Let x = lesser of two numbers.
$x + 29$ = greater of two numbers.
$x + (x + 29) = 135$

35. Let n = Irene's age now.
$n + 26$ = Irene's mother's age now.
$n + (n + 26) = 58$

36. Let w = Willie's age now.
$w + 4$ = Ellen's age now.
$(w + 10) + (w + 4 + 10) = 54$

37. Let x = weight of Seth's car in pounds.
$x + 250$ = weight of Ramón's car in pounds.
$x + (x + 250) = 7140$

38. Let h = Manuel's height in inches.
$h + 7$ = Sven's height in inches.
$2h + (h + 7) = 193$

39. Let x = the number of students present in Mr. Wyatt's class.
$x + 5 = 33$

40. Let n = the number of students in Alan's class.
$n - 7$ = the number of students in Jim's class.
$n + (n - 7) = 189$

41. Let m = Soto's height in centimeters.
$m - 31$ = Reggie's height in centimeters.
$m + 2(m - 31) = 502$

42. Let n = number of nickels that Yvette has.
$n - 17$ = number of pennies that Yvette has.
$n + (n - 17) = 63$

43. Let x = distance that Cheryl drove east.
$x + 80$ = distance that Bonnie drove west.
$x + (x + 80) = 520$

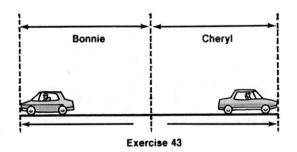

Exercise 43

44. Let p = number of pages in Ayani's book.
$p - 27$ = number of pages in Danny's book.
$p + (p - 27) = 873$

45. Let n = number of games that Greg won.
$n - 12$ = number of games that Alex won.
$n = 2(n - 12)$

46. Let x = Carol's height in inches 5 years ago.
$x + 7$ = Carol's height in inches now.
$x + 3$ = Vanesa's height in inches now.
$(x + 7) + (x + 3) = 124$

47. Let n = number of \$5 bills that Emilio has.
$n - 7$ = number of \$10 bills that Emilio has.
$n + 3$ = number of \$1 bills that Emilio has.
$n + (n + 3) + (n - 7) = 29$

48. Let x = Doug's age now.
$x - 4$ = Christy's age now.
$(x - 5) + (x - 4 - 5) = 48$

Write a problem based on the given information.

49. Let d = number of dimes that Teresa has.
$d + 4$ = number of nickels that Teresa has.
$d - 9$ = number of quarters that Teresa has.
$d + (d + 4) + (d - 9) = 37$

50. Let x = number of albums that Anna has.
$x - 3$ = number of albums that Barbara has.
$x + 4$ = number of albums that Chris has.
$x + (x - 3) + (x + 4) = 43$

mini-review

Determine whether each statement is *true* or *false*.

1. $5^2 = 5 + 5$ **2.** $7(0) = 7$ **3.** $1(4.2) = 4.2$ **4.** $0 + 18 = 18$

5. $3(4)(1) = 3(4) + 3(1)$ **6.** $7.2 + (3 + 2.8) = (7.2 + 2.8) + 3$

Answer each of the following.

7. Find the perimeter and area of a rectangle if its width is 6 inches and its length is 12 inches.

8. Find the perimeter and area of a square if the length of a side is 6 inches.

9. Pedro drives $1\frac{1}{2}$ hours to work each day. How far does he drive if his average speed is 40 miles per hour?

Define the variable, then write an equation for each problem.

10. A number increased by 17 is 42. Find the number.

11. A number decreased by 34 is 237. Find the number.

12. How old is Pete if three times his age decreased by 13 is 38?

13. Thirty-one years ago Clarice was 22 years old. How old is she now?

14. Each week for eight weeks the price of a chair was reduced $12.50. The price on the chair now is $189.50. What was the original price of the chair?

Write a problem based on the given information.

15. Let x = Sara's age now.
$x + 23 = 45$

16. Let x = the large number.
Let $x - 5$ = the smaller number.
$x + (x - 5) = 34$

Excursions in Algebra   History

The use of the term *associative* is due to Sir William Rowan Hamilton (1805–1865). Hamilton was a mathematician who studied and taught at Trinity College in Dublin, Ireland. He worked in astronomy and physics as well as higher forms of algebra. Some of his algebraic work is used in exactly the same form today. At the age of 30, Hamilton was knighted. He holds the honor of being the first foreign associate named to the United States Academy of Sciences.

Formulas

The formula for the area of a rectangle is $A = \ell w$. To find the area of a rectangle with length 51 cm and width 34 cm, you assign the variable ℓ the value 51 and the variable w the value 34. The area of the rectangle is (51)(34) or 1734 cm^2.

In the **BASIC** computer language, **READ** and **DATA** statements are used to assign values to variables. Consider the following program.

```
10   READ L,W          The computer locates the data and reads
20   DATA 51,34         them in the order in which they are listed.
30   LET A = L * W      In line 30, the variable A is assigned the
40   PRINT A            value of (51)(34) or 1734.
```

When this program is executed, the value of A, 1734, is printed.

READ and **DATA** statements are very useful when working with many calculations. For example, the following program can be used to help complete the chart for Exercises 5-10 on page 24.

```
10    PRINT "L","W","H","S"
20    PRINT
30    READ L,W,H              When the value of L is zero,
40    IF L = 0 THEN 110        the program will end.
50    DATA 5,8,6,18,10,4,20.5,3,8    Notice that more than one set of data
60    DATA 5.5,12,3.5,12.9,11,4.6     can be included in each DATA statement.
70    DATA 21.8,6.5,9.7,0,0,0
80    LET S = 2 * (L * H + W * H + L * W)    A LET statement is used in line 80
90    PRINT L,W,H.S           to write the formula as a BASIC
100   GOTO 30                 statement.
110   END
```

Enter and run this program on your computer. Compare the output with the chart that you have already completed.

Exercises

Write each formula as a BASIC statement.

1. $A = s^2$

2. $P = 2(\ell + w)$

3. $I = P \cdot r \cdot t$

4. $A = \frac{1}{2}bh$

5. $D = rt$

6. $A = \frac{1}{2}h(a + b)$

7. Write a program similar to the one above to print a chart for Exercises 11–18 on page 24. Enter and run the program.

8. Write a program to find the area for each of the figures described in Exercises 19–26 on page 25. Enter and run the program.

9. Write a program to find the distance in Exercises 39–42 on page 25. Enter and run the program.

Vocabulary

variable (3)
expression (3)
factor (3)
product (3)
dividend (3)
divisor (3)
quotient (3)
base (4)
exponent (4)
open sentence (9)
solution (9)
equals sign (9)
equation (9)
additive identity (12)
multiplicative identity (12)
Multiplicative Property of
  Zero (12)
Distributive Property (15)

properties of equality:
  Reflexive (12)
  Symmetric (12)
  Transitive (12)
  Substitution (13)
term (15)
like terms (16)
simplest form (16)
coefficient (16)
Commutative Property of
  Addition (19)
Commutative Property of
  Multiplication (19)
Associative Property of
  Addition (19)
Associative Property of
  Multiplication (19)
formula (23)

Chapter Summary

1. Variables are symbols that are used to represent unspecified numbers. (3)

2. An algebraic expression consists of one or more numbers and variables along with one or more arithmetic operations. (3)

3. In a multiplication expression, the quantities being multiplied are called factors and the result is called the product. In a division expression, the dividend is divided by the divisor. The result is called the quotient. (3)

4. A raised dot or parentheses are often used instead of × to indicate multiplication. When variables are used to represent factors, the multiplication sign is usually omitted. (3)

5. An exponent indicates the number of times the base is used as a factor. (4)

6. Order of Operations:
  1. Evaluate all powers.
  2. Do all multiplications and divisions from left to right.
  3. Do all additions and subtractions from left to right. (6)

7. Parentheses, brackets, and the fraction bar are used to clarify or change the order of operations. Evaluate expressions within parentheses or brackets first. When a fraction bar is used, the numerator and denominator are each treated as a single value. (7)

8. In an open sentence, the variable must be replaced in order to determine if the sentence is true or false. (9)

9. Finding the replacements for the variable that make a sentence true is called solving the open sentence. Each replacement is called a solution of the open sentence. (9)

10. An open sentence that contains an equals sign, =, is called an equation. (9)

11. The following properties of equality are true for any numbers a, b, and c.
    Reflexive: $a = a$ (12)
    Symmetric: If $a = b$, then $b = a$. (12)
    Transitive: If $a = b$ and $b = c$, then $a = c$. (12)
    Substitution: If $a = b$, then a may be replaced by b. (13)

12. Like terms are terms that contain the same variables, with corresponding variables raised to the same power. (16)

13. The coefficient, or numerical coefficient, is the numerical part of a term. (16)

14. The following properties are true for any numbers a, b, and c. (20)

	Addition	Multiplication
Commutative	$a + b = b + a$	$ab = ba$
Associative	$(a + b) + c = a + (b + c)$	$(ab)c = a(bc)$
Identity	0 is the identity. $a + 0 = 0 + a = a$	1 is the identity. $a \cdot 1 = 1 \cdot a = a$
Zero		$a \cdot 0 = 0 \cdot a = 0$
Distributive: $a(b + c) = ab + ac$ and $(b + c)a = ba + ca$		

15. A formula is an equation that states a rule for the relationship between certain quantities. (23)

16. To understand a verbal problem, you should explore the problem situation by asking and answering questions. (26)

17. A four-step problem-solving plan is given below. (30)
    1. Explore the problem.
    2. Plan the solution.
    3. Solve the problem.
    4. Examine the solution.

Chapter Review

1–1 Write each of the following as an expression using exponents.

    1. $a \cdot a \cdot a \cdot a$ **2.** $15 \cdot x \cdot x \cdot x \cdot y \cdot y$

Write an algebraic expression for each verbal expression. Use x as the variable.

    3. twice a number decreased by 17   **4.** a number increased by 14

    5. the product of 8 and a number   **6.** twice the cube of a number

1–2 Evaluate each expression if $a = 5$, $b = 8$, $c = \frac{2}{3}$, $d = \frac{1}{2}$, and $e = 0.3$.

    7. ab^2 **8.** $3ac - bd$ **9.** $(2a - b)^2$

    10. bcd **11.** $5e^2$ **12.** $b + c + 2d$

1–3 Solve each equation.

    13. $a = 29 - 5^2$ **14.** $5(6) - 3(5) = y$ **15.** $8 - 2 \cdot 3 = b$

    16. $r = (29 - 5)^2$ **17.** $w = (0.2)(8 + 3)$ **18.** $m = \frac{2}{3}\left(3 - \frac{1}{2}\right)$

1–4 State the property shown.

    19. $7 + 0 = 7$ **20.** $2(1) = 2$ **21.** $11 \cdot 0 = 0$

    22. If $a + b = 5$ then $5 = a + b$.   **23.** $r = r$

    24. If $a = 7$ and $7 = 9 - 2$, then $a = 9 - 2$.

1–5 Simplify.

    25. $10x + x$ **26.** $9a - 7a$ **27.** $5b + 3(b + 2)$ **28.** $9(r + s) - 2s$

1–6 State the property shown.

    29. $5(a + c) = 5(c + a)$ **30.** $10(ab) = (10a)b$

    31. $4 + (x + y) = (4 + x) + y$ **32.** $a(b + c) = (b + c)a$

Simplify.

    33. $6a + 7b + 8a + 2b$ **34.** $\frac{3}{4}a^2 + \frac{2}{3}ab + ab$

1–7 Translate each sentence into an equation.

    35. Eighteen decreased by the square of d is equal to f.

    36. The number c equals the cube of the product of 2 and x.

1–8 **Two cans of vegetables together cost $1.08. One of them costs 10¢ more than the other.**

    37. Would 2 cans of the less expensive vegetable cost more or less than $1.08?

    38. How much would 3 cans of each cost?

1–9 Define the variable, then write an equation for each problem.

    39. Minal weighs 8 pounds less than Claudia. Together they weigh 182 pounds. How much does Minal weigh?

    40. Three times a number decreased by 21 is 57. Find the number.

Chapter Test

Write an algebraic expression for each verbal expression. Use x as the variable.

1. a number increased by 17

2. twice the square of a number

3. the sum of a number and its cube

4. twice Nica's age decreased by 23 years

Evaluate.

5. $(12 - 10)^4$

6. $13 + 4 \cdot 5^2$

7. $0.7(1.4 + 0.6)$

8. $\frac{3}{4}(8 + 28)$

9. $23 - 12(1.5)$

10. $6 + 3(3.4)$

Evaluate if $m = 8$, $n = 3$, $p = \frac{3}{4}$, $q = \frac{2}{3}$, and $r = 0.5$.

11. $(mn)^2$

12. pq^2

13. $n + r^2$

Solve each equation.

14. $v = \frac{6^2 - 2^3}{7}$

15. $8(0.03) - 0.05 = y$

16. $\frac{3}{4} - \left(\frac{1}{2}\right)^2 = k$

State the property shown.

17. $t + 0 = t$

18. $7(m + 2n) = 7(2n + m)$

19. $34 \cdot 1 = 34$

20. $3(2a + b) = 6a + 3b$

21. If $m = a + b$ then $a + b = m$.

22. $3(2) = 2(3)$

23. $a + (2b + 5c) = (a + 2b) + 5c$

24. If $6 = 2a$ and $a = 3$, then $6 = 2 \cdot 3$.

Simplify.

25. $n + 5n$

26. $2.5x - x + y + 3.5y$

27. $3(a + 2) + 5a$

28. $4an + \frac{2}{3}am + 8an + \frac{1}{3}am$

Translate each sentence into an equation.

29. The product of π and the square of r is A.

30. The sum of a, b, and c is equal to P.

Define the variable, then write an equation for each problem.

31. Toshi is two years younger than Seymour. The sum of their ages is 68. How old is Toshi?

32. 79 decreased by 5 times a number is 49. Find the number.

CHAPTER 2

Adding and Subtracting Rational Numbers

Water freezes at 0° Celsius. Carbon dioxide becomes dry ice at a temperature 78.5° lower than the freezing point of water. This temperature can be written as −78.5°. The symbol "−" is used in expressing numbers less than 0. In this chapter, you will learn about these numbers.

2-1 Integers on the Number Line

The figure below is part of a **number line**. It is drawn by choosing a starting position on a line, usually 0, and marking off equal distances from that point. On a number line, the distances marked to the right of 0 are named by *members* of the set of **whole numbers**. The number line can be used to observe operations of addition or subtraction.

Remember, the whole numbers are the set of numbers 0, 1, 2, ...

Consider these subtraction sentences.

$$5 - 1 = 4 \qquad 6 - 2 = 4$$
$$5 - 3 = 2 \qquad 6 - 4 = 2$$
$$5 - 5 = 0 \qquad 6 - 6 = 0$$
$$5 - 7 = \underline{\ ?\ } \qquad 6 - 8 = \underline{\ ?\ }$$

There are no answers to $5 - 7$ and $6 - 8$ among the set of whole numbers. Thus, the whole numbers lack *closure* for subtraction. That is, the answer to a subtraction problem involving whole numbers is not always a whole number.

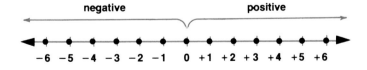

This number line shows how to find $5 - 1$, $5 - 3$, $5 - 5$, and $5 - 7$.

The number line shows that the answer for $5 - 7$ should be *two less than zero*. "Less than" is like taking away, or subtraction. You can write the number *two less than zero* as -2. This is an example of a **negative number**.

To include negative numbers on a number line, extend the line to the left of zero and mark off equal distances. To avoid confusion, name the points to the right of zero using the *positive sign* (+) and to the left of zero using the *negative sign* (−).

0 is neither positive nor negative.

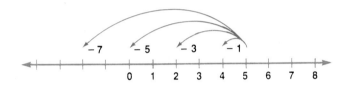

Read "-3" as *negative* 3.   Read "$+4$" as *positive* 4.

Any nonzero number written without a sign is understood to be positive.

The set of numbers used to name the points marked on the number line above is called the set of **integers**. This set can be written $\{. . . , -5, -4, -3, -2, -1, 0, 1, 2, 3, 4, 5, . . .\}$ where . . . means continued indefinitely.

Integers are used to describe many real-life situations.

Football: A loss of 5 yards is −5.
           A gain of 3 yards is +3.

Business: A profit of $540 is +540.
           A loss of $130 is −130.

Weather: 20° above zero is +20.
           8° below zero is −8.

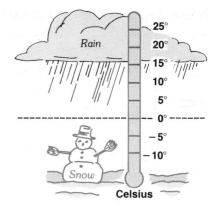

To graph a set of numbers means to locate the points named by those numbers on the number line. The number that corresponds to a point on the number line is called the **coordinate** of the point.

Examples

1 **Name the coordinate of point E.**

The coordinate of E is −1.

2 **Name the set of numbers graphed.**

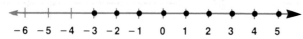

The bold arrow means that the graph continues indefinitely in that direction.

The set of numbers graphed is $\{-3, -2, -1, 0, 1, 2, 3, 4, \ldots\}$.

3 **Graph the set of all integers less than 3.**

Less than 3 means that 3 is not included.

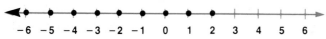

The set of numbers graphed is $\{\ldots, -3, -2, -1, 0, 1, 2\}$.

Exploratory Exercises

State the coordinate of each point.

1. J	2. H	3. A	4. D	5. K	6. L
7. C	8. F	9. B	10. M	11. E	12. I

Name the set of numbers graphed.

13.

14.

15.

16.

17.

18.

Name an integer to describe each situation.

19. 3 yard loss

20. 12 yard gain

21. 650 meters above sea level

22. 189 meters below sea level

23. up 12 floors

24. down 3 floors

25. $450 loss

26. $325 profit

27. 37° above zero

28. 25° below zero

Written Exercises

Name the set of numbers graphed.

29.

30.

31.

32.

33.

34.

35.

36.

Graph each set of numbers on a number line.

37. $\{-1, 1, 3, 5\}$

38. $\{0, 2, 6\}$

39. $\{-2, 2, 4\}$

40. $\{-2, -4, -6\}$

41. $\{-2, 1, 4\}$

42. $\{-3, 0, 2\}$

43. $\{-4, -3, -2, 1\}$

44. $\{-3, -2, 2, 3\}$

45. $\{\ldots, -3, -2, -1, 0\}$

46. $\{-3, -2, -1, \ldots\}$

47. $\{4, 5, 6 \ldots\}$

48. $\{\ldots, -1, 0, 1\}$

49. {all integers less than -2}

50. {all integers greater than 3}

Find the next three numbers in each pattern.

51. 33, 25, 17, ___, ___, ___

52. $-6, -3, 0,$ ___, ___, ___

53. $-21, -16, -11,$ ___, ___, ___

54. 37, 26, 15, ___, ___, ___

55. 1, 3, 9, 27, ___, ___, ___

56. $-8, -4, -2,$ ___, ___, ___

Challenge Exercises

Find the next three numbers in each pattern.

57. 1, 4, 9, 16, ___, ___, ___

58. $1, \frac{1}{2}, \frac{1}{4},$ ___, ___, ___

59. $0, 1, -2, 3, -4, 5,$ ___, ___, ___

60. 1, 1, 2, 3, 5, 8, ___, ___, ___

2-2 Addition on the Number Line

In Saturday's football game the Brookfield Warriors lost 3 yards on the first play. On the next play they gained 7 yards. The diagram at the right shows that the net gain was 4 yards.

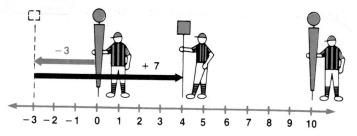

Similarly, a number line is often used to show addition of integers. For example, to find the sum of 4 and −6, follow these steps.

Step 1 Draw an arrow, starting at 0 and going to 4.

Step 2 Starting at 4, draw an arrow 6 units to the left, or −6 units.

Step 3 The second arrow points to the sum, −2.

Note that parentheses are used in the expression $4 + (-6)$ so that the sign of the number is not confused with the operation symbol.

Examples

1 **Find −4 + 6 on a number line.**

$$-4 + 6 = 2$$

2 **Find −2 + (−5) on a number line.**

$$-2 + (-5) = -7$$

3 **Find 4 + (−3) on a number line.**

$$4 + (-3) = 1$$

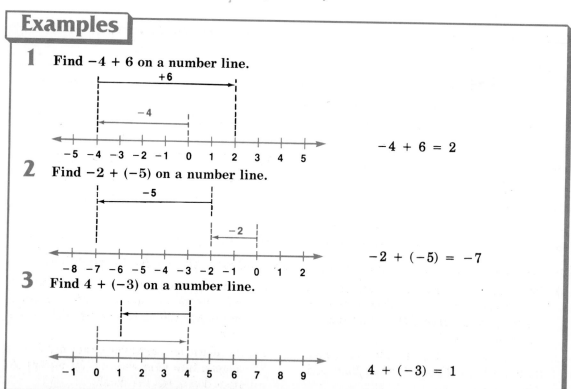

Exploratory Exercises

State the corresponding addition sentence for each diagram.

1.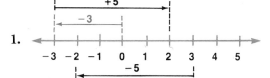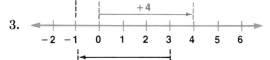

2.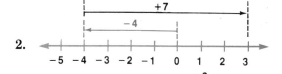

3.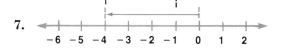

4.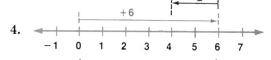

5.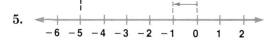

6.

7.

8.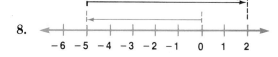

9.

10.

11. How is the meaning of the parentheses in $4 + (-6)$ different from those used in $4 + (5 - 3)$?

Written Exercises

Find each sum. If necessary, use a number line.

12. $7 + 6$	**13.** $9 + 7$	**14.** $-9 + (-7)$	**15.** $-8 + (-11)$
16. $-5 + 0$	**17.** $0 + (-9)$	**18.** $-9 + 4$	**19.** $6 + (-11)$
20. $4 + (-4)$	**21.** $-11 + 11$	**22.** $9 + (-5)$	**23.** $-3 + 9$
24. $2 + (-7)$	**25.** $-11 + 10$	**26.** $9 + (-4)$	**27.** $-14 + 6$
28. $-3 + 12$	**29.** $8 + 3$	**30.** $4 + (-7)$	**31.** $-11 + 5$
32. $7 + (-7)$	**33.** $-491 + 491$	**34.** $0 + 12$	**35.** $-13 + 0$
36. $-6 + (-11)$	**37.** $-13 + (-9)$	**38.** $-15 + (-11)$	**39.** $-18 + (-24)$

For each problem, write an open sentence using addition. Then solve the problem.

40. One night in Burlington, Vermont, the temperature was $-24°C$. During the day the temperature rose $12°C$. What was the temperature then?

41. A traffic helicopter ascended to a height of 400 meters to monitor traffic. Then it descended 300 meters. How high was it then?

42. An elevator went up to the 36th floor. Then it came down 11 floors. At what floor was it then?

43. A scuba diver was exploring at a depth of 75 meters. She then went up 30 meters. At what depth was she then?

2-3 Adding Integers

It would be inconvenient to draw a number line to find sums such as $-713 + 425$. A mathematical concept called *absolute value* can be used to find sums of integers without using a number line.

Looking at -4 and 4 on the number line, you can see that they are different numbers. However, they are the same number of units from 0. The numbers -4 and 4 have the same **absolute value**.

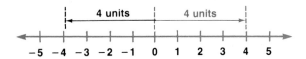

The absolute value of a number is the number of units it is from 0 on the number line.	*Definition of Absolute Value*

$|-4|$ represents the absolute value of -4.

$$|-4| = 4$$

$|4|$ represents the absolute value of 4.

$$|4| = 4$$

Absolute value can be used to find the sum of integers. Consider the following facts learned from the number line.

$$6 + 7 = 13 \qquad\qquad -6 + (-7) = -13$$

Notice that the sign of each addend is positive. The sum is positive.

Notice that the sign of each addend is negative. The sum is negative.

These and other similar examples suggest the following rule.

To add integers with the same sign, add their absolute values. Give the sum the same sign as the addends.	*Adding Integers with the Same Sign*

Example

1 Add $-5 + (-7)$.

$$\begin{aligned}
-5 + (-7) &= -(|-5| + |-7|) &&\textit{Both numbers are negative.}\\
&= -(5 + 7) &&\textit{Add absolute values.}\\
&= -12 &&\textit{The sum is negative.}
\end{aligned}$$

Study these facts found using a number line.

$$4 + (-9) = -5 \qquad\qquad -5 + 8 = 3$$

$$|4| = 4 \text{ and } |-9| = 9. \qquad\qquad |-5| = 5 \text{ and } |8| = 8.$$

Notice that $9 - 4$ is 5. Which integer, 4 or -9, has the greater absolute value? The sum of 4 and -9 has the same sign as this addend.

Notice that $8 - 5$ is 3. Which integer, -5 or 8, has the greater absolute value? The sum of -5 and 8 has the same sign as this addend.

These and other similar examples suggest the following rule.

To add integers with different signs, subtract the lesser absolute value from the greater absolute value. Give the result the same sign as the addend with the greater absolute value.	*Adding Integers with Different Signs*

Examples

2 **Add $3 + (-7)$.**

$$3 + (-7) = -(|-7| - |3|) \qquad \text{-7 has the greater absolute value.}$$
$$= -(7 - 3) \qquad\qquad \text{Subtract absolute values.}$$
$$= -4 \qquad\qquad\qquad \text{The sum is negative.}$$

3 **Evaluate $k + t$ if $k = -9$ and $t = 16$.**

$$k + t = -9 + 16 \qquad\qquad \text{Substitute -9 for k and 16 for t.}$$
$$= +(|16| - |-9|) \qquad \text{$+16$ has the greater absolute value.}$$
$$= +(16 - 9) \qquad\qquad \text{Subtract absolute values.}$$
$$= 7 \qquad\qquad\qquad\quad \text{The sum is positive.}$$

Exploratory Exercises

State the absolute value of each integer.

1. $+8$
2. $+14$
3. -6
4. -24
5. 17
6. 321
7. -21
8. -271
9. 0
10. 59

State the sign of each sum.

11. $-6 + (-11)$
12. $-14 + (-15)$
13. $+8 + (+9)$
14. $13 + 21$
15. $-8 + (+16)$
16. $-11 + (+7)$
17. $+8 + (-35)$
18. $-7 + (+27)$
19. $18 + (-3)$
20. $17 + (-38)$
21. $-17 + 12$
22. $-27 + 31$

Written Exercises

Find each sum.

23. $+7 + (+9)$ **24.** $18 + 22$ **25.** $-6 + (-13)$ **26.** $-13 + (-8)$

27. $-3 + (+16)$ **28.** $14 + (-9)$ **29.** $-5 + 31$ **30.** $-10 + 4$

31. $-18 + (-11)$ **32.** $-23 + (-47)$ **33.** $82 + (-78)$ **34.** $43 + (-67)$

35. $38 + (-47)$ **36.** $-25 + 47$ **37.** $63 + (-47)$ **38.** $-21 + 52$

39. $102 + (-12)$ **40.** $-104 + 16$ **41.** $-93 + 39$ **42.** $97 + (-79)$

Evaluate each expression if $a = -5$, $k = 3$, and $m = -6$.

43. $a + 13$ **44.** $-5 + k$ **45.** $15 + m$ **46.** $8 + a$

47. $k + (-18)$ **48.** $m + (-31)$ **49.** $m + 6$ **50.** $m + 17$

51. $|m|$ **52.** $|a|$ **53.** $|m + 4|$ **54.** $|7 + a|$

55. $|-8 + k|$ **56.** $|a + k|$ **57.** $|k + m|$ **58.** $|k| + |m|$

59. $-|3 + a|$ **60.** $-|k + 8|$ **61.** $-|-24 + m|$ **62.** $-|a + (-11)|$

Find each sum.

63. $+374 + 239$ **64.** $521 + 124$ **65.** $-374 + (-165)$ **66.** $-179 + (-826)$

67. $-582 + 379$ **68.** $573 + (-336)$ **69.** $1982 + (-1482)$ **70.** $-1492 + 876$

71. $|-285 + (-641)|$ **72.** $|-931| + (-643)$ **73.** $|871 + (-284)|$

74. $-|-423 + (-148)|$ **75.** $-197 + |-483|$ **76.** $-||-843| + |-231||$

Complete each statement.

77. If $n > 0$, then $|n| = \underline{\ ?\ }$.   **78.** If $n < 0$, then $|n| = \underline{\ ?\ }$.

79. If $n > 0$, then $|-n| = \underline{\ ?\ }$.   **80.** If $n < 0$, then $|-n| = \underline{\ ?\ }$.

---

mini-review

Find each sum. Use a number line if necessary.

1. $-13 + (-23)$ **2.** $-842 + 842$ **3.** $44 + (+29)$ **4.** $-161 + 182$

5. Graph {all integers less than 2} on a number line.

6. State the property expressed in the equation $5(p + 2q) = 5p + 10q$.

Evaluate each expression if $a = -3$, $d = 5$, and $r = -7$.

7. $d^2 - 7|a| + 2r^2$ **8.** $-7 + |r|$ **9.** $r + 6d \div 3$

10. $|a + r|$ **11.** $3(a + r + 2d) - d^3$ **12.** $|r| + |a|$

13. Write an open sentence using addition and solve. *An elevator in a hotel went up to the 47th floor and then came down 23 floors. At what floor did the elevator stop?*

14. Write an algebraic expression for *five times the sum of the square of* x *and the cube of* y.

2-4 Adding Rational Numbers

Some points on a number line cannot be named by integers. For example, the number line below is separated into fourths to show a sample of numbers that appear in the form of common fractions and decimals. The numbers shown on this number line are all **rational numbers**.

$$-\frac{6}{4} \quad -\frac{5}{4} \quad -1 \quad -\frac{3}{4} \quad -\frac{2}{4} \quad -\frac{1}{4} \quad 0 \quad \frac{1}{4} \quad \frac{2}{4} \quad \frac{3}{4} \quad 1 \quad \frac{5}{4} \quad \frac{6}{4}$$

$$-1.5 \quad -1.25 \quad -1 \quad -0.75 \quad -0.5 \quad -0.25 \quad 0 \quad 0.25 \quad 0.5 \quad 0.75 \quad 1 \quad 1.25 \quad 1.5$$

A rational number is a number that can be expressed in the form $\frac{a}{b}$, where a and b are integers and b is not equal to zero.	*Definition of a Rational Number*

Examples of rational numbers expressed in the form $\frac{a}{b}$ are shown in this chart.

Rational Number	Form $\frac{a}{b}$
3	$\frac{3}{1}$
$-2\frac{3}{4}$	$-\frac{11}{4}$
0.125	$\frac{1}{8}$
0	$\frac{0}{1}$
$0.33\overline{3}$	$\frac{1}{3}$

Notice that terminating decimals and repeating decimals can be expressed as fractions.

Addition of rational numbers can be represented on a number line by following the same steps as addition of integers.

Consider the addition $\frac{1}{4} + \left(-\frac{3}{8}\right)$. First replace $\frac{1}{4}$ with $\frac{2}{8}$. Then add $\frac{2}{8}$ and $-\frac{3}{8}$.

Remember, to add fractions, the denominators must be the same.

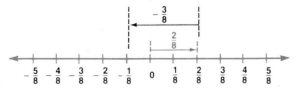

The sum of $\frac{1}{4}$ and $-\frac{3}{8}$ is $-\frac{1}{8}$.

Using the number line to find sums such as $\frac{1}{4} + \left(-\frac{3}{8}\right)$ is inconvenient. However, the rules used to add integers can also be used to add rational numbers.

Examples

1 Add $-\frac{3}{8} + \left(-\frac{5}{16}\right)$.

$-\frac{3}{8} + \left(-\frac{5}{16}\right) = -\frac{6}{16} + \left(-\frac{5}{16}\right)$ *Replace $-\frac{3}{8}$ with $-\frac{6}{16}$.*

$= -\left(\left|-\frac{6}{16}\right| + \left|-\frac{5}{16}\right|\right)$ *Since the numbers have the same sign, add their absolute values.*

$= -\left(\frac{6}{16} + \frac{5}{16}\right)$

$= -\frac{11}{16}$ *The sum is negative.*

2 Add $1.354 + (-0.765)$.

$1.354 + (-0.765) = +(|1.354| - |-0.765|)$ *1.354 has the greater absolute value.*

$= +(1.354 - 0.765)$ *Since the addends have different signs, find the difference of their absolute values. The sum is positive.*

$= 0.589$

3 Add $-1\frac{2}{3} + 4\frac{3}{4}$.

$-1\frac{2}{3} + 4\frac{3}{4} = -1\frac{8}{12} + 4\frac{9}{12}$ *Replace $-1\frac{2}{3}$ with $-1\frac{8}{12}$ and $4\frac{3}{4}$ with $4\frac{9}{12}$.*

$= +\left(\left|4\frac{9}{12}\right| - \left|-1\frac{8}{12}\right|\right)$ *$4\frac{9}{12}$ has the greater absolute value.*

$= +\left(4\frac{9}{12} - 1\frac{8}{12}\right)$ *Since the addends have different signs, find the difference of their absolute values.*

$= 3\frac{1}{12}$ *The sum is positive.*

---

1. If two numbers have the same sign, add their absolute values. Give the sum the same sign as the addends.

2. If two numbers have different signs, subtract the lesser absolute value from the greater absolute value. Give the result the same sign as the addend with the greater absolute value.

Summary of Rules for Addition of Rational Numbers

Exploratory Exercises

State the absolute value of each number.

1. $+\frac{3}{4}$ **2.** $+\frac{9}{8}$ **3.** -1.76 **4.** -2.53 **5.** $-\frac{3}{11}$

6. -16.7 **7.** 1.82 **8.** $-\frac{7}{16}$ **9.** $+\frac{3}{82}$ **10.** 1.48

State the sign of each sum.

11. $-\frac{3}{4} + \frac{7}{8}$

12. $\frac{15}{16} + \left(-\frac{13}{16}\right)$

13. $-1.354 + 1.265$

14. $37.42 + (-45.36)$

15. $-\frac{12}{31} + \frac{13}{62}$

16. $-\frac{9}{13} + \left(-\frac{3}{13}\right)$

17. $-394.1 + 427.6$

18. $-85.32 + 76.3$

19. $-\frac{5}{12} + \left(-\frac{5}{24}\right)$

20. $\frac{2}{5} + \left(-\frac{2}{7}\right)$

21. $-0.0034 + 0.034$

22. $-0.0759 + 0.4$

Written Exercises

Find each sum.

23. $-\frac{11}{9} + \left(-\frac{7}{9}\right)$

24. $\frac{17}{21} + \left(-\frac{13}{21}\right)$

25. $\frac{5}{11} + \left(-\frac{6}{11}\right)$

26. $-\frac{8}{13} + \left(-\frac{11}{13}\right)$

27. $-\frac{7}{12} + \frac{5}{12}$

28. $-\frac{9}{28} + \frac{13}{28}$

29. $4.57 + (-3.69)$

30. $-4.8 + 3.2$

31. $-1.7 + (-3.9)$

32. $-2.31 + 7.62$

33. $-38.9 + 24.2$

34. $-0.007 + 0.06$

35. $\frac{3}{4} + \left(-\frac{7}{12}\right)$

36. $-\frac{2}{7} + \frac{3}{14}$

37. $-\frac{3}{8} + \frac{5}{24}$

38. $-\frac{5}{6} + \left(-\frac{7}{12}\right)$

39. $-\frac{1}{8} + \left(-\frac{5}{2}\right)$

40. $\frac{2}{3} + \left(-\frac{2}{9}\right)$

41. $-1.543 + 2.165$

42. $-3.948 + 4.826$

43. $-0.376 + (-0.289)$

44. $-0.006 + 0.0052$

45. $-0.0005 + (-0.3)$

46. $0.0007 + (-0.001)$

47. $-\frac{3}{5} + \frac{5}{6}$

48. $\frac{3}{8} + \left(-\frac{7}{12}\right)$

49. $-\frac{4}{15} + \frac{3}{4}$

50. $-\frac{9}{4} + \left(-\frac{4}{3}\right)$

51. $\frac{17}{5} + \left(-\frac{13}{4}\right)$

52. $-\frac{7}{15} + \left(-\frac{5}{12}\right)$

Write an addition sentence. Then solve.

53. Jenelle bought a pair of shoes that cost $12.87 with tax. If she gave the clerk a twenty-dollar bill, how much would she receive in change?

54. Ted invested his savings in 50 shares of stock. Over the course of three days, the value of the stock gained $3\frac{1}{8}$ points, lost $\frac{3}{4}$ of a point, and gained another $\frac{1}{8}$ of a point. What was the net gain in stock value?

Challenge Exercises

Find each sum. Express each result in decimal form.

55. $-\frac{4}{5} + (-3.8)$

56. $-0.37 + \left(-\frac{21}{8}\right)$

57. $-5\frac{3}{4} + 6.25$

58. $-8.66 + 6\frac{7}{8}$

59. $7.43 + \left(-\frac{9}{4}\right)$

60. $4\frac{7}{8} + (-3.754)$

61. $-11\frac{7}{16} + 7.225$

62. $-3.276 + \left(-\frac{15}{8}\right)$

Problem Solving

Guess and Check

Many different strategies can be used to solve problems. One strategy is to write an equation. Another important strategy is called *guess and check*. To use this strategy, guess the answer to the problem, then check whether the guess is correct. If the first guess is incorrect, guess again until you find the right answer. Often, the results of one guess can help you make a better guess. Always keep a record of your guesses so you do not make the same guess twice.

Sometimes there will not be one right answer. You may have to find the best answer.

Example   **Insert parentheses so that a true equation results.**
$$10 - 4 \cdot 2 - 1 = 3$$

Copy the expression to the left of the equals sign.
Then insert parentheses and evaluate the expression.

$10 - 4 \cdot (2 - 1) = 6$ *Try again since the value does not equal 3.*
$(10 - 4) \cdot 2 - 1 = 11$ *Try again.*
$10 - (4 \cdot 2 - 1) = 3$ *Correct!*

Exercises

Use the guess-and-check strategy to solve each problem.

1. When a certain number is multiplied by itself the product is 2916. Find the number.

2. Find an odd number between 10 and 25 that is divisible by 3, but is not divisible by 5.

3. Using each of the digits 1, 2, . . . , 6 only once, find two whole numbers whose product is as large as possible.

4. If it costs a nickel each time you cut and weld a link, what is the minimum cost to make a chain out of 5 links?

5. Paper plates can be purchased in packages of 15 or 25. Joe bought 7 packages and got 125 plates. How many packages of 25 did he buy?

6. The cube of a certain whole number is close to 4000. Find the number.

7. Copy the figure at the right. Then fill in the digits 1, 2, . . . , 8 in such a way that no two consecutive numbers are in boxes that touch at a point or side.

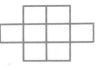

8. Gina and Jim raise cats and birds. They counted all the heads and got 10. They counted all the feet and got 34. How many birds and cats do they have?

For each of the following, insert parentheses so that a true equation results.

9. $4 \cdot 5 - 2 + 7 = 19$

10. $25 - 4 \cdot 2 + 3 = 5$

11. $3 + 5 \cdot 8 - 2 = 48$

12. $3 + 6 \cdot 4 \cdot 2 = 54$

2-5 More About Addition

Previously you have added pairs of numbers. To add three or more numbers, first group the numbers in pairs. Use the Commutative and Associative Properties to rearrange the addends if necessary.

Examples

1 **Add $(-4) + 5 + (-6)$.**

$$(-4) + 5 + (-6) = (-4 + 5) + (-6) \qquad \text{Group two of the addends.}$$
$$= 1 + (-6)$$
$$= -5 \qquad \text{Do you get the same result if you find the sum } -4 + [5 + (-6)]?$$

2 **Add $-\frac{4}{3} + \frac{5}{8} + \left(-\frac{7}{3}\right)$.**

$$-\frac{4}{3} + \frac{5}{8} + \left(-\frac{7}{3}\right) = \left[-\frac{4}{3} + \left(-\frac{7}{3}\right)\right] + \frac{5}{8} \qquad \text{Commutative and Associative Properties of Addition}$$
$$= -\frac{11}{3} + \frac{5}{8}$$
$$= -\frac{88}{24} + \frac{15}{24}$$
$$= -\frac{73}{24}$$
$$= -3\frac{1}{24}$$

3 **Add $28.32 + (-56.17) + 32.41 + (-75.13)$.**

Group the positive numbers and group the negative numbers.

$$28.32 + (-56.17) + 32.41 + (-75.13) = 28.32 + 32.41 + (-56.17) + (-75.13)$$
$$= (28.32 + 32.41) + [(-56.17) + (-75.13)]$$
$$= 60.73 + (-131.30)$$
$$= -70.57$$

You can use the Distributive Property and what you know about rational numbers to simplify expressions that have like terms.

Example

4 **Simplify $-5x + a + 7x + (-4a)$.**

$$-5x + a + 7x + (-4a) = (-5x + 7x) + [1a + (-4a)] \qquad \text{Group the like terms.}$$
$$= (-5 + 7)x + [1 + (-4)]a \qquad \text{Use the Distributive Property.}$$
$$= 2x + (-3a)$$

Exploratory Exercises

Find each sum.

1. $2 + (-6) + 4$

2. $5 + 2 + (-5)$

3. $0.6 + (-0.3) + (-0.4)$

4. $-0.2 + 0.4 + (-0.7)$

5. $\frac{1}{2} + \left(-\frac{1}{4}\right) + \frac{3}{4}$

6. $-\frac{1}{5} + \frac{3}{5} + \left(-\frac{4}{5}\right)$

7. $-4m + 8m$

8. $-8x + (-10x)$

9. $r + (-2r) + 5r$

10. $-k + 6k + (-3k)$

11. $-b + y + 4b + (-5y)$

12. $s + (-5t) + 9t + 8s$

Written Exercises

Find each sum.

13. $4 + (-12) + (-18)$

14. $7 + (-11) + 32$

15. $8 + (-15) + 13$

16. $-41 + (-78) + 51$

17. $83 + (-19) + 16$

18. $16 + (-9) + (-94)$

19. $-3a + 12a + (-14a)$

20. $5y + (-12y) + (-21y)$

21. $14b + (-21b) + 37b$

22. $5x + (-21x) + 29x$

23. $-3z + (-17z) + (-18z)$

24. $9m + 43m + (-16m)$

25. $-\frac{3}{4} + \frac{5}{12} + \left(-\frac{5}{4}\right)$

26. $\frac{7}{3} + \left(-\frac{5}{6}\right) + \left(-\frac{2}{3}\right)$

27. $-\frac{2}{7} + \frac{3}{14} + \frac{3}{7}$

28. $-\frac{3}{5} + \frac{6}{7} + \left(-\frac{2}{35}\right)$

29. $\frac{5}{8} + \left(-\frac{3}{16}\right) + \left(-\frac{7}{24}\right)$

30. $\frac{3}{4} + \left(-\frac{5}{8}\right) + \frac{3}{32}$

31. $12 + (-17) + 36 + (-45)$

32. $81 + (-31) + (-9) + 62$

33. $-75 + 47 + 32 + (-16)$

34. $28 + (-56) + 32 + (-75)$

35. $6.7 + (-8.1) + (-7.3)$

36. $-7.6 + 1.8 + (-3.5)$

37. $-37.12 + 42.18 + (-12.6)$

38. $-4.13 + (-5.18) + 9.63$

39. $-7.9 + (-5.3) + (-4.2)$

40. $-9.7 + 5.3 + 4.2$

41. $-14a + 36k + 12k + (-83a)$

42. $16b + (-22xy) + 31b + (-36xy)$

43. $13mp + (-3ps) + 76mp + (-21ps)$

44. $-17px + 22bg + 35px + (-37bg)$

45. $\begin{array}{r} 12a \\ +(-9a) \\ \hline \end{array}$

46. $\begin{array}{r} -15m \\ +\quad 6m \\ \hline \end{array}$

47. $\begin{array}{r} -23w \\ +\quad 47w \\ \hline \end{array}$

48. $\begin{array}{r} -13c \\ +(-28c) \\ \hline \end{array}$

49. $\begin{array}{r} 5.8k \\ +(-3.6k) \\ \hline \end{array}$

50. $\begin{array}{r} -7.9s \\ +(-3.8s) \\ \hline \end{array}$

51. $\begin{array}{r} -0.23x \\ +(-0.5\ x) \\ \hline \end{array}$

52. $\begin{array}{r} 0.81h \\ +(-0.93h) \\ \hline \end{array}$

Solve each problem.

53. Julie Thomas shot rounds in a recent golf tournament in which her scores relative to par were -3, $+2$, -4, and -1. What was her score for the tournament relative to par?

54. Last week, the following day-to-day changes in the noon temperatures were recorded: Sun., $+3$; Mon., -7; Tues., -5; Wed., $+4$; Thurs., $+6$; Fri., -3; Sat., -8. What was the net change for the week?

55. On Tuesday, Johnny Lomax wrote checks for $35.76 and $41.32. On Wednesday, he deposited $135.59. Friday he wrote a check for $63.17. What was the net increase or decrease in his account for the week?

56. Last week, the following day-to-day changes for the Dow-Jones stock averages were recorded: Mon., $+5\frac{3}{8}$; Tues., $-6\frac{1}{4}$; Wed., $+11\frac{1}{8}$; Thurs., $+3\frac{5}{8}$; Fri., $-7\frac{1}{2}$. What was the net change for the week?

2-6 Subtraction

Symmetry can be seen in many things in nature. Symmetry occurs when a line can be drawn through an object so that one side of the figure is like a mirrored reflection of the other side. An oak leaf is an example of symmetry in nature. Note how points on one side of the leaf can be paired with points opposite the central vein of the leaf. The idea of pairing opposites is helpful in understanding subtraction.

Notice on the number line that 3 and -3 can be paired opposite each other with respect to 0. Both numbers are the same distance from 0.

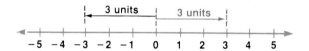

What is the result when you add two numbers such as 3 and -3?

$$3 + (-3) = 0 \qquad -7 + 7 = 0 \qquad -\tfrac{4}{7} + \tfrac{4}{7} = 0 \qquad 19.3 + (-19.3) = 0$$

If the sum of two numbers is 0, the numbers are called **additive inverses** or **opposites**.

-3 is the additive inverse, or opposite, of 3.

7 is the additive inverse, or opposite, of -7.

$\tfrac{4}{7}$ is the additive inverse, or opposite, of $-\tfrac{4}{7}$.

-19.3 is the additive inverse, or opposite, of 19.3.

Is 3 the additive inverse of -3?

Zero is its own opposite.

These and other similar examples suggest the following property.

For every number a, $a + (-a) = 0$.	*Additive Inverse Property*

Additive inverses are used in the subtraction of rational numbers. Observe the following patterns.

If a is negative, then $(-a)$ is positive.

Subtraction	*Addition*	*Subtraction*	*Addition*

additive inverses

$$7 - 3 = 4 \qquad\qquad 7 + (-3) = 4$$

same result

additive inverses

$$8 - (-2) = 10 \qquad\qquad 8 + 2 = 10$$

same result

It appears that subtracting a number is equivalent to adding its additive inverse. These and other similar examples suggest the following rule for subtracting rational numbers.

> **To subtract a rational number, add its additive inverse. For rational numbers a and b, $a - b = a + (-b)$.**

Subtraction Rule

Examples

1 **Subtract $5 - (-8)$.**

$$5 - (-8) = 5 + (+8) \qquad \textit{To subtract } -8, \textit{ add } +8.$$
$$= 13$$

2 **Subtract $6 - 11$.**

$$6 - 11 = 6 + (-11) \qquad \textit{To subtract } 11, \textit{ add } -11.$$
$$= -5$$

3 **Subtract $-4 - (-7)$.**

$$-4 - (-7) = -4 + (+7) \qquad \textit{To subtract } -7, \textit{ add } +7.$$
$$= 3$$

4 **Subtract $7.32 - (-6.85)$.**

$$7.32 - (-6.85) = 7.32 + (+6.85) \qquad \textit{To subtract } -6.85, \textit{ add } +6.85.$$
$$= 14.17$$

5 **Subtract $-\frac{7}{8} - \left(-\frac{3}{16}\right)$.**

$$-\frac{7}{8} - \left(-\frac{3}{16}\right) = -\frac{7}{8} + \frac{3}{16} \qquad \textit{To subtract } -\frac{3}{16}, \textit{ add } +\frac{3}{16}.$$
$$= -\frac{14}{16} + \frac{3}{16}$$
$$= -\frac{11}{16}$$

The Distributive Property is used to simplify expressions that contain like terms as shown in the next example.

Example

6 Simplify $7x - 12x$.

$$7x - 12x = 7x + (-12x) \qquad \textit{To subtract 12x, add } -12x.$$
$$= [7 + (-12)]x \qquad \textit{Distributive Property}$$
$$= -5x$$

Exploratory Exercises

State the additive inverse of each of the following.

1. $+6$
2. -6
3. 5
4. -14
5. -13
6. a
7. $-a$
8. $-2y$
9. 0
10. $4x$
11. 3.7
12. -7.4
13. $\frac{3}{7}$
14. $-\frac{4}{9}$
15. $-\frac{8}{17}$

State the number named.

16. $-(-5)$
17. $-(+7)$
18. $-(4)$
19. $-(-13)$
20. $-(-36)$
21. $-(-7)$
22. $-(18)$
23. $-(-16)$
24. $-(-43)$
25. $-(+56)$
26. $-(-1.8)$
27. $-(+7.1)$
28. $-\left(\frac{8}{11}\right)$
29. $-\left(-\frac{3}{4}\right)$
30. $-(6.5)$

State an addition expression.

31. $8 - 13$
32. $-18 - 7$
33. $-17 - 8$
34. $11 - (-2)$
35. $-1.7 - (-1.5)$
36. $\frac{3}{8} - \left(-\frac{2}{3}\right)$
37. $9y - 3y$
38. $-8a - (-7a)$

Written Exercises

Find each difference.

39. $27 - 19$
40. $47 - 32$
41. $52 - 37$
42. $8 - 13$
43. $13 - 31$
44. $27 - 43$
45. $17 - (-23)$
46. $29 - (-25)$
47. $18 - (-34)$
48. $-21 - (-14)$
49. $-23 - (-12)$
50. $-47 - 35$
51. $\frac{3}{5} - \left(-\frac{1}{5}\right)$
52. $\frac{1}{6} - \frac{2}{3}$
53. $-\frac{1}{2} - \left(-\frac{3}{4}\right)$
54. $3.81 - (-4.65)$
55. $0 - 21$
56. $69 - (-95)$
57. $7.3 - (-4.2)$
58. $-4.5 - 8.6$
59. $-67.1 - (-38.2)$
60. $89.3 - (-14.2)$
61. $-72.5 - 81.3$
62. $-5\frac{7}{8} - 2\frac{3}{4}$
63. $\begin{array}{r} 19 \\ -36 \\ \hline \end{array}$
64. $\begin{array}{r} 61 \\ -(-43) \\ \hline \end{array}$
65. $\begin{array}{r} -5.2 \\ -(-3.8) \\ \hline \end{array}$
66. $\begin{array}{r} -1.7 \\ -(+3.9) \\ \hline \end{array}$
67. $19m - 12m$
68. $8h - 23h$
69. $-18p - 4p$
70. $-21a - 3a$
71. $24b - (-9b)$
72. $-31x - (-33x)$
73. $-52z - (-17z)$
74. $41y - (-41y)$

Evaluate each expression.

75. $m - 7$, if $m = 5$

76. $a - 12$, if $a = -8$

77. $p - 14$, if $p = 72$

78. $y - 0.5$, if $y = -0.8$

79. $a - (-7)$, if $a = 1.9$

80. $b - (-0.5)$, if $b = -13$

81. $h - (-1.3)$, if $h = -18$

82. $k - (-12)$, if $k = 2.7$

83. $w - 3.7$, if $w = -1.8$

84. $\frac{3}{5} - n$, if $n = -\frac{7}{5}$

85. $\frac{11}{2} - m$, if $m = -\frac{5}{2}$

86. $\frac{8}{13} + a$, if $a = +\frac{9}{13}$

87. $\frac{11}{4} - x$, if $x = \frac{27}{8}$

88. $-\frac{8}{5} - y$, if $y = \frac{12}{5}$

89. $-\frac{12}{7} - z$, if $z = \frac{16}{21}$

90. $-\frac{13}{6} - k$, if $k = -\frac{11}{12}$

91. $-\frac{18}{17} - q$, if $q = \frac{31}{34}$

92. $-\frac{33}{38} - r$, if $r = -\frac{16}{19}$

mini-review

Find each sum or difference.

1. $\frac{3}{8} + \left(-\frac{7}{11}\right)$

2. $-\frac{4}{15} + \frac{3}{5}$

3. $-\frac{9}{2} + \left(-\frac{3}{4}\right)$

4. $-0.005 + 0.023$

5. $-0.34 + (-0.034)$

6. $-0.236 + 0.1234$

7. $11v - 3v - 4v$

8. $6d - (-6d) - 5d$

9. $27f - 43f - 5g$

Evaluate each expression.

10. $a - 2$, if $a = -3$

11. $13 - |b|$, if $b = -\frac{7}{3}$

12. $\left|\frac{2}{3}\right| - t$, if $t = -\frac{2}{3}$

13. $|r| - 0.123$, if $r = 3.478$

14. $ab - c^2 + 3bc \div a$, if $a = 3$, $b = 4$, and $c = 5$

15. $t(m + n)^2$, if $m = 7$, $n = 6$, and $t = 2$

Write an open sentence using addition and solve.

16. A submarine descended to a depth of 432 meters and then rose 189 meters. How far below the surface of the water was the submarine?

17. Courtney was trying to save money for a ten-speed bicycle. She opened a savings account with money she earned baby-sitting. Her first deposit was $75. She then withdrew $37 for a pair of shoes. What is her new balance?

18. Write an algebraic expression for the verbal expression *the difference of a and b divided by the product of a and b.*

19. Write a verbal expression for the algebraic expression $3 \cdot x \cdot x \cdot y(5x + 2y)$.

1. Write a mathematical expression for twice a number decreased by 17.

2. Write a mathematical expression for the square of a number increased by 21 times the number.

3. Use exponents to rewrite the expression $(3)(3)(3)(a)(a)(b)(b)(b)$.

4. Write $4^4 a^5$ without using exponents.

Evaluate.

5. $3 + 2(4) - \frac{10}{2}$

6. $\frac{(4 + 8)^2}{4(2)} + 3(2)$

7. $4(5a + 3b^2)$, if $a = 6$ and $b = 4$

8. $3ab - 4c$, if $a = 2$, $b = 3$, and $c = \frac{1}{2}$

Solve.

9. $x = \frac{3 + 7}{10}$

10. $t = 3 \cdot 4 - 2$

11. What number is called the additive identity?

State the property shown.

12. $(3 + 8)4 = 11(4)$

13. $(7x)a = (x7)a$

14. $y + 8 = y + 8$

15. $6(7a + 3) = 42a + 18$

16. $y \cdot 0 = 0$

Simplify.

17. $4mn + 12mn$

18. $16x + 12xy - 13x$

19. $3(a + 0.2b) - 0.4b$

20. $4(x + 3y) + 2(2x + y)$

21. $3a + 4b + 17a + 5b$

22. $5(a + 2b) + 6a + 7b$

Find each absolute value.

23. $|-215|$ 24. $-|-13 + 4|$

Write each sentence as an equation.

25. The sum of y and the cube of n is equal to x.

26. The square of a side (s) is equal to the area (A).

27. Name the set of numbers graphed.

28. $\{-3, -1, 2, 5\}$

29. $\{\ldots, -6, -5, -4\}$

30. Find $4 + (-6)$ on a number line.

Find each sum or difference.

31. $36 + (-73)$

32. $-8 + (-21)$

33. $-9.3 + 4.7$

34. $-\frac{1}{2} + \left(-\frac{1}{6}\right) + \left(-\frac{2}{3}\right)$

35. $-21 + (-28)$

36. $3.6 - 7.9$

37. $-3a + 5m + 16m + (-25a)$

38. $-7t - 16t$

39. $-4t - 3t + 7a - 5a$

40. $4(3r + 5y) - 6r$

41. $13h - 12j - 34h$

42. $23t - 67y + 42y$

43. $6(3r - 4t) + 3(4t - 3r)$

44. $17y - 4t - 8y - 3t$

45. The formula for finding the Celsius temperature (C) when you know the Fahrenheit temperature (F) is $C = \frac{5}{9}(F - 32)$. Find the Celsius temperature when the Fahrenheit temperature is 59°.

Define the variable and write an equation.

46. Eighty-five decreased by one-half of a number is 57.

2-7 Solving Equations Involving Addition

A scale, as shown below on the left, is in balance when each side holds equal weights. If you add weight to only one side, as shown on the right, then the scale is no longer in balance.

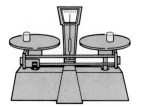 

However, if you add the same weight to each side, the scale will balance.

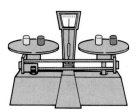

Think of an equation as a scale in balance. If the same number is added to each side of the equation, then the result is an equivalent equation. **Equivalent equations** are equations that have the same solution.

$11 = x + 3$	*The solution to this equation is 8.*
$11 + 5 = (x + 3) + 5$	*Add 5 to each side.*
$16 = x + 8$	*The solution to this equation is also 8.*

The property that is used to add the same number to each side of an equation is called the **Addition Property of Equality**.

For any numbers a, b, and c, if $a = b$, then $a + c = b + c$.	***Addition Property of Equality***

Also, if the same number is *subracted* from each side of an equation, the result is an equivalent equation.

$$11 = x + 3$$
$$11 - 3 = (x + 3) - 3 \qquad \textit{Subtract 3 from each side.}$$
$$8 = x \qquad \textit{Why is this equation equivalent to 11 = x + 3?}$$

The property that is used to subtract the same number from each side of an equation is called the **Subtraction Property of Equality.**

For any numbers *a*, *b*, and *c*, if *a* = *b*, then *a* − *c* = *b* − *c*.	*Subtraction Property of Equality*

To *solve an equation* means to isolate the variable having a coefficient of one on one side of the equation. Both the Addition and Subtraction Properties of Equality can be used to solve equations involving addition.

Remember, x means 1 · x. The coefficient of x is 1.

Examples

1 **Solve $r + 16 = -7$.**

$$r + 16 = -7 \qquad \textit{Use the Addition Property of Equality and the additive inverse.}$$
$$r + 16 + (-16) = -7 + (-16) \qquad \textit{Add −16 to each side.}$$
$$r + 0 = -23 \qquad \textit{The sum of 16 and −16 is 0.}$$
$$r = -23$$

To check that −23 is the solution, substitute −23 for r in r + 16 = −7.

Check: $r + 16 = -7$
$$-23 + 16 \overset{?}{=} -7$$
$$-7 = -7 \quad ✔ \qquad \textit{The check verifies that the solution to the equation is −23.}$$

2 **Solve $k + 15 = -6$.**

$$k + 15 = -6 \qquad \textit{Use the Subtraction Property of Equality.}$$
$$k + 15 - 15 = -6 - 15 \qquad \textit{Subtract 15 from each side.}$$
$$k + 0 = -21$$
$$k = -21$$

Check: $k + 15 = -6$
$$-21 + 15 \overset{?}{=} -6$$
$$-6 = -6 \quad ✔ \qquad \textit{The check verifies that the solution to the equation is −21.}$$

3 Solve $m + \left(-\frac{3}{4}\right) = -\frac{1}{2}$.

$$m + \left(-\frac{3}{4}\right) + \frac{3}{4} = -\frac{1}{2} + \frac{3}{4}$$

$$m + 0 = -\frac{2}{4} + \frac{3}{4}$$

$$m = \frac{1}{4}$$

Check: $m + \left(-\frac{3}{4}\right) = -\frac{1}{2}$

$$\frac{1}{4} + \left(-\frac{3}{4}\right) \stackrel{?}{=} -\frac{1}{2}$$

$$-\frac{1}{2} = -\frac{1}{2} \quad \checkmark$$

The solution is $\frac{1}{4}$.

Exploratory Exercises

State the number you add to each side of the equation to solve it.

1. $y + 21 = -7$ **2.** $13 + x = -16$ **3.** $y + (-5) = 11$

4. $z + (-9) = 34$ **5.** $-10 + k = 34$ **6.** $y + 13 = 45$

State the number you subtract from each side of the equation to solve it.

7. $m + 16 = 14$ **8.** $k + 9 = -16$ **9.** $t + 5 = 8$

10. $y + 9 = -53$ **11.** $z + (-3) = -8$ **12.** $x + (-4) = -37$

Written Exercises

Solve and check each equation.

13. $m + 10 = 7$ **14.** $y + 16 = 7$ **15.** $5 + a = -14$

16. $9 = x + 13$ **17.** $k + 11 = -21$ **18.** $b + 15 = -32$

19. $-11 = a + 8$ **20.** $18 + m = -57$ **21.** $14 + c = -5$

22. $y + 3 = -15$ **23.** $p + 12 = -4$ **24.** $w + 42 = -51$

25. $r + (-8) = 7$ **26.** $z + (-17) = 0$ **27.** $z + (-18) = 34$

28. $-12 + b = 12$ **29.** $-15 + d = 13$ **30.** $x + (-7) = 36$

31. $r + (-11) = -21$ **32.** $y + (-13) = -27$ **33.** $h + (-13) = -5$

34. $d + (-6) = -9$ **35.** $-11 = k + (-5)$ **36.** $-23 = -19 + n$

37. $-4.1 = m + (-0.5)$ **38.** $0 = t + (-1.4)$ **39.** $y + 2.3 = 1.5$

40. $x + 4.2 = 1.5$ **41.** $2.4 = m + 3.7$ **42.** $4.4 = b + 6.3$

43. $y + \frac{7}{16} = -\frac{5}{8}$ **44.** $x + \frac{4}{9} = -\frac{2}{27}$ **45.** $-\frac{7}{6} + k = \frac{5}{6}$

46. $-\frac{5}{7} + w = \frac{5}{7}$ **47.** $m + \frac{5}{9} = -\frac{3}{5}$ **48.** $m + \left(-\frac{7}{8}\right) = \frac{5}{12}$

Write an equation and solve. Check each solution.

49. A number increased by 5 is equal to 34. Find the number.

50. Eighty-two increased by some number is -34. Find the number.

51. A number increased by -45 is 77. Find the number.

52. A number increased by -56 is -82. Find the number.

2-8 Solving Equations Involving Subtraction

Recall that the Subtraction Property of Equality is used to *undo* addition.

$$x + 7 = 10$$
$$x + 7 - 7 = 10 - 7$$
$$x = 3$$

Likewise, many equations involving subtraction may be solved by using the Addition Property of Equality. Add the same number to each side of an equation to *undo* subtraction.

Addition and subtraction are inverse operations.

Example

1 **Solve $m - 9 = -13$.**

$$m - 9 = -13 \qquad \text{\textit{9 is being subtracted from m.}}$$
$$m - 9 + 9 = -13 + 9 \qquad \text{\textit{Add 9 to each side.}}$$
$$m = -4$$

Check:
$$m - 9 = -13$$
$$-4 - 9 \stackrel{?}{=} -13$$
$$-13 = -13 \quad \checkmark$$

The solution is -4.

Sometimes an equation can be solved more easily if it is first rewritten in a different form. Recall that subtracting a number is the same as adding its inverse. For example, the equation $b - (-8) = 23$ may be rewritten as $b + 8 = 23$. Also, $y + (-7.5) = -12.2$ may be rewritten as $y - 7.5 = -12.2$.

Examples

2 **Solve $b - (-8) = 23$.**

This equation is equivalent to $b + 8 = 23$.

$$b + 8 = 23 \qquad \text{\textit{8 is being added to b.}}$$
$$b + 8 - 8 = 23 - 8 \qquad \text{\textit{Subtract 8 from each side.}}$$
$$b = 15$$

Check:
$$b - (-8) = 23$$
$$15 - (-8) \stackrel{?}{=} 23$$
$$23 = 23 \quad \checkmark \qquad \text{The solution is 15.}$$

3 **Solve $y + (-7.5) = -12.2$.**

This equation is equivalent to $y - 7.5 = -12.2$.

$$y - 7.5 = -12.2$$
$$y - 7.5 + 7.5 = -12.2 + 7.5 \qquad \text{\textit{Add 7.5 to each side.}}$$
$$y = -4.7 \qquad \text{\textit{Check this result.}}$$

The solution is -4.7.

Solving some equations requires two steps. Study the following example.

Example

4 **Solve $-8 - k = 13$.**

$$-8 - k = 13$$
$$-8 + 8 - k = 13 + 8 \quad \text{\textit{Add 8 to each side.}}$$
$$-k = 21 \quad \text{\textit{The opposite of k is positive 21.}}$$
$$k = -21 \quad \text{\textit{Therefore, k is negative 21.}}$$

Check:
$$-8 - k = 13$$
$$-8 - (-21) \stackrel{?}{=} 13$$
$$13 = 13 \quad \text{✓} \qquad \text{The solution is } -21.$$

In solving equations, it is important to check solutions. It is sometimes helpful to verify a solution by using a calculator. The solution of $x + (-7.3) = -9.7$ is -2.4. This means that $-2.4 + (-7.3)$ should equal -9.7.

Check:
$$x + (-7.3) = -9.7$$
$$-2.4 + (-7.3) \stackrel{?}{=} -9.7$$

To simplify this expression, a change-sign key, labeled $\boxed{+/-}$, is used.

When this key is pressed, the calculator changes the sign of the number in the display. The change-sign key can be used to enter negative numbers. For example, to enter -2.4, press 2.4 and then press the change-sign key. For $-2.4 + (-7.3)$, press the following sequence of keys.

ENTER: 2.4 $\boxed{+/-}$ $\boxed{+}$ 7.3 $\boxed{+/-}$ $\boxed{=}$

Because -9.7 is shown in the calculator display screen, -2.4 is verified as the solution.

Example

5 **Using Calculators**

Solve $t + (-3.289) = -17.565$.

$$t + (-3.289) = -17.565$$
$$t + (-3.289) + 3.289 = -17.565 + 3.289$$

$$t = -17.565 + 3.289$$

Use a calculator to simplify.

ENTER: 17.565 $\boxed{+/-}$ $\boxed{+}$ 3.289 $\boxed{=}$

DISPLAY: *17.565   -17.565   3.289   -14.276*

The solution is -14.276.

Check: $t + (-3.289) = -17.565$

$-14.276 + (-3.289) \overset{?}{=} -17.565$ *Substitute -14.276 for t.*

ENTER: 14.276 $\boxed{+/-}$ $\boxed{+}$ 3.289 $\boxed{+/-}$ $\boxed{=}$ **DISPLAY:** *-17.565* ✔

Exploratory Exercises

Rename each expression by using its inverse operation.

1. $m + (-8)$ **2.** $r + (-12)$ **3.** $y - (-11)$ **4.** $k - (-12)$

5. $z + (-31)$ **6.** $s - (-18)$ **7.** $p - (-47)$ **8.** $q + (-62)$

Written Exercises

Solve each equation.

9. $a - 15 = -32$ **10.** $h - 26 = -29$ **11.** $y - 7 = -32$

12. $r - 21 = -37$ **13.** $y + (-7) = -19$ **14.** $x + (-8) = -31$

15. $b + (-14) = 6$ **16.** $k + (-13) = 21$ **17.** $d - (-27) = 13$

18. $m - (-13) = 37$ **19.** $r - (-31) = 16$ **20.** $t - (-16) = 9$

21. $x - (-33) = 14$ **22.** $-12 + z = -36$ **23.** $x - 13 = 45$

24. $d - 27 = -63$ **25.** $y + (-18) = 7$ **26.** $w - (-37) = 28$

27. $13 - m = 41$ **28.** $16 - y = 37$ **29.** $65 = 12 - x$

30. $41 = 32 - r$ **31.** $-7 = -16 - k$ **32.** $-27 = -6 - p$

33. $-14 - a = -21$ **34.** $-27 - b = -7$ **35.** $-19 - s = 41$

36. $r - 6.5 = -9.3$ **37.** $y - 7.3 = 5.1$ **38.** $p - (-1.3) = -7.1$

39. $s - (-7.1) = 2.4$ **40.** $-1.43 + w = 0.89$ **41.** $-0.0056 + z = 0.065$

42. $m - \frac{6}{7} = \frac{3}{14}$ **43.** $a - \frac{2}{3} = -\frac{4}{9}$ **44.** $b + \left(-\frac{3}{4}\right) = \frac{4}{5}$

45. $j + \left(-\frac{3}{5}\right) = \frac{5}{3}$ **46.** $x - \left(-\frac{5}{6}\right) = \frac{2}{7}$ **47.** $h - \left(-\frac{7}{8}\right) = \frac{5}{9}$

48. $4.891 + x = -13.228$ **49.** $-5.367 = -3.001 + z$

50. $67.992 + p = -5.429$ **51.** $a + 2888.5 = -649.03$

Write an equation and solve.

52. A number decreased by 14 is -46. Find the number.

53. The difference of a number and -23 is 35. Find the number.

54. What number decreased by 45 is -78?

55. What number decreased by -67 is -34?

56. Twenty-three minus a number is 42. Find the number.

57. A number less 135 is -59. Find the number.

Anna Lopez is a meteorologist. When forecasting the weather, Anna predicts the high and low temperatures for each day. In addition, she predicts the equivalent temperatures due to the *windchill factor*, which depends on the actual temperature and the speed of the wind.

The windchill factor is an estimate of the cooling effect the wind has on a person in cold weather. Increased wind velocity causes the body to lose heat more rapidly. Therefore, the chilling effect of cold increases as the speed of the wind increases. Anna uses a chart similar to this to predict the windchill factor.

Windchill Chart

Wind speed in mph	Actual temperature (°Fahrenheit)								
	50	40	30	20	10	0	−10	−20	−30
	Equivalent temperature (° Fahrenheit)								
0	50	40	30	20	10	0	−10	−20	−30
5	48	37	27	16	6	−5	−15	−26	−36
10	41	28	16	4	−9	−21	−33	−46	−58
15	36	22	9	−4	−18	−31	−45	−58	−72
20	33	18	4	−10	−24	−39	−53	−67	−81
25	30	15	1	−14	−29	−44	−59	−73	−88
30	28	13	−2	−17	−32	−48	−63	−78	−93

In Chicago on January 11, the low temperature was −10°F. If the wind speed was 15 mph, the windchill factor made the temperature equivalent to −45°F.

Exercises

Use the windchill chart to find each equivalent temperature.

1. 30°F, 10 mph

2. 20°F, 20 mph

3. −30°F, 25 mph

4. −20°F, 5 mph

5. 0°F, 15 mph

6. 40°F, 30 mph

7. Find the difference between the equivalent temperatures in Problems 1 and 2.

8. Find the difference between the equivalent temperatures in Problems 3 and 4.

9. Find the difference between the equivalent temperatures in Problems 5 and 6.

10. Find the difference between the equivalent temperatures in Problems 2 and 6.

2-9 Problem Solving: Rational Numbers

Before solving a problem, you must study the problem carefully. Consider the following problem.

Death Valley is about 86 meters below sea level. The top of Mt. McKinley is 6194 meters above sea level. What is the difference in these elevations?

Now follow the four-step plan for solving this problem.

explore

Read the problem to find out what is asked. Identify important facts from the problem.

The problem asks for the difference in elevations.

Death Valley	86 m below sea level
Mt. McKinley	6194 m above sea level

Then think about how the facts are related. Sometimes it is helpful to draw a chart or diagram.

Elevation above sea level can be represented with a positive number. Elevation below sea level can be represented with a negative number.

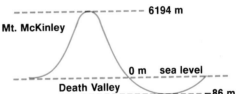

plan

Decide how to solve the problem.

To solve the problem, find the difference between 6194 and −86. Recall that difference is the result of subtraction. Find 6194 − (−86).

solve

Do the computation and answer the problem.

6194 − (−86) = 6280
The difference in elevations is 6280 m.

examine

See if the solution makes sense for the given problem. If not, try another way to solve the problem.

Since Mt. McKinley is about 6200 m above sea level and Death Valley is about 90 m below sea level, the distance between elevations should be approximately 6290 m. Therefore, the solution 6280 m makes sense.

Sometimes a problem can be solved by translating it into a mathematical equation. Study this example.

Example

1 Write an equation and solve the problem.

Jerry Flaherty has bench pressed 90 kilograms. He is working toward a goal of pressing 115 kilograms. How many more kilograms does he need to reach his goal?

explore   *Read the problem to find out what is asked, and define a variable.*
The problem asks for how many more kilograms are needed to reach the goal.
Let x = the number of kilograms.

plan   *Write an equation.*
An amount must be added to 90 kg to reach the goal of 115 kg.
$90 + x = 115$

solve   *Solve the equation and answer the problem.*
$$90 + x = 115$$
$$90 - 90 + x = 115 - 90$$
$$x = 25$$

Jerry needs to press 25 kg more.

examine   *Examine the solution.*
If Jerry presses 25 more kilograms, he will have pressed 90 + 25, or 115 kilograms.

Exploratory Exercises

State the answer to each question.

1. Frieda traveled 110 kilometers by boat up the Rhine River last summer. She then traveled 320 kilometers down along the river by train. How far downstream was she from her starting point?
   a. What is asked?
   b. How far up the river did Frieda travel?
   c. How far down the river did she travel?
   d. How can these distances be represented using positive and negative numbers?
   e. Define a variable. Then state an equation for the problem.

2. The perimeter of a triangle is $35\frac{3}{8}$ inches. One side measures $10\frac{1}{8}$ inches. A second side measures $12\frac{3}{16}$ inches. What does the third side measure?
   a. What is the perimeter of the triangle?
   b. What are the measures of the known sides?

c. Draw a diagram to represent the problem.

d. What is unknown?

e. What do you need to know about the perimeter to solve this problem?

f. Define a variable. Then state an equation for the problem.

3. The sum of two integers is −23. One of the integers is +9. What is the other integer?

    a. What is asked?

    b. What is one of the integers?

    c. What is the sum of the integers?

    d. What mathematical operation is indicated in the problem?

    e. Define a variable. Then state an equation for the problem.

4. An English assignment covers 182 pages. Kristina has 79 pages yet to read. How many pages has she read?

    a. What is asked?

    b. How many pages does the assignment cover?

    c. How many pages has Kristina yet to read?

    d. Define a variable. Then state an equation for the problem.

For each problem, define a variable. Then state an equation.

5. An elevator started at the first floor and went up 14 floors. Then it went down 9 floors. At what floor was it then located?

6. Fairmeadow High School has 1283 students. On Friday 1116 were present. How many were absent?

7. Thirteen subtracted from a number is −5. Find the number.

8. The sum of a number and 12 is equal to 47. Find the number.

9. In a mid-season slump the Yankees scored only 17 runs in 9 games. Their opponents scored 41 runs. By how many runs did their opponents outscore them?

10. A traffic helicopter descended 160 meters to observe road conditions. It leveled off at 225 meters. What was its original altitude?

Written Exercises

11-20. Solve the problems in Exploratory Exercises 1-10.

Read and explore each problem. Make a diagram if necessary. Then define a variable, write an equation, and solve the problem.

21. Jeff Simons sold 27 cars last month. This amount is 36 less than the same time period one year ago. What were his sales one year ago?

22. A rancher lost 47 cattle because of the summer heat. His herd now numbers 396. How large was the herd before the summer heat?

23. Agnes Graham bought a jacket on sale and saved $27.35. The regular price was $81.79. What was the sale price?

24. The total of Barbara Orfanello's gas bill and electric bill for January was $210.87. Her electric bill was $95.25. How much was her gas bill?

25. Gary Carson skied down the slalom run in 131.3 seconds. This was 21.7 seconds faster than his sister. What was her time?

26. Lisa Thorson skied down the slalom run in 139.8 seconds. This was 13.7 seconds slower than her best time. What is her best time?

27. The temperature at mid-afternoon was 12°C. By early evening the temperature was −7°C. What was the temperature change?

28. On February 2 the highest temperature in the United States was 87°F in Miami, Florida. The lowest temperature was −19°F in Rice Lake, Wisconsin. What was the difference between the high and low temperatures?

29. Thirteen less than some number is 64. What is the number?

30. Some number subtracted from 161 yields 87. What is the number?

31. The sum of two integers is −32. One of the integers is +6. What is the other integer?

32. The difference of two integers is 26. The lesser integer is −11. What is the greater integer?

33. A scuba diver swam to a depth of 53 meters below sea level. Then he saw a shark 10 meters above him. At what depth was the shark?

34. The entrance to a South Dakota gold mine is 3273 feet above sea level. The mine is 4386 feet deep, measured from the entrance. What is the elevation of the bottom of the mine?

35. Eagle's Bluff is about 143 meters above sea level. From the peak of the bluff to the floor of Sulphur Springs Canyon is 217 meters. How far below sea level is Sulphur Springs Canyon?

36. A 4-sided city lot has a perimeter of 340.8 feet. Two sides have lengths of 60.23 feet. A third side has a length of 109.13 feet. What is the length of the fourth side?

37. Four cave explorers descended to a depth of 112 meters below the cave entrance. They discovered a large cavern whose ceiling was 27 meters above them. At what depth below the cave entrance was the cavern ceiling?

38. The area of a triangular courtyard is 1520.2 square feet. The area occupied by a circular fountain in the middle is 132.7 square feet. A walkway covers 253.6 square feet. If the remaining area is used for gardens, how much area is set aside for gardens?

39. Shares of stock in Olympia Motors were listed at $37\frac{3}{4}$ per share. The shares then dropped $2\frac{1}{8}$ points. What was the new listing?

40. A triangle has sides of $16\frac{3}{8}$, $14\frac{1}{16}$, and $17\frac{3}{4}$ inches. What is its perimeter?

Solve each equation.

1. $y + (-34) = 35$
2. $f + 35 = -98$
3. $125 + z = -76$
4. $27 = 45 + r$
5. $-23 = w + (-12)$
6. $-57 = 22 + t$
7. $x - 43 = 56$
8. $y - 67 = -29$
9. $32 - t = 89$
10. $12 - w = -34$
11. $-18 - a = -38$
12. $-58 = e - (-27)$

Write an equation and solve.

13. Thirteen subtracted from a number is -6. Find the number.

14. The sum of a number and -35 is 98. Find the number.

15. A number added to 258 is -77. Find the number.

16. A number decreased by -11 is -176. Find the number.

17. An elevator started at the first floor and went up to the twenty-third floor. It then came down seven floors. At what floor was it then located?

18. Joe skied down the mountain in 230.5 seconds. That was 42.7 seconds faster than his brother. What was his brother's time?

19. A triangle has a perimeter of 36.8 meters. If the lengths of two sides are 11.5 meters and 12.7 meters, find the length of the third side.

20. Mick purchased stock in Chesco, Inc. for $25\frac{1}{8}$ points per share. The shares dropped $1\frac{1}{4}$ points and then climbed $2\frac{3}{4}$ points. Find the new listing.

Excursions in Algebra   Magic Squares

Magic squares are square arrays of numbers. The sum of the numbers in each row, column, and diagonal is the same.

1	2	−3
−4	0	4
3	−2	−1

The sum of the numbers in each row, column, and diagonal of this magic square is 0.

New magic squares can be formed by adding the same number to each entry in the magic square above. In the squares below, how was each entry obtained? Find each new magic square.

3	4	−1

−2		
	−3	
		−4

	−7	

Temperature

Temperature is commonly measured in Celsius degrees or Fahrenheit degrees. Celsius and Fahrenheit thermometers are shown at the right.

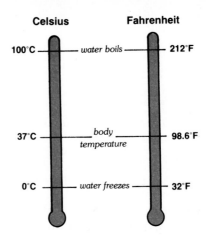

The following BASIC program converts a Fahrenheit reading to the equivalent Celsius reading.

```
10  INPUT F
20  LET C = 5/9*(F-32)     Note how the formula C = ⁵⁄₉(F − 32) is written in line 20.
30  PRINT ''F = '';F,''C = '';C
40  END
```

When you input the Fahrenheit degrees shown on the thermometer, the corresponding Celsius degrees will be printed.

There is one temperature at which the value of F is the same as C. The program can be modified so that the computer will evaluate all integer values of F from −50 to 50 and print only the value of F that is the same as C.

```
10  FOR F = −50 TO 50
20  LET C = 5/9*(F−32)
25  IF C< >F THEN 35     In line 25, the symbol < > represents not equal.
30  PRINT ''F = '';F,''C = '';C
35  NEXT F
40  END
```

Exercises

1. At what temperature is the value of F the same as C?

2. If line 25 was removed from the program, how would the output change?

3. Change lines 10 and 25 of the program to determine if there is a Fahrenheit temperature that is 100 more than the corresponding Celsius temperature.

4. Modify the program to determine if there is a Fahrenheit temperature that is 50 more than the corresponding Celsius temperature.

whole numbers   (41)
negative   (41)
integers   (41)
coordinate   (42)
absolute value   (46)
rational numbers   (49)

additive inverse   (55)
opposite   (55)
equivalent equations   (60)
Addition Property of Equality   (60)
Subtraction Property of Equality   (61)

Chapter Summary

1. On a number line, points to the right of zero are named with a positive sign (+) and points to the left of zero are named with a negative sign (−). (41)

2. The set of integers can be written {. . . , −5, −4, −3, −2, −1, 0, 1, 2, 3, 4, 5, . . .} where . . . means continued indefinitely.   (41)

3. To graph a set of numbers means to locate the points named by those numbers on the number line.   (42)

4. A number line can be used to show addition of integers. Move to the right for positive integers and to the left for negative integers.   (44)

5. Definition of Absolute Value: The absolute value of a number is the number of units it is from 0 on the number line.   (46)

6. Adding Integers with the Same Sign: To add two integers with the same sign, add their absolute values. Give the sum the same sign as the addends. (46)

7. Adding Integers with Different Signs: To add two integers with different signs, subtract the lesser absolute value from the greater absolute value. Give the result the same sign as the addend with the greater absolute value.   (47)

8. A rational number is a number that can be expressed in the form $\frac{a}{b}$, where a and b are integers and b is not equal to zero.   (49)

9. Summary of Rules for Addition of Rational Numbers: The rules for adding two integers with the same sign and adding two integers with different signs can also be used to add rational numbers.   (50)

10. If the sum of two numbers is 0, the numbers are called additive inverses or opposites.   (55)

11. Additive Inverse Property: For every number a, $a + (-a) = 0$.   (55)

12. Subtraction Rule: To subtract a rational number, add its additive inverse. For rational numbers a and b, $a - b = a + (-b)$.   (56)

13. Addition Property of Equality: For any numbers a, b, and c, if $a = b$, then $a + c = b + c$.   (60)

14. Subtraction Property of Equality: For any numbers a, b, and c, if $a = b$, then $a - c = b - c$.   (61)

15. To solve an equation means to isolate the variable having a coefficient of one on one side of the equation. Both the Addition and Subtraction Properties of Equality can be used to solve equations involving addition.   (61)

16. Equations involving subtraction may be solved by using the Addition Property of Equality.   (63)

17. Before solving a problem, you must explore the problem and plan carefully. It is often helpful to draw a chart or diagram. Sometimes a problem can be solved by translating it into a mathematical equation.   (67)

Chapter Review

2–1   **Name the set of numbers graphed.**

1.    2.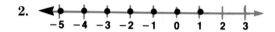

3.    4.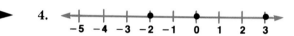

Graph each set of numbers on a number line.

5. $\{-2, 1, 4\}$

6. $\{5, 3, -1, -3\}$

7. $\{-3, -2, -1, 0, \ldots\}$

8. $\{\ldots, 3, 4, 5\}$

2–2   **Use a number line to find each sum.**

9. $4 + (-5)$ **10.** $-3 + 6$ **11.** $-2 + (-4)$ **12.** $-4 + 1$

2–3   **Find each sum.**

13. $17 + (-9)$ **14.** $-9 + (-12)$ **15.** $-12 + 8$

16. $13 + (-51)$ **17.** $-74 + 21$ **18.** $-17 + (-31)$

2–4   **Find each sum.**

19. $\frac{6}{7} + \left(-\frac{13}{7}\right)$ **20.** $-\frac{5}{8} + \frac{3}{16}$ **21.** $-7.5 + 10.2$

22. $-0.37 + (-0.812)$ **23.** $3.707 + (-0.058)$ **24.** $-4.375 + 5.432$

2–5 **Find each sum.**

25. $84 + (-31) + 13$

26. $65 + (-13 + 28)$

27. $-4b + 17b + (-36b)$

28. $a + 9a + (-12a)$

29. $\frac{7}{3} + \frac{5}{12} + \left(-\frac{5}{6}\right)$

30. $-\frac{4}{3} + \frac{5}{6} + \left(-\frac{7}{3}\right)$

31. $-6.7 + 8.1 + (-5.3)$

32. $-5 + (-3.7) + 9.4$

33. $-13pq + 41k + 12k + (-38pq)$

34. $61mb + (-51pg) + (-48mb) + 13pg$

2–6 **Find each difference.**

35. $14 - 36$

36. $8 - (-5)$

37. $-7 - (-11)$

38. $-13 - 16$

39. $3.72 - (-8.65)$

40. $0.702 - 0.98$

41. $-13x - (-7x)$

42. $-4.5y - 8.1y$

43. $\quad 36$
$\quad \underline{-19}$

44. $\quad -61$
$\quad \underline{-(-43)}$

45. $\quad -2.8$
$\quad \underline{-(+3.9)}$

46. $\frac{4}{5} - \left(-\frac{9}{5}\right)$

47. $-\frac{4}{7} - \frac{3}{14}$

48. $-\frac{3}{2} - \frac{5}{4}$

2–7 **Solve each equation.**

49. $k + 13 = 5$

50. $19 = y + 7$

51. $z + 15 = -9$

52. $m + (-5) = -17$

53. $p + (-7) = 31$

54. $19 = -8 + d$

2–8 **Solve each equation.**

55. $x - 16 = 37$

56. $15 - y = 9$

57. $r - (-5) = -8$

58. $m - (-4) = 21$

59. $-13 = 6 - k$

60. $y + (-9) = -35$

2–9 **For each problem, define a variable. Then use an equation to solve the problem.**

61. Some number added to -16 is equal to 39. What is the number?

62. When 9 is subtracted from another integer, the difference is -13. What is the other integer?

63. The temperature at sunrise was $-2°$F. At noon the temperature was $15°$F. What was the temperature change?

64. Darlene's score at the end of the first half of a game was -7.8. At the end of the game her score was 19.2. How many points did she score during the second half of the game?

Chapter Test

Graph each set on a number line.

1. $\{1, 2, -5\}$

2. $\{\ldots, -2, -1, 0, 1\}$

Find each sum.

3. $-11 + (-13)$

4. $-15 + 23$

5. $36 + (-42)$

6. $1.654 + (-2.367)$

7. $\frac{3}{7} + \left(-\frac{9}{7}\right)$

8. $-\frac{1}{8} + \frac{3}{4}$

9. $12x + (-21x)$

10. $-7 + (-6) + 37$

11. $[4 + (-13)] + (-12)$

12. $\frac{5}{8} + \left(-\frac{3}{16}\right) + \left(-\frac{3}{4}\right)$

13. $-7.9 + 3.5 + 2.4$

14. $18b + 13xy + (-46b)$

State the additive inverse of each value.

15. $\frac{3}{8}$

16. -32.7

17. xy

18. $-14b$

Find each difference.

19. $12 - 19$

20. $-14 - 3$

21. $21 - (-7)$

22. $-41 - (-52)$

23. $-\frac{7}{16} - \frac{3}{8}$

24. $6.32 - (-7.41)$

Evaluate each expression.

25. $|x|$, if $x = -2.1$

26. $5 - y$, if $y = 20$

27. $-x - 38$, if $x = -2$

28. $\left|-\frac{1}{2} + z\right|$, if $z = \frac{1}{4}$

29. $-|a| + |b|$, if $a = 6$ and $b = -2$

30. $d - (-3.8)$, if $d = 0$

Solve each equation.

31. $m + 13 = -9$

32. $k + 16 = -4$

33. $y + (-3) = 14$

34. $x + (-6) = 13$

35. $k - (-3) = 28$

36. $-5 - k = 14$

37. $r - (-1.2) = -7.3$

38. $b - \frac{2}{3} = -\frac{5}{6}$

Solve each problem.

39. The sum of two integers is -11. One integer is 8. Find the other integer.

40. Joe's golf score was 68. This was 4 less than Marie's golf score. What was Marie's score?

The test questions on this page deal with number concepts and basic operations. The information at the right may help you with some of the questions.

Directions: Choose the best answer.
Write A, B, C, or D.

1. Which of the following numbers is *not* a prime number?

   (A) 17 **(B)** 23 **(C)** 37 **(D)** 87

2. How many integers between 325 and 400, inclusive, are divisible by 4?

   (A) 18 **(B)** 19 **(C)** 20 **(D)** 24

3. How many integers between 99 and 201 are divisible by 2 or 5?

   (A) 60 **(B)** 61 **(C)** 70 **(D)** 71

4. How many integers are between, but not including, 5 and 1995?

   (A) 1988   **(B)** 1989
   (C) 1990   **(D)** 2000

5. A person is standing in line, thirteenth from the front and eleventh from the back. How many people are in the line?

   (A) 22 **(B)** 23 **(C)** 24 **(D)** 25

6. What number is missing from the sequence 2, 5, 15, 18, 54, 171, 174, 522?

   (A) 36 **(B)** 56 **(C)** 57 **(D)** 398

7. What digit is represented by △ in this subtraction problem?

   $$\begin{array}{r} 80\square \\ -60\,2 \\ \hline \triangle 98 \end{array}$$

   (A) 1 **(B)** 2 **(C)** 3 **(D)** 4

8. Find the smallest positive integer which gives a remainder of 3 when divided by 4, 6, or 8.

   (A) 51 **(B)** 27 **(C)** 15 **(D)** 26

---

1. A number is *divisible* by another number if it can be divided exactly with no remainder. A number is divisible by each of its factors.

2. A number is *prime* if it has no factors except itself and 1.

   7 is a prime number.
   8 is not *prime* since it has factors other than itself and 1.
   2 and 4 are factors of 8.

9. 8(916) + 916 =
   (A) 4(916) + 3(916)
   (B) 5(916) + 4(916)
   (C) 6(916) + 4(916)
   (D) 3(916) + 4(916)

10. Alan owes Bruce $4, Bruce owes Carl $12, and Carl in turn owes Alan some money. If all three debts could be settled by having Bruce pay $3 to Alan and $5 to Carl, how much does Carl owe Alan?

    (A) $4 **(B)** $5 **(C)** $6 **(D)** $7

11. A person has 100 green, 100 orange, and 100 yellow jelly beans. How many jars can be filled if each jar must contain 8 green, 5 orange, and 6 yellow jelly beans?

    (A) 12 **(B)** 15 **(C)** 16 **(D)** 25

12. A patient must be given medication every 5 hours, starting at 10 A.M. Thursday. What is the first day on which the patient will receive medication at noon?

    (A) Thursday   **(B)** Friday
    (C) Saturday   **(D)** Sunday

13. Which is the difference of two consecutive prime numbers less than 30?

    (A) 5 **(B)** 6 **(C)** 7 **(D)** 8

14. The sum of five consecutive integers is always divisible by

    (A) 2 **(B)** 4 **(C)** 5 **(D)** 10

CHAPTER 3

Multiplying and Dividing Rational Numbers

Steve Bernard earns 3% commission on sales of all new cars. To earn $850.00 per week in commission, what must his weekly sales be? In this chapter, you will learn to use equations to solve problems such as this.

3-1 Multiplying Rational Numbers

How can you multiply positive and negative numbers? One way is to consider multiplication as repeated addition. For example, $4(-2)$ may be expressed as $(-2) + (-2) + (-2) + (-2)$.

$$(-2) + (-2) + (-2) + (-2) = -8$$
$$\text{Therefore, } 4(-2) = -8.$$

The product of *positive* 4 and *negative* 2 is *negative* 8.

Another way to find the product of a positive number and a negative number is to consider an example such as the following.

$0 = 7(0)$	*Multiplication Property of Zero*
$0 = 7[5 + (-5)]$	*Substitution and Additive Inverse Properties*
$0 = 7(5) + 7(-5)$	*Distributive Property*
$0 = 35 + 7(-5)$	*Substitution Property*

Using the Additive Inverse Property, 0 is equal to $35 + (-35)$. Therefore, $7(-5)$ must equal -35. Using the Commutative Property, $(-5)(7)$ must also equal -35.

These and other similar examples suggest the following rule.

The product of two numbers that have different signs is negative.	*Multiplying Two Numbers with Different Signs*

Examples

1 Multiply $9(-4)$.

$9(-4) = -36$

2 Simplify $(-3a)(4b)$.

$(-3a)(4b) = (-3)(4)ab$ *Commutative and Associative Properties*
$\qquad\qquad\;\; = -12ab$

3 Multiply $(0.5)(-6)$.

$(0.5)(-6) = -3$

You know that the product of two positive numbers is positive. What is the sign of the product of two negative numbers?

$0 = -6(0)$	*Multiplicative Property of Zero*
$0 = -6[8 + (-8)]$	*Substitution and Additive Inverse Properties*
$0 = -6(8) + (-6)(-8)$	*Distributive Property*
$0 = -48 + (-6)(-8)$	*Substitution Property*

The additive inverse of -48 is 48. Thus, $(-6)(-8)$ must equal 48. The product of *negative* 6 and *negative* 8 is *positive* 48.

These and other similar examples suggest the following rule.

The product of two numbers that have the same sign is positive.	*Multiplying Two Numbers with the Same Sign*

Examples

4 **Multiply $(-4)(-13)$.**

$(-4)(-13) = 52$ *Both factors are negative. So the product is positive.*

5 **Multiply $\left(-\frac{2}{3}\right)\left(-\frac{1}{5}\right)$.**

$\left(-\frac{2}{3}\right)\left(-\frac{1}{5}\right) = \frac{2}{15}$ *Why is the product positive?*

6 **Simplify $6x(-7y) + (-3x)(-5y)$.**

$$6x(-7y) + (-3x)(-5y) = -42xy + 15xy$$
$$= (-42 + 15)xy \quad \textit{Distributive Property}$$
$$= -27xy$$

What is the result if a number is multiplied by -1?

$$-1(3) = -3 \qquad \tfrac{4}{5}(-1) = -\tfrac{4}{5} \qquad -1(-4) = 4 \qquad (-0.3)(-1) = 0.3$$

These examples are summarized in the Multiplicative Property of -1.

The product of any number and -1 is its additive inverse. $$-1(a) = -a \text{ and } a(-1) = -a$$	*Multiplicative Property of -1*

To find the product of three or more numbers, first group the numbers in pairs.

Example

7 **Multiply $(-2)(-3)(-1)(4)$.**

$$(-2)(-3)(-1)(4) = 6(-1)(4) \qquad \textit{First multiply } -2 \textit{ by } -3.$$
$$= -6(4) \qquad\qquad \textit{Then multiply 6 by } -1.$$
$$= -24$$

Recall that the product of *two* negative factors is positive. Why are the following statements true?

The product of an *even* number of negative factors is positive.
The product of an *odd* number of negative factors is negative.

Exploratory Exercises

Determine whether each product is positive or negative.

1. $3(-2)$
2. $(-2)(-8)$
3. $3(-5)$
4. $(-6)(-5)$

5. $(-9)(7)$
6. $(-11)(10)$
7. $\left(\frac{7}{3}\right)\left(\frac{7}{3}\right)$
8. $\left(-\frac{7}{9}\right)\left(\frac{9}{14}\right)$

9. $(-3)(4)(-2)$
10. $(-6)(4)(-3)$
11. $\left(\frac{3}{5}\right)\left(\frac{2}{3}\right)(-3)$
12. $\left(-\frac{4}{5}\right)\left(-\frac{1}{5}\right)(-5)$

13. Under what conditions is ab positive?
14. Under what conditions is ab negative?
15. Under what conditions is abc positive?
16. If a^2 is positive, what can you conclude about a?
17. If a^3 is positive, what can you conclude about a?
18. If a^3 is negative, what can you conclude about a?
19. If ab is negative, what is the sign of the inverse of ab?
20. Under what conditions is ab equal to zero?

Written Exercises

21–32. Find each product for Exploratory Exercises 1–12.

Find each product.

33. $5(12)$
34. $(-6)(11)$
35. $\left(\frac{3}{5}\right)\left(-\frac{4}{7}\right)$

36. $\left(-\frac{7}{8}\right)\left(-\frac{1}{3}\right)$
37. $4\left(-\frac{7}{8}\right)$
38. $(-5)\left(-\frac{2}{5}\right)$

39. $\frac{3}{5}(5)(-2)\left(-\frac{1}{2}\right)$
40. $\frac{2}{11}(-11)(-4)\left(-\frac{3}{4}\right)$
41. $\left(-\frac{1}{3}\right)\left(-\frac{3}{4}\right)\left(-\frac{4}{5}\right)$

42. $\left(-\frac{7}{12}\right)\left(\frac{6}{7}\right)\left(-\frac{3}{4}\right)$
43. $(-4)(0)(-2)(-3)$
44. $(4)(-2)(-1)(-3)$

45. $(4.265)(-1.83)$
46. $(-2.93)(-0.003)$
47. $(-6.8)(-5.415)(3.1)$

Simplify.

48. $3(-4) + 2(-7)$
49. $-3(2) + (-2)(-3)$
50. $-4(4) + (-2)(-3)$

51. $(-7)(8) + 8(-7)$
52. $5(-2) - 3(-8)$
53. $4(-1) - 2(-6)$

54. $4(7) - 3(11)$
55. $8(9) - 6(3)$
56. $\frac{2}{3}\left(\frac{1}{2}\right) - \left(-\frac{3}{2}\right)\left(\frac{2}{3}\right)$

57. $\left(-\frac{1}{2}\right)\left(-\frac{1}{2}\right) - \frac{2}{3}\left(\frac{3}{2}\right)$
58. $\left(\frac{5}{8}\right)\left(-\frac{1}{2}\right) - \left(\frac{5}{8}\right)\left(\frac{1}{2}\right)$
59. $\frac{5}{6}\left(\frac{6}{7}\right) - \left(\frac{5}{6}\right)\left(-\frac{6}{7}\right)$

60. $5x(-6y) + 2x(-8y)$
61. $(-2x)(-4y) - (-9x)(10y)$

62. $(0.5)(2.632) + (1.203)(3.3)$
63. $(-1.9)(3.04) - (-5.425)(-11.6)$

64. $5(3t - 2t) + 2(4t - 3t)$
65. $3(5x + 3x) - 4(2x + 6x)$

66. $4[3x + (-2x)] - 5(3x + 2x)$
67. $6(3x + 7x) + 7[8x + (-4x)]$

68. $\frac{1}{2}(6x + 8x) - \frac{1}{3}(6x + 9x)$
69. $-\frac{2}{7}(21x + 35a) + \frac{4}{7}(35x - 21a)$

70. $\frac{3}{4}(4a - 12b) + \frac{1}{8}(16a + 48b)$
71. $\frac{5}{6}(-24a + 36b) + \left(-\frac{1}{3}\right)(60a - 42b)$

72. $\frac{1}{2}\left(-\frac{1}{3}a + \frac{2}{3}b\right) + \frac{2}{3}\left(\frac{1}{2}a - \frac{3}{4}b\right)$
73. $\frac{1}{2}(-3a - 4b) + \left(-\frac{2}{9}\right)(-4a - 7b)$

74. $(-0.25)(-4x - 2y) + 3(11x - 4.9y)$
75. $1.2(4x - 5y) - 0.2(-1.5x + 8y)$

Write a multiplication sentence for each problem.

Suppose a balloon is ascending at a rate of 3 meters per minute.

76. Ten minutes from now the balloon will be 30 meters higher.

77. Ten minutes ago the balloon was 30 meters lower.

Suppose a balloon is descending at the rate of 3 meters per minute.

78. Ten minutes from now the balloon will be 30 meters lower.

79. Ten minutes ago the balloon was 30 meters higher.

Challenge Exercises

Evaluate each expression if $a = -8$, $b = -12$, and $c = \frac{1}{3}$.

80. $9(2ab - 3c) - (4ab - 6c)$

81. $\frac{1}{4}\left(-6a + \frac{2}{3}b\right) + \frac{3}{4}\left(\frac{1}{2}b - 10c\right)$

82. $\frac{2}{3}[8b + (-18c)] - \frac{1}{4}\left[-9c + \left(-\frac{3}{4}a\right)\right]$

83. $2.1a[15c + (-0.4) + 4.3b(1.8a - 2b)]$

mini-review

Find each sum or difference.

1. $-3 - (-4)$

2. $51 + (-93)$

3. $-0.187 + (-8.15)$

4. $-5 + (-4.8) + 10.4$

5. $-3.2x - 8.4x$

6. $\frac{3}{5}x + \frac{2}{15}x - \frac{x}{3}$

Solve each equation.

7. $w - (-34) = 89$

8. $23 - w = 34$

9. $w - (-42) = -28$

10. $61 = -56 - w$

11. $11.2 - w = 49.1$

12. $a - (-2.34) = -4.56$

Solve each problem.

13. Twenty-two less than some number is 64. What is the number?

14. A number decreased by -27 is 43. Find the number.

15. Two cave explorers descended to a depth of 123 meters. Here they discovered a cavern whose ceiling was 45 meters above them. How far was the ceiling below the cave entrance?

Find each product.

16. $(-12)(-3)(-2)$

17. $23(-1)(-1)$

18. $(2a)(-3a)(-4b)$

Simplify.

19. $4(-3) + 5(-4)$

20. $7[8s + (-3y)] + (-4s)$

21. $\left(-\frac{2}{3}\right)\left(-\frac{5}{4}\right) - \left(\frac{1}{2}\right)\left(\frac{7}{6}\right)$

22. $-\frac{3}{8}(40xy + 24) + \frac{2}{3}(60xy - 42)$

3-2 Dividing Rational Numbers

Study the following examples.

$\frac{60}{6} \cdot 6 = 10 \cdot 6$ *Multiplying by 6* $\frac{38 \cdot 2}{2} = \frac{76}{2}$ *Dividing by 2 undoes*
  undoes dividing by 6.   *multiplying by 2.*
$= 60$ $= 38$

These and other examples show that multiplication and division are *inverse operations.* You can use this fact to derive rules for dividing positive and negative numbers.

$$3 \cdot 4 = 12 \quad \text{so} \quad 12 \div 4 = 3$$

same   *positive*
signs   *quotient*

$$5 \cdot (-6) = -30 \quad \text{so} \quad -30 \div (-6) = 5$$

The quotient of two numbers with the same sign is *positive.*

What happens if the numbers have different signs?

$$-7 \cdot 9 = -63 \quad \text{so} \quad -63 \div 9 = -7$$

different   *negative*
signs   *quotient*

$$-8 \cdot (-6) = 48 \quad \text{so} \quad 48 \div (-6) = -8$$

The quotient of two numbers with different signs is *negative.*

The quotient of two numbers is positive if the numbers have the same sign. The quotient of two numbers is negative if the numbers have different signs.

*Dividing
Rational
Numbers*

Notice that the rule for finding the sign of the quotient of two numbers is the same as the rule for finding the sign of their product.

Examples

1 **Divide $-36 \div 3$.**

Since the numbers have different signs, the quotient is negative.
$-36 \div 3 = -12$

2 **Simplify $\frac{-45}{-9}$.**

$\frac{-45}{-9} = 5$ *The fraction bar indicates division.*
      The quotient is positive.

Recall that it is possible to subtract a number by adding its additive inverse. In a similar manner, it is possible to divide a number by multiplying by its multiplicative inverse. Two numbers whose product is 1 are **multiplicative inverses** or **reciprocals**.

The reciprocal of $\frac{4}{9}$ is $\frac{9}{4}$ because $\frac{4}{9} \cdot \frac{9}{4} = 1$.

The reciprocal of -5 is $-\frac{1}{5}$ because $-5\left(-\frac{1}{5}\right) = 1$.

The reciprocal of 1 is 1 because $1 \cdot 1 = 1$.

Zero has no reciprocal because the product of zero and any rational number is zero, not one.

These examples can be summarized in the Multiplicative Inverse Property.

For every nonzero number a, there is exactly one number $\frac{1}{a}$, such that $a\left(\frac{1}{a}\right) = \frac{1}{a}(a) = 1$.	*Multiplicative Inverse Property*

To divide $\frac{2}{3}$ by $\frac{3}{4}$, multiply by the multiplicative inverse of $\frac{3}{4}$.

$$\frac{2}{3} \div \frac{3}{4} = \frac{2}{3} \cdot \frac{4}{3} \qquad \text{\small $\frac{3}{4}$ and $\frac{4}{3}$ are reciprocals.}$$
$$= \frac{8}{9}$$

In general, *any* division expression can be changed to an equivalent multiplication expression. To divide by any nonzero number, multiply by the reciprocal of that number.

For all numbers a and b, with $b \neq 0$, $$a \div b = \frac{a}{b} = a\left(\frac{1}{b}\right) = \frac{1}{b}(a).$$	*Division Rule*

To divide by b, multiply by $\frac{1}{b}$.

Examples

3 **Divide $-\frac{3}{4} \div 8$.**

$-\frac{3}{4} \div 8 = -\frac{3}{4} \cdot \frac{1}{8}$ *Multiply by the reciprocal of 8.*

$= -\frac{3}{32}$

4 **Simplify $\frac{4a + 32}{4}$.**

$\frac{4a + 32}{4} = (4a + 32) \div 4$

$= (4a + 32)\left(\frac{1}{4}\right)$

$= 4a\left(\frac{1}{4}\right) + 32\left(\frac{1}{4}\right)$

$= a + 8$

Exploratory Exercises

State the reciprocal of each number.

1. 3
2. −5
3. 0
4. −1
5. $\frac{2}{3}$
6. $\frac{1}{15}$

7. $\frac{4}{3}$
8. $\frac{7}{9}$
9. $-\frac{3}{11}$
10. $-\frac{1}{11}$
11. $\frac{10}{7}$
12. $\frac{21}{5}$

13. $-\frac{3}{5}$
14. $-\frac{8}{15}$
15. $3\frac{1}{4}$
16. $2\frac{1}{8}$
17. $-2\frac{3}{7}$
18. $-6\frac{5}{11}$

Written Exercises

Simplify.

19. $\frac{-30}{-5}$
20. $\frac{-48}{8}$
21. $\frac{30}{-6}$
22. $\frac{-55}{11}$
23. $\frac{70}{5}$
24. $\frac{80}{20}$

25. $\frac{-40}{8}$
26. $\frac{-48}{6}$
27. $\frac{-96}{-16}$
28. $\frac{-84}{-7}$
29. $\frac{-36}{-9}$
30. $\frac{42}{-6}$

31. $\frac{-38}{2}$
32. $\frac{54}{-9}$
33. $\frac{-200x}{50}$
34. $\frac{-450n}{10}$
35. $\frac{-36a}{-6}$
36. $\frac{45b}{9}$

37. $\frac{63a}{-9}$
38. $\frac{77b}{-11}$
39. $49 \div 7$
40. $-16 \div 8$
41. $65 \div (-13)$
42. $\frac{75}{-15}$

43. $\frac{-\frac{5}{6}}{8}$
44. $\frac{-\frac{3}{4}}{9}$
45. $\frac{\frac{7}{8}}{-10}$
46. $\frac{\frac{3}{11}}{-6}$
47. $\frac{\frac{1}{3}}{4}$
48. $\frac{\frac{3}{8}}{6}$

49. $\frac{7}{-\frac{2}{5}}$
50. $\frac{11}{-\frac{5}{6}}$
51. $\frac{-6}{-\frac{4}{7}}$
52. $\frac{-5}{\frac{2}{7}}$
53. $\frac{-6}{-\frac{4}{9}}$
54. $\frac{-9}{-\frac{10}{17}}$

55. $\frac{3a + 9}{3}$
56. $\frac{6a + 24}{6}$
57. $\frac{7a + 35}{-7}$
58. $\frac{14a + 56}{-7}$

59. $\frac{20a + 30b}{-2}$
60. $\frac{-5x + (-10y)}{-5}$
61. $\frac{60a - 30b}{-6}$
62. $\frac{70x - 30y}{-5}$

Excursions in Algebra
Reciprocals

The key labeled $\frac{1}{x}$ on your calculator is the reciprocal key. When this key is pressed, the calculator replaces the number on display with its reciprocal.

Example   Evaluate $\frac{1}{4}$.

ENTER: 4 $\boxed{1/x}$ DISPLAY: 4   0.25

Exercises
Use a calculator to evaluate each expression. Round each result to the nearest thousandth.

1. $\frac{1}{2.5}$
2. $\frac{1}{7}$
3. $\frac{1}{3(5)}$
4. $\frac{1}{3} \cdot \frac{1}{5}$
5. $\frac{1}{0.618}$
6. $\frac{1}{5 - 2(0.8)}$

7. Enter a number then press the reciprocal key twice. What happens? Predict what will happen if you press the key n times.

8. Enter 0 and then press the reciprocal key. What happens? Why?

9. Enter 7.328 $\boxed{1/x}$ $\boxed{\times}$ 7.328 $\boxed{\div}$. What is the result? Why?

3-3 Solving Equations Using Multiplication and Division

A small basket holds $\frac{1}{3}$ of the number of apples in a larger basket. There are 12 apples in the small basket. About how many apples would the larger basket hold?

An equation can be used to solve this problem. If x is the number of apples in the larger basket, then the equation is $\frac{1}{3}x = 12$.

To solve this equation, you would use the **Multiplication Property of Equality**. It states that if each side of an equation is multiplied by the same nonzero number, the result is an equivalent equation.

Recall that equivalent equations have the same solution.

For any numbers a, b, and c, with $c \neq 0$, **if $a = b$, then $ac = bc$.**	*Multiplication Property of Equality*

Study how the Multiplication Property of Equality is used below.

Example

1 Solve $\frac{1}{3}x = 12$.

$\frac{1}{3}x = 12$　　　*The coefficient of x is $\frac{1}{3}$.*

$3\left(\frac{1}{3}x\right) = 3(12)$　　*Multiply each side by 3, the reciprocal of $\frac{1}{3}$.*

$1x = 36$

$x = 36$

Check:　　$\frac{1}{3}x = 12$

$\frac{1}{3}(36) \stackrel{?}{=} 12$　　*Replace x with 36.*

$12 = 12$　✔

The solution is 12.

In the equation $\frac{1}{3}x = 12$, the variable x represented the number of apples in the larger basket. Thus, there are about 36 apples in the larger basket.

Examples

2 **Solve $\frac{k}{7} = -45$.**

$$\frac{k}{7} = -45$$

$$\frac{1}{7}k = -45 \qquad \textit{Substitute } \frac{1}{7}k \textit{ for } \frac{k}{7}.$$

$$7 \cdot \frac{1}{7}k = 7(-45) \qquad \textit{Multiply each side by 7.}$$

$$k = -315 \qquad \textit{Check this result.}$$

The solution is -315.

3 **Solve $\left(2\frac{1}{3}\right)m = 3\frac{1}{9}$.**

$$\left(2\frac{1}{3}\right)m = 3\frac{1}{9}$$

$$\frac{7}{3}m = \frac{28}{9} \qquad \textit{Rewrite the mixed numbers as improper fractions.}$$

$$\frac{3}{7}\left(\frac{7}{3}m\right) = \frac{3}{7}\left(\frac{28}{9}\right) \qquad \textit{Multiply each side by } \frac{3}{7}.$$

$$m = \frac{4}{3} \qquad \textit{Check this result.}$$

The solution is $\frac{4}{3}$.

4 **Solve $24 = -2a$.**

$$24 = -2a$$

$$-\frac{1}{2}(24) = -\frac{1}{2}(-2a) \qquad \textit{Multiply each side by } -\frac{1}{2}.$$

$$-12 = a \qquad \textit{Check this result.}$$

The solution is -12.

The equation $24 = -2a$ was solved by multiplying each side by $-\frac{1}{2}$. The same result could have been obtained by dividing each side by -2. In general, dividing each side of an equation by the same nonzero number results in an equivalent equation. This is called the **Division Property of Equality**.

The Division Property of Equality is a special case of the Multiplication Property of Equality.

For any numbers a, b, and c, with $c \neq 0$,

if $a = b$, then $\frac{a}{c} = \frac{b}{c}$.

Division Property of Equality

Example

5 Solve $-6x = 11$.

$$-6x = 11$$

$$\frac{-6x}{-6} = \frac{11}{-6} \qquad \textit{Divide each side by } -6.$$

$$x = -\frac{11}{6} \qquad \textit{Check this result.}$$

The solution is $-\frac{11}{6}$.

Exploratory Exercises

State the number by which each side can be multiplied to solve each equation.

1. $\frac{b}{3} = -6$ **2.** $\frac{b}{5} = 10$ **3.** $\frac{3}{4}n = 30$ **4.** $\frac{5}{7}t = -10$ **5.** $\frac{4}{9}n = -24$

6. $-\frac{5}{9}x = 15$ **7.** $-8n = 24$ **8.** $3n = -18$ **9.** $1 = \frac{k}{9}$ **10.** $-14 = \frac{k}{2}$

State the number by which each side can be divided to solve each equation.

11. $4x = 24$ **12.** $28 = 7x$ **13.** $35 = 4y$ **14.** $-36 = 4z$ **15.** $-7x = 21$

16. $-5x = 14$ **17.** $-8x = -9$ **18.** $-14x = -30$ **19.** $-6x = -36$ **20.** $-10x = 40$

Written Exercises

Solve.

21. $-4r = -28$ **22.** $-7w = -49$ **23.** $-8t = 56$ **24.** $5t = -45$ **25.** $6x = -42$

26. $11y = -77$ **27.** $-5s = -85$ **28.** $-7h = -91$ **29.** $9x = 40$ **30.** $-3y = 52$

31. $3w = -11$ **32.** $5c = 8$ **33.** $434 = -31y$ **34.** $17b = -391$ **35.** $42.51x = 8$

36. $0.132s = 6.41$ **37.** $6.205x = 2.83$ **38.** $0.49x = 6.277$ **39.** $\frac{k}{8} = 6$ **40.** $11 = \frac{k}{5}$

41. $-10 = \frac{b}{-7}$ **42.** $-13 = \frac{b}{-8}$ **43.** $\frac{h}{11} = -25$ **44.** $\frac{h}{12} = -24$ **45.** $\frac{d}{3} = -30$

46. $\frac{d}{4} = -5$ **47.** $-65 = \frac{f}{29}$ **48.** $-70 = \frac{f}{25}$ **49.** $\frac{c}{-8} = -14$ **50.** $\frac{c}{-5} = -24$

51. $\frac{2}{5}t = -10$ **52.** $\frac{4}{9}t = 72$ **53.** $\frac{4}{7}r = 20$ **54.** $\frac{3}{4}x = -12$ **55.** $\frac{4}{9}x = -9$

56. $\frac{9}{11}y = 20$ **57.** $-\frac{3}{5}y = -50$ **58.** $-\frac{2}{3}x = -41$ **59.** $\frac{2}{3}x = -6$ **60.** $-\frac{11}{8}x = 42$

61. $\frac{3}{5}x = 9$ **62.** $-\frac{13}{5}y = -22$ **63.** $\frac{5}{2}x = -25$ **64.** $\frac{9}{2}r = -30$ **65.** $3x = 4\frac{2}{3}$

66. $-5x = -3\frac{2}{3}$ **67.** $\left(5\frac{1}{2}\right)x = 33$ **68.** $\left(-4\frac{1}{2}\right)x = 36$ **69.** $\left(3\frac{1}{2}\right)x = 1\frac{1}{6}$ **70.** $\left(2\frac{1}{3}\right)x = 4\frac{2}{3}$

Complete.

71. If $3x = 15$, then $9x = \underline{\ ?\ }$.

72. If $4x = 9$, then $16x = \underline{\ ?\ }$.

73. If $10y = 46$, then $5y = \underline{\ ?\ }$.

74. If $8y = 14$, then $2y = \underline{\ ?\ }$.

88 *Multiplying and Dividing Rational Numbers*

75. If $2a = -10$, then $6a = \underline{\ ?\ }$.

76. If $-7a = 13$, then $35a = \underline{\ ?\ }$.

77. If $12b = -1$, then $4b = \underline{\ ?\ }$.

78. If $-16b = 22$, then $4b = \underline{\ ?\ }$.

79. If $4k + h = 6$, then $8k + 2h = \underline{\ ?\ }$.

80. If $7k - 5 = 4$, then $21k - 15 = \underline{\ ?\ }$.

Solve.

81. Fourteen times a number is 70. What is the number?

82. Eight times a number is 216. What is the number?

83. Negative twelve times a number is -156. What is the number?

84. Negative seven times a number is 1.476. What is the number?

85. One fourth of a number is -16.325. What is the number?

86. Four thirds of a number is 4.82. What is the number?

87. Joyce Conners paid \$47.50 for five football tickets. What is the cost of each ticket?

88. Ling Ho paid \$8.10 for 7.5 gallons of gasoline. What is the cost per gallon of gasoline?

Challenge Exercises

89. A store sells a six-pack of ginger ale in aluminum cans for \$2.28. Each time Ellen buys ginger ale, she sells the empty cans to a recycler at a rate of 10 cans for 40 cents. How many cans of ginger ale did Ellen buy if the final net total of her expenditures was \$7.48?

mini-review

Write each of the following as an expression using exponents.

1. $6 \cdot 6 \cdot 6 \cdot 6$ **2.** $\frac{1}{3} \cdot x \cdot x \cdot y \cdot y \cdot y$ **3.** $4(n)(n)$ **4.** $5 \cdot 5 \cdot t \cdot t \cdot t \cdot t$

State the property shown.

5. $x + 13 = x + 13$ **6.** $(7 - 4) + n = 3 + n$

Simplify.

7. $25 + (-42)$ **8.** $-\frac{2}{7} - \frac{3}{14}$ **9.** $-16.32 + 0.359$

10. $\left(\frac{3}{5}\right)\left(-\frac{5}{7}\right)$ **11.** $12(-4) + 12(3)$ **12.** $\frac{54a}{-9}$

13. $\frac{\frac{3}{11}}{-6}$ **14.** $-5(4d + 5h) + 8(-5d)$ **15.** $\frac{-11x - (-22y)}{-11}$

Solve.

16. $m + 15 = -21$ **17.** $p - (-3.6) = 4.8$ **18.** $\frac{7}{3}y = -\frac{10}{9}$

Define the variable. Then write an equation for each problem. Do not solve.

19. Alberto scored 6 points in a volleyball game. This was 3 points less than the previous game. How many points did Alberto score in both games?

20. Each week for 8 weeks a store reduced the price of a bicycle by \$9.99. The final reduced price was \$135.07. What was the original price?

3-4 Solving Equations Using More Than One Operation

Each of these equations can be solved by performing the inverse of the operation indicated in the equation.

$$x + 2 = 5 \qquad 3x = 4.8 \qquad \frac{x}{2} = -11$$

Note that each of the following equations contains more than one operation.

$$\frac{x}{4} + 7 = 10 \qquad \frac{a + 10}{3} = 6 \qquad -5x - 7 = 28$$

These equations can be solved using the same methods that were used to solve the first set of equations. However, more than one step is involved.

Recall the order of operations for evaluating an expression. To solve an equation with more than one operation, undo the operations in reverse order.

Examples

1 **Solve $3x + 7 = 13$.**

$3x + 7 = 13$	*Addition of 7 is indicated.*
$3x + 7 - 7 = 13 - 7$	*Therefore, subtract 7 from each side.*
$3x = 6$	*Multiplication by 3 is also indicated.*
$\frac{3x}{3} = \frac{6}{3}$	*Therefore, divide each side by 3.*
$x = 2$	

Check:
$$3x + 7 = 13$$
$$3(2) + 7 \overset{?}{=} 13$$
$$6 + 7 \overset{?}{=} 13$$
$$13 = 13 \quad \checkmark$$

The solution is 2.

2 **Solve $\frac{x}{4} + 9 = 6$.**

$\frac{x}{4} + 9 = 6$	
$\frac{x}{4} + 9 - 9 = 6 - 9$	*First subtract 9 from each side. Why?*
$\frac{x}{4} = -3$	
$4\left(\frac{x}{4}\right) = 4(-3)$	*Then multiply each side by 4. Why?*
$x = -12$	*Check this result.*

The solution is -12.

Examples

3 Solve $7 = 14 - 3x$.

$$7 = 14 - 3x$$
$$7 - 14 = 14 - 3x - 14$$
$$7 - 14 = (14 - 14) - 3x$$
$$-7 = -3x$$
$$\frac{-7}{-3} = \frac{-3x}{-3}$$
$$\frac{7}{3} = x \qquad \textit{Check this result.}$$

The solution is $\frac{7}{3}$.

4 Solve $\frac{d-4}{3} = 5$.

$$\frac{d-4}{3} = 5$$
$$3\left(\frac{d-4}{3}\right) = 3(5)$$
$$d - 4 = 15$$
$$d - 4 + 4 = 15 + 4$$
$$d = 19 \qquad \textit{Check this result.}$$

The solution is 19.

Exploratory Exercises

State the steps you would use to solve each equation.

1. $3x - 7 = 2$
2. $5n - 4 = -7$
3. $2x + 5 = 13$
4. $8 + 3x = 5$

5. $\frac{a+2}{5} = 10$
6. $\frac{a-7}{15} = -30$
7. $\frac{3}{11}x + 2 = 5$
8. $-\frac{4}{13}y - 7 = 6$

Written Exercises

Solve and check each equation.

9. $4t - 7 = 5$
10. $6 = 4n + 2$
11. $4 + 7x = 39$
12. $7n - 4 = 17$

13. $5n + 3 = 9$
14. $34 = 8 - 2t$
15. $5 - 3x = 32$
16. $2 - 7s = -19$

17. $-3x - 7 = 18$
18. $-4y + 2 = 29$
19. $7 = \frac{x}{2} + 5$
20. $14 = \frac{m}{3} - 6$

21. $\frac{y}{3} + 6 = -45$
22. $\frac{b}{6} - 7 = 13$
23. $\frac{b}{3} + 2 = -21$
24. $\frac{c}{-4} - 8 = -42$

25. $\frac{c}{-3} - 1 = 26$
26. $\frac{c}{-9} + 18 = 40$
27. $\frac{d+5}{3} = -9$
28. $\frac{3+n}{7} = -5$

29. $5 = \frac{m-5}{4}$
30. $8 = \frac{t-6}{7}$
31. $16 = \frac{s-8}{-7}$
32. $7 = \frac{r-8}{6}$

33. $\frac{4d+5}{7} = 7$
34. $\frac{4r+8}{16} = 7$
35. $\frac{7n+(-1)}{8} = 8$
36. $\frac{-3n-(-4)}{-6} = -9$

37. $0.2n + 3 = 8.6$
38. $0.5x + 1.5 = 12$
39. $8 - 1.2s = -1.6$

40. $4.3 + 2.306y = -5.25$
41. $0.8 = 4.11n - 24.831$
42. $-0.0007 = 2.6t + 8.06$

43. $\frac{3}{4}n - 3 = 9$
44. $\frac{1}{3}x + 4 = 6$
45. $7 = 3 - \frac{n}{3}$

46. $8 = 5 + \frac{t}{4}$
47. $5 - \frac{2}{3}n = 7$
48. $8 + \frac{3}{4}n = 26$

Challenge Exercises

Solve.

49. $2[x + 3(x - 1)] = 18$
50. $4(2x - 7) + 3(x - 1) = 46$

51. $-6 = \frac{1}{4}\left[\frac{3}{2}(x - 2) + 2x\right]$
52. $-\frac{4}{5} = -\frac{1}{2}\left[4 - 3\left(x + \frac{1}{5}\right)\right]$

Applications in Physics

Suppose an automobile is moving at a speed of 50 mph. Wishing to pass another car, the driver increases the speed to 55 mph in 10 seconds. The driver has accelerated the car.

Acceleration is the rate at which speed is changing with respect to time. To compute acceleration (a) you divide the change in speed by the time (t) needed to make the change. The change in speed is the difference between the final speed (f) and the starting speed (s). This can be summarized in the following formula.

$$a = \frac{f - s}{t}$$

Example **A race car goes from 44 m/s to 77 m/s in 11 seconds. Find the change in speed per second.**   *m/s is the symbol for meters per second.*

$$a = \frac{f - s}{t}$$

$$a = \frac{77 - 44}{11} \text{ or } 3 \qquad \textit{Substitute 77 for f, 44 for s, and 11 for t.}$$

The car accelerates 3 meters per second each second or 3 m/s².

Consider what happens when the racing car slows down from 77 m/s to 44 m/s during the 11 seconds. The same equation can be used. Subtracting the starting speed from the final speed gives a negative value. This means that the acceleration is negative.   *Negative acceleration is <u>deceleration</u>.*

$$a = \frac{44 - 77}{11} = -3$$

The acceleration is −3 m/s².   *The deceleration is 3 m/s².*

Exercises

Find the acceleration for each problem.

1. A motorcycle goes from 2 m/s to 14 m/s in 6 seconds.

2. A skateboard goes from 5 m/s to 0 m/s in 1 second.

3. A train is traveling for 1 hour at a constant speed of 50 mph.

4. A car starts from a standstill. It accelerates to 42.5 mph in 11.2 seconds.

5. A jet plane decreases its speed from 500 km/h to 350 km/h in 29.6 seconds.

6. A car traveling at 36.4 mph is stopped in 5.75 seconds.

3-5 More Equations

Many equations contain variables on each side. To solve such equations, first use the Addition or Subtraction Property of Equality to write an equivalent equation that has all the variables on one side. Then finish solving the equation by using the methods shown previously in this chapter.

Examples

1 Solve $5n - 7 = 3n + 2$.

$$5n - 7 = 3n + 2$$
$$5n - 3n - 7 = 3n - 3n + 2 \qquad \text{Subtract 3n from each side.}$$
$$2n - 7 = 2$$
$$2n - 7 + 7 = 2 + 7 \qquad \text{Add 7 to each side.}$$
$$2n = 9$$
$$\frac{2n}{2} = \frac{9}{2} \qquad \text{Divide each side by 2.}$$
$$n = \frac{9}{2}$$

Check:
$$5n - 7 = 3n + 2$$
$$5\left(\frac{9}{2}\right) - 7 \stackrel{?}{=} 3\left(\frac{9}{2}\right) + 2$$
$$\frac{45}{2} - \frac{14}{2} \stackrel{?}{=} \frac{27}{2} + \frac{4}{2}$$
$$\frac{31}{2} = \frac{31}{2} \quad \checkmark$$

The solution is $\frac{9}{2}$.

2 Solve $\frac{3}{5}x + 3 = \frac{1}{5}x - 7$.

$$\frac{3}{5}x + 3 = \frac{1}{5}x - 7$$
$$\frac{3}{5}x - \frac{1}{5}x + 3 = \frac{1}{5}x - \frac{1}{5}x - 7 \qquad \text{Subtract } \tfrac{1}{5}x \text{ from each side.}$$
$$\frac{2}{5}x + 3 = -7$$
$$\frac{2}{5}x + 3 - 3 = -7 - 3 \qquad \text{Subtract 3 from each side.}$$
$$\frac{2}{5}x = -10$$
$$\frac{5}{2}\left(\frac{2}{5}x\right) = \frac{5}{2}(-10) \qquad \text{Multiply each side by } \tfrac{5}{2}.$$
$$x = -25 \qquad \text{Check this result.}$$

The solution is -25.

Many equations also contain grouping symbols. When solving equations of this type, first use the Distributive Property to remove the grouping symbols.

Example

3 **Solve $5 - 2(3 - r) = 2(2r + 5) + 1$.**

$$5 - 2(3 - r) = 2(2r + 5) + 1$$

$5 - 6 + 2r = 4r + 10 + 1$ *Use the Distributive Property.*

$-1 + 2r = 4r + 11$ *Simplify.*

$-1 + 2r - 2r = 4r - 2r + 11$

$-1 = 2r + 11$

$-1 - 11 = 2r + 11 - 11$

$-12 = 2r$

$\frac{-12}{2} = \frac{2r}{2}$

$-6 = r$ *Check this result.*

The solution is -6.

Some equations may have *no* solution. Other equations may have *all numbers* in their solution sets. An equation that is true for every value of the variable is called an **identity**.

Examples

4 **Solve $2x + 5 = 2x - 3$.**

$$2x + 5 = 2x - 3$$

$$2x + 5 - 2x = 2x - 3 - 2x$$

$$2x - 2x + 5 = 2x - 2x - 3$$

$5 = -3$ *This is a false statement.*

This equation has no solution.

There is no value of x that will make the equation 2x + 5 = 2x − 3 true.

5 **Solve $3(x + 1) - 5 = 3x - 2$.**

$$3(x + 1) - 5 = 3x - 2$$

$$3x + 3 - 5 = 3x - 2$$

$$3x - 2 = 3x - 2$$

$3x - 2 + 2 = 3x - 2 + 2$ *Add 2 to each side.*

$3x = 3x$

$\frac{3x}{3} = \frac{3x}{3}$ *Divide each side by 3.*

$x = x$ *This statement is true for all values of x.*

This equation is an identity. The solution is the set of all numbers.

Exploratory Exercises

State the steps you would use to solve each equation.

1. $3x + 2 = 4x - 1$ **2.** $6x - 8 = 9 + 2x$ **3.** $4y - 3 = -5y + 7$ **4.** $2y + 7 = -3y$

5. $8y - 10 = -3y + 2$ **6.** $-2y - 5 = 10 + 9y$ **7.** $3(x + 1) = 7$ **8.** $4(x - 1) = 5$

9. $4(3 + 5y) = -11$ **10.** $-7(8 - y) = 12$ **11.** $6(2x - 3) = -9$ **12.** $-7(x - 3) = -4$

Written Exercises

Solve and check each equation.

13. $6x + 7 = 8x - 13$

14. $3x - 5 = 7x + 7$

15. $3 - 4x = 10x + 10$

16. $13 - 8x = 5x + 2$

17. $17 + 2x = 21 + 2x$

18. $-5x - 1 = -5x - 1$

19. $6x + 3 = 6x + 3$

20. $18 - 4x = 42 - 4x$

21. $14x - 6 = -2x + 8$

22. $32 - 15x = -10 + 26x$

23. $28x + 1 = 1 + 28x$

24. $43x - 7 = -9 + 43x$

25. $3(x + 2) = 12$

26. $0.003(x - 5.25) = -6.367$

27. $7.225(x - 3.1) - 2 = 0.554$

28. $3x - 2(x + 3) = x$

29. $7 + 2(x + 1) = 2x + 9$

30. $6(x + 2) - 4 = -10$

31. $6 = 3 + 5(x - 2)$

32. $-2(2x - 3) = 6 - 4x$

33. $4(x - 2) = 4x$

34. $-5 = 4 - 2(x - 5)$

35. $6.25(x + 0.333) = 4.8(x - 1.1)$

36. $8.1(x - 9.25) = -8.1(x + 5.399)$

37. $-3(x + 5) = 3(x - 1)$

38. $5x - 7 = 5(x - 2) + 3$

39. $7 - 3x = x - 4(2 + x)$

40. $4(2x - 1) = -10(x - 5)$

41. $5(x - 14) = 3(12 - 6x)$

42. $-8(4 + 9x) = 7(-2 - 11x)$

43. $2(x - 3) + 5 = 3(x - 1)$

44. $5x + 4 = 7(x + 1) - 2x$

45. $2x - 4(x - 5) = 2(10 - x)$

46. $6(x - 2) = 5(x - 11) - 21$

47. $\frac{1}{3}(x + 6) = 7$

48. $5 - \frac{1}{2}(x - 6) = 4$

49. $3(x - 5) = \frac{1}{5}(10x - 25)$

50. $4(2x - 8) = \frac{1}{7}(49x + 70)$

51. $\frac{2}{3}n + 8 = \frac{1}{3}n - 2$

52. $\frac{3}{4}n + 16 = 2 - \frac{1}{8}n$

53. $3 + \frac{2}{5}x = 11 - \frac{2}{5}x$

54. $\frac{1}{2}t + 6 = \frac{1}{3}t$

55. $2.1k - 50 = 7.3k - 3.2$

56. $18 - 3.8x = 7.36 - 1.9x$

57. $8n - 11.35 = 1.2n + 17.21$

58. $3.1x - 2.78 = 7.22 - 1.9x$

59. $3(4x - 8) = -7(2x - 3)$

60. $16(2x - 5) = -2(5x - 18)$

61. $3(5 - t) = 14$

62. $3(2x + 1) = 6(x - 1)$

63. $10\left(t - \frac{3}{5}\right) = 8$

64. $5\left(2w - \frac{3}{5}\right) = w - 3$

65. $8.002 + 10.39x = 5x + 1.111$

66. $12.35(4.26 - y) = 5.7008$

67. $9(2 + w) = 33$

68. $2(3v - 1) = 6v - 2$

69. $-3(2n - 5) = \frac{1}{2}(-12n + 30)$

70. $2(5 - 8n) = -4(3 + 4n)$

Write an algebraic expression for each verbal expression.

1. a number increased by 23

2. forty-two less than a number

3. a number decreased by 67

4. the product of 5 and the cube of a number

Evaluate each expression.

5. $3 + \frac{8}{2} - 6$

6. $\frac{5(3 + 2)}{5} + \frac{24}{6}$

7. $8(3.6 + 0.9) - 2.3$

Evaluate each expression if $a = 4$, $b = -6$, and $c = 8$.

8. $a^2 - b + c$

9. $c(a - b) + b$

10. $\frac{5ab^2}{b - c}$

Simplify.

11. $7a + 14a$

12. $16ay - 9ay$

13. $7ab^2 + 8a^2b - 5ab^2$

14. $x^2 + \frac{7}{8}x - \frac{x}{8}$

15. $5a + 8c + 12a$

16. $6a + 5b + 7c + 8a$

Translate each sentence into an equation. Do not solve.

17. The number y equals the square of the sum of a and b.

18. A number decreased by 23 is 45.

19. Seventy-eight decreased by twice a number is 50.

20. The sum of t and r equals the square of x.

Find each sum or difference.

21. $-3 + 16$

22. $-5 - (-6)$

23. $63 + (-23)$

24. $\left(-\frac{8}{15}\right) + \left(-\frac{4}{15}\right)$

25. $315 + |-71|$

26. $|-43 - (-10)| - 8$

27. $7 + (-12) + (-34)$

28. $\left(-\frac{3}{4}\right) + \left(-\frac{4}{3}\right) + \frac{7}{4}$

29. $6.7 + 8.4 - 9.3$

30. $11.4 + (-18.9) + 34.2$

31. $-12a + 5b + 23a$

32. $11a + (-3a) + 4b + (-12b)$

Solve.

33. $4 + a = -17$

34. $a - 5 = -36$

35. $y - 2.3 = 7.8$

36. $1.7 - y = -5.6$

37. $-4r = 76$

38. $\frac{k}{6} = 9$

39. $\frac{3}{7}y = 5$

40. $8a - 6 = 42$

41. $\frac{m - 8}{4} = 6$

42. $0.2n + 3 = 8.4$

43. $3n - 8 = 5n + 8$

44. $7 - 2n = 9 - 4n$

Simplify.

45. $-5(8) + (-4)(7)$

46. $\frac{72a}{15}$

47. $\frac{6x - 8}{-4}$

48. $5(3t - 4b) + 7(4t + 5b)$

49. $-\frac{2}{7}(35x + 42y) + \frac{3}{7}(49x)$

50. Solve $0.02x - 7.39 = 16.25 - 0.8x$.

3-6 Still More Equations

You can solve equations containing fractions and decimals by first using the Multiplication Property of Equality to eliminate the fractions and decimals.

Examples

1 Solve $\frac{2x}{5} + \frac{x}{4} = \frac{26}{5}$.

$$\frac{2x}{5} + \frac{x}{4} = \frac{26}{5}$$ *The least common denominator is 20.*

$$20\left(\frac{2x}{5} + \frac{x}{4}\right) = 20\left(\frac{26}{5}\right)$$ *Multiply each side of the equation by 20.*

$$20\left(\frac{2x}{5}\right) + 20\left(\frac{x}{4}\right) = 20\left(\frac{26}{5}\right)$$ *Use the Distributive Property.*

$$8x + 5x = 104$$ *The fractions are eliminated.*

$$13x = 104$$

$$x = 8$$

Check:
$$\frac{2x}{5} + \frac{x}{4} = \frac{26}{5}$$

$$\frac{2(8)}{5} + \frac{8}{4} \overset{?}{=} \frac{26}{5}$$

$$\frac{16}{5} + \frac{8}{4} \overset{?}{=} \frac{26}{5}$$

$$\frac{64}{20} + \frac{40}{20} \overset{?}{=} \frac{104}{20}$$

$$\frac{104}{20} = \frac{104}{20} \quad \checkmark$$

The solution is 8.

2 Solve $2.1x + 45.2 = -7.3 - 8.4x$.

$$2.1x + 45.2 = -7.3 - 8.4x$$ *Each decimal involves tenths.*

$$10(2.1x + 45.2) = 10(-7.3 - 8.4x)$$ *Multiply each side of the equation by 10.*

$$10(2.1x) + 10(45.2) = 10(-7.3) - 10(8.4x)$$

$$21x + 452 = -73 - 84x$$ *The decimals are eliminated.*

$$21x + 84x + 452 = -73 - 84x + 84x$$

$$105x + 452 - 452 = -73 - 452$$

$$105x = -525$$

$$x = -5$$ *Check this result.*

The solution is -5.

Some equations contain more than one variable. It is often necessary to solve such equations for a specific variable.

Example

3 Solve for x in $ax + b = dx + c$.

$$ax + b = dx + c$$
$$ax + b - b = dx + c - b$$
$$ax = dx + c - b$$
$$ax - dx = dx - dx + c - b$$
$$(a - d)x = c - b$$
$$\frac{(a - d)x}{a - d} = \frac{c - b}{a - d}$$
$$x = \frac{c - b}{a - d}$$

Division by zero is undefined.
Therefore $a - d \neq 0$, or $a \neq d$.

Exploratory Exercises

State the number by which each side can be multiplied to eliminate the fractions or decimals. Then rewrite each equation.

1. $\frac{3}{4}x - 7 = 8 + \frac{2}{3}x$

2. $\frac{2}{5}x = 7 - \frac{3}{4}x$

3. $4t - 7 = \frac{5t}{2} - 3$

4. $1.2s + 8.1 = 3.5 - 2s$

5. $5.2z = 3 + 1.7z$

6. $8.17y = 4.2 - 3.7y$

Written Exercises

Solve and check each equation.

7. $\frac{y + 5}{3} = 7$

8. $\frac{x - 7}{5} = 12$

9. $\frac{3n - 2}{5} = \frac{7}{10}$

10. $0.2x + 1.7 = 3.9$

11. $5.3 - 0.3x = -9.4$

12. $1.9s + 6 = 3.1 - s$

13. $\frac{3}{4}x - 4 = 7 + \frac{1}{2}x$

14. $\frac{3}{8} - \frac{1}{4}x = \frac{1}{2}x - \frac{3}{4}$

15. $\frac{5}{8}x + \frac{3}{5} = x$

16. $3y - \frac{4}{5} = \frac{1}{3}y$

17. $\frac{1}{3}y + 3 = \frac{1}{2}y$

18. $\frac{2}{5}x - 1 = \frac{1}{4}x$

19. $\frac{4y + 3}{7} = \frac{9}{14}$

20. $\frac{4 - x}{5} = \frac{1}{5}x$

21. $\frac{7 + 3t}{4} = -\frac{1}{8}t$

22. $1.033x - 4 = 2.15x + 8.7$

23. $0.882 - 0.25z = 0.301z - 4$

24. $2.15y - 7 = 1.285y - 2$

25. $\frac{2}{3}x + \frac{1}{2} = \frac{1}{3}x + 4$

26. $\frac{3y}{2} - y = 4 + \frac{1}{2}y$

27. $\frac{x}{2} - \frac{1}{3} = \frac{x}{3} - \frac{1}{2}$

Solve for x.

28. $5x = y$

29. $x + r = 2d$

30. $\frac{x + a}{3} = c$

31. $\frac{d + x}{e} = f$

32. $ax + b = c$

33. $ex - 2y = 3z$

Solve for y.

34. $ay - b = c$

35. $c - 2y = d$

36. $ay + z = am - n$

37. $a(y + 1) = b$

38. $\frac{x}{y} = 4$

39. $\frac{3}{5}y + a = b$

Solve.

40. Find a number that when increased by 6 is 12 less than its opposite.

41. Find a number that when decreased by 30 is 12 more than twice its opposite.

42. Five more than two-thirds of a number is the same as three less than one-half of the number. Find the number.

43. One-fifth of a number plus five times that number is equal to seven times the number less 18. Find the number.

Challenge Exercises

44. The sum of two numbers is 25. Twelve less than four times one of the numbers is 16 more than twice the other number. Find both numbers.

45. The difference of two numbers is 12. Two-fifths of one of the numbers is six more than one-third of the other number. Find both numbers.

mini-review

Evaluate each expression.

1. $4 - 10 \div 2 + 18$

2. $\frac{3^3 - 6^3 + 9^2}{3 \cdot 5}$

3. $\frac{3}{4}(18) + \frac{3}{5}(35)$

4. $2692 + (-1368)$

5. $|296.5 - 413|$

6. $|-32| - |48.66|$

Simplify.

7. $\frac{\frac{16}{4}}{9}$

8. $0.2(x + 3y) + 1.9x$

9. $\frac{n}{4} - 5\left(\frac{n}{3} - n^2\right)$

10. $-\frac{6}{11} + \frac{7}{3} + (-9)$

11. $-18.33mn + 39ab + (-263mn)$

Solve.

12. $-38 = -16 - x$

13. $\frac{a}{-6} = -8$

14. $\frac{4a - 3}{3} = -1$

15. $-2 + \frac{2}{5}w = -6$

16. $-6(2h - 8) = -7(3h + 5)$

Excursions in Algebra   Mental Computation

The Distributive Property can be used to find products mentally.

$35(22) = 35(20 + 2)$

  Rename 22 as 20 + 2.

$= 35(20) + 35(2)$

$= 700 + 70$

$= 770$ *35(22) = 770*

$75(2.6) = 75(3.0 - 0.4)$

  Rename 2.6 as 3.0 - 0.4.

$= 75(3.0) - 75(0.4)$

$= 225 - 30$

$= 195$ *75(2.6) = 195*

Exercises

Find each product using the strategy shown above.

1. $7(14)$

2. $9(102)$

3. $(42)(18)$

4. $(35)(33)$

5. $(2.4)(27)$

6. $(1.5)(340)$

7. $3000(63)$

8. $\left(3\frac{1}{2}\right)(\$12.50)$

3-7 Problem Solving: Using Equations

In cross-country, a team's score is found by adding the place numbers of the first five finishers. For example, five runners for a team finished 2nd, 6th, 8th, 9th, and 13th. Their team's score is found as follows.

$$2 + 6 + 8 + 9 + 13 = 38$$

Jeff and Raul ran in an invitational meet. Jeff finished 3 places ahead of Raul. Their combined score was 27. In what place did each runner finish?

Use the problem solving plan to find the solution.

explore

Define a variable.

Let x = Jeff's place number.
Then $x + 3$ = Raul's place number.

plan

Write an equation.

The combined score was 27.

$$x + (x + 3) \qquad = \quad 27$$

solve

Solve the equation and answer the problem.

$$
\begin{aligned}
x + (x + 3) &= 27 \\
2x + 3 &= 27 \\
2x + 3 - 3 &= 27 - 3 \\
2x &= 24 \\
\frac{2x}{2} &= \frac{24}{2} \\
x &= 12
\end{aligned}
$$

Since x represents Jeff's place number, Jeff finished in 12th place. Since $x + 3$ represents Raul's place number, Raul finished in 15th place.

examine

Check to see if the answer makes sense.

Read the problem again and compare your answer to the statements given in the problem. For the problem above you could ask the following questions.

1. If Jeff finished 12th and Raul finished 15th, did Jeff finish 3 places ahead of Raul?
2. Was their combined score 27?

Example

1 **Use an equation to solve the problem. Then examine the solution.**

Megan wants to buy a 10-speed bicycle that costs $117. This is $12 more than three times the amount that she saved last month. How much did she save last month?

explore   Let x = amount of money saved last month.

plan   Cost is $12 more than 3 times amount saved last month.

$$117 \;=\; 12 \quad\quad + \quad\quad\quad\quad\quad\quad 3x$$

solve
$$117 = 12 + 3x$$
$$117 - 12 = 12 - 12 + 3x$$
$$105 = 3x$$
$$\frac{105}{3} = \frac{3x}{3}$$
$$35 = x$$

Megan saved $35 last month.

examine   Is $117 equal to $12 more than three times $35?
$$117 \overset{?}{=} 3(35) + 12$$
$$117 \overset{?}{=} 105 + 12$$
$$117 = 117 \quad ✔$$

Since this is a true statement, $35 is the correct answer.

Consecutive numbers are numbers in counting order such as 3, 4, 5. Beginning with an even integer and counting by two gives *consecutive even integers*. For example, $-6, -4, -2, 0, 2, 4$ are consecutive even integers. Beginning with an odd integer and counting by two gives *consecutive odd integers*. For example, $-3, -1, 1, 3, 5$ are consecutive odd integers.

Example

2 **Find three consecutive even integers whose sum is -12.**

Let x = the least even integer.
$x + 2$ = the next greater even integer.
$x + 4$ = the greatest of the three even integers.

$$x + (x + 2) + (x + 4) = -12 \quad\quad \textit{The sum of the integers is } -12.$$
$$3x + 6 = -12$$
$$3x = -18$$
$$x = -6$$

Therefore, $x + 2 = -4$ and $x + 4 = -2$. The integers are $-6, -4,$ and -2.

Exploratory Exercises

Solve each problem.

1. State three consecutive integers if the greatest one is 4.

2. State three consecutive integers if the least one is −2.

3. State three consecutive even integers if the least one is −4.

4. State three consecutive even integers if the greatest one is 10.

5. State four consecutive odd integers if the least one is 13.

6. State four consecutive odd integers if the least one is −7.

For each sentence, define a variable. Then state an equation.

7. The sum of two consecutive integers is 17.

8. The sum of three consecutive even integers is 48.

9. The sum of two consecutive odd integers is −36.

10. The sum of four consecutive integers is −46.

11. Three times a number increased by 4 is −11.

12. Seventeen decreased by twice a number is 5.

Written Exercises

For each problem, define a variable. Then use an equation to solve the problem. Some problems may have no solution.

13. Find three consecutive integers whose sum is 87.

14. Find three consecutive integers whose sum is 114.

15. Find four consecutive integers whose sum is 130.

16. Find four consecutive integers whose sum is 278.

17. Find two consecutive even integers whose sum is 115.

18. Find two consecutive even integers whose sum is 138.

19. Find two consecutive odd integers whose sum is 64.

20. Find two consecutive odd integers whose sum is 35.

21. Find three consecutive odd integers whose sum is 99.

22. Find three consecutive even integers whose sum is 72.

23. Twice a number increased by 20.065 is 62.885. Find the number.

24. Three times a number decreased by 11.36 is 58.309. Find the number.

25. Karen has 6 more than twice as many newspaper customers as when she started. She now has 98 customers. How many did she have when she started?

26. Bonnie sold some stock for $42 a share. This was $10 a share more than twice what she paid for it. At what price did she buy the stock?

27. Last year Marc Ames sold 7 sedans more than twice the number of vans Lola Franco sold. Marc sold 83 sedans. How many vans did Lola sell?

28. One season Reggie Walker scored 9 more runs than twice the number of runs he batted in. He scored 117 runs that season. How many runs did he bat in?

29. A rectangular playground is 60 meters longer than it is wide. It can be enclosed by 920 meters of fencing. Find its length.

30. A soccer field is 75 yards shorter than 3 times its width. Its perimeter is 370 yards. Find its dimensions.

31. The captain of the cross-country team placed second in a meet. The other four members placed in consecutive order, but farther behind. The team score was 40. In what places did the other members finish?

32. One member of the cross-country team placed fourth in a meet. The other four members placed in consecutive order, but farther behind. The team score was 70. In what places did the other members finish?

33. Namid bought a used bike for $8 more than half its original price. Namid paid $40 for the bike. What was the original price?

34. Heather bought a winter coat for $6 less than half its original price. Heather paid $65 for the coat. What was the original price?

35. The lengths of the sides of a triangle are consecutive odd integers. The perimeter is 27 m. What are the lengths of the sides?

36. The lengths of the sides of a four-sided figure are consecutive even integers. The perimeter is 156 ft. What are the lengths of the sides?

37. Twice a number increased by 4 times the number is 96. Find the number.

38. Twice a number increased by 12 is 31 less than three times the number. Find the number.

39. A number increased by 5.255 less than twice itself is 415.33. Find the number.

40. Five times a number decreased by 4.002 more than twice the number is 119.34. Find the number.

41. Five-eights of a number is three more than half of the number. Find the number.

42. One-half of a number increased by 16 is four less than two-thirds of the number. Find the number.

43. Find four consecutive even integers such that twice the least increased by the greatest is 96.

44. Find four consecutive odd integers such that the sum of the first and twice the second is 175.

45. Twice the greater of two consecutive odd integers is 13 less than three times the lesser. Find the integers.

46. Three times the greatest of 3 consecutive even integers exceeds twice the least by 38. Find the integers.

47. Find two consecutive integers such that 3 times the first integer plus 2 times the second integer is equal to 107.

48. Find two consecutive integers such that 2 times the first integer plus 4 times the second integer is equal to 256.

Challenge Exercises

49. The length of a rectangle is 40 m less than 2 times its width. Its perimeter is 220 m. Find its dimensions.

50. The length of a rectangle is 35 m more than 3 times its width. Its perimeter is 390 m. Find its dimensions.

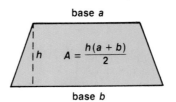

base a

$$A = \frac{h(a + b)}{2}$$

base b

51. A trapezoid has an area of 117 sq ft and a height of 9 ft. One base is 4 ft shorter than the other base. How long is each base?

Problem Solving

Many problems can be solved more easily if you draw a picture or diagram to represent the situation. Sometimes a picture will help you decide how to work a problem. At other times the picture will show the answer to the problem.

Example   **The length of a rectangle is twice its width. The perimeter is 39 meters. Find the length and width.**

Draw one side and mark it x.

Then draw an adjacent side twice as long as the first side.

Finish drawing the rectangle.

Label all four sides.

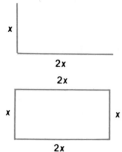

You can use the diagram to write an equation for the problem.

$$x + 2x + x + 2x = 39$$
$$6x = 39$$
$$x = 6.5$$

The width is 6.5 meters and the length is twice the width, or 13 meters.

Exercises

Solve each problem. Use pictures and diagrams.

1. Two sides of a triangle have the same length. The third side is 2 m longer. If the perimeter of the triangle is 20 m, find the lengths of the sides.

2. You can cut a pizza into 7 pieces with only 3 straight cuts. What is the greatest number of pieces you can make with 5 straight cuts?

3. A row of 20 seats has some red seats and some blue seats. The first two are red, the next two are blue, the next two are red, and so on. What color is the 14th seat?

4. Three spiders are on a 9-ft wall. Susie spider is 4 feet from the top. Sam spider is 7 feet from the bottom. Shirley spider is 3 feet below Sam spider. Which spider is nearest the top of the wall?

5. A row of coins is arranged in the following pattern: 1 dime, followed by 3 nickels, then 5 dimes, 7 nickels, and so on. What is the 40th coin in the row?

6. An ant is climbing a 30-ft flagpole. Each day he climbs up 7 feet. Each night he slips back 4 feet. How many days will it take to reach the top of the flagpole?

7. A staircase with 3 steps was built with 6 blocks. Can a staircase be built with 28 blocks? How many steps would it have?

8. There were 9 people at a party. Everyone shook hands with everyone else exactly once. How many handshakes occurred?

3-8 Ratio and Proportion

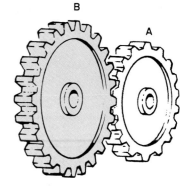

As gear A revolves 4 times, it will cause gear B to revolve 3 times. Hence, we say that the gear ratio of A to B is 4 to 3.

In mathematics, a **ratio** is a comparison of two numbers by division. The gear ratio above can be expressed in the following ways.

$$4 \text{ to } 3 \qquad 4{:}3 \qquad \frac{4}{3}$$

A ratio is most commonly expressed as a fraction in simplest form.

Examples

1 **What is the ratio of 10 meters to 6 meters?**

The ratio is $\frac{10}{6}$ or $\frac{5}{3}$.

2 **What is the ratio of 20 inches to 4 feet?**

The units should be the same.

$$4 \text{ feet} = 48 \text{ inches} \qquad \textit{Change feet to inches.}$$

The ratio is $\frac{20}{48}$ or $\frac{5}{12}$.

An equation of the form $\frac{a}{b} = \frac{c}{d}$ that states that two ratios are equal is called a proportion.

Definition of Proportion

Every proportion consists of four terms.

first   *third*

$$\frac{a}{b} = \frac{c}{d}$$

second fourth

The first and fourth terms, a and d, are called the **extremes**.

The second and third terms, b and c, are called the **means**.

In a proportion, the product of the extremes is equal to the product of the means.

$$\text{If } \frac{a}{b} = \frac{c}{d}, \text{ then } ad = bc.$$

Means-Extremes Property of Proportions

You can use this property to solve some equations. The equation must be in the form of a proportion.

3 Solve $\frac{21}{27} = \frac{x}{18}$.

$$\frac{21}{27} = \frac{x}{18}$$

$21(18) = 27x$ *Means-Extremes Property*

$378 = 27x$

$14 = x$ *Check this result.*

The solution is 14.

4 **Using Calculators**

Use a calculator to solve the proportion $\frac{6}{2.56} = \frac{9.32}{m}$.   *By the Means-Extremes Property, $6m = (2.56)(9.32)$.*

ENTER: 2.56 $\boxed{\times}$ 9.32 $\boxed{\div}$ 6 $\boxed{=}$ *Multiply the values shown diagonally. Then divide by the third value.*

Your calculator display should read 3.9765333. When rounded to the nearest thousandth, the solution is 3.977.

5 Solve $\frac{x}{5} = \frac{x+3}{10}$.

$$\frac{x}{5} = \frac{x+3}{10}$$

$10x = 5(x + 3)$ *Means-Extremes Property*

$10x = 5x + 15$ *Distributive Property*

$5x = 15$

$x = 3$ *Check this result.*

The solution is 3.

A ratio of two measurements having different units of measure is called a *rate*. For example, 20 miles per gallon is a rate. Proportions are often used to solve problems involving rate.

Example

6 **A trip of 96 miles required 6 gallons of gasoline. At that rate, how many gallons would be required for a 152-mile trip?**

    Let x = gallons required for 152-mile trip.

plan   Write a proportion for the problem.

$$\frac{96 \text{ miles}}{6 \text{ gallons}} = \frac{152 \text{ miles}}{x \text{ gallons}}$$ *Note that both ratios compare miles to gallons.*

solve

$$\frac{96}{6} = \frac{152}{x}$$

$$96x = 912$$

$$x = 9\frac{1}{2} \text{ or } 9.5 \quad \textit{Check this result.}$$

A trip of 152 miles would require $9\frac{1}{2}$ or 9.5 gallons of gasoline.

examine A trip of approximately 100 miles required 6 gallons of gasoline. A trip of 150 miles would require about $1\frac{1}{2}$ times the amount of gasoline or 9 gallons. Therefore, a solution of $9\frac{1}{2}$ gallons is reasonable.

Exploratory Exercises

Write each ratio as a fraction in simplest form.

1. 3 grams to 11 grams
2. 7 feet to 3 feet
3. 21 meters to 16 meters
4. 16 cm to 5 cm
5. 12 ounces to 6 ounces
6. 15 km to 5 km
7. 8 feet to 28 inches
8. 4 pounds to 100 ounces
9. 16 cm to 40 mm
10. 72 mm to 90 mm
11. 35 minutes to 2 hours
12. 5 hours to 52 minutes

Written Exercises

Solve each proportion.

13. $\frac{3}{4} = \frac{x}{8}$
14. $\frac{3}{15} = \frac{b}{45}$
15. $\frac{2}{10} = \frac{1}{y}$
16. $\frac{10}{a} = \frac{20}{28}$

17. $\frac{6}{8} = \frac{7}{x}$
18. $\frac{x}{9} = \frac{7}{16}$
19. $\frac{9}{m} = \frac{15}{10}$
20. $\frac{6}{5} = \frac{18}{t}$

21. $\frac{5}{n} = \frac{4}{1.6}$
22. $\frac{8.6}{25.8} = \frac{1}{n}$
23. $\frac{1}{0.19} = \frac{12}{n}$
24. $\frac{n}{8} = \frac{0.21}{2}$

25. $\frac{x}{4.085} = \frac{5}{16.33}$
26. $\frac{2.405}{3.67} = \frac{g}{1.88}$
27. $\frac{s}{9.65} = \frac{7}{1.066}$
28. $\frac{19.25}{a} = \frac{7.055}{2.94}$

29. $\frac{x + 2}{5} = \frac{7}{5}$
30. $\frac{6}{14} = \frac{7}{x - 3}$
31. $\frac{3}{5} = \frac{x + 2}{6}$
32. $\frac{14}{10} = \frac{5 + x}{x - 3}$

33. $\frac{9}{x - 8} = \frac{4}{5}$
34. $\frac{4 - x}{3 + x} = \frac{16}{25}$
35. $\frac{x - 12}{6} = \frac{x + 7}{-4}$
36. $\frac{x + 9}{-5} = \frac{x - 10}{11}$

Use a proportion to solve each problem.

37. Stewart earns $97 in 4 days. At that rate, how many days will it take him to earn $485?

38. Peggy saves $18 in 4 weeks. How long will it take her to save $81 at the same rate?

39. Shan Wong used 25 gallons of gasoline in traveling 350 miles. How much gasoline will he use in traveling 462 miles?

40. Andrea Jones drove 244 kilometers in 4 hours. At that rate, how long will it take her to drive 366 kilometers?

41. The scale on a map is 1 centimeter to 57 kilometers. Fargo and Bismarck are 4.7 centimeters apart on the map. What is the actual distance between these cities?

42. The scale on a map is 2 centimeters to 5 kilometers. Dove Creek and Kent are 15.75 kilometers apart. How far apart are they on the map?

43. One gear has 36 teeth. The ratio of this gear to a second gear is 4 to 3. How many teeth does the second gear have?

44. One gear has 45 teeth. The ratio of this gear to a second gear is 3 to 2. How many teeth does the second gear have?

Challenge Exercises

Find the ratio of a to b.

45. $10a = 25b$

46. $4a + 2b = 16a$

47. $\frac{3a}{5} = \frac{12b}{7}$

48. $\frac{6a + 2b}{9} = \frac{3a - b}{4}$

49. $\frac{11a - 2b}{12} = \frac{2a + 5b}{-10}$

50. $\frac{2(3a + 6b)}{-4} = \frac{-3(12a - 5b)}{8}$

mini-review

Write an algebraic expression for each verbal expression. Use x as the variable.

1. a number decreased by 11

2. one-third the cube of a number

3. twice a number increased by 6

4. five-fourths of the square of a number

Find each sum or difference.

5. $67 + (-28) + 16$

6. $\frac{7}{5} + \frac{2}{15} + \left|-\frac{4}{3}\right|$

7. $-18.6x - 15.3x$

Solve.

8. $r - (-42) = -53$

9. $\frac{4}{3} = \frac{7}{6} - k$

10. $3(x + 2) = 13$

11. $4x + 12 = -5x + 30$

12. $\frac{3}{4}x + 5 = 3 - \frac{1}{2}x$

13. $\frac{2}{7}y + 5 = \frac{5}{7}y$

14. $\frac{y}{3.9} = \frac{2.505}{6.1}$

15. $\frac{5}{12} = \frac{2 - x}{3 + x}$

Simplify.

16. $\left(\frac{1}{4}\right)\left(-\frac{2}{3}\right) \div (6)$

17. $2a(4a^2x - 5a + 7)$

Solve for x.

18. $3x = 4a$

19. $4x + 2a = 12b$

20. $\frac{\ell + x}{y} = 33$

Solve.

21. Alma rides her bicycle for three-fourths of an hour every day. Find the distance she rides if she averages 13.65 miles per hour.

22. Find three consecutive even integers such that twice the first plus the second is 74.

The basic unit of length in the metric system is the meter. A unit of capacity, the **liter**, is defined in terms of the meter.

The large cube shown on this page is 10 centimeters along each edge. A cube this size has a volume of 1000 cubic centimeters (cm^3) and a capacity of one liter (L). The usual metric prefixes are used with liter.

1 cm^3

A **milliliter** is $\frac{1}{1000}$ of a liter. One milliliter of water will fill a cube that is 1 cm along each edge. A thimble holds about 1 milliliter (1 mL).

There is also a relationship between capacity and mass in the metric system. A milliliter of water has a mass of one gram. A liter of water has a mass of one kilogram.

Exercises

Would you use milliliter or liter to measure each quantity?

1. gasoline for a car
2. eye drops
3. a glass of soda pop
4. vanilla for a cake
5. a bottle of liquid laundry detergent
6. water for the garden

Complete.

7. 3000 mL = ____ L
8. 2 L = ____ mL
9. 4.6 L = ____ mL
10. 2100 mL = ____ L
11. 4200 mL = ____ L
12. 5.06 L = ____ mL
13. 0.54 L = ____ mL
14. 236 mL = ____ L
15. 20 mL = ____ L
16. 3 mL = ____ L
17. 3.05 L = ____ mL
18. 17.98 L = ____ mL

Challenge Exercises

19. How many liters of water will fill a cube that has a volume of 1 cubic meter?

3-9 Percent

Les is the quarterback for the JFK High School football team. In last week's game he completed 17 out of 25 passes. What is his rate per 100?

You can solve this problem by using ratios. Write a ratio that is equivalent to $\frac{17}{25}$ and has a denominator of 100.

$$\frac{17}{25} = \frac{17 \cdot 4}{25 \cdot 4} = \frac{68}{100}$$

Les completed passes at a rate of 68 per 100 or 68 percent. The word **percent** means per hundred, or hundredths. The symbol for percent is %.

$$68\% = \frac{68}{100} = 0.68$$

You can use a proportion to change a fraction to a percent, as shown in the following examples.

Examples

1 **Change $\frac{3}{5}$ to a percent.**

$$\frac{3}{5} = \frac{n}{100}$$

$300 = 5n$ *Means-Extremes Property*

$60 = n$

Thus, $\frac{3}{5}$ is equal to $\frac{60}{100}$ or 60%.

2 **Using Calculators**

Change $\frac{7}{8}$ to a percent.

$$\frac{7}{8} = \frac{n}{100}$$

$$\left(\frac{7}{8}\right)(100) = n$$

ENTER: 7 $\boxed{\div}$ 8 $\boxed{\times}$ 100 $\boxed{=}$

DISPLAY: 7   8   0.875   100   87.5

Thus, $\frac{7}{8}$ is equal to $87\frac{1}{2}\%$ or 87.5%.

Proportions can also be used to solve percent problems. One of the ratios in such proportions is always a comparison of two numbers called the **percentage** and the **base**. The other ratio is formed by expressing the percent as a fraction, called the **rate**.

percentage → $\frac{17}{25} = \frac{r}{100}$ ← *rate*   *In the percent proportion,*
base →   *the rate is $\frac{r}{100}$.*

$$\frac{\text{Percentage}}{\text{Base}} = \text{Rate} \quad \text{or} \quad \frac{\text{Percentage}}{\text{Base}} = \frac{r}{100}$$

Percent Proportion

Example

3 **50 is what percent of 60?**

$$\frac{\text{Percentage}}{\text{Base}} = \frac{r}{100}$$

$$\frac{50}{60} = \frac{r}{100} \qquad \textit{The number 50 is being compared to 60.}$$

$$5000 = 60r$$

$$83\tfrac{1}{3} = r$$

Thus, 50 is $83\tfrac{1}{3}\%$ of 60. The decimal solution is $83.\overline{3}\%$.

Sometimes it is easier to solve a percent problem by translating the problem into a mathematical equation. Recall that the word "of" suggests multiplication, and the word "is" suggests equality. The problem in Example 3 can be translated into an equation as follows. The percent is represented by $\frac{x}{100}$.

50 is what percent of 60?

$$50 = \qquad \frac{x}{100} \qquad \cdot \quad 60 \qquad \textit{Compare this equation to the one in Example 3.}$$

Examples

4 **What number is 36% of 150?**

$$x \qquad = \frac{36}{100} \cdot 150$$

$$x = 54$$

Thus, 54 is 36% of 150.

5 **40% of what number is 30?**

$$\frac{40}{100} \cdot \qquad x \qquad = 30$$

$$40x = 3000$$

$$x = 75$$

Thus, 40% of 75 is 30.

Most calculators have a key labeled $\boxed{\%}$. This key can be used to find percents of numbers. The way this key is used varies for different types of calculators. Two possible key sequences are shown in the following example.

Example

6 **Using Calculators**

Find 18% of 46.3.

a. **ENTER:** 18 $\boxed{\%}$ $\boxed{\times}$ 46.3 $\boxed{=}$

   DISPLAY: *18 0.18 46.3 8.334*

b. **ENTER:** 46.3 $\boxed{\times}$ 18 $\boxed{\%}$ $\boxed{=}$

   DISPLAY: *46.3 18 0.18 8.334*

Thus, 18% of 46.3 is 8.334.

There are many applications of percent, such as sales tax, discounts, and commission. Example 7 involves commission.

Example

7 **Jim Byars earns 3% commission on sales of all new cars. If he earned $861 in commission last week, what was the dollar amount of his total sales?**

explore Let x = total sales in dollars.

plan 3% of total sales equals $861.

$$\frac{3}{100} \cdot x = 861$$

solve
$$\frac{3}{100}x = 861$$
$$3x = 86{,}100$$
$$x = 28{,}700 \qquad \textit{Check this result.}$$

The total sales amount was $28,700.

examine 3% of $28,700 is $861.

Exploratory Exercises

Change each ratio or fraction to a percent.

1. $\frac{31}{100}$
2. $\frac{9}{100}$
3. $\frac{3}{10}$
4. $\frac{1}{25}$
5. $\frac{4}{5}$

6. $\frac{7}{20}$
7. $\frac{3}{8}$
8. $\frac{1}{3}$
9. $\frac{7}{4}$
10. $\frac{9}{5}$

Written Exercises

Use a proportion to solve each problem.

11. Six is what percent of 15?

12. Eighteen is what percent of 60?

13. What percent of 50 is 35?

14. What percent of 17 is 34?

15. Five is what percent of 40?

16. What percent of 75 is 225?

Use an equation to solve each problem.

17. What number is 40% of 80?

18. Thirty-five is 50% of what number?

19. 17.65 is 25% of what number?

20. What number is 0.3% of 62.7?

21. Fourteen is what percent of 56?

22. What percent of 72 is 12?

Solve.

23. What is 40% of 60?

24. Seventy-five is what percent of 250?

25. Twenty-one is 35% of what number?

26. Find 37.5% of 80.

27. Fifty-two is what percent of 80%

28. Thirty-six is 45% of what number?

29. Find 7.5% of 405.

30. Find 81% of 32.

31. Twenty-eight is 20% of what number?

32. Sixteen is 40% of what number?

33. Nineteen is what percent of 76?

34. Thirty-seven is what percent of 296?

35. Find 4% of $6070.

36. Find 6% of $9.40

37. Fifty-five is what percent of 88?

38. Eighty-eight is what percent of 55?

39. $7030.50 is 107.5% of how many dollars?

40. $54,000 is 108% of how many dollars?

41. Find 112% of $500.

42. Find 113.4% of $1000.

43. 96.027 is what percent of 60?

44. 84.312 is what percent of 93.55?

45. Eight is 20% of what number?

46. Ninety is 60% of what number?

47. Find 0.12% of $5200.75.

48. What is 98.5% of $140.32?

49. A theater was filled to 75% of capacity. How many of the 720 seats were filled?

50. The sales tax on a $20 purchase was $0.90. What was the rate of sales tax?

51. Janice scored 85% on the last test. She answered 34 questions correctly. How many questions were on the test?

52. In a 180-kilogram sample of ore, there was 3.2% metal. How many kilograms of metal were in the sample?

53. Suppose 6% of 8000 people polled regarding an election expressed no opinion. How many people expressed no opinion?

54. Bob Stapleton earns 2% of sales on all truck bodies he sells. Last week he earned $974. What were his sales?

55. Henri Rici paid $7230 for a car. After one year, its value had decreased by $1622. By what percent had the car depreciated in value?

56. June Carlos earns $125.48 per week in salary and 8.25% commission on all sales. How much must she sell in order to earn $200 per week?

Challenge Exercises

57. Brad Gee works as a salesman in a men's clothing store. One day, he began with $75 in the cash register. At the end of the day he had $1422.79. If he charged 6% sales tax on all items sold, what were his total sales for the day?

3-10 Problem Solving: Percents

A jacket that cost $50 last year is now priced at $60. The price increased by $10 since last year. You can write a ratio that compares the amount of increase to the price last year. The ratio can be changed to a percent.

$$\frac{\textit{amount of increase}}{\textit{price last year}} \qquad \frac{10}{50} = \frac{20}{100} \quad \text{or} \quad 20\% \qquad \textit{A percent is a ratio of a number to 100.}$$

The amount of increase is 20% of the price last year. Therefore, you could say that the price of the jacket *increased* by 20% since last year. The **percent of increase** was 20%.

The **percent of decrease** can be found in a similar manner, as shown in the following example.

Example

1 **A sweater that originally cost $35 is now on sale for $28. Find the percent of decrease.**

explore The price decreased from $35 to $28. The amount of decrease was $7.

plan You need to write a ratio that compares the amount of decrease with the original price. Then, express the ratio as a percent.

$$\frac{\textit{amount of decrease}}{\textit{original price}} \qquad \frac{7}{35} = \frac{r}{100} \qquad \textit{To express } \tfrac{7}{35} \textit{ as a percent, change it to the ratio of some number to 100.}$$

solve
$$\frac{7}{35} = \frac{r}{100}$$
$$700 = 35r$$
$$\frac{700}{35} = r$$
$$20 = r \qquad \textit{The percent of decrease was 20\%.}$$

examine Check the solution with the words of the problem. The original price of the sweater was $35. Find 20% of $35 and subtract the result from $35.

$$20\% \text{ of } 35 = \frac{20}{100}(35)$$
$$= \frac{1}{5}(35)$$
$$= 7 \qquad \text{The amount of decrease was \$7.}$$

Since $35 - $7 = $28, the solution is correct. ✔

Sometimes an increase or decrease is given as a percent, and you must find the amount. The following problem involves a discount. A discount of 35% means that the price is *decreased* by 35%.

In advertisements, discounts are often stated as %-off.

Example

2 **Amy bought a television set that had an original price of $495.50. She received a 35% discount. What was the discount price?**

explore The original price was $495.50, and the discount was 35%.

plan You need to find the amount of discount, then subtract that amount from $495.50. The result is the discount price.

solve 35% of 495.50 = 0.35(495.50)   *Note 35% = $\frac{35}{100}$ or 0.35.*
$\quad\quad\quad\quad\quad\quad\quad = 173.425$

Subtract this amount from the original price.
$\quad\quad 495.50 - 173.425 = 322.075$
The discount price was $322.08.

examine Here is another way to solve the problem. The discount was 35%, so the discount price was 65% of the original price.
Find 65% of $495.50.
$\quad\quad 0.65 \ (495.50) = 322.075$
This method produces the same discount price, $322.08.

The sales tax on a purchase is a *percent* of the purchase price. To find the total price, you must calculate the *amount* of sales tax and add it to the purchase price.

Example

3 **Frank Orfanello purchased a new tennis racket for $31.78. He also had to pay a sales tax of 5%. Find the amount of tax and the total price.**

explore The price is $31.78 and the tax rate is 5%.

plan First find 5% of $31.78. Then add the result to $31.78.

solve 5% of 31.78 = 0.05(31.78)   *Note 5% = 0.05.*
$\quad\quad\quad\quad\quad\quad = 1.589$ *Round 1.589 to 1.59.*
The amount of tax is $1.59. Now find the total price.
$\quad\quad$ purchase price + amount of tax = total price
$\quad\quad\quad\quad$ 31.78 $\quad\quad + \quad\quad$ 1.59 $\quad\quad = 33.37$
The total price is $33.37.

examine Since 100 + 5 = 105, the total price is 105% of the purchase price. Find 105% of $31.78 and compare the result to $33.37.
$\quad\quad 1.05(31.78) = 33.369$ The total price of $33.37 is correct. ✔

Sometimes it is helpful to use equations to solve percent problems.

Example

4 Jane Shriver works as a sales person in a department store. One of her benefits is a 20% discount on all items she buys for herself. She paid $54 for a new dress. What was the price before the discount?

explore
Let x = price of dress before discount.
Then $0.20x$ = amount of discount. *Note 20% = 0.20 or 0.2.*

plan
$$\underbrace{\text{price}}_{x} - \underbrace{\text{discount}}_{0.20x} = \underbrace{\text{discount price}}_{54}$$

solve
$$x - 0.20x = 54$$
$$(1 - 0.20)x = 54 \quad \text{Distributive Property}$$
$$0.80x = 54$$
$$x = 67.5$$

The price before the discount was $67.50.

examine
If the price before the discount was $67.50 and Jane paid $54 for the dress, the amount of discount was $13.50. Since $13.50 is 20% of $67.50, the solution is correct.

Exploratory Exercises

Solve.

1. What is 50% of 20?
2. What is 50% of 350?
3. What is 25% of 32?
4. What is 25% of 60?
5. What is 75% of 60?
6. What is 75% of 120?
7. Eighteen is 50% of what number?
8. Ninety-two is 50% of what number?
9. Ten is 25% of what number?
10. Thirty is 75% of what number?

Written Exercises

Solve.

11. What number increased by 40% equals 14?
12. What number decreased by 20% is 16?
13. What number increased by $33\frac{1}{3}\%$ equals 52?
14. What number decreased by $66\frac{2}{3}\%$ is 18?
15. Fourteen is 50% less than what number?
16. Twenty is 20% less than what number?
17. 55.394 is 10.5% more than what number?
18. 60.065 is 1.65% more than what number?
19. The price in dollars (p) plus 5% tax is equal to $3.15. Find p.
20. The price in dollars (p) minus a 15% discount is $3.40. Find p.

Copy and complete the following chart. The first line is given as a sample.

	Earlier Amount	Later Amount	Did the amount increase (I) or decrease (D)?	Amount of Increase or Decrease	Percent of Increase or Decrease
	$50	$70	I	$20	40%
21.	$100	$94			
22.	$100	$108			
23.	$200		D		14%
24.	$313.49		I		14.5%
25.		$60.10	I	$12.21	
26.		$36	D	$36	

Notice that 20 is 40% of 50.

Solve each problem.

27. What is 30% more than 30?

28. What is 75% less than 80?

29. What percent of 11.295 is 6.35?

30. What percent of 80.95 is 11.612?

31. A price decreased from $50 to $40. Find the percent of decrease.

32. A price increased from $40 to $50. Find the percent of increase.

33. An item priced $36 has a 25% discount. Find the discount price.

34. Sales tax of 6% is added to a purchase of $11. Find the total price.

Find the customer price for each problem. When there is a discount and sales tax, compute the discount price first.

35. Stereo Set: $345.00
Discount: 12%

36. Ten-speed Bike: $148.00
Discount: 18%

37. Clothing: $74.00
Sales Tax: 6.5%

38. Books: $38.50
Sales Tax: 6%

39. Shoes: $44.00
Discount: 10%
Sales Tax: 4%

40. Auto Tires: $154.00
Discount: 20%
Sales Tax: 5%

Solve each problem.

41. Wilma paid $92.04 for new school clothes. This included 4% sales tax. What was the cost of the clothes before taxes?

42. Chapa paid $45.10, including $7\frac{1}{2}$% tax, for a pair of jeans. What was the cost of the jeans before taxes?

43. Jason paid $13.96 for a new shirt, after receiving a 22.333% discount. What was the price before the discount?

44. Ben received a discount of $4.50 on a new radio. The discount price was $24.65. What was the percent of discount?

45. A group of 25 people share equally in the profit from a bake sale. What percent does each person receive?

46. The original selling price of a stove was $550. This price was increased by 20%. The increased price was then discounted by 10%. What was the new selling price?

Successive Discounts

A store may give a second discount on an item that is already being sold at a discount. Perhaps the item is out of style or slightly damaged. Consider this example.

Example The regular price of a sofa is $825. What is the final sale price if an additional 20% discount is given on this sofa that is already selling at a discount of 10%?

Regular price	$ 825	$742.50	*A 20% discount is the same as*
A 10% discount	× 0.90	× 0.80	*80% of $742.50.*
is the same as	$742.50	$594.00	**Sale price**
90% of the			
regular price.			

The two successive discounts can be summarized as follows.

$$(\$825 \times 0.90) \times 0.80 = \$594$$

The following program will help you determine whether two successive discounts or one combined discount will give a lower price.

```
10   PRINT "ENTER ORIGINAL PRICE AND TWO DISCOUNTS
        AS DECIMALS"
20   INPUT P,X1,X2
30   LET S1 = P * (1 − X1) * (1 − X2)   This computes sale price with successive discounts.
40   LET S2 = P * (1 − (X1 + X2))   This computes sale price with one combined discount.
50   PRINT "TWO SUCCESSIVE DISCOUNTS", "$";S1
60   PRINT "COMBINED DISCOUNT", "$";S2
70   IF S2 < S1 THEN PRINT "ONE COMBINED DISCOUNT HAS A LOWER
        SALE PRICE"
80   IF S1 < S2 THEN PRINT "SUCCESSIVE DISCOUNTS HAVE A LOWER
        SALE PRICE"
90   IF S1 = S2 THEN PRINT "THERE IS NO DIFFERENCE"
100  END
```

Exercises

Find the sale price of each item using successive discounts and one combined discount.

1. Price, $49.00; Discounts, 20% and 15%

2. Price, $185; Discounts, 25% and 10%

3. Price, $12.50; Discounts, 30% and $12\frac{1}{2}$%

4. Price, $156.95; Discounts, $33\frac{1}{3}$% and 10%

5. What is the relationship between the sale price using successive discounts and one combined discount?

6. Modify the program to determine whether the order in which the discounts are applied affects the sale price.

7. Forty percent of the students who auditioned for the school play were cut in the first week. Twenty percent of the remaining students were chosen for the play. If 12 students were chosen, how many students were in the original audition?

8. If a is 180% of b, then b is what percent of a?

Vocabulary

multiplicative inverse (84)	proportion (105)	base (110)
reciprocal (84)	extremes (105)	rate (110)
identity (94)	means (105)	percent of increase (114)
consecutive numbers (101)	percent (110)	percent of decrease (114)
ratio (105)	percentage (110)	

Chapter Summary

1. The product of two numbers that have different signs is negative. (79)
2. The product of two numbers that have the same sign is positive. (80)
3. The product of any number and -1 is its additive inverse.
$$-1(a) = -a \text{ and } a(-1) = -a \quad (80)$$
4. The quotient of two numbers is positive if the numbers have the same sign. The quotient of two numbers is negative if the numbers have different signs. (83)
5. Multiplicative Inverse Property: For every nonzero number a, there is exactly one number $\frac{1}{a}$, such that $a\left(\frac{1}{a}\right) = \frac{1}{a}(a) = 1$. (84)
6. Division Rule: For all numbers a and b, with $b \neq 0$, $a \div b = \frac{a}{b} = a\left(\frac{1}{b}\right) = \frac{1}{b}(a)$. (84)
7. Multiplication Property of Equality: For any numbers a, b, and c, with $c \neq 0$, if $a = b$, then $ac = bc$. (86)
8. Division Property of Equality: For any numbers a, b, and c, with $c \neq 0$, if $a = b$, then $\frac{a}{c} = \frac{b}{c}$. (87)
9. An equation that is true for every value of the variable is called an identity. (94)
10. Consecutive numbers are numbers in counting order such as 3, 4, 5. (101)

11. An equation of the form $\frac{a}{b} = \frac{c}{d}$ that states that two ratios are equal is called a proportion. (105)

12. Means-Extremes Property of Proportions: In a proportion, the product of the extremes is equal to the product of the means. If $\frac{a}{b} = \frac{c}{d}$, then $ad = bc$. (105)

13. Proportions can be used to solve percent problems. (110)

14. Percent Proportion: $\frac{\text{Percentage}}{\text{Base}} = \text{Rate}$ or $\frac{\text{Percentage}}{\text{Base}} = \frac{r}{100}$ (111)

15. Percent of Increase: $\frac{\text{amount of increase}}{\text{original amount}}$ (114)

16. Percent of Decrease: $\frac{\text{amount of decrease}}{\text{original amount}}$ (114)

Chapter Review

3–1 **Find each product.**

    1. $(-11)(9)$ **2.** $(-8)(-12)$ **3.** $\frac{3}{5}\left(-\frac{5}{7}\right)$

    Simplify each expression.

    4. $-3(7) + (-8)(-9)$ **5.** $\frac{1}{2}(6a + 8b) - \frac{2}{3}(12a + 24b)$

3–2 **Simplify.**

    6. $\frac{-54}{6}$ **7.** $\frac{63b}{-7}$ **8.** $\frac{\frac{4}{5}}{-7}$ **9.** $\frac{-12}{-\frac{2}{3}}$ **10.** $\frac{33a + 66}{-11}$

3–3 **Solve.**

    11. $-7r = -56$ **12.** $23y = 1035$ **13.** $\frac{3}{4}x = -12$ **14.** $\frac{x}{5} = 7$

3–4   **15.** $3x - 8 = 22$ **16.** $-4y + 2 = 32$ **17.** $0.5n + 3 = -6$

    18. $-6 = 3.1t + 6.4$ **19.** $\frac{x}{-3} + 2 = -21$ **20.** $\frac{r - 8}{-6} = 7$

3–5   **21.** $5a - 5 = 7a - 19$ **22.** $-3(x + 2) = -18$

    23. $4(2y - 1) = -10(y - 5)$ **24.** $11.2n + 6 = 5.2n$

3–6   **25.** $\frac{2}{3}x + 5 = \frac{1}{2}x + 4$ **26.** $2.9m + 1.7 = 3.5 + 2.3m$ **27.** $\frac{3t + 1}{4} = \frac{3}{4}t - 5$

    Solve for x.

    28. $x + r = q$ **29.** $\frac{x + y}{c} = d$

    Solve for y.

    30. $5(2a + y) = 3b$ **31.** $\frac{2y - a}{3} = \frac{a + 3b}{4}$

3–7   **32.** Find two consecutive even integers whose sum is 94.   **33.** Find three consecutive odd integers whose sum is 81.

    34. Four times a number decreased by twice the number is 100. What is the number?

3–8 **Solve each proportion.**

35. $\frac{6}{15} = \frac{n}{45}$ **36.** $\frac{4}{8} = \frac{11}{t}$ **37.** $\frac{5}{6} = \frac{n-2}{4}$

3–9 **Change each fraction to a percent.**

38. $\frac{6}{10}$ **39.** $\frac{7}{8}$ **40.** $\frac{17}{100}$

Solve.

41. Nine is what percent of 15?

42. What number is 60% of 80?

43. Twenty-one is 35% of what number?

44. Eighty-four is what percent of 96?

3–10 **45.** What number increased by 20% is equal to 42?

46. Bernice bought a dress on sale for $64.77. This price included a 15% discount. How much was the discount?

Chapter Test

Simplify each expression.

1. $\frac{8(-3)}{2}$ **2.** $(-5)(-2)(-2) - (-6)(-3)$ **3.** $\frac{2}{3}\left(\frac{1}{2}\right) - \left(-\frac{3}{2}\right)\left(-\frac{2}{3}\right)$

4. $\frac{3}{4}(8x + 12y) - \frac{5}{7}(21x - 35y)$ **5.** $\frac{70a - 42b}{-14}$ **6.** $\frac{\frac{11}{5}}{-6}$

Solve each equation.

7. $-3y = 63$ **8.** $\frac{3}{4}y = -27$ **9.** $3x + 1 = 16$

10. $5.2n + 0.7 = 2.8 + 2.2n$ **11.** $5(8 - 2n) = 4n - 2$ **12.** $3(n + 5) - 6 = 3n + 9$

13. $7x + 9 = 3(x + 3)$ **14.** $-2(3n - 5) + 3n = 2 - n$ **15.** $\frac{3}{4}n - \frac{2}{3}n = 5$

16. $\frac{t - 7}{4} = 11$ **17.** $\frac{2r - 3}{-7} = 5$ **18.** $8r - \frac{r}{3} = 46$

Solve for x.

19. $\frac{x + y}{b} = c$ **20.** $yx + a = c$

Solve each proportion.

21. $\frac{7}{8} = \frac{5}{t}$ **22.** $\frac{9}{11} = \frac{x - 3}{x + 5}$ **23.** $\frac{y + 2}{8} = \frac{7}{5}$ **24.** $\frac{2}{5} = \frac{x - 3}{-2}$

Solve each problem.

25. Find 6.5% of 80.

26. Forty-two is what percent of 126?

27. Eighty-four is 60% of what number?

28. Find three consecutive odd integers whose sum is 93.

29. Twice a number increased by 12 is 31 less than three times the number. Find the number.

30. Find two consecutive integers such that twice the lesser integer, increased by the greater integer, is 50.

31. A price decreased from $60 to $45. Find the percent of decrease.

CHAPTER 4

Inequalities

Nancy Lopez, a leading professional golfer, has scores of 68, 72, and 69 for the first three rounds in a golf tournament. Experts figure that it will take an average score of less than 70 to win the tournament. What must Nancy score on the last round of golf? In this chapter, you will learn how to solve problems such as this using inequalities.

4-1 Inequalities and Their Graphs

You know that 3 is less than 8. This can be shown in two ways.

$3 < 8$ means 3 *is less than* 8.

$8 > 3$ means 8 *is greater than* 3.

Note that the $<$ and $>$ point to the lesser number.

A mathematical sentence containing $<$ or $>$ is called an **inequality**.

The sentence $3 < 8$ is true. Is $3 > 8$ true? Is $3 = 8$ true? Note that only one of the three sentences is true. This can be summarized by the following property.

For any two numbers a and b, exactly one of the following sentences is true. $a < b$ $\quad$ $a = b$ $\quad$ $a > b$	*Comparison Property*

Examples

1 Is $4 < 5\frac{1}{2}$ *true* or *false*?

$4 < 5\frac{1}{2}$ means 4 *is less than* $5\frac{1}{2}$. This sentence is true.

The inequalities $4 < 5\frac{1}{2}$ and $5\frac{1}{2} > 4$ have the same meaning.

2 Is $9 > 4 + 3 + 2 + 1$ *true* or *false*?

$9 \overset{?}{>} 4 + 3 + 2 + 1$ $\qquad$ *Is 9 greater than the sum of 4, 3, 2, and 1?*

$9 \overset{?}{>} 10$ $\qquad\qquad\qquad$ *Simplify the right side of the inequality.*

Since 9 is not greater than 10, the sentence is false.

Recall that there are three properties of equality. Equality is reflexive, symmetric, and transitive. Do inequalities have these same properties?

Determine whether each sentence below is true or false.

Reflexive: $\quad$ $6 > 6$

Symmetric: $\quad$ If $4 > 3$, then $3 > 4$.

Transitive: $\quad$ If $6 > 2$ and $2 > 1$, then $6 > 1$.

Determine whether the Transitive Property is true for other examples.

The relation $>$ is not reflexive or symmetric. However, it is transitive. Explore some similar examples to verify that $<$ is transitive.

For all numbers a, b, and c, 1. If $a < b$ and $b < c$, then $a < c$. 2. If $a > b$ and $b > c$, then $a > c$.	*Transitive Property of Order*

Example

3 **Determine whether the following sentence is *true* or *false*.**

If $-3 < 1$ and $1 < 4$, then $-3 < 4$.

The sentence is true by the Transitive Property of Order.

The symbols $\neq$, $\leq$, and $\geq$ can also be used when comparing numbers. The following chart shows several inequality symbols and their meanings.

Symbol	Meaning
$<$	is less than
$>$	is greater than
$\neq$	is not equal to
$\leq$	is less than or equal to
$\geq$	is greater than or equal to

Example

4 **Is $5.1 \leq 7 + 0.9$ *true* or *false*?**

$5.1 \overset{?}{\leq} 7 + 0.9$ *Is 5.1 less than or equal to 7 + 0.9?*

$5.1 \leq 7.9$

Since 5.1 is less than 7.9, the sentence is true.

Consider the graphs of -3, -1, $2\frac{1}{2}$, and 4.5, shown on the number line below.

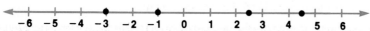

The following statements can be made about the numbers and their graphs.

The graph of -3 is to the left of the graph of -1.	$-3 < -1$
The graph of -1 is to the left of the graph of $2\frac{1}{2}$.	$-1 < 2\frac{1}{2}$
The graph of $2\frac{1}{2}$ is to the right of the graph of -3.	$2\frac{1}{2} > -3$
The graph of 4.5 is to the right of the graph of $2\frac{1}{2}$.	$4.5 > 2\frac{1}{2}$

If a and b represent any numbers and the graph of a is to the left of the graph of b, then $a < b$. If the graph of a is to the right of the graph of b, then $a > b$.

Comparing Numbers on the Number Line

Recall that an equation such as $x + 6 = 7$ is an open sentence. An inequality such as $x < 5$ is also considered an open sentence. The set of all replacements for the variable that makes an open sentence true is called the **solution set** of the sentence.

To find the solution set of $x < 5$, determine what replacements for x make $x < 5$ true. All numbers less than 5 make the inequality true. This can be shown by the solution set {all numbers less than 5}, read *the set of all numbers less than 5*. Study the graph of this solution set in Example 5.

Examples

5 **Graph the solution set of $x < 5$.**

The heavy arrow shows that all numbers to the left of 5 are included.

The circle indicates that 5 is not included in the solution set.

6 **Graph the solution set of $y \neq -1.5$.**

The solution set is {all numbers except -1.5}.

The circle indicates that -1.5 is not included in the solution set.

7 **Graph the solution set of $n \geq -\frac{3}{2}$.**

The solution set is $\left\{-\frac{3}{2}\text{ and all numbers greater than }-\frac{3}{2}\right\}$.

The dot at $-\frac{3}{2}$ indicates that $-\frac{3}{2}$ is included in the solution set.

Exploratory Exercises

Determine whether each sentence is *true* or *false*.

1. $7 < 4$ **2.** $-3 < 3$ **3.** $2 > 0$ **4.** $-9 > -4$

5. $-\frac{8}{3} \leq \frac{8}{3}$ **6.** $-5\frac{2}{3} \neq -6\frac{2}{3}$ **7.** $-4 < 2 - 8$ **8.** $0.6 \leq 1 - 0.4$

9. If $4 > 3$ and $3 < 5$, then $4 > 5$.   **10.** If $1 < 3$, then $1 > 3$.

State the solution set of each inequality.

11. $n > 3$ **12.** $y < -5$ **13.** $z < 6$ **14.** $y > -1$ **15.** $y \geq -4$ **16.** $m \neq 0$

Determine whether each number is included on the graph below.

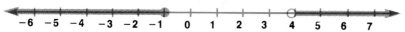

17. -1 **18.** 4 **19.** 3.9 **20.** -4.5 **21.** $-1\frac{1}{2}$ **22.** $2\frac{1}{2}$

Written Exercises

Write an inequality for each graph.

23.   **24.**

25.   **26.**

27.   **28.**

29.   **30.**

Replace each _?_ with <, >, or = to make each sentence true.

31. $-5 \underline{\ ?\ } 7$ **32.** $-2 \underline{\ ?\ } -3$ **33.** $6 \underline{\ ?\ } 4 + 2$ **34.** $-7 - 2 \underline{\ ?\ } -9$

35. $-5 \underline{\ ?\ } 0 - 3$ **36.** $3 \underline{\ ?\ } \frac{15}{3}$ **37.** $10 \underline{\ ?\ } \frac{27}{3}$ **38.** $12 \underline{\ ?\ } 15 - 27$

39. $8 \underline{\ ?\ } 4.1 + 3.9$ **40.** $5 \underline{\ ?\ } 8.4 - 1.5$ **41.** $-7 \underline{\ ?\ } -\frac{3.6}{0.6}$ **42.** $\frac{5.4}{18} \underline{\ ?\ } -4 + 1$

43. $(-7.502)(0.511) \underline{\ ?\ } -3.115$ **44.** $(-6.01)(-4.122) \underline{\ ?\ } 25.005$ **45.** $\frac{4}{3}(6) \underline{\ ?\ } 4\left(\frac{3}{2}\right)$

46. $8\left(\frac{3}{4}\right) \underline{\ ?\ } 6\left(\frac{2}{3}\right)$ **47.** $3 + 1.077 \underline{\ ?\ } \frac{9.624}{2.2}$ **48.** $\frac{27.155}{3} \underline{\ ?\ } 2(4.459)$

49. If $2\frac{2}{3} > -1\frac{1}{2}$ and $-1\frac{1}{2} > -3$, then $-3 \underline{\ ?\ } 2\frac{2}{3}$.   **50.** If $4.5 > 2.1$ and $2.1 > -3.6$, then $4.5 \underline{\ ?\ } -3.6$.

Graph the solution set of each inequality on a number line.

51. $n > 5$ **52.** $x > -3$ **53.** $m < 1$ **54.** $y < -2$ **55.** $x \neq -1$ **56.** $y \neq 3$

57. $y \leq 6$ **58.** $n \leq -2$ **59.** $a \geq -3$ **60.** $x \geq -2$ **61.** $x < -10$ **62.** $m < 3$

Challenge Exercises

63. If $a < b$ and $a < c$, what conclusion can be made about b and c?

64. Consider $3 < 4$ and $3 < 5$. Is the conclusion $4 < 5$ valid?

In algebra a compound statement consists of two simple statements that are connected by the words *and* or *or*. You must read compound statements very carefully. Understanding the meaning of the words *and* and *or* will help you determine whether a compound statement is true or false. Study the statements below.

triangle

A triangle has three sides *and* a pentagon has four sides.

For this compound statement to be true, both parts must be true. Since triangles have three sides, the first part is true. However, pentagons do not have four sides, so the second part is false. Therefore, the compound statement is false.

pentagon

A triangle has three sides *or* a pentagon has four sides.

For this compound statement to be true, at least one part must be true. You know that triangles have three sides, so the compound statement is true.

hexagon

The compound statements described above can be defined as follows.

octagon

A compound statement formed by joining two statements with the word *and* is called a conjunction. For a conjunction to be true, both statements must be true.	*Conjunction*

A compound statement formed by joining two statements with the word *or* is called a disjunction. For a disjunction to be true, at least one statement must be true.	*Disjunction*

Exercises

Determine whether each statement is a *conjunction* or a *disjunction*. Then, write *true* or *false*.

1. A hexagon has six sides and an octagon has nine sides.

2. An octagon has eight sides or a pentagon has six sides.

3. A pentagon has five sides and a hexagon has six sides.

4. A triangle has four sides and an octagon does not have seven sides.

5. A pentagon has six sides or an octagon has ten sides.

6. A triangle has three sides or a hexagon has six sides.

7. $3 < 4$ or $7 < 6$

8. $-4 > 0$ and $2 < 5$

9. $6 > 0$ and $-6 < 0$

10. $2 = 0$ or $-2 > -3$

11. $7 \neq 7$ and $6 > 4$

12. $0 > -4$ or $2 < -2$

4-2 Solving Inequalities Using Addition and Subtraction

Kristina had $267.23 in her savings account and Mike had $134.53 in his account. They each received $10 from their grandmother and deposited it into their accounts. Whose account has more money?

Kristina   **Mike**

$267.23 > $134.53

$267.23 + $10.00 $134.53 + $10.00

$277.23 > $144.53

Kristina had more money at the beginning and at the end. Notice that adding the same number to each side of the inequality did not change the truth of the inequality.

Suppose Kristina and Mike each withdrew, or subtracted, $25 from their accounts. Kristina would still have more money than Mike.

You know that the sentence $-1 < 3$ is true. Is $-1 + 1 < 3 + 1$ a true sentence? Is $-1 - 2 < 3 - 2$ a true sentence? These examples suggest the following rules that can be used in solving inequalities.

For all numbers a, b, and c,
1. If $a > b$, then $a + c > b + c$ and $a - c > b - c$.
2. If $a < b$, then $a + c < b + c$ and $a - c < b - c$.

Addition and Subtraction Properties for Inequalities

---

1 Solve $m + 7 > 12$.

$$m + 7 > 12$$
$$m + 7 - 7 > 12 - 7 \quad \textit{Subtract 7 from each side.}$$
$$m > 5$$

The solution set is {all numbers greater than 5}.

Check: First check whether 5 is the correct *boundary*. Is 5 the value of m in $m + 7 = 12$?

$$m + 7 = 12$$
$$5 + 7 \stackrel{?}{=} 12$$
$$12 = 12 \quad \text{✓}$$

Next substitute one or two numbers greater than 5, such as 6 and 10, into the inequality. For numbers greater than 5, the inequality should be true.

$m + 7 > 12$	$m + 7 > 12$
$6 + 7 \stackrel{?}{>} 12 \quad \textit{Try 6.}$	$10 + 7 \stackrel{?}{>} 12 \quad \textit{Try 10.}$
$13 > 12 \quad$ ✓	$17 > 12 \quad$ ✓

2 Solve $8y + 3 > 9y - 14$.

$$8y + 3 > 9y - 14$$
$$8y - 8y + 3 > 9y - 8y - 14 \quad \textit{Subtract 8y from each side.}$$
$$3 > y - 14$$
$$3 + 14 > y - 14 + 14 \quad \textit{Add 14 to each side.}$$
$$17 > y \text{ or } y < 17$$

The solution set is {all numbers less than 17}. $\quad$ *Try checking 10 and 16.*

The solution of an inequality is expressed as a set. In Example 2, the solution set is {all numbers less than 17}. Another way of writing such a solution set is by using **set-builder notation**. The solution set for Example 2 in set-builder notation is $\{y \mid y < 17\}$. This is read *the set of all numbers y such that y is less than 17.*

Exploratory Exercises

State the number you would add to each side to solve each inequality.

1. $r + 7 < 21$ **2.** $y + 2 < -16$ **3.** $m - 13 \geq 41$ **4.** $y - 18 \geq -3$

5. $4y < 3y + 12$ **6.** $7x < 6x - 12$ **7.** $6z > 5z - 21$ **8.** $7n > 6n - 4$

9. $12 > m - (-3)$ **10.** $16 \leq k - (-9)$ **11.** $14 < b - 7$ **12.** $9 > p - 21$

Written Exercises

Solve each inequality.

13. $a + 2 < 10$

14. $x + 3 < -17$

15. $r - 9 \geq 23$

16. $y - 15 \geq -2$

17. $5b < 4b + 8$

18. $9z > 8z - 8$

19. $3m > 2m - 19$

20. $6y < 5y - 11$

21. $y + 2 \leq -6$

22. $x + 7 \leq 10$

23. $n - 3 > 8$

24. $-7 + n \leq -16$

25. $6 + n > 40$

26. $3 + y \geq -4$

27. $x - 5 > 3$

28. $-8 + y < 10$

29. $n - (-5) < -6$

30. $y - (-4) > -7$

31. $-12 < s - (-3)$

32. $-21 > r - (-8)$

33. $t + 13.1 \leq 47.7$

34. $y + 18.7 < 81.6$

35. $m - 1.73 > 4.65$

36. $p - 7.35 \geq 6.81$

37. $\frac{3}{4} + r > \frac{7}{4}$

38. $r + \frac{5}{3} \leq \frac{2}{3}$

39. $x - \frac{1}{3} < \frac{1}{2}$

40. $x + \left(-\frac{1}{4}\right) \geq -\frac{7}{4}$

41. $9f - 6 > 10f$

42. $8x - 31 > 7x$

43. $6n > 5n + 6$

44. $3m < 2m - 7$

45. $14 + 7x > 8x$

46. $-3 + 14z < 15z$

47. $6a + 4 \geq 5a$

48. $9y - 6 > 10y$

49. $2r - 2.1 < -8.7 + r$

50. $5s - 6.5 < -13.4 + 4s$

51. $12.37 + 4z > 181.3 + 3z$

52. $16.803 + p < 14.615 + 2p$

53. $17.42 - 8.025t \leq 15.667 - 9.025t$

54. $2r - \frac{3}{4} < r + \frac{5}{3}$

55. $3t - \frac{2}{5} \geq 2t - \frac{1}{4}$

56. $7d - \frac{3}{2} < 6d + \frac{5}{8}$

Solve each inequality. Express the solution set in set-builder notation.

57. $y + \frac{3}{8} \leq \frac{7}{24} + 2y$

58. $7x + \frac{3}{16} < -\frac{11}{16} + 6x$

59. $y + \frac{16}{9} > 2y + \frac{5}{6}$

mini-review

Evaluate if $x = 7$, $y = -2$, and $z = \frac{3}{4}$.

1. $3x^2y$

2. $(4z - 2y)^2$

3. $x + 5y - 3z$

Simplify.

4. $13b - 5(b + 2)$

5. $\frac{-11}{\frac{-4}{5}}$

6. $\frac{63x - 18}{9}$

Solve.

7. $r = -0.25(8 + 14)$

8. $p + 26 = -32$

9. $2.5 = a - (-3.6)$

10. $\frac{6}{5}x = -300$

11. $-4 = 2.5m + 8.62$

12. $\frac{5t + 1}{2} = \frac{5}{2}t - 6$

13. $-5(x - 1) = 45$

14. $\frac{6}{17} = \frac{4}{n}$

15. $\frac{2}{11} = \frac{x - 5}{x + 3}$

Replace each $\underline{\ ?\ }$ with <, >, or = to make each sentence true.

16. $\frac{7}{9} \underline{\ ?\ } 0.7$

17. $-2.125 \underline{\ ?\ } -2\frac{1}{8}$

18. $(-4.3)(0.7) \underline{\ ?\ } -3$

Solve.

19. The sum of two integers is -55. One integer is -16. Find the other integer.

20. Fifteen is 40% of what number?

4-3 Solving Inequalities Using Multiplication and Division

You know that the sentence $-4 < 6$ is true. Suppose each side of the inequality is multiplied by the same positive number. Study the following examples. Are the resulting sentences true?

$$-4 < 6$$
$$-4(3) \overset{?}{<} 6(3)$$
$$-12 \overset{?}{<} 18 \qquad true$$

$$-4 < 6$$
$$-4\left(\tfrac{1}{2}\right) \overset{?}{<} 6\left(\tfrac{1}{2}\right)$$
$$-2 \overset{?}{<} 3 \qquad true$$

Notice that dividing each side by 2 would have given the same result as multiplying by $\tfrac{1}{2}$.

The inequalities $-12 < 18$ and $-2 < 3$ are true. Thus, if each side of a true inequality is multiplied by the same positive number, the result is also true.

What happens if each side of an inequality is multiplied by the same negative number?

$$-4 < 6$$
$$-4(-3) \overset{?}{<} 6(-3)$$
$$12 \overset{?}{<} -18 \qquad false$$

$$-4 < 6$$
$$-4\left(-\tfrac{1}{2}\right) \overset{?}{<} 6\left(-\tfrac{1}{2}\right)$$
$$2 \overset{?}{<} -3 \qquad false$$

The inequality $12 < -18$ is false, but $12 > -18$ is true. Also, $2 < -3$ is false, but $2 > -3$ is true. Thus, when each side of an inequality is multiplied by the same negative number, the direction of the inequality symbol must be reversed.

For all numbers a, b, and c,	

1. If c is positive and $a < b$, then $ac < bc$.
   If c is positive and $a > b$, then $ac > bc$.
2. If c is negative and $a < b$, then $ac > bc$.
   If c is negative and $a > b$, then $ac < bc$.
	Multiplication Property for Inequalities

This property can be used to solve inequalities.

Example

1 Solve $\frac{k}{4} > 13$.

$$\frac{k}{4} > 13$$

$$4\left(\frac{k}{4}\right) > 4(13) \qquad \text{Multiply each side by 4.}$$

$$k > 52$$

The solution set is {all numbers greater than 52}.

2 **Solve $-\frac{a}{7} > -2$.** *Another way to write $-\frac{a}{7}$ is $\frac{a}{-7}$.*

$$\frac{a}{-7} > -2$$

$$-7\left(\frac{a}{-7}\right) < -7(-2) \qquad \text{\textit{Multiply each side by -7 and reverse}}$$
$$\qquad \text{\textit{the direction of the inequality symbol.}}$$

$$a < 14$$

The solution set is {all numbers less than 14} or $\{a | a < 14\}$.

3 **Solve $\frac{4x}{3} \leq -12$.** *Another way to write $\frac{4x}{3}$ is $\frac{4}{3}x$.*

$$\frac{4}{3}x \leq -12$$

$$\frac{3}{4}\left(\frac{4}{3}x\right) \leq \frac{3}{4}(-12) \qquad \text{\textit{The reciprocal of $\frac{4}{3}$ is $\frac{3}{4}$. Multiply each side by $\frac{3}{4}$.}}$$

$$x \leq -9$$

The solution set is {all numbers less than or equal to -9} or $\{x | x \leq -9\}$.

Recall that $\frac{a}{b}$ (or $a \div b$) is equivalent to $a\left(\frac{1}{b}\right)$ for all numbers when b is *not* zero. Thus, the Multiplication Property for Inequalities can also apply to division. When solving inequalities, you can multiply (or divide) each side by the same positive number. You can also multiply (or divide) each side by the same negative number if you *reverse* the direction of the inequality symbol.

For all numbers a, b, and c,

1. If c is positive and $a < b$, then $\frac{a}{c} < \frac{b}{c}$.

   If c is positive and $a > b$, then $\frac{a}{c} > \frac{b}{c}$.

2. If c is negative and $a < b$, then $\frac{a}{c} > \frac{b}{c}$.

   If c is negative and $a > b$, then $\frac{a}{c} < \frac{b}{c}$.

Division Property for Inequalities

The following example shows two methods to solve the inequality $-72 \geq -6m$. *Choose the method that is easier for you.*

Example

4 **Solve $-72 \geq -6m$.**

a.
$$-72 \geq -6m$$
$$\left(-\frac{1}{6}\right)(-72) \leq \left(-\frac{1}{6}\right)(-6m)$$
$$12 \leq m$$

Multiply each side by $-\frac{1}{6}$. Change $\geq$ to $\leq$.

b.
$$-72 \geq -6m$$
$$\frac{-72}{-6} \leq \frac{-6m}{-6}$$
$$12 \leq m$$

Divide each side by -6. Change $\geq$ to $\leq$.

The solution set is $\{m | 12 \leq m\}$. *This can also be expressed as $\{m | m \geq 12\}$.*

Example

5 Solve $(-0.4)x < 0.6$.

$(-0.4)x < 0.6$

$\dfrac{(-0.4)x}{-0.4} > \dfrac{0.6}{-0.4}$ *Divide each side by -0.4 and reverse the direction of the inequality symbol.*

$x > -1.5$

The solution set is $\{x \mid x > -1.5\}$.

Exploratory Exercises

State the number by which to multiply each side to solve each equation. Then indicate if the direction of the inequality symbol reverses.

1. $\frac{r}{3} > -4$
2. $\frac{s}{-6} < -11$
3. $\frac{d}{-8} < -9$
4. $\frac{d}{-11} > 3$

5. $3y < 21$
6. $-6z < 18$
7. $-4x > 8$
8. $5a < -20$

9. $\frac{1}{6}j \geq -9$
10. $-\frac{1}{8}h \leq 10$
11. $-\frac{s}{11} < 2$
12. $-\frac{t}{12} < -4$

13. $\frac{4}{3}k > 16$
14. $\frac{2}{7}m < 12$
15. $-\frac{5n}{3} \leq -10$
16. $-\frac{3r}{8} \geq 9$

Written Exercises

Solve each inequality. Express the solution set in set-builder notation.

17. $14p < 84$
18. $-16q > -128$
19. $-17s > 119$
20. $4z < -6$

21. $-2r \geq 35$
22. $23b > 276$
23. $-8s < -34$
24. $5m \leq -17$

25. $\frac{h}{-18} > -25$
26. $\frac{k}{-32} \leq 50$
27. $-16 \leq \frac{b}{8}$
28. $-35 < \frac{a}{-7}$

29. $7r < -4.9$
30. $-\frac{1}{4}m \geq 19$
31. $-\frac{1}{5}y < -\frac{2}{3}$
32. $-3n \leq -6.6$

33. $-\frac{5s}{8} \geq \frac{15}{4}$
34. $\frac{2}{5}z < \frac{4}{3}$
35. $-\frac{3}{4}k > \frac{6}{7}$
36. $\frac{4z}{7} > -\frac{2}{5}$

37. $-5.1m < -3.57$
38. $6.1g < 3.66$
39. $-0.998 \leq 0.715t$
40. $4.008 \leq -1.26x$

41. $-13z > -1.04$
42. $\frac{3b}{4} < \frac{2}{5}$
43. $-\frac{5x}{6} < \frac{2}{3}$
44. $1.8z > -54$

45. $\frac{13}{18} \leq \frac{2}{3}w$
46. $-2.58 > 4.3n$
47. $51.3 < -5.7a$
48. $-\frac{3}{14} \geq -\frac{5m}{7}$

49. $8.7x < 40$
50. $0.07x \geq 0.93$
51. $-0.471y \leq 2.556$
52. $-2070a > 30.2$

53. $843a \geq 25$
54. $-3.106y > -4.259$
55. $-50.4 \leq -1525b$
56. $3x < 5x$

Challenge Exercises

Determine the conditions under which each sentence is *true*.

57. If $x > y$, then $x^2 > y^2$.
58. If $x < y$, then $x^2 < y^2$.

4-4 Inequalities with More Than One Operation

You can solve inequalities involving more than one operation by applying the methods you have already used to solve equations.

Examples

1 **Solve $16 - 5b > 29$.**

$$16 - 5b > 29$$
$$16 - 16 - 5b > 29 - 16 \qquad \text{Subtract 16 from each side.}$$
$$-5b > 13$$
$$\frac{-5b}{-5} < \frac{13}{-5} \qquad \text{Divide each side by } -5 \left(\text{or multiply by } -\frac{1}{5}\right).$$
$$\qquad\qquad\qquad \text{Change } > \text{ to } <.$$
$$b < -\frac{13}{5}$$

The solution set is $\left\{b \mid b < -\frac{13}{5}\right\}$.

2 **Solve $9x + 4 < 13x - 7$.**

$$9x + 4 < 13x - 7$$
$$9x - 13x + 4 < 13x - 13x - 7 \qquad \text{Subtract 13x from each side.}$$
$$-4x + 4 < -7$$
$$-4x + 4 - 4 < -7 - 4 \qquad \text{Subtract 4 from each side.}$$
$$-4x < -11$$
$$\frac{-4x}{-4} > \frac{-11}{-4} \qquad \text{Divide each side by } -4 \left(\text{or multiply by } -\frac{1}{4}\right).$$
$$\qquad\qquad\qquad \text{Change } < \text{ to } >.$$
$$x > \frac{11}{4}$$

The solution set is $\left\{x \mid x > \frac{11}{4}\right\}$.

Use the Distributive Property to eliminate grouping symbols.

Example

3 **Solve $(0.7)(m + 3) \leq (0.4)(m + 5)$.**

$$(0.7)(m + 3) \leq (0.4)(m + 5)$$
$$0.7m + 2.1 \leq 0.4m + 2.0 \qquad \text{Use the Distributive Property.}$$
$$0.7m + 2.1 - 2.1 \leq 0.4m + 2.0 - 2.1 \qquad \text{Subtract 2.1 from each side.}$$
$$0.7m \leq 0.4m - 0.1$$
$$0.7m - 0.4m \leq 0.4m - 0.4m - 0.1 \qquad \text{Subtract 0.4m from each side.}$$
$$0.3m \leq -0.1$$
$$\frac{0.3m}{0.3} \leq \frac{-0.1}{0.3} \qquad \text{Divide each side by 0.3.}$$
$$m \leq -\frac{1}{3}$$

The solution set is $\left\{m \mid m \leq -\frac{1}{3}\right\}$.

Example

4

Solve $-3(2x - 7) \geq 4x - (x - 3)$.

$$-3(2x - 7) \geq 4x - (x - 3)$$
$$-3(2x - 7) \geq 4x + (-1)(x - 3) \qquad \text{\textit{To subtract, add the inverse.}}$$
$$-3(2x) + (-3)(-7) \geq 4x + (-1)x + (-1)(-3) \qquad \text{\textit{Use the Distributive Property.}}$$
$$-6x + 21 \geq 4x + (-x) + 3 \qquad \text{\textit{Simplify.}}$$
$$-6x + 21 \geq 3x + 3 \qquad \text{\textit{Combine 4x and} } -x.$$
$$-6x - 3x + 21 \geq 3x - 3x + 3 \qquad \text{\textit{Subtract 3x from each side.}}$$
$$-9x + 21 \geq 3$$
$$-9x + 21 - 21 \geq 3 - 21 \qquad \text{\textit{Subtract 21 from each side.}}$$
$$-9x \geq -18$$
$$\frac{-9x}{-9} \leq \frac{-18}{-9} \qquad \text{\textit{Divide each side by} } -9.$$
$$\text{\textit{Change} } \geq \text{ \textit{to} } \leq.$$
$$x \leq 2$$

The solution set is $\{x | x \leq 2\}$.

Exploratory Exercises

State the steps you would use to solve each inequality.

1. $3x - 1 > 14$
2. $9x + 2 > 20$
3. $4y - 7 > 21$
4. $-7y + 6 < 48$
5. $32 + 14t < 4$
6. $5 + 9a > -67$
7. $-12 + 11y \leq 54$
8. $-9 + 6r \leq -33$
9. $\frac{z}{4} + 7 \geq -5$
10. $\frac{b}{-3} + 5 \leq -13$
11. $\frac{d}{-4} - 5 < 23$
12. $\frac{m}{3} - 7 > 11$
13. $13 - 2a \leq 15$
14. $5 - 6g > -19$
15. $2k + 7 > k - 10$
16. $5y + 4 > y + 1$

Written Exercises

Solve each inequality. Express the solution in set-builder notation.

17. $13r - 11 > 7r + 37$
18. $6a + 9 < -4a + 29$
19. $10p - 14 < 8p - 17$
20. $9q + 2 \leq 7q - 25$
21. $3y + 7 \leq 4y + 8$
22. $5 + 10b > 12b + 10$
23. $0.1x < 0.2x - 8$
24. $0.3x + 6.8 \geq 2.0x$
25. $7(x + 8) < 3(x + 12)$
26. $-5(y + 5) \geq 5(y - 1)$
27. $3n - 8n + 21 > 0$
28. $9d - 5 + d < -8$
29. $0.6(y + 7) \geq 0.7(2y + 6)$
30. $0.3(m + 4) \leq 0.4(2m + 3)$
31. $8c - (c - 5) < c + 17$
32. $10x - 2(x - 4) \leq 0$
33. $1.393x - 12 < 0.959x + 4.1$
34. $4.2x - 13.075 > 2.751x$
35. $-3(2x - 8) < 2(x + 14)$
36. $5x \leq 10 + 3(2x + 4)$
37. $-4.65(2x - 3.099) + 5.5(2.061x + 10) \geq 0$
38. $y - 3.025(y + 1.75) \leq 5.982y - 10$
39. $6 - (3y + 5) > 4 - (2y + 7)$
40. $3y - 2(8y - 11) > 5 - (2y + 6)$

Challenge Exercises

41. $\frac{3x + 8}{12} < \frac{5}{12}$
42. $\frac{5y - 4}{3} > \frac{y + 5}{3}$
43. $\frac{2n + 1}{7} \geq \frac{n + 4}{5}$
44. $\frac{c + 8}{4} \leq \frac{-c + 5}{9}$

4-5 Comparing Rational Numbers

You know that $\frac{6}{9}$ is less than $\frac{7}{9}$ because the denominators are the same and 6 is less than 7. How can you compare $\frac{3}{8}$ and $\frac{4}{11}$? If you change both fractions so that they have a common denominator of 88, you can see that $\frac{33}{88}$ is greater than $\frac{32}{88}$. Thus, $\frac{3}{8} > \frac{4}{11}$.

In comparing the fractions, if the denominators are the same, compare the numerators.

A shortcut for comparing two rational numbers is to use cross products. Study the property stated below.

For any rational numbers $\frac{a}{b}$ and $\frac{c}{d}$, with $b > 0$, $d > 0$,	
1. If $\frac{a}{b} < \frac{c}{d}$, then $ad < bc$.	*Comparison Property for Rational Numbers*
2. If $ad < bc$, then $\frac{a}{b} < \frac{c}{d}$.	

This property also holds if $<$ is replaced by $>$, $\leq$, $\geq$, or $=$.

The following example shows how this property is used to compare $\frac{3}{8}$ and $\frac{4}{11}$.

You can use the Multiplication Property for Inequalities to prove the Comparison Property.

Find the product of the extremes. $\frac{3}{8} \overset{?}{-} \frac{4}{11}$ *Find the product of the means.*

$$3(11) \overset{?}{-} 8(4)$$
$$33 > 32$$

Since $33 > 32$, you can conclude that $\frac{3}{8} > \frac{4}{11}$.

Examples

1 Replace the $\overset{?}{-}$ with $<$, $>$, or $=$ to make $\frac{6}{11} \overset{?}{-} \frac{7}{12}$ a true sentence.

$\frac{6}{11} \overset{?}{-} \frac{7}{12}$ *Find the cross products.*

$$6(12) \overset{?}{-} 11(7)$$
$$72 < 77$$

The true sentence is $\frac{6}{11} < \frac{7}{12}$.

2 Replace the $\overset{?}{-}$ with $<$, $>$, or $=$ to make $-\frac{1}{4} \overset{?}{-} -\frac{1}{5}$ a true sentence.

Rewrite $-\frac{1}{4}$ as $\frac{-1}{4}$ and $-\frac{1}{5}$ as $\frac{-1}{5}$. Then find the cross products.

$-\frac{1}{4} \overset{?}{-} -\frac{1}{5}$ *Find the cross products.*

$$-5 < -4$$

The true sentence is $-\frac{1}{4} < -\frac{1}{5}$.

The Comparison Property can help you determine which of two items is a better buy. Study the following example.

Example

3 **Super Saver Mart advertised a 9.4-ounce tube of toothpaste for \$1.61. Is this a better buy than another brand weighing 6 ounces that sells for 95 cents?**

Compare the cost per ounce or unit cost of each tube. If the quality of the two items is the same, the item with the smaller unit cost is the better buy. Be sure to express costs using the same unit.

$$\text{unit cost} = \frac{\text{total cost}}{\text{number of units}} \text{ or } \frac{\text{total cents}}{\text{number of ounces}}$$

$$\text{unit cost of first brand} = \frac{161}{9.4} \qquad \text{unit cost of second brand} = \frac{95}{6}$$

In each case the unit cost is expressed as cents per ounce.

Compare $\frac{161}{9.4}$ and $\frac{95}{6}$ by finding the cross products.

$$\frac{161}{9.4} \; \overset{?}{=} \; \frac{95}{6}$$

$$966 > 893$$

Thus, $\frac{161}{9.4} > \frac{95}{6}$.

The lesser number is $\frac{95}{6}$. So, 6 ounces for 95¢ is the better buy.

A property that is true for rational numbers but is not true for integers is called the **Density Property**.

| Between every pair of distinct rational numbers, there is another rational number. | *Density Property for Rational Numbers* |

If a and b are rational numbers, then one way to find a rational number between them is to take the average of a and b. Thus, $\frac{1}{2}(a + b)$ will be a rational number.

Example

4 **Find a rational number between $\frac{3}{4}$ and $\frac{11}{7}$.**

Take the average of the two rational numbers.

$$\frac{1}{2}\left(\frac{3}{4} + \frac{12}{7}\right) = \frac{1}{2}\left(\frac{21}{28} + \frac{44}{28}\right)$$
$$= \frac{1}{2}\left(\frac{65}{28}\right)$$
$$= \frac{65}{56} \qquad \text{A rational number between } \frac{3}{4} \text{ and } \frac{11}{7} \text{ is } \frac{65}{56}.$$

Exploratory Exercises

State which ratio in each pair is greater.

1. $\frac{3}{4}, \frac{4}{5}$

2. $\frac{11}{12}, \frac{7}{8}$

3. $\frac{9}{10}, \frac{10}{11}$

4. $\frac{7}{8}, \frac{8}{9}$

5. $\frac{6}{5}, \frac{7}{6}$

6. $\frac{11}{9}, \frac{12}{10}$

7. $-\frac{1}{3}, -\frac{1}{4}$

8. $-\frac{6}{5}, -\frac{7}{6}$

9. $-\frac{9}{7}, -\frac{7}{5}$

10. $-\frac{3}{4}, -\frac{2}{3}$

11. $-\frac{13}{11}, -\frac{15}{13}$

12. $\frac{0.06}{0.4}, \frac{0.9}{5}$

Determine the ratio for the unit cost of each item.

13. 10 ounces of coffee for $4.59

14. $3.79 for 8 ounces of coffee

Written Exercises

Replace each $\underline{\ ?\ }$ with <, >, or = to make each sentence true.

15. $\frac{6}{7} \underline{\ ?\ } \frac{7}{8}$

16. $\frac{8}{7} \underline{\ ?\ } \frac{9}{8}$

17. $\frac{7}{19} \underline{\ ?\ } \frac{6}{17}$

18. $\frac{8}{15} \underline{\ ?\ } \frac{9}{16}$

19. $-\frac{2}{3} \underline{\ ?\ } -\frac{3}{5}$

20. $-\frac{1}{3} \underline{\ ?\ } -\frac{2}{7}$

21. $-\frac{7}{6} \underline{\ ?\ } -\frac{21}{18}$

22. $\frac{5}{14} \underline{\ ?\ } \frac{25}{70}$

23. $\frac{0.4}{3} \underline{\ ?\ } \frac{1.2}{2}$

24. $\frac{1.1}{4} \underline{\ ?\ } \frac{2.2}{5}$

25. $-\frac{3.26}{0.965} \underline{\ ?\ } -\frac{5.83}{2.136}$

26. $-\frac{6.908}{4.01} \underline{\ ?\ } -\frac{5.355}{2.4}$

Find a number between the given numbers.

27. $\frac{1}{2}$ and $\frac{6}{7}$

28. $\frac{4}{7}$ and $\frac{9}{4}$

29. $-\frac{2}{9}$ and $-\frac{8}{11}$

30. $-\frac{19}{30}$ and $-\frac{31}{45}$

31. $-\frac{3}{8}$ and $\frac{9}{10}$

32. $-\frac{8}{17}$ and $\frac{1}{9}$

33. $-\frac{11}{19}$ and $\frac{5}{14}$

34. $\frac{0.12}{5}$ and $\frac{1.31}{6}$

Write a ratio for the unit cost of each item. Then compare the ratios to determine which of the two items is the better buy. Express all ratios in the form cents/units.

35. a 21-ounce can of baked beans for 79¢ or a 28-ounce can for 97¢

36. a 10-ounce jar of coffee for $4.27 or an 8-ounce jar for $3.64

37. a 184-gram can of peanuts for 91¢ or a 340-gram can for $1.89

38. a half-pound bag of cashews for $2.93 or a $\frac{3}{4}$-pound bag for $4.19

39. a six-pack of cola containing 2.1 liters for $1.79, or a six-pack containing 1.9 liters for $1.69

40. three liters of soda for $2.25 or two liters for $1.69

41. a 27-ounce loaf of bread for 93¢ or a 20-ounce loaf for 79¢

42. a dozen extra-large eggs weighing 27 ounces for $1.09, or a dozen large eggs weighing 24 ounces for 99¢

Solve each problem.

43. At Whittaker's Market a head of lettuce is 79¢. A head of lettuce weighs approximately $\frac{3}{4}$-pound. Dudley's Market sells lettuce at 44¢ for $\frac{1}{2}$-pound. Which store has the better price on lettuce?

44. Haloke wanted to buy soda for a party. Eight 1-liter bottles of Brand X cost $3.92. Six 1-liter bottles of Brand Y cost $3.08. A 2-liter bottle of Brand Z costs $1.09. Which soda is least expensive per liter?

State the property shown.

1. $5(a - 2b) = 5a - 10b$

2. $7(2 + x) = (2 + x)7$

Simplify.

3. $4.26 + (-0.5)$

4. $-\frac{4}{9} - \frac{13}{18}$

5. $\frac{36a}{-4}$

Solve.

6. $-48 = 15 - m$

7. $\frac{2x + 5}{3} = 14$

8. $3n - \frac{n}{4} = 50$

Solve each inequality. Express the solution set in set-builder notation.

9. $5t < -3 + 4t$

10. $-8k < 102$

11. $0.4z + 5.88 \geq 8.8z$

Excursions in Algebra

Comparing Numbers

You can use a calculator to compare rational numbers. First change each fraction to a decimal by dividing the numerator by the denominator. Then compare the decimals.

Example Change the fractions $\frac{6}{17}$, $\frac{17}{49}$, and $\frac{35}{99}$ to decimals. Then write the fractions in order from least to greatest.

$$\frac{6}{17} = 0.3529412$$

$$\frac{17}{49} = 0.3469388$$

$$\frac{35}{99} = 0.3535354$$

It is not always necessary to copy all the decimal places shown in the calculator display. Usually, you can round results to the nearest thousandth.

Therefore $\frac{17}{49} < \frac{6}{17} < \frac{35}{99}$.

Exercises

Write the fractions in order from least to greatest.

1. $\frac{17}{21}, \frac{20}{27}, \frac{19}{24}$

2. $\frac{17}{19}, \frac{32}{35}, \frac{45}{49}$

3. $\frac{3}{14}, \frac{5}{23}, \frac{9}{43}$

Use a calculator to determine which of the two items is the better buy.

4. a dozen oranges for $1.59 or half a dozen for 85¢

5. five pounds of green beans for $3.50 or 2 pounds for $1.38

6. a 48-ounce bottle of dish soap for $2.39 or a 22-ounce bottle for $1.09

7. a 1-pound package of lunch meat for $1.98 or a 12-ounce package for $1.80

4-6 Compound Sentences

Megan Peroni works part-time. Last year she paid $629 in federal income tax. The tax table that she used is shown at the right. According to the table, since she is single, her taxable income must have been at least $4650 but less than $4700.

Let I represent her income. Then the two inequalities below describe the amount of her income.

$$I \geq 4650 \quad \text{and} \quad I < 4700$$

When considered together, these inequalities form a **compound sentence**. A compound sentence containing *and* is true only if *both* inequalities are true.

Another way of writing $I \geq 4650$ and $I < 4700$ without using *and* is shown below.

$$4650 \leq I < 4700$$

Recall that $4650 \leq I$ means $I \geq 4650$.

This sentence is read, *I is greater than or equal to 4650 and less than 4700.*

If 1040A, line 17, OR 1040EZ, line 7 is—		And you are—			
At least	But less than	Single (and 1040EZ filers)	Married filing jointly	Married filing sepa- rately	Head of a house- hold
		Your tax is—			
4,000	4,050	532	484	544	504
4,050	4,100	539	491	551	511
4,100	4,150	547	499	559	519
4,150	4,200	554	506	566	526
4,200	4,250	562	514	574	534
4,250	4,300	569	521	581	541
4,300	4,350	577	529	589	549
4,350	4,400	584	536	596	556
4,400	4,450	592	544	604	564
4,450	4,500	599	551	611	571
4,500	4,550	607	559	619	579
4,550	4,600	614	566	626	586
4,600	4,650	622	574	634	594
4,650	4,700	629	581	641	601
4,700	4,750	637	589	649	609
4,750	4,800	644	596	656	616
4,800	4,850	652	604	664	624
4,850	4,900	659	611	671	631
4,900	4,950	667	619	679	639
4,950	5,000	674	626	686	646

Examples

1 Write the compound sentence $x > -5$ and $x < 1$ without using *and*.

$x > -5$ is the same as $-5 < x$. Therefore $x > -5$ and $x < 1$ can be written $-5 < x < 1$.

Since $x < 1$ can be written $1 > x$, the compound sentence can also be written $1 > x > -5$.

2 Write the compound sentence $y \geq 0$ and $y \leq 5$ without using *and*.

$y \geq 0$ and $y \leq 5$ can be written $0 \leq y \leq 5$ or $5 \geq y \geq 0$.

The graph of a compound sentence containing *and* is the *intersection* of the graphs of the two inequalities.

3 **Graph the solution set of $x > -5$ and $x < 1$.**

Graph $x > -5$.

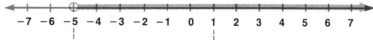

Graph $x < 1$.

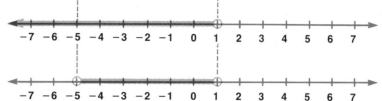

Find the intersection, that is, where the graphs overlap.

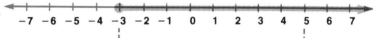

Why are 1 and -5 not included in the solution?

4 **Graph the solution set of $-3 \le x \le 5$.**

*$-3 \le x$ means $x \ge -3$.
Graph $x \ge -3$.*

Graph $x \le 5$.

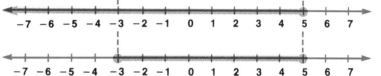

Find the intersection.

Notice that -3 and 5 are included in the solution.

5 **Graph the solution set of $x < 3$ and $x > 5$.**

Graph $x < 3$.

Graph $x > 5$.

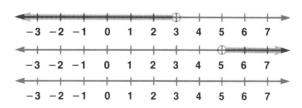

Find the intersection.

There are *no* points in the intersection of the two graphs. That is, there are no numbers that are *both* less than 3 and greater than 5. Therefore, the intersection is the empty or null set. The symbol $\emptyset$ or { } is used to denote this set.

A compound sentence may contain *or* instead of *and*. A compound sentence containing *or* is true if *either* or *both* inequalities are true. The solution of an *or* sentence is the *union* of the solution sets of each inequality.

6 **Graph the solution set of $x \geq 3$ or $x < -2$.**

Graph $x \geq 3$.

Graph $x < -2$.

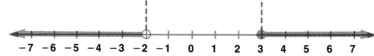

Find the union. That is, combine both graphs.

7 **Solve $2y > y - 3$ or $3y < y + 6$. Graph the solution set.**

First solve each inequality for y.

$$2y > y - 3 \qquad \text{or} \qquad 3y < y + 6$$
$$2y - y > y - y - 3 \qquad\qquad 3y - y < y - y + 6$$
$$y > -3 \qquad\qquad\qquad\qquad 2y < 6$$
$$\frac{2y}{2} < \frac{6}{2}$$
$$y < 3$$

Then graph the solution sets.

Graph $y > -3$.

Graph $y < 3$.

The solution set is {all numbers}.

Exploratory Exercises

Determine whether each compound sentence is *true* or *false*.

1. $8 < 3$ and $8 > 3$
2. $13 > 9$ and $13 > 12$
3. $5 < 7$ and $-8 < -6$
4. $9 \neq 0$ and $12 < 17$
5. $5 \geq -4$ and $11 \leq 7$
6. $5 > -3$ and $-5 > -1$
7. $7 > 4$ or $5 < 6$
8. $13 > 6$ or $0 < -2$
9. $-11 < -19$ or $3 > 0$
10. $8 \leq 8$ or $7 > 15$
11. $8 \neq 9$ or $16 < -7$
12. $-21 > -19$ or $-7 > 0$

Written Exercises

Write each compound sentence without using *and*.

13. $0 \le m$ and $m < 9$

14. $0 < y$ and $y \le 12$

15. $p > \frac{3}{4}$ and $p \le \frac{11}{9}$

16. $r > -\frac{1}{2}$ and $r < \frac{8}{3}$

17. $z > -\frac{4}{5}$ and $z < \frac{2}{3}$

18. $y \le \frac{4}{9}$ and $y \ge -\frac{4}{3}$

19. $m < -\frac{6}{5}$ and $m > -\frac{13}{7}$

20. $r > -\frac{3}{4}$ and $r \le -\frac{1}{10}$

21. $a \le -2.4$ and $a \ge -4.9$

22. $m \ge 0.35$ and $m \le 0.99$

Graph the solution set of each compound sentence.

23. $m < -7$ or $m \ge 0$

24. $x \ge -2$ and $x \le 5$

25. $n \le -5$ and $n \ge -1$

26. $r > 2$ or $r \le -2$

27. $b > 5$ or $b \le 0$

28. $p < -3$ and $p > 3$

29. $x > -5$ and $x < 0$

30. $d \ge -6$ and $d \le -3$

31. $q \ge -5$ and $q \le 1$

32. $w > -3$ or $w < 1$

33. $d > 0$ or $d < 4$

34. $s \le 8$ or $s \ge 3$

35. $a > 8$ or $a < 5$

36. $r > -4$ or $r \le 0$

37. $p \le 6$ and $p \ge -1$

Solve each inequality and graph the solution set.

38. $3 + x < -4$ or $3 + x > 4$

39. $-1 + b > -4$ or $-1 + b < 3$

40. $2 > 3t + 2$ and $3t + 2 > 14$

41. $9 - 2m > 11$ and $5m < 2m + 9$

42. $-2 \le x + 3$ and $x + 3 < 4$

43. $7 + 3q < 1$ or $-12 < 11q - 1$

44. $x \ne 6$ and $3x + 1 > 10$

45. $2x + 4 \le 6$ or $x \ge 2x - 4$

46. $5(x - 3) + 2 < 7$ and $5x > 4(2x - 3)$

47. $2 - 5(2x - 3) > 2$ or $3x < 2(x - 8)$

Write the compound sentence whose solution set is graphed.

48.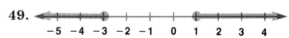

49.

50.

51.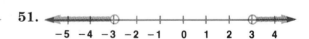

52.

53.

Challenge Exercises

Solve each inequality and graph the solution set.

54. $2.135x + 0.55 < 6$ and $3.5x + 3.205 > -1.28$

55. $-3.25x - 0.667 \ge 2.7$ and $-4x - \frac{5}{6} \le \frac{1}{2}$

56. $\frac{5}{x} + 3 > 0$

57. $\frac{3}{x} + 2 > 0$

58. $-5 < 4 - 3x < 13$

59. $-3 - x < 2x < 3 + x$

60. $2x - 1 < 2x + 8 < 2x + 4$

61. $x - 1 < 2x + 3 < x + 4$

Cumulative Review

1. Write a mathematical expression for the square of a number increased by twice the number.

2. Evaluate $\frac{c^2 - a^2}{(c - a)^2}$ if $a = 5$ and $c = 7$.

3. Solve $x = \frac{2}{3} \cdot \frac{3}{4} \div \frac{3}{5}$.

4. State the property shown by $0 + 96 = 96$.

Simplify.

5. $13a + 21a + 14b - 11b$

6. $5x^2 + 2 + 3x - 4x^2$

Write each sentence as an equation.

7. Twice x decreased by y is z.

8. The product of a, b, and the square of c is f.

9. Graph $\{-1, 0, 1, \ldots\}$ on a number line.

Find each sum or difference.

10. $12 + (-6)$

11. $-11 + (-5)$

12. $-38.6 + 42.73$

13. $11a + (-3a)$

14. $13 - (-18)$

Solve.

15. $y + (-9) = 19$

16. $p - 13 = 27$

17. $-18.3 = x - 11.6$

Simplify.

18. $(-9)(7)(2)$

19. $\frac{3}{4}(12a - 72b) + \frac{2}{5}(10a + 15b)$

20. $\frac{-36}{-12}$

21. $\frac{20x + 30y}{-10}$

Solve.

22. $-\frac{a}{3} = 42$

23. $4x + 2 = -10$

24. $3(7m - 2) = 4(-7m + 2)$

25. $\frac{y - 3}{6} = \frac{2}{3}$

26. $\frac{4}{9} = \frac{8}{r}$

27. $\frac{3}{5}x - 6 = \frac{1}{5}x + 1$

28. What is 35% of 70?

29. State an inequality for the graph.

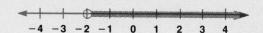

Solve each inequality.

30. $y - 9 > 4$ 31. $-\frac{3}{8}x \geq 6$

32. $7 + 3t \leq 14$

33. $4(a + 6) > 7a - 8$

34. Replace $\underline{\ ?\ }$ with $<$, $>$, or $=$ to make the sentence $-\frac{7}{8} \underline{\ ?\ } -\frac{28}{32}$ true.

Solve the inequality and graph the solution set. Express the solution set in set-builder notation.

35. $\frac{3}{2}x + \frac{7}{2} < 2$

Solve each problem.

36. The difference of two integers is -117. One of the integers is -68. What is the other integer?

37. Two meters of copper wire weigh 0.25 kg. How much will 50 meters of wire weigh?

38. The sales tax on a $44 purchase is $2.31. What is the rate of sales tax?

4-7 Open Sentences Involving Absolute Value

Remember that the *absolute value* of a number is the number of units it is from 0 on the number line. An open sentence that involves absolute value must be interpreted carefully. Study the graph and interpretation of each open sentence given below.

$|x| = 2$

The distance from 0 to x is 2 units.
Therefore, $x = -2$ or $x = 2$.
The solution set is $\{-2, 2\}$.
This can also be written as $\{x|x = -2$ or $x = 2\}$.

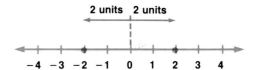

$|x| < 2$

The distance from 0 to x is less than 2 units.
Therefore, $x > -2$ *and* $x < 2$.
The solution set is $\{x|-2 < x < 2\}$.

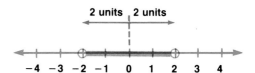

$|x| > 2$

The distance from 0 to x is greater than 2 units. Therefore, $x < -2$ *or* $x > 2$.
The solution set is $\{x|x < -2$ or $x > 2\}$.

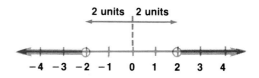

Notice that an open sentence involving absolute value should be interpreted as a compound sentence.

Example

1 **Solve $|3x - 1| = 5$.**

$|3x - 1| = 5$ means $3x - 1 = 5$ *or* $3x - 1 = -5$.

Solve each equation to find the solution set.

$$3x - 1 = 5 \qquad \text{or} \qquad 3x - 1 = -5$$
$$3x - 1 + 1 = 5 + 1 \qquad\qquad 3x - 1 + 1 = -5 + 1$$
$$3x = 6 \qquad\qquad\qquad 3x = -4$$
$$\frac{3x}{3} = \frac{6}{3} \qquad\qquad\qquad \frac{3x}{3} = \frac{-4}{3}$$
$$x = 2 \qquad \text{or} \qquad x = -\frac{4}{3}$$

The solution set is $\left\{x|x = 2 \text{ or } x = -\frac{4}{3}\right\}$.

Examples

2 Solve $|2x + 1| < 8$ and graph the solution set.

$|2x + 1| < 8$ means $2x + 1 > -8$ *and* $2x + 1 < 8$.

$2x + 1 > -8$	and	$2x + 1 < 8$
$2x + 1 - 1 > -8 - 1$		$2x + 1 - 1 < 8 - 1$
$2x > -9$		$2x < 7$
$\frac{2x}{2} > \frac{-9}{2}$		$\frac{2x}{2} < \frac{7}{2}$
$x > -\frac{9}{2}$	and	$x < \frac{7}{2}$

The solution set is $\left\{x \mid x > -\frac{9}{2} \text{ and } x < \frac{7}{2}\right\}$.

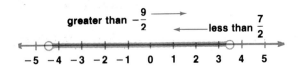

greater than $-\frac{9}{2}$ ⟶
⟵ less than $\frac{7}{2}$

-5 -4 -3 -2 -1 0 1 2 3 4 5

This solution set can also be written as $\left\{x \mid -\frac{9}{2} < x < \frac{7}{2}\right\}$.

3 Solve $|13.04 - a| \geq 2.34$ and graph the solution set.

$|13.04 - a| \geq 2.34$ means $13.04 - a \geq 2.34$ or $13.04 - a \leq -2.34$.

$13.04 - a \geq 2.34$	or	$13.04 - a \leq -2.34$
$13.04 - 13.04 - a \geq 2.34 - 13.04$		$13.04 - 13.04 - a \leq -2.34 - 13.04$
$-a \geq -10.7$		$-a \leq -15.38$
$a \leq 10.7$	or	$a \geq 15.38$

The solution set is $\{a \mid a \leq 10.7 \text{ or } a \geq 15.38\}$.

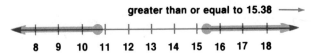

⟵ less than or equal to 10.7

greater than or equal to 15.38 ⟶

8 9 10 11 12 13 14 15 16 17 18

Exploratory Exercises

State each open sentence as a compound sentence.

1. $|x| = 4$ **2.** $|y| = 7$ **3.** $|y| > 3$ **4.** $|x| > 5$

5. $|y| < \frac{5}{2}$ **6.** $|x| < \frac{7}{2}$ **7.** $|x + 2| > 3$ **8.** $|y + 1| > 0$

9. $|x + 2| < 3$ **10.** $|y + 1| < 4$ **11.** $|2x - 5| \geq 3$ **12.** $|2x + 3| \geq 5$

13. $|7 - x| = 4$ **14.** $|7 + x| = 2$ **15.** $|3x + 1| > 6$ **16.** $|3x - 1| < 6$

Written Exercises

Solve each open sentence. Graph each solution set for Exercises 17–40.

17. $|y + 1| > 4$ **18.** $|x + 1| > -2$ **19.** $|y - 1| < 4$ **20.** $|x - 7| < 2$

21. $|2 - y| \le 1$ **22.** $|6 - x| \le 2$ **23.** $|9 - y| \ge -3$ **24.** $|7 - x| \ge 4$

25. $|2y - 10| \ge 6$ **26.** $|12 - 3x| \ge 12$ **27.** $|3 - 3x| = 0$ **28.** $|14 - 2y| = 16$

29. $|4x + 4| \le 14$ **30.** $|6x + 6| \le 28$ **31.** $|10x + 10| \ge 90$ **32.** $|9x + 9| \ge -72$

33. $|2y - 5| \ge 4$ **34.** $|2y - 7| \ge -6$ **35.** $|4 - 3x| = 0$ **36.** $|13 - 2y| = 8$

37. $\left|3t - \frac{1}{2}\right| < \frac{7}{2}$ **38.** $\left|\frac{1}{2} - 3t\right| \ge \frac{11}{2}$ **39.** $|y| + 6 = 8$ **40.** $|y| - 7 = 4$

41. $|2a - 0.06| \le 2.8$ **42.** $|26 + 94.8y| \ge 3.9$ **43.** $|x + 2.54| > 7.9$ **44.** $|8.36 - r| < 29.4$

45. $|194t + 2.36| \le 14.8$ **46.** $|88.6 - 433m| > 849$ **47.** $|4.96x + 8.9| \le 10.8$

48. $|16.25 - 0.097x| \ge 2.008$ **49.** $|52b + 0.268| > 1.225$ **50.** $|12.8 + 14a| < 16.982$

Write an open sentence involving absolute value for each problem. Then graph its solution set.

51. The temperature was within 8 degrees Fahrenheit of 0°F.

52. The temperature was more than 5 degrees Fahrenheit from 0°F.

53. The board was supposed to be 1.5 meters long, plus or minus 0.005 meters.

54. Karen's desired golf score was 90. Her actual score was within 6 strokes of her desired score.

Challenge Exercises

For each graph write an open sentence involving absolute value.

55.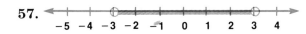

56.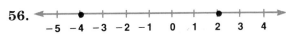

57.

58.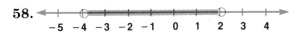

59.

60.

Solve each problem.

61. Find all integer solutions of $|x| \le 2$.

62. Find all integer solutions of $|x| < 4$.

63. If $|x| \le a$, how many integer solutions exist?

64. If $|x| < a$, how many integer solutions exist?

65. Under what conditions is $-|a|$ negative?

66. Under what conditions is $-|a|$ positive?

67. Suppose $x < 0$, $y > 0$, and $x + y = 0$. Is $|x| > |y|$?

Parts used in automobiles must have very precise measurements. Otherwise, they will not work properly. However, it is impossible to produce parts with exact dimensions. Thus, the dimensions of the parts must be between specified limits.

For example, a certain ball bearing that is 1 centimeter in diameter will only work if it is slightly larger or slightly smaller. The diameter may not differ from 1 centimeter by more than 0.001 centimeter. The 0.001 centimeter is called the tolerance of the ball bearing. The diameter of the ball bearing must be 1 ± 0.001 centimeter. That is, the diameter can vary between $1 + 0.001$ centimeter and $1 - 0.001$ centimeter. The acceptable diameter can be represented by the following inequality.

$$1 - 0.001 \leq x \leq 1 + 0.001$$
$$0.999 \leq x \leq 1.001$$

The tolerance interval is 0.999 cm to 1.001 cm.

That is, the least possible diameter is 0.999 cm, and the greatest possible diameter is 1.001 cm.

Exercises

Write each expression as an inequality.

1. $x = 3 \pm 0.01$
2. $x = 5 \pm 0.003$
3. $x = 7 \pm 0.0002$
4. $y = 6 \pm 0.0015$
5. $y = 1 \pm 0.15$
6. $y = 2 \pm 0.003$
7. $r = 0.5 \pm 0.0001$
8. $r = 1.5 \pm 0.001$
9. $d = \frac{1}{2} \pm 0.0035$

Solve each problem.

10. A chemical supply company guarantees the precision weighing of its products. They advertise that a certain product weighs 8 oz $\pm$ 0.03 oz. What is the tolerance interval?

11. A pane of glass should be 26 inches wide by 32 inches long. The tolerance is $\frac{3}{16}$ inch. Find the tolerance interval for each dimension.

4-8 Problem Solving: Inequalities

Problems containing the phrases *greater than* or *less than* can often be solved using inequalities. Other phrases that suggest inequalities are *at least*, *at most*, and *between*. Study the inequalities that correspond to each statement below.

The number x is *at least* 8.	$x \geq 8$	*At least 8 means 8 or greater.*
The number y is *at most* 4.	$y \leq 4$	*At most 4 means 4 or less.*
The number z is *between* 7 and 10.	$7 < z < 10$	*Between 7 and 10 means greater than 7 and less than 10.*
The number w is *between* 7 and 10 *inclusive*.	$7 \leq w \leq 10$	*Between 7 and 10, inclusive means greater than or equal to 7 and less than or equal to 10.*

Examples

1 **Jerry wishes to spend at most $12.50 on new equipment for his model railroad. He has chosen a new railroad car that costs $7.98. How much can he spend on other equipment?**

explore   At most $12.50 means $12.50 or less.
Let x = the amount that Jerry can spend on other equipment.

plan   Total to spend ≤ 12.50
$7.98 + x \leq 12.50$

solve $7.98 - 7.98 + x \leq 12.50 - 7.98$
$x \leq 4.52$
Jerry can spend $4.52 or less on other equipment.

examine   Jerry can spend at most $4.52 since $7.98 + $4.52 = $12.50.
The answer given above is correct.

2 **Roberto has scores of 9.3, 9.2, 9.7, and 8.9 in a figure skating competition. He has one more trial. Roberto will win the competition if his total score is greater than 46.3. What must Roberto's fifth score be?**

explore   You must find Roberto's fifth score so that his total score will be greater than 46.3.
Let x = Roberto's fifth score.

plan $\underbrace{\text{Roberto's total score}} > \underbrace{\text{score to beat}}$
$9.3 + 9.2 + 9.7 + 8.9 + x > \qquad 46.3$

solve $37.1 + x > 46.3$
$37.1 - 37.1 + x > 46.3 - 37.1$
$x > 9.2$
Roberto's final score must be greater than 9.2.   *Examine this solution.*

Written Exercises

Use an inequality to solve each problem.

1. Five times a number increased by 12 is at least 37. What is the number?

2. Seven times a number decreased by 0.005 times the number is less than 3250. What is the number?

3. The sum of two consecutive positive odd integers is at most 18. What are the integers?

4. The sum of two consecutive positive even integers is at most 22. What are the integers?

5. The Two Rivers Tribune pays 5¢ per paper to the carrier. How many papers must the carrier deliver to earn at least $3.50 per day?

6. A bookstore makes a profit of $4.30 on each two-volume set of books sold. How many sets must the store sell to make a profit of at least $175.00?

7. A stove and a freezer weigh at least 260 kg. The stove weighs 115 kg. What is the weight of the freezer?

8. Bill and his father spent at least $110.00 while shopping. Bill spent $47.32. How much did his father spend?

9. Duane earns $12,000 per year in salary and 6% commission on his sales. How much were his sales if his annual income was between $21,000 and $27,000?

10. Sheila bought between 10 and 18 gallons of paint for her house. The paint cost $9.75 per gallon. What was the total cost of the paint?

11. George plans to spend at most $40 for shirts and ties. He bought 2 shirts for $13.95 each. How much can he spend for ties?

12. Diego plans to spend at most $10.00 on model airplanes and supplies. If he buys two models for $3.49 each, how much will he have left for supplies?

13. If 4.05 times an integer is increased by 3.116, the result is betwen 13 and 25. What is the integer?

14. If 9 less than 6 times a number lies between 31 and 37, what is the number?

15. Cecilla has scores of 8.7, 9.3, 8.8, and 9.4 in a figure skating competition. She has one more trial. The leading opponent has a total score of 45.9. What must Cecilla's fifth score be if she is to at least tie for first place?

16. Jenny has a total score of 45.9 in five trials of a skating competition. The leading opponent has scores of 9.1, 8.7, 9.5, and 9.3, and has one more trial. What can Jenny's opponent score to assure Jenny of first place?

17. Keith has scores of 9.1, 9.3, 9.6, and 8.7 in a pommel horse competition. He has one more trial. The leading opponent has completed all five trials and has an average score of 9.22. What must Keith score on his final trial if he is to win the competition?

18. Mali has scores of 9.9, 9.5, 8.9, and 8.7 on the uneven parallel bars. She has one more trial. The leading opponent has completed the competition with an average score of 9.12. What must Mali score on her final trial if she is to win the competition?

19. The sum of two consecutive even integers is greater than 75. Find the pair with the least sum.

20. Find all sets of three consecutive positive even integers whose sum is less than 30.

Comparisons

In the BASIC language conditional program branches are like railroad branches that are controlled by switches. Whether a train takes a branch depends on whether the switch is open or closed. Whether a computer takes a conditional branch depends on whether a statement (the condition) is true or false.

A conditional branch uses an IF–THEN statement. In the following program, IF–THEN statements are used to compare two rational numbers.

```
10   INPUT "ENTER THE FIRST NUMERATOR AND DENOMINATOR ";A,B.
20   INPUT "ENTER THE SECOND NUMERATOR AND DENOMINATOR ";C,D
30   IF A/B > C/D THEN 60
40   IF A/B < C/D THEN 70
50   PRINT A;"/";B;"=";C;"/";D:GOTO 80
60   PRINT A;"/";B;">";C;"/";D:GOTO 80
70   PRINT A;"/";B;"<";C;"/";D:GOTO 80
80   END
```

A colon is used to connect two statements.

Notice how the Comparison Property is used in the program. By this property, you know that either $\frac{A}{B} > \frac{C}{D}, \frac{A}{B} < \frac{C}{D},$ or $\frac{A}{B} = \frac{C}{D}$. Line 30 checks the first possibility. If the condition is true, the computer branches to line 60. If not, the second possibility is checked in line 40. The branching continues until the appropriate PRINT statement is executed and the program ends.

Exercises

Compare each pair of rational numbers. If a rational number is negative, enter the negative sign with the numerator.

1. $\frac{9}{11}, \frac{15}{19}$

2. $\frac{7}{8}, \frac{13}{15}$

3. $-\frac{7}{8}, -\frac{13}{15}$

4. $\frac{16}{17}, \frac{17}{18}$

5. $-\frac{7}{6}, -\frac{21}{18}$

6. $\frac{-6}{-5}, \frac{5}{6}$

7. $\frac{-6}{5}, \frac{5}{6}$

8. $-\frac{15}{33}, -\frac{22}{45}$

9. Determine which of the following two items is a better buy: a 16-ounce box of crackers for $1.69 or a 9-ounce box for $1.09.

10. Determine which of the following two items is a better buy: $1\frac{1}{4}$ pounds of cheese for $3.90 or $1\frac{3}{4}$ pounds for $4.25.

Vocabulary

inequality (123)
Comparison Property (123)
Transitive Property of Order (124)
solution set (125)
set-builder notation (129)
compound sentence (140)

properties for inequalities:
   Addition (128)
   Subtraction (128)
   Multiplication (131)
   Division (132)
Comparison Property for Rational Numbers (136)
Density Property for Rational Numbers (137)

Chapter Summary

1. A mathematical sentence containing $<$ or $>$ is called an inequality. (123)

2. Comparison Property: For any two numbers a and b, exactly one of the following sentences is true.

$$a < b \qquad a = b \qquad a > b \qquad (123)$$

3. Transitive Property of Order:
   For all numbers a, b, and c,
   1. If $a < b$ and $b < c$, then $a < c$.
   2. If $a > b$ and $b > c$, then $a > c$. (124)

4. The symbols $\neq$, $\leq$, and $\geq$ can also be used when comparing numbers. The symbol $\neq$ means *is not equal to*, $\leq$ means *is less than or equal to*, and $\geq$ means *is greater than or equal to*. (124)

5. If a and b represent any numbers and the graph of a is to the left of the graph of b, then $a < b$. If the graph of a is to the right of the graph of b, then $a > b$. (125)

6. Addition and Subtraction Properties for Inequalities:
   For all numbers a, b, and c,
   1. If $a > b$, then $a + c > b + c$ and $a - c > b - c$.
   2. If $a < b$, then $a + c < b + c$ and $a - c < b - c$. (128)

7. Multiplication and Division Properties for Inequalities:
   For all numbers a, b, and c,
   1. If c is positive and $a < b$, then $ac < bc$ and $\frac{a}{c} < \frac{b}{c}$.

      If c is positive and $a > b$, then $ac > bc$ and $\frac{a}{c} > \frac{b}{c}$.

   2. If c is negative and $a < b$, then $ac > bc$ and $\frac{a}{c} > \frac{b}{c}$.

      If c is negative and $a > b$, then $ac < bc$ and $\frac{a}{c} < \frac{b}{c}$.
      (131–132)

8. Comparison Property for Rational Numbers:
   For any rational numbers $\frac{a}{b}$ and $\frac{c}{d}$ with $b > 0$, $d > 0$,

   1. If $\frac{a}{b} < \frac{c}{d}$, then $ad < bc$.   2. If $ad < bc$, then $\frac{a}{b} < \frac{c}{d}$.
   This property also holds if $<$ is replaced by $>$, $\leq$, $\geq$, or $=$. (136)

9. Density Property for Rational Numbers: Between every pair of distinct rational numbers, there is another rational number. (137)

10. A compound sentence containing *and* is true only if both inequalities are true. (140)

11. A compound sentence containing *or* is true if either or both inequalities are true. (141)

12. An open sentence involving absolute value should be interpreted as a compound sentence. (145)

4-1 State whether each sentence is *true* or *false*.

    1. $4 < -5$ **2.** $13 > 4$ **3.** $7 - 8 \le 0$ **4.** $\frac{6}{2} - 3 \ge 0$

Replace each $\underset{=}{?}$ with $<$, $>$, or $=$ to make each sentence true.

    5. $-9 \ \underset{=}{?} \ -11$ **6.** $5 \ \underset{=}{?} \ \frac{13}{3}$ **7.** $(4.1)(0.2) \ \underset{=}{?} \ 8.2$

Graph the solution set of each inequality on a number line.

    8. $y \ge -4$ **9.** $x \le 5$ **10.** $n < 4$ **11.** $r \ne -2$

Solve each inequality.

4-2   **12.** $m - 4 < 9$ **13.** $5y - 6 \ge 4y$ **14.** $r - 2.3 \ge -7.8$

        15. $n + 8 < -3$ **16.** $2x + 7 < 3x$ **17.** $y + \frac{5}{8} < \frac{11}{24}$

4-3   **18.** $\frac{h}{-8} \le 13$ **19.** $6a \le -24$ **20.** $\frac{4}{3}k > 16$

        21. $-\frac{3}{8}r > 9$ **22.** $-7c \le 91$ **23.** $0.7t < -0.98$

        24. $\frac{2}{3}k > \frac{2}{15}$ **25.** $-0.3x < 4.5$ **26.** $\frac{4}{7}z > -\frac{2}{5}$

4-4   **27.** $7m - 12 < 30$ **28.** $2r - 0.5 > 3.1$ **29.** $14a - 11 > 6a + 37$

        30. $7y + 8 \le 4y - 11$ **31.** $4(m - 1) < 7m + 8$ **32.** $4z - 11 > 7.3z + 22$

4-5 Replace each $\underset{=}{?}$ with $<$, $>$, or $=$ to make each sentence true.

    33. $\frac{3}{8} \ \underset{=}{?} \ \frac{4}{11}$ **34.** $\frac{10}{11} \ \underset{=}{?} \ \frac{11}{12}$ **35.** $-\frac{3}{4} \ \underset{=}{?} \ -\frac{7}{9}$

Find a number between the given numbers.

    36. $\frac{2}{9}$ and $\frac{5}{8}$ **37.** $-\frac{3}{5}$ and $-\frac{7}{12}$ **38.** $-\frac{1}{2}$ and $\frac{7}{11}$

    39. Which is a better buy: 0.75 liter of soda for 89¢ or 1.25 liters of soda for $1.31?

4-6 Write each compound sentence without using *and*.

    40. $-4 < y$ and $y \le 7$ **41.** $-3 \le x$ and $x \le 17$

Graph the solution set of each compound sentence.

    42. $y > -1$ and $y \le 3$ **43.** $x \le -3$ or $x > 0$

    44. $2m + 5 \le 7$ or $2m \ge m - 3$ **45.** $4r \ge 3r + 7$ and $3r < 33$

4-7 Solve each inequality. Then graph its solution set.

    46. $|y| > 2$ **47.** $|m - 1| \le 5$ **48.** $\left| 2t - \frac{1}{2} \right| > \frac{9}{2}$

4-8 Use an inequality to solve each problem.

    49. If 8 times a number is decreased by 2, the result is between 5 and 15. What is the result?

    50. Linda plans to spend at most $85 on jeans and shirts. She bought 2 shirts for $15.30 each. How much can she spend on jeans?

Chapter Test

State whether each sentence is _true_ or _false_.

1. $4 > -8$

2. $-2 > 0$

3. $0.3 < -0.6$

Replace each ? with >, <, or = to make each sentence true.

4. $2 \underline{\ ?\ } -7$

5. $-4 \underline{\ ?\ } -3$

6. $\frac{7}{6} \underline{\ ?\ } \frac{13}{12}$

State an inequality for each graph.

7.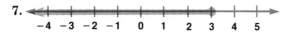

8.

Graph the solution set of each inequality on a number line.

9. $y \leq 2$

10. $-3 \leq x < 4$

11. $y \neq 4$

Solve each inequality.

12. $y - 2 > 11$

13. $7y > 6y - 11$

14. $\frac{r}{5} > -4$

15. $\frac{a}{-3} > -8$

16. $3z \leq 4.2$

17. $5p \geq -19$

18. $\frac{3}{4}r + 7 \leq -8$

19. $\frac{2}{3}r > -\frac{7}{12}$

20. $8x + 3 < 13x - 7$

21. $0.3(m + 4) \leq 0.5(m - 4)$

Find a number between the given numbers.

22. $\frac{5}{11}$ and $\frac{13}{7}$

23. $-\frac{2}{3}$ and $-\frac{9}{14}$

24. $-\frac{13}{2}$ and $\frac{12}{7}$

Graph the solution set of each compound sentence.

25. $x > -3$ and $x < 2$

26. $x \geq 7$ or $x \leq -1$

Solve each inequality. Then graph its solution set.

27. $|n| < 5$

28. $|n| > 3$

29. $|2x - 1| < 5$

30. $|3x - 5| \geq 1$

Solve.

31. At Carl's Market, a 5-pound bag of flour costs 89¢. A 2-pound bag of flour at Kirby's Market costs 41¢. Which store has the better price on flour?

32. The average of four consecutive positive odd integers is less than 20. What are the greatest integers that will satisfy this condition?

33. Seven less than twice a number is less than 83. What is the number?

The test questions on this page deal with fraction concepts. The information at the right may help you with some of the questions. Study the examples.

Directions: Choose the best answer. Write A, B, C, or D.

1. In a class of 27 students, six are honor students. What part of the class is *not* honor students?

   (A) $\frac{7}{11}$ (B) $\frac{7}{9}$ (C) $\frac{2}{7}$ (D) $\frac{2}{9}$

2. Which of the following fractions is less than $\frac{1}{5}$?

   (A) $\frac{3}{14}$ (B) $\frac{21}{100}$ (C) $\frac{2}{11}$ (D) $\frac{101}{501}$

3. Which fraction is greater than $\frac{1}{4}$ but less than $\frac{1}{3}$?

   (A) $\frac{4}{13}$ (B) $\frac{1}{5}$ (C) $\frac{5}{12}$ (D) $\frac{3}{4}$

4. Which of the following is the least?

   (A) 0.77 (B) $\frac{7}{9}$ (C) $\frac{8}{11}$ (D) $\frac{3}{4}$

5. Which of the following is the greatest?

   (A) $\frac{1}{2}$ (B) $\frac{8}{15}$ (C) $\frac{4}{9}$ (D) $\frac{7}{13}$

6. The months from April through December, inclusive, are what fractional part of a year?

   (A) $\frac{7}{12}$ (B) $\frac{3}{4}$ (C) $\frac{5}{12}$ (D) $\frac{2}{3}$

7. What fractional part of a week is one-sixth of a day?

   (A) $\frac{1}{42}$ (B) $\frac{1}{30}$ (C) $\frac{4}{7}$ (D) $\frac{7}{168}$

8. The difference between $42\frac{3}{8}$ minutes and $41\frac{2}{3}$ minutes is approximately how many seconds?

   (A) 18 (B) 63 (C) $22\frac{1}{2}$ (D) 43

1. The fraction $\frac{a}{b}$ means $a \div b$. The *numerator* is a and the *denominator* is b.

2. As the numerator of a positive fraction increases, the value of the fraction increases.

   Example: $\frac{1}{5} < \frac{2}{5} < \frac{3}{5} < \frac{10}{5} < \frac{13}{5}$

3. As the denominator of a positive fraction increases, the value of the fraction decreases.

   Example $\frac{3}{2} > \frac{3}{3} > \frac{3}{4} > \frac{3}{8} > \frac{3}{11}$

4. A fraction may be used to indicate part of a group.

   $$\frac{\text{part}}{\text{whole}} = \text{fractional part}$$

   Example: If there are 4 boys in a class of 9 students, then boys represent $\frac{4}{9}$ of the class.

9. Which group of numbers is arranged in descending order?

   (A) $\frac{5}{7}, \frac{7}{12}, \frac{6}{11}, \frac{3}{13}$ (B) $\frac{7}{12}, \frac{5}{7}, \frac{6}{11}, \frac{3}{13}$

   (C) $\frac{5}{7}, \frac{6}{11}, \frac{7}{12}, \frac{3}{13}$ (D) $\frac{3}{13}, \frac{7}{12}, \frac{6}{11}, \frac{5}{7}$

10. Which group of numbers is arranged from greatest to least?

    (A) $3, \frac{1}{4}, -1, -0.5$ (B) $-1, -0.5, \frac{1}{4}, 3$

    (C) $3, \frac{1}{4}, -0.5, -1$ (D) $-0.5, \frac{1}{4}, -1, 3$

11. A car was driven twice as many miles in July as in each of the other 11 months of the year. What fraction of the total mileage for the year occurred in July?

    (A) $\frac{2}{11}$ (B) $\frac{2}{13}$ (C) $\frac{1}{6}$ (D) $\frac{1}{7}$

CHAPTER 5

Powers

About one-fifth of the world's population lives in China. In 1987, the population of China was estimated to be 1,085,000,000. The number 1,085,000,000 can be expressed more concisely by using powers of ten. You will learn more about powers in this chapter.

5-1 Multiplying Monomials

A **monomial** is a number, a variable, or a product of a number and one or more variables. Some examples of monomials are -9, y, $7a$, $3y^3$, and $\frac{1}{2}abc^5$. Monomials such as -9 and $\frac{5}{3}$, which do not contain variables, are called **constant monomials**, or **constants**.

The following expressions are *not* monomials.

$$m + n \qquad \frac{x}{y} \qquad 3 - 4ab \qquad \frac{1}{x^2}$$

Why is each expression not a monomial?

Consider each of the following products.

$$4 \cdot 16 = 64 \qquad 8 \cdot 16 = 128 \qquad 8 \cdot 32 = 256$$

Each number can be expressed as a power of 2.

$$2^2 \cdot 2^4 = 2^6 \qquad 2^3 \cdot 2^4 = 2^7 \qquad 2^3 \cdot 2^5 = 2^8$$

Recall that an expression of the form x^n is a power. The base is x and the exponent is n.

Consider the exponents only. How do you obtain the exponent 6 from the exponents 2 and 4? How do you obtain 7 from 3 and 4? How do you obtain 8 from 3 and 5?

Think about $a^2 \cdot a^3$.

a^2 means $a^1 \cdot a^1$ and a^3 means $a^1 \cdot a^1 \cdot a^1$.

$a^2 \cdot a^3$ means $(a^1 \cdot a^1) \cdot (a^1 \cdot a^1 \cdot a^1)$.

$$\underbrace{a^1 \cdot a^1}_{2} \cdot \underbrace{a^1 \cdot a^1 \cdot a^1}_{3} = a^5$$
$$2 \quad + \quad 3 \quad = 5$$
$$a^2 \cdot a^3 = a^{2+3} = a^5$$

These and many similar examples suggest that you can multiply powers that have the same base by adding exponents.

For any number a and any integers m and n, $$a^m \cdot a^n = a^{m+n}.$$	*Product of Powers*

Examples

1 Simplify $x^3 \cdot x^4$.

$x^3 \cdot x^4 = x^{3+4}$ *To multiply powers that have the same base, write the common base, then add the exponents.*

$\quad\quad\quad = x^7$

2 Simplify $(3a^2)(4a)$.

$(3a^2)(4a) = 3(4)(a^2)(a)$ *Use the Commutative and Associative Properties.*

$\quad\quad\quad\quad = 12a^{2+1}$ *Recall $a = a^1$.*

$\quad\quad\quad\quad = 12a^3$

Examples

3 Simplify $(a^3b^2)(a^2b^4)$.

$$(a^3b^2)(a^2b^4) = (a^3 \cdot a^2)(b^2 \cdot b^4)$$
$$= a^{3+2} \cdot b^{2+4}$$
$$= a^5b^6$$

4 Simplify $(-5x^2)(3x^3y^2)\left(\frac{2}{5}xy^4\right)$.

$$(-5x^2)(3x^3y^2)\left(\frac{2}{5}xy^4\right)$$
$$= \left(-5 \cdot 3 \cdot \frac{2}{5}\right)(x^2 \cdot x^3 \cdot x)(y^2 \cdot y^4)$$
$$= -6x^6y^6$$

The variables in a monomial are usually arranged in alphabetical order.

Exploratory Exercises

Verify each product by multiplication.

1. $3^1 \cdot 3^2 = 3^3$

2. $2^3 \cdot 2^2 = 2^5$

3. $4^2 \cdot 4^2 = 4^4$

4. $2^5 \cdot 2^2 = 2^7$

5. $5^2 \cdot 5 = 5^3$

6. $3^2 \cdot 3^4 = 3^6$

Simplify.

7. $x^3 \cdot x^5$

8. $y^7 \cdot y^7$

9. $a^4 \cdot a^4$

10. $b^5 \cdot b^2$

11. $m \cdot m^3$

12. $d \cdot d^6$

Written Exercises

Simplify.

13. $a^2(a^5)$

14. $b^4 \cdot b^2$

15. $m^5 \cdot m$

16. $m^2(m^2)$

17. $t^2(t^2)(t^2)$

18. $r^2(r^3)(r^3)$

19. $a^5(a)(a^7)$

20. $(b^4)(b^4)(b)$

21. $(a^2b)(ab^4)$

22. $(x^3y)(xy^3)$

23. $(m^3n)(mn^2)$

24. $(r^3t^4)(r^4t^4)$

25. $(3a^2)(4a^3)$

26. $(-5x^3)(4x^4)$

27. $(-10x^3y)(2x^2)$

28. $(3y^3z)(7y^4)$

29. $(y^3z^4)(y^2)$

30. $m^4(m^3b^2)$

31. $(3a^2b)(2ab^5)$

32. $(4x^2y^3)(2xy^6)$

33. $(3x^4)(-2x^4y^3)$

34. $(-8x^3y)(2x^4)$

35. $(3x^4y)(4x^2y^2)$

36. $(-2x^2y)(-6x^4y^7)$

37. $(-3x^5y)(2x^4)$

38. $(-2n^4y^3)(3ny^4)$

39. $(3x^2y^2z)(2x^2y^2z^3)$

40. $(r^2xy)(-2r^3x)$

41. $(5a^2b^2c)(-7a^3)$

42. $(2am^3n)(-3am^4)$

43. $(2am)(3a^2m^2n)$

44. $(4r^2st)(-6s^2t^2)$

45. $(ab)(ac)(bc)$

46. $(m^2n)(am)(an^2)$

47. $\frac{3}{4}a(12b^2)$

48. $-\frac{5}{6}c(12a^3)$

49. $\left(\frac{1}{2}a^2\right)(6ab^2)$

50. $(-27ay^3)\left(-\frac{1}{3}ay^3\right)$

51. $\left(-\frac{1}{8}a\right)\left(-\frac{1}{6}\right)(b)(48c)$

52. $(bc)\left(\frac{2}{3}b\right)\left(\frac{2}{3}a\right)$

53. $ab\left(\frac{1}{2}a\right)\left(\frac{1}{2}b\right)\left(\frac{1}{2}c\right)$

54. $\left(-\frac{1}{3}c^2b^3a\right)(18a^2b^2c^3)$

Challenge Exercises

Simplify.

55. $y^2 \cdot y^b$

56. $3^x \cdot 3^2$

57. $a^{x-2} \cdot a^5$

58. $2^{4a} \cdot 2^{5a}$

59. $x^{2a} \cdot x^{4a}$

60. $(2^{7x+6})(2^{3x-4})$

61. $(y^{3a+1})(y^{a-6})$

62. $(x+3)^a \cdot (x+3)^6$

63. $(x+3)^{2a} \cdot (x+3)^{b-a}$

Another problem-solving strategy is to look for a pattern. When using this strategy, it is important to organize information about the problem. Study the following example.

Example Pentagonal dot numbers are named by the number of dots used to form a five-sided figure. There is the same number of dots on each side. For the pentagonal dot number 35, determine the number of dots on each side of its dot figure.

To find the number of dots per side of the pentagonal dot number 35, it is helpful to examine other dot numbers and *look for a pattern.*

The dot figures for pentagonal dot numbers 5, 10, and 15 are shown. After studying these figures, try to solve the problem. You should consider how the number of dots on each side increases as the total number of dots increases.

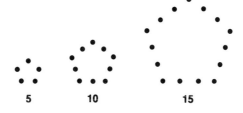

5   **10**   **15**

The total number of dots is a multiple of five.

$5 = 1 \cdot 5$ There are $1 + 1$ or 2 dots per side.
$10 = 2 \cdot 5$ There are $2 + 1$ or 3 dots per side.
$15 = 3 \cdot 5$ There are $3 + 1$ or 4 dots per side.
$\vdots$
$35 = 7 \cdot 5$ There are $7 + 1$ or 8 dots per side of the pentagonal dot number 35.

The total number of dots per side is one more than the 5-multiple for that figure.

Exercises

Solve each problem.

1. There were 10 people at a party. Each person shook hands with each of the other people exactly once. How many handshakes occurred?

2. How many squares are shown at the right? (Hint: There are more than 16.)

3. How many rectangles are shown below?

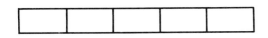

4. How many triangles are shown at the right?

5. Imagine 25 lockers, all closed, and 25 people. Suppose the first person opens every locker. Then the second person closes every second locker. Next the third person changes the state of every third locker. (If it's open, she closes it. If it's closed, she opens it.) Suppose this procedure is continued until the 25th person changes the state of the 25th locker. Which lockers will be open at the end of the procedure?

5-2 Powers of Monomials

Gina wanted to rewrite the expressions $(5^2)^4$ and $(x^6)^2$ using a single exponent. She wrote the expressions correctly as follows.

$$(5^2)^4 = (5^2)(5^2)(5^2)(5^2) \qquad (x^6)^2 = (x^6)(x^6)$$
$$= 5^{2+2+2+2} \qquad\qquad\quad = x^{6+6}$$
$$= 5^8 \qquad\qquad\qquad\quad = x^{12}$$

How can you obtain the exponent 8 from the exponents 2 and 4? How can you obtain 12 from 6 and 2? These and other similar examples suggest that you can find the power of a power by multiplying exponents.

For any number a and any integers m and n, $$(a^m)^n = a^{mn}.$$	*Power of a Power*

Next Gina wanted to find $(xy)^3$ and $(ab)^4$. She used the Associative and Commutative Properties of Multiplication as follows.

$$(xy)^3 = (xy)(xy)(xy) \qquad\quad (ab)^4 = (ab)(ab)(ab)(ab)$$
$$= (x \cdot x \cdot x)(y \cdot y \cdot y) \qquad = (a \cdot a \cdot a \cdot a)(b \cdot b \cdot b \cdot b)$$
$$= x^3 y^3 \qquad\qquad\qquad\quad = a^4 b^4$$

These examples suggest that the power of a product is the product of the powers.

For all numbers a and b and any integer m, $$(ab)^m = a^m b^m.$$	*Power of a Product*

Examples

1 Simplify $[(-4)^3]^2$.

$$[(-4)^3]^2 = (-4)^{3 \cdot 2}$$
$$= (-4)^6 \text{ or } 4096$$

2 Simplify $(x^2 y^4)^3$.

$$(x^2 y^4)^3 = (x^2)^3 \cdot (y^4)^3$$
$$= x^{2 \cdot 3} y^{4 \cdot 3}$$
$$= x^6 y^{12}$$

Example 2 shows the combined use of the rules for the power of a power and the power of a product. This can be stated as follows.

For all numbers a and b and any integers m, n, and p, $$(a^m b^n)^p = a^{mp} b^{np}.$$	*Power of a Monomial*

Examples

3 Simplify $\left(-\frac{2}{3}a^2x^3\right)^3$.

$\left(-\frac{2}{3}a^2x^3\right)^3 = \left(-\frac{2}{3}\right)^3 (a^2)^3 (x^3)^3$

$= -\frac{8}{27}a^6x^9$

4 Simplify $(9b^4y)^2[(-b)^2]^3$.

$(9b^4y)^2[(-b)^2]^3 = 9^2(b^4)^2y^2(b^2)^3$

$= 81b^8y^2b^6$

$= 81b^{14}y^2$

Some calculators have a *power key* labeled y^x. The following example shows how you can use this key to find 7^8.

Example

5 **Using Calculators**

Evaluate 7^8.

ENTER: 7 $\boxed{y^x}$ 8 $\boxed{=}$ DISPLAY: $7 \quad 8 \quad 5764801$

Therefore, $7^8 = 5{,}764{,}801$.

Exploratory Exercises

Complete each exercise.

1. Evaluate $(4y)^2$ and $4y^2$ if $y = 2$.
2. For all numbers a and b and any integer m, is $(a + b)^m = a^m + b^m$ a true sentence? (Hint: Substitute numbers and evaluate.)
3. For all numbers a and b and any integer m, is $\left(\frac{a}{b}\right)^m = \frac{a^m}{b^m}$ a true sentence?
4. For any number a and any integers m and n, is $(a^m)^n = a^{mn}$ a true sentence?

Written Exercises

Simplify.

5. $(5^3)^3$
6. $(10^2)^2$
7. $[(-5)^2]^3$
8. $[(-4)^2]^2$
9. $(40.28)^3$
10. $(13.75)^7$
11. $[(2^3)^2]^4$
12. $[(3^2)^4]^2$
13. $(x^4)^3$
14. $(m^2)^5$
15. $(10y)^2$
16. $(-7z)^3$
17. $\left(\frac{1}{2}c\right)^2$
18. $\left(\frac{2}{5}d\right)^2$
19. $(0.4d)^2$
20. $(0.6d)^3$
21. $(ab^2)^3$
22. $(a^3x^2)^4$
23. $(2a^2b)^2$
24. $(-3a^2b^5)^2$
25. $4(a^2b^3)^3$
26. $-3(ax^3y)^2$
27. $\left(\frac{1}{2}xy^2\right)^3$
28. $\left(\frac{3}{4}x^2y^5\right)^2$
29. $(-0.1x^2)^2$
30. $(-0.2x^3y)^3$
31. $(0.2a^2)^3$
32. $(0.3x^3y^2)^2$
33. $(4xy)^2(-3x)$
34. $(-3ab)^3(2b^2)$
35. $(2a)^3(-3b)$
36. $(2a)^2(3y)$
37. $(2x^2)^2\left(\frac{1}{2}y^2\right)^2$
38. $\left(\frac{3}{10}y^2\right)^2(10y^2)^3$
39. $10^2 \cdot 10^3$
40. $4^2 \cdot 4^3 \cdot 4^4$
41. $(3a^2)^3 + (5a^2)^3$
42. $(5y)^2 + (3y)(7y)$
43. $(-3x^3y)^3 - 3(x^2y)^2(x^5y)$

5-3 Dividing Monomials

Consider each of the following quotients.

$$\frac{81}{27} = 3 \qquad \frac{27}{3} = 9 \qquad \frac{243}{9} = 27$$

Each number can be expressed as a power of 3.

$$\frac{3^4}{3^3} = 3^1 \qquad \frac{3^3}{3^1} = 3^2 \qquad \frac{3^5}{3^2} = 3^3$$

Consider the exponents only. How do you obtain the exponent 1 from the exponents 4 and 3? How do you obtain 2 from 3 and 1? How do you obtain 3 from 5 and 2?

You can use your knowledge of exponents to simplify $\frac{b^5}{b^2}$, $b \neq 0$.

$$\frac{b^5}{b^2} = \frac{\cancel{b} \cdot \cancel{b} \cdot b \cdot b \cdot b}{\cancel{b} \cdot \cancel{b}} \qquad \text{Notice that } \frac{b \cdot b}{b \cdot b} \text{ is equal to 1.}$$

$$= b \cdot b \cdot b \qquad \text{The quotient has } (5 - 2) \text{ or 3 factors.}$$

$$= b^3$$

Notice that you can divide powers that have the same base by subtracting exponents.

For all integers m and n, and any nonzero number a, $$\frac{a^m}{a^n} = a^{m-n}.$$	*Dividing Powers*

Why is 0 not an acceptable replacement for a?

Examples

1 Simplify $\frac{x^8}{x^5}$.

$$\frac{x^8}{x^5} = x^{8-5}$$

$$= x^3$$

2 Simplify $\frac{a^4 b^3}{ab^2}$.

$$\frac{a^4 b^3}{ab^2} = \left(\frac{a^4}{a^1}\right)\left(\frac{b^3}{b^2}\right) \qquad \text{Group the powers that have the same base}$$

$$= a^{4-1} b^{3-2}$$

$$= a^3 b$$

Study the two ways shown below to simplify $\frac{a^3}{a^3}$, $a \neq 0$.

$$\frac{a^3}{a^3} = \frac{a \cdot a \cdot a}{a \cdot a \cdot a} \qquad \frac{a^3}{a^3} = a^{3-3} \qquad \text{Use the rule for dividing powers.}$$

$$= 1 \qquad = a^0$$

Since $\frac{a^3}{a^3}$ cannot have two different values, you can conclude that $a^0 = 1$. In general, *any* nonzero number raised to the zero power is equal to 1.

0^0 is not defined.

For any nonzero number a, $a^0 = 1$.	*Zero Exponent*

Study the two ways shown below to simplify $\frac{k^2}{k^7}$, $k \neq 0$.

$$\frac{k^2}{k^7} = \frac{\cancel{k} \cdot \cancel{k}}{\cancel{k} \cdot \cancel{k} \cdot k \cdot k \cdot k \cdot k \cdot k} \qquad\qquad \frac{k^2}{k^7} = k^{2-7}$$

$$= \frac{1}{k \cdot k \cdot k \cdot k \cdot k} \qquad\qquad = k^{-5} \qquad \textit{Use the rule for dividing powers.}$$

$$= \frac{1}{k^5}$$

Since $\frac{k^2}{k^7}$ cannot have two different values, you can conclude that $k^{-5} = \frac{1}{k^5}$. This example suggests the following rule.

For any nonzero number a and any integer n, $a^{-n} = \frac{1}{a^n}$.	*Negative Exponents*

To simplify a quotient of monomials, write an equivalent expression that has positive exponents and no powers of powers. Also, each base should appear only once and all fractions should be in simplest form.

Examples

3 Simplify $\frac{-6r^3s^5}{18r^{-7}s^5t^{-2}}$.

$$\frac{-6r^3s^5}{18r^{-7}s^5t^{-2}} = \left(\frac{-6}{18}\right)\left(\frac{r^3}{r^{-7}}\right)\left(\frac{s^5}{s^5}\right)\left(\frac{1}{t^{-2}}\right)$$

$$= -\frac{1}{3}r^{10}s^0t^2$$

$$= -\frac{r^{10}t^2}{3}$$

4 Simplify $\frac{(4a^{-1})^{-2}}{(2a^4)^2}$.

$$\frac{(4a^{-1})^{-2}}{(2a^4)^2} = \frac{4^{-2}a^2}{4a^8} \qquad \textit{Power of a Product Rule}$$

$$= (4^{-2})\left(\frac{1}{4}\right)a^{-6}$$

$$= \left(\frac{1}{16}\right)\left(\frac{1}{4}\right)\left(\frac{1}{a^6}\right)$$

$$= \left(\frac{1}{64}\right)\left(\frac{1}{a^6}\right)$$

$$= \frac{1}{64a^6}$$

Exploratory Exercises

Evaluate.

1. 5^0 **2.** $(-3)^0$ **3.** 4^{-1} **4.** $(-8)^{-1}$ **5.** 10^{-2} **6.** $(-2)^{-3}$

7. $(5^{-1})^2$ **8.** $(3^{-2})^3$ **9.** $\frac{4^{-2}}{4}$ **10.** $\frac{2}{2^{-4}}$ **11.** $\left(\frac{1}{3}\cdot\frac{1}{6}\right)^{-1}$ **12.** $(2^0 3^{-2})^{-2}$

Simplify. Remember to express the results with positive exponents. Assume no denominator is equal to zero.

13. $m^{-5}n^0$ **14.** $f^6 g^0$ **15.** $c^0 d^{-2} e^{-1}$ **16.** $x^5 y^0 z^{-5}$ **17.** $\frac{k^9}{k^4}$ **18.** $\frac{5n^5}{n^8}$

19. $\frac{a^2 b^7}{a^4}$ **20.** $\frac{b^9}{b^4 c^3}$ **21.** $\frac{1}{x^{-1}}$ **22.** $\frac{1}{r^{-4}}$ **23.** $\frac{a^{-4}}{b^{-3}}$ **24.** $\frac{r^{-5}}{k^{-1}}$

Written Exercises

Simplify. Assume no denominator is equal to zero.

25. $\frac{n^8}{n^5}$ **26.** $\frac{w^9}{w^2}$ **27.** $\frac{x^2}{x^3}$ **28.** $\frac{b^6}{b^7}$ **29.** $\frac{a^0}{a^{-2}}$ **30.** $\frac{1}{r^{-3}}$

31. $\frac{k^{-2}}{k^4}$ **32.** $\frac{m^2}{m^{-4}}$ **33.** $\frac{an^6}{n^5}$ **34.** $\frac{xy^7}{y^4}$ **35.** $\frac{an^3}{n^5}$ **36.** $\frac{kn^2}{n^4}$

37. $\frac{b^6 c^5}{b^3 c^2}$ **38.** $\frac{(-a)^4 b^8}{a^4 b^7}$ **39.** $\frac{(-x)^3 y^3}{x^3 y^6}$ **40.** $\frac{a^2 b^2}{a^4 b^5}$

41. $\frac{12b^5}{4b^4}$ **42.** $\frac{48a^8}{12a}$ **43.** $\frac{12b^4}{60b}$ **44.** $\frac{10m^4}{30m}$

45. $\frac{x^3 y^6}{x^3 y^3}$ **46.** $\frac{a^6 b^3}{a^2 b^9}$ **47.** $\frac{a^3 b^4}{a^2 b^2}$ **48.** $\frac{b^6 c^5}{b^{14} c^2}$

49. $\frac{w^5 t^7}{w^3 t^{12}}$ **50.** $\frac{(-r)^5 s^8}{r^5 s^2}$ **51.** $\frac{30x^4 y^7}{-6x^{13} y^2}$ **52.** $\frac{24a^3 b^6}{-2a^2 b^2}$

53. $\frac{16b^4 c}{-4bc^3}$ **54.** $\frac{-8a^3 b^7}{a^2 b^6}$ **55.** $\frac{22a^2 b^5 c^7}{-11abc^2}$ **56.** $\frac{24x^2 y^7 z^3}{-6x^2 y^3 z^1}$

57. $\frac{9xyz^5}{x^4}$ **58.** $\frac{ab^5 c}{ac}$ **59.** $\frac{7x^3 z^5}{4z^{15}}$ **60.** $\frac{27a^4 b^6 c^9}{15a^3 c^{15}}$

61. $\frac{(a^7 b^2)^2}{(a^{-2}b)^{-2}}$ **62.** $\frac{r^{-5} s^{-2}}{(r^2 s^5)^{-1}}$ **63.** $\frac{(5r^{-1}s)^3}{(s^2)^3}$ **64.** $\frac{(r^{-4}k^2)^2}{(5k^2)^2}$

65. $\left(\frac{3m^2 n^2}{6m^{-1}k}\right)^0$ **66.** $\frac{(-b^{-1}c)^0}{4a^{-1}c^2}$ **67.** $\left(\frac{7m^{-1}n^3}{n^2 r^{-2}}\right)^{-1}$ **68.** $\left(\frac{2xy^{-2}z^4}{3xyz^{-1}}\right)^{-2}$

Challenge Exercises

Simplify.

69. $\frac{x^{y+2}}{x^{y-3}}$ **70.** $\frac{y^{x-3}}{y^{x+4}}$ **71.** $\frac{(a^{x+2})^2}{(a^{x-3})^2}$ **72.** $\frac{a^b}{a^{a-b}}$

Solve each equation for k.

73. $x^{k-12} = (x^2)^{-k-3}$ **74.** $m^k \cdot m^{-15} = (m^3)^{k+2}$ **75.** $x^{2k} \cdot x^{3k} = x^{15}$

76. $x^{2k+1} = (x^5)^{k-4}$ **77.** $a^k \cdot a^3 = \frac{1}{a^{2k}}$ **78.** $4^k \cdot (4^2)^k = 4^6$

5-4 Scientific Notation

Astronomers use large numbers when measuring the distance from a star to Earth. Sometimes it is not desirable to express these numbers in *decimal notation.*

For example, the earth is about 93,000,000 miles away from the sun.

$$93{,}000{,}000 = 9.3 \times 10{,}000{,}000$$
$$= 9.3 \times 10^7$$

The decimal point was moved 7 places to the left. $10{,}000{,}000 = 10^7$

This form of expressing numbers is called **scientific notation.** To write a number in scientific notation, express it as the product of a number greater than or equal to 1, but less than 10, and a power of 10.

A number is expressed in scientific notation when it is in the form $$a \times 10^n$$ **where $1 \le a < 10$ and n is an integer.**	*Definition of Scientific Notation*

Examples

1 **Express 38,245 in scientific notation.**

38,245   *To get a number greater than or equal to 1 and less than 10, the decimal point must be moved 4 places to the left.*

$$38{,}245 = 3.8245 \times 10^4$$

2 **Express 5093.4 in scientific notation.**

5093.4   *The decimal point must be moved 3 places to the left.*

$$5093.4 = 5.0934 \times 10^3$$ *The exponent in 10^3 tells how many places the decimal point was moved.*

3 **Express 2.6×10^5 in decimal notation.**

$$2.6 \times 10^5 = 2.6 \times 100{,}000$$
$$= 260{,}000$$ *Notice that the decimal point was moved 5 places to the right.*

Scientific notation is also used to express very small numbers.

$$0.000034 = 3.4 \times 0.00001$$ *The decimal point must be moved 5 places to the right.*
$$= 3.4 \times \frac{1}{100{,}000}$$
$$= 3.4 \times \frac{1}{10^5}$$
$$= 3.4 \times 10^{-5}$$ *The exponent of 10 is −5.*

When numbers between zero and one are written in scientific notation, the exponent of 10 is negative.

4 **Express 0.00319 in scientific notation.**

0.00319   *To get a number greater than or equal to 1 and less than 10, the decimal point must be moved to the right.*

$0.00319 = 3.19 \times 10^{-3}$ *How many places was the decimal point moved?*

5 **Express 3.2×10^{-7} in decimal notation.**

$3.2 \times 10^{-7} = 3.2 \times 0.0000001$
$= 0.00000032$ *Notice that the decimal point was moved 7 places to the left.*

You can find products or quotients of numbers that are expressed in scientific notation.

Examples

6 **Evaluate $\dfrac{5.6 \times 10^{-8}}{8.0 \times 10^{-3}}$. Express the result in scientific notation and**

decimal notation.

$\dfrac{5.6 \times 10^{-8}}{8.0 \times 10^{-3}} = \dfrac{5.6}{8.0} \times \dfrac{10^{-8}}{10^{-3}}$ *Note: $-8 - (-3) = -8 + 3 = -5$*

$= 0.7 \times 10^{-5}$ *Express 0.7×10^{-5} in scientific notation.*

$= 7 \times 10^{-6}$ or 0.000007

7 **Use scientific notation to find the product of 0.000008 and 3,500,000,000. Express the result in scientific notation and decimal notation.**

$(0.000008)(3,500,000,000) = (8 \times 10^{-6})(3.5 \times 10^{9})$ *Express in scientific notation.*

$= (8)(3.5)(10^{-6})(10^{9})$ *Use the Commutative and Associative Properties.*

$= 28(10^{3})$

$= 2.8 \times 10^{4}$ or 28,000

Scientific notation can be used to enter very large or very small numbers into many calculators. These numbers are entered using the key labeled $\boxed{\text{EE}}$ or $\boxed{\text{EXP}}$.

Example

8 **Using Calculators**

Express 230×10^{15} in scientific notation.

ENTER: 230 $\boxed{\text{EE}}$ 15 $\boxed{=}$

DISPLAY: *230   230 00   230 15   2.3 17*

The calculator has expressed 230×10^{15} in scientific notation as 2.3×10^{17}.

The exponent of 10 is 17. That is, 2.3   17 means 2.3×10^{17}.

Exploratory Exercises

Express each number in the second column in decimal notation. Express each number in the third column in scientific notation.

	Planet	Diameter (km)	From Sun (km)
1.	Mercury	5.0×10^3	57,900,000
2.	Venus	1.218×10^4	108,230,000
3.	Earth	1.276×10^4	149,590,000
4.	Mars	6.76×10^3	227,720,000
5.	Jupiter	1.427×10^5	778,120,000
6.	Saturn	1.21×10^5	1,428,300,000

Written Exercises

Express each number in scientific notation.

7. 4293   **8.** 5280   **9.** 240,000   **10.** 389,500

11. 0.000319   **12.** 0.004296   **13.** 0.000000092   **14.** 0.00000000317

15. 32×10^5 **16.** 284×10^3 **17.** 0.76×10^7 **18.** 0.0031×10^3

Evaluate. Express each result in scientific notation and decimal notation.

19. $\frac{4.8 \times 10^3}{1.6 \times 10^1}$ **20.** $\frac{5.2 \times 10^5}{1.3 \times 10^2}$ **21.** $\frac{7.8 \times 10^{-5}}{1.3 \times 10^{-7}}$

22. $\frac{8.1 \times 10^2}{2.7 \times 10^{-3}}$ **23.** $\frac{1.32 \times 10^{-6}}{2.4 \times 10^2}$ **24.** $\frac{2.31 \times 10^{-2}}{3.3 \times 10^{-9}}$

25. $(2 \times 10^5)(3 \times 10^{-8})$ **26.** $(4 \times 10^2)(1.5 \times 10^6)$ **27.** $(3.1 \times 10^{-2})(2.1 \times 10^5)$

28. $(3.1 \times 10^4)(4.2 \times 10^{-3})$ **29.** $(78 \times 10^6)(0.01 \times 10^3)$ **30.** $(0.2 \times 10^5)(31 \times 10^{-6})$

31. $(0.000003)(70,000)$ **32.** $(86,000,000)(0.005)$

33. $24,000 \div 0.00006$ **34.** $0.0000039 \div 650,000$

mini-review

Simplify.

1. $\frac{7}{8} + \left(-\frac{13}{16}\right) + \frac{5}{32}$ **2.** $0.26mn + 4.32m^2n + (-1.08mn)$

3. $\frac{1}{2}(10m + 6n) - \frac{3}{4}(8m - 12n)$ **4.** $3(-26) + (-114) - 5(-21)$

5. $\frac{42x}{-7}$ **6.** $\left(-\frac{1}{4}zy^3x^2\right)(-12xy^4z^2)$ **7.** $\frac{(2a^{-1}b^2)^2}{4a^{-3}b}$

Solve.

8. $t - (-56) = -18$ **9.** $-2(x - 5) = 11$ **10.** $\frac{6}{9} = \frac{n}{12}$

11. $\frac{3(t + 2)}{5} = \frac{6t}{2}$ **12.** $6.3 \leq -0.3t$ **13.** $|p + 6| > 8$

14. 43.2 is what percent of 120?   **15.** 165 is 55% of what number?

The Consumer Price Index is used to keep track of the change of prices of food, clothing, shelter, fuels, transportation, charges for medical and dental services, and other goods and services that people buy for day-to-day living. Current prices are compared to average prices during 1967. For example, an index of 310 points means that goods and services that cost

$100 in 1967 would now cost $310. Changes in the CPI are often expressed as percents. An index of 310 means that the prices average 210% higher than in 1967.

Example   The CPI was 327.5 in February, 1986, and 334.4 in February, 1987. Find the annual percent of increase.

$$334.4 - 327.5 = 6.9$$ *Subtract the earlier value from the later value.*

$$\frac{6.9}{327.5} = \frac{x}{100}$$ *Express the difference as a percent of the earlier value.*

$$6.9(100) = 327.5x$$ *Find the cross products.*

$$\frac{690}{327.5} = \frac{327.5x}{327.5}$$ *Use a calculator to simplify.*

$$2.1 = x$$ *Round to the nearest tenth.*

The annual percent of increase was 2.1%.

Exercises

Compute the monthly percent of increase in the CPI to the nearest tenth.

1. March 245.1
   April 247.8

2. Jan. 179.3
   Feb. 180.7

3. June 234.8
   July 236.4

4. August 294.3
   Sept. 295.4

5. Oct. 301.4
   Nov. 306.1

6. April 280.6
   May 281.8

Compute the yearly percent of increase in the CPI to the nearest tenth. The CPI for two successive years is given.

7. Jan. 195.3
   Jan. 209.4

8. July 223.7
   July 251.4

9. Dec. 245.1
   Dec. 283.8

10. Aug. 291.4
    Aug. 310.8

11. May 263.1
    May 295.6

12. Nov. 236.4
    Nov. 275.2

5-5 Problem Solving: Age Problems

To solve verbal problems you must analyze each sentence carefully.
Decide what is asked and explore how the given facts are related. Many
problems contain extra information. Other problems do not contain
enough information. Study the following problems.

Examples

1 **Maria is 8 years older than Jose. Carla is 2 years younger than Maria.
The sum of Maria's age and Jose's age is 38. How old is Jose?**

explore   This problem asks for Jose's age. What facts are given?

Maria is 8 years older than Jose.
The sum of Maria's age and Jose's age is 38.

Why is Carla's age not stated in the facts given?

Define variables, then use an equation to solve the problem.

Let a = Jose's age now.
Then $a + 8$ = Maria's age.

plan $\underbrace{\text{Jose's age}}_{a} + \underbrace{\text{Maria's age}}_{(a + 8)} = \underbrace{38}_{= 38}$

solve
$$a + (a + 8) = 38$$
$$2a + 8 = 38$$
$$2a + 8 - 8 = 38 - 8$$
$$2a = 30$$
$$a = 15$$

examine   If Jose is 15 years old, then Maria must be 23 years old. Since the sum of
their ages is 38, the solution is correct.

2 **Six years ago, Mr. Winters was five times as old as his son Mark. How old is
Mark now if his age is one-third of his father's present age?**

explore   Let a = Mark's age now.
$3a$ = Mr. Winters' age now.   *Why?*
$a - 6$ = Mark's age 6 years ago.
$3a - 6$ = Mr. Winters' age 6 years ago.

plan $\underbrace{\text{Mr. Winters' age 6 years ago}}_{3a - 6} = \underbrace{\text{5 times}}_{= 5 \ \times} \underbrace{\text{Mark's age 6 years ago}}_{(a - 6)}$

| **solve** | $3a - 6 = 5(a - 6)$ |

$$3a - 6 = 5(a - 6)$$
$$3a - 6 = 5a - 30$$
$$-2a = -24$$
$$a = 12$$

Mark is now 12 years old.   *Examine this solution.*

Exploratory Exercises

State an expression for each of the following.

1. Marilyn is now n years old. Represent her age:
   a. 5 years ago.
   b. in 11 years.
   c. x years ago.

2. Don is now x years old. Represent his age:
   a. 7 years ago.
   b. in 12 years.
   c. in d years.

3. Paul is now $(n + 7)$ years old. Represent his age:
   a. in 7 years.
   b. 4 years ago.
   c. n years ago.
   d. in $(n + 2)$ years.
   e. $(n - 4)$ years ago.
   f. in $5n$ years.

4. Helene is now $(2n + 8)$ years old. Represent her age:
   a. 5 years ago.
   b. in 6 years.
   c. n years ago.
   d. in $(n - 3)$ years.
   e. $(n + 1)$ years ago.
   f. in $3n$ years.

5. Mary is m years old. Represent John's age if he is:
   a. twice as old as Mary.
   b. one-third as old as Mary.
   c. 5 years younger than Mary.
   d. 8 years older than Mary.

Written Exercises

Use an equation to solve each problem. If there is not enough information given, write *not enough information.*

6. Abe is 3 years older than Mindy. The sum of their ages is 39. What are their ages now?

7. Charlie is 14 years younger than Jack. The sum of their ages is 74. What are their ages now?

8. Nica is five times as old as his sister. The sum of their ages is 12. How old is Nica's sister?

9. Joshua is 7 years younger than his brother, and 30 years younger than his father. How old is Joshua?

10. Erin is 25 years younger than her father. The sum of their ages is 75. How old is Erin?

11. Matt is 5 years older than his sister, and 23 years younger than his mother. If the sum of Matt's age and his mother's age is 35, how old is Matt?

12. Lisa is 6 years older than her brother Tom. Their father's age is twice the sum of their ages. How old is Lisa if her father is 32?

13. Todd is 12 years older than Delores. In 2 years, Todd will be twice as old as Delores. How old is Todd now?

14. Twelve years ago, Thea was 7 years old. In 8 years, Thea will be three times as old as Nancy. Nancy is two years younger than Cindy. How old is Nancy now?

15. The sum of David's age and Ben's age is 40. Ann is 3 years older than David. David is 4 years older than Ben. How old is David?

5-6 Problem Solving: Mixtures

Sue Murphy sold tickets for the basketball game. Each adult ticket costs $3.00 and each student ticket costs $1.50. The total number of tickets sold was 105 and the total income was $250.50. How many of each kind of ticket were sold?

Equations can be used to solve problems such as this. It is often helpful to use a chart to organize the information.

Example

1 **Use an equation to solve the problem above.**

explore Let a = the number of adult tickets sold.
Then $105 - a$ = the number of student tickets sold.

plan

	Number of Tickets	Price per Ticket	Total Price
Adult	a	3.00	$3.00a$
Student	$105 - a$	1.50	$1.50(105 - a)$

$$\underbrace{\text{Income from}}_{3.00a} + \underbrace{\text{Income from}}_{1.50(105 - a)} = \underbrace{\text{Total}}_{250.50}$$
Adult Tickets   Students Tickets   Income

solve
$$3.00a + 1.50(105 - a) = 250.50$$
$$3.00a + 157.50 - 1.50a = 250.50$$
$$3.00a - 1.50a = 250.50 - 157.50$$
$$1.50a = 93.00$$
$$a = 62$$

There were 62 adult tickets sold. There were $105 - 62$, or 43 student tickets sold.

examine If 62 adult tickets were sold, the total income from adult tickets was $3.00(62) or $186. If 43 student tickets were sold, the total income from student tickets was $1.50(43) or $64.50.
Since $186 + $64.50 = $250.50, the solution is correct.

Example

2 Mr. Leibovitch owns "The Coffee Pot," which is a specialty coffee store. He wishes to create a special mix of two coffees priced at $6.40 a pound and $7.28 a pound. How many pounds of the $7.28 coffee must he mix with 9 pounds of the $6.40 coffee to sell the mixture for $6.95 per pound?

explore Let n = number of pounds of the $7.28 coffee.

plan

	Number of Pounds	Total Price
$6.40/lb	9	6.40(9)
$7.28/lb	n	7.28n
$6.95/lb	9 + n	6.95(9 + n)

$$\underbrace{\text{Total Cost of \$6.40/lb Coffee}}_{6.40(9)} + \underbrace{\text{Total Cost of \$7.28/lb Coffee}}_{7.28n} = \underbrace{\text{Total Cost of Mixture}}_{6.95(9 + n)}$$

solve

$$6.40(9) + 7.28n = 6.95(9 + n)$$
$$57.60 + 7.28n = 62.55 + 6.95n$$
$$7.28n - 6.95n = 62.55 - 57.60$$
$$0.33n = 4.95$$
$$n = 15$$

Mr. Leibovitch must mix 15 pounds of coffee at $7.28 a pound with 9 pounds of coffee at $6.40 a pound to obtain the desired mixture.

Examine this solution.

Exploratory Exercises

Copy and complete the chart for each problem. Then state an equation.

1. Bonnie bought 16 paperback books for $10.95. Some cost 60¢ each and the rest cost 75¢ each. How many of each did she buy?

	Number	Total Price
60¢ books	x	
75¢ books		

2. Rodolfo has $2.55 in dimes and quarters. He has eight more dimes than quarters. How many quarters does he have?

	Number	Total Value
Quarters	x	
Dimes		

3. Peanuts sell for $3.00 a pound. Cashews sell for $6.00 a pound. How many pounds of cashews should be mixed with 12 pounds of peanuts to obtain a mixture selling for $4.20 a pound?

	Pounds	Total Price
$3.00 peanuts	12	
$6.00 cashews		
$4.20 mixture		

4. At the Granada Theater, tickets for adults cost $4.75 and tickets for children cost $2.25. How many tickets of each were purchased if 26 tickets cost $83.50?

	Number of Tickets	Total Value
Adults	n	
Children		

Written Exercises

Solve each problem.

5. Adrienne has 5 more dimes than nickels. In all, she has $2.30. How many nickels does she have?

6. Lois has 27 coins in nickels and dimes. In all she has $1.90. How many of each does she have?

7. Jeanne Ramos has 53 coins in pennies and nickels with a value of $1.37. How many coins of each does she have?

8. Tim has twice as much money in nickels as in dimes. He has 30 coins in all. How many of each coin does he have?

9. How many pounds of apples costing 64¢ per pound must be added to 30 pounds of apples costing 49¢ per pound to create a mixture that would sell for 58¢ per pound?

10. Ground chuck sells for $1.75 a pound. How many pounds of ground round selling for $2.45 a pound should be mixed with 20 pounds of ground chuck to obtain a mixture that sells for $2.05 a pound?

11. Drew has twice as many quarters as nickels and three more dimes than nickels. He has $4.20 in all. How many of each does he have?

12. Gary has 42 coins in nickels, dimes, and quarters. If he has 8 more nickels than dimes and has $7.15 in all, how many of each does he have?

13. On the first day of school, 264 notebooks were sold. Some sold for 95¢ each and the rest sold for $1.25 each. How many of each were sold if the total sales were $297.00?

14. Fred Suter bought tickets to the Olympic Games. Some tickets cost $5 and the others cost $8. He paid $101 for 16 tickets. How many of each did he buy?

15. How much coffee that costs $3 a pound should be mixed with 5 pounds of coffee that costs $3.50 a pound to obtain a mixture that costs $3.25 a pound?

16. Theatre tickets cost $2.50 for children and $3.50 for adults. The price for 8 tickets was $23.00. How many adult tickets were purchased?

17. The Kulichs are going to Cedar Island, an amusement park, for a family outing. The total cost of tickets for a family of two adults and three children is $61.75. If a child's ticket costs $6.00 less than an adult ticket, find the cost of each.

5-7 Problem Solving: Mixtures Using Percents

As shown in the previous lesson, mixture problems involve a blend of two or more parts. Sometimes these parts are expressed as percents as in the problem below.

Jim Mallinson invested a portion of $28,000 at 8% interest and the balance at 10% interest. How much did he invest at each rate if his total income from both investments is $2550?

Example

1 **Use an equation to solve the problem above.**

explore

Let n = amount invested at 8%.
Then $28,000 - n$ = amount invested at 10%.

plan

Amount Invested	Rate	Annual Income
n	0.08	$0.08n$
$28,000 - n$	0.10	$0.10(28,000 - n)$

Income from 8% Investment	+	Income from 10% Investment	=	Total Income
$0.08n$	+	$0.10(28,000 - n)$	=	2550

solve

$$0.08n + 0.10(28,000 - n) = 2550$$
$$8n + 10(28,000 - n) = 255,000$$
$$8n + 280,000 - 10n = 255,000$$
$$-2n = -25,000$$
$$n = 12,500$$
$$28,000 - n = 15,500$$

To eliminate the decimals, multiply each term by 100.

Jim invested $12,500 at 8% interest and $15,500 at 10% interest.

examine

$12,500 at 8% = $1000 and $15,500 at 10% = $1550.
Since $1000 + $1550 = $2550, the solution is correct.

2 Hal is doing a chemistry experiment that calls for a 30% solution of copper sulfate. Hal has 40 mL of 25% solution. How many milliliters of a 60% solution should Hal add to obtain the required 30% solution?

explore Let x = amount of 60% solution to be added.

plan

	Amount of Solution (mL)	Amount of Copper Sulfate
25% solution	40	0.25(40)
60% solution	x	0.60x
30% solution (mixture)	40 + x	0.30(40 + x)

Amount of Copper Sulfate in 25% Solution + Amount of Copper Sulfate in 60% Solution = Amount of Copper Sulfate in Mixture

$$0.25(40) \quad + \quad 0.60x \quad = \quad 0.30(40 + x)$$

solve

$$0.25(40) + 0.60x = 0.30(40 + x)$$
$$10 \quad + \quad 0.6x = 12 + 0.3x$$
$$0.3x = 2$$
$$x = 6.667 \qquad \textit{Round to the nearest thousandth.}$$

Hal should add 6.667 mL of the 60% solution to the 40 mL of 25% solution. *Examine this solution.*

Exploratory Exercises

Copy and complete the chart for each problem. Then state an equation.

1. An advertisement for an orange drink claims that the drink contains 10% orange juice. How much pure orange juice would have to be added to 5 quarts of the drink to obtain a mixture containing 40% orange juice?

	Quarts	Amount of Orange Juice
10% juice	5	
Pure juice		
40% juice		

2. The amount of $6000 is being put into two accounts, part at 4.5% and the remainder at 6%. If the total annual interest earned from the two accounts is $279, how much was deposited at each rate?

	Amount	Yearly Interest
At 4.5%	x	
At 6%		

3. Janice wishes to create 30 pounds of a 20% copper alloy by mixing a 10% alloy with a 25% alloy. How many pounds of each would be needed?

	Pounds	Amount of Copper
10% alloy	n	
25% alloy		
20% alloy		

4. How much whipping cream (9% butterfat) should be added to 1 gallon of milk (4% butterfat) to obtain a 6% butterfat mixture?

	Gallons	Amount of Butterfat
9% butterfat	x	
4% butterfat		
6% butterfat		

3, 5, 7

Written Exercises

Solve each problem.

5. A chemist has 2.5 liters of a solution that is 70% acid. How much water should be added to obtain a 50% acid solution?

6. How much pure copper must be added to 50.255 kg of an alloy containing 12% copper to raise the copper content to 21%?

7. How many grams of salt must be added to 40 grams of a 28% salt solution to obtain a 40% salt solution?

8. A liter of cream has 9.2% butterfat. How much skim milk containing 2% butterfat should be added to the cream to obtain a mixture with 6.4% butterfat?

9. Marcos has 50.075 mL of a 60% solution of silver nitrate. How many milliliters of a 30% silver nitrate solution should be added to obtain a 50% solution?

10. A pharmacist has 150 dL of a 25% solution of peroxide in water. How many deciliters of peroxide should be added to obtain a 40% solution?

11. An aluminum alloy containing 48% aluminum is to be made by combining 30% and 60% alloys. How many pounds of the 60% alloy must be added to 24 pounds of the 30% alloy to produce the desired alloy?

12. A car radiator has a capacity of 16 quarts and is filled with a 25% antifreeze solution. How much must be drained off and replaced with pure antifreeze to obtain a 40% antifreeze solution?

13. Yolanda invested a portion of $12,500 at 6.2% interest and the balance at 8.6% interest. How much did she invest at each rate if her total income from both investments is $967?

14. Setsu Yamamoto invested $33,600, part at 5% interest and the remainder at 8% interest. If she earned twice as much from her 5% investment as her 8% investment, how much did she invest at each rate?

15. A solution is 50% alcohol. If 10 liters are removed from the solution and replaced with 10 liters of alcohol, the resulting solution is 75% alcohol. How many liters of the solution are there?

16. A stockbroker has invested part of $25,000 at 10.5% interest and the rest at $12\frac{1}{4}$% interest. If the annual income earned from these investments is $2,843.75, find the amount invested at each rate.

Cumulative Review

Complete.

1. 28% of what number is 50.4?

2. Find 19% of 47.

3. Graph $-3 < 2x + 1$ and $2x + 1 \leq 7$.

4. Write 82,100,000 in scientific notation.

5. Write 6.359×10^{-4} in decimal notation.

6. Can m and $-m$ be consecutive even integers?

7. Can m and $-m$ be consecutive odd integers?

8. Can m and $-m$ be consecutive integers?

Simplify.

9. $-6.5x + 3.7y - 1.9x - 5y$

10. $6(2m - 3n) - 5(3m + 2n - 1)$

11. $ab^2(3a^3b)^3$

12. $2xy(-8x^3y)$

13. $x^{3a}(x^{2a})^3$

14. $\frac{-24x^3y^2z}{8xy^3z^2}$

15. $(0.3a^2)^3$

16. $(2a^3b^2c^{-1}d)^3$

17. $(8t^3)(4t^5)^2$

Solve.

18. $x - 19 = 37$

19. $\frac{5}{7}x = 10$

20. $3\frac{2}{3}n = \frac{1}{2}$

21. $8t - 7 = 3t + 8$

22. $3x - 5 < 13$

23. $7 - 3d \geq d + 9$

24. $|3x| > 18$

25. $|2x + 7| \leq 11$

26. $\frac{n}{6} = \frac{-2}{11}$

27. $\frac{x + 2}{x - 3} = \frac{4}{9}$

28. Solve for x in $5x - 2a = 5b + x$.

Solve each problem.

29. Find 38% of 650.

30. 90 is 45% of what number?

31. 5.825 is what percent of 46.6?

32. The sum of three consecutive odd integers is 141. Find the numbers.

33. The sum of two numbers is 37. The difference of these same numbers is 11. Find the numbers.

34. Justine earned $85.10 for 23 hours of work at a fast-food restaurant. Find her hourly earnings.

35. Nick traveled 264 miles on 12 gallons of gasoline. How many miles per gallon does his car average?

36. Blaine bought a stereo for $476.79. This price was 20% less than the original price. What was the original price?

37. A chemist has 4 mL of a 65% solution of silver nitrate. How many milliliters of a 20% silver nitrate solution should be added to obtain a 25% solution?

38. An airline issued round-trip first class tickets from Orlando to Dallas at $255 and round-trip coach tickets at $198. If 167 tickets were purchased for a total fare of $40,191, find the number of first class tickets sold.

5-8 Direct Variation

Diego delivers newspapers. He is paid 3¢ for each newspaper he delivers. The following table relates the number of newspapers Diego delivers (x) and his income (y).

x	20	50	100	120	150
y	$0.60	$1.50	$3.00	$3.60	$4.50

Diego's income depends *directly* on the number of newspapers he delivers. The relationship between newspapers delivered and income is shown by the equation $y = 0.03x$. Such an equation is called a **direct variation**. We say that *y varies directly as x.*

A direct variation is described by an equation of the form $y = kx$, where $k \neq 0$.	*Definition of Direct Variation*

In a direct variation, k is called the **constant of variation**. To find the constant of variation, divide each side of $y = kx$ by x.

$$\frac{y}{x} = k \quad \text{or} \quad k = \frac{y}{x}$$

Example

1 The weight of an object on the moon varies directly as its weight on Earth. A certain astronaut weighs 168 pounds on Earth and 28 pounds on the moon. Kristina weighs 108 pounds on Earth. What would she weigh on the moon?

Let x = weight on Earth.
Let y = weight on the moon.
Then $y = kx$. Find the value of k.

$k = \frac{y}{x}$

$k = \frac{28}{168}$ *Substitute the astronaut's weights for x and y.*

$k = \frac{1}{6}$ *The constant of variation is $\frac{1}{6}$.*

Next find Kristina's weight on the moon. Find y when x is 108.

$y = kx$

$y = \frac{1}{6}(108)$ *Substitute 108 for x and $\frac{1}{6}$ for k.*

$y = 18$

Thus, Kristina would weigh 18 pounds on the moon.

Direct variations are also related to proportions. From the table on Diego's newspapers and income, many true proportions can be formed. Two proportions are shown below.

number of papers
$$\frac{20}{60} = \frac{50}{150}$$
income

number of papers $\displaystyle \frac{20}{50} = \frac{60}{150}$ income

Two general forms for proportions such as these can be derived from the equation $y = kx$. Let (x_1, y_1) be a solution for $y = kx$. Let a second solution be (x_2, y_2). Then $y_1 = kx_1$ and $y_2 = kx_2$. A proportion can be written using x_1, x_2, y_1, and y_2. Study the following.

x_1 is read x sub 1.

$y_1 = kx_1$ *This equation describes a direct variation.*

$\dfrac{y_1}{y_2} = \dfrac{kx_1}{kx_2}$ *Use the Division Property of Equality. Since y_2 and kx_2 are equivalent, you can divide the left side by y_2 and the right side by kx_2.*

$\dfrac{y_1}{y_2} = \dfrac{x_1}{x_2}$ *Simplify.*

Study how another proportion can be formed from this proportion.

$\dfrac{y_1}{y_2} = \dfrac{x_1}{x_2}$

$x_2 y_1 = x_1 y_2$ *Find the cross products.*

$\dfrac{x_2 y_1}{y_1 y_2} = \dfrac{x_1 y_2}{y_1 y_2}$ *Divide each side by $y_1 y_2$.*

$\dfrac{x_2}{y_2} = \dfrac{x_1}{y_1}$ *Simplify.*

You can use proportions to solve problems involving direct variation.

Be consistent when using proportions: compare x's and y's; or compare the first two elements and the second two elements.

Example

2 **If y varies directly as x and $y = 27$ when $x = 6$, find x when $y = 45$.**

Use the following proportion to solve the problem.

$\dfrac{y_1}{y_2} = \dfrac{x_1}{x_2}$

Since y is 27 when x is 6, let $y_1 = 27$ and $x_1 = 6$. Let $y_2 = 45$ and solve for x_2.

$\dfrac{27}{45} = \dfrac{6}{x_2}$ *Substitute.*

$27x_2 = 45(6)$ *Find the cross products.*

$27x_2 = 270$

$x_2 = 10$ Thus, $x = 10$ when $y = 45$.

Exploratory Exercises

State which equations are direct variations. For each direct variation, state the constant of variation.

1. $y = 3x$
2. $ab = 8$
3. $5x^2 = 6y$
4. $-3a = b$
5. $d = 7t$
6. $r = \frac{1}{3}m$
7. $\frac{1}{6}y = \frac{5}{3}$
8. $z = \frac{3}{p}$

Written Exercises

Solve each problem. Assume that y varies directly as x.

9. If $y = 12$ when $x = 3$, find y when $x = 7$.

10. If $y = -8$ when $x = 2$, find y when $x = 10$.

11. If $y = 3$ when $x = 15$, find y when $x = -25$.

12. If $y = -7$ when $x = -14$, find y when $x = 20$.

13. If $y = -6$ when $x = 9$, find x when $y = -4$.

14. If $y = -8$ when $x = -3$, find x when $y = 6$.

15. If $y = 12$ when $x = 15$, find x when $y = 21$.

16. If $y = 1.765$ when $x = 2.204$, find x when $y = 13$.

17. If $y = 11.723$ when $x = 2.06$, find x when $y = 3.459$.

18. If $y = 2\frac{2}{3}$ when $x = \frac{1}{4}$, find y when $x = 1\frac{1}{8}$.

19.–28. **For Exercises 9–18, find the constant of variation. Then write an equation of the form $y = kx$ for each variation.**

Solve.

29. If 2.5 m of copper wire weigh 0.325 kg, how much will 75 m weigh?

30. If 6 pounds of sugar cost $2.00, how much will 40 pounds cost?

31. Hugo's wages vary directly as the time he works. If his wages for 4 days are $110, how much will they be for 17 days?

32. Deidre's wages vary directly as the time she works. If her wages for six days are $121, what are her wages for 20 days?

33. A car uses 5 gallons of gasoline to travel 143 miles. How much gasoline will the car use to travel 200 miles?

34. A car uses 9.055 liters of gasoline to travel 88.333 km. How much gasoline will the car use to travel 215 km?

35. Charles' Law says that the volume of a gas is directly proportional to its temperature. If the volume of a gas is 2.5 cubic feet at 150° (absolute temperature), what is the volume of the same gas when the temperature is 200° (absolute temperature)?

36. In an electrical transformer, voltage is directly proportional to the number of turns on the coil. If 110 volts comes from 55 turns, what would be the voltage produced by 66 turns?

Challenge Exercises

Solve.

37. y varies directly as the square of x. If $y = 14$ when $x = 4$, find y when $x = 9$.

38. y varies directly as $x - 10$. If $y = -12$ when $x = -3$, find x when $y = 7$.

5-9 Inverse Variation

Anne Marie plans to drive on a 600-mile trip. Her time, in hours, for the trip will be determined by her average rate in miles per hour. The following table shows the time (t) for various rates (r).

r	30	40	45	50	60
t	20	15	$13\frac{1}{3}$	12	10

The formula $d = rt$ is used to find values for t.

Notice that as the rate *increases*, the time to make the trip *decreases*. As the rate *decreases*, the time *increases*. We say that the *rate varies inversely as the time*.

> **An inverse variation is described by an equation of the form $xy = k$, where $k \neq 0$.**

Definition of Inverse Variation

Sometimes an **inverse variation** is written in the form $y = \frac{k}{x}$. We say that *y varies inversely as x*.

Example

1 If y varies inversely as x and $y = 6$ when $x = 18$, find x when $y = -3$.

First find the value of k.

$xy = k$

$18(6) = k$ *Substitute 18 for x and 6 for y.*

$108 = k$ *The constant of variation is 108.*

Then find x when $y = -3$.

$xy = k$

$x(-3) = 108$ *Substitute -3 for y and 108 for k.*

$x = -36$

Thus, $x = -36$ when $y = -3$.

Consider again the table of rates and times for Anne Marie's trip. There are many ways to form true mathematical statements from the data. Compare the following mathematical statements.

$$50 \cdot 12 = 60 \cdot 10 \qquad \frac{50}{60} = \frac{10}{12}$$

Two general forms for mathematical statements such as these can be derived from the equation $xy = k$.

Let (x_1, y_1) be a solution of an inverse variation, $xy = k$. Let (x_2, y_2) be a second solution. Then $x_1 y_1 = k$ and $x_2 y_2 = k$. Study the following.

$$x_1 y_1 = k$$
$$\boxed{x_1 y_1 = x_2 y_2} \qquad \textit{You can substitute } x_2 y_2 \textit{ for } k \textit{ because } x_2 y_2 = k.$$

The equation $x_1 y_1 = x_2 y_2$ is called the *product rule for inverse variations*. Study how it can be used to form a proportion.

$$x_1 y_1 = x_2 y_2$$
$$\frac{x_1 y_1}{x_2 y_1} = \frac{x_2 y_2}{x_2 y_1} \qquad \textit{Divide each side by } x_2 y_1.$$

$$\frac{x_1}{x_2} = \frac{y_2}{y_1} \qquad \textit{Notice that this proportion is different from}$$
$$\textit{the proportions for direct variation on page 179.}$$

The product rule or the proportion above can be used to solve problems involving inverse variation.

Example

2 **If y varies inversely as x and $y = 3$ when $x = 12$, find x when $y = 4$.**

Let $x_1 = 12$, $y_1 = 3$, and $y_2 = 4$. Solve for x_2.

a. Use the product rule.

$$x_1 y_1 = x_2 y_2$$
$$12 \cdot 3 = x_2 \cdot 4$$
$$\frac{36}{4} = x_2$$
$$9 = x_2$$

b. Use the proportion.

$$\frac{x_1}{x_2} = \frac{y_2}{y_1}$$
$$\frac{12}{x_2} = \frac{4}{3}$$
$$36 = 4x_2$$
$$9 = x_2$$

Thus, $x = 9$ when $y = 4$.

Exploratory Exercises

State which equations are inverse variations. For each inverse variation, state the constant of variation.

1. $ab = 6$

2. $c = 3.14d$

3. $\frac{50}{y} = x$

4. $\frac{1}{5}a = d$

5. $\frac{-13}{a} = b$

6. $14 = ab$

7. $xy = 1$

8. $y = \frac{1}{x}$

9. $bh = 40$

10. $a = \frac{7}{b}$

11. $d = 4t^2$

12. $s = 3t$

Written Exercises

Solve each problem. Assume that y varies inversely as x.

13. If $y = 24$ when $x = 8$, find y when $x = 4$.

14. If $y = -6$ when $x = -2$, find y when $x = 5$.

15. If $y = 99$ when $x = 11$, find x when $y = 11$.

16. If $y = 7$ when $x = \frac{2}{3}$, find y when $x = 7$.

17. If $x = 2.7$ when $y = 8.1$, find y when $x = 3.6$.

18. If $x = 6.099$ when $y = 4.351$, find x when $y = 6.405$.

Solve each problem.

19. Susan drove 3.25 hours at a rate of 50 mph. How long would the same distance take Susan if she drove at 45 mph?

20. Gary drove 4 hours at a rate of 80 km/h. How long would the same distance take Gary if he drove at 75 km/h?

Boyle's law states that the volume of a gas (V) varies inversely with applied pressure (P). This is shown by the formula $P_1 V_1 = P_2 V_2$. Use the formula to solve each problem.

21. Pressure acting on 60 m³ of a gas is raised from 1 atmosphere to 2 atmospheres. What new volume does the gas occupy?

22. A helium-filled balloon has a volume of 16 m³ at sea level. The pressure at sea level is 1 atmosphere. The balloon rises to a point in the atmosphere where the pressure is 0.75 atmosphere. What is its volume?

In sound and harmonics, the frequency of a vibrating string is inversely proportional to its length. Use this information to solve each problem.

23. A violin string 10 inches long vibrates at a frequency of 512 cycles per second. Find the frequency of an 8-inch string.

24. A piano string 36 inches long vibrates at a frequency of 480 cycles per second. Find the frequency of the string if it were shortened to 24 inches.

mini-review

Solve each equation.

1. $t - 652 = -325$

2. $4\frac{2}{5} = p - \left(-1\frac{3}{10}\right)$

3. $n - (-0.02) = -3.5$

4. $5x + 3 = 2(x - 1)$

5. $7r - \frac{r}{4} = 22$

6. $\frac{x+1}{x-5} = \frac{7}{4}$

Solve each inequality.

7. $4n > 5n + 6$

8. $-\frac{4}{5}q \le -\frac{2}{15}$

9. $|2x + 3| \le 5$

Simplify. Assume no denominator is equal to zero.

10. $(3ab^2c)(2a^3b^3)^2$

11. $\frac{16a^4b^7}{-12a^{-2}b^3}$

12. $\frac{(-m)^3 n^{-4}}{(3m^2 n)^{-2}}$

5-10 Lever Problems

Laura and Jason are on a seesaw. They want the seesaw to balance. Laura weighs 132 pounds and Jason weighs 108 pounds. Which person should sit closer to the fulcrum (the pivot point)?

If you have observed people on a seesaw, you may have noticed that the heavier person must sit closer to the fulcrum to balance the seesaw. This is an example of an inverse variation. A seesaw is a type of lever.

All lever problems involve inverse variation.

Suppose weights w_1 and w_2 are placed on a lever at distances d_1 and d_2, respectively, from the fulcrum. The lever is balanced when $w_1 d_1 = w_2 d_2$.	***Property of Levers***

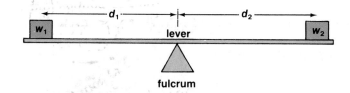

The property of levers is illustrated at the right.

Examples

1   The fulcrum of a 16-ft seesaw is placed in the middle. Jason, who weighs 108 pounds is seated 8 feet from the fulcrum. How far from the fulcrum should Laura sit if she weighs 132 pounds?

Use the property of levers, $w_1 d_1 = w_2 d_2$.
Let $w_1 = 108$, $d_1 = 8$, and $w_2 = 132$. Solve for d_2.

$$w_1 d_1 = w_2 d_2$$
$$108 \cdot 8 = 132 \cdot d_2$$
$$864 = 132 d_2$$
$$6\frac{6}{11} = d_2$$

Laura should sit $6\frac{6}{11}$ or about 6 ft 7 in. from the fulcrum.

2   An 8-ounce weight is placed at one end of a yardstick. A 10-ounce weight is placed at the other end. Where should the fulcrum be placed to have the yardstick balanced?

Let d = distance from fulcrum to 8-ounce weight, in inches.
Then $36 - d$ = distance from fulcrum to 10-ounce weight, in inches.

$$8d = 10(36 - d)$$
$$8d = 360 - 10d$$
$$18d = 360$$
$$d = 20$$

The fulcrum should be placed 20 inches from the 8-ounce weight.

3 Patti and Cathy are seated on the same side of a seesaw. Patti is 6 feet from the fulcrum and weighs 115 pounds. Cathy is 8 feet from the fulcrum and weighs 120 pounds. Jack is seated on the other side of the seesaw, 10 feet from the fulcrum. If the seesaw is balanced, how much does Jack weigh?

Draw a diagram.

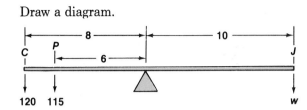

Let w = Jack's weight.

$$120(8) + 115(6) = 10w$$
$$960 + 690 = 10w$$
$$1650 = 10w$$
$$165 = w$$

Thus, Jack weighs 165 pounds.

Exploratory Exercises

For each of the following, suppose the two people are on a seesaw. For the seesaw to balance, which person must sit closer to the fulcrum?

1. Jorge, 168 pounds
   Emilio, 220 pounds

2. Shawn, 114 pounds
   Shannon, 97 pounds

3. Juanita, 49 pounds
   Lucia, 49 pounds

4. Soto, 52 kg
   Helen, 55 kg

5. Jack, 50 kg
   Betty, 58 kg

6. Jimmy, 72 pounds
   Jeff, 68 pounds

Written Exercises

Solve.

7. Mary Jo weighs 120 pounds and Doug weighs 160 pounds. They are seated at opposite ends of a seesaw. Doug and Mary Jo are 14 feet apart, and the seesaw is balanced. How far is Mary Jo from the fulcrum?

8. Grant, who weighs 150 pounds, is seated 8 feet from the fulcrum of a seesaw. Mariel is seated 10 feet from the fulcrum. If the seesaw is balanced, how much does Mariel weigh?

9. Weights of 100 pounds and 115 pounds are placed on a lever. The two weights are 15 feet apart, and the lever is balanced. How far from the fulcrum is the 100-pound weight?

10. A lever has a 140-pound weight on one end and a 160-pound weight on the other end. The lever is balanced, and the 140-pound weight is exactly one foot farther from the fulcrum than the 160-pound weight. How far from the fulcrum is the 160-pound weight?

11. Macawi, who weighs 108 pounds, is seated 5 feet from the fulcrum of a seesaw. Barbara is seated on the same side of the seesaw, two feet farther from the fulcrum than Macawi. Barbara weighs 96 pounds. The seesaw is balanced when Sue, who weighs 101 pounds, sits on the other side. How far is Sue from the fulcrum?

12. Christy, who weighs 112 pounds, is seated 6 feet from the fulcrum of a seesaw. Sooyun, who weighs 124 pounds, is seated on the same side of the seesaw, 8 feet from the fulcrum. The other side of the seesaw is 9 feet long. Is it possible for Shim, who weighs 180 pounds, to balance the seesaw?

Powers and E Notation

In the BASIC language, powers are written using the symbols ↑ or ∧. For example, $(2ab^2)^3$ can be written $(2*A*B \uparrow 2) \uparrow 3$. The following program evaluates this expression if $a = 9$ and $b = 10$.

```
10   READ A,B
20   DATA 9,10
30   PRINT (2 * A * B ↑ 2) ↑ 3
40   END

RUN
5.832E + 09
```

Lines 10 and 20 assign the variables A the value 9 and B the value 10.

Notice that the output for this program is in E notation. This is the computer equivalent of scientific notation. Thus, 5.832E + 09 means 5.832×10^9. The computer used E notation because there were more than six significant digits in the output. (Some computers allow nine significant digits.)

Expressions such as $\frac{3.8 \times 10^{-9}}{1.3 \times 10^{-4}}$ can be evaluated using E notation.

This can be done by entering the following command into a computer.

```
PRINT (3.8E-09)/(1.3E-04)
```

The answer will be printed.

```
2.92307692 E-05
```

Exercises

Write a program that will evaluate each expression if $a = 4$, $b = 6$, and $c = 8$.

1. $a^2b^3c^4$

2. $(-2a)^2(4b)^3$

3. $(4a^2b^4)^3$

4. $(-6ab^3c^2)^4$

5. $\left(\frac{1}{2}ab\right)^2(2c)^2$

6. $(-5a^2c^3)^2(2b^2)^2$

7. $(ac)^3 + (3b)^2$

8. $(2a^3)^2 + (ab^2)^2$

Write PRINT commands that will evaluate each expression.

9. $\frac{4.2 \times 10^8}{2.3 \times 10^{15}}$

10. $\frac{3.4 \times 10^{-10}}{1.2 \times 10^{10}}$

11. $\frac{4.24 \times 10^{-12}}{3.1 \times 10^{-9}}$

12. $(5 \times 10^{19})(6 \times 10^{-8})$

13. $(2 \times 10^{12})(1.3 \times 10^7)$

14. $(1.4 \times 10^{-10})(2.7 \times 10^{-8})$

Vocabulary

monomial   (157)

constant   (157)

scientific notation   (165)

direct variation   (178)

constant of variation   (178)

inverse variation   (181)

Chapter Summary

1. A monomial is a number, a variable, or a product of a number and one or more variables.   (157)

2. Monomials that do not contain variables are called constants.   (157)

3. Product of Powers Rule: For any number a and any integers m and n,
$$a^m \cdot a^n = a^{m+n}. \quad (157)$$

4. Power of a Power Rule: For any number a and any integers m and n,
$$(a^m)^n = a^{mn}. \quad (160)$$

5. Power of a Product Rule: For all numbers a and b and any integer m,
$$(ab)^m = a^m b^m. \quad (160)$$

6. Power of a Monomial Rule: For all numbers a and b and any integers m, n, and p,
$$(a^m b^n)^p = a^{mp} b^{np}. \quad (160)$$

7. Dividing Powers: For all integers m and n, and any nonzero number a,
$$\frac{a^m}{a^n} = a^{m-n}. \quad (162)$$

8. Zero Exponent: For any nonzero number a, $a^0 = 1$.   (163)

9. Negative Exponents: For any nonzero number a and any integer n,
$$a^{-n} = \frac{1}{a^n}. \quad (163)$$

10. A number is expressed in scientific notation when it is in the form $a \times 10^n$ where $1 \le a < 10$ and n is an integer.   (165)

11. Direct Variation is described by an equation of the form $y = kx$, where $k \ne 0$.   (178)

12. Inverse Variation is described by an equation of the form $xy = k$, where $k \ne 0$.   (181)

13. Property of Levers: Suppose weights w_1 and w_2 are placed on a lever at distances d_1 and d_2, respectively, from the fulcrum. The lever is balanced when $w_1 d_1 = w_2 d_2$.   (184)

Chapter Review

5–1 **Simplify.**

1. $b^2 \cdot b^7$

2. $y^3 \cdot y^3 \cdot y$

3. $(a^2b)(a^2b^2)$

4. $(a^2)(a^2b)$

5. $(3ab)(-4a^2b^3)$

6. $(-4a^2x)(-5a^3x^4)$

5–2

7. $(7a)^2$

8. $(4a^2b)^3$

9. $\left(\frac{1}{3}b^2\right)^2$

10. $(-0.5)^3a^4$

11. $(-3xy)^2(4x)^3$

12. $(5a^2)^3 + 7(a^6)$

5–3 **Simplify. Assume no denominator is equal to zero.**

13. $\frac{y^{10}}{y^6}$

14. $\frac{(3y)^0}{6a}$

15. $\frac{42b^7}{14b^4}$

16. $\frac{27b^{-2}}{14b^{-3}}$

17. $\frac{(3a^3bc^2)^2}{18a^2b^3c^4}$

18. $\frac{-16a^3b^2c^4d}{-48a^4bcd^3}$

5–4 **Express each number in scientific notation.**

19. 240,000

20. 0.000314

Evaluate. Express each result in scientific notation.

21. $(2 \times 10^5)(3 \times 10^6)$

22. $(3 \times 10^3)(1.5 \times 10^6)$

23. $\frac{5.4 \times 10^{-3}}{0.9 \times 10^4}$

24. $\frac{8.4 \times 10^{-6}}{1.4 \times 10^{-9}}$

Solve.

5–5 25. Joe is 10 years older than Jim. The sum of their ages is 64. What are their ages?

26. Mary is 20 years old. Her sister, Sara, is 5 years old. In how many years will Mary be twice as old as Sara?

5–6,
5–7 27. Jan has 32 coins in nickels and dimes. He has 10 more nickels than dimes and has $2.15 in all. How many of each kind of coin does he have?

28. A salt solution is 8% salt. How much water needs to be added to 100 mg of this solution to lower the concentration to 5%?

5–8,
5–9 29. If x varies directly as y and $x = 15$ when $y = 1.5$, find x when $y = 9$.

30. If y varies inversely as x and y is 24 when x is 30, find x when y is 10.

5–10 31. Lee and Rena sit on opposite ends of a 12-foot teetertotter. Lee weighs 180 pounds and Rena weighs 108 pounds. Where should the fulcrum be placed in order for them to be balanced?

32. Chip O'Hall needs to move a rock weighing 1050 pounds. If he has a 6 foot long pinchbar and uses a fulcrum placed one foot from the rock, will he be able to lift the rock if he weighs 205 pounds?

Simplify. Assume no denominator is equal to zero.

1. $y^2 \cdot y^{13}$

2. $a^2 \cdot a^3 \cdot b^4 \cdot b^5$

3. $(a^2b)(a^3b^2)$

4. $(-12abc)(4a^2b^3)$

5. $(9a)^2$

6. $(-3a)^4(a^5b)$

7. $\left(\frac{1}{5}r^2\right)^2$

8. $(-0.3a)^3$

9. $(-5a^2)(-6b)^2$

10. $(5a)^2b + 7a^2b$

11. $\frac{y^{11}}{y^6}$

12. $\frac{y^3x}{yx}$

13. $\frac{63a^2bc}{9abc}$

14. $\frac{48a^2bc^5}{(3ab^3c^2)^2}$

15. $\frac{14ab^{-3}}{21a^2b^{-5}}$

16. $\frac{10a^2bc}{20a^{-1}b^{-1}c}$

Express in scientific notation.

17. 5280

18. 0.00378

Evaluate. Express each result in scientific notation.

19. $(3 \times 10^3)(2 \times 10^4)$

20. $(4 \times 10^{-3})(2 \times 10^{16})$

21. $\frac{25 \times 10^3}{5 \times 10^{-3}}$

22. $\frac{91 \times 10^{18}}{13 \times 10^{14}}$

Solve each problem.

23. Jim is 5 years older than Les. The sum of their ages is 105. What are their ages?

24. If m varies inversely as n and $n = 42$ when $m = 28$, find m if $n = 56$.

25. Ahmed's wages vary directly as the time he works. If his wages for 5 days are $26, how much will they be for 12 days?

26. Dan has twice as much money in dimes as in quarters. He has 42 coins in all. How many of each coin does he have?

27. How many ounces of a 6% iodine solution needs to be added to 12 ounces of a 10% solution to create a 7% solution?

28. Marce weighs 133 pounds and sits 6 feet from the fulcrum of a seesaw. She balances with Suzanne if Suzanne sits 7 feet from the fulcrum. How much does Suzanne weigh?

CHAPTER 6

Polynomials

Poly-, a prefix meaning many or several, is used in some mathematical words. There are many *poly*gons, plane figures with many angles, in the photograph above. In algebra, expressions consisting of one or more monomials are called *poly*nomials. In this chapter, you will learn to add, subtract, and multiply polynomials.

6-1 Polynomials

A **polynomial** is a monomial or a sum of monomials. Recall that a monomial is a number, a variable, or a product of numbers and variables. A **binomial** is a polynomial with two terms, and a **trinomial** is a polynomial with three terms.

A monomial has one term.

Examples of each of these types of polynomials are given in the following chart.

monomial	binomial	trinomial
$5x^2$	$3x + 2$	$5x^2 - 2x + 7$
$4abc$	$4x + 5y$	$a^2 + 2ab + b^2$
-7	$3x^2 - 8xy$	$4a + 2b^2 - 3c$

Recall that $5x^2 - 2x + 7$ is equivalent to $5x^2 + (-2x) + 7$.

The expression $x + \frac{3}{b}$ is not a polynomial because $\frac{3}{b}$ is not a monomial.

Examples

State whether each expression is a polynomial. If the expression is a polynomial, identify it as either a monomial, binomial, or trinomial.

1 $8x^2 - 3xy$

The expression $8x^2 - 3xy$ can be written as $8x^2 + (-3xy)$. Therefore, $8x^2 - 3xy$ is a polynomial because it can be written as the sum of monomials, $8x^2$ and $-3xy$. Since it has two terms, $8x^2 - 3xy$ is a binomial.

2 $\frac{5}{2y^2} + 7y + 6$

The expression $\frac{5}{2y^2} + 7y + 6$ is not a polynomial because $\frac{5}{2y^2}$ is not a monomial.

The **degree** of a monomial is the sum of the exponents of its variables.

monomial	degree
$5x^2$	2
$4ab^3c^4$	$1 + 3 + 4 = 8$
-9	0

It is understood that a has an exponent of 1.

What is the degree of $-\frac{1}{2}x$?

To find the *degree of a polynomial*, first find the degree of each of its terms. The degree of the polynomial is the greatest of the degrees of its terms.

State the degree of each polynomial.

3 $8x^3 - 2x^2 + 7$

The polynomial has three terms, $8x^3$, $-2x^2$, and 7. Their degrees are 3, 2, and 0. The greatest degree is 3. Therefore, the degree of $8x^3 - 2x^2 + 7$ is 3.

4 $6x^2y + 5x^3y^2z - x + x^2y^2$

The polynomial has four terms, $6x^2y$, $5x^3y^2z$, $-x$, and x^2y^2. Their degrees are 3, 6, 1, and 4. Therefore, the degree of $6x^2y + 5x^3y^2z - x + x^2y^2$ is 6.

The terms of a polynomial are usually arranged so that the powers of one variable are in either ascending or descending order.

Ascending Order	Descending Order
$3 + 5a - 8a^2 + a^3$	$a^3 - 8a^2 + 5a + 3$
(in x) $5xy + x^3y^2 - x^4 + x^5y^2$	(in x) $x^5y^2 - x^4 + x^3y^2 + 5xy$
(in y) $x^3 - 3x^2y + 4x^2y^2 - y^3$	(in y) $-y^3 + 4x^2y^2 - 3x^2y + x^3$

Exploratory Exercises

State whether each expression is a polynomial. If the expression is a polynomial, identify it as either a monomial, binomial, or trinomial.

1. $5x^2y + 3xy + 7$ **2.** $\frac{5}{k} - k^2y$ **3.** 0

4. $3a^2x - 5a$ **5.** $\frac{a^3}{3}$ **6.** $x^2 - \frac{x}{2} + \frac{1}{3}$

State the degree of each monomial.

7. $100x$ **8.** -18 **9.** $29xyz$ **10.** $8x^2$

11. $17x^2y$ **12.** $-14x^3z$ **13.** 0 **14.** $51x^5yz$

15. $28xst^3$ **16.** $14m^2n^3$ **17.** $4x$ **18.** $-36xw^4$

Written Exercises

Find the degree of each polynomial.

19. $12s + 21t$ **20.** 19 **21.** $-2a^2b^{10} + 3a^9b$

22. $22x + 5y^4$ **23.** $13s^2t^2 + 4st^2 - 5s^5t$ **24.** $7x^3 + 4xy + 3xz^3$

25. $17r^2t + 3r + t^2$ **26.** $29x + x^{29}$ **27.** $17x^3y^4 - 11xy^5$

28. $11x^2$ **29.** $5r - 3s + 7t$ **30.** $x + 2y^2 + 3z^3$

31. $n - 2p^2$ **32.** $27x^4 - 3x^3yz$ **33.** $12x^4y^5 + xy^6$

34. $29n^2 + 17n^2t^2$

35. $11wxyz - 9w^4$

36. $2xy^2z + 5xyz^5 + x^4$

37. $5x^2 - 2x^5$

38. 2^2r^2

39. $3^3xy + 5x$

40. $2x^3yz - 5xy^3z + 11z^5$

41. $32xyz + 11x^2y + 17xz^2$

42. $4xy + 9xz^2 + 17rs^3$

43. $-4yzw^4 + 10x^4z^2w - 2z^2w^3$

44. $n^2 + r^3 + s^{14} + z^2$

45. $6mn^4t - 3m^4nt^2 + mn^6$

Arrange the terms of each polynomial so that the powers of x are in ascending order.

46. $3 + x^4 + 2x^2$

47. $1 + x^3 + x^5 + x^2$

48. $2x^2 + 5ax + a^3$

49. $-2x^2y + 3xy^3 + x^3$

50. $17bx^2 + 11b^2x - x^3$

51. $21p^2x + 3px^3 + p^4$

52. $5b + b^3x^2 + \frac{2}{3}bx$

53. $\frac{1}{3}s^2x^3 + 4x^4 - \frac{2}{5}s^4x^2 + \frac{1}{4}x$

Arrange the terms of each polynomial so that the powers of x are in descending order.

54. $-6x + x^5 + 4x^3 - 20$

55. $5x^2 - 3x^3 + 2x + 7$

56. $4x^3y + 3xy^4 - x^2y^3 + y^4$

57. $11x^2 + 7ax^3 - 3x + 2a$

58. $\frac{3}{4}x^3y - x^2 + 4 + \frac{2}{3}x$

59. $7a^3x - 8a^3x^3 + \frac{1}{5}x^5 + \frac{2}{3}x^2$

60. $0.2mx^4 - 1.3x^5 + 0.4m^2 + 2.6x^3$

61. $1.7t^3x + 2.4tx^5 - 0.3tx^2 + 5.1x^3$

mini-review

Simplify.

1. $6(5) - 4(-9)$

2. $\dfrac{10}{\frac{-8}{9}}$

3. $(2x^5y^2)(-3xy^4)$

Solve and check each equation.

4. $4x + 2 = 4(x - 8)$

5. $2.5(3n - 1) = -0.2(12 - 2n)$

Solve each inequality.

6. $n - 6 > 8.2$

7. $4m < 2m + 8$

8. $\frac{1}{2} - 2r > \frac{3}{4} - 4r$

Express each number in scientific notation.

9. $42{,}350$

10. 0.00000628

11. 28.5×10^6

12. 0.005×10^3

Solve each problem.

13. A rectangular garden is 15 feet longer than it is wide. It can be enclosed by 170 feet of fencing. Find its dimensions.

14. Football tickets cost $3 for students and $4.50 for adults. The price for 10 tickets was $39. How many student tickets were purchased?

15. The price of a suit was marked down $37.50. The original price was $125. What was the rate of discount?

16. How many grams of salt must be added to 65 grams of a 15% salt solution to obtain a 35% salt solution?

6-2 Adding and Subtracting Polynomials

You have used various properties to simplify expressions.

$$4a + 3a - 2 = (4 + 3)a - 2 \qquad \textit{Distributive Property}$$
$$= 7a - 2 \qquad\qquad \textit{Substitution Property of Equality}$$
$$5x^2 + 3y + 2x^2 - y = 5x^2 + 2x^2 + 3y - y \qquad \textit{Commutative Property for Addition}$$
$$= (5 + 2)x^2 + (3 - 1)y \qquad \textit{Distributive Property}$$
$$= 7x^2 + 2y \qquad\qquad \textit{Substitution Property of Equality}$$

Suppose you want to add the polynomials $(3x + 2y)$ and $(8x + 3y)$.
You can use the same properties to find such sums.

Examples

1 **Add $(3x + 2y) + (8x + 3y)$.**

To add polynomials, add their like terms.

$$(3x + 2y) + (8x + 3y) = (3x + 8x) + (2y + 3y) \qquad \textit{Associative and Commutative Properties for Addition}$$
$$= (3 + 8)x + (2 + 3)y \qquad \textit{Distributive Property}$$
$$= 11x + 5y \qquad\qquad \textit{Substitution Property of Equality}$$

2 **Add $(-3x^2 + 2x + 7) + (6x^2 - 5x - 3)$.**

$$(-3x^2 + 2x + 7) + (6x^2 - 5x - 3) = (-3x^2 + 6x^2) + [2x + (-5x)] + [7 + (-3)]$$
$$= (-3 + 6)x^2 + [2 + (-5)]x + [7 + (-3)]$$
$$= 3x^2 + (-3x) + 4$$
$$= 3x^2 - 3x + 4$$

Sometimes it is convenient to find the sum by writing the polynomials in column form.

Examples

3 **Add $(5x^2 - 3xy + 7y^2) + (4x^2 + xy - 3y^2)$ in column form.**

$$\begin{array}{l} 5x^2 - 3xy + 7y^2 \\ \underline{4x^2 + xy - 3y^2} \qquad \textit{Notice that like terms are aligned.} \\ 9x^2 - 2xy + 4y^2 \end{array}$$

4 **Add $(3y^2 + 5y - 6) + (7y^2 - 9)$ in column form.**

$$\begin{array}{l} 3y^2 + 5y - 6 \\ \underline{7y^2 - 9} \qquad \textit{Notice that there is no term in the y-column.} \\ 10y^2 + 5y - 15 \end{array}$$

Recall that you can subtract a rational number by adding its additive inverse or opposite. Similarly, you can subtract a polynomial by adding its additive inverse.

$$a - b = a + (-b)$$

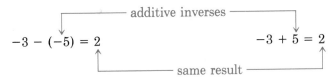

Subtraction   **Addition**

additive inverses

$-3 - (-5) = 2$ $-3 + 5 = 2$

same result

To find the additive inverse of a polynomial, replace each term by its additive inverse.

Another way to find the additive inverse is to multiply each term by -1.

Polynomial	Additive Inverse
$x + 2y$	$-x - 2y$
$2x^2 - 3x + 5$	$-2x^2 + 3x - 5$
$-8x + 5y - 7z$	$8x - 5y + 7z$
$3x^3 - 2x^2 - 5x$	$-3x^3 + 2x^2 + 5x$

Examples

5 **Subtract $(5x^2 - 2x + 7) - (2x^2 + 4x - 3)$.**

Add the additive inverse of $(2x^2 + 4x - 3)$.

$$
\begin{aligned}
(5x^2 - 2x + 7) - (2x^2 + 4x - 3) &= (5x^2 - 2x + 7) + [-(2x^2 + 4x - 3)] \\
&= (5x^2 - 2x + 7) + (-2x^2 - 4x + 3) \\
&= 5x^2 - 2x + 7 - 2x^2 - 4x + 3 \\
&= (5x^2 - 2x^2) + (-2x - 4x) + (7 + 3) \\
&= 3x^2 - 6x + 10
\end{aligned}
$$

6 **Subtract $\left(\frac{2}{3}x^2 - \frac{3}{5}xy + 5y^2\right) - \left(\frac{1}{3}x^2 - \frac{3}{5}xy - 2y^2\right)$.**

$$
\begin{aligned}
\left(\frac{2}{3}x^2 - \frac{3}{5}xy + 5y^2\right) - \left(\frac{x^2}{3} - \frac{3}{5}xy - 2y^2\right) &= \left(\frac{2}{3}x^2 - \frac{3}{5}xy + 5y^2\right) + \left[-\left(\frac{1}{3}x^2 - \frac{3}{5}xy - 2y^2\right)\right] \\
&= \left(\frac{2}{3}x^2 - \frac{3}{5}xy + 5y^2\right) + \left(-\frac{1}{3}x^2 + \frac{3}{5}xy + 2y^2\right) \\
&= \left(\frac{2}{3}x^2 - \frac{1}{3}x^2\right) + \left(-\frac{3}{5}xy + \frac{3}{5}xy\right) + (5y^2 + 2y^2) \\
&= \frac{1}{3}x^2 + 7y^2
\end{aligned}
$$

7 **Subtract $(4x^2 - 3y^2 + 5xy) - (8xy + 6x^2 + 3y^2)$ in column form.**

Reorder terms so that the powers of x are in descending order.

$$
\begin{array}{r}
4x^2 + 5xy - 3y^2 \\
(-)\ 6x^2 + 8xy + 3y^2 \\
\hline
\end{array}
$$

Add the additive inverse. →

$$
\begin{array}{r}
4x^2 + 5xy - 3y^2 \\
(+)\ -6x^2 - 8xy - 3y^2 \\
\hline
-2x^2 - 3xy - 6y^2
\end{array}
$$

Example

8 Subtract $(5a^3 + 2a^2 - 7) - (3a^3 + 8a^2 - 2a + 4)$ and check by addition.

$$
\begin{array}{l}
5a^3 + 2a^2 \quad\quad - 7 \\
(-)\underline{3a^3 + 8a^2 - 2a + 4}
\end{array}
\quad \text{Add the additive inverse.} \Rightarrow
\begin{array}{l}
5a^3 + 2a^2 \quad\quad - 7 \\
(+)\underline{-3a^3 - 8a^2 + 2a - 4} \\
2a^3 - 6a^2 + 2a - 11
\end{array}
$$

Check:
$$
\begin{array}{ll}
2a^3 - 6a^2 + 2a - 11 & \textit{difference} \\
(+)\ \underline{3a^3 + 8a^2 - 2a + \ 4} & \textit{second polynomial} \\
5a^3 + 2a^2 \quad\quad - \ 7 & \textit{first polynomial}
\end{array}
$$

Exploratory Exercises

Simplify.

1. $-(5a + 9b)$
2. $-(7a - 6b)$
3. $-(4x^2 - 5)$
4. $-(x^2 + 7xy + y^2)$
5. $-(2y^2 - 7y + 12)$
6. $-(6a^2 - 3ab + b^2)$
7. $-(-10r^2 + 4rs - 6s)$
8. $-(-6a + 5b - c)$
9. $-(-x^3 - x - 1)$

State the additive inverse of each polynomial.

10. 29
11. $-15x^2y$
12. $3x + 2y$
13. $-8m + 7n$
14. $-11x + 13y$
15. $x^2 + 3x + 7$
16. $-4h^2 - 5hk - k^2$
17. $-3ab^2 + 5a^2b - b^3$
18. $x^3 + 5x^2 - 3x - 11$

Name the like monomials in each group.

19. $5m, 4mn, -3m, 2n, -mn, 8n$
20. $2x^3, 5xy, -x^2y, 14xy, 12xy$
21. $-7ab^2, 8a^2b, 11b^2, 16a^2b, -2b^2$
22. $3p^3q, -2p, 10p^3q, 15pq, -p$
23. $6x^3, 6x^2, 7y^3, -8x^2, 3y^3$
24. $14cd^2, 11c^2d, -7c^2, cd^2, 12c^2$

Written Exercises

Add or subtract.

25. $(5x + 6y) + (2x + 8y)$
26. $(4a + 6b) + (2a + 3b)$
27. $(7n + 11m) - (4m + 2n)$
28. $(7y + 9x) - (6x + 5y)$
29. $(3x - 7y) + (3y + 4x)$
30. $(3x - 2y) + (5x + 8y)$
31. $(5a - 6m) - (2a + 5m)$
32. $(3s - 5t) - (8t + 2s)$
33. $(5m + 3n) + 8m$
34. $(12x + 7y) + 8y$
35. $(13x + 9y) - 11y$
36. $(9x + 3y) - 9y$
37. $(n^2 + 5n + 3) + (2n^2 + 8n + 8)$
38. $(5a^2 + 7a + 5) + (a^2 + 6a + 3)$
39. $(3 + 2a + a^2) - (5 + 8a + a^2)$
40. $(3x^2 - 7x + 4) - (2x^2 + 8x - 6)$
41. $\left(\frac{1}{2}x^2 + \frac{5}{3}x - \frac{1}{4}\right) + \left(\frac{1}{2}x^2 - \frac{1}{3}x + \frac{1}{2}\right)$
42. $\left(\frac{3}{4}x^2 + \frac{2}{3}x - 7\right) + \left(\frac{5}{4}x^2 - 6x + 4\right)$
43. $(5ax^2 + 3a^2x - 5x) + (2ax^2 - 5ax + 7x)$
44. $\left(\frac{5}{7}a^2 - \frac{3}{4}a + \frac{1}{2}\right) - \left(\frac{3}{7}a^2 + \frac{1}{2}a - \frac{1}{2}\right)$
45. $(2xy + 6xy^2 + y^2) - (3x^2y + 2xy + 3y^2)$
46. $\left(\frac{3}{8}m^2 - 4m + \frac{2}{3}\right) + \left(\frac{5}{8}m^2 - 2m + \frac{1}{3}\right)$

47. $(3mn^2 + 3mn - n^3) - (5mn^2 + n + 2n^3)$

48. $(x^3 - 3x^2y + 4xy^2 + y^3) - (7x^3 + x^2y - 9xy^2 + y^3)$

Add.

49. $4a + 5b - 6c + d$
$\underline{3a - 7b + 2c + 8d}$
$2a - b + 7d$

50. $2x^2 - 5x + 7$
$5x^2 + 7x - 3$
$\underline{x^2 - x + 11}$

51. $5ax^2 + 3a^2x - 7a^3$
$\underline{2ax^2 - 8a^2x + 4}$

52. $a^3 - b^3$
$\underline{3a^3 + 2a^2b - b^2 + 2b^3}$

Subtract.

53. $6x^2y^2 - 3xy - 7$
$\underline{5x^2y^2 + 2xy + 3}$

54. $5x^2 - 4$
$\underline{3x^2 + 8x + 4}$

55. $11m^2n^2 + 2mn - 11$
$\underline{5m^2n^2 - 6mn + 17}$

56. $2a - 7$
$\underline{5a^2 + 8a - 11}$

Subtract and check by addition.

57. $11m^2n^2 + 4mn - 6$
$\underline{5m^2n^2 - 6mn + 17}$

58. $7z^2 + 4$
$\underline{3z^2 + 2z - 6}$

59. $4y - 11$
$\underline{3y^2 + 4y - 11}$

Find the measure of the third side of each triangle. P is the measure of the perimeter.

60. $P = 3x + 3y$

$x + y$ $x + y$

61. $P = 7x + 2y$

$2x + y$ $3x - 5y$

62. $P = 11x^2 - 29x + 10$

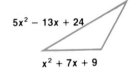

$5x^2 - 13x + 24$

$x^2 + 7x + 9$

Challenge Exercises

Two angles are complementary if the sum of their degree measures is 90. Find the degree measure of the complement of each angle having the given degree measure.

63. 85   **64.** $3x$ **65.** $x - 2$ **66.** $2x + 40$

Two angles are supplementary if the sum of their degree measures is 180. Find the degree measure of the supplement of each angle having the given degree measure.

67. 45   **68.** $2x$ **69.** $8x - 10$ **70.** $5x + 130$

The sum of the degree measures of the three angles of a triangle is 180. Find the degree measure of the third angle of each triangle given the degree measures of the other two angles.

71. 80; 70   **72.** $3x; x + 2$ **73.** $x - 1; x + 5$ **74.** $2x - 3; 3x - 8$

75. $5 - 2x; 7 + 8x$ **76.** $x^2 - 8x + 2; x^2 - 3x - 1$

77. $3x^2 - 5; 4x^2 + 2x + 1$ **78.** $4 - 2x; x^2 - 1$

Powers containing the exponent 2 or 3 can be read in two ways.

x^2 is read "x squared" or "x to the second power."

x^3 is read "x cubed" or "x to the third power."

Powers containing numerical exponents other than 2 or 3 are usually read as follows.

x^0 is read "x to the zero power."

x^6 is read "x to the sixth power."

x^n is read "x to the nth power."

Recall that an exponent indicates the number of times that the base is used as a factor. Suppose you are to write each of the following in symbols.

Words	Symbols
three times x squared	$3x^2$
three times x the quantity squared	$(3x)^2$

In the second expression, parentheses are used to show that the expression $3x$ is used as a factor twice.

$$(3x)^2 = (3x)(3x)$$

The phrase *the quantity* is used to indicate parentheses when reading expressions.

Exercises

State how you would read each expression.

1. 4^2 **2.** 3^3 **3.** a^5 **4.** m^9

5. $5b^2$ **6.** $(12r)^5$ **7.** $9x^3y$ **8.** $(x + 2y)^2$

9. $4m^2n^4$ **10.** $(6a^2b)^4$ **11.** $a - b^3$ **12.** $(2a + b)^4$

Decide if the expressions are equivalent. Write *yes* or *no*.

13. $12 \cdot x^3$ and $12 \cdot x \cdot x \cdot x$ **14.** $3xy^5$ and $3(xy)^5$

15. $(2a)^3$ and $8a^3$ **16.** $(ab)^2$ and $a^2 \cdot b^2$

17. xy^3 and x^3y^3 **18.** $4(x^3)^2$ and $4x^6$

6-3 Multiplying a Polynomial by a Monomial

The Distributive Property can be used to multiply a polynomial by a monomial.

Examples

1 **Multiply $5a(3a^2 + 4)$.**

$5a(3a^2 + 4) = 5a(3a^2) + 5a(4)$ *Use the Distributive Property.*
$= 15a^3 + 20a$

2 **Multiply $2m^2(5m^2 - 7m + 8)$.**

$2m^2(5m^2 - 7m + 8) = 2m^2(5m^2) + 2m^2(-7m) + 2m^2(8)$
$= 10m^4 - 14m^3 + 16m^2$ *Use the Product of Powers Rule.*

3 **Multiply $-3xy(2x^2y + 3xy^2 - 7y^3)$.**

$-3xy(2x^2y + 3xy^2 - 7y^3) = -3xy(2x^2y) + (-3xy)(3xy^2) + (-3xy)(-7y^3)$
$= -6x^3y^2 - 9x^2y^3 + 21xy^4$

4 **Simplify $2a(5a^2 + 3a - 2) - 8(3a^2 - 7a + 1)$.**

$2a(5a^2 + 3a - 2) - 8(3a^2 - 7a + 1)$
$= 2a(5a^2) + 2a(3a) + 2a(-2) + (-8)(3a^2) + (-8)(-7a) + (-8)(1)$
$= 10a^3 + 6a^2 - 4a - 24a^2 + 56a - 8$
$= 10a^3 - 18a^2 + 52a - 8$ *Combine like terms.*

Exploratory Exercises

Multiply.

1. $-5a(12a^2)$
2. $4x^2(7x^3)$
3. $2(5x - 3)$
4. $8(3a + 5)$
5. $7a(5a + 8)$
6. $3m(8m + 7)$
7. $3ab(5a - 3)$
8. $2mn(m - 7)$
9. $-8a^2(3a^2 + 7a)$
10. $-4m^3(5m^2 + 2m)$
11. $3xy(3xy + 2x)$
12. $7ab(5ab^2 + b^2)$

Written Exercises

Multiply.

13. $5(3a + 7)$
14. $8(7m + 2)$
15. $-3(8x + 5)$
16. $-7(5x + 8)$
17. $\frac{1}{2}x(8x + 6)$
18. $\frac{2}{3}a(6a + 15)$
19. $3b(5b + 8)$
20. $7a(8a + 11)$
21. $-2x(5x + 11)$
22. $-5y(7y + 12)$
23. $1.1a(2a + 7)$
24. $2.6b(5b - 1)$
25. $7a(3a^2 - 2a)$
26. $5b(8b^2 - 7b)$
27. $3st(5s^2 + 2st)$
28. $5mn(5m^2 + 3mn)$
29. $7xy(5x^2 - y^2)$
30. $4ab(3a^2 - 7b^2)$

31. $2a(5a^3 - 7a^2 + 2)$

32. $5a(a^2 + 9a - 3)$

33. $7x^2y(5x^2 - 3xy + y)$

34. $8cd^2(7c^2d - cd + d^2)$

35. $5y(8y^3 + 7y^2 - 3y)$

36. $8v(7v^3 + v^2 - 7v)$

37. $-4x(7x^2 - 4x + 3)$

38. $-8a(5a^2 + 8a - 3)$

39. $5x^2y(3x^2 - 7xy + y^2)$

40. $7a^2b^2(a^4 - 5a^2b + 6b^2)$

41. $4m^2(9m^2n + mn - 5n^2)$

42. $2x^2(9x^2y - 7xy + y^2)$

43. $-8xy(4xy + 7x - 14y^2)$

44. $-7ab(ab + 11a^2b - 11b^2)$

45. $-\frac{1}{3}x(9x^2 + x - 5)$

46. $\frac{2}{5}a(10a^2 - 15a + 8)$

47. $-2mn(8m^2 - 3mn + n^2)$

48. $-7am(2a^2m^2 - 7am + 11)$

49. $-\frac{3}{4}ab^2\left(\frac{1}{3}b^2 - \frac{4}{9}b + 1\right)$

50. $-\frac{1}{3}xy\left(12x^2 + 8xy - \frac{2}{3}y^2\right)$

Simplify.

51. $2a(a^3 - 2a^2 + 7) + 5(a^4 + 5a^3 - 3a + 5)$

52. $6m(m^2 - 11m + 4) - 7(m^3 + 8m - 11)$

53. $5m^2(m + 7) - 2m(5m^2 - 3m + 7) + 2(m^3 - 8)$

54. $6a^2(3a - 4) + 5a(7a^2 - 6a + 5) - 3(a^2 + 6a)$

55. $3a^2(a - 4) + 6a(3a^2 + a - 7) - 4(a - 7)$

56. $8r^2(r + 8) - 3r(5r^2 - 11) - 9(3r^2 - 8r + 1)$

57. $2.5t(8t - 12) + 5.1(6t^2 + 10t - 20)$

58. $3.2a(5a + 1.1) + 9.3(6a^2 - 1.1a + 3.5)$

59. $\frac{3}{4}m(8m^2 + 12m - 4) + \frac{3}{2}(8m^2 - 9m)$

60. $\frac{1}{2}a(a^2 + 7a - 8) + \frac{1}{4}(5a^3 + 8a^2 - 3a)$

Challenge Exercises

61. $\frac{3}{4}a(a^2 + 0.79a + 0.345) - \frac{4}{5}a\left(\frac{1}{4}a^2 + 0.295a + \frac{1}{2}\right)$

62. $\frac{2}{5}x^2(x^3 + 0.86x^2 - 0.49) - 0.67x\left(0.34x^4 - \frac{2}{5}x^3 - 0.631\right)$

Excursions in Algebra	History

Amalie Emmy Noether (1882–1935) was a German mathematician who worked with the structure of algebra. She studied noncommutative algebras. In the period 1930–1933, she was the center of mathematical activity at the University of Göttingen with her research program and her influence on many students.

Noether's strength as a mathematician lay in her ability to operate abstractly with concepts. In her hands the axiomatic method (using axioms or properties) became a powerful tool of mathematical research. Albert Einstein paid her a great tribute in 1935: "In the realm of algebra . . . , she discovered methods which have proved of enormous importance in the development of the present day younger generation of mathematicians."

6-4 Multiplying Polynomials

The Distributive Property can be used to multiply polynomials as well as to multiply a monomial by a polynomial. To find the product of two binomials, use the Distributive Property twice.

Example

1 **Multiply $(2x + 3)(5x + 8)$.**

$(2x + 3)(5x + 8) = 2x(5x + 8) + 3(5x + 8)$	*Use the Distributive Property.*
$= 2x(5x) + 2x(8) + 3(5x) + 3(8)$	*Use the Distributive Property again.*
$= 10x^2 + 16x + 15x + 24$	*Simplify each term.*
$= 10x^2 + 31x + 24$	*Combine like terms.*

Although two binomials can always be multiplied as shown above, the following shortcut called the **FOIL method** is used frequently.

	$2x \cdot 5x$	
Multiply the First terms.	$(2x + 3)\ (5x + 8)$	$10x^2$
	$2x \cdot 8$	$+$
Multiply the Outer terms.	$(2x + 3)\ (5x + 8)$	$16x$
	$3 \cdot 5x$	$+$
Multiply the Inner terms.	$(2x + 3)\ (5x + 8)$	$15x$
	$3 \cdot 8$	$+$
Multiply the Last terms.	$(2x + 3)\ (5x + 8)$	24

$$(2x + 3)(5x + 8) = 10x^2 + 16x + 15x + 24$$
$$= 10x^2 + 31x + 24 \quad \text{\textit{Combine the like terms}}$$
16x and 15x.

To multiply two binomials, find the sum of the products of	
F the first terms,	
O the outer terms,	*FOIL Method*
I the inner terms, and	*for Multiplying*
L the last terms.	*Two Binomials*

2 **Multiply $(x + 5)(x + 7)$.**

$(x\rangle x + 7$ $(5) x + 7$
$x^2 + 7x + 5x + 35$

$$
\underset{\text{F}}{}\underset{\text{L}}{}\quad \underset{\text{F}}{}\quad \underset{\text{O}}{}\quad \underset{\text{I}}{}\quad \underset{\text{L}}{}
$$

$(x + 5)(x + 7) = x \cdot x + x \cdot 7 + 5 \cdot x + 5 \cdot 7$

$$
\begin{aligned}
&= x^2 + 7x + 5x + 35 \\
&= x^2 + 12x + 35 \quad \textit{Combine like terms.}
\end{aligned}
$$

3 **Multiply $(3x - 5)(5x + 2)$.**

$$\text{F} \qquad \text{O} \qquad \text{I} \qquad \text{L}$$

$(3x - 5)(5x + 2) = 3x(5x) + 3x(2) + (-5)(5x) + (-5)(2)$

$$
\begin{aligned}
&= 15x^2 + 6x - 25x - 10 \\
&= 15x^2 - 19x - 10
\end{aligned}
$$

4 **Multiply $\left(\frac{1}{3}a - \frac{2}{3}\right)\left(\frac{1}{2}a + \frac{1}{3}\right)$.**

$$
\begin{aligned}
\left(\frac{1}{3}a - \frac{2}{3}\right)\left(\frac{1}{2}a + \frac{1}{3}\right) &= \left(\frac{1}{3}a\right)\left(\frac{1}{2}a\right) + \left(\frac{1}{3}a\right)\left(\frac{1}{3}\right) + \left(-\frac{2}{3}\right)\left(\frac{1}{2}a\right) + \left(-\frac{2}{3}\right)\left(\frac{1}{3}\right) \\
&= \frac{1}{6}a^2 + \frac{1}{9}a - \frac{1}{3}a - \frac{2}{9} \\
&= \frac{1}{6}a^2 + \frac{1}{9}a - \frac{3}{9}a - \frac{2}{9} \\
&= \frac{1}{6}a^2 - \frac{2}{9}a - \frac{2}{9}
\end{aligned}
$$

The Distributive Property can be used to multiply any two polynomials.

5 **Multiply $(2x + 5)(3x^2 - 5x + 4)$.**

$$
\begin{aligned}
(2x + 5)(3x^2 - 5x + 4) &= (2x)(3x^2 - 5x + 4) + (5)(3x^2 - 5x + 4) \\
&= 2x(3x^2) + 2x(-5x) + 2x(4) + 5(3x^2) + 5(-5x) + 5(4) \\
&= 6x^3 - 10x^2 + 8x + 15x^2 - 25x + 20 \\
&= 6x^3 + 5x^2 - 17x + 20
\end{aligned}
$$

6 **Multiply $(x^2 - 5x + 4)(2x^2 + x - 7)$.**

$$
\begin{aligned}
&(x^2 - 5x + 4)(2x^2 + x - 7) \\
&= (x^2)(2x^2 + x - 7) + (-5x)(2x^2 + x - 7) + (4)(2x^2 + x - 7) \\
&= x^2(2x^2) + x^2(x) + x^2(-7) + (-5x)(2x^2) + (-5x)(x) + (-5x)(-7) + 4(2x^2) + 4(x) + \\
&\quad 4(-7) \\
&= 2x^4 + x^3 - 7x^2 - 10x^3 - 5x^2 + 35x + 8x^2 + 4x - 28 \\
&= 2x^4 - 9x^3 - 4x^2 + 39x - 28
\end{aligned}
$$

6-5 Some Special Products

Study the following diagram. Here are two ways to find the area of the large square.

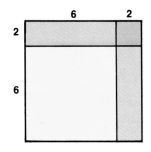

Method 1: The length of each side is 6 + 2 units. Find the square of (6 + 2).

$$(6 + 2)^2 = 8^2$$
$$= 64$$

Method 2: The area of the large square is the sum of the areas of the smaller parts.

$$6^2 + 6 \cdot 2 + 6 \cdot 2 + 2^2$$
$$= 36 + 12 + 12 + 4$$
$$= 64$$

Notice that $(6 + 2)^2 = 6^2 + 6 \cdot 2 + 6 \cdot 2 + 2^2$.

Using a similar procedure, the general form $(a + b)^2$ can be simplified as follows.

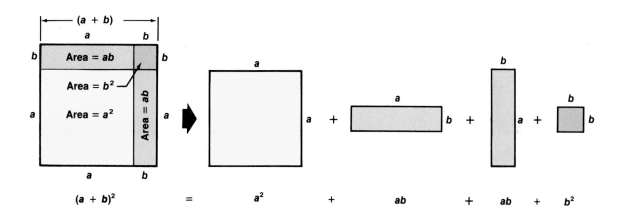

$$(a + b)^2 = a^2 + ab + ab + b^2$$
$$= a^2 + 2ab + b^2$$

Check this result using the FOIL method.

$$(a + b)^2 = (a + b)(a + b) = a^2 + 2ab + b^2$$

Square of a Sum

Examples

1 **Multiply $(x + 3)^2$.**

Use the square of a sum, $(a + b)^2 = a^2 + 2ab + b^2$.

$$(x + 3)^2 = x^2 + 2(x)(3) + (3)^2 \qquad \textit{Replace a with x and b with 3.}$$
$$= x^2 + 6x + 9$$

2 **Multiply $(5m + 3n)^2$.**

Use $(a + b)^2 = a^2 + 2ab + b^2$.

$$(5m + 3n)^2 = (5m)^2 + 2(5m)(3n) + (3n)^2 \qquad \textit{Replace a with 5m and b with 3n.}$$
$$= 25m^2 + 30mn + 9n^2$$

To find $(a - b)^2$, express $(a - b)$ as $[a + (-b)]$ and square it as a sum.

$$(a - b)^2 = [a + (-b)]^2$$
$$= a^2 + 2(a)(-b) + (-b)^2$$
$$= a^2 - 2ab + b^2$$

$(a - b)^2 = (a - b)(a - b) = a^2 - 2ab + b^2$	***Square of a Difference***

Compare the square of a sum and the square of a difference. How do they vary?

Examples

3 **Multiply $(c - 2)^2$.**

Use the square of a difference, $(a - b)^2 = a^2 - 2ab + b^2$.

$$(c - 2)^2 = c^2 - 2(c)(2) + (2)^2 \qquad \textit{Replace a with c and b with 2.}$$
$$= c^2 - 4c + 4$$

4 **Multiply $(3x - 2y)^2$.**

Use $(a - b)^2 = a^2 - 2ab + b^2$.

$$(3x - 2y)^2 = (3x)^2 - 2(3x)(2y) + (2y)^2$$
$$= 9x^2 - 12xy + 4y^2 \qquad \textit{You can use the FOIL method to check this.}$$

You can use the FOIL method to find the product of a sum and a difference of the same two numbers.

$$(a + b)(a - b) = a(a) + a(-b) + b(a) + b(-b)$$
$$= a^2 - ab + ba - b^2$$
$$= a^2 - b^2$$

This product is called the **difference of squares**.

$(a + b)(a - b) = a^2 - b^2$	***Product of a Sum and a Difference***

Examples

5 **Multiply $(x + 5)(x - 5)$.**

Use $(a + b)(a - b) = a^2 - b^2$.

$(x + 5)(x - 5) = x^2 - 5^2$ *Replace a with x and b with 5.*
$\qquad\qquad\quad = x^2 - 25$

6 **Multiply $(5a + 6b)(5a - 6b)$.**

Use $(a + b)(a - b) = a^2 - b^2$.

$(5a + 6b)(5a - 6b) = (5a)^2 - (6b)^2$
$\qquad\qquad\qquad\quad = 25a^2 - 36b^2$

Exploratory Exercises

Multiply.

1. $(a + 2b)^2$

2. $(a - 3b)^2$

3. $(2x + y)^2$

4. $(3x - 2y)^2$

5. $(3m^2 + 2n)^2$

6. $(3m - 4x)^2$

7. $(a^2 + b)^2$

8. $(2a + 3)(2a - 3)$

9. $(3a - 8b)^2$

10. $(4x - 3)(4x + 3)$

11. $(5a - 3b)(5a - 3b)$

12. $(4x^2 + y)^2$

13. $(3x - y)^2$

14. $(5m^2 + 7n)^2$

15. $(2x^2 - 3y^3)^2$

Written Exercises

Multiply.

16. $(4x + y)^2$

17. $(5x + y)^2$

18. $(2a - b)^2$

19. $(5a - b)^2$

20. $(6m + 2n)^2$

21. $(7x + 3y)^2$

22. $(4x - 9y)^2$

23. $(5r - 7s)^2$

24. $(5a - 12b)^2$

25. $(6a - 5b)^2$

26. $(5x + 6y)^2$

27. $(11m + 7n)^2$

28. $\left(\frac{1}{2}a + b\right)^2$

29. $\left(\frac{1}{3}x + y\right)^2$

30. $\left(\frac{4}{3}x - 2y\right)^2$

31. $\left(\frac{3}{4}a - 2b\right)^2$

32. $(5 - x)^2$

33. $(1 + x)^2$

34. $(1.1x + y)^2$

35. $(3.1 + 2y)^2$

36. $(a^2 - 3b^3)^2$

37. $(x^3 - 5y^2)^2$

38. $(3x + 5)(3x - 5)$

39. $(8a + 2b)(8a - 2b)$

40. $(3.2x^2 - 1.4y)^2$

41. $(1.8a^2 - 2.3y^2)^2$

42. $\left(\frac{1}{3}a + \frac{3}{2}b\right)\left(\frac{1}{3}a - \frac{3}{2}b\right)$

43. $\left(\frac{4}{3}x^2 - y\right)\left(\frac{4}{3}x^2 + y\right)$

44. $(7a^2 - b)^2$

45. $(8x^2 - 3y)^2$

46. $(3a^2 + b)^2$

47. $(4x^3 - 3y^2)^2$

48. $(6a^3 - a^2)^2$

49. $(x - 3)(x + 4)(x + 3)(x - 4)$

50. $(x + 2)(x - 2)(2x + 5)$

51. $(4x - 1)(4x + 1)(x - 4)$

52. $(1.3x^2 + 2.1y^3)^2$

Challenge Exercises

Multiply.

53. $(a + b)^4$

54. $(2x - y)^4$

55. $(3m + 2n)^4$

56. $(a + b)^5$

57. $(a - b)^5$

58. $(x - 2y)^3$

59. $(2x - 3y)^4$

60. $\left(\frac{1}{2}x + y\right)^3$

61. $(0.3m - n)^4$

62. $(a^{2n} - b^{5n})(a^{2n} - b^{5n})$

63. $(x^{2n} + y^n)(x^{2n} - y^n)$

64. $(a^{3n} - b^{5n})(a^{3n} + b^{5n})$

65. $(x^{3n} + y^{5n})(x^{3n} + y^{5n})$

Excursions in Algebra   The Store and Recall Keys

Many calculators have keys labeled STO and RCL. The key labeled STO is the *store* key. When you enter a number and then press this key, the number is *stored* in the memory of the calculator.

The key labeled RCL is the *recall* key. When you press this key, the calculator retrieves a stored number from the memory and displays it.

The store and recall keys are helpful when evaluating expressions where the same value will be used repeatedly.

Example   Evaluate $5 - 4x^2$ when x is 7.29.

First, store 7.29.

ENTER: 7.29

Some calculators have different labels for store and recall keys.

DISPLAY: 7.29

Then evaluate the expression.

ENTER: 5 4 $\boxed{x^2}$ $\boxed{=}$

DISPLAY: 5 4 7.29 53.1441 -207.5764

When x is 7.29, the value of $5 - 4x^2$ is -207.5764.

Exercises

Evaluate each expression if the value of x is -1.06.

1. $2 - 3x$

2. $-6 - 7x$

3. $2x + 5x - 7$

4. $4x + x - 9$

5. $(7 - x)^2$

6. $7 - x$

Evaluate if $a = 6$, $b = -4$, and $y = -\frac{4}{3}$.

1. $3b^3y$

2. $9ay - b$

3. $\frac{b^2}{6a^2}$

4. $y(a - b)^2$

State the property shown.

5. $12x^2 + 4xy = 4x(3x + y)$

6. $(4 + x) + 6 = 4 + (x + 6)$

Add or subtract.

7. $-21 + 13$

8. $\frac{3}{5} + \left(-\frac{3}{10}\right)$

9. $-21 - 5.4 - 18.2$

Simplify.

10. $4a + 3(a + 4)$

11. $-3(-6) - 6(-5)$

12. $(2r^2s)(8rs^3)$

13. $(4a^2b^3)^3$

14. y^{-4}

15. $\frac{-\frac{3}{5}}{9}$

16. $\frac{52m^8}{14m^5}$

17. $(a + 4b) - (3a - b)$

18. $\frac{1}{2}y\left(2y + \frac{1}{y}\right)$

19. $(2x + 5)(3x - 8)$

20. $(3x - 2)(3x + 2)$

21. $(3m + 4)^2$

Solve.

22. $5 + a = 17$

23. $-13 = -42 - p$

24. $\frac{3}{5}x = -8$

25. $7a - 5 = 30$

26. $7 = 3 + 4(m - 1)$

27. $1.3a - 2.3 = 4.5a + 4.1$

28. What is 75% of 92?

29. 48 is what percent of 144?

Solve each inequality. Then graph its solution set.

30. $2t < t + 3$

31. $-3(y + 4) \leq -5$

32. Write the following compound sentence without *and*: $x > 4$ and $x \leq 7$.

33. Express 983,000,000 in scientific notation.

34. Find the degree of the polynomial $3^2a^3b^7c$.

35. Find the degree of the polynomial $27x^2 + 4xy - 13y^3$.

Solve each problem.

36. Which is a better buy, a gallon of milk for $1.67 or a half-gallon of milk for $0.96?

37. Find three consecutive integers whose sum is 117.

38. Six times an integer decreased by 5 is greater than 67. What is the least integer that will satisfy this condition?

39. Maggie is 3 years older than Agnes. The sum of their ages is 35. What are their ages now?

40. An item priced at $65 has a 20% discount. Find the discount price.

41. Mike has 400 mL of 75% solution of silver nitrate. How many milliliters of a 30% solution should be added to obtain a 50% solution?

6-6 Solving Equations

Many equations contain polynomials that must be added, subtracted, or multiplied before the equation can be solved.

Examples

1 **Solve $-8 - (9 - 4m) = 15$.**

$$-8 - (9 - 4m) = 15$$
$$-8 - 9 + 4m = 15 \qquad \text{\textit{Replace each term in the parentheses}}$$
$$-17 + 4m = 15 \qquad \text{\textit{with its additive inverse.}}$$
$$-17 + 17 + 4m = 15 + 17$$
$$4m = 32$$
$$\frac{4m}{4} = \frac{32}{4}$$
$$m = 8$$

The solution is 8.   *Check this result.*

2 **Solve $5(x - 3) + 3x = 8(9 - x) + 21$.**

$$5(x - 3) + 3x = 8(9 - x) + 21$$
$$5x - 15 + 3x = 72 - 8x + 21 \qquad \text{\textit{Simplify each side of the equation.}}$$
$$8x - 15 = 93 - 8x \qquad \text{\textit{Combine like terms.}}$$
$$8x + 8x - 15 + 15 = 93 + 15 - 8x + 8x$$
$$16x = 108$$
$$\frac{16x}{16} = \frac{108}{16}$$
$$x = \frac{27}{4} \text{ or } 6.75$$

The solution is $\frac{27}{4}$ or 6.75.   *Check this result.*

3 **Solve $x(x - 3) + 4x - 3 = 8x + 4 + x(3 + x)$.**

$$x(x - 3) + 4x - 3 = 8x + 4 + x(3 + x)$$
$$x^2 - 3x + 4x - 3 = 8x + 4 + 3x + x^2$$
$$x^2 + x - 3 = x^2 + 11x + 4$$
$$x - 3 = 11x + 4 \qquad \text{\textit{Subtract } } x^2 \text{\textit{ from each}}$$
$$x - 11x - 3 + 3 = 11x - 11x + 4 + 3 \qquad \text{\textit{side of the equation.}}$$
$$-10x = 7$$
$$\frac{-10x}{-10} = \frac{7}{-10}$$
$$x = -\frac{7}{10} \text{ or } -0.7$$

The solution is $-\frac{7}{10}$ or -0.7.   *Check this result.*

Exploratory Exercises

Simplify.

1. $3x + 4x - 5 - 6$
2. $12 - 3y + 12y - 23$
3. $3y + 5y - 8 - 34$
4. $17 - (3a + 4) - 6a$
5. $6y - 8y + 34 - 34y$
6. $8r + 34 + 39r - 8r$
7. $12(q + 4) - 2q$
8. $5(x - 3) + 22 - 8x$
9. $-3(x - 12) + 5(x - 23)$
10. $-9(9 - y) - 8(12 - y) + 23$

Written Exercises

Solve each equation.

11. $3x + 8x - 7 = 21x - 5$
12. $9x - 8 + 4x = 7x + 16$
13. $4y + 8y - 7 = 30y + 19$
14. $8y + 16 + 8y = 21 - 9y$
15. $-8 + 12a = 14a - 23 + 5a$
16. $7a + 23 - 12a = 34 + 23a$
17. $-9a + 34 - 2a = 28 - 5a + 27$
18. $9a - 11 = 23 + 24a - 27a$
19. $11(a - 3) + 5 = 2a + 44$
20. $-3(2a - 12) + 48 = 3a - 3$
21. $26 + 4(3a - 5) = 14a - 15$
22. $-13 + 3(3a + 11) = -33 - 3a$
23. $57 + 2a = 3a - 5(a - 9)$
24. $29 - 3a = 2(3a - 4) + 3$
25. $3a - 35 = 4(a + 12)$
26. $13 - 3a = 23(2a - 3)$
27. $2(5w - 12) = 6(-2w + 3) + 2$
28. $-11(3a - 4) = -3(-4a - 14)$
29. $-6(12 - 2w) = 7(-2 - 3w)$
30. $15(3a - 4) = -4(3a - 3)$
31. $-5(2x - 13) + 12 = 3(2x - 5)$
32. $7(x - 12) = 13 + 5(3x - 4)$
33. $\frac{1}{2}(2x - 34) = \frac{2}{3}(6x - 27)$
34. $\frac{3}{4}(8z - 12) = \frac{5}{6}(12z - 18)$
35. $19 - (2y + 3) = 2(y + 3) + y$
36. $2(a + 2) + 3a = 13 - (2a + 2)$
37. $x(x + 2) + 3x = x(x - 3)$
38. $w(w + 12) = w(w + 14) + 12$
39. $a(a - 6) + 2a = 3 + a(a - 2)$
40. $q(2q + 3) + 20 = 2q(q - 3)$
41. $x(x + 8) - x(x + 3) - 23 = 3x + 11$
42. $y(y - 12) + y(y + 2) + 25 = 2y(y + 5) - 15$
43. $4x - \frac{1}{2}(2x - 4) = \frac{1}{3}(x + 6) - 5$

Challenge Exercises

Solve each equation for x or y.

44. $2x = 5a$
45. $4aby = 8a^2bc$
46. $30b^2c^2d = 10bc^2y$
47. $y - 2a = 0$
48. $4n + 3y = 2m$
49. $ax + 3 = b$
50. $2x - 7 = 8 + a$
51. $2y + b = 3 - 5y$
52. $7ax - 2d = 3ax + 6d$
53. $4x + 5mn = 7mn + 2x$
54. $by - c = a - d$
55. $3(y - 2a) = 24a$
56. $a(x + b) = 3ab + 5$
57. $2(3a - 2x) = 5a + x$
58. $4(y + 3a) = 4a - y$

6-7 Problem Solving: Perimeter and Area

Recall that a polygon is a closed figure formed by line segments. The **perimeter** of a polygon is the sum of the lengths of its sides.

Example

1   **In a certain isosceles triangle, the third side is 3 inches shorter than either of the congruent sides. If the perimeter is 69 inches, find the lengths of the sides.**

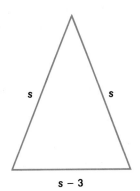

An isosceles triangle has at least two congruent sides.

explore   Let the following expressions represent the lengths of the sides.

s *length of each congruent side*

$(s - 3)$ *length of third side*

plan   Recall that the perimeter of a triangle is the sum of the lengths of its sides. To find these lengths, solve the following equation.

$$\frac{first}{side} + \frac{second}{side} + \frac{third}{side} = \frac{perimeter\ of}{the\ triangle}$$

solve

$$s + s + (s - 3) = 69$$
$$3s - 3 = 69$$
$$3s = 72$$
$$s = 24$$

Therefore, the length of each congruent side is 24 inches. The length of the third side is $24 - 3$ or 21 inches.

examine   The sum of the lengths of the three sides is $24 + 24 + 21$ or 69 inches.

The **area** of a polygon is the measurement of the region bounded by the polygon. Area is measured in square units. Some common units of area are square inches, square feet, square yards, square centimeters, and square meters.

Example

2 Maria Coulson has a rectangular garden that is 10 feet longer than it is wide. A sidewalk that is 3 feet in width surrounds the garden. The total area of the sidewalk is 396 square feet. What are the dimensions of the garden?

explore

Let x = width of garden.
$(x + 10)$ = length of garden

$(x + 6)$ = width of garden and sidewalk

$(x + 16)$ = length of garden and sidewalk

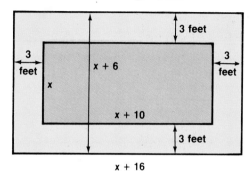

Since the area of a rectangle is the product of its length and width, the following expressions represent area.

$$x(x + 10) \quad \textit{area of garden}$$
$$(x + 6)(x + 16) \quad \textit{area of garden and sidewalk}$$

plan

To find the dimensions of the garden, solve the following equation.

$$\underset{\textit{and sidewalk}}{\textit{area of garden}} - \underset{\textit{garden}}{\textit{area of}} = \underset{\textit{sidewalk}}{\textit{area of}}$$

solve

$$(x + 6)(x + 16) - x(x + 10) = 396$$
$$x^2 + 22x + 96 - x^2 - 10x = 396$$
$$12x + 96 = 396$$
$$12x = 300$$
$$x = 25$$

Therefore, the width is 25 feet. The length is $(x + 10)$ or 35 feet.

examine

To examine the solution, compute the total area of the garden and sidewalk in two ways and compare the results.

$$\textit{area of garden} + \textit{area of sidewalk}$$
$$25 \times 35 \quad + \quad 396 \quad = 1271$$

$$\textit{total length} \times \textit{total width}$$
$$(35 + 6) \times (25 + 6)$$
$$41 \quad \times \quad 31 \quad = 1271$$

Exploratory Exercises

Write an expression that represents each problem.

1. The length of a rectangle is 5 units more than the width. Find the length if the measure of the width is: **a.** n **b.** $2t$ **c.** $2n + 3$ **d.** $t - 1$

2. The length of a rectangle is 2 units less than twice the width. Find the length if the measure of the width is: **a.** t **b.** $2x$ **c.** $3a + 2$ **d.** $2t + 1$

Written Exercises

Use an equation to solve each problem.

3. The area of a rectangle is 360 square meters. Its length is 24 meters. What is its width?

4. The area of a rectangle is 440 square inches. Its width is 20 inches. What is its length?

5. The length of a rectangle is 4 feet more than twice the width. The perimeter is 116 feet. Find the dimensions of the rectangle.

6. The perimeter of a football field is 1040 feet. The length of the field is 120 feet less than 3 times the width. What are the dimensions of the field?

7. A certain triangle has two congruent sides. The third side is 17 cm shorter than either of the equal sides. If the perimeter is 91 cm, what is the length of the third side?

8. The sides of a triangle are $4x$, $2x + 6$, and $7x - 9$. If the perimeter of the triangle is 62 inches, find the lengths of the three sides.

9. To get a square photograph to fit into a square frame, Linda LaGuardia had to trim a 1-inch strip from each side of the photo as shown below. In all, she trimmed off 40 square inches. What were the original dimensions of the photograph?

10. A rectangular garden is 5 feet longer than twice its width. It has a sidewalk 3 feet wide on two of its sides, as shown below. The area of the sidewalk is 213 square feet. Find the dimensions of the garden.

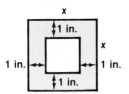

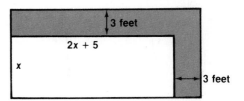

11. The second side of a triangle is twice the length of the first. The third side is 3 cm less than the second side. What are the lengths of the sides if the perimeter is 37 cm?

12. The three sides of a triangle have measures that are consecutive odd numbers. What are the lengths of the sides if the perimeter is 87 m?

13. A trapezoid has an area of 162 m² and a height of 12 m. The lower base is 6 m more than twice the upper base. Find the length of the upper base. Use $A = \frac{1}{2}h(a + b)$.

14. A trapezoid has an area of 81 ft² and a height of 9 ft. The upper base is 14 ft less than 3 times the lower base. Find the length of the lower base. Use $A = \frac{1}{2}h(a + b)$.

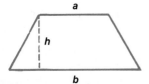

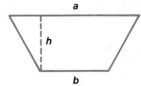

15. The length of a rectangle is 20 yards greater than the width. If the length was decreased by 5 yards, and the width increased by 4 yards, the area would remain unchanged. Find the original dimensions of the rectangle.

16. The length of a rectangle is 7 cm less than twice its width. If the length was increased by 11 cm and the width decreased by 6 cm, the area would be decreased by 40 square centimeters. Find the original dimensions of the rectangle.

Challenge Exercise

17. Mr. Herrera had a concrete sidewalk built on three sides of his yard as shown at the right. The yard measures 24 by 42 feet. The longer walk is 3 feet wide. The price of the concrete was $22 per square yard, and the total bill was $902. What is the width of the walk on the remaining two sides?

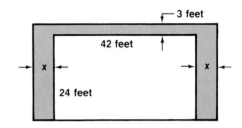

---

mini-review

Simplify.

1. $1.5(4x - 2y) + (0.1)(6.5x)$ **2.** $(ab)\left(\frac{4}{5}b\right)\left(\frac{4}{5}a\right)$ **3.** $\dfrac{(5a^{-3}b)^2}{(5a)^3}$

Graph the solution set of each compound sentence.

4. $x < 5$ or $x \geq 8$ **5.** $d > 0$ and $d < 4$ **6.** $p \leq 3$ or $p \geq -2$

Solve each problem. Assume that y varies inversely as x.

7. If $y = 34$ when $x = 2$, find y when $x = 17$.

8. If $x = 3.6$ when $y = 4.5$, find y when $x = 7.2$.

Solve each problem.

9. At Dawson's Market, a bunch of broccoli is 85¢. A bunch of broccoli weighs approximately $\frac{3}{4}$ pound. Ruwe's Market sells broccoli at 70¢ for $\frac{1}{2}$ pound. Which store has the better price on broccoli?

10. Weights of 50 pounds and 75 pounds are placed on a lever. The two weights are 16 feet apart, and the lever is balanced. How far from the fulcrum is the 50-pound weight?

11. A car uses 8 gallons of gasoline to travel 264 miles. How much gasoline will the car use to travel 500 miles?

12. The three sides of a triangle have measures that are consecutive even integers. What are the lengths of the sides if the perimeter is 126 m?

6-8 Problem Solving: Simple Interest

Sue Bailey deposited an amount of money in the bank at 7% annual interest. After 6 months, she received $52.50 interest. How much money had Sue deposited in the bank?

The formula $I = Prt$ is used to solve **simple interest** problems such as the one above. In the formula, I represents interest, P represents the amount of money invested or principal, r represents the annual interest rate, and t represents time in years.

Examples

1 Solve the problem above to find out how much money Sue had deposited in the bank.

> **explore**   Let P = amount of money Sue deposited.

> **plan** $I = Prt$
> $52.50 = P(0.07)\left(\frac{1}{2}\right)$ *Change 7% to 0.07. Change 6 months to $\frac{1}{2}$ year.*

> **solve** $52.50 = 0.035P$
> $1500 = P$
>
> Sue had deposited $1500 in the bank.

> **examine**   When P is 1500, I is $(1500)(0.07)\left(\frac{1}{2}\right)$ or $52.50.

2 Marilyn Mallinson invested $30,000, part at 6% annual interest and the balance at $7\frac{1}{2}$% annual interest. Last year she earned $1995 in interest. How much money did she invest at each rate?

Let n = amount of principal invested at 6%.
Then $(30,000 - n)$ = amount of principal invested at $7\frac{1}{2}$%.

Since $I = Prt$, the following expressions represent the interest earned on each portion of the investment.

interest on 6% investment   *interest on $7\frac{1}{2}$% investment*

$\quad I = n(0.06)(1) \qquad\qquad I = (30,000 - n)(0.075)(1)$
$\qquad = 0.06n \qquad\qquad\qquad\quad = 2250 - 0.075n$

Use the following equation to find the amount of money invested at each rate.

interest at 6% + interest at $7\frac{1}{2}$% = total interest

$$0.06n + (2250 - 0.075n) = 1995$$
$$-0.015n = -255$$
$$n = 17,000$$

Therefore, Marilyn invested $17,000 at 6% and $(30,000 - n)$ or $13,000 at $7\frac{1}{2}$%.

Exploratory Exercises

Use $I = Prt$ to find the missing quantity.

1. Find I, if $P = \$8000$, $r = 6\%$, and $t = 1$ year.

2. Find I, if $P = \$5000$, $r = 12\frac{1}{2}\%$, and $t = 5$ years.

3. Find t, if $I = \$1890$, $P = \$6000$, and $r = 9\%$.

4. Find t, if $I = \$2160$, $P = \$6000$, and $r = 8\%$.

5. Find r, if $I = \$2430$, $P = \$9000$, and $t = 2$ years, 3 months.

6. Find r, if $I = \$780$, $P = \$6500$, and $t = 1$ year.

7. Find P, if $I = \$756$, $r = 9\%$, and $t = 3\frac{1}{2}$ years.

8. Find P, if $I = \$196$, $r = 10\%$, and $t = 7$ years.

9. Find r, if $I = \$3487.50$, $P = \$6000$, and $t = 3$ years, 9 months.

10. Find t, if $I = \$3528$, $P = \$8400$, and $r = 10\frac{1}{2}\%$.

Written Exercises

Solve each problem.

11. The selling price of a house includes the 7% sales commission for the real estate company. A house sold for $160,000. How much did the owner receive and how much did the real estate company receive?

12. Walking shoes were on sale at the Sport Shop at a 15% discount. What did Cecilia pay for a pair of walking shoes that were originally $52.50?

13. Gladys deposited an amount of money in the bank at 6.5% annual interest. After 6 months, she received $7.80 interest. How much money had Gladys deposited in the bank?

14. Isaac deposited an amount of money in the bank at 7% annual interest. After 3 months he received $14 interest. How much money had Isaac deposited in the bank?

15. Michele Limotta invested $10,000 for one year, part at 8% annual interest and the balance at 12% annual interest. Her total interest for the year was $944. How much money did she invest at each rate?

16. Steve Devine invested $7200 for one year, part at 10% annual interest and the balance at 14% annual interest. His total interest for the year was $960. How much money did he invest at each rate?

17. Fred Ferguson invested $5000 for one year, part at 9% annual interest and the balance at 12% annual interest. The interest from the investment at 9% was $198 more than the interest from the investment at 12%. How much money did he invest at 9%?

18. Angela Raimondi wants to invest $8500, part at 14% annual interest and part at 12% annual interest. If she wants to earn the same amount of interest from each investment, how much should she invest at 14%? (Answer to the nearest cent.)

19. In one year, Cholena earned the same interest from an investment at 8% annual interest as an investment at 12% annual interest. She had invested $1500 more in the 8% account. How much money did she have invested at 12%?

20. John and Inger Johnson have $8000 to invest. They want to earn $850 interest for the year. Money can be invested at annual interest rates of either 10% or 11%. What is the minimum amount at 11%, with the balance at 10%, that will give $850 in interest?

21. Carlos and Bonita Díaz have invested $2500 at 10% annual interest. They have $6000 more to invest. At what rate must they invest the $6000 to have a total annual interest of $940?

22. Ken Bauman invested $9450, part at 8% annual interest and the balance at 11% annual interest. He earned twice as much interest from the 11% investment as from the 8% investment. How much money did he have invested at 11%?

Challenge Exercises

23. If an investor had earned one-fourth of a percent more in annual interest on an investment, the interest for one year would have been $45 greater. How much was invested at the beginning of the year?

24. A student deposited $1000 in a savings account at 8% interest compounded semi-annually. If no withdrawals or additional deposits were made, what was the amount in the account at the end of eighteen months?

Excursions in Algebra   Squaring Special Numbers

Study the following list of squares.

$$15^2 = 225 \qquad 25^2 = 625 \qquad 35^2 = 1225$$

Notice that each number ends in five and the square of each number ends in 25 (5^2). Notice that the numbers preceding 25 in each square follow a pattern.

 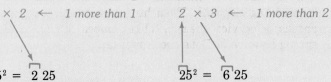 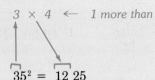

This pattern can be used to mentally square a number that ends in five.

Find 65^2.

The square will end in 25. ____25

Multiply 6 by the next whole number. $6 \times 7 = 42$
4225

So $65^2 = 4225$.

Use the above strategy to find each square.

1. 45^2 **2.** 55^2 **3.** 75^2 **4.** 85^2 **5.** 95^2 **6.** 105^2

Applications in Banking

Checking Account

Del Vargas has a checking account. He records his checks and deposits in the check register below. To find his balance, Del adds any deposits to his previous balance and subtracts the checks he has written.

CHECK NO.	DATE	CHECK ISSUED TO	AMOUNT OF DEPOSIT	AMOUNT OF CHECK	BALANCE
					35.72	←to previous balance
—	11/10	Deposit	115.00		150.72	←add deposits
428	11/12	S.E. Electric		25.50	125.22	←subtract checks
429	11/12	Star Oil Co.		21.35	
430	11/19	Kelly's Casuals		42.14	
431	11/23	Photo Journal Ass		6.50	
432	11/25	Palmer's Market		29.83	
—	11/25	Deposit	119.75		
433	11/28	Central Loan		90.00	

Each month Del receives a bank statement along with his canceled checks and deposit slips. The statement shows which checks and deposits the bank has processed. Unprocessed checks and deposits are said to be outstanding. Parts of the statement are shown below.

ENDING BALANCE
$31.90

CHECK NO.	AMOUNT
CHECKS	
428	25.50
429	21.35
430	42.14
432	29.83
DEPOSITS	
	115.00

Del must bring his statement and register ending balances into agreement. To the bank's ending balance, he adds any outstanding deposits. Then he subtracts any outstanding checks or service charges. This number should equal the ending balance in Del's check register.

Exercises

1-6. In Del's check register, find the balance after each check or deposit is made.

7. What checks and deposits are outstanding?

8. Bring the ending balance (from Exercise 6) into agreement with the bank balance.

6-9 Problem Solving: Uniform Motion

When an object moves at a constant speed, or rate, it is said to be in **uniform motion**. The formula **$d = rt$** is used to solve uniform motion problems. In the formula, d represents distance, r represents rate, and t represents time. When solving uniform motion problems, it is often helpful to draw a diagram and to organize relevant information in a chart.

In the formula, r can represent an <u>average</u> rate instead of a constant rate.

Examples

1 **Manuel Fernandez rides his bicycle at a speed of 8 mph (miles per hour). How long will it take him to ride 28 miles?**

explore Let t = the time it takes Manuel to ride 28 miles.

plan
$$d = rt$$
$$28 = 8t$$

solve
$$3\tfrac{1}{2} = t$$
Manuel will take $3\tfrac{1}{2}$ hours to ride 28 miles.

examine When t is $3\tfrac{1}{2}$, d is $8 \cdot 3\tfrac{1}{2}$ or 28 miles.

2 **Dan and Donna Wyatt leave their home in Chatsworth at the same time. They travel in opposite directions. Dan travels at 80 km/h (kilometers per hour) and Donna travels at 72 km/h. In how many hours will they be 760 km apart?**

760 km

Dan   Donna

80 km/h   72 km/h

Draw a diagram to help analyze the problem.

Let t represent the number of hours.

	r	t	d
Dan	80	t	$80t$
Donna	72	t	$72t$

Dan travels 80t km.
Donna travels 72t km.

They travel a total of 760 km.

Organize the information in a chart.

Dan's distance + Donna's distance = total distance

$$80t \quad + \quad 72t \quad = 760$$
$$152t = 760$$
$$t = 5$$

In 5 hours, Dan and Donna will be 760 km apart.   *Examine this solution.*

220 *Polynomials*

Example

3 At 8:00 A.M. Peggy leaves home driving at 35 mph. A half hour later, Doug discovers that she left her briefcase. He drives 50 mph to catch up with her. If Doug is delayed 15 minutes with a flat tire, at what time will he catch up to Peggy?

Let x = the time Peggy travels until Doug arrives.

	r	$\cdot$	t	$=$	d
Peggy	35		x		$35x$
Doug	50		$\left(x - \frac{3}{4}\right)$		$50\left(x - \frac{3}{4}\right)$

Peggy travels x hours.

Doug starts $\frac{1}{2}$ hour later and is delayed for $\frac{1}{4}$ hour (flat tire).

Doug travels $\left(x - \frac{3}{4}\right)$ hours.

Peggy and Doug travel the same distance.

$$35x = 50\left(x - \frac{3}{4}\right)$$
$$35x = 50x - \frac{75}{2}$$
$$-15x = -\frac{75}{2}$$
$$x = 2\frac{1}{2}$$

Peggy has been traveling for $2\frac{1}{2}$ hours when Doug catches up to her.

Doug catches up to Peggy at 8 A.M. $+ 2\frac{1}{2}$ hours or 10:30 A.M.

Exploratory Exercises

Use $d = rt$ to answer each question.

1. Pat is driving 80 km/h. How far will she travel: **a.** in 2 hours? **b.** in 6 hours? **c.** in h hours?

2. Makya is driving 40 mph. How far will he travel: **a.** in 3 hours? **b.** in $4\frac{1}{2}$ hours? **c.** in 15 minutes? **d.** in k minutes?

3. Marilyn traveled 240 miles. What was her rate if she made the trip:
a. in 6 hours? **b.** in t hours?

4. Juan traveled 270 miles. What was his rate if he made the trip:
a. in 5 hours? **b.** in x hours?

5. Rudy rode his bicycle 72 km. How long did it take him if his rate was:
a. 9 km/h? **b.** 18 km/h?

6. Patsy traveled 360 miles. How long did it take her if her rate was:
a. 40 mph? **b.** 30 mph? **c.** x mph?

Written Exercises

Solve each problem. Use charts and diagrams if necessary.

7. Two trains leave Bridgeport at the same time, one traveling north, the other south. The first train travels at 40 mph and the second at 30 mph. In how many hours will the trains be 245 miles apart?

8. Brad and Scott leave home traveling on their bicycles in opposite directions. Scott travels 10 km/h and Brad travels 12 km/h. In how many hours will they be 110 kilometers apart?

9. Rosita drives from Boston to Cleveland, a distance of 616 miles. Her rest, gasoline, and food stops amount to 2 hours. What was her rate if the trip took 16 hours?

10. At the same time Kris leaves Washington, D.C., for Detroit, Amy leaves Detroit for Washington, D.C. The distance between the cities is 510 miles. Amy drives 5 mph faster than Kris. How fast is Kris driving if they pass each other in 6 hours?

11. Two cyclists are traveling in the same direction on the same course. One travels 20 mph and the other 14 mph. After how many hours will they be 15 miles apart?

12. Hana leaves at 10:00 A.M., traveling at 50 mph. At 11:30 A.M., Yong starts in the same direction at 45 mph. When will they be 100 miles apart?

13. Boat *A* leaves the pier at 9:00 A.M., at 8 knots (nautical miles per hour). A half hour later, Boat *B* leaves the same pier in the same direction traveling at 10 knots. At what time will Boat *B* overtake Boat *A*?

14. Bob is driving 40 mph. After Bob has driven 30 miles, Jack starts driving in the same direction. At what rate must Jack drive to catch up to Bob in 5 hours?

15. An express train travels 80 km/h from Wheaton to Whitfield. A passenger train, traveling 48 km/h, takes 2 hours longer for the same trip. How far apart are Wheaton and Whitfield?

16. Kenny Brown drives to town at 36 mph and returns at 48 mph. If his total driving time is $3\frac{1}{2}$ hours, how far is his home from town?

17. Two airplanes leave Dallas at the same time and fly in opposite directions. One plane travels 80 mph faster than the other. After three hours, they are 2940 miles apart. What is the rate of each plane?

18. Jackson runs a 440 yard run in 55 seconds and Rey runs it in 88 seconds. To have Jackson and Rey finish at the same time, how many yards headstart should Rey be given?

19. At 1:30 P.M., a plane leaves Tucson for Baltimore, a distance of 2240 miles. The plane flies 280 mph. A second plane leaves Tucson at 2:15 P.M., and is scheduled to land in Baltimore 15 minutes before the first plane. At what rate must the second plane travel to arrive on schedule?

20. Sherry leaves Columbus, Ohio, on a train traveling 70 mph to Dallas. Twelve hours later, Soto leaves Columbus by plane traveling 500 mph to Dallas. If Columbus and Dallas are 1050 miles apart, will Soto arrive before or after Sherry?

Challenge Exercise

21. Two trains are 240 miles apart traveling toward each other on parallel tracks. One travels 35 mph and the other travels 45 mph. At the front of the faster train is a bee that can fly 75 mph. Assume that the bee can change direction instantaneously. It flies from one train to the other until the trains pass each other. When the trains pass, what distance has the bee flown?

Compound Interest

Suppose you have deposited $1000 in a savings account that pays 7.5% interest compounded annually. If you do not withdraw or deposit any additional money to the account, in how many years will your money double?

The formula $I = Prt$ is used to solve this problem.

$$I = Prt$$
$$I = 1000(0.075)(1)$$
$$I = 75$$

The interest earned after 1 year is $75. The new principal is $1075. This procedure is now repeated using $1075 as the principal and continues until the principal is greater than $2000.

Certainly you could solve the problem with a calculator, but a computer has the advantage of performing the operations very quickly. Use this BASIC program to determine the number of years it will take your money to double.

```
10   PRINT "ENTER THE PRINCIPAL, THE RATE AS A DECIMAL,
        AND THE TIME IN YEARS"
20   INPUT P,R,T        When interest is compounded annually, T = 1.
30   LET DEP = P        DEP is the amount deposited.
40   LET Y = 0
50   PRINT "YEAR", "PRINCIPAL"
60   LET I = P * R * T  Line 60 calculates the interest.
70   LET P = P + I      Line 70 adds interest to the principal.
80   LET Y = Y + T      Line 80 keeps a count of the years.
90   PRINT Y, "$";P
100  IF P > 2 * DEP THEN 120
110  GOTO 60
120  PRINT "IT TOOK ";Y;" YEARS TO DOUBLE YOUR MONEY"
130  END
```

Exercises

1. If $P = \$1000$, $r = 7.5\%$, and $t = 1$, in how many years will your money double?

2. If $P = \$1000$, $r = 10\%$, and $t = 1$, in how many years will your money double?

3. Run the program several times using $P = \$1000$ and $t = 1$, but vary the rate of interest. What is the relationship between the interest rate and the number of years it takes to double the principal?

4. If interest is compounded semiannually, $t = 0.5$. In how many years would your money double if $P = 1000$, $r = 10\%$ and $t = 0.5$?

5. Run the program several times using $P = \$1000$ and $r = 10\%$, but vary the time, t. What is the relationship between t and the number of years it takes to double the principal?

Vocabulary

polynomial (191)
binomial (191)
trinomial (191)
degree (191)
ascending order (192)
descending order (192)
FOIL method (201)

square of a sum (205)
square of a difference (206)
product of a sum and a difference (207)
perimeter (212)
area (212)
simple interest (216)
uniform motion (220)

Chapter Summary

1. A polynomial is a monomial or a sum of monomials. A binomial is a polynomial with two terms. A trinomial is a polynomial with three terms. (191)

2. The degree of a monomial is the sum of the exponents of its variables. The degree of a nonzero constant is 0. (191)

3. The degree of a polynomial is the greatest of the degrees of its terms. (192)

4. The terms of a polynomial are usually arranged so that the powers of one variable are in either ascending or descending order. (192)

5. The same properties used to simplify expressions can be used to add polynomials. (194)

6. To subtract a polynomial from a polynomial, add its additive inverse. (195)

7. To multiply two binomials, find the sum of the products of the **F**irst terms, the **O**uter terms, the **I**nner terms, and the **L**ast terms. This method is called FOIL. (201)

8. Square of a Sum:
$$(a + b)^2 = (a + b)(a + b) = a^2 + 2ab + b^2 \quad (205)$$

9. Square of a Difference:
$$(a - b)^2 = (a - b)(a - b) = a^2 - 2ab + b^2 \quad (206)$$

10. Product of a Sum and a Difference:
$$(a + b)(a - b) = a^2 - b^2 \quad (207)$$

11. The formula $I = Prt$ is used to solve simple interest problems. (216)

12. The formula $d = rt$ is used to solve uniform motion problems. (220)

6–1 Arrange the terms of each polynomial so that the powers of x are in descending order.

1. $3x^4 - x + x^2 - 5$

2. $ax^2 - 5x^3 + a^2x - a^3$

6–2 Add or subtract.

3. $(2x^2 - 5x + 7) - (3x^3 + x^2 + 2)$

4. $(x^2 - 6xy + 7y^2) + (3x^2 + xy - y^2)$

5. $(x^2 - 5x + 3) - (3x - 4)$

6. $\left(-\frac{1}{2}x^2 - \frac{1}{3}x + 2\right) + \left(\frac{3}{2}x^2 + \frac{1}{3}x - 4\right)$

6–3 Multiply. Simplify when possible.

7. $7xy(x^2 + 4xy - 8y^2)$

8. $x(3x - 5) + 7(x^2 - 2x + 9)$

9. $4x^2(x + 8) - 3x(2x^2 - 8x + 3) + 3(x^3 - 7x + 1)$

6–4 Multiply.

10. $(x + 5)(3x - 2)$

11. $(4x - 3)(x + 4)$

12. $(x - 4)(x^2 + 5x - 7)$

13. $\left(\frac{1}{2}a + 1\right)\left(\frac{3}{4}a^2 - 2a + 4\right)$

6–5 Multiply.

14. $(x - 6)(x + 6)$

15. $(5x - 3y)(5x + 3y)$

16. $(4x + 7)^2$

17. $(8x - 5)^2$

6–6 Solve each equation.

18. $4a + 9a - 7 = 14a + 6$

19. $4 + 6x = 2(4x - 2)$

Solve each problem.

6–7 **20.** Bill McClure is making a picture frame whose length will be 4 in. greater than its width. The frame will have a uniform width of 2 in. If the area of the frame will be 192 square inches, what will the inside dimensions of the picture frame be?

21. The sum of two congruent sides of a certain triangle is 22 inches greater than the base. If the perimeter is 50 inches, what is the length of each congruent side?

6–8 **22.** Marie Espinoza invested $8000 for one year, part at 8% and the remainder at 12%. Her total interest for the year was $744. How much money did she invest at each rate?

23. Francie earned $786 interest on a one year investment earning 15% annual interest. How much money did Francie invest?

6–9 **24.** At 8:00 A.M., Alan drove west at 35 mph. At 9:00 A.M., Reiko drove east from the same point at 42 mph. When will they be 266 miles apart?

25. Karen leaves home driving 32 mph. A half hour later, her sister Gail drives 40 mph to catch up to her. When will Gail catch up to Karen?

Chapter Test

Arrange the terms of each polynomial so that the powers of x are in descending order.

1. $5x^2 - 3 + x^3 + 5x$

2. $5 - xy^3 + x^3y^2 - x^2$

Simplify.

3. $(a + 5)(a - 5)$

4. $(2x - 5)(7x + 3)$

5. $(3a^2 + 3)[2a - (-6)]$

6. $3x^2y^3(2x - xy^2)$

7. $-4xy(5x^2 - 6xy^3 + 2y^2)$

8. $(4x^2 - y^2)(4x^2 + y^2)$

9. $\left(\frac{1}{2}x + 4\right)\left(\frac{1}{2}x + 3\right)$

10. $(0.2a - 0.4)^2$

11. $0.3b(0.4b^2 - 0.7b + 4)$

12. $(2a^2b + b^2)^2$

13. $x^2(x - 8) - 3x(x^2 - 7x + 3) + 5(x^3 - 6x^2)$

14. $a^2(a + 5) + 7a(a^2 + 8) - 7(a^3 - a + 2)$

Solve each equation.

15. $5y - 8 - 13y = 12y + 6$

16. $3(4m + 5) = 6m - 9$

Solve each problem.

17. The length of a rectangle is 3 inches less than twice the width. The perimeter is 84 inches. Find the dimensions of the rectangle.

18. The length of a rectangle is eight times its width. If the length was decreased by 10 meters and the width was decreased by 2 meters, the area would be decreased by 162 square meters. Find the original dimensions.

19. Patricia invested $5000 for one year. Maureen also invested $5000 for one year. Maureen's account earned interest at a rate of 10% per year. At the end of the year, Maureen's account earned $125 more than Patricia's account. What was the rate of interest on Patricia's account?

20. The second side of a triangle is twice the length of the first side. The length of the third side is 12 inches more than the length of the first side. If the perimeter is 304 inches, what is the length of each side?

21. Two cyclists start toward each other from two towns that are 96 miles apart. One cyclist rides at 18 mph and the other rides at 14 mph. In how many hours will they meet?

22. Luis invested $5000 for one year at 9% interest. Harry invested some money at the same time at 8% interest. At the end of the year, Luis and Harry together had earned $810 in interest. How much money did Harry invest?

The test questions on this page deal with ratios, proportions, and percents. The information at the right may help you with some of the questions.

Directions: Choose the best answer. Write A, B, C, or D.

1. 0.7% is the ratio of 7 to

   (A) 100 **(B)** $\frac{1}{10}$

   (C) $\frac{1}{1000}$ **(D)** 1000

2. 37.5% of a pound is equivalent to what fractional part of a pound?

   (A) $\frac{37}{100}$ **(B)** $\frac{3}{8}$ **(C)** $\frac{37}{99}$ **(D)** $37\frac{1}{2}$

3. Five-sixths is how many sevenths?

   (A) $5\frac{5}{7}$ **(B)** $8\frac{2}{5}$ **(C)** $5\frac{5}{6}$ **(D)** $4\frac{2}{7}$

4. How many elevenths is 75%?

   (A) $8\frac{1}{4}$ **(B)** $\frac{3}{4}$ **(C)** $6\frac{9}{11}$ **(D)** $\frac{12}{11}$

5. 90% of 270 is 2.7% of

   (A) 729 **(B)** 900

   (C) 2000 **(D)** 9000

6. A box contains 60 red, blue, and green pens. If 35% are red and 9 are blue, what percent are green?

   (A) 40 **(B)** 50 **(C)** 60 **(D)** 65

7. Sharon wishes to run a certain distance in 20% less time than she usually takes. By what percent must she increase her overall average speed?

   (A) 5% **(B)** 20% **(C)** 25% **(D)** 50%

8. 150% of 5c is b. What percent of 2b is c?

   (A) $5\frac{1}{2}$ **(B)** $6\frac{2}{3}$ **(C)** 15 **(D)** 75

9. Two-fifths times five-sevenths is equal to what number times six-elevenths?

   (A) $\frac{11}{21}$ **(B)** $\frac{55}{63}$ **(C)** $\frac{20}{77}$ **(D)** $\frac{22}{43}$

1. A ratio is an expression that compares two quantities.

   Example: If a container holds 4 pens and 7 pencils, the ratio of pencils to pens is 7:4 or $\frac{7}{4}$.

2. Two equivalent ratios form a proportion. The cross products of a proportion are equal. That is, if $\frac{a}{b} = \frac{c}{d}$, then $ad = bc$.

3. *Percent* means *hundredths*.

   7% means $\frac{7}{100}$ or 7 out of 100.

10. A woman owns $\frac{3}{4}$ of a business. She sells half of her share for $30,000. What is the total value of the business?

    (A) $45,000 **(B)** $75,000
    (C) $80,000 **(D)** $112,000

11. On a map, a line segment 1.75 inches long represents 21 miles. What distance in miles is represented by a segment 0.6 inches long?

    (A) 6 **(B)** 7.2 **(C)** 12.6 **(D)** 24.15

12. A cake recipe that serves 6 people calls for 2 cups of flour. How many cups of flour would be needed to bake a smaller cake that serves 4 people?

    (A) $\frac{3}{4}$ **(B)** 1 **(C)** $1\frac{1}{4}$ **(D)** $1\frac{1}{3}$

13. The price of an item was reduced by 10%, then later reduced by 20%. The two reductions were equivalent to a single reduction of

    (A) 15% **(B)** 28% **(C)** 30% **(D)** 70%

14. The discount price for a $50 item is $40. What is the discount price for a $160 item if the rate of discount is $1\frac{1}{2}$ times the rate of discount for the $50 item?

    (A) 20 **(B)** 48 **(C)** 96 **(D)** 112

CHAPTER 7

Factoring

   Suppose the measure of the area of a certain square garden is
represented by the polynomial $y^2 - 16y + 64$. What is the measure of
one side of the garden? To answer questions such as this one, you can
use *factoring*. In this chapter, you will learn to factor certain types of
polynomials.

7-1 Factors, Greatest Common Factors

If two or more numbers are multiplied, each number is a **factor** of the product. For example, you can express 18 as the product of different pairs of positive integers.

$$18 = 1 \cdot 18 \qquad 18 = 2 \cdot 9 \qquad 18 = 3 \cdot 6$$

The integers 1, 18, 2, 9, 3, and 6 are factors of 18.

Some integers have only two whole-number factors, the integer itself and 1. These integers are called **prime numbers**.

A prime number is an integer, greater than 1, whose only whole number factors are 1 and itself.	*Definition of Prime Number*

The prime numbers less than 50 are 2, 3, 5, 7, 11, 13, 17, 19, 23, 29, 31, 37, 41, 43, and 47.

A composite number is an integer, greater than 1, that is not prime.	*Definition of Composite Number*

The first few **composite numbers** are 4, 6, 8, 9, 10, and 12.

The number 9 is a factor of 18, but 9 is not a **prime factor** because it is not a prime number. When a composite number is expressed as a product of factors that are all prime, the expression is called the **prime factorization** of the number. The prime factorization of 18 is $2 \cdot 3 \cdot 3$ or $2 \cdot 3^2$.

The number 1 is neither prime nor composite.

Example

1 **Find the prime factorization of 84.**

Begin by dividing 84 by the least prime number that is a factor. Then use each prime in order as many times as it is a factor.

$$84 = 2 \cdot 42$$

$$2 \cdot 2 \cdot 21$$

$$2 \cdot 2 \cdot 3 \cdot 7$$

The least prime number that is a factor of 84 is 2.

When all factors are prime, the process stops.

The expression $2 \cdot 2 \cdot 3 \cdot 7$ or $2^2 \cdot 3 \cdot 7$ is the prime factorization of 84.

A negative integer may be expressed as a product of a positive integer and -1. For example, to find the factors of -525, first write it as the product of -1 and 525. Then find the prime factors of 525.

$$-525 = -1 \cdot 525$$
$$= -1 \cdot 3 \cdot 5^2 \cdot 7$$

$$525 = 3 \cdot 5^2 \cdot 7$$

A monomial such as $20a^2b$ may be expressed in factored form as a product of prime numbers and variables with no exponent greater than 1.

$$20a^2b = 2 \cdot 10 \cdot a \cdot a \cdot b$$
$$= 2 \cdot 2 \cdot 5 \cdot a \cdot a \cdot b$$

Two or more integers may have some common factors. Consider the prime factorizations of 90 and 105.

$$90 = 2 \cdot 3 \cdot 3 \cdot 5 \qquad 105 = 3 \cdot 5 \cdot 7$$

The integers 90 and 105 have the common prime factors 3 and 5. The product of these prime factors, $3 \cdot 5$ or 15, is called the **greatest common factor (GCF)** of 90 and 105.

The number 1 is always a common factor.

The greatest common factor of two or more integers is the product of the prime factors common to each integer.	*Definition of Greatest Common Factor*

The GCF of two or more monomials is the product of their common prime factors, when each monomial is in prime factored form.

Examples

2 **Find the GCF of 54, 63, and 180.**

$$54 = 2 \cdot ③ \cdot ③ \cdot 3$$
$$63 = ③ \cdot ③ \cdot 7$$
$$180 = 2 \cdot 2 \cdot ③ \cdot ③ \cdot 5$$

Factor each integer. Then circle the common factors.

The GCF of 54, 63, and 180 is $3 \cdot 3$ or 9.

3 **Find the GCF of $8a^2b$ and $18a^2b^2c$.**

$$8a^2b = ② \cdot 2 \cdot 2 \cdot ⓐ \cdot ⓐ \cdot ⓑ$$
$$18a^2b^2c = ② \cdot 3 \cdot 3 \cdot ⓐ \cdot ⓐ \cdot ⓑ \cdot b \cdot c$$

The GCF of $8a^2b$ and $18a^2b^2c$ is $2a^2b$.

Exploratory Exercises

State whether the second integer is a factor of the first.

1. 16; 8 **2.** 15; 3 **3.** 41; 6 **4.** 55; 11 **5.** 72; 36

State whether each number is prime or composite. If the number is composite, find its prime factorization.

6. 89 **7.** 39 **8.** 24 **9.** 91 **10.** 53

Find the GCF of each pair of numbers.

11. 4, 12   **12.** 10, 15   **13.** 9, 36   **14.** 15, 5   **15.** 11, 22

16. 15, 45   **17.** 20, 30   **18.** 18, 35   **19.** 36, 72   **20.** 16, 18

Written Exercises

Find the prime factorization of each number.

21. 21   **22.** 28   **23.** 60   **24.** 51   **25.** 39

26. 63   **27.** 34   **28.** 72   **29.** 112   **30.** 150

31. 304   **32.** 216   **33.** 300   **34.** 1000   **35.** 1540

Factor each expression. Do not use exponents.

36. -64 **37.** -26 **38.** -240 **39.** -500 **40.** -231

41. $98a^2b$ **42.** $44ab^2c^3$ **43.** $196b^2$ **44.** $756(ab)^3$ **45.** $-102x^3y$

Find the GCF of the given monomials.

46. 16, 60   **47.** 15, 50   **48.** -80, 45

49. 29, -58 **50.** 305, 55   **51.** 252, 126

52. 128, 245   **53.** 95, 304   **54.** $7y^2$, $14y^3$

55. $17a$, $34a^2$ **56.** $-12ab$, $4a^2b^2$ **57.** $4xy$, $-6x$

58. 6, 8, 4   **59.** 5, 10, 15   **60.** 16, 24, 30

61. 18, 30, 54   **62.** 24, 84, 168   **63.** 16, 22, 30

64. $50n^4$, $40n^2p^2$ **65.** $20a^2$, $24ab^5$ **66.** $60x^2y^2$, $35xz^3$

67. $12an^2$, $40a^4$ **68.** $56x^3y$, $49ax^2$ **69.** $12mn$, $10mn$, $15mn$

70. $6a^2$, $18b^2$, $9b^3$ **71.** $8b^4$, $5c$, $3a^2b$ **72.** $15abc$, $35a^2c$, $105a$

73. $14a^2b^2$, $18ab$, $2a^3b^3$ **74.** $18a^2b^2$, $6b$, $42a^2b^3$ **75.** $24a^2b$, $28axy$, $36ay$

Excursions in Algebra   Sieve of Eratosthenes

Eratosthenes was an early Greek (c. 200 B.C.) who was one of the first to calculate the circumference of Earth. He is also remembered in mathematics for the "sieve of Eratosthenes," a method for finding prime numbers.

To use his method, list the positive integers beginning with 2. Each multiple of a prime is to be crossed out. For example, cross out every second number after 2, every third number after 3, every fifth number after 5, and so on. When the process is finished, the numbers not crossed out are prime. *Begin by crossing out every second number after 2.*

2   3   4̸   5   6̸   7   8̸   9   10̸   11   12̸   13   14̸   15   16̸   17

18̸   19   20̸   21   22̸   23   24̸   25   26̸   27   28̸   29   30̸

Exercises

1. Copy and complete the sieve of Eratosthenes above for integers to 30.

2. List the positive integers from 2 to 100. Use the sieve of Eratosthenes to find the prime numbers between 2 and 100.

7-2 Factoring Using the Distributive Property

The Distributive Property has been used to multiply a polynomial by a monomial. It can also be used to express a polynomial in factored form.

Multiplying	**Factoring**
$3(a + b) = 3a + 3b$	$3a + 3b = 3(a + b)$
$x(y - z) = xy - xz$	$xy - xz = x(y - z)$
$3y(4x + 2) = 3y(4x) + 3y(2)$	$12xy + 6y = 3y(4x) + 3y(2)$
$\qquad\qquad = 12xy + 6y$	$\qquad\qquad = 3y(4x + 2)$

Examples

1 **Use the Distributive Property to express $10y^2 + 15y$ in factored form.**

First, find the greatest common factor of $10y^2$ and $15y$.

$10y^2 = 2 \cdot ⑤ \cdot ⓨ \cdot y$

$15y = 3 \cdot ⑤ \cdot ⓨ$ $\qquad$ *The GCF is 5y.*

Then, express each term as a product of the GCF and its remaining factors.

$10y^2 + 15y = 5y(2y) + 5y(3)$
$\qquad\qquad\quad = 5y(2y + 3)$ $\qquad$ *Use the Distributive Property.*

2 **Factor $21ab^2 - 33a^2bc$.**

$21ab^2 = ③ \cdot 7 \cdot ⓐ \cdot ⓑ \cdot b$
$33a^2bc = ③ \cdot 11 \cdot ⓐ \cdot a \cdot ⓑ \cdot c$ $\qquad$ *The GCF is 3ab.*

Express the terms as products.

$21ab^2 - 33a^2bc = 3ab(7b) - 3ab(11ac)$
$\qquad\qquad\qquad\quad = 3ab(7b - 11ac)$ $\qquad$ *Use the Distributive Property.*

3 **Factor $12a^5b + 8a^3 - 24a^3c$.**

$12a^5b = ② \cdot ② \cdot 3 \cdot ⓐ \cdot ⓐ \cdot ⓐ \cdot a \cdot a \cdot b$
$8a^3 = ② \cdot ② \cdot 2 \cdot ⓐ \cdot ⓐ \cdot ⓐ$
$24a^3c = ② \cdot ② \cdot 2 \cdot 3 \cdot ⓐ \cdot ⓐ \cdot ⓐ \cdot c$ $\qquad$ *The GCF is $4a^3$.*
$12a^5 b + 8a^3 - 24a^3c = 4a^3(3a^2b) + 4a^3(2) - 4a^3(6c)$
$\qquad\qquad\qquad\qquad\quad = 4a^3(3a^2b + 2 - 6c)$

4 **Factor $6x^3y^2 + 14x^2y + 2x^2$.**

$6x^3y^2 = ② \cdot 3 \cdot ⓧ \cdot ⓧ \cdot x \cdot y \cdot y$
$14x^2y = ② \cdot 7 \cdot ⓧ \cdot ⓧ \cdot y$
$2x^2 = ② \cdot ⓧ \cdot ⓧ$ $\qquad$ *The GCF is $2x^2$.*
$6x^3y^2 + 14x^2y + 2x^2 = 2x^2(3xy^2 + 7y + 1)$ $\qquad$ *Remember $2x^2 = 2x^2(1)$.*

Exploratory Exercises

Find the GCF of the terms in each expression.

1. $3y^2 + 12$

2. $4a + 2b$

3. $5y - 9y^2$

4. $6a + 3b$

5. $9b + 5c$

6. $3a + 6a$

7. $9y^2 + 3y$

8. $11a + 10q$

9. $8x^2 - 4x$

10. $12a^2b + 6a$

11. $7ac - 21a^2$

12. $16c - 21b$

13. $21xyz - xy$

14. $14b^2 - 42c$

15. $42a - 13r$

Written Exercises

Factor each expression.

16. $16x + 4y$

17. $24x^2 + 12y^2$

18. $12xy + 12x^2$

19. $5a^2b + 10ab$

20. $11x + 44x^2y$

21. $16x^2 + 8x$

22. $25a^2b^2 + 30ab^3$

23. $36a^2b^2 - 12ab$

24. $14xy - 18xy^2$

25. $27a^2b + 9b^3$

26. $3a^2b - 6a^2b^2$

27. $18xy^2 - 24x^2y$

28. $14xy^2 + 2xy$

29. $15xy^3 + y^4$

30. $x^5y - x$

31. $29xy - 3x$

32. $17a - 41a^3b$

33. $a + a^4b^3$

34. $3x^2y + 9y^2 + 6$

35. $5a^2 + 10ab - 15b^2$

36. $2a^3b^2 - 16a^2b^3 + 8ab$

37. $3x^3y + 9xy^2 + 36xy$

38. $24x^2y^2 + 12xy + x$

39. $28a^2b^2c^2 + 21a^2bc^2 - 14abc$

40. $12ax + 20bx + 32cx$

41. $a + a^2b + a^3b^3$

42. $ax^3 + 5bx^3 + 9cx^3$

43. $14a^3x + 19a^3y + 11a^3z$

44. $6x^2 - 9xy + 24x^2y^2$

45. $42abc - 12a^2b^2 + 3a^2c^2$

46. $24abx + 12ax^2 + 6x^3$

47. $x^5 + 5x^4 + 3x^2 + 2x$

48. $\frac{1}{2}x^2 - \frac{1}{4}ax$

49. $\frac{2}{3}x + \frac{1}{3}y$

50. $\frac{4}{5}x^2y + \frac{3}{5}y^2$

51. $\frac{2}{5}a - \frac{2}{5}b + \frac{4}{5}c$

Excursions in Algebra   Divisibility Rules

Some simple rules can help you determine if a number is divisible by 2, 3, 5, or 10.

If the ones digit is 0, 2, 4, 6, or 8, the number is divisible by 2.
If the ones digit is 0 or 5, the number is divisible by 5.
If the ones digit is 0, the number is divisible by 10.
If the *sum* of the digits is divisible by 3, the number is divisible by 3.

For example, the sum of the digits of 741 is 7 + 4 + 1 or 12.
741 is divisible by 3 because 12 is divisible by 3.

Exercises

Determine whether each number is divisible by 2, 3, 5, or 10.

1. 44   **2.** 75   **3.** 110   **4.** 123   **5.** 405   **6.** 570

Applications in Geometry Area

You can simplify equations involving area by using the Distributive Property and factoring. Consider this example.

A deck 3 meters wide is to be built around a swimming pool that is shown in the diagram at the right. In order to determine the amount of material that is needed to build the deck, it is necessary to calculate the area of the deck.

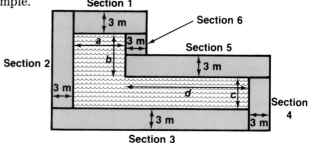

To find the area of the deck, find the sum of the areas of the rectangular regions.

$$\underbrace{}_{\text{Section 1}}\ \underbrace{}_{\text{Section 2}}\ \underbrace{}_{\text{Section 3}}\ \underbrace{}_{\text{Section 4}}\ \underbrace{}_{\text{Section 5}}\ \underbrace{}_{\text{Section 6}}$$

$$\text{Area} = 3(a + 3) + 3(b + c + 3) + 3(a + d + 3) + 3(c + 3) + 3(d + 3) + 3(b - 3)$$

Use the Distributive Property.
$$A = 3a + 9 + 3b + 3c + 9 + 3a + 3d + 9 + 3c + 9 + 3d + 9 + 3b - 9$$
Combine like terms.
$$A = 6a + 6b + 6c + 6d + 36$$
Factor the polynomial.
$$A = 6(a + b + c + d + 6)$$
The area of the deck can be represented by $A = 6(a + b + c + d + 6)$.

Exercises

For each diagram, write an equation that represents the area of the shaded region.

1.

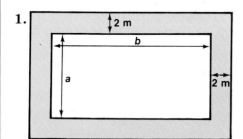

2.

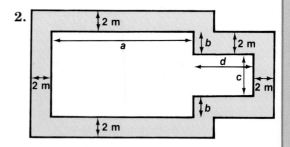

3.

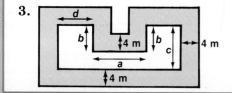

4.

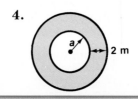

7-3 Factoring Differences of Squares

Consider the two squares that are shown. The area of the larger square is a^2, and the area of the smaller square is b^2.

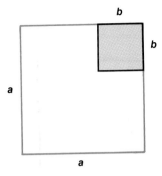

The area $a^2 - b^2$ can be found by subtracting the area of the smaller square from the area of the larger square.

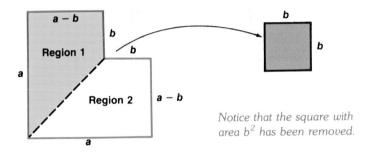

Notice that the square with area b^2 has been removed.

Rearranging these regions as shown below, you can see that $a^2 - b^2$ is equal to the product of $(a - b)$ and $(a + b)$.

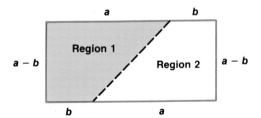

The area of the rectangle is $(a - b)(a + b)$. This is the same as $a^2 - b^2$.

$$a^2 - b^2 = (a - b)(a + b) = (a + b)(a - b)$$

Difference of Squares

You can use the difference of squares to factor binomials of the form $a^2 - b^2$.

1 **Factor $a^2 - 64$.**

$a^2 - 64 = (a)^2 - (8)^2$ *$a \cdot a = a^2$ and $8 \cdot 8 = 64$.*

$\qquad\quad = (a - 8)(a + 8)$ *Use the difference of squares.*

2 **Factor $9x^2 - 100y^2$.**

$9x^2 - 100y^2 = (3x)^2 - (10y)^2$ *$3x \cdot 3x = 9x^2$ and $10y \cdot 10y = 100y^2$.*

$\qquad\qquad\quad = (3x - 10y)(3x + 10y)$ *Use the difference of squares.*

3 **Factor $\frac{1}{4}y^2 - \frac{4}{9}p^2$.**

$\frac{1}{4}y^2 - \frac{4}{9}p^2 = \left(\frac{1}{2}y\right)^2 - \left(\frac{2}{3}p\right)^2$ *Why?*

$\qquad\qquad = \left(\frac{1}{2}y - \frac{2}{3}p\right)\left(\frac{1}{2}y + \frac{2}{3}p\right)$ *Check your answers using FOIL.*

Sometimes the terms of a binomial have common factors. If so, the GCF should always be factored out before factoring the difference of squares. Occasionally, factoring the difference of squares needs to be used more than once in a problem.

Examples

4 **Factor $8x^2 - 18y^2$.**

$8x^2 - 18y^2 = 2(4x^2 - 9y^2)$ *2 is the GCF of $8x^2$ and $18y^2$.*

$\qquad\qquad = 2(2x - 3y)(2x + 3y)$ *$2x \cdot 2x = 4x^2$ and $3y \cdot 3y = 9y^2$.*

5 **Factor $162a^4 - 32y^8$.**

$162a^4 - 32y^8 = 2(81a^4 - 16y^8)$ *Why?*

$\qquad\qquad = 2(9a^2 - 4y^4)(9a^2 + 4y^4)$ *$9a^2 + 4y^4$ cannot be factored.*

$\qquad\qquad = 2(3a - 2y^2)(3a + 2y^2)(9a^2 + 4y^4)$

Exploratory Exercises

State whether each binomial can be factored as the difference of squares.

1. $x^2 - y^2$ **2.** $a^2 + b^2$ **3.** $b^2 - 25$

4. $a^2 - 4b^4$ **5.** $9y^2 + a^2$ **6.** $9x^2 - 5$

7. $a - 9$ **8.** $8c^2 - 7$ **9.** $0.04m^2 - 0.09n^2$

10. $1.44r^2 - 0.49t^2$ **11.** $\frac{4}{9}a^2 - \frac{1}{4}$ **12.** $\frac{16}{25}a^2 + 1$

Written Exercises

Factor each polynomial completely.

13. $a^2 - 9$

14. $x^2 - 49$

15. $4x^2 - 9y^2$

16. $x^2 - 36y^2$

17. $a^2 - 4b^2$

18. $1 - 9y^2$

19. $16a^2 - 9b^2$

20. $3m^2 - 6n^2$

21. $16a^2 - 25$

22. $49 - 4a^2b^2$

23. $2a^2 + 18a$

24. $6x^2 + 12$

25. $8x^2 - 12y^2$

26. $2z^2 - 98$

27. $12a^2 - 48$

28. $8x^2 - 18$

29. $45x^2 - 20z^2$

30. $25y^2 - 49z^4$

31. $9x^4 - 25y^4$

32. $17 - 68a^2$

33. $0.01n^2 - 1.69r^2$

34. $a^2x^2 - 0.64y^2$

35. $1.69p^2 - 2.56q^2$

36. $0.0004a^2 - b^2c^2$

37. $4x^2 - 64y^2$

38. $12a^2 - 12$

39. $9x^4 - 16y^2$

40. $36x^2 - 81y^4$

41. $28x^2 - 7$

42. $81 - 9x^2$

43. $-16 + 9x^2$

44. $-9 + 4y^2$

45. $15x^2 - 60y^2$

46. $x^4 - y^4$

47. $a^4 - b^2$

48. $x^4 - 1$

49. $\frac{1}{4}x^2 - 16$

50. $\frac{1}{16}x^2 - 25$

51. $\frac{9}{2}x^2 - \frac{49}{2}y^2$

52. $\frac{2}{3}x^2 - \frac{8}{3}$

53. $x^8 - 1$

54. $(a + b)^2 - m^2$

55. $(x - y)^2 - y^2$

56. $p^2 - (m + n)^2$

57. $a^2 - (b - c)^2$

58. $9x^2 - (y + z)^2$

59. $(x - y)^2 - 0.36z^2$

60. $y^4 - 81$

61. $x^4 - 16$

62. $256 - n^4$

63. $162a^5 - 32ab^4$

64. $98x^6 - 128y^8$

65. $48x^5 - 3xy^4$

66. $48s^5t - 243st^5$

Challenge Exercises

67. Find positive integers m and n such that $m^2 - n^2 = 11$.

68. Find positive integers x and y such that $x^2 - y^2 = 15$.

69. Find positive integers a and b such that $a^2 - b^2 = 45$.

70. Find positive integers p and q such that $p^2 - q^2 = 105$.

mini-review

Solve each inequality. Express the solution set in set-builder notation.

1. $-6n \geq -72$

2. $\frac{5}{4}k < \frac{11}{12}$

3. $6.3m > -16.9$

Simplify. Assume no denominator is equal to zero.

4. $(4x^2y)(12xy^4z)$

5. $\left(\frac{3}{5}y\right)^2(-5xy^3)^3$

6. $\frac{-16a^3bc^2}{28ab^5}$

Multiply.

7. $8b(11b^2 + 6a^3)$

8. $(2x + 5y)(3x - y)$

9. $(8a + 3b)^2$

Factor each polynomial completely.

10. $10a^4 - 6a^3b - 8a^2b^2$

11. $a^2 - 16b^2$

12. $\frac{2}{3}x^2 + \frac{5}{3}$

In the metric system, a unit of mass is the gram (g). One gram is defined as the mass of one cubic centimeter of water at its maximum density.

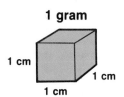

1 gram

1 cm
1 cm
1 cm

You can think of a gram as about the mass of two paper clips.

1 g

A gram is too small to use conveniently. Therefore, the kilogram (kg) is used as the basic unit for mass. There are 1000 grams in a kilogram. The mass of this textbook is about one kilogram.

1 kg

A metric ton (t) equals 1000 kilograms. Shippers measure grain in metric tons. A metric ton is about the mass of a small automobile.

1 t

The milligram (mg) is used to measure very small quantities. One gram equals 1000 milligrams.

Exercises

Would you use gram or kilogram to measure the mass of each object?

1. bag of groceries
2. pencil
3. mouse
4. wrestler
5. suitcase
6. can of pepper

Complete each exercise.

7. 3 kg = ▓ g
8. 8.5 kg = ▓ g
9. 5000 g = ▓ kg
10. 6700 g = ▓ kg
11. 8 g = ▓ mg
12. 6.5 g = ▓ mg
13. 3000 mg = ▓ g
14. 3900 mg = ▓ g
15. 9.52 kg = ▓ g
16. 3852 g = ▓ kg
17. 3598 mg = ▓ g
18. 6.257 g = ▓ mg
19. 38.575 kg = ▓ g
20. 875 g = ▓ kg
21. 575 mg = ▓ g
22. 5 t = ▓ kg
23. 5300 kg = ▓ t
24. 88,000 kg = ▓ t
25. 40 mg = ▓ kg
26. 260 mg = ▓ kg
27. 2 kg = ▓ mg
28. 1.93 kg = ▓ mg
29. 10 mg = ▓ t
30. 10t = ▓ mg

7-4 Perfect Squares and Factoring

Recall that $1^2 = 1$, $2^2 = 4$, $3^2 = 9$, $4^2 = 16$, and so on. Numbers such as 1, 4, 9, and 16 are called **perfect squares**. Products of the form $(a + b)^2$ and $(a - b)^2$ are also called perfect squares and these expansions are called **perfect square trinomials**.

You should recognize these patterns.

$$(a + b)^2 = a^2 + 2ab + b^2$$
$$(a - b)^2 = a^2 - 2ab + b^2$$

Perfect Square Trinomials

These patterns can help you factor trinomials such as $y^2 + 16y + 64$ and $4x^2 - 20xy + 25y^2$.

Finding a Product

$(y + 8)^2 = y^2 + 2(y)(8) + (8)^2$
$\qquad = y^2 + 16y + 64$

$(2x - 5y)^2$
$\quad = (2x)^2 - 2(2x)(5y) + (5y)^2$
$\quad = 4x^2 - 20xy + 25y^2$

Factoring

$y^2 + 16y + 64 = (y)^2 + 2(y)(8) + (8)^2$
$\qquad\qquad\quad = (y + 8)^2$

$4x^2 - 20xy + 25y^2$
$\quad = (2x)^2 - 2(2x)(5y) + (5y)^2$
$\quad = (2x - 5y)^2$

To determine whether a trinomial can be factored in this way, first decide if it is a perfect square. In other words, decide if it can be written in either the form $a^2 + 2ab + b^2$ or $a^2 - 2ab + b^2$.

Example

1   **Determine whether $x^2 + 22x + 121$ is a perfect square. If it is, factor it.**

To determine whether $x^2 + 22x + 121$ is a perfect square, answer each question.

a. Is x^2 a perfect square? $x^2 \overset{?}{=} (x)^2$ *yes*
b. Is 121 a perfect square? $121 \overset{?}{=} (11)^2$ *yes*
c. Is the middle term twice
    the product of x and 11? $22x \overset{?}{=} 2(x)(11)$ *yes*

Since all three answers are "yes," the trinomial $x^2 + 22x + 121$ is a perfect square. It can be factored as follows.

$x^2 + 22x + 121 = (x)^2 + 2(x)(11) + (11)^2$
$\qquad\qquad\qquad\quad = (x + 11)^2$

Examples

2 **Determine whether $16a^2 + 72a + 81$ is a perfect square. If it is, factor it.**

    a. Is the first term a square? $16a^2 \overset{?}{=} (4a)^2$ *yes*

    b. Is the last term a square? $81 \overset{?}{=} (9)^2$ *yes*

    c. Is the middle term twice the
        product of $4a$ and 9? $72a \overset{?}{=} 2(4a)(9)$ *yes*

    $16a^2 + 72a + 81$ is a perfect square.

$$16a^2 + 72a + 81 = (4a)^2 + 2(4a)(9) + (9)^2$$
$$= (4a + 9)^2$$

3 **Determine whether $15 + 4a^2 - 20a$ is a perfect square. If it is, factor it.**

To determine whether $15 + 4a^2 - 20a$ is a perfect square, first arrange the terms so that the powers of a are in descending order.

$$15 + 4a^2 - 20a = 4a^2 - 20a + 15$$

    a. Is the first term a square? $4a^2 \overset{?}{=} (2a)^2$ *yes*

    b. Is the last term a square? $15 \overset{?}{=} (?)^2$ *no*

    $15 + 4a^2 - 20a$ is *not* a perfect square.

4 **Determine whether $16x^2 - 26x + 49$ is a perfect square. If it is, factor it.**

    a. Is the first term a square? $16x^2 \overset{?}{=} (4x)^2$ *yes*

    b. Is the last term a square? $49 \overset{?}{=} (7)^2$ *yes*

    c. Is the middle term twice
        the product of $4x$ and 7? $26x \overset{?}{=} 2(4x)(7)$ *no*

    $16x^2 - 26x + 49$ is *not* a perfect square.

5 **Factor $9x^2 - 12xy + 4y^2$.**

    $9x^2 - 12xy + 4y^2$ is a perfect square.   *Why?*

$$9x^2 - 12xy + 4y^2 = (3x)^2 - 2(3x)(2y) + (2y)^2$$
$$= (3x - 2y)^2$$

Exploratory Exercises

Determine whether each trinomial is a perfect square trinomial. If it is, factor it.

  1. $a^2 + 4a + 4$ **2.** $b^2 + 14b + 49$ **3.** $x^2 - 10x - 100$

  4. $y^2 + 10y - 100$ **5.** $n^2 - 13n + 36$ **6.** $p^2 + 12p + 36$

  7. $y^2 - 8y + 10$ **8.** $x^2 + 17x + 21$ **9.** $4x^2 - 4x + 1$

10. $9y^2 - 10y + 4$ **11.** $9b^2 - 6b + 1$ **12.** $4a^2 - 20a + 25$

Written Exercises

Factor each trinomial, if possible.

13. $a^2 + 12a + 36$

14. $b^2 + 10b + 25$

15. $x^2 + 16x + 64$

16. $y^2 + 14y + 49$

17. $n^2 - 8n + 16$

18. $m^2 - 22m + 121$

19. $4a^2 + 4a + 1$

20. $64b^2 + 16b + 1$

21. $1 - 10a + 25a^2$

22. $1 - 12y + 36y^2$

23. $x^2 + 6x - 9$

24. $9a^2 + 6a + 2$

25. $50x^2 + 40x + 8$

26. $18a^2 - 48a + 32$

27. $121y^2 + 22y + 1$

28. $81n^2 + 36n + 4$

29. $9x^2 + 42x + 49$

30. $64m^2 + 160m + 100$

31. $25b^2 - 30b + 9$

32. $25x^2 - 120x + 144$

33. $64b^2 - 72b + 81$

34. $49m^2 - 126m + 81$

35. $m^2 + 16mn + 64n^2$

36. $40a + 9 + 25a^2$

37. $9x^2 + 24xy + 16y^2$

38. $16p^2 - 40pq + 25q^2$

39. $144p^2 + 72p + 9$

40. $144n^2 + 168n + 49$

41. $64x^2 - 80x + 25$

42. $2x^2 - 20x + 50$

43. $3x^2 + 36x + 108$

44. $3x^2 + 18x + 48$

45. $24 + 6x^2 + 18x$

46. $3b^2 - 18bc + 27c^2$

47. $4x^2 + 4xz^2 + z^4$

48. $4x^2 - 28xy^2 + 49y^4$

49. $\frac{1}{4}a^2 + 3a + 9$

50. $\frac{4}{9}b^2 - \frac{16}{3}b + 16$

51. $m^4 + 12m^2n^2 + 36n^4$

52. $16x^2y^2 - 72xy + 81$

53. $9a^2 + \frac{24}{5}a + \frac{16}{25}$

54. $a^2 - \frac{4}{3}a + \frac{16}{36}$

55. $x^2 + \frac{5}{7}x + \frac{9}{49}$

56. $4n^2 - \frac{7}{9}n + \frac{1}{81}$

Find the replacement for c that would make a perfect square trinomial.

57. $9x^2 + cxy + 49y^2$

58. $16m^2 + cmp + 25p^2$

59. $cx^2 + 28x + 49$

60. $cy^2 + 50y + 25$

61. $36 + cy^2 - 36y$

62. $cn^2 - 220n + 121$

63. $4x^2 + 4xy + c$

64. $9a^2 + 24ab + c$

65. $9x^2 - 12xy + c$

66. $64x^2 - 16xy + c$

Challenge Exercises

67. The measure of the area of a square is $x^2 - 20x + 100$. What is the measure of the length of a side?

68. The measure of the area of a square is $400 + y^2 + 40y$. What is the measure of its perimeter?

69. The measure of the area of a circle is $(4a^2 - 36ab^2 + 81b^4)\pi$. What is the measure of its diameter?

Find the GCF of the given monomials.

1. $81, 108$

2. $38ab^3, -74a^3b$

3. $4xy^2z^4, 64x^2z^2, 32x^2yz^3$

4. $55m^3n, 275m^5n^2, 165m^2n^3$

5. Write the compound sentence whose solution set is graphed.

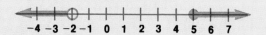

Simplify.

6. $(8x^2 - 7x + 3) - (6x^2 + 8)$

7. $(5x^2y)^3$

Multiply.

8. $2ab(5a^2 - 3ab + 7b^2)$

9. $(8t - 3)(2t + 5)$

10. $(2n^2 - 7m)^2$

11. $(4x + 5y^2)^2$

Factor.

12. $5x^2 - 80y^4$

13. $9a^2 - 30ab^2 + 25b^4$

Find the degree of each polynomial.

14. $13abc + 26a^2b^3$

15. $12n^4xy^3 - 4n^2x^3 + 2nxy^5$

Solve.

16. $\frac{1}{4}(12x - 8) = \frac{3}{2}(4 + 10x)$

17. $5(2x - 3) - 4x = 6x + 5$

18. $5.3x - 8.5 = -2.9x + 40.7$

19. $|3n + 27| \le 99$

20. $2x(x - 7) - 9 = (x - 3)(2x + 1) + 1$

21. $5t - (t + 3) < 6t + 7$

22. Solve $a\left(\frac{n}{3} - b\right) = 7 - 2ab$ for n.

23. Find the replacement for c that would make $25x^2 + 5cxy + 64y^2$ a perfect square trinomial.

Evaluate. Express each result in scientific notation and decimal notation.

24. $\frac{7.8 \times 10^4}{2.6 \times 10^2}$

25. $\frac{8.5 \times 10^{-3}}{1.7 \times 10^{-8}}$

Solve.

26. The area of a rectangle is 1287 square yards. Its length is 33 yards. What is its width?

27. Brian saves $26 in four weeks. At that rate, how much money can he save in 14 weeks?

28. If four times a number is decreased by 11, the result is between -15 and 29. What is the number?

29. Antonio is five years less than twice as old as Rebecca. The sum of their ages is 46 years. How old is each person?

30. Sue Gordon invested $11,700, part at 5% interest and the balance at 7% interest. If her annual earnings from both investments is $733, how much is invested at each rate?

31. Mr. and Mrs. Owens took a trip to Lake Teragram. On the way there, their average speed was 42 miles per hour. On the way home, their average speed was 56 miles per hour. If their total travel time was 7 hours, find the distance to the lake.

7-5 Factoring $x^2 + bx + c$

Recall that if two numbers are multiplied, each number is a *factor* of the product. Similarly, if two binomials are multiplied, each binomial is a *factor* of the product.

Use the FOIL method to find the product $(x + 3)(x + 7)$.

$$\begin{array}{cccc} \mathbf{F} & \mathbf{O} & \mathbf{I} & \mathbf{L} \end{array}$$

$$\begin{aligned} (x + 3)(x + 7) &= x \cdot x + x \cdot 7 + 3 \cdot x + 3 \cdot 7 \\ &= x^2 + 7x + 3x + 3 \cdot 7 \\ &= x^2 + (7 + 3)x + 3 \cdot 7 \\ &= x^2 + 10x + 21 \end{aligned}$$

21 is the product of 3 and 7.
10 is the sum of 3 and 7.

The binomials $(x + 3)$ and $(x + 7)$ are factors of $x^2 + 10x + 21$.

In general, the product of the binomials $(x + m)$ and $(x + n)$ can be found by using FOIL.

$$\begin{aligned} (x + m)(x + n) &= x^2 + nx + mx + mn \\ &= x^2 + (m + n)x + mn \end{aligned}$$

This pattern can be used to factor certain types of trinomials. For example, to factor $x^2 + 7x + 10$ into the form $(x + m)(x + n)$, find the factors of 10 whose sum is 7. Consider only positive factors of 10 since $(m + n)$ and mn are both positive.

$$mn = 10$$
$$m + n = 7$$

Factors of 10	Sum of Factors
1, 10	$1 + 10 = 11$
2, 5	$2 + 5 = 7$

Since the sum must be 7, the factors are 2 and 5.

$$x^2 + 7x + 10 = (x + 2)(x + 5) \qquad \textit{Check by using FOIL.}$$

Example

1 **Factor $y^2 + 8y + 12$.**

$$y^2 + 8y + 12 = y^2 + (m + n)y + mn \qquad m + n = 8 \text{ and } mn = 12$$

Find the factors of 12 whose sum is 8. Consider only positive integers.

Factors of 12	Sum of Factors
1, 12	$1 + 12 = 13$
2, 6	$2 + 6 = 8$
3, 4	$3 + 4 = 7$

The factors are 2 and 6. Therefore, $y^2 + 8y + 12 = (y + 2)(y + 6)$.

Check that $(y + 2)(y + 6) = y^2 + 8y + 12$ by using FOIL.

Example

2 **Factor $a^2 - 9a + 18$.**

$a^2 - 9a + 18 = a^2 + (m + n)a + mn$ $m + n = (-9)$ and $mn = 18$
Since $(m + n)$ is negative and mn is positive, the factors of 18 must *both* be *negative*. Do you see why?

Factors of 18	Sum of Factors
$-1, -18$	$-1 + (-18) = -19$
$-2, -9$	$-2 + (-9) = -11$
$-3, -6$	$-3 + (-6) = -9$

Therefore, $a^2 - 9a + 18 = (a - 3)(a - 6)$.   *Check by using FOIL.*

Notice that in the trinomial $x^2 + 5x - 6$ the value of c is *negative*. Since $x^2 + 5x - 6$ can be expressed as $x^2 + 5x + (-6)$, the same pattern can be used to factor trinomials in which c is negative.

Examples

3 **Factor $x^2 + 5x - 6$.**

$x^2 + 5x - 6 = x^2 + (m + n)x + mn$ where $m + n = 5$ and $mn = -6$. Since $(m + n)$ is positive and mn is negative, exactly one factor of each pair of factors below must be negative. Do you see why?

Factors of -6	Sum of Factors
$1, -6$	$1 + (-6) = -5$
$-1, 6$	$-1 + 6 = 5$
$2, -3$	$2 + (-3) = -1$
$-2, 3$	$-2 + 3 = 1$

Select the factors -1 and 6. Therefore, $x^2 + 5x - 6 = (x - 1)(x + 6)$.
Check by using FOIL.

4 **Factor $a^2 - 3a - 10$.**

$a^2 - 3a - 10 = a^2 + (m + n)a + mn$ $m + n = -3$ and $mn = -10$
$2 + (-5) = -3$ and $2(-5) = -10$.

Therefore, $a^2 - 3a - 10 = (a + 2)(a - 5)$.   *Check by using FOIL.*

When the value of c is negative, the factors of a trinomial are a *difference* and a *sum*. For $x^2 + 5x - 6$, the difference is $(x - 1)$ and the sum is $(x + 6)$. For $a^2 - 3a - 10$, the difference is $(a - 5)$ and the sum is $(a + 2)$.

Example

5 **Factor $14 + 13x + x^2$.**

The polynomial $14 + 13x + x^2$ may be written as $x^2 + 13x + 14$.
$x^2 + 13x + 14 = x^2 + (m + n)x + mn$ $m + n = 13$ and $mn = 14$

Factors of 14	Sum of Factors
1, 14	$1 + 14 = 15$
2, 7	$2 + 7 = 9$

Why should you test only positive factors?

There are no factors m and n such that $m + n = 13$ and $mn = 14$.
Therefore, $x^2 + 13x + 14$ cannot be factored using integers.

A polynomial that *cannot* be written as a product of two polynomials is called a **prime polynomial**. In Example 5 the trinomial $x^2 + 13x + 14$ is prime.

Example

6 **Find all values of k so that the trinomial $x^2 + kx + 8$ can be factored using integers.**

For $x^2 + kx + 8$, the value of k is the sum of the factors of 8.

Factors of 8	Sum of Factors
1, 8	$1 + 8 = 9$
2, 4	$2 + 4 = 6$
$-1, -8$	$-1 + (-8) = -9$
$-2, -4$	$-2 + (-4) = -6$

The values of k are 9, 6, -9, and -6.

Exploratory Exercises

For each polynomial, find integers m and n whose product is the last term and whose sum is the coefficient of the second term.

1. $x^2 + 15x + 14$
2. $y^2 + 9y + 14$
3. $b^2 - 8b + 12$
4. $x^2 - 19x + 34$
5. $p^2 + 5p - 36$
6. $r^2 - 10r - 24$
7. $y^2 - 7y - 30$
8. $g^2 + 5g - 24$
9. $x^2 + 18x + 45$
10. $t^2 - 19t + 48$
11. $c^2 - 17c - 60$
12. $a^2 + 13a - 48$

Complete the factoring.

13. $p^2 + 9p - 10 = (p + \blacksquare)(p - 1)$
14. $y^2 + 3y - 28 = (y + 7)(y \blacksquare 4)$
15. $x^2 - 2x - 35 = (x + 5)(x - \blacksquare)$
16. $a^2 - 4a - 21 = (a + \blacksquare)(a - 7)$
17. $h^2 + 6h - 16 = (h - 2)(h \blacksquare \blacksquare)$
18. $x^2 - 9x - 36 = (x - 12)(x \blacksquare \blacksquare)$
19. $a^2 + 4am - 21m^2 = (a - 3m)(a \blacksquare \blacksquare)$
20. $p^2 + 16rp + 63r^2 = (p \blacksquare \blacksquare)(p + 9r)$

Written Exercises

Factor each trinomial, if possible. If the trinomial cannot be factored using integers, write *prime*.

21. $y^2 + 12y + 27$

22. $a^2 + 22a + 21$

23. $c^2 + 2c - 3$

24. $x^2 + 2x - 15$

25. $y^2 - 8y + 15$

26. $a^2 - 12a + 35$

27. $m^2 - m - 20$

28. $x^2 - 5x - 24$

29. $h^2 + 5h + 8$

30. $c^2 + 3c + 6$

31. $z^2 - 10z - 39$

32. $y^2 - 7y + 60$

33. $r^2 - 10r - 24$

34. $y^2 - 9y - 36$

35. $y^2 + 8y - 20$

36. $n^2 + 12n - 45$

37. $p^2 + 10p + 25$

38. $g^2 + 15g + 26$

39. $b^2 - 11b + 28$

40. $x^2 - 9x + 14$

41. $15 + 16m + m^2$

42. $36 + 15a + a^2$

43. $y^2 + 11y + 28$

44. $x^2 + 14x + 33$

45. $a^2 - 18a - 40$

46. $b^2 + 13b - 30$

47. $x^2 - 3x - 18$

48. $y^2 - 7y - 30$

49. $36 - 16m + m^2$

50. $36 - 13y + y^2$

51. $m^2 - 6m - 55$

52. $h^2 - 3h - 108$

53. $66 - 17j + j^2$

54. $64 - 16c + c^2$

55. $42 - 23a + a^2$

56. $50 - 27x + x^2$

57. $c^2 - 2cd - 8d^2$

58. $x^2 - 4xy - 5y^2$

59. $a^2 + 2ab - 3b^2$

60. $n^2 + 4an - 32a^2$

61. $(a + b)^2 - 5(a + b) - 6$

62. $(c + q)^2 - 8(c + q) - 9$

Find all values of k so that each trinomial can be factored using integers.

63. $x^2 + kx + 10$

64. $m^2 + km + 6$

65. $r^2 + kr - 13$

66. $y^2 + ky - 4$

67. $n^2 + kn + 7$

68. $p^2 + kp + 17$

69. $a^2 + ka - 5$

70. $b^2 + kb - 16$

71. $t^2 + kt + 21$

72. $c^2 + kc + 12$

73. $s^2 + ks - 14$

74. $y^2 + ky - 15$

Excursions in Algebra   Factoring Cubes

You know how to factor the difference of two squares. Can you factor the difference of two cubes? Consider $a^3 - b^3$.

$$a^3 - b^3 = a^3 - a^2b + a^2b - b^3 \qquad \textit{Notice that } -a^2b + a^2b = 0.$$
$$= a^2(a - b) + b(a^2 - b^2)$$
$$= a^2(a - b) + b(a - b)(a + b)$$
$$= (a - b)[a^2 + b(a + b)]$$
$$= (a - b)(a^2 + ab + b^2)$$

Example   **Factor $y^3 - 27$.**

y^3 is the cube of y. 27 is the cube of 3. Now use the pattern above.

$$y^3 - 27 = (y - 3)(y^2 + 3y + 9)$$

Exercises

Factor using the pattern above.

1. $x^3 - 64$

2. $8x^3 - y^3$

3. $27a^3 - 64b^3$

4. Find a pattern for factoring $a^3 + b^3$ by using a method similar to the one above.

7-6 Factoring by Grouping

Some polynomials have four terms. It may be possible to factor these polynomials by first grouping the terms in pairs and factoring a monomial from each group. Then use the Distributive Property again to factor the common binomial.

Examples

1 **Factor $3xy - 21y + 5x - 35$.**

$3xy - 21y + 5x - 35 = (3xy - 21y) + (5x - 35)$ *Group pairs of terms that have a common monomial factor.*

$= 3y(x - 7) + 5(x - 7)$ *Factor the GCF from each group.*

$= 3y(x - 7) + 5(x - 7)$ *Notice that $(x - 7)$ is a common factor.*

$= (3y + 5)(x - 7)$ *Use the Distributive Property.*

Check by using **FOIL**.

$\qquad\qquad\qquad$ F $\qquad$ O $\qquad$ I $\qquad$ L

$(3y + 5)(x - 7) = 3y(x) + 3y(-7) + 5(x) + 5(-7)$

$\qquad\qquad\quad = 3xy - 21y + 5x - 35$ ✔

2 **Factor $8x^2y - 5x - 24xy + 15$.**

$8x^2y - 5x - 24xy + 15 = (8x^2y - 5x) + (-24xy + 15)$ *Why are the terms grouped this way?*

$= x(8xy - 5) + (-3)(8xy - 5)$ *What is the common factor?*

$= (x - 3)(8xy - 5)$

Check: $(x - 3)(8xy - 5) = x(8xy) + x(-5) + (-3)(8xy) + (-3)(-5)$

$\qquad\qquad\qquad\qquad\quad = 8x^2y - 5x - 24xy + 15$ ✔

Sometimes you can group the terms in more than one way. The following example shows another way to factor the polynomial in Example 2 above.

Example

3 **Factor $8x^2y - 5x - 24xy + 15$.**

$8x^2y - 5x - 24xy + 15 = (8x^2y - 24xy) + (-5x + 15)$

$= 8xy(x - 3) + (-5)(x - 3)$

$= (8xy - 5)(x - 3)$ *The result is the same as in Example 2.*

Recognizing binomials that are additive inverses is often helpful in factoring. For example, the binomials $(3 - a)$ and $(a - 3)$ are additive inverses since $(3 - a)$ and $-1(a - 3)$ are equivalent. What is the additive inverse of $(5 - y)$?

$-1(a - 3)$
$= (-1)(a) + (-1)(-3)$
$= -a + 3$
$= 3 - a$

Example

4 Factor $15x - 3xy + 4y - 20$.

$15x - 3xy + 4y - 20 = (15x - 3xy) + (4y - 20)$

$\qquad = 3x(5 - y) + 4(y - 5)$

$\qquad = -3x(y - 5) + 4(y - 5)$

$\qquad = (-3x + 4)(y - 5)$

The binomials $(5 - y)$ and $(y - 5)$ are additive inverses.

$(5 - y) = -1(y - 5)$.

Check: $(-3x + 4)(y - 5) \overset{?}{=} (-3x)(y) + (-3x)(-5) + 4(y) + 4(-5)$

$\qquad\qquad\qquad\quad \overset{?}{=} -3xy + 15x + 4y - 20$

$\qquad\qquad\qquad\quad = 15x - 3xy + 4y - 20$ ✔

Exploratory Exercises

Express each polynomial in factored form.

1. $k(r + s) - m(r + s)$

2. $y(m + n) - 4(m + n)$

3. $4b(x - y) + y(x - y)$

4. $t(t - s) + s(t - s)$

5. $7rp(r - 3p) - 4q(r - 3p)$

6. $3ab(a - 4) - 8(a - 4)$

7. $8m(x + y) + (x + y)$

8. $p(p - 3r) - (p - 3r)$

9. $5mp(2m + 3p) - 3bc(2m + 3p)$

10. $5z(5x^2 - 3y^2) + m^2(5x^2 - 3y^2)$

11. $4m(y - 5) + 3p(5 - y)$

12. $5y(z - 6) - 3m(6 - z)$

13. $7x(a + b) + 3m(a + b) - 4p(b + a)$

14. $3a(x - y) - 4b(y - x) + 5c(x - y)$

15. $a(8r - 3y) + b(8r - 3y) + c(3y - 8r)$

16. $x(4a^2 + 3b^2) - 3y(4a^2 + 3b^2) + z(3b^2 + 4a^2)$

Written Exercises

Find the common factor for the grouped binomials.

> **Sample** $(2a^2 + ab) + (4ac + 2bc)$
>
> $(2a^2 + ab) + (4ac + 2bc) = a(2a + b) + 2c(2a + b)$
>
> The common factor is $(2a + b)$.

17. $(bx + by) + (3ax + 3ay)$

18. $(3mx + 2my) + (3kx + 2ky)$

19. $(a^2 + 3ab) + (2ac + 6bc)$

20. $(rx + 2ry) + (kx + 2ky)$

21. $(a^2 - 4ac) + (ab - 4bc)$

22. $(10x^2 - 6xy) + (15x - 9y)$

23. $(4m^2 - 3mp) + (3p - 4m)$

24. $(16k^2 - 28kp) + (7p^2 - 4kp)$

25. $(6x^3 + 7x^2y) + (6x + 7y)$

26. $(5a + 2b) + (10a^2 + 4ab)$

27. $(3a^2b + 2ab^3) + (6ab + 4b^3)$

28. $(15a^2 - 21ab) + (20ab - 28b^2)$

Factor each polynomial. Check by using FOIL.

29. $2ax + 6xc + ba + 3bc$

30. $6mx - 4m + 3rx - 2r$

31. $ay + 4p + 4a + yp$

32. $3my - ab + am - 3by$

33. $2my + 7x + 7m + 2xy$

34. $zr + 6q + rq + 6z$

35. $ay - ab + cb - cy$

36. $3ax - 6bx + 8b - 4a$

37. $a^2 - 2ab + a - 2b$

38. $2ab + 2am - b - m$

39. $4ax + 3ay + 4bx + 3by$

40. $x^2 + 5xy + ax + 5ay$

41. $3m^2 - 5m^2p + 3p^2 - 5p^3$

42. $10x^2 - 14xy + 5xy - 7y^2$

43. $5a^2 - 4ab + 12b^3 - 15ab^2$

44. $4ax - 14bx + 35by - 10ay$

45. $6a^2 - 6ab + 3cb - 3ca$

46. $5a^2 + 5ab + 5cd + 5c^2$

47. $ax + a^2x - a - 2a^2$

48. $5xy + 15x - 6y - 18$

49. $7mp^2 + 2np^2 - 7mq^2 - 2nq^2$

50. $5a^3 + 3a^2b - 5ab^2 - 3b^3$

51. $x^3 + 2x^2 - x - 2$

52. $y^3 + y^2 - y - 1$

53. $a^3 - a^2b + ab^2 - b^3$

54. $2a^3 - 5ab^2 - 2a^2b + 5b^3$

55. $7m(a^2 - b^2) + 5y(b^2 - a^2)$

56. $3x(k^2 - m^2) - 4a(m^2 - k^2)$

57. $a^2(m - 2) + 5a(m - 2) + 6(m - 2)$

58. $x^2(a + 1) - 2x(a + 1) - 15(a + 1)$

59. $2a^2x + 3a^2y - 14ax - 21ay + 24x + 36y$

60. $ax^2 - 3bx^2 - 7ax + 21bx - 18a + 54b$

Excursions in Algebra   Grouping Three Terms

In the following example, the first three terms have been grouped to form a perfect square trinomial.

Factor $a^2 + 4a + 4 - 9b^2$.

$$a^2 + 4a + 4 - 9b^2 = (a^2 + 4a + 4) - 9b^2$$

The perfect square trinomial $a^2 + 4a + 4$ can be factored as $(a + 2)^2$.

$$= (a + 2)^2 - 9b^2$$

$$= [(a + 2) - 3b][(a + 2) + 3b]$$ *Use the difference of squares.*

$$= (a + 2 - 3b)(a + 2 + 3b)$$

Exercises

Factor each polynomial.

1. $x^2 + 4x + 4 - 25y^2$

2. $y^2 + 8y + 16 - 36z^2$

3. $x^2 + 2xy + y^2 - r^2$

4. $m^2 - k^2 + 6k - 9$

5. $9 - 9y^2 - x^2 + 6xy$

6. $c^2 - 10ab - a^2 - 25b^2$

7. $4a^2 + 12ab + 9b^2 - 16x^2$

8. $4a^2 + 4ax + x^2 - b^2$

Reading Algebra

When you read mathematics, do not expect to read rapidly. Mathematics is a concise language. Try reading aloud each of these statements.

Liquids have no definite shape.

$$x^2 + 5x + 4 = 0$$

The equation takes longer to read because the mathematical symbols are compact forms of words and concepts. In words, it would be read as:

x squared plus five x plus four equals zero.

The following concepts are involved.

x is a variable.

The raised 2 in x^2 means x is used as a factor twice.

$5x$ means 5 multiplied by x.

You must also know the meaning of +, =, 5, 4, and 0.

For understanding, you need to translate the symbols into words and concepts.

Suppose you read a section in this book. First read it rapidly to get a general idea of the content. Then reread it slowly so that you understand all of the ideas in the section. Turn to page 256 and read Lesson 7-8 rapidly to find the main ideas in that section.

The main ideas in the section are listed here.

The section gives a summary of the methods used to factor polynomials.

If there is a greatest common factor, factor it out first.

Continue factoring until each of the remaining factors is prime.

Now reread Lesson 7-8 carefully to find out how to factor a polynomial completely.

Exercises

Write each expression in words.

1. $3x^2 + 14x + 15$

2. $(3x + 5)(x + 3)$

3. $3xy - 6x + 5y - 10$

4. $(3x + 5)(y - z)$

5. $4x(x + 8) = 16$

6. $8x^3 - 27 = 0$

7. List the concepts contained in Exercise 1.

8. List the concepts contained in Exercise 2.

9. List the methods of factoring polynomials that are discussed in Lesson 7-8.

7-7 Factoring $ax^2 + bx + c$

In the trinomial $x^2 - 3x - 18$, the coefficient of x^2 is 1. To factor this trinomial, you find the factors of -18 whose sum is -3.

Consider the trinomial $2y^2 + 7y + 6$. In this trinomial the coefficient of y^2 is not 1. To factor a trinomial such as this one, another method is needed.

What is the coefficient of y^2 in $2y^2 + 7y + 6$?

Study how **FOIL** is used to multiply $(3x + 4)$ by $(2x + 7)$.

$$\overset{\textbf{F} \quad \textbf{O} \quad \textbf{I} \quad \textbf{L}}{(3x + 4)(2x + 7) = 6x^2 + 21x + 8x + 28}$$

$$= 6x^2 + (21 + 8)x + 28$$

$$21 \cdot 8 = 168$$

$$6 \cdot 28 = 168$$

Notice that the *product* of 6 and 28 is the same as the *product* of 21 and 8. You can use this method to factor the trinomial $2y^2 + 7y + 6$.

$2y^2 + 7y + 6$

$2y^2 + (\blacksquare + \blacksquare)y + 6$

The product of 2 and 6 is 12. You need to find two integers whose *product is 12* and whose *sum is 7*.

Factors of 12	Sum of Factors
1, 12	$1 + 12 = 13$
2, 6	$2 + 6 = 8$
3, 4	$3 + 4 = 7$

$2y^2 + (3 + 4)y + 6$ Select the factors 3 and 4.

$2y^2 + 3y + 4y + 6$

$(2y^2 + 3y) + (4y + 6)$ Group pairs of terms that have a common monomial factor.

$y(2y + 3) + 2(2y + 3)$ Factor the GCF from each group.

$(y + 2)(2y + 3)$ Use the Distributive Property.

Therefore, $2y^2 + 7y + 6 = (y + 2)(2y + 3)$.   *Check by using FOIL.*

The following example illustrates that the order in which the factors are selected does not affect the result.

Examples

1 **Factor $3x^2 + 11x + 6$.**

$3x^2 + 11x + 6$ *The product of 3 and 6 is 18.*

Factors of 18	Sum of Factors
1, 18	$1 + 18 = 19$
2, 9	$2 + 9 = 11$

You can stop listing factors when you find a pair that works.

$3x^2 + (\blacksquare + \blacksquare)x + 6$ *Select the factors 2 and 9.*

$3x^2 + (2 + 9)x + 6$	or $3x^2 + (9 + 2)x + 6$
$3x^2 + 2x + 9x + 6$	$3x^2 + 9x + 2x + 6$
$(3x^2 + 2x) + (9x + 6)$	$(3x^2 + 9x) + (2x + 6)$
$x(3x + 2) + 3(3x + 2)$	$3x(x + 3) + 2(x + 3)$ *Factor the GCF from each group.*
$(x + 3)(3x + 2)$	$(3x + 2)(x + 3)$ *Use the Distributive Property.*

Notice that $(x + 3)(3x + 2) = (3x + 2)(x + 3)$.

Therefore, $3x^2 + 11x + 6 = (x + 3)(3x + 2)$.   *Check by using FOIL.*

2 **Factor $5x^2 - 17x + 14$.**

$5x^2 - 17x + 14$ *The product of 5 and 14 is 70.*
Both factors of 70 must be negative. Why?

Factors of 70	Sum of Factors
$-1, -70$	$-1 + (-70) = -71$
$-2, -35$	$-2 + (-35) = -37$
$-5, -14$	$-5 + (-14) = -19$
$-7, -10$	$-7 + (-10) = -17$

$5x^2 + (\blacksquare + \blacksquare)x + 14$
$5x^2 + [-10x + (-7)x] + 14$
$(5x^2 - 10x) + (-7x + 14)$
$5x(x - 2) + (-7)(x - 2)$ *What is the common factor?*
$(5x - 7)(x - 2)$

Therefore, $5x^2 - 17x + 14 = (5x - 7)(x - 2)$.   *Check by using FOIL.*

3 **Factor $8y^2 - 6y - 9$.**

$$8y^2 - 6y - 9$$

The product of 8 and -9 is -72.
Exactly one factor of -72 in each pair must be negative.
Why? Since the sum is negative, the factor with the
greater absolute value is negative. The product of a positive
number and a negative number is negative.

Factors of -72	Sum of Factors
$1, -72$	$1 + (-72) = -71$
$2, -36$	$2 + (-36) = -34$
$3, -24$	$3 + (-24) = -21$
$4, -18$	$4 + (-18) = -14$
$6, -12$	$6 + (-12) = -6$

$8y^2 + (\blacksquare + \blacksquare)y - 9$
$8y^2 + [6y + (-12)y] - 9$
$(8y^2 + 6y) + (-12y - 9)$
$2y(4y + 3) + (-3)(4y + 3)$
$(2y - 3)(4y + 3)$

Therefore, $8y^2 - 6y - 9 = (2y - 3)(4y + 3)$. *Check by using FOIL.*

Exploratory Exercises

For each trinomial $ax^2 + bx + c$, supply the missing numbers so that the sum will equal b and the product will equal ac.

1. $3y^2 + 11y + 6 = 3y^2 + (\blacksquare + \blacksquare)y + 6$

2. $3c^2 + 14c + 8 = 3c^2 + (\blacksquare + \blacksquare)c + 8$

3. $4y^2 + 11y + 6 = 4y^2 + (\blacksquare + \blacksquare)y + 6$

4. $6b^2 + 31b + 5 = 6b^2 + (\blacksquare + \blacksquare)b + 5$

5. $4x^2 - 8x + 3 = 4x^2 + (\blacksquare + \blacksquare)x + 3$

6. $2y^2 - 11y + 15 = 2y^2 + (\blacksquare + \blacksquare)y + 15$

7. $2k^2 - 9k + 10 = 2k^2 + (\blacksquare + \blacksquare)k + 10$

8. $4a^2 - 25a + 6 = 4a^2 + (\blacksquare + \blacksquare)a + 6$

9. $5c^2 + 6c - 8 = 5c^2 + (\blacksquare + \blacksquare)c - 8$

10. $2x^2 + x - 21 = 2x^2 + (\blacksquare + \blacksquare)x - 21$

11. $5r^2 - 13r - 6 = 5r^2 + (\blacksquare + \blacksquare)r - 6$

12. $6a^2 + 7a - 3 = 6a^2 + (\blacksquare + \blacksquare)a - 3$

13. $3m^2 + 11m - 20 = 3m^2 + (\blacksquare + \blacksquare)m - 20$

14. $3d^2 - 5d - 28 = 3d^2 + (\blacksquare + \blacksquare)d - 28$

Written Exercises

Factor each trinomial, if possible. If the trinomial cannot be factored using integers write prime.

15. $3y^2 + 8y + 5$

16. $7a^2 + 22a + 3$

17. $3x^2 + 8x + 4$

18. $3a^2 + 14a + 15$

19. $8m^2 - 10m + 3$

20. $2y^2 - 7y + 3$

21. $2h^2 - h - 3$

22. $3y^2 + 5y - 2$

23. $3k^2 + 7k - 6$

24. $3m^2 - 7m - 6$

25. $6p^2 - p - 2$

26. $4b^2 + 5b - 6$

27. $2a^2 + 3a - 14$

28. $2x^2 + 5x - 12$

29. $6t^2 + 5t - 6$

30. $7n^2 - 22n + 3$

31. $2y^2 - 5y + 3$

32. $3x^2 + 4x - 15$

33. $2q^2 - 9q - 18$

34. $6y^2 - 11y + 4$

35. $6m^2 + 19m + 10$

36. $4y^2 - 17y - 15$

37. $6x^2 - 19x - 11$

38. $10n^2 - 19n + 7$

39. $12r^2 - 11r + 3$

40. $9k^2 - 12k + 4$

41. $8p^2 - 18p + 9$

42. $5b^2 - 13b - 10$

43. $15p^2 + 14p - 8$

44. $3t^2 - 32t + 20$

45. $6y^2 - 19y + 15$

46. $10k^2 - 11k - 6$

47. $6s^2 + 7s - 20$

48. $12b^2 + 17b + 6$

49. $18x^2 + 55x + 25$

50. $16m^2 + 14m - 15$

51. $18c^2 + 41c - 10$

52. $15y^2 + 17y - 18$

53. $18r^2 - 19r - 12$

54. $15x^2 - 13xy + 2y^2$

55. $8m^2 - 14mn + 3n^2$

56. $3s^2 - 10st - 8t^2$

57. $16x^2 - 16xy - 5y^2$

58. $16a^2 - 38ab - 5b^2$

59. $20p^2 + 11pq - 4q^2$

60. $25r^2 + 25rs + 6s^2$

61. $9k^2 + 30km + 25m^2$

62. $36a^2 + 9ab - 10b^2$

63. $20s^2 + 17st - 24t^2$

64. $3x^2 - 30xy + 56y^2$

65. $8x^2 - 42xq + 27q^2$

66. $12r^2 - 16rs - 11s^2$

67. $14c^2 + 41cd + 15d^2$

68. $21x^2 + 52xy + 32y^2$

69. $30a^2 + 47a - 88$

70. $99a^2 + 23a - 42$

71. $24y^2 + 23y + 6$

72. $6 + 29b + 35b^2$

73. $2 - 21y + 27y^2$

74. $100 - 100x + 21x^2$

75. $48a^2 - 26ab + 3b^2$

76. $24x^2 - 61xy + 35y^2$

77. $10a^2 - 34ab + 27b^2$

78. $6a^3 + 13a^2 + 5a$

79. $40x^4 - 116x^3 + 84x^2$

80. $16y^3 + 4y^2 - 6y$

81. $60t^3 - 65t^2 - 70t$

82. $12m^3n - 22m^2n^2 - 144mn^3$

83. $20a^4b - 59a^3b^2 + 42a^2b^3$

Challenge Exercises

Solve.

84. The measure of the area of a triangle is $7.5x^2 + 15.5x - 12$. What is the measure of its base if the measure of its height is $5x - 3$?

85. If the measure of the area of a circle is $(4x^2 + 28x + 49)\pi$, what is the measure of its diameter?

mini-review

Write each compound sentence without using *and*.

1. $-2.5 < k$ and $k < -1$

2. $m < 16$ and $m \geq -\frac{12}{5}$

Solve each problem if y varies directly as x.

3. If $y = 14$ when $x = 2$, find y when $x = 3$.

4. If $y = 6.25$ when $x = -8$, find x when $y = 13.5$.

Find the replacement for c that would make a perfect square trinomial.

5. $4x^2 + cx + 25$

6. $cm^2 + 84mn + 49n^2$

Solve each problem.

7. Deb Rogers invested $5500 for one year, part at 8% annual interest and the balance at 12% annual interest. Her total interest for the year was $580. How much money did she invest at each rate?

8. Jose Martinez needs to move a rock that weighs 725 pounds. If he has a 5-foot crow bar and uses a fulcrum placed one foot from the rock, will Jose be able to lift the rock if he weighs 182 pounds?

You can use a calculator to solve and check equations such as $0.35x + 6.789x = 15$. First, find the sum of 0.35 and 6.789 and store it.

ENTER: .35 [+] 6.789 [=] [STO]

DISPLAY: *0.35*   *6.789*   *7.139*   *7.139*

When you enter .35, the display shows 0.35.

Then, divide 15 by 7.139.   *You are dividing each side of the equation by 7.139.*

ENTER: 15 [÷] [RCL] [=]

DISPLAY: *15*   *7.139*   *2.1011346*

The solution is about 2.101.   *Use your calculator to check this solution.*

You can also use your calculator to check whether $\frac{1}{18}$ is a solution of $243y^2 = \frac{3}{4}$. First enter $\left(\frac{1}{18}\right)^2$ by using the reciprocal and square keys. Then store it.

ENTER: 18 [¹/ₓ] [x²] [STO]

DISPLAY: *18*   *.05555556*   *.00308642*   *.00308642*

Then, evaluate $243y^2$ for $y = \frac{1}{18}$ and compare it to the value of $\frac{3}{4}$.

ENTER: 243 [×] [RCL] [=]

DISPLAY: *243*   *.00308642*   *0.75*

ENTER: 3 [÷] 4 [=]

DISPLAY: *3*   *4*   *0.75*

Since $0.75 = 0.75$, the number $\frac{1}{18}$ is a solution of $243y^2 = \frac{3}{4}$.

Exercises

Solve and check each equation.

1. $0.02y + 3.87y = 45$
2. $6.09x + 0.089x = 9$
3. $35 = 56.7a - 60.02a$
4. $7.8y + 9.2y - 0.06y = -20$
5. $75 = 0.008c - 4.2c + 9.01c$
6. $2.5x + 7.3x - 5.06x = -1$
7. $4.6m + 56 = 9.3m - 2.05$
8. $15.9 - 56.7t = 28.05t + 91$
9. $79.5n = 35 - 0.006n + 13.8n$
10. $-9.1x - 4.78x - 63 = 53x$

Use a calculator to check whether the given number is a solution of the equation.

11. $128x^2 = 2; \frac{1}{8}$
12. $108y^2 = 3; \frac{1}{5}$
13. $144a^2 - 16 = 0; -\frac{1}{3}$
14. $256b^2 - 4 = 0; -\frac{1}{8}$
15. $0 = 169 - 100m^2; \frac{13}{10}$
16. $0 = 25y - 625y^3; -\frac{2}{5}$
17. $25x^2 - 30x = -9; \frac{3}{5}$
18. $49c^2 = -112c - 64; -\frac{8}{7}$
19. $4a^2 = -20a - 25; \frac{2}{5}$
20. $25 + 90x = -81x^2; -\frac{4}{9}$

7-8 Summary of Factoring

You have used many methods to factor polynomials. This table can help you decide which method to use.

Check for:	Number of Terms		
	Two	Three	Four or More
greatest common factor	✔	✔	✔
difference of squares	✔		✔
perfect square trinomial		✔	
trinomial that has two binomial factors		✔	
pairs of terms that have a common monomial factor			✔

If there is a GCF, factor it out first. Then, check the appropriate factoring methods in the order shown in the table. Using these methods, factor until each of the remaining factors is prime. When you follow this procedure, you are *factoring a polynomial completely.*

Examples

1 **Factor $3x^2 - 27$.**

Since $3x^2 - 27$ has two terms, first check for the GCF. Then check for the difference of squares.

$3x^2 - 27 = 3(x^2 - 9)$ *3 is the GCF.*
$\qquad\quad\;\; = 3(x - 3)(x + 3)$ *$x^2 - 9$ is the difference of squares.*

Check: $3(x - 3)(x + 3) = 3(x^2 - 9)$
$\qquad\qquad\qquad\qquad\quad = 3x^2 - 27$ ✔

The binomial $3x^2 - 27$ is completely factored as $3(x - 3)(x + 3)$.

2 **Factor $6y^2 - 20y - 16$.**

Since $6y^2 - 20y - 16$ has three terms, check for the GCF, a perfect square trinomial, and a trinomial that has two binomial factors.

$6y^2 - 20y - 16 = 2(3y^2 - 10y - 8)$ *2 is the GCF.*

The trinomial $3y^2 - 10y - 8$ is not a perfect square. Why?   *$3y^2$ and 8 are not*
Try factoring $3y^2 - 10y - 8$ using two binomial factors.   *perfect squares.*

$6y^2 - 20y - 16 = 2(3y^2 - 10y - 8)$
$\qquad\qquad\qquad\;\; = 2(y - 4)(3y + 2)$

The trinomial $6y^2 - 20y - 16$ is completely factored as $2(y - 4)(3y + 2)$.

Example

3 Factor $2m^3 + 3m^2n - 8m - 12n$.

Since $2m^3 + 3m^2n - 8m - 12n$ has four terms, first check for the GCF. Then, check for pairs of terms that have a common monomial factor.

$$2m^3 + 3m^2n - 8m - 12n = (2m^3 + 3m^2n) + (-8m - 12n) \quad \textit{The GCF is 1.}$$
$$= m^2(2m + 3n) + (-4)(2m + 3n)$$
$$= (m^2 - 4)(2m + 3n) \quad \textit{m}^2 - 4 \textit{ is the}$$
$$= (m - 2)(m + 2)(2m + 3n) \quad \textit{difference of squares.}$$

The polynomial $2m^3 + 3m^2n - 8m - 12n$ is completely factored as $(m - 2)(m + 2)(2m + 3n)$. *Check by multiplying the factors.*

Exploratory Exercises

Factor. Check by using FOIL or the Distributive Property.

1. $3x^2 + 15$
2. $4y^2 - 24$
3. $5ax + 6ay$
4. $8mn^2 - 13m^2n$
5. $12ax^2 + 18ay^2$
6. $4c^2d + 16bc^2$
7. $5x^2 - 10x^2y$
8. $4m^2 - 20m^2n$
9. $a^2 - 9b^2$
10. $2x^2y + 8x^2y^2 + 10x^2y^3$
11. $3a^2b + 6ab + 9ab^2$
12. $9x^2 - 16y^2$

Written Exercises

Factor. Check by using FOIL or the Distributive Property.

13. $2a^2 - 72$
14. $3y^2 - 147$
15. $m^3 + 6m^2 + 9m$
16. $18y + 12y^2 + 2y^3$
17. $4a^3 - 36a$
18. $3x^3 - 27x$
19. $m^4 - p^2$
20. $b^4 - 16$
21. $2k^2 + 3k + 1$
22. $6r^2 + 13r + 6$
23. $6y^2 - 24x^2$
24. $2m^2 - 32n^2$
25. $3y^2 + 21y - 24$
26. $6x^2 + 27x - 15$
27. $20y^2 + 34y + 6$
28. $5a^2 + 7a + 3$
29. $2b^2 + 6b + 2$
30. $3b^2 - 36$
31. $m^2 + 8mn + 16n^2$
32. $4a^2 + 12ab + 9b^2$
33. $6p^2 + 9p - 105$
34. $12c^2 + 10c - 42$
35. $3ax^2 + 16ax + 21a$
36. $4y^3 - 12y^2 + 8y$
37. $9a^3 + 66a^2 - 48a$
38. $2m^2 - mb - 10b^2$
39. $3x^2 - 9xy - 30y^2$
40. $a^2b^3 - 25b$
41. $m^3n^2 - 49m$
42. $2a^2 - 4ab - 70b^2$
43. $3x^2 + 24xy - 99y^2$
44. $3y^4 - 48$
45. $4x^4 - 324$
46. $m^2p + m^2 - 36p - 36$
47. $9y^4 + 8y^2 - 1$
48. $27k^2m + 27km^2 - 12m^3$
49. $60r^3 - 54r^2s + 12rs^2$
50. $8y^4 + 14y^2 - 4$
51. $x^2y^2 - z^2 - y^2 + x^2z^2$

Challenge Exercises

Factor by first factoring out a common decimal or fraction.

52. $0.3x^2 - 4.8$
53. $0.7y^2 - 6.3$
54. $\frac{1}{4}a^2 + a + 1$
55. $\frac{1}{3}b^2 + 2b + 3$
56. $y^2 + \frac{5}{12}y - \frac{1}{6}$
57. $\frac{1}{4}r^2 + \frac{3}{2}r + 2$
58. $0.4r^2 + 1.2r + 0.8$
59. $0.7y^2 + 3.5y + 4.2$
60. $m^2 - \frac{5}{12}m - \frac{1}{4}$

Bits and Bytes

The decimal numeration system is a base 10 system. The ten digits 0, 1, 2, 3, 4, 5, 6, 7, 8, and 9 are used to name numbers.

Powers of Two

$2^0 = 1$	$2^4 = 16$
$2^1 = 2$	$2^5 = 32$
$2^2 = 4$	$2^6 = 64$
$2^3 = 8$	$2^7 = 128$

Internally, a digital computer uses only two digits, 0 and 1. These two digits (bits) are used in a base two, or binary, numeration system.

In this system the position of each digit gives its place value as a power of 2. If 0 and 1 are used, then 11101 in base two means 1 sixteen, 1 eight, 1 four, 0 two's, and 1 one.

binary numeral $\rightarrow$

$$11101 = 1 \cdot 2^4 + 1 \cdot 2^3 + 1 \cdot 2^2 + 0 \cdot 2^1 + 1$$
$$= 16 + 8 + 4 + 0 + 1$$
$$= 29 \quad \leftarrow \quad \textit{decimal numeral}$$

To show which base is used, you can write $11101_{two} = 29_{ten}$.

The circuits of a computer store numbers in base 2. The diagram shows how a computer would store 1010_{two}.

A color spot indicates a circuit that conducts electricity. It represents 1.

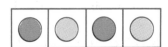

A grey spot indicates a circuit that does not conduct electricity. It represents 0.

Bits are the alphabet of machine language. The pattern formed by a group of bits stands for a number, and that number itself may stand for a character or even a whole word. Usually, eight bits make up one **byte**. Computer size is often stated in bytes of memory.

In mathematics and business $1K = 1000$; in a computer, $1K = 1024(2^{10})$. Many small personal computers have 64K bytes of memory, but some have 128K bytes or more.

Exercises

Change each binary numeral to a decimal numeral.

1. 1101_{two} **2.** 10110_{two} **3.** 11111_{two} **4.** 10101_{two}

Draw a diagram that shows how a computer would store each of the following.

5. 1100_{two} **6.** 1001_{two} **7.** 19_{ten} **8.** 33_{ten}

Vocabulary

factor   (229)

prime number   (229)

composite number   (229)

prime factor   (229)

prime factorization   (229)

greatest common factor, GCF   (230)

difference of squares   (235)

perfect square trinomial   (239)

prime polynomial   (245)

Chapter Summary

1. A prime number is an integer, greater than 1, whose only whole number factors are 1 and itself.   (229)

2. A composite number is an integer, greater than 1, that is not prime.   (229)

3. When a composite number is expressed as a product of factors that are all prime, the expression is called the prime factorization of the number. (229)

4. The greatest common factor (GCF) of two or more integers is the product of the prime factors common to each integer.   (230)

5. The Distributive Property may be used to express a polynomial in factored form.   (232)

6. Differences of Two Squares:
   $a^2 - b^2 = (a - b)(a + b) = (a + b)(a - b)$ (235)

7. The result of squaring an integer is called a perfect square.   (239)

8. The square of a binomial is a perfect square trinomial.   (239)

9. Factoring Perfect Square Trinomials:
   $a^2 + 2ab + b^2 = (a + b)^2$
   $a^2 - 2ab + b^2 = (a - b)^2$ (239)

10. The pattern $(x + m)(x + n) = x^2 + (m + n)x + mn$ can be used to factor trinomials of the form $x^2 + bx + c$.   (243)

11. When the value of c is negative, the factors of $x^2 + bx + c$ are a difference and a sum.   (244)

12. A polynomial that cannot be written as a product of two polynomials is called a prime polynomial.   (245)

13. It may be possible to factor a polynomial with four terms by first grouping the terms in pairs and factoring a monomial from each group. Then use the Distributive Property again to factor the common binomial.   (247)

14. To factor a trinomial of the form $ax^2 + bx + c$, first find two factors whose product is $a \cdot c$ and whose sum is b. Then, rename b as the sum of these factors. Continue to factor by grouping the terms in pairs and factoring the GCF from each group.   (251)

15. Summary of Factoring   (256)

Check for:	Number of Terms		
	Two	Three	Four or More
greatest common factor	✔	✔	✔
difference of squares	✔		✔
perfect square trinomial		✔	
trinomial that has two binomial factors		✔	
pairs of terms that have a common monomial factor			✔

Chapter Review

7–1 **Find the prime factorization of each monomial.**

1. 42

2. 66

3. 80

4. $96y^3x^2$

5. $121(ab)^4$

6. $16a^3b$

Find the GCF of the given monomials.

7. 35, 50

8. 12, 18, 40

9. $16abc$, $30a^2b$

Factor each polynomial. If a polynomial cannot be factored using integers, write prime.

7–2 **10.** $13x + 26y$

11. $6x^2y + 12xy + 6$

12. $24a^2b^2 - 18ab$

13. $24ab + 18ac + 32a^2$

14. $\frac{3}{5}a - \frac{3}{5}b + \frac{6}{5}c$

15. $\frac{3}{4}x + \frac{1}{4}y$

7–3 **16.** $b^2 - 16$

17. $25 - 9y^2$

18. $16a^2 - 81b^4$

19. $2y^2 - 128$

20. $\frac{1}{4}y^2 - \frac{9}{16}z^2$

21. $x^4 - 1$

7–4 **22.** $a^2 + 18a + 81$

23. $16x^2 - 8x + 1$

24. $9x^2 - 12x + 4$

25. $32x^2 - 80x + 50$

26. $6a^2 - 24a + 6$

27. $y^2 - \frac{3}{2}y + \frac{9}{16}$

7–5   **28.** $y^2 + 7y + 12$ **29.** $b^2 - 8b + 15$ **30.** $y^2 - 9y - 36$

        31. $b^2 + 5b - 6$ **32.** $r^2 + 23r + 132$ **33.** $a^2 - 3a - 28$

        34. $x^2 - 13x + 30$ **35.** $s^2 + 10s - 39$ **36.** $x^2 + 11xy + 24y^2$

        37. $m^2 - mn - 42n^2$ **38.** $a^2 - 10ab + 9b^2$ **39.** $r^2 - 8rs - 65s^2$

7–6   **40.** $a^2 - 4ac + ab - 4bc$ **41.** $4m^2 - 3mp + 3p - 4m$

        42. $2rs + 6ps + rm + 3mp$ **43.** $bm + 7r + rm + 7b$

        44. $10x^2 - 6xy + 15x - 9y$ **45.** $16k^2 - 4kp - 28kp + 7p^2$

7–7   **46.** $3y^2 + 14y + 8$ **47.** $4m^2 + 11m + 6$ **48.** $4y^2 - 7y + 3$

        49. $2r^2 - 3r - 20$ **50.** $3a^2 - 13a + 14$ **51.** $4m^2 - 8m + 3$

        52. $6b^2 + 7b + 3$ **53.** $3x^2 + 11x - 20$ **54.** $12x^2 + 31x + 20$

7–8   **55.** $3x^2 - 12$ **56.** $12mx + 3xb + 4my + by$ **57.** $15ay^2 + 37ay + 20a$

        58. $2r^3 - 18r^2 + 28r$ **59.** $mx^2 + bx^2 - 49m - 49b$ **60.** $4y^2 + 2y - 6$

Chapter Test

Find the prime factorization of each monomial.

  1. 95   **2.** 300   **3.** $18a^2b$

Find the GCF of the given monomials.

  4. 36, 20   **5.** $18a^2b, 28a^3b^2$ **6.** $6x^2y^3, 12x^2y^2z, 15x^2y$

Factor each polynomial. If a polynomial cannot be factored using integers, write *prime*.

  7. $25y^2 - 49w^2$ **8.** $w^2 - 16w + 64$ **9.** $x^2 + 14x + 24$

10. $36a^2 - 48a + 16$ **11.** $y^2 - 8y + 7$ **12.** $a^2 - 11ab + 18b^2$

13. $28m^2 + 18m$ **14.** $m^2 + 7m + 10$ **15.** $5x^2 + 20y^2$

16. $h^2 - 3h - 10$ **17.** $8m^2 - 72n^2$ **18.** $r^2 + 3r - 18$

19. $x^2 + 18x + 81$ **20.** $a^2 - 12a - 13$ **21.** $r^2 - rs - 72s^2$

22. $36m^2 - 24mn + 4n^2$ **23.** $3x^2 + 19x + 28$ **24.** $5x^2 - 19x + 14$

25. $36m^2 + 60mn + 25n^2$ **26.** $192a^2 - 75b^2$ **27.** $6p^2 + 7p - 3$

28. $16a^2b^2 - c^2d^2$ **29.** $2m^2 + 11m + 12$ **30.** $36a^2b^3 - 45ab^4$

31. $4m^2 - 196$ **32.** $6x^3 + 15x^2 - 9x$

33. $3a^2b - 8a - 12ab - 32$ **34.** $4my - 20m + 15p - 3py$

35. $15a + 6b + 10a^2 + 4ab$ **36.** $y^2m + y^2n - 4m - 4n$

37. $16a^4 - 1$ **38.** $3x^2 + 15x + 5$

CHAPTER 8

Applications of Factoring

A certain fireworks rocket is set off at an average velocity of 440 feet per second. This type of fireworks is designed to explode at a height of 3000 feet. How many seconds after it is set off will it reach 3000 feet and explode?

In this chapter, you will learn how factoring can be used to solve problems such as this.

8-1 Solving Equations Already in Factored Form

Consider the following products.

$$3(0) = 0 \qquad 0(-8) = 0 \qquad \left[\tfrac{1}{2} + \left(-\tfrac{1}{2}\right)\right](0) = 0 \qquad (0)(x + 2) = 0$$

The above products all equal zero. Notice that in each case *at least one* of the factors is zero.

For all numbers a and b, if $ab = 0$, then $a = 0$, $b = 0$, or both a and b equal 0.	***Zero Product Property***

You can use this property to solve equations already written in factored form.

Examples

1 **Solve $2x(x - 5) = 0$. Check the solution.**

If $2x(x - 5) = 0$, then $2x = 0$ or $x - 5 = 0$.   *Zero Product Property*
$2x = 0 \qquad$ or $\qquad x - 5 = 0$ *Solve each equation.*
$x = 0 \qquad$ or $\qquad x = 5$

Check: $\qquad\qquad 2x(x - 5) = 0$
$\qquad 2(0)(0 - 5) \overset{?}{=} 0 \qquad$ or $\qquad 2(5)(5 - 5) \overset{?}{=} 0 \qquad$ *Substitute 0 and 5 for*
$\qquad\qquad 0(-5) \overset{?}{=} 0 \qquad\qquad\qquad\quad 10(0) \overset{?}{=} 0 \qquad$ *x into the original equation.*
$\qquad\qquad\qquad 0 = 0 \ \checkmark \qquad\qquad\qquad\qquad 0 = 0 \ \checkmark$

The solutions of $2x(x - 5) = 0$ are 0 and 5. The solution set is $\{0, 5\}$.

2 **Solve $(y + 2)(3y + 5) = 0$. Check the solution.**

If $(y + 2)(3y + 5) = 0$, then $y + 2 = 0$ or $3y + 5 = 0$.
$y + 2 = 0 \qquad$ or $\qquad 3y + 5 = 0 \qquad$ *Zero Product Property*
$\qquad y = -2 \qquad\qquad\qquad 3y = -5$
$\qquad\qquad\qquad\qquad\qquad\qquad\quad y = -\tfrac{5}{3}$

Check: $\qquad\qquad (y + 2)(3y + 5) = 0$
$(-2 + 2)[3(-2) + 5] \overset{?}{=} 0 \quad$ or $\quad \left(-\tfrac{5}{3} + 2\right)\left[3\left(-\tfrac{5}{3}\right) + 5\right] \overset{?}{=} 0$

$\qquad 0[3(-2) + 5] \overset{?}{=} 0 \qquad\qquad\quad \left(-\tfrac{5}{3} + 2\right)(-5 + 5) \overset{?}{=} 0$

$\qquad\qquad\qquad 0 = 0 \ \checkmark \qquad\qquad\qquad\qquad \left(-\tfrac{5}{3} + 2\right)(0) \overset{?}{=} 0$

$\qquad\qquad\qquad\qquad\qquad\qquad\qquad\qquad\qquad\qquad 0 = 0 \ \checkmark$

The solution set is $\left\{-2, -\tfrac{5}{3}\right\}$.

Example

3   Maria gave this puzzle to her friends. "The product of 4 times my age and 45 less than 3 times my age is 0. How old am I?" Find Maria's age now.

explore   This problem can be solved by using an equation.
Let y = Maria's age now.

plan $4y(3y - 45) = 0$

solve $4y = 0$ or $3y - 45 = 0$
$\quad\quad y = 0$ $\quad\quad\quad\quad\quad 3y = 45$
$\quad\quad\quad\quad\quad\quad\quad\quad\quad\quad y = 15$

examine   Although 0 is a solution of $4y(3y - 45) = 0$, Maria cannot be 0 years old. Therefore, disregard 0 as a solution. Maria is 15 years old.

Exploratory Exercises

State the conditions under which each equation will be true.
Sample: $x(x + 7) = 0$ will be true if $x = 0$ or $x + 7 = 0$.

1. $x(x + 3) = 0$
2. $y(y - 12) = 0$
3. $3r(r - 4) = 0$
4. $7a(a + 6) = 0$
5. $3t(4t - 32) = 0$
6. $2x(5x - 10) = 0$
7. $(x - 6)(x + 4) = 0$
8. $(b - 3)(b - 5) = 0$
9. $(a + 3)(3a - 12) = 0$
10. $(2y + 22)(y - 1) = 0$
11. $(2y + 8)(3y + 24) = 0$
12. $(4x + 4)(2x + 6) = 0$
13. $(x - 3)(x - 3) = 0$
14. $(3x - 9)(5x - 15) = 0$
15. $(3x + 2)(x - 7) = 0$
16. $(x - 8)(2x + 7) = 0$
17. $(4x - 7)(3x + 5) = 0$
18. $(3x - 5)(4x - 7) = 0$
19. $(2x + 3)(x + 7) = 0$
20. $\left(4x - \frac{1}{3}\right)\left(3x + \frac{1}{8}\right) = 0$

Written Exercises

21–40. Solve each equation in Exploratory Exercises 1–20. Check the solutions.

Solve each problem. Disregard unreasonable solutions.

41. Ramón told Jenny, "The product of twice my age decreased by 32 and 5 times my age is 0. How old am I?" Find Ramón's age.

42. The product of a certain positive number decreased by 5 and the same number increased by 7 is 0. What is the number?

43. The product of a certain negative number increased by 2 and the same number decreased by $\frac{3}{4}$ is 0. What is the number?

44. The result is 0 when 7 is multiplied by the sum of 6 and a certain number. What is the number?

Challenge Exercises

Write an equation in factored form that would have each given solution set.

45. $\{-3, 5\}$
46. $\{0, -6\}$
47. $\{-9, -11\}$
48. $\left\{2, -\frac{2}{3}\right\}$
49. $\left\{-\frac{3}{4}, -\frac{5}{6}\right\}$
50. $\left\{0, -7, -\frac{1}{3}\right\}$

A useful problem-solving strategy is to organize relevant information by using a table. Study the following problem.

> **The cost to mail a small package includes a certain amount for the first ounce and 17¢ for each additional ounce. It costs 90¢ to mail a 5-ounce package. How much would it cost to mail a 10-ounce package?**

You know that it costs 90¢ to mail a 5-ounce package. Write this specific case in your table. You also know that it costs 17¢ more for each additional ounce. Use this information to generate more cases.

Number of Ounces	1	2	3	4	5	6	7	8	9	10
Cost (in dollars)					0.90	1.07	1.24	1.41	1.58	1.75

Therefore, it would cost $1.75 to mail a 10-ounce package.

Using the pattern in the table, you could find the cost of mailing a 1-ounce package.

It costs 0.90 − 0.17 or 0.73 to mail a 4-ounce package.

Number of Ounces	1	2	3	4	5	6	7	8	9	10
Cost (in dollars)	0.22	0.39	0.56	0.73	0.90	1.07	1.24	1.41	1.58	1.75

It costs 22¢ to mail a 1-ounce package. The following expression represents the cost of mailing an n-ounce package.

cost of first ounce + (number of ounces − first ounce) 0.17

$$0.22 + (n - 1)0.17$$

Exercises

Use a table to solve each problem.

1. Bob Hartschorn was charged $2.12 for a 20-minute call to Chicago. If Bob was charged $2.32 for a 22-minute call, how much should he be charged for an hour call?

2. Find the sum of the first three consecutive odd integers. What is the sum of the first five consecutive odd integers? What is the sum of the first fifty consecutive odd integers?

3. Find pairs of positive integers whose sum is 45 and whose difference is less than 9.

4. Find the positive integers less than 50 that have an odd number of factors.

5. To determine a grade point average, four points are given for an A, three for a B, two for a C, one for a D, and zero for an F. If Greg has a total of 13 points for 5 classes, what combinations of grades could he have?

6. Mei tore a sheet of paper in half. Then she tore each of the resulting pieces in half. If she continued this process 15 more times, how many pieces of paper would she have at the end?

8-2 Factoring Review

Equations can be easily solved if the equation is factorable. You have used several methods to factor polynomials. A summary of these is given in the following chart.

Check for:	Number of Terms		
	Two	Three	Four or More
greatest common factor	✔	✔	✔
difference of squares	✔		✔
perfect square trinomial		✔	
trinomial that has two binomial factors		✔	
pairs of terms that have a common monomial factor			✔

Examples

1 Factor $2x^3 - 50x$.

$2x^3 - 50x = 2x(x^2 - 25)$ *2x is the GCF.*

$\qquad\qquad = 2x(x - 5)(x + 5)$ *$x^2 - 25$ is the difference of squares.*

2 Factor $5a^2x^2 - 30ax^2 + 45x^2$.

$5a^2x^2 - 30ax^2 + 45x^2 = 5x^2(a^2 - 6a + 9)$ *$5x^2$ is the GCF.*

$\qquad\qquad\qquad = 5x^2(a - 3)(a - 3)$ *$a^2 - 6a + 9$ is a perfect square trinomial.*

$\qquad\qquad\qquad = 5x^2(a - 3)^2$

3 Factor $6m^3 - 11m^2n - 10mn^2$.

$6m^3 - 11m^2n - 10mn^2 = m(6m^2 - 11mn - 10n^2)$ *m is the GCF.*

$\qquad\qquad\qquad = m(2m - 5n)(3m + 2n)$

4 Factor $4ax^2 - 36ay^2 - 5bx^2 + 45by^2$.

$4ax^2 - 36ay^2 - 5bx^2 + 45by^2 = (4ax^2 - 36ay^2) - (5bx^2 - 45by^2)$

$\qquad\qquad\qquad = 4a(x^2 - 9y^2) - 5b(x^2 - 9y^2)$

$\qquad\qquad\qquad = (x^2 - 9y^2)(4a - 5b)$

$\qquad\qquad\qquad = (x - 3y)(x + 3y)(4a - 5b)$

Exploratory Exercises

Indicate which method of factoring you would apply first to each polynomial. Use the table from the previous page.

1. $x^3 - 5x^2$

2. $64c^2 - 25p^2$

3. $9x^2 - 12xy + 4y^2$

4. $16m^4 - 64n^2$

5. $35z^2 + 13z - 12$

6. $x^2 + y^2$

7. $8x^2 - 72x + 162$

8. $a^2 + 8a + 16$

9. $1 - 49k^2$

10. $a^5 - a^3b^2$

11. $x^2 - 5x + 6$

12. $x^2 - (y + z)^2$

13. $4x^4 - 3x^3y - 12x^2 + 9xy$

14. $6a^2 + a - 35$

15. $9a^2 - 30a + 25$

16. $3x^2 - xy - 6x + 2y$

17. $4m^2 + 40m + 100$

18. $6a^2 - 24b^2$

19. $2x^2 + 4xy - x - 2y$

20. $28m^2 + 17mn - 3n^2$

Written Exercises

21–40. Factor each polynomial in Exploratory Exercises 1–20.

Factor each polynomial.

41. $6x^2 + 10xy$

42. $35ab - 15b^3$

43. $4a^2 - 9b^2$

44. $25v^4 - 36w^4$

45. $x^2 + 12x + 36$

46. $a^2 - 8a + 16$

47. $12y^2 + 19y - 21$

48. $20m^2 + 17mn - 24n^2$

49. $x^3 + 5x^2 - 9x - 45$

50. $4x^3 + 12x^2 - x - 3$

51. $m^4 - m^2n^2$

52. $x^6 - x^4y^4$

53. $5x^2 + 90x + 405$

54. $6x^2 - 96x + 384$

55. $20a^2c^2 + 60a^2c + 45a^2$

56. $48a^2m^2 - 24a^2m + 3a^2$

57. $15x^3y - 6x^2y^3 + 3x^2yz$

58. $5ab^2c + 5ab^4 - 15a^3b^2$

59. $12x^2 - 3y^2$

60. $45x^4 - 5y^4$

61. $8x^3 - 32xy^4$

62. $48a^5y - 147ay^7$

63. $x^3 + 2x^2 + 8x + 16$

64. $4a^3 + 3a^2 + 8a + 6$

65. $12x^2 - 12xy - 72y^2$

66. $10ab^2 + 20abc - 80ac^2$

67. $(x + y)^2 - (a - b)^2$

68. $(3a + 2b)^2 - 9(x + 3y)^2$

69. $6x^2 + 2xy - 4y^2$

70. $10x^2 + 29xy + 10y^2$

71. $\frac{x^4}{9} - 49y^2$

72. $36a^2 - \frac{b^4}{100}$

73. $\frac{m^2}{4} - \frac{9}{25}$

74. $\frac{16}{49} - \frac{a^2}{9}$

75. $6ax - 21ay + 10bx - 35by$

76. $20ax - 4ay - 15bx + 3by$

77. $2x^2 - 7x - 8$

78. $2x^2 - 7x - 4$

79. $x^2 + 6x + 9 - y^2$

80. $a^4 - b^4 - 2b^2c^2 - c^4$

81. $0.01x^2 - 0.06xy + 0.09y^2$

82. $0.15x^4 - 0.019x^2y + 0.0006y^2$

83. $(x + 1)^2 - 3(x + 1) + 2$

84. $(2x - 3)^2 - 4(2x - 3) - 5$

Challenge Exercises

Factor each polynomial.

85. $0.12m^3n^2 - 0.001m^2n$

86. $\frac{3}{5}a^2b - \frac{4}{5}a^3b^2$

87. $\frac{1}{4}x^2 - \frac{3}{5}xy + \frac{9}{25}y^2$

88. $\frac{m^2}{3} + \frac{2}{3}m - \frac{35}{3}$

89. $x^4 - 2x^2y + 3x^3 - 6xy$

90. $4p^2m - 5p^2n - 8pm + 10pn$

mini-review

Simplify.

1. $[(-2^3)]^4$

2. $\left(\frac{4}{3}x^3y^0\right)^2$

3. $(3n)^3 + (6n)(-5n^2)$

Multiply.

4. $5x^2y(3xy^3 + 6z)$

5. $(x + 2)(2x^2 + 5x - 1)$

6. $(2x + 7)^2$

Factor each polynomial.

7. $y^2 - 6y - 72$

8. $9x^4 - 25$

9. $12x^2 - 22x - 144$

Solve each problem.

10. How many pounds of peaches costing 88¢ per pound must be added to 15 pounds of peaches costing 65¢ per pound to create a mixture that would sell for 75¢ per pound?

11. Shannon weighs 126 pounds and Jacob weighs 154 pounds. They are seated at opposite ends of a seesaw. Shannon and Jacob are 16 feet apart, and the seesaw is balanced. How far is Shannon from the fulcrum?

Excursions in Algebra   Factoring

Use a calculator to find the prime factorization of each number.

1. 8658
2. 18060
3. 24,030
4. 240,075
5. 1,337,050
6. 4,952,920

Use a calculator to factor each polynomial.

7. $1764x^2 - 121$

8. $2450k^2 - 392$

9. $11,532m^3 + 32,364m^2 + 22,707m$

10. $0.070225t^2 - 0.06519t + 0.015129$

Use a calculator to find the replacement for c that would make a perfect square trinomial.

11. $784x^2 + cxy + 81y^2$

12. $1681m^2 + 3034m + c$

13. $ck^2 - 1584k + 1089$

14. $-1638xy + cx^2 + 169y^2$

Applications in Banking

Many formulas are used daily in business and science. One such formula is used in finding **compound interest**. Compound interest is interest that is added to an original investment at specified time intervals. Thus, the interest becomes part of a new principal.

Suppose you deposited $500 in a savings account for 3 years at a rate of 6% annual interest.

	amount deposited +	interest	= new principal
End of first year	$500 +	$500(0.06)	= $530
End of second year	$530 +	$530(0.06)	= $561.80
End of third year	$561.80 +	$561.80(0.06)	= $595.51 *This result is approximate.*

To write a general formula, let P represent the amount deposited, or principal, and let r represent the annual interest rate. Then, $P \cdot r$ represents the amount of interest earned.

End of first year
$$P + P \cdot r = P(1 + r)^1 \qquad \textit{Factor out P.}$$

End of second year
$$P(1 + r) + P(1 + r) \cdot r = P(1 + r)(1 + r) \qquad \textit{Factor out P(1 + r).}$$
$$= P(1 + r)^2$$

End of third year
$$P(1 + r)^2 + P(1 + r)^2 \cdot r = P(1 + r)^2(1 + r) \qquad \textit{Factor out P(1 + r)}^2.$$
$$= P(1 + r)^3$$
$$\vdots$$

End of t years
$$P(1 + r)^t$$

Therefore, at the end of t years the total amount of savings including interest could be presented by $P(1 + r)^t$.

Suppose that interest is compounded biannually for 1 year at an annual interest rate of 6%. The interest would be computed twice per year at 3% each time. Thus the total amount in savings could be presented by:

$$P\left(1 + \frac{0.06}{2}\right)^{2 \cdot 1} \longleftarrow \textit{interest computed twice per year for 1 year}$$
$$\underbrace{} \textit{interest computed every six months at } \tfrac{0.06}{2} \textit{ or 0.03}$$

Exercises

Use a calculator to solve each problem.

1. Alonso invested $750 for 1 year at 8% annual interest. If interest is compounded biannually, how much will Alonso have at the end of the year?

2. Ana invested $600 in a savings account for 2 years. If the annual interest rate is 12% compounded biannually, what is the total amount Ana will have at the end of 2 years?

3. For interest compounded quarterly for 2 years at 12% annual interest the expression is $P\left(1 + \frac{0.12}{4}\right)^{4 \cdot 2}$. If Jim invested $600 in such an account, how much will he have at the end of 2 years?

4. In Exercises 2 and 3, Ana's money was compounded biannually and Jim's was compounded quarterly. Who had more money at the end of 2 years? How much more?

8-3 Solving Some Equations by Factoring

If a polynomial equation can be written in the form $ab = 0$, then the Zero Product Property can be applied to solve the equation.

Examples

1 **Solve $x^2 - 9 = 0$. Check the solution.**

$$x^2 - 9 = 0 \qquad \textit{Factor } x^2 - 9 \textit{ using the difference of squares.}$$
$$(x - 3)(x + 3) = 0$$
$$x - 3 = 0 \qquad \text{or} \qquad x + 3 = 0 \qquad \textit{Use the Zero Product Property.}$$
$$x = 3 \qquad\qquad\qquad x = -3$$

Check $\qquad\qquad\qquad x^2 - 9 = 0$

$$(3)^2 - 9 \stackrel{?}{=} 0 \qquad \text{or} \qquad (-3)^2 - 9 \stackrel{?}{=} 0$$
$$9 - 9 \stackrel{?}{=} 0 \qquad\qquad\qquad 9 - 9 \stackrel{?}{=} 0$$
$$0 = 0 \quad \checkmark \qquad\qquad\qquad 0 = 0 \quad \checkmark$$

The solution set is $\{3, -3\}$.

2 **Solve $y^2 = -7y$. Check the solution.**

To use the Zero Product Property, one side of the equation must be zero.

$$y^2 = -7y$$
$$y^2 + 7y = 0$$
$$y(y + 7) = 0 \qquad \textit{Factor out the GCF, y, to write the equation in the form ab = 0.}$$
$$y = 0 \qquad \text{or} \qquad y + 7 = 0$$
$$y = -7$$

Check: $\qquad\qquad y^2 = -7y$

$$(0)^2 \stackrel{?}{=} -7(0) \qquad \text{or} \qquad (-7)^2 \stackrel{?}{=} -7(-7)$$
$$0 = 0 \quad \checkmark \qquad\qquad\qquad 49 = 49 \quad \checkmark$$

The solution set is $\{0, -7\}$.

3 **Solve $\frac{2}{3}x^3 - 3x = 0$.**

$$\frac{2}{3}x^2 - 3x = 0$$
$$x\left(\frac{2}{3}x - 3\right) = 0$$
$$x = 0 \qquad \text{or} \qquad \frac{2}{3}x - 3 = 0$$
$$\frac{2}{3}x = 3$$
$$x = \frac{9}{2} \qquad \textit{Check the result.}$$

The solution set is $\left\{0, \frac{9}{2}\right\}$.

Examples

4 **Solve $a^2 + 64 = -16a$.**

To use the Zero Product Property, one side of the equation must be zero.

$$a^2 + 64 = -16a$$
$$a^2 + 16a + 64 = 0 \qquad \text{\textit{Use the fact that $a^2 + 16a + 64$ is a perfect square}}$$
$$(a + 8)^2 = 0 \qquad \text{\textit{trinomial to write the equation in the form ab = 0.}}$$
$$(a + 8)(a + 8) = 0$$
$$a + 8 = 0 \qquad \text{or} \qquad a + 8 = 0$$
$$a = -8 \qquad\qquad\qquad a = -8$$

The solution set is $\{-8\}$.   *Why?*

5 **The product of twice a number increased by five and three times the number decreased by two is zero. Find the number.**

explore   Let $n =$ the number.

plan $(2n + 5)(3n - 2) = 0$

solve
$$2n + 5 = 0 \qquad \text{or} \qquad 3n - 2 = 0$$
$$2n = -5 \qquad\qquad\qquad 3n = 2$$
$$n = -\tfrac{5}{2} \qquad\qquad\qquad n = \tfrac{2}{3}$$

examine   *Check* $(2n + 5)(3n - 2) = 0$

$$\left[2\left(-\tfrac{5}{2}\right) + 5\right]\left[3\left(-\tfrac{5}{2}\right) - 2\right] \stackrel{?}{=} 0 \quad \text{or} \quad \left[2\left(\tfrac{2}{3}\right) + 5\right]\left[3\left(\tfrac{2}{3}\right) - 2\right] \stackrel{?}{=} 0$$
$$(-5 + 5)\left(-\tfrac{15}{2} - 2\right) \stackrel{?}{=} 0 \qquad\qquad \left(\tfrac{4}{3} + 5\right)(2 - 2) \stackrel{?}{=} 0$$
$$0\left(-\tfrac{19}{2}\right) \stackrel{?}{=} 0 \qquad\qquad\qquad \left(\tfrac{19}{3}\right)(0) \stackrel{?}{=} 0$$
$$0 = 0 \ \checkmark \qquad\qquad\qquad 0 = 0 \ \checkmark$$

Both solutions check. The number is $-\tfrac{5}{2}$ or $\tfrac{2}{3}$.

Exploratory Exercises

Express each equation in factored form so that the Zero Product Property can be applied. Do not solve.

1. $n^2 - 3n = 0$

2. $y^2 - 6y = 0$

3. $8c^2 + 32c = 0$

4. $9d^2 + 36d = 0$

5. $3x^2 - \tfrac{3}{4}x = 0$

6. $x^2 + \tfrac{5}{3}x = 0$

7. $7y^2 = 14y$

8. $8y^2 = -4y$

9. $-13y = -26y^2$

10. $x^2 = 9$

11. $y^2 - 16 = 0$

12. $x^2 + 4x + 4 = 0$

13. $y^2 - 16y + 64 = 0$

14. $x^2 - 8x = -16$

15. $y^2 = 10y - 25$

Written Exercises

Solve each equation. Check the solution.

16. $x^2 - 6x = 0$

17. $m^2 + 36m = 0$

18. $2x^2 + 4x = 0$

19. $4x^2 - 9 = 0$

20. $a^2 - 81 = 0$

21. $16x^2 = 64$

22. $t^2 - 15 = 10$

23. $x^2 + 22 = 58$

24. $2x^2 - 205 = 37$

25. $4s^2 = -36s$

26. $y^2 = 7y$

27. $x^2 = -8x$

28. $6y = -3y^2$

29. $x^2 = 36$

30. $y^2 = 64$

31. $y^2 = -5y$

32. $\frac{1}{2}y^2 - \frac{1}{4}y = 0$

33. $3y^2 - \frac{4}{3} = 0$

34. $\frac{2}{3}y = \frac{1}{3}y^2$

35. $5x^2 = 45$

36. $2y^2 - 98 = 0$

37. $4y^2 - 16 = 0$

38. $m^2 - 24m + 144 = 0$

39. $y^2 + 10y = -25$

40. $25x^2 + 20x = -4$

41. $81n^2 + 36n = -4$

42. $9y^2 = -42y - 49$

43. $y^2 - \frac{16}{81} = 0$

44. $\frac{5}{6}x^2 = \frac{2}{3}x$

45. $\frac{3}{4}x^2 - \frac{1}{8}x = 0$

Use an equation to solve each problem. Disregard any unreasonable solutions.

46. The square of a number decreased by 144 is zero. Find the number.

47. The square of a number added to 6 times the number is zero. Find the number.

48. Eleven less than the square of a number is 38. Find the number.

49. The square of a number is equal to 10 times the number decreased by 25. Find the number.

50. Jake is as many years less than 30 as Jackie is more than 30. The product of their ages is 884. How old is each?

51. The product of Fernando's age and 6 less than his age is 135. Find Fernando's age.

52. A certain number increased by 12 is multiplied by the same number decreased by 12. The product is -63. Find the number.

53. A certain number decreased by 7 is multiplied by the same number increased by 7. The product is 51. Find the number.

54. The square of a number increased by 1070 is 1359. Find the number.

55. If the square of a number is subtracted from 900, the result is 275. Find the number.

Challenge Exercises

56. The height h in meters of a rocket with initial velocity v, t seconds after blast-off, is given by the formula $h = vt - 5t^2$. A toy rocket is launched with an initial velocity of 80 meters/second. How many seconds after blast-off will it hit the ground?

57. The measure of the area of a square is $4x^2 - 28x + 49$. Find the value of x if the measure of the perimeter of the square is 60.

58. A tinsmith has a rectangular piece of tin with a 3-inch square cut from each corner. After folding up the sides to form a box, she notices the box is twice as long as it is wide and has a volume of 1350 in³. Find the length and width of the box.

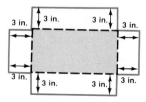

Evaluate each expression if $x = 4$, $y = -5$, and $z = \frac{2}{3}$.

1. $2xy^2$
2. $4x - 12yz$
3. $(3x + y)^2$

State the property shown.

4. $6 \cdot 0 = 0$
5. $3(a + 5) = 3a + 15$
6. $7 + (3 + 4) = (7 + 3) + 4$

Simplify.

7. $2a + 5(a - 4)$
8. $\frac{1}{2}(4a + 6b) - \frac{4}{3}(27a - 15b)$
9. $\frac{24}{-\frac{4}{5}}$
10. $(4a^2b)(-3b^3)$
11. $-2(ax^3y)^3$
12. $\frac{-64a^3b^5c^4d}{-16a^2b^6c^3d}$

Factor completely.

13. $25x^4 + 5y^2$
14. $121x^4 - 81y^2$
15. $b^2 - b - 20$
16. $12x^2 + 26x + 12$
17. $8x^2 + 6x - 35$
18. $x^3 + 3x^2 - x - 3$
19. $2a^2 + 13ab + 15b^2$

Solve each equation.

20. $m - (-12) = 24$
21. $\frac{x}{-4} + 3 = -11$
22. $2|5x - 3| = 28$
23. $(3y + 2)(y - 5) = 0$
24. $5m^2 + 5m = 0$
25. $a^3 + 6a^2 + 9a = 0$
26. $n^2 + n = 2$
27. Solve $5y - 7a = 3b$ for y.

Solve each inequality. Then graph its solution set.

28. $m + 14 < 6$
29. $5x - 1 \geq 3x + 7$
30. $-8r > 88$
31. $|y| > 4$
32. $|3t + 1| \leq 5$

Solve.

33. When 7 is subtracted from another integer, the difference is -16. What is the other integer?

34. Susan plans to spend at most $50 on shorts and tops. She bought 2 pairs of shorts for $14.20 each. How much can she spend on tops?

35. Dolores earned $340 in 4 days. At that rate how long will it take her to earn $935?

36. Ogima paid $96.70 for a special computer program. The price included $7\frac{1}{2}\%$ sales tax. What was the price without the tax?

37. Sid Sitkoff invested $12,200, part at 5% interest and the remainder at $6\frac{1}{2}\%$ interest. His total annual earnings from both investments were $670. How much did he invest at each rate?

38. The length of a rectangle is 12 meters greater than the width. If the length was decreased by 5 meters and the width increased by 2 meters, the area would be decreased by 55 square meters. Find the original dimensions of the rectangle.

39. Rosemary Villa drove from home to New York City at 48 miles per hour and returned at 40 miles per hour. The round trip took her $5\frac{1}{2}$ hours. How far is her home from New York City?

40. Sue is 6 years younger than Becky. The sum of their ages is 42. What are their ages?

8-4 Solving More Equations by Factoring

Some equations of second degree or higher can be written as a product of factors equal to zero. The Zero Product Property can be applied to solve these equations.

Example

1 **Solve $2x^2 = 11x - 12$. Check the solution.**

$$2x^2 = 11x - 12$$

$2x^2 - 11x + 12 = 0$	*Rewrite the equation.*
$(2x - 3)(x - 4) = 0$	*Factor $2x^2 - 11x + 12$.*
$2x - 3 = 0$ or $x - 4 = 0$	*Use the Zero Product Property.*
$x = \dfrac{3}{2}$ or $x = 4$	

Check:
$$2x^2 = 11x - 12$$

$$2\left(\tfrac{3}{2}\right)^2 \overset{?}{=} 11\left(\tfrac{3}{2}\right) - 12 \qquad or \qquad 2(4)^2 \overset{?}{=} 11(4) - 12$$

$$2\left(\tfrac{9}{4}\right) \overset{?}{=} \tfrac{33}{2} - \tfrac{24}{2} \qquad\qquad\qquad 2(16) \overset{?}{=} 44 - 12$$

$$\tfrac{9}{2} = \tfrac{9}{2} \; ✔ \qquad\qquad\qquad\qquad 32 = 32 \; ✔$$

The solution set of $2x^2 = 11x - 12$ is $\left\{\tfrac{3}{2}, 4\right\}$.

You may have to use more than one method of factoring to solve an equation.

Example

2 **Solve $x^3 - 24x = 5x^2$. Check the solution.**

$$x^3 - 24x = 5x^2$$

$x^3 - 5x^2 - 24x = 0$	*Arrange terms in descending order.*
$x(x^2 - 5x - 24) = 0$	*Factor out the GCF, x.*
$x(x - 8)(x + 3) = 0$	*Factor $x^2 - 5x - 24$.*
$x = 0$ or $x - 8 = 0$ or $x + 3 = 0$	
$x = 0$ or $x = 8$ or $x = -3$	

Check:
$$x^3 - 24x = 5x^2$$

$$0^3 - 24(0) \overset{?}{=} 5(0)^2 \qquad 8^3 - 24(8) \overset{?}{=} 5(8)^2 \qquad (-3)^3 - 24(-3) \overset{?}{=} 5(-3)^2$$

$$0 - 0 \overset{?}{=} 0 \qquad\qquad 512 - 192 \overset{?}{=} 5(64) \qquad\qquad -27 + 72 \overset{?}{=} 5(9)$$

$$0 = 0 \; ✔ \qquad\qquad\quad 320 = 320 \; ✔ \qquad\qquad\qquad 45 = 45 \; ✔$$

The solution set of $x^3 - 24x = 5x^2$ is $\{0, 8, -3\}$.

In Example 2, if you divide each side of $x(x^2 - 5x - 24) = 0$ by x, the result would not be an equivalent equation. This follows since zero is one of the solutions. Dividing by x would mean dividing by zero and division by zero is not defined. For this reason, when solving equations, do not divide by expressions that contain a variable.

Exploratory Exercises

Solve each equation.

1. $(a + 7)(a - 3) = 0$

2. $(x - 6)(x + 5) = 0$

3. $(r - 8)(r - 2) = 0$

4. $(b + 12)(b - 3) = 0$

5. $(2y + 3)(y + 1) = 0$

6. $(3z - 1)(z + 2) = 0$

7. $(3x - 4)(2x - 1) = 0$

8. $(2m - 1)(5m + 3) = 0$

9. $7(k - 5)(k + 3) = 0$

10. $4(s + 7)(s - 6) = 0$

11. $x(2x + 7)(3x + 4) = 0$

12. $z(5z + 4)(3z - 11) = 0$

Written Exercises

Solve each equation. Check the solutions.

13. $x^2 + 13x + 36 = 0$

14. $y^2 + 19y + 34 = 0$

15. $a^2 + 4a - 21 = 0$

16. $b^2 - 8b - 33 = 0$

17. $s^2 + 2s - 63 = 0$

18. $y^2 - 64 = 0$

19. $r^2 - 49 = 0$

20. $c^2 + 17c + 72 = 0$

21. $x^2 - 12x = 0$

22. $k^2 - 13k = 0$

23. $p^2 = 5p + 24$

24. $r^2 = 18 + 7r$

25. $2t^2 + 7t = 15$

26. $3y^2 - 7y = 20$

27. $6z^2 + 5 = -17z$

28. $12m^2 + 3 = -20m$

29. $2x^2 + 13x = 24$

30. $3y^2 + 16y = 35$

31. $m^3 - 81m = 0$

32. $4b^3 - 36b = 0$

33. $r^3 - 6r^2 + 8r = 0$

34. $s^3 + 2s^2 - 35s = 0$

35. $5b^3 + 34b^2 = 7b$

36. $2k^3 + 5k^2 = 42k$

37. $2y^3 - 15y = y^2$

38. $6x^3 - 7x = 11x^2$

39. $\frac{x^2}{12} - \frac{2}{3}x - 4 = 0$

40. $x^2 - \frac{1}{6}x - \frac{35}{6} = 0$

41. $0.03x^2 - x + 3 = 0$

42. $0.3x^2 - 1.06x + 0.2 = 0$

43. $(x + 8)(x + 1) = -12$

44. $(r - 1)(r - 1) = 36$

45. $(3y + 2)(y + 3) = y + 14$

46. $(y + 4)(3y - 2) = -y - 14$

47. $h^3 + h^2 - 4h - 4 = 0$

48. $y^3 - y^2 - y + 1 = 0$

49. $9a^3 - 18a^2 - a + 2 = 0$

50. $2m^3 + 5m^2 - 18m - 45 = 0$

51. $y^4 - 8y^2 + 16 = 0$

52. $m^4 - 2m^2 + 1 = 0$

53. $xy + 4x - 3y - 12 = 0$

54. $2my + 5m + 8y + 20 = 0$

55. $3rs - 4r + 6s - 8 = 0$

56. $4pz - z + 12p - 3 = 0$

Challenge Exercises

Write an equation with integral coefficients in the form $ax^2 + bx + c = 0$ that has the given solutions.

57. $\{2, 3\}$

58. $\{-3, 5\}$

59. $\left\{\frac{2}{3}, -1\right\}$

60. $\{-2, 2, 5\}$

8-5 Problem Solving: Integer Problems

Some types of integer problems can be solved by using factoring.

Examples

1 **Find two consecutive integers whose product is 72.**

explore This problem can be solved by using an equation.
Let x = one integer.
Then $x + 1$ = the next greater integer.

plan $x(x + 1) = 72$

solve
$$x(x + 1) = 72$$
$$x^2 + x = 72$$
$$x^2 + x - 72 = 0$$
$$(x + 9)(x - 8) = 0$$

$x + 9 = 0$ or $x - 8 = 0$
     $x = -9$ or $x = 8$

If $x = -9$, then $x + 1 = -8$.
If $x = 8$, then $x + 1 = 9$.

examine Since $-9(-8) = 72$ and $8 \cdot 9 = 72$, the two consecutive integers can be -9 and -8 or 8 and 9.

2 **The sum of the squares of two consecutive odd integers is 130. Find the integers.**

explore Let x = one odd integer.
Then $x + 2$ = the next greater odd integer.

plan $x^2 + (x + 2)^2 = 130$

solve
$$x^2 + (x + 2)^2 = 130$$
$$x^2 + x^2 + 4x + 4 = 130$$
$$2x^2 + 4x - 126 = 0$$
$$2(x^2 + 2x - 63) = 0 \qquad \textit{Factor out the GCF, 2.}$$
$$x^2 + 2x - 63 = 0 \qquad \textit{Divide each side by 2.}$$
$$(x + 9)(x - 7) = 0$$

$x + 9 = 0$ or $x - 7 = 0$
     $x = -9$ or $x = 7$

If $x = -9$, then $x + 2 = -7$.
If $x = 7$, then $x + 2 = 9$.

examine Since $(-9)^2 + (-7)^2 = 130$ and $(7)^2 + (9)^2 = 130$, the two consecutive odd integers can be -9 and -7 or 7 and 9.

Exploratory Exercises

Define a variable and state an equation for each sentence.

1. The product of two consecutive integers is 110.

2. The product of two consecutive integers is 156.

3. The product of two consecutive even integers is 168.

4. The product of two consecutive odd integers is 143.

5. The sum of two integers is 15 and their product is 44.

6. The sum of two integers is 22 and their product is 117.

7. The sum of the squares of two consecutive integers is 181.

8. The sum of the squares of two consecutive even integers is 100.

9. The difference of the squares of two consecutive integers is 17.

10. The difference of the squares of two consecutive even integers is 52.

Written Exercises

11–20. Solve each problem in Exploratory Exercises 1–10.

For each problem below, define a variable. Then use an equation to solve the problem.

21. Find two consecutive even integers whose product is 120.

22. Find two consecutive even integers whose product is 360.

23. Find two consecutive positive odd integers whose product is 195.

24. Find two consecutive odd integers whose product is 399.

25. Find two consecutive positive integers whose product is 182.

26. Find two consecutive positive integers whose product is 272.

27. Find two integers whose sum is 11 and whose product is 24.

28. Find two integers whose sum is 19 and whose product is 60.

29. Find two integers whose difference is 3 and whose product is 88.

30. Find two integers whose difference is 23 and whose product is -120.

31. The sum of the squares of two consecutive positive odd integers is 202. Find the integers.

32. The sum of the squares of two consecutive positive integers is 113. Find the integers.

33. The sum of two integers is 13. The sum of their squares is 97. Find the integers.

34. The sum of two integers is 3. The sum of their squares is 185. Find the integers.

35. When one integer is added to the square of the next consecutive integer, the sum is 41. Find the integers.

36. When one integer is added to the square of the next consecutive integer, the sum is 55. Find the integers.

37. When the square of the second of two consecutive even integers is added to twice the first integer, the sum is 76. Find the integers.

38. When the square of the second of two consecutive even integers is added to twice the first integer, the sum is 116. Find the integers.

39. Find 3 consecutive odd integers if the difference of the squares of the least and greatest is 120.

40. Lyle and Connie Fisher's ages are consecutive odd integers. The product of their ages is 399. If Lyle is older, find their ages.

8-6 Problem Solving: Velocity and Area

If an object is launched from ground level, it reaches its maximum height in the air at a time halfway between launch and impact times. Its height above the ground after t seconds is given by the formula $h = vt - 16t^2$. In this formula, h represents the height of the object in feet, and v represents the initial velocity in feet per second. Thus, the height of an object with an initial velocity of 144 feet per second is given by the formula $h = 144t - 16t^2$.

Example

1 **A flare is launched from a life raft with an initial velocity of 144 feet per second. How many seconds will it take for the flare to return to the sea?**

explore   Use the formula $h = vt - 16t^2$.

The variable h represents the height of the flare in feet when it returns to the sea. Thus, $h = 0$.

The variable v represents the initial velocity of the flare, in feet per second.
Thus, $v = 144$.

plan   Substitute the appropriate values into the formula.

solve
$$0 = 144t - 16t^2 \qquad \text{\textit{t is the time in seconds.}}$$
$$0 = 16t(9 - t)$$
$$16t = 0 \quad \text{or} \quad 9 - t = 0$$
$$t = 0 \quad \text{or} \quad 9 = t$$

examine   The flare returns to the sea in 9 seconds. The answer 0 is not reasonable since it represents the time when the flare is launched.

You can also use factoring to solve some area problems. Recall that the area of a rectangle is equal to the product of its length and width.

2 Pat Bing has a photograph that is 8 cm long and 6 cm wide. Pat wants to reduce the length and width of the photo by the same amount. She also wants the reduced photo to have half the area of the original photo. By what amount should she reduce the length and width?

Let x = the amount the length and width should be reduced.

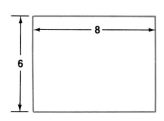

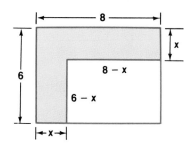

The dimensions of the reduced photo will be $(8 - x)$ cm and $(6 - x)$ cm.

The area of the photo is $8 \cdot 6$, or 48 cm².

The area of the reduced photo will be $\frac{48}{2}$, or 24 cm².

$$\text{length} \cdot \text{width} = \text{area}$$
$$(8 - x)(6 - x) = 24 \qquad \textit{Substitute the appropriate values into the formula.}$$
$$48 - 8x - 6x + x^2 = 24 \qquad \textit{Solve the equation.}$$
$$x^2 - 14x + 48 = 24$$
$$x^2 - 14x + 24 = 0$$
$$(x - 12)(x - 2) = 0$$
$$x - 12 = 0 \quad \text{or} \quad x - 2 = 0$$
$$x = 12 \quad \text{or} \quad x = 2 \qquad \textit{Why is 12 cm not a reasonable answer?}$$

Pat should reduce the length and width by 2 cm each.
The dimensions of the reduced photo will be $8 - 2$, or 6 cm and $6 - 2$, or 4 cm.

Written Exercises

For each problem, define a variable. Then use an equation to solve the problem.

1. The area of Jane Redfern's living room is 40 m². The length of the room is 3 m more than the width. What are its dimensions?

2. The length of Mrs. Boland's garden is 5 yards more than its width. The area of the garden is 234 square yards. What are its dimensions?

3. A rectangle is 4 in. wide and 7 in. long. When the length and width are increased by the same amount, the area is increased by 26 in². What are the dimensions of the new rectangle?

4. A photo is 8 cm wide and 12 cm long. The length and width are increased by an equal amount in order to double the area of the photo. What are the dimensions of the new photo?

5. Mr. Steinborn wants to double the area of his garden by adding a strip of uniform width along each of the four sides. The original garden is 10 ft by 15 ft. How wide a strip must be added?

6. A strip of uniform width is plowed along both sides and both ends of a garden 120 ft by 90 ft. How wide is the strip if the garden is half plowed?

Use $h = vt - 16t^2$ to find the missing quantity.

7. Find v, if $t = 5$ seconds and $h = 480$ feet.

8. Find v, if $t = 8$ seconds and $h = 32$ feet.

9. Find h, if $t = 7$ seconds and $v = 1700$ feet per second.

10. Find h, if $t = 2$ seconds and $v = 110$ feet per second.

11. Find two values for t, if $v = 160$ feet per second and $h = 336$ feet.

12. Find two values for t, if $v = 120$ feet per second and $h = 224$ feet.

Use the formula $h = vt - 16t^2$ to solve each problem.

13. A flare is launched from a life raft with an initial velocity of 192 feet per second. How many seconds will it take for the flare to return to the sea?

14. A golf ball is hit into the air with an initial velocity of 64 feet per second. How many seconds will it take for the golf ball to hit the ground?

15. A missile is fired with an initial velocity of 2320 feet per second. When will it be 40,000 feet high?

16. A rocket is fired with an initial velocity of 1920 feet per second. When will it be 32,000 feet high?

17. A rocket is fired with an initial velocity of 1640 feet per second. When will it be 816 feet high?

18. A flare is launched with an initial velocity of 128 feet per second. How many seconds will it take the flare to return to the sea?

mini-review

Simplify.

1. $(5x^2 + 3x - 16) + (-2x^2 - 16x + 7)$

2. $\left(-\frac{1}{4}x^2 + \frac{1}{3}x + 5\right) - \left(\frac{3}{4}x^2 + \frac{2}{3}x - 7\right)$

Factor each polynomial.

3. $4x^3 + 12xy$

4. $9a^2 + 42a + 49$

5. $72a^3 - 2a$

Solve.

6. $10a + 6a - 13 = 14a + 9$

7. $6 + 11x = 5(2 - 3x)$

8. $x^2 = 49$

9. $x^2 - 7x + 12 = 0$

10. $(b - 10)(b - 7) = -2$

Solve.

11. A car uses 8.5 liters of gasoline to travel 95 km. How much gasoline will the car use to travel 250 km?

12. Find two consecutive positive odd integers if the difference of their squares is 72.

Use the formula $h = vt - 16t^2$ to solve each problem.

13. Find the height of a rocket 4 seconds after launch if its initial velocity is 760 feet per second.

14. A ball is tossed directly upward with an initial velocity of 120 feet per second. How many seconds will it take for the ball to hit the ground?

Excursions in Algebra

When you solve a polynomial equation by factoring, the Zero Product Property is used.

If $ab = 0$, then $a = 0$, $b = 0$, or a and b both equal 0.

Now consider the inequality $ab > 0$. For what values of a and b will this inequality be true? It will be true only when a and b are either both positive or both negative.

If $ab > 0$, then either $a > 0$ and $b > 0$, or $a < 0$ and $b < 0$.

Example   Solve $(x + 6)(x - 2) > 0$ and graph the solution set.

$x + 6 > 0$ and $x - 2 > 0$ or $x + 6 < 0$ and $x - 2 < 0$

$x > -6$ and $x > 2$ or $x < -6$ and $x < 2$

$x > 2$ or $x < -6$ *Why?*

The solution set is $\{x \mid x > 2 \text{ or } x < -6\}$.

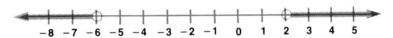

Consider the inequality $ab < 0$. Either a must be positive and b negative or b must be positive and a negative.

If $ab < 0$, then either $a > 0$ and $b < 0$, or $a < 0$ and $b > 0$.

Example   Solve $(x + 7)(x - 8) < 0$ and graph the solution set.

$x + 7 > 0$ and $x - 8 < 0$ or $x + 7 < 0$ and $x - 8 > 0$

$x > -7$ and $x < 8$ or $\underbrace{x < -7 \text{ and } x > 8}$

This can never be true.

The compound sentence $x > -7$ and $x < 8$ can be written $-7 < x < 8$.
The solution set is $\{x \mid -7 < x < 8\}$.

Exercises

Solve each inequality and graph the solution set.

1. $x(x + 4) > 0$
2. $x(x - 3) > 0$
3. $(x + 2)(x - 3) \geq 0$
4. $x(x - 5) < 0$
5. $(x + 6)(x - 3) \leq 0$
6. $x^2 + 5x + 6 > 0$
7. $x^2 - 2x - 15 \geq 0$
8. $x^2 + 6x + 5 < 0$
9. $y^2 < 9$
10. $y^2 + 2y > 35$
11. $x^2 > 36$
12. $x^2 - 3x \leq -2$

8-7 Simplifying Rational Expressions

Factoring polynomials is also a useful tool for simplifying **algebraic fractions**, or **rational expressions**.

Rational numbers and algebraic fractions are similar. Recall that any rational number can be expressed as the quotient of two integers.

Rational expressions contain variables and can be expressed as the quotient of two polynomials. The expressions $\frac{36a^2bc}{24b^2}$, $\frac{2x}{x-5}$, and $\frac{p^2-25}{p^2+4p+1}$ are examples of rational expressions.

Assume that the denominator does not equal zero.

To simplify a rational number such as $\frac{6}{21}$, first factor the numerator and denominator. Then divide each by the greatest common factor.

$$\frac{6}{21} = \frac{2 \cdot 3}{7 \cdot 3}$$ *The greatest common factor is 3.*

$$= \frac{2 \cdot 3}{7 \cdot 3} \quad \text{or} \quad \frac{2}{7} \qquad \text{Notice that } \tfrac{3}{3} = 1.$$

Note that the product of the common factors is the greatest common factor. (GCF)

The same procedure can be used to simplify rational expressions.

Examples

1 Simplify $\frac{42y^2}{18xy}$.   *Assume the denominator does not equal zero.*

$$\frac{42y^2}{18xy} = \frac{2 \cdot 3 \cdot 7 \cdot y \cdot y}{2 \cdot 3 \cdot 3 \cdot x \cdot y} \qquad \text{Factor the numerator and denominator.}$$

$$= \frac{\overset{1}{2} \cdot \overset{1}{3} \cdot 7 \cdot y \cdot \overset{1}{y}}{\underset{1}{2} \cdot \underset{1}{3} \cdot 3 \cdot x \cdot \underset{1}{y}} \qquad GCF = 2 \cdot 3 \cdot y \text{ or } 6y$$

$$= \frac{7y}{3x}$$

2 Simplify $\frac{y^2-9}{y^2-y-6}$.

$$\frac{y^2-9}{y^2-y-6} = \frac{(y+3)(y-3)}{(y-3)(y+2)}$$

$$= \frac{(y+3)\overset{1}{(y-3)}}{\underset{1}{(y-3)}(y+2)}$$

$$= \frac{y+3}{y+2}$$

What is the GCF?

3 Simplify $\frac{a+2}{a^2-4}$.

$$\frac{a+2}{a^2-4} = \frac{(a+2) \cdot 1}{(a+2)(a-2)}$$

$$= \frac{\overset{1}{(a+2)} \cdot 1}{\underset{1}{(a+2)}(a-2)}$$

$$= \frac{1}{a-2}$$

What is the GCF?

Exploratory Exercises

For each expression, find the greatest common factor (GCF) of the numerator and denominator. Then simplify. Assume that no denominator is equal to zero.

1. $\frac{24}{72}$

2. $\frac{99}{132}$

3. $\frac{42y}{18xy}$

4. $\frac{13x}{39x^2}$

5. $\frac{-3x^2y^5}{18x^5y^2}$

6. $\frac{14y^2z}{49yz^3}$

7. $\frac{x(y+1)}{x(y-2)}$

8. $\frac{4a}{a(a+7)}$

9. $\frac{m+5}{2(m+5)}$

10. $\frac{(a+b)(a-b)}{(a-b)(a-b)}$

11. $\frac{z+3}{z^2-9}$

12. $\frac{y-4}{y^2-16}$

Written Exercises

Simplify. Assume that no denominator is equal to zero.

13. $\frac{38a^2}{42ab}$

14. $\frac{79a^2b}{158a^3bc}$

15. $\frac{a+b}{a^2-b^2}$

16. $\frac{x+y}{x^2-y^2}$

17. $\frac{c^2-4}{(c+2)^2}$

18. $\frac{y^2-81}{(y-9)^2}$

19. $\frac{a^2-a}{a-1}$

20. $\frac{m^2-2m}{m-2}$

21. $\frac{x^2+4}{x^4-16}$

22. $\frac{y^2+6}{y^4-36}$

23. $\frac{2x^2+6}{2x+6}$

24. $\frac{3n^2-9}{3n-9}$

25. $\frac{-4y^2}{2y^2-4y^3}$

26. $\frac{3a^3}{3a^3+6a^2b}$

27. $\frac{-4a^2}{a^3+6a^2b+9ab}$

28. $\frac{2a^3-8a}{am^3+am^2+am}$

29. $\frac{x+y}{x^2+2xy+y^2}$

30. $\frac{x+6}{x^2+7x+6}$

31. $\frac{x-3}{x^2+x-12}$

32. $\frac{x+5}{x^2+7x+10}$

33. $\frac{m^2-16}{m^2+8m+16}$

34. $\frac{n^2-y^2}{n^2-2ny+y^2}$

35. $\frac{y^2-9}{y^2+6y+9}$

36. $\frac{y^2+8y-20}{y^2-4}$

37. $\frac{x^2+8x+15}{x+3}$

38. $\frac{2m^2-13m+15}{2m-3}$

39. $\frac{6x^2+24x}{x^2+8x+16}$

40. $\frac{8m^2-16m}{m^2-4m+4}$

41. $\frac{4k^2-25}{4k^2-20k+25}$

42. $\frac{x^2-x-20}{x^2+7x+12}$

43. $\frac{3-x}{6-17x+5x^2}$

44. $\frac{2x-14}{x^2-4x-21}$

45. $\frac{5x^2+10x+5}{3x^2+6x+3}$

46. $\frac{3x^2+3x+18}{7x^2+7x+42}$

47. $\frac{m^2-2m-8}{m^2-m-6}$

48. $\frac{b^2+b-12}{b^2+2b-15}$

49. $\frac{p^2-4p+4}{p^2+4p-12}$

50. $\frac{b^2-5b+6}{b^4-13b^2+36}$

51. $\frac{2x^2-5x+3}{3x^2-5x+2}$

52. $\frac{3m^2+8m-3}{6m^2+17m-3}$

53. $\frac{2y^2-18}{2y-6}$

54. $\frac{8x^2-x^3}{16-2x}$

55. $\frac{2x^2+8x}{x^2-11x}$

56. $\frac{6y^3-9y^2}{2y^2+5y-12}$

57. $\frac{x^2-x^2y}{x^3-x^3y}$

Sums of Integers

One advantage of using a computer in problem solving is that the computer can perform many operations very quickly. Consider this problem.

The sum of the first 5 positive even integers is 30. Find the sum of the first 100 positive even integers.

The following BASIC program can be used to find the sum.

```
10  INPUT "HOW MANY EVEN INTEGERS? ";N
20  LET SUM = 0
30  FOR I = 2 TO 2 * N STEP 2
40  LET SUM = SUM + I          Line 40 calculates the sum.
50  NEXT I
60  PRINT "THE SUM OF THE FIRST ";N;" EVEN INTEGERS IS ";SUM
70  END

] RUN
HOW MANY EVEN INTEGERS? 100
THE SUM OF THE FIRST 100 EVEN INTEGERS IS 10100
```

Exercises

1. Run the program several times. Input 1, 2, 3, . . . 10 for the value of N. Write the corresponding sum.

2. Modify lines 10, 30, and 60 to find the sum of the first N positive odd integers.

3. For N = 1, 2, 3, . . . 10, write the sum of the first N positive odd integers.

4. Modify the program to find the sum of the first N positive integers. Run the program for N = 1, 2, 3, . . . 10.

Vocabulary

Zero Product Property   (263)
algebraic fractions   (282)

rational expressions   (282)

Chapter Summary

1. The Zero Product Property: For all numbers a and b, if $ab = 0$, then $a = 0$, $b = 0$, or both a and b equal 0.   (263)

2. If an equation of second degree or higher can be written as a product of factors equal to zero, the Zero Product Property can be applied to solve the equation. (273)

3. To simplify a rational expression, first factor the numerator and denominator. Then eliminate common factors.   (282)

8–1 **Solve each equation. Check the solutions.**

1. $y(y + 11) = 0$ **2.** $7a(a - 7) = 0$

3. $4t(2t - 10) = 0$ **4.** $(2y - 9)(2y + 9) = 0$

5. $(3y - 2)(4y + 7) = 0$ **6.** $(y - 2)(y + 7) = 0$

7. The product of a positive number decreased by 6 and the same number increased by 11 is zero. What is the number?

8–2 **Factor.**

8. $50x^2 - 98y^2$ **9.** $6m^3 + m^2 - 15m$

10. $56a^2 - 93a + 27$ **11.** $24am - 9an + 40bm - 15bn$

12. $81x^8 - 16c^4$ **13.** $49m^2 + 70mn + 25n^2$

14. $28x^2 + 13x - 6$ **15.** $8a^2 + 32b^2$

Solve. Check the solutions.

8–3 **16.** $y^2 - 9y = 0$ **17.** $y^2 = -7y$

18. $2y^2 - 98 = 0$ **19.** $\frac{3}{4}y = \frac{1}{4}y^2$

20. $a^2 + 10a + 25 = 0$ **21.** $25y^2 + 20y = -4$

22. The square of a number added to seven times the number is zero. What is the number?

8–4 **23.** $y^2 + 40 = -13y$ **24.** $x^2 + 12x = -35$

25. $2m^2 + 13m = 24$ **26.** $(x + 6)(x - 1) = 78$

27. $6p^2 - 11p + 4 = 0$ **28.** $6x^3 + 29x^2 + 28x = 0$

For each problem, define a variable. Then use an equation to solve the problem.

8–5 **29.** The product of two consecutive odd integers is 143. Find the integers.

30. The sum of the squares of two consecutive odd integers is 202. Find the integers.

8–6 **31.** A rectangle is 11 inches wide and 13 inches long. When its length and width are decreased by the same amount, its area is decreased by 63 square inches. What are the dimensions of the new rectangle?

32. A baseball travels upward with an initial velocity of 80 feet per second. How long does it take to return to the ground? Use the formula $h = vt - 16t^2$.

8–7 **Simplify.**

33. $\frac{3x^2y}{12xy^3z}$ **34.** $\frac{x + y}{x^2 + 3xy + 2y^2}$

35. $\frac{x^2 - 9}{x^2 - 6x + 9}$ **36.** $\frac{x^2 + 10x + 21}{x^3 + x^2 - 42x}$

Chapter Test

Factor.

1. $9x^2 - 36y^4$

2. $12x^2 + 23x - 24$

3. $25a^2 - 20a + 4$

4. $15ac - 21ad + 10bc - 14bd$

5. $3a^2b - 5ab + 11ab^2$

6. $50a^2 - 25b^2$

7. $25m^2n^2 + 90mnp + 81p^2$

8. $x^3 - 5x^2 - 9x + 45$

9. $x^8 - y^8$

10. $98a^2 - 140ab + 50b^2$

Simplify.

11. $\frac{21x^2y}{28ax}$

12. $\frac{x^2 + 7x - 18}{x^2 + 12x + 27}$

13. $\frac{7x^2 - 28}{5x^3 - 20x}$

Solve each equation.

14. $(3w - 21)(w + 8) = 0$

15. $18s^2 + 72s = 0$

16. $x^2 - 6x + 9 = 0$

17. $6w(4w - 28) = 0$

18. $t^2 - 121 = 0$

19. $4x^2 = 36$

20. $(3m + 5)(2m - 3) = 0$

21. $\frac{4}{9} = x^2$

22. $x^2 + 25 = 10x$

23. $a^2 - 9a - 52 = 0$

24. $3x^2 - 7x - 20 = 0$

25. $x + 6 = 12x^2$

26. $10m^2 + 7m = 3$

27. $x^3 - 5x^2 - 66x = 0$

28. $x^3 + 7x^2 - 25x - 175 = 0$

For each problem, define a variable. Then use an equation to solve the problem.

29. The product of 5 times Jóse's salary and 97,000 less than 4 times Jóse's salary is zero. What is Jóse's salary?

30. The sum of two integers is 21 and their product is 104. Find the integers.

31. When one integer is added to the square of the next consecutive integer, the sum is 55. Find the integers.

32. A photograph is 8 cm wide and 12 cm long. The length and width are increased by an equal amount in order to double the area of the photograph. What are the dimensions of the new photograph?

For problems 33 and 34, use $h = vt - 16t^2$.

33. An arrow is shot directly up with an initial velocity of 144 ft/sec. How long does it take to return to the ground?

34. A rocket is fired directly up at an initial velocity of 2240 ft/sec. How high will the rocket be after 70 seconds?

The questions on this page involve comparing two quantities, one in Column A and one in Column B. In certain questions, information related to one or both quantities is centered above them. All variables used stand for real numbers.

Directions:

Write A if the quantity in Column A is greater.

Write B if the quantity in Column B is greater.

Write C if the quantities are equal.

Write D if there is not enough information to determine the relationship.

Column A	Column B
$x \geq 13$	
1. $\frac{37}{3}$	x
2. $\frac{0.4}{20}$	0.002
3. $10 + 5 \div 4 - 3$	$10 - 5 \cdot 4 + 3$
$b > c + 1$	
4. b	c
$b < c + 1$	
5. b	c
6. The number of which 6 is 20%.	10% of 310
$a = b$	
7. $-5(a - b)$	$7(3b - 3a)$
8. $\frac{3}{4}(5 + 1)\left(\frac{6-6}{2}\right)$	$\frac{1}{6}(1 + 3)(12 \div 2)$
9. 25% of $\frac{4}{7}$	$\frac{4}{7}$ of $\frac{1}{4}$
10. $0.02 \div 0.2$	$0.2 \cdot 0.02$
11. The value of $2y + 7$ when $y = -2$.	The value of $2x - 8$ when $x = 6$.

Examples

Column A	Column B
1. $\frac{x}{y}$	$\frac{y}{x}$

The answer is D because it is not known whether x and y are positive or negative.

The values of $\frac{x}{y}$ and $\frac{y}{x}$ depend on the choices for x and y.

$$a > 0$$

2. $4a - 3a \qquad\qquad 4 \cdot 3a$

The answer is B because $12a$ is greater than a for all positive values of a.

Column A	Column B
12. 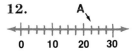	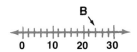
The number indicated by arrow A on the number line above.	The number indicated by arrow B on the number line above.

13. The price of a stereo increased by 20% and then decreased by 20%.

The original price of the stereo.	The new price of the stereo.
$a:b = c:d$	
14. ad	bc
$3x - 3y = 24$	
15. x	y
$r < 0 < s$	
16. r^2	$\frac{s}{2}$

CHAPTER 9

Functions and Graphs

Barb and Tim have reserved seats F-3 and F-4 for a concert. This means they will sit in row F, seats 3 and 4.

Suppose the letter F was replaced by 6 because it is the sixth row. The reserved seats could be identified by 6-3 and 6-4, or by the ordered pairs (6, 3) and (6, 4). Would (3, 6) and (4, 6) identify the same seats? Why is the order of the numbers important?

In this chapter, you will learn to graph ordered pairs. You will also learn more about functions and their graphs.

9-1 Ordered Pairs

In mathematics, **ordered pairs** of numbers are used to locate points in a plane. The points are in reference to two *perpendicular* number lines, lines that meet to form 90° angles. Notice that the lines intersect at their zero points. This point of intersection is called the **origin**. The ordered pair that names this point is (0, 0). The two number lines are called the **x-axis** and the **y-axis**. The first component of an ordered pair corresponds to a number on the horizontal, or *x*-axis. The second component of the ordered pair corresponds to a number on the vertical, or *y*-axis.

To find the ordered pair for point *A*, think of a horizontal line and vertical line through point *A*. Notice where these lines intersect the axes. The number on the *x*-axis that corresponds to *A* is −3. The number on the *y*-axis that corresponds to *A* is 4. Thus, the ordered pair for point *A* is (−3, 4).

The first component, −3, is called the **x-coordinate** of point *A*. The second component, 4, is called the **y-coordinate** of point *A*. The plane is called the **coordinate plane**.

The first component is sometimes called the <u>abscissa</u>. The second component is sometimes called the <u>ordinate</u>.

Example

1 **Name the ordered pairs for points *R*, *S*, *T*, and *U*.**

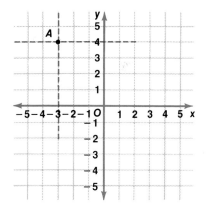

Think of a horizontal line and a vertical line passing through each point.

Be sure to express the x-coordinate first in each ordered pair.

The *x*-coordinate of point *R* is 4. The *y*-coordinate is −3. Thus, the ordered pair for *R* is (4, −3).

The *x*-coordinate of point *S* is 1. The *y*-coordinate is 4. Thus, the ordered pair for *S* is (1, 4).

The ordered pair for *T* is (−3, −5) and the ordered pair for *U* is (−5, 3). Do you see why?

Sometimes a point is named not only by a letter but also by its location. For example, $R(4, 3)$ means that the location of point R is $(4, 3)$.

To **graph** an ordered pair means to place a dot at the point that is named by the ordered pair. This is sometimes called *plotting* the point. When graphing an ordered pair, you should start at the origin. The x-coordinate gives the number of units to move right or left. The y-coordinate gives the number of units to move up or down. Study these examples.

Examples

2 **Graph point $C(3, 1)$.**

Start at O. Move 3 units to the right. Then move 1 unit up and place a dot. Label the dot with the letter C.

> *Check to make sure the dot corresponds to 3 on the x-axis and 1 on the y-axis.*

3 **Graph point $D(-3, -2)$.**

Start at O. Move 3 units to the left. Do you see why the move was to the left? Then move 2 units down and place a dot. Do you see why the move was down? Label the dot with the letter D.

> *Check to make sure the dot corresponds to −3 on the x-axis and −2 on the y-axis.*

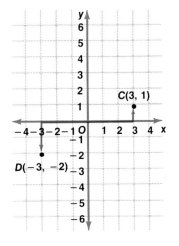

When plotting points, the following is true.

1. **Exactly one point in the plane is named by a given ordered pair of numbers.**
2. **Exactly one ordered pair of numbers names a given point in the plane.**

Completeness Property for Points in the Plane

The x-axis and y-axis separate the plane into four regions, called **quadrants**. The quadrants are numbered as shown at the right. In which quadrant is the graph of $(5, -4)$? In which quadrant is the graph of $(-2, -7)$? In which quadrant is the graph of $(2, 0)$? The axes are not in any quadrant. Can you think of a reason why?

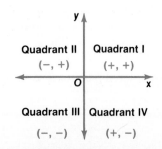

Exploratory Exercises

State the ordered pair for each point.

1. A
2. B
3. C
4. D
5. E
6. F
7. G
8. H
9. I
10. J
11. K
12. L
13. M
14. N
15. P
16. Q
17. R
18. S
19. T
20. U

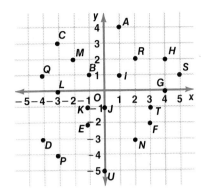

State the quadrant in which the point for each ordered pair is located.

21. $(5, 2)$
22. $(-3, -1)$
23. $(-2, 3)$
24. $(6, 0)$
25. $(0, -2)$
26. $(-6, -2)$
27. $(4, -3)$
28. $(5, 1)$
29. $(3, -2)$
30. $(3, 6)$
31. $(-4, 2)$
32. $(-3, 0)$
33. $(-3, 1)$
34. $(-4, 0)$
35. $(-1, -3)$
36. $(5, -1)$

Written Exercises

Name the quadrant for the graph of (x, y), given each condition.

37. $x > 0, y < 0$
38. $x < 0, y < 0$
39. $x = 0, y > 0$
40. $x < 0, y > 0$
41. $x > 0, y > 0$
42. $x > 0, y = 0$
43. $x = -1, y < 0$
44. $x < 0, y = 3$

Graph each point.

45. $A(5, -2)$
46. $B(3, 5)$
47. $C(-6, 0)$
48. $D(-3, 4)$
49. $E(-3, -3)$
50. $F(-5, 1)$
51. $G(2, -1)$
52. $H(4, 0)$
53. $I(3, -4)$
54. $J(-3, 0)$
55. $K(-5, -2)$
56. $L(4, 2)$
57. $M(-5, -4)$
58. $N(1, 4)$
59. $P(2, -3)$
60. $Q(0, -2)$
61. $R(-2, 1)$
62. $S(-3, -1)$
63. $T(-3, 6)$
64. $U(6, 2)$

Graph each point. Then connect the points in alphabetical order and identify the figure.

65. $A(4, 6)$, $B(3, 7)$, $C(2, 6)$, $D(1, 7)$, $E(0, 6)$, $F(1, 8)$, $G(4, 9)$, $H(7, 8)$, $I(8, 6)$, $J(7, 7)$, $K(6, 6)$, $L(5, 7)$, $M(4, 6)$, $N(4, 0)$, $P(5, 0)$, $Q(5, 1)$.

66. $A\left(-3, \frac{1}{2}\right)$, $B(1, 0)$, $C(4, 0)$, $D(7, -4)$, $E(8, -4)$, $F(7, 0)$, $G(10, 0)$, $H(11, -2)$, $I(12, -2)$, $J\left(11\frac{1}{2}, 0\right)$, $K\left(13, \frac{1}{2}\right)$, $L\left(11\frac{1}{2}, 1\right)$, $M(12, 3)$, $N(11, 3)$, $P(10, 1)$, $Q(7, 1)$, $R(8, 5)$, $S(7, 5)$, $T(4, 1)$, $U(1, 1)$, $V\left(-3, \frac{1}{2}\right)$.

9-2 Relations

Consider the ordered pairs $(2, 2)$, $(-2, 3)$, and $(0, -1)$. In these examples, 2 is paired with 2, -2 is paired with 3, and 0 is paired with -1. When expressed as $\{(2, 2), (-2, 3), (0, -1)\}$, the set is called a **relation**.

A relation is a set of ordered pairs.	*Definition of Relation*

The first components of the ordered pairs given above are 2, -2, and 0. The **domain** of the relation is $\{2, -2, 0\}$.

The domain of a relation is the set of all *first* components from each ordered pair.	*Definition of Domain*

The second components are 2, 3, and -1. The **range** of the relation is $\{2, 3, -1\}$.

The range of a relation is the set of all *second* components from each ordered pair.	*Definition of Range*

A relation can also be shown as a table, a mapping, or a graph. A *mapping* pairs a member of the domain of a relation with a member in the range.

Table

x	y
2	2
-2	3
0	-1

Mapping

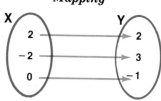

Graph

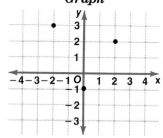

The **inverse** of any relation is obtained by switching the components of the ordered pairs. The inverse of the relation $\{(2, 2), (-2, 3), (0, -1)\}$ is $\{(2, 2), (3, -2), (-1, 0)\}$. Notice the domain of the relation becomes the range of the inverse and the range of the relation becomes the domain of the inverse.

	Relation Q is the inverse of Relation S if and only if for every ordered pair (a, b) in S, there is an ordered pair (b, a) in Q.	*Definition of Inverse of a Relation*

Examples

1 State the set of ordered pairs shown by the table. State the domain and range of the relation. Then state the set of ordered pairs in the inverse.

x	y
0	5
2	3
1	−4
−3	3
−1	−2

The set of ordered pairs in the relation is $\{(0, 5), (2, 3), (1, -4), (-3, 3), (-1, -2)\}$.

The domain is the set of first components or $\{0, 2, 1, -3, -1\}$.

The range is the set of second components or $\{5, 3, -4, -2\}$.

The set of ordered pairs in the inverse is $\{(5, 0), (3, 2), (-4, 1), (3, -3), (-2, -1)\}$.

2 State the relation shown by the graph. State the domain and range. Then state the inverse of the relation.

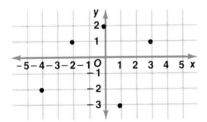

The relation is $\{(3, 1), (0, 2), (-2, 1), (-4, -2), (1, -3)\}$.

The domain is $\{3, 0, -2, -4, 1\}$.

The range is $\{1, 2, -2, -3\}$.

The inverse is $\{(1, 3), (2, 0), (1, -2), (-2, -4), (-3, 1)\}$.

3 State the relation for the mapping. Then state the domain and range.

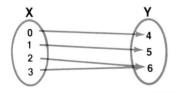

The relation is $\{(0, 4), (1, 5), (2, 6), (3, 6)\}$.

The domain is $\{0, 1, 2, 3\}$.

The range is $\{4, 5, 6\}$.

Exploratory Exercises

State the domain and range of each relation.

1. $\{(0, 2), (1, -2), (2, 4)\}$

2. $\{(-4, 2), (-2, 0), (0, 2), (2, 4)\}$

3. $\{(-3, 1), (-2, 0), (-1, 1), (0, 2)\}$

4. $\{(5, 2), (0, 0), (-9, -1)\}$

5. {(7, 5), (−2, −3), (4, 0), (5, −7), (−9, 2)}

6. {(1, 1), (2, 4), (−2, 4), (3, 9)}

7. {(−3, 0), (−2, −5), (−2, 6), (3, 7), (0, −17)}

8. {(−5, 2), (−4, 5), (−2, 1), (−4, 4)}

9. {(3.1, −1), (−4.7, 3.9), (2.4, −3.6), (−9, 12.12)}

10. {(4.7, −2.3), (5.1, 2.8), (−3.5, 1)}

11. $\left\{\left(\frac{1}{2}, \frac{1}{4}\right), \left(1\frac{1}{2}, -\frac{2}{3}\right), \left(-3, \frac{2}{5}\right), \left(-5\frac{1}{4}, -7\frac{2}{7}\right)\right\}$

12. $\left\{(0, 0), \left(\frac{1}{2}, \frac{1}{8}\right), \left(-\frac{1}{2}, -\frac{1}{8}\right)\right\}$

State each relation as a set of ordered pairs. Then state the domain, range, and inverse of each.

13.

x	y
1	5
2	7
3	9
4	11

14.

x	y
1	3
2	2
4	9
6	5

15.

x	y
−4	1
−2	3
0	1
2	3
4	1

16.

x	y
1	−2
3	−4
5	−6
9	−4
11	−2

Written Exercises

Write the relation for each mapping as a set of ordered pairs.

17.

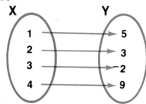

18.

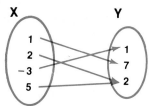

19.

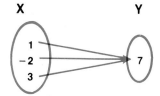

20.

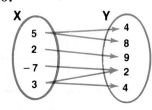

21.

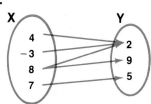

22.

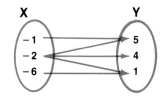

Draw a mapping for each relation. Then write the relation and the inverse as sets of ordered pairs.

23.

x	y
4	2
1	3
3	3
6	4

24.

x	y
1	4
3	−2
4	4
6	−2

25.

x	y
8	1
7	3
6	5
2	−2

26.	x	y
	1	3
	2	4
	3	5
	4	6
	5	7

27.	x	y
	1	3
	2	5
	1	-7
	2	9

28.	x	y
	1	-2
	$2\frac{1}{2}$	-3
	3	8
	4	-3

Write each relation as a set of ordered pairs. Then state the domain and range of each.

29.

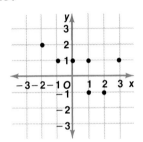

30.

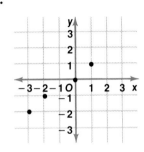

31.

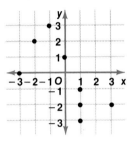

32.

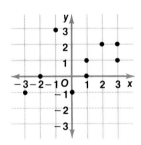

33.

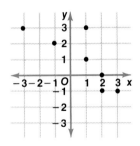

34.

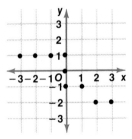

Draw a mapping for each relation.

35. $\{(1, 3), (2, 5), (8, 2), (5, -3)\}$

36. $\{(2, 3), (3, 2)\}$

37. $\{(8, 1), (4, 2), (6, -4), (5, -3), (6, 0)\}$

38. $\{(0, 1)\}$

39. $\{(2, 2), (3, 3), (4, 4)\}$

40. $\{(1, 3), (2, 7), (4, 1), (-3, 3), (3, 3)\}$

41. $\{(-3, 3), (-2, 2), (-1, 1), (0, 0), (1, -1)\}$

42. $\{(-6, 0), (-1, 2), (-3, 4)\}$

43. $\left\{\left(4\frac{1}{2}, \frac{1}{2}\right), \left(2\frac{1}{2}, 1\right), \left(-1\frac{1}{2}, 2\right)\right\}$

44. $\left\{\left(\frac{1}{3}, \frac{3}{4}\right), \left(\frac{2}{3}, -\frac{1}{2}\right), \left(-\frac{3}{4}, \frac{2}{3}\right), \left(\frac{2}{3}, \frac{3}{4}\right)\right\}$

9-3 Equations as Relations

Anita works part-time in a bookstore. She earns $5 per hour. She made the following chart to show her earnings for different numbers of hours worked. She let x represent the number of hours worked per week and y represent her total pay.

x(hours)	$5x$	y(pay)	(x, y)
1	5(1)	5	(1, 5)
5	5(5)	25	(5, 25)
8	5(8)	40	(8, 40)
10	5(10)	50	(10, 50)
12	5(12)	60	(12, 60)
15	5(15)	75	(15, 75)
20	5(20)	100	(20, 100)

The equation $y = 5x$ describes her pay (y) for any number of hours worked (x). Anita could have selected other values of x. Generally, however, these values are restricted. For example, Anita could not work less than 0 hours or more than 168 hours in a week. The ordered pairs in the chart above are *solutions* to the equation $y = 5x$.

If a true statement results when an ordered pair is substituted into an equation in two variables, the ordered pair is a solution of the equation.	*Definition of the Solution of an Equation in Two Variables*

Since the solutions to an equation in two variables are ordered pairs, such an equation describes a relation. The set of values of x is the domain of the relation. The set of corresponding values of y is the range.

1 **Solve $y = 2x + 3$ if the domain is $\{-5, -3, -1, 0, 1, 3, 5, 7, 9\}$.**

Make a table. The values of x come from the domain. Now substitute the given values of x to determine the corresponding values of y.

x	$2x + 3$	y	(x, y)
-5	$2(-5) + 3$	-7	$(-5, -7)$
-3	$2(-3) + 3$	-3	$(-3, -3)$
-1	$2(-1) + 3$	1	$(-1, 1)$
0	$2(0) + 3$	3	$(0, 3)$
1	$2(1) + 3$	5	$(1, 5)$
3	$2(3) + 3$	9	$(3, 9)$
5	$2(5) + 3$	13	$(5, 13)$
7	$2(7) + 3$	17	$(7, 17)$
9	$2(9) + 3$	21	$(9, 21)$

The solution set is
$\{(-5, -7), (-3, -3), (-1, 1), (0, 3), (1, 5), (3, 9), (5, 13), (7, 17), (9, 21)\}$.

2 **Solve $2y + 4x = 8$ if the domain is $\{-4, -3, -2, 2, 3, 4\}$.**

Values of y are usually easier to determine if the equation is first solved for y in terms of x.

$$2y + 4x = 8$$
$$2y = 8 - 4x$$
$$\frac{2y}{2} = \frac{8 - 4x}{2}$$
$$y = 4 - 2x$$

Now substitute the given values of x to determine the corresponding values of y.

x	$4 - 2x$	y	(x, y)
-4	$4 - 2(-4)$	12	$(-4, 12)$
-3	$4 - 2(-3)$	10	$(-3, 10)$
-2	$4 - 2(-2)$	8	$(-2, 8)$
2	$4 - 2(2)$	0	$(2, 0)$
3	$4 - 2(3)$	-2	$(3, -2)$
4	$4 - 2(4)$	-4	$(4, -4)$

The solution set is $\{(-4, 12), (-3, 10), (-2, 8), (2, 0), (3, -2), (4, -4)\}$.

When variables other than x and y are used, assume that the values of the letters that come first in the alphabet are from the domain.

Example

3 Solve $3r + 2s = 11$ if the domain is $\{-3, 0, 1, 2, 5\}$.

The values for r come from the domain. Therefore, first solve the equation so that s is expressed in terms of r.

$$3r + 2s = 11$$
$$2s = 11 - 3r$$
$$\frac{2s}{2} = \frac{11 - 3r}{2}$$
$$s = \frac{11 - 3r}{2}$$

Now substitute the given values of r to determine the corresponding values of s.

r	$\dfrac{11 - 3r}{2}$	s	(r, s)
-3	$\frac{11 - 3(-3)}{2}$	10	$(-3, 10)$
0	$\frac{11 - 3(0)}{2}$	$\frac{11}{2}$	$\left(0, \frac{11}{2}\right)$
1	$\frac{11 - 3(1)}{2}$	4	$(1, 4)$
2	$\frac{11 - 3(2)}{2}$	$\frac{5}{2}$	$\left(2, \frac{5}{2}\right)$
5	$\frac{11 - 3(5)}{2}$	-2	$(5, -2)$

The solution set is $\left\{(-3, 10), \left(0, \frac{11}{2}\right), (1, 4), \left(2, \frac{5}{2}\right), (5, -2)\right\}$.

Exploratory Exercises

Copy each table. Then find the solutions for each equation for the domain indicated.

1. $y = 3x$

2. $y = 4x - 3$

3. $n = \dfrac{2m + 5}{3}$

x	y	(x, y)
-4		
-2		
0		
1		
2		
3		

x	y	(x, y)
-3		
-2		
-1		
0		
2		
4		

m	n	(m, n)
-4		
-2		
0		
1		
3		

Written Exercises

Solve each equation for the variable indicated.

4. $x + y = 5$, for y

5. $3x + y = 7$, for y

6. $b - 5a = 3$, for b

7. $4m + n = 7$, for n

8. $8x + 2y = 6$, for y

9. $6x + 3y = 12$, for y

10. $4a + 3b = 7$, for b

11. $6r + 5s = 2$, for s

12. $6x = 3y + 2$, for y

13. $3a = 7b + 8$, for b

14. $4p = 7 - 2q$, for q

15. $7q = 4 + 5m$, for q

State which of the ordered pairs given are solutions of the equation.

16. $3x + y = 8$ **a.** $(2, 2)$ **b.** $(3, 1)$ **c.** $(4, -4)$ **d.** $(8, 0)$

17. $2x + 3y = 11$ **a.** $(3, 1)$ **b.** $(1, 3)$ **c.** $(-2, 5)$ **d.** $(4, -1)$

18. $2m - 5n = 1$ **a.** $(-2, -1)$ **b.** $(2, 1)$ **c.** $\left(-\frac{3}{2}, -\frac{1}{2}\right)$ **d.** $(-7, -3)$

19. $3r = 8s - 4$ **a.** $\left(\frac{2}{3}, \frac{3}{4}\right)$ **b.** $\left(0, \frac{1}{2}\right)$ **c.** $(4, 2)$ **d.** $(2, 4)$

20. $3y = x + 7$ **a.** $(2, 4)$ **b.** $(2, -1)$ **c.** $(2, 3)$ **d.** $(-1, 2)$

21. $4x = 8 - 2y$ **a.** $(2, 0)$ **b.** $(0, 2)$ **c.** $\left(\frac{1}{2}, -3\right)$ **d.** $(1, -2)$

Solve each equation if the domain is $\{-2, -1, 0, 2, 5\}$.

22. $y = 2x + 1$

23. $y = 5x - 3$

24. $x + y = 7$

25. $x - y = 4$

26. $2x + y = 7$

27. $5x + y = 4$

28. $2a + 3b = 13$

29. $4r + 3s = 16$

30. $3x = 5 + 2y$

31. $4y = 3 + 2x$

32. $5a = 8 - 4b$

33. $2t = 3 - 5s$

34. $a - 6b = -32$

35. $5x - y = -3$

36. $6x + 5y = 11$

37. $4m + 3n = 10$

mini-review

Multiply.

1. $2.3a(3a - 8)$

2. $(6a + 1)(2a^2 - 1)$

3. $(2a - 9b)(2a + 9b)$

Solve each equation.

4. $5x^2 = 20$

5. $18y^3 = 66y^2 + 24y$

6. $(k + 4)(k + 2) = -1$

Factor, if possible.

7. $12x^3 - 147x$

8. $64m^2 + 48m + 9$

9. $6x^2 + 11x - 7$

Complete the following for $\{(3, 0), (-2, 5), (2, 7), (-4, -4)\}$.

10. State the domain.

11. State the range.

12. State the inverse.

Solve each problem.

13. The selling price of a house includes the 6.5% sales commission for the real-estate company. A house sold for $145,000. How much did the owner receive and how much did the real-estate company receive?

14. A rectangle is 5 cm wide and 16 cm long. When the length and width are decreased by the same amount, the area is decreased by 54 cm². What are the dimensions of the new rectangle?

Graphs are often used to help visualize the relationship between two variables. To fully understand the information presented in the graph, ask yourself the following three questions.

1. What does the title indicate will be represented in the graph?
2. What variable is represented along each axis?
3. What units are used along each axis?

Apply each of the three questions to the graph below. The bar graph shows information from a survey.

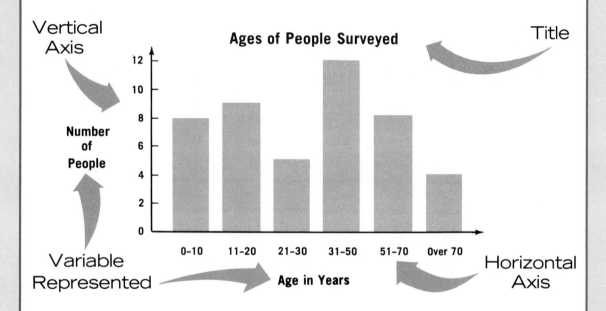

On the graph, the title indicates that the graph shows the number of people of each age group who were surveyed. The vertical axis shows the number of people in each age group. The horizontal axis shows age groups in years. Each unit on the vertical axis represents two people. The horizontal axis is labeled to show what age groups were used. Notice that the units used for each axis do not have to be the same.

Exercises

Turn to page 315 *Misleading Graphs*. Answer the three questions at the top of this page about the graphs on page 315.

9-4 Graphing Relations

Some of the solutions of $y = 2x - 1$ are shown in the table. The solutions can also be shown by a graph.

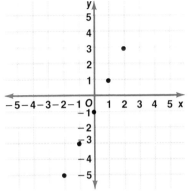

x	$2x - 1$	y	(x, y)
-2	$2(-2) - 1$	-5	$(-2, -5)$
-1	$2(-1) - 1$	-3	$(-1, -3)$
0	$2(0) - 1$	-1	$(0, -1)$
1	$2(1) - 1$	1	$(1, 1)$
2	$2(2) - 1$	3	$(2, 3)$

If the domain of $y = 2x - 1$ is the set of all numbers, an infinite number of ordered pairs are solutions to the equation.

Suppose you draw a line connecting the points in the graph above. All points for the solutions of $y = 2x - 1$ lie on this line. The coordinates of any other point on this line satisfy the equation. Hence, the line is called the *graph* of $y = 2x - 1$.

Remember, the values of the first components in the ordered pair of a relation represent its domain.

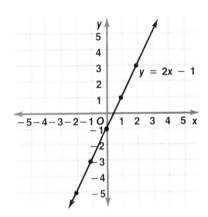

The equation $y = 2x - 1$ can be changed to equivalent forms such as $-y = 1 - 2x$ or $2x - y = 1$. All of these forms will have the same graph. An equation whose graph is a straight line is called a **linear equation**.

A linear equation is an equation that can be written in the form $Ax + By = C$, where A, B, and C are any numbers, and A and B are not both 0.

Definition of Linear Equation

The equations $4x + 3y = 7$, $2x = 8 + y$, $5m - n = \frac{1}{2}$, $x = 3$, and $y = 7$ are linear equations. The equations $3x + y^2 = 7$ and $\frac{1}{x} + y = 4$ are *not* linear equations. Why?

Examples

1 **Draw the graph of $3x - y = 3$.**

First solve the equation so that y is expressed in terms of x.

$3x - y = 3$

$\qquad -y = 3 - 3x$

$\qquad y = 3x - 3$

Now set up a table of values for x and y. Graph the ordered pairs and connect them with a line.

x	$3x - 3$	y	(x, y)
-2	$3(-2) - 3$	-9	$(-2, -9)$
-1	$3(-1) - 3$	-6	$(-1, -6)$
0	$3(0) - 3$	-3	$(0, -3)$
1	$3(1) - 3$	0	$(1, 0)$
2	$3(2) - 3$	3	$(2, 3)$

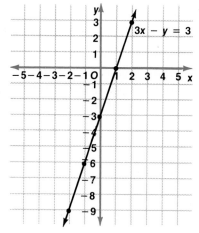

2 **Draw the graph of $3x + 2y = 4$.**

$3x + 2y = 4$

$\qquad 2y = 4 - 3x$

$\qquad y = \dfrac{4 - 3x}{2}$

x	$\dfrac{4 - 3x}{2}$	y	(x, y)
-2	$\dfrac{4 - 3(-2)}{2}$	5	$(-2, 5)$
-1	$\dfrac{4 - 3(-1)}{2}$	$\dfrac{7}{2}$	$\left(-1, \dfrac{7}{2}\right)$
0	$\dfrac{4 - 3(0)}{2}$	2	$(0, 2)$
1	$\dfrac{4 - 3(1)}{2}$	$\dfrac{1}{2}$	$\left(1, \dfrac{1}{2}\right)$
2	$\dfrac{4 - 3(2)}{2}$	-1	$(2, -1)$

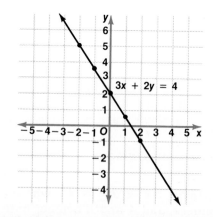

Exploratory Exercises

State whether each equation is a linear equation.

1. $5x + 2y = 7$

2. $3x^2 + 2y = 4$

3. $x + \frac{1}{y} = 7$

4. $\frac{1}{x} - \frac{1}{y} = \frac{1}{2}$

5. $\frac{3}{4}x + 2y = 3$

6. $\frac{3}{5}x - \frac{2}{3}y = 5$

7. $x = y^2$

8. $x = y$

9. $\frac{3}{5x} = y$

10. $\frac{3}{5}x = y$

11. $\frac{1}{3}x + \frac{1}{3y} = 4$

12. $5x - y = 4$

13. $xy^2 = y$

14. $xy = 4$

15. $\frac{y}{2} = 2$

16. $5m + 2n = 8$

17. $2w = 3z - 1$

18. $\frac{5}{3}x + \frac{2}{3}y = 7$

Written Exercises

Solve each equation for y.

19. $x + 5y = 16$

20. $8x - y = 16$

21. $6x + 7 = -14y$

22. $3x + 4y = 12$

23. $4x - \frac{3}{8}y = 1$

24. $\frac{1}{2}x - \frac{2}{3}y = 10$

State whether each equation is a linear equation.

25. $xy = 2$

26. $3x = 2y$

27. $x = y^2$

28. $y^2 = x^2$

29. $3m - 2n = 8$

30. $3mn^2 = 1$

31. $2x - 3 = y^2$

32. $5x - 7 = 3y$

33. $8a - 7b = 2a$

34. $3m - 2n = 0$

35. $4x^2 - 3x = y$

36. $2m + 5m = 7n$

Graph each equation.

37. $3m + n = 4$

38. $2x - y = 8$

39. $b = 5a - 7$

40. $y = 3x + 1$

41. $4x + 3y = 12$

42. $5x + 2y = 10$

43. $5x + 3y = 8$

44. $2x + 7y = 9$

45. $3x - 2y = 12$

46. $4x - 3y = 24$

47. $x = 4$

48. $y = -2$

49. $\frac{1}{2}x + y = 8$

50. $x + \frac{1}{3}y = 6$

51. $\frac{2}{3}x - \frac{1}{3}y = 2$

52. $\frac{3}{4}x + \frac{1}{2}y = 6$

53. $x = -\frac{5}{2}$

54. $y = \frac{4}{3}$

55. $\frac{2}{5}x - \frac{3}{5}y = 1$

56. $\frac{3}{4}x - \frac{2}{3}y = 7$

57. $\frac{3}{5}x = 6$

58. $\frac{3}{4}y = 6$

59. $\frac{5}{2}x + \frac{2}{3}y = 1$

60. $\frac{4}{3}x - \frac{3}{4}y = 1$

Challenge Exercises

Graph each equation.

61. $y = x^2$

62. $y = x^2 + 2$

63. $y = -(x^2) + 3$

64. $x = 2y^2$

65. $y = -(x^2) - 3$

66. $y - 2x^2 = 3$

67. $y = x^3$

68. $y = 2x^3 + 1$

69. $x = y^3 - 1$

Cumulative Review

1. Arrange the terms of the polynomial $8 - xy^3 - 2x^3y^2 - 5x^2$ so the powers of x are in descending order.

State the degree of each polynomial.

2. $5x^2y^3 + 3x^4 - 7x^5y + x^3y$

3. $6 - x^3y + x^5 - x^2y^2 + 7xy^3$

Multiply.

4. $5a^2(3a^4 - 2a^2 + 3)$

5. $(5x - 3y)(2x + y)$

6. $(3a - 4b)^2$

7. $(0.2x - 1.3y^2)^2$

8. Simplify $\dfrac{15x^2 + 7xy - 2y^2}{25x^2 - y^2}$.

9. State the domain, range and inverse for $\{(5, 1), (-3, 2), (4, 2), (5, 0), (2, 2)\}$.

10. Graph $4x - 3y = 8$.

Factor.

11. $8m^2 - 72n^2$

12. $y^2 - \dfrac{3}{2}y + \dfrac{9}{16}$

13. $18x^2 - 50y^2$

14. $20a^2 - 3ab - 56b^2$

15. $3x^2y + 12x - 5xy - 20$

16. $20p^2 + 11p - 4$

17. $24a^3 + 15a^2 - 32a - 20$

18. Find the area of a square whose perimeter is $(6x + 12)$ meters.

19. Solve $3x - y = 8$ if the domain is $\{-3, -1, 0, 2\}$.

Solve.

20. $(2m + 7)(5m - 3) = 0$

21. $a^2 + 2a = 63$

22. $3(x + 5) - 6x = 5(14 + x) + 1$

23. $5x^2 = 3x$

24. $x^2 - 16 = 6x$

Solve each problem.

25. Find two consecutive integers if their product is 552.

26. Barbara Martinelli is going on a 240-mile trip. She leaves home and drives 40 miles per hour for 3 hours. What speed must she now drive for the remainder of the trip to complete the trip in $2\frac{1}{2}$ more hours?

27. The sum of a number and its square is 182. Find the number.

28. An open market is selling three kinds of oranges: tangelos at 43¢ per pound, valencias at 39¢ per pound, and navels at 59¢ per pound. Yesterday three times as many pounds of tangelos sold as valencias, and 6 pounds more of navels sold as valencias. If the oranges sold for a total of $35.32, how many pounds of each were sold?

29. A chemist has 800 mL of a solution that is 25% acid. How much pure (100%) acid must be added to the 25% solution to obtain a 40% acid solution?

30. A price increased from $85 to $110. Find the percent of increase.

9-5 Functions

The following graphs and accompanying tables describe two different relations.

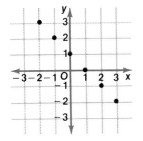

x	y
-2	3
-1	2
0	1
1	0
2	-1
3	-2

For each value of x, there is exactly one value of y.

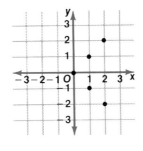

x	y
0	0
1	-1
1	1
2	-2
2	2

For x = 1, there are two values of y, 1 and −1. For x = 2, there are two values, 2 and −2.

The first relation is an example of a special type of relation called a **function**.

A function is a relation in which each element of the domain is paired with *exactly* one element of the range.	*Definition of Function*

Examples

1 Is $\{(5, -2), (3, 2), (4, -1), (-2, 2)\}$ a function? State the inverse of the relation. Is the inverse a function?

Since each element of the domain is paired with exactly one element of the range, the relation *is* a function.

The inverse is $\{(-2, 5), (2, 3), (-1, 4), (2, -2)\}$.
The element 2 of the domain is paired with more than one element of the range, namely 3 and -2. Therefore, the relation *is not* a function.

2 Is $\{(-1, 3), (0, 5), (2, 3), (5, -2), (2, 4)\}$ a function?

The element 2 of the domain is paired with more than one element of the range, namely 3 and 4. Therefore, the relation *is not* a function.

3 **Two relations are described by the following mappings. Which are functions?**

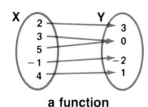

a function   *not* **a function**

The relation at the left is a function. For each element in the domain of the relation, there is *only one* corresponding element in the range. Why is the relation at the right not a function?

4 **Is the relation described by $x + 2y = 8$ a function?**

Try substituting a value for x. For example, try 2. What is the corresponding value for y? Can there be more than one value for y when $x = 2$? No. If $x = 2$, the only value for y is 3. Can you find any values for x that will give more than one value for y? No. There are none. Therefore, $x + 2y = 8$ is a function. Specifically, we say that y is a function of x.

Suppose you graph the relation $x + 2y = 8$.

x	2	0	8
y	3	4	0

Now place your pencil at the left of the graph to represent a vertical line. Slowly move the pencil to the right across the graph.

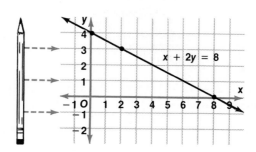

For each value of x, the vertical line passes through no more than one point of the graph. This is true for every function.

If any vertical line passes through no more than one point of the graph of a relation, then the relation is a function.	*Vertical Line Test for a Function*

5 Use the vertical line test to determine if the relation graphed is a function.

a.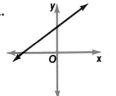
b.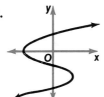
c.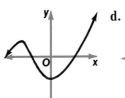
d.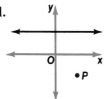

The relations in **a** and **c** are functions since any vertical line passes through no more than 1 point of the graph of the relation. The relation in **b** is *not* a function, since a vertical line near the *y*-axis passes through *three* points. The relation in **d** is *not* a function. A vertical line through point *P* also intersects the graph in another point.

Equations that describe functions are often written in a special way. The equation $y = 2x + 1$ can be written $f(x) = 2x + 1$. The symbol $f(x)$ is read "*f* of *x*" and represents the range of a function whose domain is represented by *x*. Specific values of the domain can then be substituted for *x* in the equation. For example, if $x = 3$, then $f(3) = 2(3) + 1$ or 7. Note $f(3)$ is a convenient way of referring to the value of *y* corresponding to $x = 3$. We say $f(3)$ is the **functional value** at $x = 3$.

Letters other than f are sometimes used for names of functions.

Examples

6 If $f(x) = 2x - 7$, find $f(2)$, $f(5)$, and $f(-3)$.

$$f(x) = 2x - 7 \qquad f(x) = 2x - 7 \qquad f(x) = 2x - 7$$
$$f(2) = 2(2) - 7 \qquad f(5) = 2(5) - 7 \qquad f(-3) = 2(-3) - 7$$
$$= 4 - 7 \qquad\qquad = 10 - 7 \qquad\qquad = -6 - 7$$
$$= -3 \qquad\qquad\;\; = 3 \qquad\qquad\;\; = -13$$

7 If $g(x) = x^2 - 2x + 1$, find $g(6a)$.

$$g(x) = x^2 - 2x + 1$$
$$g(6a) = (6a)^2 - 2(6a) + 1 \qquad \textit{Substitute 6a for x.}$$
$$= 36a^2 - 12a + 1$$

8 If $f(t) = 100t - 5t^2$, find $4[f(3)]$.

$$f(t) = 100t - 5t^2$$
$$4[f(3)] = 4[100(3) - 5(3)^2] \qquad \textit{4[f(3)] means 4 times the value of f(3).}$$
$$= 4[300 - 45]$$
$$= 4 \cdot 255$$
$$= 1020$$

Exploratory Exercises

State whether each relation is a function.

1. $\{(3, 4), (5, 4), (-2, 3), (5, 3)\}$

2. $\{(8, 4), (5, -2), (6, 3), (2, 3)\}$

3. $3m + 5n = 4$

4. $3s^2 + 2t^2 = 7$

5. $5a^2 - 7 = b$

Use the vertical line test to determine whether each graph represents a function.

6.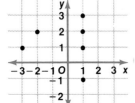

7.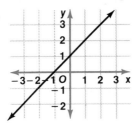

8.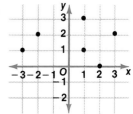

9.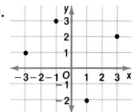

10.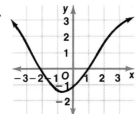

11.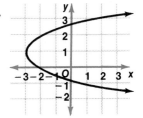

Given $g(x) = 2x - 1$, determine each value.

12. $g(2)$

13. $g(-4)$

14. $g(-6)$

15. $g(0)$

16. $g\left(\frac{1}{2}\right)$

17. $g\left(\frac{5}{2}\right)$

Written Exercises

Determine whether each relation is a function. Then state the inverse relation and determine whether it is a function.

18. $\{(3, 1), (5, 1), (7, 1)\}$

19. $\{(1, 3), (1, 5), (1, 7)\}$

20. $\{(-2, 4), (1, 3), (5, 2), (1, 4)\}$

21. $\{(-3, 3), (-2, 2), (-1, 1), (0, 0)\}$

22. $\{(6, -1), (1, 4), (2, 3), (6, 1)\}$

23. $\{(5, 4), (6, 1), (-2, 3), (0, 3)\}$

24. $\{(5, 4), (-6, 5), (4, 5), (0, 4)\}$

25. $\{(3, -2), (4, -1), (-2, 5), (4, 5)\}$

Determine whether each relation is a function.

26. $3x + 5y = 7$

27. $y = 2$

28. $4x - 7y = 3$

29. $x^2 + y = 7$

30. $x + y^2 = 7$

31. $x = -3$

32. $\frac{1}{x} = y$

33. $3 = x - y$

34. $4x = 5y$

35. $x = -2$

36. $\frac{2}{3}x = \frac{3}{2}y + 1$

37. $x^2 - y^2 = 3$

Given $f(x) = 3x - 5$ and $g(x) = x^2 - x$, determine each value.

38. $f(2)$

39. $f(-3)$

40. $g(-4)$

41. $g(3)$

42. $f\left(\frac{1}{2}\right)$

43. $f\left(\frac{2}{3}\right)$

44. $g\left(-\frac{1}{2}\right)$

45. $g\left(\frac{1}{3}\right)$

46. $f(2a)$

47. $g(2b)$

48. $3[f(5)]$

49. $2[g(-2)]$

50. $f(5.5)$

51. $1.5[f(-3)]$

52. $f(a + 3)$

53. $g(b - 3)$

Applications in Surveying

Many land regions have irregular shapes. Aerial surveyors often use coordinates when finding areas of such regions. The coordinate method described in the steps below can be used to find the area of *any* polygonal region. Study how this method is used to find the area of the region at the right.

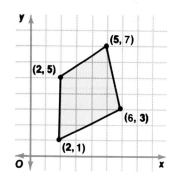

Step 1 List the ordered pairs for the vertices in counter-clockwise order, repeating the first ordered pair at the bottom of the list.

Step 2 Find D, the sum of the downward diagonal products (from left to right).

$$D = (5 \cdot 5) + (2 \cdot 1) + (2 \cdot 3) + (6 \cdot 7)$$
$$= 25 + 2 + 6 + 42 \text{ or } 75$$

(5, 7)

(2, 5)

(2, 1)

(6, 3)

(5, 7)

Step 3 Find U, the sum of the upward diagonal products (from left to right).

$$U = (2 \cdot 7) + (2 \cdot 5) + (6 \cdot 1) + (5 \cdot 3)$$
$$= 14 + 10 + 6 + 15 \text{ or } 45$$

Step 4 Use the formula $A = \frac{1}{2}(D - U)$ to find the area.

$$A = \frac{1}{2}(D - U)$$
$$= \frac{1}{2}(75 - 45)$$
$$= \frac{1}{2}(30) \text{ or } 15$$

The area is 15 square units. Count the number of square units enclosed by the polygon. Does this result seem reasonable?

Exercises

Use the coordinate method to find the area of each region.

1.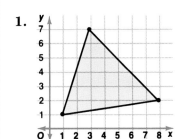

2.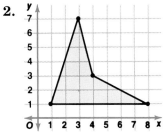

3.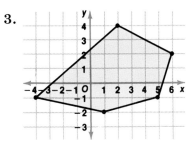

Graphing Calculators: **Absolute Value Function**

Compare the graphs of $y = |x|$ shown in blue at the right, and $y = |x| + 3$ shown in red at the right. Notice that the graph of $y = |x| + 3$ looks like the graph of $y = |x|$, but it is moved, or translated, upward 3 units.

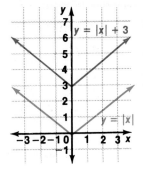

Note that the lowest, or minimum, point on the graph of $y = |x|$ is (0, 0). What are the coordinates of the minimum point of the graph of $y = |x| + 3$?

A graphing calculator can be used to explore the relationship between the graph of the absolute value function $y = |ax + b| + c$, and the values of a, b, and c.

On most graphing calculators, the GRAPH key is used to start the graphics function. After this key is pressed, the function to be graphed must be entered. Then the EXE key is used to graph the function on the graphics screen.

Some calculators use the DRAW *key to graph the function.*

Examples

1 **Use the graphing function to graph $y = |x|$.**

    A. Set the range values as desired.

    B. ENTER: GRAPH SHIFT ABS ALPHA X EXE

One suggestion for the range values is as follows.

    Xmin: -10 Xmax: 10   Xscl: 1
    Ymin: -10 Ymax: 10   Yscl: 1

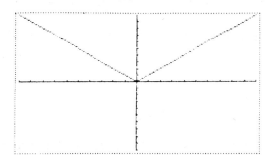

The size of the graphing screen varies among graphing calculators. The graphing screens shown are for a 95×63 dot screen.

2 **Without clearing the screen, graph $y = |2x| - 4$.**

    ENTER: GRAPH SHIFT ABS 2 ALPHA X $-$ 4 EXE

The graph of $y = |2x| - 4$ looks like the graph of $y = |x|$, but it is translated downward 4 units and stretched vertically so that it is narrower.

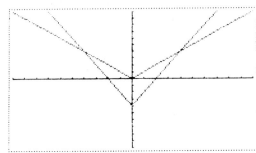

Example

3 Graph $y = |3x + 6| - 7$ on the same set of axes as the previous example.

ENTER: GRAPH SHIFT ABS (3 ALPHA X + 6) − 7 EXE

This graph looks like the graph of $y = |x|$. However, it has been translated so that its minimum point is now at $(-2, -7)$. It has also been stretched vertically so that it is narrower than the other two graphs.

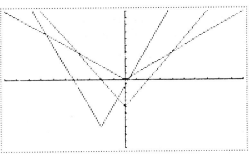

The equation $y = |3x + 6| - 7$ is equivalent to $y = 3|x + 2| - 7$. Notice, then, the relationship between the coordinates of the minimum point, $(-2, -7)$, and the equation $y = 3|x + 2| - 7$.

Written Exercises

Graph each pair of equations on the same set of axes on a graphing calculator. Then compare the graphs.

1. $y = |x|; y = |x| + 5$
2. $y = |x|; y = |x| - 6$
3. $y = |x|; y = |x| + 1.5$
4. $y = |x|; y = |2x|$
5. $y = |x|; y = |4x|$
6. $y = |x|; y = |0.5x|$
7. $y = |3x|; y = -3|x|$
8. $y = \frac{1}{2}|x|; y = -\frac{1}{2}|x|$
9. $y = |x|; y = |x - 5| - 1$
10. $y = |x|; y = |5 - x| - 4$
11. $y = |x|; y = |2x + 2|$
12. $y = |x|; y = -0.5|x - 3|$
13. $y = |4x|; y = |4x + 8| + 6$
14. $y = \left|\frac{1}{3}x\right|; y = -\frac{1}{3}|x + 9| - 4$

State the letter corresponding to the graph of each equation using the figure at the right.

15. $y = |x - 3|$
16. $y = |x| + 1$
17. $y = -|x| - 3$
18. $y = |3x + 9|$
19. $y = |x + 3| - 4$
20. $y = -\frac{1}{2}|x - 4|$

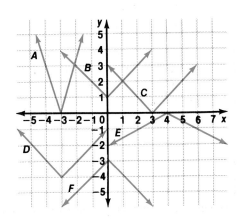

9-6　Graphing Inequalities in Two Variables

The graph of the equation $x = 3$ is a line that separates the coordinate plane into two regions. One region is shaded blue. The other is shaded yellow. Each region is called a **half-plane**. The line for $x = 3$ is called the **boundary**, or **edge**, for each half-plane.

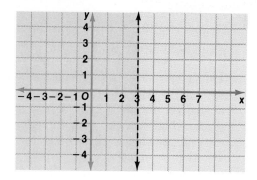

Choose a value for x that appears in the yellow region, namely 5. Since $5 > 3$, we see that the inequality $x > 3$ describes the yellow region. The x-coordinate of every point in that region is greater than 3. The inequality $x < 3$ describes the blue region. Points on the line for $x = 3$ are in neither region.

Consider the graphs of $y > x + 1$ and $y \le x + 1$. The boundary line in both is the line for $y = x + 1$.

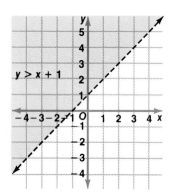

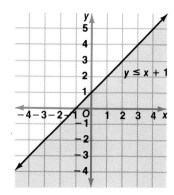

In the graph of $y > x + 1$, the boundary is *not* part of the graph. Therefore, the boundary is shown as a broken line. However, all points above the broken line are part of the graph. The graph is called an **open half-plane**.

The inequality $y \le x + 1$ means $y < x + 1$ or $y = x + 1$. Thus, the boundary line, $y = x + 1$, *is* part of the graph and is shown as a solid line. The graph also contains all points below the line. The graph is called a **closed half-plane**.

1 **Graph $y > -4x - 3$.**

First graph $y = -4x - 3$. Draw it as a broken line since this boundary is *not* part of the graph. Test a point on either side of the boundary. This will tell you which half-plane is the graph. The origin is an easy point to check. The coordinates of the origin, $(0, 0)$, satisfy the inequality $y > -4x - 3$. So, the coordinates of all the points on that side of the boundary satisfy the inequality.

Check: Choose a point on the opposite side of the boundary, say $(-2, -2)$. Test the point in the inequality. This ordered pair does not satisfy the inequality $y > -4x - 3$.

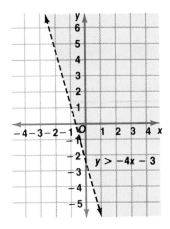

2 **Graph $-2y + 3x \geq 8$.**

First solve for y.

$$-2y + 3x \geq 8$$
$$-2y \geq -3x + 8$$
$$y \leq \frac{3}{2}x - 4$$

Then graph $y = \frac{3}{2}x - 4$. Draw it as a solid line since this boundary *is* part of the graph. Then test a point to determine which half-plane is also part of the graph. The origin is not part of the graph. Therefore, the graph is the right half-plane and the boundary.

Check this solution.

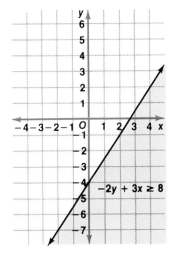

3 **Graph $y \leq 3x$.**

First graph $y = 3x$. Draw it as a solid line since this boundary *is* part of the graph. Then test a point to determine which half-plane is also part of the graph. The origin is on the boundary, so you must test a point other than the origin. Try $(1, 1)$. Since $(1, 1)$ satisfies the inequality, the graph is the right half-plane and the boundary.

Check this solution.

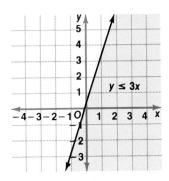

Exploratory Exercises

State which of the ordered pairs given are solutions of the inequality. Then tell whether the boundary line would be included in the graph of the inequality.

1. $x + 2y \geq 3$ **a.** $(-2, 2)$ **b.** $(4, -1)$ **c.** $(3, 1)$
2. $2x - 3y \leq 1$ **a.** $(2, 1)$ **b.** $(5, -1)$ **c.** $(1, 1)$
3. $5x + 2 > 3y$ **a.** $(2, 3)$ **b.** $(4, 7)$ **c.** $(-2, -3)$
4. $4y - 8 \geq 0$ **a.** $(0, 2)$ **b.** $(2, 5)$ **c.** $(-2, 0)$
5. $3x - 2y \leq 6$ **a.** $(4, -3)$ **b.** $(2, 0)$ **c.** $(0, 3)$
6. $-2x < 8 - y$ **a.** $(5, 10)$ **b.** $(3, 6)$ **c.** $(-4, 0)$
7. $3x + 4y < 7$ **a.** $(1, 1)$ **b.** $(2, -1)$ **c.** $(-2, 4)$
8. $y > x - 1$ **a.** $(5, 8)$ **b.** $(3, 3)$ **c.** $(-2, -3)$

Use $(0, 0)$ as a test point to determine whether it satisfies each inequality. Then describe which half-plane is part of each graph.

9.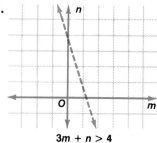

$3m + n > 4$

10.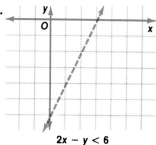

$2x - y < 6$

11.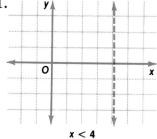

$x < 4$

12.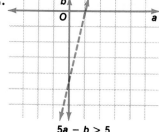

$5a - b > 5$

13.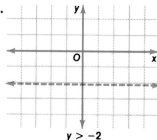

$y > -2$

14.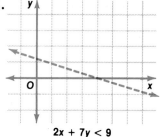

$2x + 7y < 9$

Written Exercises

Graph each inequality. Draw a separate set of axes for each.

15. $y > 3$ 16. $y \leq -2$ 17. $x \leq -1$ 18. $x > 4$
19. $x + y > 1$ 20. $x + y < 2$ 21. $x + y < -2$ 22. $x + y > 4$
23. $2x - y < 1$ 24. $3x + y > 1$ 25. $2x + 3y \geq -2$ 26. $2x - 2y \leq 2$
27. $x - 2y < 4$ 28. $4y + x < 16$ 29. $2x > 3y$ 30. $x < y$
31. $-x < -y$ 32. $-y > x$ 33. $y - x > 0$ 34. $2x < -y$
35. $y > x - 1$ 36. $y \leq x + 1$ 37. $y \leq 3x - 1$ 38. $y > 4x - 1$

Challenge Exercises

39. $y > |x|$ 40. $|y| \geq 2$ 41. $y > 2$ and $x < 3$ 42. $y > 2$ or $y < 1$

mini-review

Factor, if possible.

1. $10x^2 + 11x - 3$ **2.** $3x^3 - 42x^2 + 147x$ **3.** $81 - n^4$

4. $10a + 8a^2 - 15b - 12ab$ **5.** $4m + 4b - x^2m - x^2b$

Simplify.

6. $\frac{14a^2b^3}{56ab^4c}$ **7.** $\frac{y^2 - 49}{y^2 - 14y + 49}$ **8.** $\frac{p^2 + 8p + 12}{2p^2 - 3p - 14}$

Solve each equation for the domain given.

9. $3x - y = 2$; domain $= \{-3, -2, 1, 5\}$ **10.** $5m + 2n = 3$; domain $= \{-2, 0, 1\}$

Given $f(x) = 3x - 3$ and $g(x) = 1 + 3x^2$, determine the value of each function.

11. $f(2)$ **12.** $g(-2)$ **13.** $f(3a)$ **14.** $g(0)$

15. $f(-4)$ **16.** $g(2a)$ **17.** $f(0)$ **18.** $g(1)$

Solve each problem.

19. A certain triangle has two congruent sides. The third side is 11 in. longer than either of the congruent sides. If the perimeter is 53 in., what is the length of the third side?

20. The square of a number is equal to 5 times the number increased by 66. Find the number.

Excursions in Algebra   Misleading Graphs

Greg Friedman had received scores of 72, 74, 75, 77, and 78 on his first five algebra exams. The two graphs below were made to show how his test scores increased. Do they show the same results?

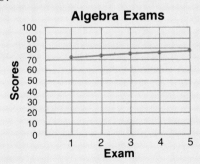

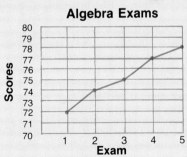

Both graphs use the same data but appear to show different results. Different scales change the appearance of the graphs. Sometimes different scales are used to make the graphs easier to read. Other times scales are chosen to make the data appear to support one point of view. Which graph do you think Greg showed his parents? Why?

9-7 Finding Equations from Relations

Suppose you purchased a number of packages of chewing gum. If each package contained five sticks of gum, you could make a chart to show the relationship between the number of packages of gum and the number of sticks. Use x for the number of packages and y for the number of sticks.

x	1	2	3	4	5	6
y	5	10	15	20	25	30

This relation can also be shown as an equation. Since y is always five times x, the equation is $y = 5x$. Another way to discover this relationship is to study the differences between successive values of x and y.

	+1	+1	+1	+1	+1	
x	1	2	3	4	5	6
y	5	10	15	20	25	30
	+5	+5	+5	+5	+5	

Notice that the differences of the y-values are exactly five times the differences of the x-values. This suggests the relation $y = 5x$.

The following chart shows a different relationship.

	+1	+1	+1	+1	+1	
a	1	2	3	4	5	6
b	7	11	15	19	23	27
	+4	+4	+4	+4	+4	

Notice that the differences of the b-values are four times the differences of the a-values. This suggests the relation $b = 4a$. However, $b = 4a$ does *not* describe the relationship given in the chart. If you substitute 1 for a in $b = 4a$, then $b = 4$, not 7, as given in the chart. To obtain a b-value of 7 when $a = 1$, you need to add 3. This suggests the relation $b = 4a + 3$. Check the relationship with other values.

If $a = 2$ and $b = 4a + 3$,
then $b = 4(2) + 3$.
$= 8 + 3$
$= 11$ ✔

If $a = 5$ and $b = 4a + 3$,
then $b = 4(5) + 3$.
$= 20 + 3$
$= 23$ ✔

1 Write an equation for the relation given in the chart.

x	1	2	3	4	5	6
y	1	4	7	10	13	16

Find the differences.

y = 3x does not correspond with the values in the chart.

$$+1 \quad +1 \quad +1 \quad +1 \quad +1$$

x	1	2	3	4	5	6
y	1	4	7	10	13	16

$$+3 \quad +3 \quad +3 \quad +3 \quad +3$$

The differences suggest $y = 3x$, but this equation does *not* describe the relation. If $x = 1$ then $y = 3$ not 1. You need to subtract 2 from $3x$. Thus, $y = 3x - 2$ will describe the relation.   *Check this relationship with other values.*

2 Write an equation for the relation given in the chart.

m	1	2	4	5	6	9
n	9	6	0	−3	−6	−15

Find the differences.

Notice that the values for m are increasing, but the values for n are decreasing.

$$+1 \quad +2 \quad +1 \quad +1 \quad +3$$

m	1	2	4	5	6	9
n	9	6	0	−3	−6	−15

$$-3 \quad -6 \quad -3 \quad -3 \quad -9$$

The differences suggest $n = -3m$. This means if $m = 1$, then $n = -3$, not 9. This equation does *not* describe the relation. For n to equal 9, you must add 12 to -3. Thus, the equation for the relation is $n = -3m + 12$, or $n = 12 - 3m$.

Exploratory Exercises

Determine whether each equation is the correct equation for the relation given in each chart. If it is, then complete the chart.

1. $y = 2x + 6$

x	3	4	5	6	7	8	9
y	12	14	16				

2. $b = 14 - 3a$

a	−4	−3	−2	−1	0	1	2
b	26	23	19				

3. $n = \frac{1}{2}m - 5$

m	2	4	6	8	10	12	14
n	−4			−1	1		

4. $d = 1 - \frac{2}{3}c$

c	3	6	9	12	15	18	21
d		−3		−7		−11	

Written Exercises

Find the missing terms in each sequence.

5. 5, 7, 9, 11, —, —, —

6. 12, 9, 6, 3, —, —, —

7. $-1, -4, -7, -10,$ —, —

8. 5, —, —, 20, 25, —, —

9. 6, 11, —, —, 26, —

10. 1, 2, 4, 8, —, —, —

11. $3, 1, \frac{1}{3},$ —, —, —

12. $4, -2, 1, -\frac{1}{2},$ —, —, —

13. 1, 0, 1, 0, —, —, —

14. 1, 11, 21, 31, —, —, —

15. $3, 1, -1, -3,$ —, —, —

16. 17, 14, —, —, 5, —

17. 3, 3.5, 4, 4.5, —, —, —

18. 11.75, 11.5, —, —, —

19. $-6.2, -5.7,$ —, —, $-4.2,$ —

20. $-8,$ —, $-8.4,$ —, —, -9

Write an equation for the relation given in each chart. Then copy and complete each chart.

21.

x	1	2	3	4	5	6
y	4	8	12	16		

22.

x	1	2	3	4	5	6
y	-3	-6	-9			-18

23.

m	-3	-2	-1	0	1	2	3
n	-5	-3	-1				

24.

a	-2	-1	0	1	2	3	4
b	-3	1	5				

25.

a	-4	-2	0	2	4	6	8
b	-13	-5	3	11			

26.

x	1	3	5	7	9	11	13
y	5	17	29	41			

27.

x	1	2	3	4	5	6	7
y	14	13	12				

28.

m	-2	-1	0	1	2	3	4
n	13	12	11	10			

29.

a	-5	-3	-1	1	2	4	7
b	28	18	8				

30.

x	-4	-2	0	2	4	6	8
y	26	22	18	14			

31.

r	-4	-2	0	2	4	6	8
s	-1	0	1				

32.

c	6	12	18	24	30	36	42
d		2	4				10

Challenge Exercises

Write an equation for each relation.

33. $\{(-2, 4), (-1, 1), (0, 0), (1, 1), (2, 4), (3, 9)\}$

34. $\{(1, 2), (2, 9), (3, 28), (4, 65), (5, 126)\}$

35. $\{(-2, 11), (-1, 14), (0, 15), (1, 14), (2, 11)\}$

36. $\{(-2, 0), (-1, -3), (0, -4), (1, -3), (2, 0)\}$

37. $\{(1, 24), (2, 12), (3, 8), (4, 6), (6, 4)\}$

38. $\{(-4, 3), (-2, 12), (-1, 48), (1, 48), (2, 12)\}$

FOR/NEXT Statements

An important step in graphing an equation is to select several values for x and then find the corresponding values for y. This means that the same expression is evaluated many times with different values for x. A computer can perform the operations easily using a loop.

FOR/ NEXT statements provide the simplest means for establishing a loop.

This BASIC program uses a loop to generate several ordered pairs that could be used to graph the equation $y = 2x - 1$.

```
10  FOR X = -2 TO 2
20  LET Y = 2 * X - 1
30  PRINT "(";X;",";Y;")"
40  NEXT X
50  END
```

What values for x will the computer read?

When the program is executed, five ordered pairs, $(-2, -5)$, $(-1, -3)$, $(0, -1)$, $(1, 1)$, and $(2, 3)$, are printed.

Exercises

1. Modify line 20 of the program to generate ordered pairs for $y = x + 2$, $y = 2x + 2$, $y = 3x + 2$, $y = 10x + 2$, and $y = 0.5x + 2$. Graph the five equations on the same axes. What effect does the coefficient of x have on the graph? At what point do the graphs intersect the y-axis?

2. Generate ordered pairs for $y = -x + 2$, $y = -2x + 2$, $y = -3x + 2$, $y = -10x + 2$, and $y = -0.5x + 2$. Graph the five equations on the same axes. If the coefficient of x is negative, describe the graphs. At what point do the graphs intersect the y-axis?

3. Generate ordered pairs for $y = 2x - 3$, $y = 2x$, and $y = 2x + 4$. Graph the three equations on the same axes. Describe the graphs. At what points do the graphs intersect the y-axis?

4. Generate ordered pairs for $y = -2x - 3$, $y = -2x$, and $y = -2x + 4$. Graph the three equations on the same axes. Describe the graphs. At what points do these graphs intersect the y-axis?

Vocabulary

ordered pairs   (289)

origin   (289)

x-axis   (289)

y-axis   (289)

x-coordinate   (289)

y-coordinate   (289)

coordinate plane   (289)

graph   (290)

quadrants   (290)

relation   (296)

domain   (296)

range   (296)

mapping   (292)

inverse of a relation   (292)

solution   (296)

linear equation   (301)

function   (305)

functional value   (307)

half-plane   (312)

boundary   (312)

edge   (312)

open half-plane   (312)

closed half-plane   (312)

Chapter Summary

1. Ordered pairs of numbers are used to locate points in the plane. The first component (*x*-coordinate) of an ordered pair corresponds to a number on the horizontal, or *x*-axis. The second component (*y*-coordinate) of the ordered pair corresponds to a number on the vertical, or *y*-axis.   (289)

2. Completeness Property for Points in the Plane: Exactly one point in the plane is located by a given ordered pair of numbers. Exactly one ordered pair of numbers locates a given point in the plane.   (290)

3. The *x*-axis and *y*-axis separate the plane into four regions, called quadrants.   (290)

4. A relation is a set of ordered pairs.   (292)

5. The domain of a relation is the set of all first components from each ordered pair.   (292)

6. The range of a relation is the set of all second components from each ordered pair.   (292)

7. Relation Q is the inverse of Relation S if and only if for every ordered pair (*a*, *b*) in S, there is an ordered pair (*b*, *a*) in Q.   (293)

8. The solutions to equations in two variables are ordered pairs.   (296)

9. A linear equation is an equation that can be written in the form $Ax + By = C$, where A, B, and C are any numbers, and A and B are not both 0.   (301)

10. A function is a relation in which each element of the domain is paired with *exactly* one element of the range.   (305)

11. Vertical Line Test for a Function: If any vertical line passes through no more than one point of the graph of a relation, then the relation is a function. (306)

12. $f(x)$ is read "*f* of *x*" and represents the range of a function whose domain is represented by x. (307)

13. When graphing inequalities, an open half-plane is used if the boundary is not part of the solution set. A closed half-plane is used if the boundary is part of the solution set. (312)

Chapter Review

9–1 State the quadrant in which the point for each ordered pair is located.

  1. $(8, -2)$ **2.** $(-3, -3)$ **3.** $(5, 6)$ **4.** $(-2, 5)$

  Graph each point.

  5. $A(6, -5)$ **6.** $B(-4, 0)$ **7.** $C(-2, -3)$ **8.** $D(3, 5)$

9–2 State the domain, range, and inverse for each relation.

  9. $\{(3, 5), (2, 6), (5, 7)\}$ **10.** $\{(4, 1), (4, -2), (4, 6), (4, -1)\}$
  11. $\{(-3, 5), (-3, 6), (4, 5), (4, 6)\}$ **12.** $\{(1, 3), (1, 5), (1, 7)\}$

  State each relation as a set of ordered pairs. Then state the domain and range.

13.

x	y
-1	4
3	4
4	6
0	-4

14.

x	y
-1	1
-2	4
-3	9
-4	6

15.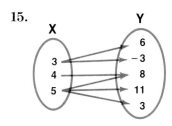

16.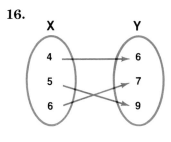

17.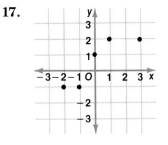

18.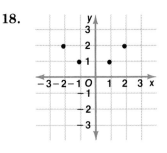

9–3   Solve each equation for y in terms of x.

19. $3x + y = 7$ $\qquad\qquad\qquad\qquad$ **20.** $4x - 3y = 9$

State which of the ordered pairs are solutions of the given equation.

21. $2x - 3y = 8$ $\qquad$ **a.** $(4, 4)$ $\qquad$ **b.** $(6, 2)$ $\qquad$ **c.** $(4, 0)$ $\qquad$ **d.** $(2, 4)$

22. $y = 3x - 6$ $\qquad$ **a.** $(0, 2)$ $\qquad$ **b.** $(2, 0)$ $\qquad$ **c.** $(6, 6)$ $\qquad$ **d.** $(-6, 0)$

Solve each equation if the domain is $\{-4, -2, 0, 2, 4\}$.

23. $y = 4x + 5$ $\qquad\qquad\qquad\qquad$ **24.** $x - y = 9$

25. $3x + 2y = 9$ $\qquad\qquad\qquad\qquad$ **26.** $4x - 3y = 0$

9–4   State whether each equation is a linear equation.

27. $3x - y^2 = 6$ $\qquad\qquad\qquad\qquad$ **28.** $xy = 8$

29. $y = 6$ $\qquad\qquad\qquad\qquad\qquad$ **30.** $2x - 8y = 5$

Graph each equation.

31. $2x - 3y = 6$ $\qquad\qquad\qquad\qquad$ **32.** $x + 5y = 4$

33. $y = 3x - 9$ $\qquad\qquad\qquad\qquad$ **34.** $\frac{1}{2}x + \frac{1}{3}y = 3$

9–5   State whether each relation is a function.

35. $\{(3, 2), (5, 3), (4, 3), (5, 2)\}$ $\qquad$ **36.** $\{(3, 8), (9, 3), (-3, 8), (5, 3)\}$

37. $3x - 4y = 7$ $\qquad\qquad\qquad\qquad$ **38.** $x^2 - y = 4$

Given $g(x) = x^2 - x + 1$, determine each functional value.

39. $g(2)$ $\qquad$ **40.** $g(-2)$ $\qquad$ **41.** $g(0)$ $\qquad$ **42.** $g\left(\frac{1}{2}\right)$ $\qquad$ **43.** $g(-1)$ $\qquad$ **44.** $g(2a)$

9–6   Graph each inequality. Draw a separate set of axes for each.

45. $x + 2y > 5$ $\qquad\qquad\qquad\qquad$ **46.** $4x - y \leq 8$

47. $3x - 2y < 6$ $\qquad\qquad\qquad\qquad$ **48.** $\frac{1}{2}y \geq x + 4$

9–7   Write an equation for the relation given in each chart.

49.

x	0	1	2	3	4
y	5	8	11	14	17

50.

x	2	4	5	7	10
y	-2	0	1	3	6

State the quadrant in which the point for each ordered pair is located.

1. $(-3, 5)$

2. $(-4, -1)$

3. $(5, -9)$

State each relation as a set of ordered pairs. Then state the domain, range, and inverse.

4.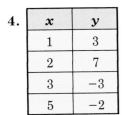

x	y
1	3
2	7
3	-3
5	-2

5.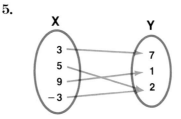

6.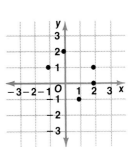

Solve each equation if the domain is $\{-2, -1, 0, 1, 3\}$.

7. $y = 3x + 10$

8. $2x - 5y = 4$

9. $2y - x = 8$

State whether each equation is a linear equation.

10. $5x = 17 - 4y$

11. $y = x^2 - y$

12. $y = \frac{1}{x}$

Graph each equation.

13. $x - 2y = 8$

14. $5x - 2y = 8$

15. $\frac{2}{3}x + \frac{3}{4}y = 1$

State whether each relation is a function.

16. $\{(2, 4), (4, 2), (5, 4), (4, 6)\}$

17. $\{(-6, 9), (4, 5), (8, -2), (0, 0)\}$

18. $8x - 3y = 7$

19. $2x = 9$

Given $f(x) = 2x - 3$, determine each functional value.

20. $f(-3)$

21. $f(7)$

22. $f(0)$

Graph each equation or inequality. Draw a separate set of axes for each.

23. $4x + 3y = 12$

24. $3x - 2y < 6$

25. $y \geq 5x + 1$

26. $4x + 2y = 9$

Write an equation for the relation given in the chart.

27.

x	1	2	3	4	7
y	3	8	13	18	33

CHAPTER 10

Lines and Slopes

Each incline of the roller coaster ride has a steepness, or slope. Some of the inclines have a greater slope than others. Slope is also used in mathematics to give information about the graphs of equations.

10-1 Slope

Jack Ferguson is a truck driver. When driving he sees many road signs similar to the one shown at the right. Suppose that the grade of a hill is 6%. This means that for every 6 feet of *rise* (vertical distance), there is a *run* (horizontal distance) of 100 feet. The ratio of *rise to run* is called **slope**. The slope of a line describes its steepness, or rate of change.

6% GRADE
3 MILES

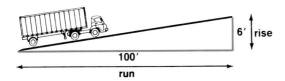

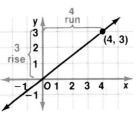

On the graph at the right, the line passes through the origin, (0, 0), and (4, 3). The rise (change in the *y*-coordinates) is 3, while the run, (change in the *x*-coordinates) is 4. Thus, the slope of this line is $\frac{3}{4}$.

The slope of a line is the ratio of the change in *y* to the corresponding change in *x*.

$$\text{Slope} = \frac{\text{change in } y}{\text{change in } x} \qquad \text{or} \qquad m = \frac{\text{change in } y}{\text{change in } x}$$

Definition of Slope

The traditional variable for slope is m.

Example

1 Determine the slope of each line.

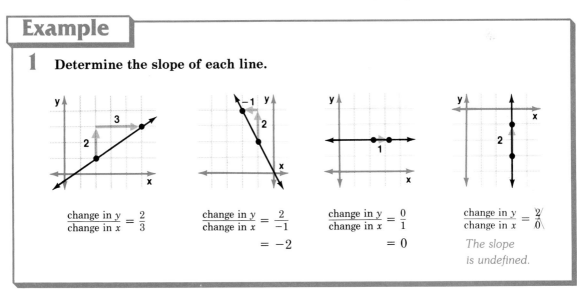

$\dfrac{\text{change in } y}{\text{change in } x} = \dfrac{2}{3}$

$\dfrac{\text{change in } y}{\text{change in } x} = \dfrac{2}{-1}$
$= -2$

$\dfrac{\text{change in } y}{\text{change in } x} = \dfrac{0}{1}$
$= 0$

$\dfrac{\text{change in } y}{\text{change in } x} = \dfrac{2}{0}$
The slope is undefined.

Notice that in **Example 1**, the line extending from lower left to upper right has a positive slope. The line extending from upper left to lower right has a negative slope. The slope of the horizontal line is zero.

For a vertical line the change in x would be zero. Since division by zero is not defined, *the slope of a vertical line is undefined.* We say that a vertical line has *no slope.*

Example

2 **Determine the slope of the line containing the points listed in the table.**

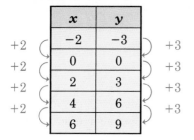

Notice that y increases 3 units for each 2 units that x increases.

$$\text{slope} = \frac{\text{change in } y}{\text{change in } x}$$

$$= \frac{3}{2}$$

The example above and other similar examples suggest that the slope of a nonvertical line can be determined from the coordinates of any two points on the line.

<table>
<tr><td>

Given the coordinates of two points on a line, (x_1, y_1) and (x_2, y_2), the slope, m, can be found as follows.

$$m = \frac{y_2 - y_1}{x_2 - x_1} \text{ where } x_2 \neq x_1$$

</td><td>

*Determining Slope
Given Two Points*

</td></tr>
</table>

y_2 is read "y sub 2." The 2 is a subscript.

Example

3 **Determine the slope of the line passing through $(3, -9)$ and $(4, -12)$.**

$$m = \frac{y_2 - y_1}{x_2 - x_1}$$

$$= \frac{-12 - (-9)}{4 - 3}$$

$$= \frac{-3}{1}$$

$$= -3$$

The difference of y-coordinates was expressed as $-12 - (-9)$. Suppose $-9 - (-12)$ had been used instead. The corresponding difference of the x-coordinates would have been $3 - 4$.

Notice that $\frac{-9 - (-12)}{3 - 4}$ is also equal to -3.

Example

4 Determine the value of r so the line through $(r, 4)$ and $(9, -2)$ has a slope of $-\frac{3}{2}$.

$$m = \frac{y_2 - y_1}{x_2 - x_1}$$

$$-\frac{3}{2} = \frac{-2 - 4}{9 - r}$$ *Replace each variable with its appropriate value.*

$$-\frac{3}{2} = \frac{-6}{9 - r}$$

$$-3(9 - r) = -6(2)$$ *Means-Extremes Property*

$$-27 + 3r = -12$$ *Solve for r.*

$$3r = 15$$

$$r = 5$$

Exploratory Exercises

State the slope of each line.

1.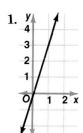

2.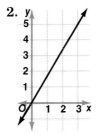

3.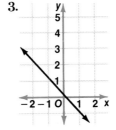

4.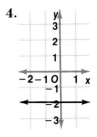

For each table, state the change in y and the change in x. Then determine the slope of the line passing through the points listed.

5.
x	y
0	0
1	2
2	4
3	6
4	8

6.
x	y
0	0
1	1
2	2
3	3
4	4

7.
x	y
-2	2
-1	1
0	0
1	-1
2	-2

8.
x	y
-2	-8
-1	-4
0	0
1	4
2	8

9.
x	y
-6	8
-3	4
0	0
3	-4
6	-8

Draw a line through the given point with the given slope.

10. $(2, 4)$, $m = \frac{1}{3}$

11. $(4, -1)$, $m = -\frac{2}{5}$

12. $(-3, 4)$, $m = 4$

13. $(-1, 5)$, $m = -2$

14. $(-3, -3)$, $m = \frac{4}{3}$

15. $(2, -4)$, $m = 0$

Written Exercises

Determine the slope of each line named below.

16. *a*

17. *b*

18. *c*

19. *d*

20. *e*

21. *f*

22. *g*

23. *h*

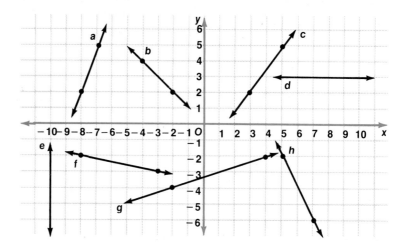

24. Look at lines *a*, *c*, and *g* above. These lines have positive slopes. What is the general direction of lines with positive slopes?

25. Look at lines *b*, *f*, and *h* above. These lines have negative slopes. What is the general direction of lines with negative slopes?

26. What is the slope of a horizontal line?

27. What is the slope of a vertical line?

Determine the slope of the line passing through each pair of points.

28. (3, 4), (4, 6)

29. (5, 2), (7, 6)

30. (−3, 6), (−5, 9)

31. (−2, 1), (−6, 3)

32. (−3, 6), (−8, 4)

33. (−1, 11), (−5, 4)

34. (7, −4), (9, −1)

35. (11, −1), (14, −6)

36. (18, −4), (6, −10)

37. (12, −3), (14, −7)

38. (14, 3), (−11, 3)

39. (−4, −6), (−3, −8)

40. $(0, 0), \left(\frac{1}{3}, \frac{1}{6}\right)$

41. $\left(\frac{3}{4}, \frac{1}{2}\right), \left(\frac{1}{2}, \frac{3}{4}\right)$

42. $\left(\frac{5}{8}, \frac{1}{3}\right), \left(\frac{1}{8}, \frac{2}{3}\right)$

43. $\left(\frac{3}{4}, 1\right), \left(\frac{3}{4}, -1\right)$

44. $\left(\frac{2}{3}, 5\right), \left(\frac{1}{3}, 2\right)$

45. $\left(3\frac{1}{2}, 5\frac{1}{4}\right), \left(2\frac{1}{2}, 6\right)$

Determine the value of *r* so the line through each pair of points has the given slope.

46. $(9, r), (6, 3), m = -\frac{1}{3}$

47. $(r, 4), (7, 3), m = \frac{3}{4}$

48. $(4, -7), (-2, r), m = \frac{8}{3}$

49. $(6, -2), (r, -6), m = -4$

50. $(r, 7), (11, r), m = -\frac{1}{5}$

51. $(4, r), (r, 2), m = -\frac{5}{3}$

Challenge Exercises

Determine the value of *r* so the line through each pair of points has the given slope.

52. $(9, r), (6, 2), m = r$

53. $(8, r^2), (3, -6), m = r$

54. $(r, 4), (3, -r), m = 2r$

55. $(1, r), (r, 8), m = \frac{1}{3}r$

10-2 Equations of Lines in Point-Slope and Standard Form

The graph at the right shows the non-vertical line, ℓ, and point R having coordinates (x_1, y_1). The slope of line ℓ is represented by m.

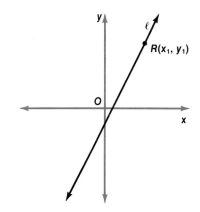

Let (x, y) represent any other point on line ℓ. Since m is the slope of the line, all points (x, y) on the line satisfy the following equation.

$$\frac{y - y_1}{x - x_1} = m \qquad \textit{Assume } x - x_1 \neq 0.$$

If each side of the equation is multiplied by $(x - x_1)$, the result is the equation $y - y_1 = m(x - x_1)$. This linear equation is said to be in **point-slope form**.

> For a given point (x_1, y_1) on a nonvertical line with slope m, the point-slope form of a linear equation is
> $$y - y_1 = m(x - x_1).$$

Point-Slope Form

If you know the slope of a line and the coordinates of one point on the line, you can write an equation of the line.

Example

1 Write the point-slope form of an equation of the line passing through $(2, -4)$ and having a slope of $\frac{2}{3}$.

Substitute in the point-slope form. Let $(x_1, y_1) = (2, -4)$ and $m = \frac{2}{3}$.

$$y - y_1 = m(x - x_1)$$
$$y - (-4) = \frac{2}{3}(x - 2)$$
$$y + 4 = \frac{2}{3}(x - 2)$$

An equation of the line is $y + 4 = \frac{2}{3}(x - 2)$.

Any linear equation can be expressed in the form $Ax + By = C$ where $A, B,$ and C are integers and A and B are not both zero. This is called the **standard form**. An equation that is written in point-slope form can be changed to standard form.

Example

2 **Write $y + 4 = \frac{2}{3}(x - 2)$ in standard form.**

$$y + 4 = \frac{2}{3}(x - 2)$$
$$3(y + 4) = 2(x - 2) \quad \textit{Multiply each side by 3 to eliminate the fractions.}$$
$$3y + 12 = 2x - 4$$
$$-2x + 3y = -16$$
$$2x - 3y = 16 \quad \textit{Multiply each side by -1 to obtain a positive coefficient for x.}$$

$2x - 3y = 16$ is in standard form.

You can also find an equation of a line if you know the coordinates of two points on the line.

Example

3 **Write the standard form of an equation of the line passing through $(5, -2)$ and $(-3, 8)$.**

First determine the slope.

$$m = \frac{y_2 - y_1}{x_2 - x_1}$$
$$= \frac{8 - (-2)}{-3 - 5} \quad \textit{Let $y_2 = 8$, $y_1 = -2$, $x_2 = -3$, and $x_1 = 5$.}$$
$$m = \frac{10}{-8} \text{ or } -\frac{5}{4}$$

Now, use the point-slope form. The slope is $-\frac{5}{4}$. Either $(5, -2)$ or $(-3, 8)$ can be substituted for (x_1, y_1).

$$y - y_1 = m(x - x_1) \quad \textit{Let $(x_1, y_1) = (5, -2)$.}$$
$$y - (-2) = -\frac{5}{4}(x - 5)$$
$$y + 2 = -\frac{5}{4}(x - 5) \quad \textit{This equation is in point-slope form.}$$

Now write the equation in standard form.

$$y + 2 = -\frac{5}{4}(x - 5)$$
$$4(y + 2) = -5(x - 5) \quad \textit{Multiply each side by 4 to eliminate the fractions.}$$
$$4y + 8 = -5x + 25$$
$$5x + 4y = 17 \quad \textit{This equation is in standard form.}$$

Check to see that you get the same result using the point $(-3, 8)$.

A horizontal line has a slope of zero. The point-slope form can be used to write the equation of horizontal lines. For example, the equation of a line through $(5, 2)$ and $(-9, 2)$ is $(y - 2) = 0(x - 5)$ or $y = 2$.

The slope of a vertical line is undefined. Therefore, the point-slope form cannot be used for vertical lines. The equation of a line through $(3, 5)$ and $(3, -9)$ is $x = 3$ since the x-coordinate of every point on the line is equal to 3. Note that the equation $x = 3$ is in standard form.

Exploratory Exercises

State the slope and a point through which the line passes for each linear equation.

1. $y - 2 = 3(x - 5)$

2. $y - 3 = 4(x - 6)$

3. $y - (-5) = -2(x + 1)$

4. $y + 4 = -2[x - (-2)]$

5. $y + 6 = -\frac{3}{2}(x + 5)$

6. $y + 7 = -\frac{3}{4}(x + 4)$

7. $2(x - 3) = y + \frac{3}{2}$

8. $-\frac{2}{3}(x + 7) = y - \frac{3}{4}$

9. $-\frac{3}{5}(x + 6) = y + \frac{3}{8}$

10. $y = 3$

11. $y = -2$

12. $x = 1$

Express each equation in standard form.

13. $y - 3 = 2\left(x + \frac{3}{2}\right)$

14. $y + 5 = -3\left(x - \frac{1}{3}\right)$

15. $y + 1 = \frac{2}{3}(x + 2)$

16. $y + \frac{3}{2} = \frac{1}{2}(x + 4)$

17. $y - \frac{2}{3} = \frac{5}{3}(x + 7)$

18. $y - 1 = \frac{5}{6}\left(x + \frac{3}{5}\right)$

Written Exercises

Write the standard form of an equation of the line passing through the given point and having the given slope.

19. $(5, 4)$, $-\frac{2}{3}$

20. $(-6, -3)$, $-\frac{1}{2}$

21. $(9, 1)$, $\frac{2}{3}$

22. $(8, 2)$, $\frac{3}{4}$

23. $(4, -3)$, 2

24. $(-2, 4)$, -3

25. $(-6, 1)$, $\frac{3}{2}$

26. $(6, -2)$, $\frac{4}{3}$

27. $(5, 7)$, 0

28. $(-2, 6)$, 0

29. $(1, 3)$, undefined

30. $(-2, 1)$, undefined

Write the standard form of an equation of the line passing through each pair of points.

31. $(5, 4)$, $(6, 3)$

32. $(9, 1)$, $(8, 2)$

33. $(6, 1)$, $(7, -4)$

34. $(8, 3)$, $(5, -1)$

35. $(4, -2)$, $(8, -3)$

36. $(6, -1)$, $(4, -2)$

37. $(-6, 1)$, $(-8, 2)$

38. $(-5, 1)$, $(6, -2)$

39. $(-8, 2)$, $(-1, -2)$

40. $(4, -2)$, $(4, 8)$

41. $(5, -4)$, $(5, 5)$

42. $(5, 3)$, $(-6, 3)$

43. $\left(\frac{3}{4}, 1\right)$, $\left(2, \frac{1}{2}\right)$

44. $\left(\frac{1}{2}, \frac{3}{4}\right)$, $\left(\frac{2}{3}, \frac{4}{5}\right)$

45. $\left(2\frac{1}{2}, \frac{1}{3}\right)$, $\left(\frac{3}{4}, 1\frac{1}{2}\right)$

46. $(4, 2)$, $(-7, 2)$

47. $\left(-8, \frac{1}{2}\right)$, $\left(9, \frac{1}{2}\right)$

48. $(3, -2)$, $(3, 5)$

49. $(-2, 7)$, $\left(-2, \frac{16}{3}\right)$

50. $\left(-\frac{2}{9}, 3\right)$, $\left(\frac{8}{9}, 3\right)$

51. $\left(-2, \frac{2}{3}\right)$, $\left(-2, \frac{2}{7}\right)$

The freezing point of water on the Celsius and Fahrenheit scales can be represented by the ordered pair (0, 32), where 0 is the Celsius temperature and 32 is the Fahrenheit temperature. Similarly, the boiling point of water can be represented by (100, 212). With these two points you can draw the graph of the line showing the relationship between Celsius and Fahrenheit.

The graph at the right can be used to quickly estimate equivalent temperatures. For a Fahrenheit temperature of 100, for example, you can estimate the corresponding Celsius temperature by first locating 100 on the Fahrenheit scale. Follow the value horizontally until you reach the line, then drop vertically to the Celsius scale. The corresponding Celsius temperature is between 35° and 40°.

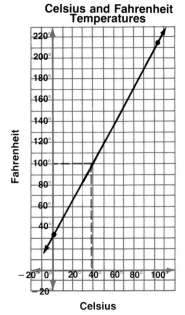

Celsius and Fahrenheit Temperatures

Speed of Sound (through water)

Many other relationships can be shown graphically. For example, the speed of sound through water is 1454 m/s, or about 1.5 km/s. This is represented on the graph by (1, 1.5). The graph at the left shows the relationship between the time in seconds and the distance in kilometers. Note that the ordered pairs (4, 6) and (6, 9) were used to draw the graph.

Exercises

Use the upper graph to estimate the Celsius or Fahrenheit equivalents.

1. 80°F
2. 65°F
3. 20°C
4. 50°C

5. Write an equation to represent the relationship between Fahrenheit and Celsius temperatures.

6. At what temperature are the Celsius and Fahrenheit values equal?

Use the lower graph to solve each problem.

7. Using sonar, a ship finds that it takes sound waves 2.5 seconds to reach the ocean floor. Estimate the depth of the ocean at this point.

8. Estimate how many seconds it would take sound waves to reach a depth of 5 km.

10-3 Slope-Intercept Form

The x-coordinate of the point where a line crosses the x-axis is called the **x-intercept** of the line. Line ℓ crosses the x-axis at $(4, 0)$. Therefore, the x-intercept is 4. Note that the corresponding y-coordinate is 0.

Similarly, the y-coordinate of the point where the line crosses the y-axis is called the **y-intercept** of the line. Line ℓ crosses the y-axis at $(0, 6)$. Therefore, the y-intercept is 6. Note that the corresponding x-coordinate is 0.

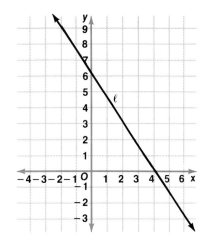

Consider the graph at the right. The line with slope m crosses the y-axis at $(0, b)$. You can write an equation for this line using the point-slope form. Let $(x_1, y_1) = (0, b)$.

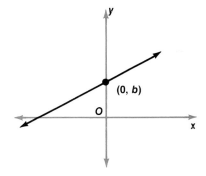

$$y - y_1 = m(x - x_1)$$
$$y - b = m(x - 0) \quad \textit{Replace } y_1 \textit{ with } b.$$
$$\textit{Replace } x_1 \textit{ with } 0.$$
$$y = mx + b$$
$$\uparrow \qquad \uparrow$$
$$\textit{slope} \quad \textit{y-intercept}$$

Given the points (x, y) and $(0, b)$ and slope m, the slope-intercept form of a linear equation is $y = mx + b$.	*Slope-Intercept Form*

Example

1   State the slope and y-intercept of the graph of $y = 5x - 3$.

$$y = mx + b$$
$$y = 5x + (-3)$$

Since $m = 5$, the slope is 5.
Since $b = -3$, the y-intercept is -3.

Compare the slope-intercept form of a linear equation with the standard form, $Ax + By = C$.

Solve for y.

$$Ax + By = C$$

$$By = -Ax + C \qquad \text{\textit{Assume } } B \neq 0.$$

$$y = -\frac{A}{B}x + \frac{C}{B} \qquad \text{\textit{This equation is in slope-}}$$
$$\qquad\qquad\qquad \text{\textit{intercept form.}}$$

$$\uparrow \qquad \uparrow$$
$$\text{\textit{slope}} \quad \text{\textit{y-intercept}}$$

If the equation is given in standard form and B is *not* zero, the slope of the line can be determined by computing $-\frac{A}{B}$ and the y-intercept can be determined by computing $\frac{C}{B}$.

Example

2 **State the slope and y-intercept of the graph of $3x + 2y = 6$.**

In the equation $3x + 2y = 6$, $A = 3$, $B = 2$, and $C = 6$.

slope: $\quad -\frac{A}{B} = -\frac{3}{2}$ $\qquad$ **y-intercept:** $\quad \frac{C}{B} = \frac{6}{2}$ or 3

Check: Write the equation in slope-intercept form.

$$3x + 2y = 6$$

$$2y = -3x + 6$$

$$y = -\frac{3}{2}x + 3 \qquad \text{\textit{The slope is } } -\frac{3}{2} \text{ \textit{and the y-intercept is 3.}}$$

Recall that the x-coordinate of the y-intercept is 0 and the y-coordinate of the x-intercept is 0. You can use these facts to find the x- and y-intercepts of the graph of a linear equation.

Example

3 **Determine the x- and y-intercepts of the graph of $3x - 6y = 12$.**

To find the x-intercept, let $y = 0$. $\qquad$ To find the y-intercept, let $x = 0$.

$$3x - 6y = 12 \qquad\qquad\qquad 3x - 6y = 12$$

$$3x - 6(0) = 12 \qquad\qquad\qquad 3(0) - 6y = 12$$

$$3x = 12 \qquad\qquad\qquad\qquad -6y = 12$$

$$x = 4 \qquad\qquad\qquad\qquad\quad y = -2$$

The x-intercept is 4. $\qquad\qquad$ The y-intercept is -2.
The graph crosses the $\qquad\qquad$ The graph crosses the
x-axis at $(4, 0)$. $\qquad\qquad\qquad$ y-axis at $(0, -2)$.

The y-intercept of an equation in the form $Ax + By = C$ is $\frac{C}{B}$. To find a formula for the x-intercept, let $y = 0$.

$$Ax + By = C$$
$$Ax + B(0) = C$$
$$Ax = C$$
$$x = \frac{C}{A}$$

The x-intercept is $\frac{C}{A}$, $A \neq 0$.

For the equation $2x - 3y = 8$, the x-intercept is $\frac{8}{2}$ or 4.

Exploratory Exercises

State the slope and y-intercept of the graph of each equation.

1. $y = 5x + 3$ **2.** $y = 2x + 1$ **3.** $y = 3x - 7$

4. $y = 6x - 8$ **5.** $y = \frac{1}{3}x$ **6.** $y = \frac{3}{4}x + 1$

7. $y = \frac{3}{5}x - \frac{1}{4}$ **8.** $y = \frac{2}{3}x$ **9.** $y = \frac{3}{5}x + \frac{1}{4}$

10. $2x + 3y = 5$ **11.** $-x + 4y = 3$ **12.** $2x - y = 1$

13. $y - 6x = 5$ **14.** $3y - 8x = 2$ **15.** $2y - 3x = 7$

16. $5y = -8x - 2$ **17.** $4y = 2x + 7$ **18.** $3y = -2x + 11$

19. State an equation of a line that has an undefined slope.

20. State an equation of a line that has no y-intercept.

Written Exercises

Write an equation in slope-intercept form of the line with the given slope and y-intercept.

21. $m = 3, b = 1$ **22.** $m = 2, b = 3$ **23.** $m = -3, b = 5$

24. $m = -4, b = 3$ **25.** $m = 4, b = -2$ **26.** $m = 2, b = -5$

27. $m = \frac{1}{2}, b = 5$ **28.** $m = \frac{2}{3}, b = 6$ **29.** $m = -\frac{5}{4}, b = 3$

30. $m = 0, b = 14$ **31.** $m = -3.1, b = 0.6$ **32.** $m = 4.2, b = -3.8$

Determine the x- and y-intercepts of the graph of each equation.

33. $3x + 2y = 6$ **34.** $5x + 3y = 15$ **35.** $5x + y = 10$

36. $2x + y = 6$ **37.** $3x + 4y = 24$ **38.** $2x - 7y = 28$

39. $3x - 2y = -5$ **40.** $2x + 5y = -11$ **41.** $8x + y = 4$

42. $6x + 2y = 3$

43. $x = -2$

44. $3y = 12$

45. $3x + \frac{1}{2}y = 8$

46. $\frac{3}{4}x - 2y = 7$

47. $\frac{1}{2}x - \frac{3}{2}y = 4$

48. $\frac{4}{3}x + \frac{5}{6}y = 10$

49. $2.2x + 0.5y = 1.1$

50. $1.8x - 2.5y = 5.4$

Determine the slope and y-intercept of the graph of each equation. Then write each equation in slope-intercept form.

51. $2x + 5y = 10$

52. $3x + 4y = 12$

53. $5x - y = 15$

54. $3x - y = 9$

55. $2x + 5y = 8$

56. $7x + 4y = 8$

57. $5x - 4y = 11$

58. $7x - 3y = 10$

59. $12x + 9y = 15$

60. $10x - 14y = 21$

61. $13x - 11y = 22$

62. $14x + 20y = 10$

63. $2x + \frac{1}{3}y = 5$

64. $3x - \frac{1}{4}y = 6$

65. $\frac{1}{2}x + \frac{1}{4}y = 3$

66. $\frac{2}{3}x + \frac{1}{6}y = 2$

67. $\frac{1}{4}x - \frac{3}{4}y = \frac{1}{2}$

68. $\frac{7}{4}x - \frac{5}{3}y = \frac{1}{6}$

69. $3x = 2y - 7$

70. $5x = 8 - 2y$

71. $3y = 2x - 7$

72. $8y = 4x + 12$

73. $1.1x - 0.2y = 3.2$

74. $0.3x - 0.5y = 1.8$

75. $5(x + 2) = y - 6x - 4$

76. $3(x - 7) = 2y + 5x + 8$

77. $y - 3x = 6(y + 7x) + 10$

78. $4y + x = 9 - 3(2y - 2x)$

79. $\frac{1}{2}x + 3y = \frac{2}{3}(x - 6y + 9)$

80. $\frac{4}{5}(2x - y) = 6x + \frac{2}{5}y - 10$

81. $\frac{3}{2}(4x + 9y) = 4(7x - \frac{1}{2}y)$

82. $-2\left(\frac{2}{3}x - y\right) = -\frac{1}{4}(3y + 8)$

83. Write an equation in slope-intercept form of the line with slope $\frac{4}{5}$ and y-intercept the same as the line whose equation is $3x + 4y = 6$.

84. Write an equation in slope-intercept form of the line with slope $-\frac{3}{5}$ and y-intercept the same as the line whose equation is $7x - 3y = 12$.

85. Write an equation in slope-intercept form of the line with y-intercept 12 and slope the same as the line whose equation is $2x - 5y - 10 = 0$.

86. Write an equation in slope-intercept form of the line with y-intercept -0.65 and slope the same as the line whose equation is $\frac{1}{2}x - \frac{3}{4}y = 6$.

87. Write an equation in slope-intercept form of the horizontal line with y-intercept 9.

88. Write an equation in slope-intercept form of the horizontal line with y-intercept -13.

Challenge Exercises

Solve each problem.

89. Find the coordinates of a point on the graph of $3x - 4y = -20$, if the y-coordinate is twice the x-coordinate.

90. Find the coordinates of a point on the graph of $4x - y = -2$, if the y-coordinate is three times the x-coordinate.

91. Find the coordinates of a point on the graph of $x + 2y = 11$, if the y-coordinate is 5 less than the x-coordinate.

92. Find the coordinates of a point on the graph of $7x + 3y = 2$, if the y-coordinate is 4 more than the x-coordinate.

mini-review

Factor, if possible.

1. $\frac{9}{25}x^2 - 49$ **2.** $18x^3 - 12x^2 + 2x$ **3.** $9k^2 + 12kp + 3kp + 4p^2$

Solve.

4. $m^2 = 11m$ **5.** $2x^2 = 13x + 7$ **6.** $8(x - 1)(2x + 1) = -9$

State the domain, range, and inverse for each relation.

7. $\{(2, 6), (2, 7), (2, -4), (2, -10)\}$ **8.** $\{(7, 3), (-2, -4), (-2, 8)\}$

State whether each relation is a function.

9. $\{(6, -3), (2, -3), (1, 7), (5, -2)\}$ **10.** $x^2 - y = 9$

Determine the slope of the line passing through each pair of points named below.

11. $(6, 3), (3, 7)$ **12.** $(2, 4), (2, -6)$ **13.** $\left(2\frac{1}{6}, 1\frac{3}{4}\right), \left(2\frac{2}{3}, 3\frac{1}{2}\right)$

Solve each problem.

14. Find two consecutive odd integers whose product is 195.

15. When one integer is added to the square of the next consecutive integer, the sum is 109. Find the integers.

Excursions in Algebra   Rise Over Run

Slope is sometimes defined as $\frac{\text{rise}}{\text{run}}$. In the coordinate plane, this is just another way of saying $\frac{\text{change in } y}{\text{change in } x}$.

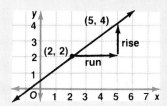

When building a stairway, a carpenter considers the ratio of riser to tread. This ratio is equivalent to $\frac{\text{rise}}{\text{run}}$ and describes the steepness of the stairs.

$$\frac{\text{riser}}{\text{tread}} = \frac{\text{rise}}{\text{run}} = \frac{8}{12} = \frac{2}{3}$$

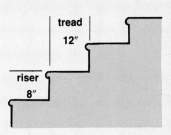

Exercises

Write the $\frac{\text{riser}}{\text{tread}}$ ratio for each problem.

	Riser	Tread
1.	9"	12"
2.	6"	$7\frac{1}{2}$"
3.	7"	9"

4. For Exercises 1–3, which stairs are the steepest?

10-4 Graphing Linear Equations

Previously you graphed a linear equation by finding any two ordered pairs that satisfied the equation. However, there are other methods that can be used. One such method is to use the x- and y-intercepts.

Example

1 **Graph $3x - 2y = 12$ using the x- and y-intercepts.**

To find the x-intercept,
let $y = 0$.

$$3x - 2y = 12$$
$$3x - 2(0) = 12$$
$$3x = 12$$
$$x = 4 \qquad \frac{C}{A} = 4$$

To find the y-intercept,
let $x = 0$.

$$3x - 2y = 12$$
$$3(0) - 2y = 12$$
$$-2y = 12$$
$$y = -6 \qquad \frac{C}{B} = -6$$

Graph $(4, 0)$ and $(0, -6)$. Then draw the line that passes through these points.

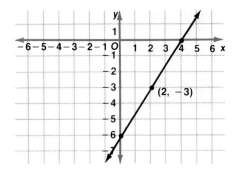

To check, choose some other point on the line and determine whether it is a solution of the linear equation. In this case, you could use $(2, -3)$.

$$3x - 2y = 12$$
$$3(2) - 2(-3) \stackrel{?}{=} 12$$
$$6 + 6 \stackrel{?}{=} 12$$
$$12 = 12 \quad ✔$$

If an equation is in standard form and A, B, or C is zero, the graph of the equation has only one intercept. In such cases, you must find some other ordered pairs that satisfy the equation.

When $A = 0$, the graph is horizontal. When $B = 0$, the graph is vertical. When $C = 0$, the graph passes through the origin.

If a linear equation is in slope-intercept form, then it is convenient to use the slope and y-intercept to draw the graph of the equation.

Example

2 **Graph $y = 3x + 2$ using the slope and y-intercept.**

The y-intercept is 2 so the graph passes through $(0, 2)$. Since the slope is 3, we know

$$3 = \frac{3}{1} = \frac{\text{change in } y}{\text{change in } x}.$$

Starting at $(0, 2)$ go up 3 units and to the right 1 unit. This will be the point $(1, 5)$. Then draw the line through $(0, 2)$ and $(1, 5)$.

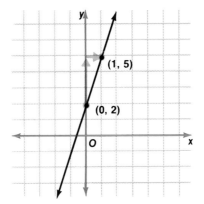

To check, you should substitute the points $(0, 2)$ and $(1, 5)$ in the equation $y = 3x + 2$ to determine whether they are solutions of the equation.

If the equation is in standard form, it is sometimes convenient to use the slope and y-intercept to draw the graph. Recall that for an equation in standard form, $-\frac{A}{B}$ is the slope of the line and $\frac{C}{B}$ is the y-intercept.

Consider the line $-4x + 3y = 12$. The slope, $-\frac{A}{B}$, is $\frac{4}{3}$. The y-intercept, $\frac{C}{B}$, is $\frac{12}{3}$ or 4. To draw the graph, use the slope and y-intercept, as in Example 2, and check the graph with at least one other ordered pair on the line.

Exploratory Exercises

State the x-intercept and y-intercept for each equation.

1. $4x + y = 8$
2. $6x + y = 6$
3. $3x + 4y = 6$
4. $5x + 2y = 20$
5. $2x - y = 8$
6. $3x - 2y = 6$
7. $7x + 2y = 10$
8. $5x + 6y = 8$
9. $5x - \frac{1}{2}y = 2$
10. $3x + \frac{1}{3}y = 2$
11. $y = 3x - 1$
12. $y = 4x + 3$
13. $2x = 5 - 3y$
14. $3x = 4y + 1$
15. $\frac{1}{2}x + \frac{1}{3}y = 0$

Given the slope and a point on a line, name two other points on the line.

16. $m = \frac{4}{3}$, $(0, 1)$
17. $m = 2$, $(0, 1)$
18. $m = \frac{5}{2}$, $(0, 3)$
19. $m = -7$, $(0, 4)$
20. $m = -\frac{1}{4}$, $(6, 1)$
21. $m = \frac{1}{2}$, $(5, 2)$
22. $m = -\frac{2}{3}$, $(-3, 2)$
23. $m = 8$, $(-3, 6)$
24. $m = 0$, $(3, 2)$

Written Exercises

Graph each equation using the x- and y-intercepts.

25. $6x - 3y = 6$
26. $4x + 5y = 20$
27. $5x - y = -10$
28. $2x + 10y = 5$
29. $2x + 5y = -10$
30. $7x - 2y = -7$
31. $y = 6x - 9$
32. $x + \frac{1}{2}y = 4$
33. $\frac{1}{2}x + \frac{2}{3}y = -3$
34. $x = 8y - 4$
35. $x = \frac{3}{4}y + 6$
36. $\frac{2}{3}y = \frac{1}{2}x + 6$

Graph each equation using the slope and y-intercept.

37. $y = \frac{2}{3}x + 3$
38. $y = \frac{3}{4}x + 4$
39. $y = -\frac{3}{5}x - 1$
40. $y = -\frac{3}{4}x + 4$
41. $y = \frac{3}{2}x - 5$
42. $y = \frac{1}{2}x + 4$
43. $-4x + y = 6$
44. $-3x + 2y = 12$
45. $-2x + y = 3$
46. $3y - 7 = 2x$
47. $5x + 2 = 7y$
48. $\frac{3}{4}x + \frac{1}{2}y = 4$

49. Compare the equations and graphs of Exercises 37 and 45. What is true about their y-intercepts? What is their point of intersection?

50. Compare the equations and graphs of Exercises 38 and 40. What is true about their slopes? What is the relationship between the lines?

51. Compare the equations and graphs of Exercises 37 and 46. What is true about their slopes? What is the relationship between the lines?

52. Compare the equations and graphs of Exercises 46 and 48. What is true about their slopes? What is the relationship between the lines?

Simplify.

1. $\dfrac{y + 3}{y^2 + y - 6}$

2. $\dfrac{x + 5}{x^2 - 4x - 45}$

3. $\dfrac{p^2 + p - 2}{p^2 - p - 6}$

Factor, if possible.

4. $4a^2 - 8ab + 12ab^2$

5. $b^2 - 25$

6. $m^2 - 6m + 9$

7. $12a^2 + 29a + 15$

8. $8x^2y - 3x - 16xy + 6$

Solve.

9. $n^2 + n = 12$

10. $(y + 2)(2y - 5) = -y - 10$

11. $5x(2x - 3) = 0$

12. $3m^2 - 108 = 0$

13. Solve $4m - 6n = 14$ for n.

14. Solve $4x + 2y = 12$ if the domain is $\{-1, 0, 1, 2, 3\}$.

15. Is $x^2 + 3x - 4 = 0$ a linear equation?

16. Is the graph of $x = 7$ a function?

17. State the quadrant in which the graph of the ordered pair $(-3, 5)$ is located.

18. State the domain, range, and inverse of the relation $\{(1, 2), (5, 7), (6, 3), (11, -2), (-3, 7)\}$.

19. Write an equation for the relationship between the variables in the chart.

x	0	2	4	6
y	-6	-9	-12	-15

20. Given $f(x) = 5x^2 - 3x + 2$, find $f(-3)$.

21. Determine the slope of the line passing through $(4, -2)$ and $(8, 1)$.

22. Graph $6x - 2y \le 10$.

23. Graph $3x - 2y = 12$ using the x- and y-intercepts.

24. Graph $2x + 3y = -6$ using the slope and y-intercept.

25. Determine the value of r so that the line through $(2, r)$ and $(5, 1)$ has a slope of $-\dfrac{5}{3}$.

26. Write the slope-intercept form of the line through $(0, -6)$ and with a slope of $\dfrac{7}{4}$.

27. Write an equation in slope-intercept form of the horizontal line with y-intercept -4.

28. Write the standard form of an equation of the line through $(4, -1)$ and $(2, 5)$.

Use the formula $h = vt - 16t^2$ to find the missing quantity.

29. Find v, if $t = 6$ seconds and $h = 28$ feet.

Solve each problem.

30. The product of a certain negative number increased by 7 and the same number decreased by $\dfrac{5}{4}$ is 0. What is the number?

31. Find two integers whose sum is 12 and whose product is 32.

32. The area of a rectangle is 64 in^2. The length is 4 more than 3 times the width. Find the length and width.

10-5 Writing Slope-Intercept Equations of Lines

Given any of the three facts below for a line, you can write an equation for that line.

> **1.** the slope and a point on the line
>
> **2.** two points on the line
>
> **3.** the x- and y-intercepts

Example

1 **Write an equation of the line whose slope is 3 and passing through (4, −2).**

$y = mx + b$	*Use slope-intercept form.*
$y = 3x + b$	*The slope m is 3.*
$-2 = 3(4) + b$	*Substitute 4 for x and −2 for y.*
$-2 = 12 + b$	*Now solve for b.*
$-14 = b$	

The slope-intercept form of the equation for the line is $y = 3x + (-14)$. This can also be expressed as $y = 3x - 14$. *In standard form, the equation is 3x − y = 14.*

The following procedure can be used to write an equation of a line when two points on the line are known.

Example

2 **Write an equation of the line that passes through (5, 1) and (8, −2).**

$$m = \frac{y_2 - y_1}{x_2 - x_1} \qquad \textit{First determine the slope.}$$

$$= \frac{-2 - 1}{8 - 5}$$

$$= \frac{-3}{3}$$

$$= -1$$

In slope-intercept form, $y = -1x + b$ or $y = -x + b$.

Now substitute the coordinates of either point into the equation and solve for b.

using (5, 1)	**or**	using (8, −2)
$y = -x + b$		$y = -x + b$
$1 = -(5) + b$		$-2 = -(8) + b$
$6 = b$		$6 = b$

The slope-intercept form of the equation is $y = -x + 6$.   *The standard form is x + y = 6.*

Example

3 Using Calculators

Write an equation of the line that passes through (7.6, 10.8) and (12.2, 93.7). Round values to the nearest thousandth.

$m = \frac{y_2 - y_1}{x_2 - x_1}$ *First determine the slope.*

ENTER: (93.7 − 10.8) ÷ (12.2 − 7.6) = STO

DISPLAY: *93.7 10.8 82.9 12.2 7.6 4.6 18.021739*

Rounded to the nearest thousandth, the slope is 18.022. In slope-intercept form, $y = 18.022x + b$.

To determine the y-intercept, solve for b. Thus, $b = y - 18.022x$. Let $(x, y) = (7.6, 10.8)$.

ENTER: 10.8 − RCL × 7.6 =

DISPLAY: *10.8 18.021739 7.6 -126.16522*

Rounded to the nearest thousandth, the y-intercept is −126.166.
Therefore, the slope-intercept form of the equation is $y = 18.022x - 126.166$.
The standard form is 18.022x − y = 126.166.

You can also use the slope and y-intercept obtained from the graph to write an equation of a line.

Example

4 Write an equation for line *PQ* whose graph is shown below.

First determine the slope. Start at P. The y-coordinate *decreases by 4* as you move from P to Q. The x-coordinate *increases by 3* as you move from P to Q.

$\text{slope} = \frac{\text{change in } y}{\text{change in } x} = \frac{-4}{3} = -\frac{4}{3}$

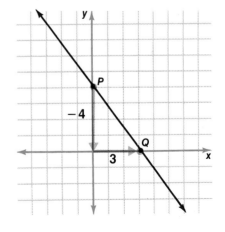

The line intersects the y-axis at $(0, 4)$. Thus, the y-intercept is 4.

Now substitute these values into the slope-intercept form.

$y = mx + 4$

$y = -\frac{4}{3}x + 4$

The equation for line *PQ* is $y = -\frac{4}{3}x + 4$. *In standard form, the equation is 4x + 3y = 12.*

Exploratory Exercises

Given the equation of a line and a point on the line, determine its y-intercept, b.

1. $y = 3x + b$, $(2, 1)$ 　　　　**2.** $y = -2x + b$, $(6, 2)$ 　　　　**3.** $y = -x + b$, $(4, 5)$

4. $y = -\frac{3}{2}x + b$, $(2, 2)$ 　　**5.** $y = \frac{3}{4}x + b$, $(8, 1)$ 　　**6.** $y = -\frac{2}{3}x + b$, $(-6, 5)$

7. $y = -\frac{5}{3}x + b$, $(1, -2)$ 　**8.** $y = \frac{5}{6}x + b$, $(3, -1)$ 　**9.** $y = \frac{4}{3}x + b$, $(2, 3)$

State the slope and y-intercept for each line. Then express an equation of the line in slope-intercept form.

10. a

11. b

12. c

13. d

14. e

15. f

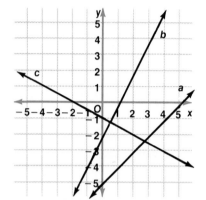
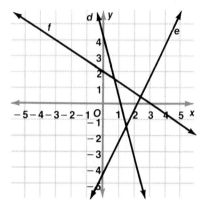

Written Exercises

Write an equation in slope-intercept form of the line having the given slope and passing through the indicated point.

16. 3; $(5, -2)$ 　　　　**17.** 2; $(4, -2)$ 　　　　**18.** $\frac{2}{3}$; $(-1, 0)$

19. $\frac{1}{2}$; $(5, 3)$ 　　　　**20.** -5; $(5, 4)$ 　　　　**21.** -3; $(3, 7)$

22. $\frac{3}{4}$; $(-2, -4)$ 　　**23.** $-\frac{5}{3}$; $(-3, -5)$ 　　**24.** $\frac{1}{4}$; $(0, 8)$

Write an equation in slope-intercept form of the line that passes through each pair of points.

25. $(-1, 7)$, $(8, -2)$ 　　**26.** $(5, 2)$, $(-7, -4)$ 　　**27.** $(6, 0)$, $(0, 4)$

28. $(4, 1)$, $(5, 2)$ 　　　　**29.** $(8, -1)$, $(7, -1)$ 　　**30.** $(1, 0)$, $(0, 1)$

31. $(6, 3)$, $(-2, 4)$ 　　　**32.** $(-4, 3)$, $(6, -5)$ 　　**33.** $(8, 0)$, $(2, 5)$

34. $(5, 7)$, $(-1, 6)$ 　　　**35.** $(-6, 2)$, $(3, -5)$ 　　**36.** $(5, 7)$, $(0, 6)$

Write an equation in slope-intercept form of the line that passes through each pair of points. Round values to the nearest thousandth.

37. $(4.67, 5.235)$, $(0.25, -1.5)$ 　　　　**38.** $(-3.2, 7.198)$, $(12.34, -0.8)$

39. $(0.4, 2.63)$, $(6.25, 12.05)$ 　　　　**40.** $(-2.1, -4.08)$, $(-0.2, -7.11)$

41. $(6.27, -0.001)$, $(4.33, 1.33)$ 　　　**42.** $(18.2, 1.008)$, $(-4.3, -11.5)$

Applications in Aeronautics

A light twin-engine aircraft has a useful carrying load of 2061 pounds. Of this a maximum of 1164 pounds of fuel can be carried. The remaining weight may be made up of people and cargo. If the plane is to carry more cargo or people, less fuel is carried and there is less flying time.

Shown below is a graph of the amount of fuel in pounds needed to fly a given number of hours.

Suppose a plane carries 575 pounds of cargo and 6 people weighing a total of 970 pounds. How many hours of flying time does the plane have?

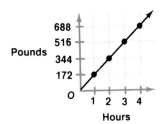

First, determine the amount of fuel the plane can carry. Then check the graph for the number of flying hours.

575	2061 *total pounds allowed*	516 pounds of fuel
+970	−1545	permit about 3 hours
1545 *total pounds of cargo and people*	516 *pounds of fuel that can be carried*	of flying time.

Exercises

Solve.

1. How many hours of flying time does a plane have if it is carrying 1000 pounds of cargo and 4 people weighing a total of 717 pounds?

2. Approximately how many hours of flying time does a plane have if it is carrying 2 people weighing a total of 305 pounds and 1061 pounds of cargo?

3. How many pounds of cargo can be carried if a plane must fly 3 hours to reach its destination? The pilot and copilot each weigh 145 pounds.

10-6 Parallel and Perpendicular Lines

The graphs below show parallel lines.

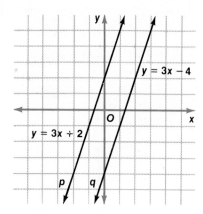

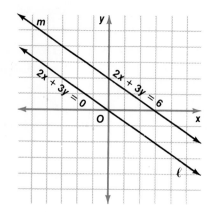

When graphed, parallel lines do not meet.

What is the slope of line p?
Line q?

What is the slope of line ℓ?
Line m?

What do you think the relationship is between the slopes of parallel lines?

If two lines have the same slope, then they are parallel. All vertical lines are parallel.

Definition of Parallel Lines

Example

1 **Find an equation of the line that is parallel to the graph of $5x - 2y = 6$ and that passes through $(4, -2)$. Use slope-intercept form.**

The slope of the graph of $5x - 2y = 6$ is $\frac{5}{2}$. $-\frac{A}{B} = \frac{5}{2}$

Therefore, the slope-intercept form of an equation whose graph is parallel to the graph of $5x - 2y = 6$ is $y = \frac{5}{2}x + b$.   *Parallel lines have the same slope.*

Now substitute $(4, -2)$ into the equation, $y = \frac{5}{2}x + b$, and solve for b.

$$-2 = \frac{5}{2}(4) + b$$

$$-2 = 10 + b$$

$$-12 = b \qquad \text{*The y-intercept is } -12.$$

The equation of the line is $y = \frac{5}{2}x - 12$.   *In standard form, the equation is $5x - 2y = 24$.*

Line ℓ is shown in the figure below at the left. Suppose line ℓ is rotated counterclockwise 90° to form line ℓ'. After the rotation, line ℓ' will be perpendicular to line ℓ.

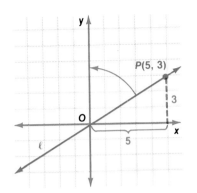 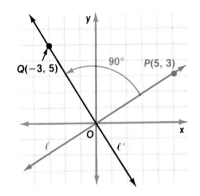 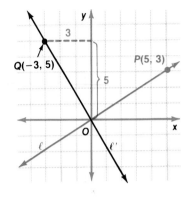

Study the slopes of ℓ and ℓ'.

slope of ℓ

$$\frac{y_2 - y_1}{x_2 - x_1} = \frac{0 - 3}{0 - 5}$$

$$= \frac{-3}{-5}$$

$$= \frac{3}{5}$$

slope of ℓ'

$$\frac{y_2 - y_1}{x_2 - x_1} = \frac{0 - 5}{0 - (-3)}$$

$$= \frac{-5}{3}$$

$$= -\frac{5}{3}$$

Compare the slopes. What is their product?

If the product of the slopes of two lines is −1, then the lines are perpendicular. In a plane, vertical lines are perpendicular to horizontal lines.	*Definition of Perpendicular Lines*

Example

2 Show that the lines given by the equations $7x + 3y = 4$ and $3x - 7y = 1$ are perpendicular.

The slope of the line for $7x + 3y = 4$ is $-\frac{7}{3}$.

The slope of the line for $3x - 7y = 1$ is $\frac{3}{7}$.

Since $-\frac{7}{3} \cdot \frac{3}{7} = -1$, the lines are perpendicular.

Example

3 Find an equation of the line that is perpendicular to the line whose equation is $2x - 5y = 3$ and that passes through $(-2, 7)$. Use slope-intercept form.

The slope of the line for $2x - 5y = 3$ is $\frac{2}{5}$. The slope of the line perpendicular to that line is $-\frac{5}{2}$. The equation in slope-intercept form for the line perpendicular to the line whose equation is $2x - 5y = 3$ is $y = -\frac{5}{2}x + b$.

Now substitute $(-2, 7)$ into the equation, $y = -\frac{5}{2}x + b$, and solve for b.

$$7 = -\frac{5}{2}(-2) + b$$
$$7 = 5 + b$$
$$2 = b$$

The equation in slope-intercept form is $y = -\frac{5}{2}x + 2$.
In standard form, the equation is $5x + 2y = 4$.

Exploratory Exercises

State the slopes of the lines parallel and perpendicular to the graph of each equation.

1. $5x - y = 7$ **2.** $3x + 4y = 2$ **3.** $2x - y = 7$

4. $6x + 3y = 4$ **5.** $2x - 3y = 7$ **6.** $7x + y = 4$

7. $x = 7$ **8.** $2x + 7y = 8$ **9.** $5x - 3y = 4$

10. $y = -4$ **11.** $y = 4x + 2$ **12.** $y = 3x - 7$

13. $3y = 2x + 5$ **14.** $4x = 2y - 7$ **15.** $3x = 4 - 3y$

Written Exercises

Write an equation of the line that is parallel to the graph of each equation and that passes through the point. Use slope-intercept form.

16. $y = -\frac{3}{5}x + 4$; $(0, -1)$ **17.** $6x + y = 4$; $(-2, 3)$ **18.** $y = \frac{3}{4}x - 1$; $(0, 0)$

19. $2x + 3y = 1$; $(4, 2)$ **20.** $5x - 2y = 7$; $(0, -4)$ **21.** $4x - 3y = 2$; $(4, 0)$

22. $y = \frac{3}{4}x - 1$; $(2, 2)$ **23.** $y = -\frac{1}{3}x + 7$; $(2, -5)$ **24.** $x = y$; $(7, -2)$

Write an equation of the line that is perpendicular to the graph of each equation and that passes through the point. Use slope-intercept form.

25. $5x - 3y = 7$; $(8, -2)$ **26.** $3x + 8y = 4$; $(0, 4)$ **27.** $y = 3x - 2$; $(6, -1)$

28. $y = -3x + 7$; $(-3, 1)$ **29.** $y = 5x - 3$; $(0, -1)$ **30.** $y = \frac{2}{3}x + 1$; $(-3, 0)$

31. $3x + 7y = 4$; $(1, -3)$ **32.** $5x + 9y = 3$; $(0, 0)$ **33.** $y = 2x - 7$; $(4, -6)$

Find the common factor for the grouped binomials.

1. $(ax + ay) + (5ax + 5ay)$
2. $(3k^2 - 4kp) + (8p - 6k)$

Solve.

3. $(x + 1)(2x - 5) = 0$
4. $2r^2 - 50 = 0$
5. $2xy - 3x + 8y - 12 = 0$

Simplify.

6. $\dfrac{30a^3bc}{165ab^4}$
7. $\dfrac{m^2 - 100}{(m + 10)^2}$
8. $\dfrac{x^2 - 2x - 3}{3x^2 - 8x - 3}$

Given $f(x) = 3x + 2$, determine the value for each function.

9. $f(3)$ **10.** $f(-4)$ **11.** $f(0)$ **12.** $f(2.5)$ **13.** $f(6a)$

Write an equation of the line that is perpendicular to the graph of each equation and that passes through the given point. Use slope-intercept form.

14. $y = \frac{5}{4}x + 2$; $(0, 6)$ **15.** $2x + y = 6$; $(2, 3)$ **16.** $x - 4y = 16$; $(-1, 1)$

Excursions in Algebra   Slopes and Grades

On page 327, the similarity between slope and grade was discussed. Slope and grade can be expressed in similar ways.

$$\text{grade} = \frac{\text{vertical change}}{\text{horizontal change}} \qquad \text{slope} = \frac{\text{rise}}{\text{run}} = \frac{\text{change in } y}{\text{change in } x}$$

Using 0.06 for 6% the grade can be expressed as follows.

$$0.06 = \frac{v}{h} \qquad \begin{array}{l} v \text{ represents the vertical change.} \\ h \text{ represents the horizontal change.} \end{array}$$

$$v = 0.06h$$

Example   **How far does a road with a 6% grade drop in 3 miles?**
Use the equation $v = 0.06h$.

$v = 0.06(3)$ *The horizontal change in miles is 3.*
$v = 0.18$

Since 1 mile equals 5280 feet, 0.18 miles equals 0.18(5280) feet, or 950.4 feet.

Thus, a road with a 6% grade drops 950.4 feet in 3 miles.

Exercises

1. The eastern entrance of the Eisenhower tunnel is at an elevation of 11,080 feet. The tunnel is 8941 feet long and has an upgrade of 0.895% toward the western end. What is the elevation of the western end of the tunnel?

2. At the Royal Gorge in Colorado, an inclined railway takes visitors down to the Arkansas River. If the grade is 175% and the vertical drop is 1015 feet, what is the horizontal change of the railway?

10-7   Midpoint of a Line Segment

The **midpoint** of a line segment is the point on that segment that separates it into two segments of equal length.

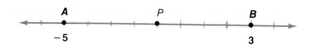

The coordinate of the midpoint, P, of line segment AB shown above can be found as follows.

Find the distance from A to B on the number line.

$$|-5 - 3| = 8$$

The distance from A to P on the number line is $\frac{1}{2}(8)$ or 4.
Add the distance, 4, to the coordinate of A, -5, to obtain the coordinate of P. Thus, the coordinate of P is $-5 + 4$ or -1.

A simpler approach is to find the average of the coordinates.

$$P = \frac{-5 + 3}{2} = \frac{-2}{2} = -1$$

The coordinate of the midpoint, P, of two points, x_1 and x_2, on a number line is $$P = \frac{x_1 + x_2}{2}.$$	*Midpoint on a Number Line*

This method can be extended to find the coordinates of the midpoint of a line segment in the coordinate plane.

The coordinates of the midpoint, (x, y), of a line segment whose endpoints are at (x_1, y_1) and (x_2, y_2) are $$(x, y) = \left(\frac{x_1 + x_2}{2}, \frac{y_1 + y_2}{2}\right).$$	*Midpoint of a Line Segment*

Example

1 **Find the coordinates of the midpoint of the line segment from $(-2, 4)$ to $(8, 6)$.**

$$(x, y) = \left(\frac{x_1 + x_2}{2}, \frac{y_1 + y_2}{2}\right)$$

$$= \left(\frac{-2 + 8}{2}, \frac{4 + 6}{2}\right) \qquad \text{Replace } x_1 \text{ with } -2, x_2 \text{ with } 8, y_1 \text{ with } 4, \text{ and } y_2 \text{ with } 6.$$

$$= (3, 5)$$

Example

2 If one endpoint of a line segment is (2, 8) and the midpoint is (−1, 4), find the coordinates of the other endpoint.

$$(x, y) = \left(\frac{x_1 + x_2}{2}, \frac{y_1 + y_2}{2}\right)$$

$$(-1, 4) = \left(\frac{2 + x_2}{2}, \frac{8 + y_2}{2}\right)$$ *Substitute (−1, 4) for (x, y).*
Substitute (2, 8) for (x₁, y₁).

Now separate this last expression into two equations.

$-1 = \dfrac{2 + x_2}{2}$ *Solve for x₂.* $4 = \dfrac{8 + y_2}{2}$ *Solve for y₂.*

$-2 = 2 + x_2$ $8 = 8 + y_2$

$-4 = x_2$ $0 = y_2$

The coordinates of the endpoint are (−4, 0).

Exploratory Exercises

State the coordinate of the point midway between each pair of points on the number line.

1. 4 and 8	**2.** 3 and 1	**3.** 10 and 18
4. −2 and 6	**5.** −3 and 13	**6.** −4 and −10
7. −3 and 6	**8.** 5 and −6	**9.** −10 and 15

State the coordinates of the midpoint of each line segment named below.

10. a	**11.** b
12. c	**13.** d
14. e	**15.** f
16. g	**17.** h

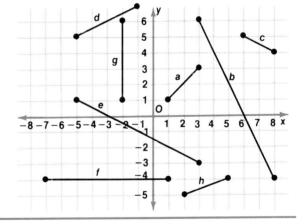

Written Exercises

Find the coordinates of the midpoint of the line segment whose endpoints are given.

18. (8, 4), (12, 2)	**19.** (9, 5), (17, 3)	**20.** (11, 4), (9, 2)
21. (17, 9), (11, −3)	**22.** (14, 4), (2, 0)	**23.** (19, −3), (11, 5)
24. (4, 2), (8, −6)	**25.** (−6, 5), (8, −11)	**26.** (−6, 5), (8, −11)
27. (5, −2), (7, 3)	**28.** (8, 2), (12, −5)	**29.** (−11, 6), (13, 4)
30. (12, −3), (5, 8)	**31.** (4, 7), (8, 0)	**32.** (2, 5), (4, 1)
33. (9, 10), (−8, 4)	**34.** (x, y), (a, b)	**35.** (2x, 3y), (6x, y)

36. $\left(\frac{1}{2}, 3\right), (1, -2)$

37. $\left(\frac{5}{6}, \frac{1}{3}\right), \left(\frac{1}{6}, \frac{1}{3}\right)$

38. $\left(3\frac{1}{2}, 2\frac{1}{4}\right), \left(5\frac{1}{3}, 1\frac{1}{2}\right)$

39. $\left(2\frac{1}{5}, 3\frac{1}{3}\right), \left(5\frac{3}{5}, 2\frac{1}{3}\right)$

If P is the midpoint of line segment AB, find the coordinates of the missing point A, B, or P.

40. $A(3, 5), P(11, 7)$

41. $A(3, 5), P(5, -7)$

42. $A(5, 9), B(-7, 3)$

43. $B(11, -4), P(3, 8)$

44. $B(5, 3), P(9, 7)$

45. $A(11, -6), B(5, -9)$

46. $P(5, -9), A(4, -11)$

47. $P(3, 9), B(-4, 1)$

48. $A(4, -7), B(-8, 1)$

49. $A(7, 4), P(9, -3)$

50. $P(3, -5), A(-3, 8)$

51. $P(5, 6), B(5, 7)$

52. The two endpoints of a diameter of a circle are $(8, -2)$ and $(4, -6)$. Find the coordinates of the center.

53. The center of a circle is $(3, -2)$ and one endpoint of a diameter is $(8, 3)$. Find the other endpoint of the diameter.

The coordinates on a map for certain cities are given below.

Los Angeles $(-6, -1)$ **Dallas $(1, -3)$**   **Chicago $(3, 3)$**

Atlanta $(5, -1)$ **Miami $(6, -4)$**   **Boston $(7, 5)$**

Find the coordinates of the point on the map midway between each pair of cities.

54. Los Angeles and Boston

55. Atlanta and Dallas

56. Chicago and Miami

57. Boston and Atlanta

58. Dallas and Boston

59. Miami and Dallas

60. Chicago and Los Angeles

61. Boston and Chicago

Challenge Exercises

Find the coordinates of P on line segment AB if P is one-fourth of the distance from A to B.

62. $A(8, 4), B(12, 12)$

63. $A(-3, 9), B(5, 1)$

64. $A(-3, 2), B(5, 4)$

65. $A(2, -6), B(9, 5)$

For quadrilateral $ABCD$, determine whether the diagonals of $ABCD$ bisect each other.

66. $A(8, 6), B(5, 5), C(4, 2), D(7, 3)$

67. $A(-2, 6), B(2, 11), C(3, 8), D(-1, 3)$

68. $A(11, 6), B(1, -2), C(-2, 4), D(3, 8)$

69. $A(8, -2), B(3, -5), C(-3, 5), D(2, 8)$

Testing Points On A Line

Each of the two graphs has three points plotted. Are the points on the same line?

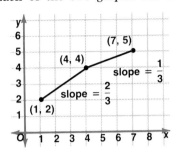

The points are not on the same line. The segments have different slopes.

The points are on the same line. The segments have the same slope and have a common point between.

You can write a computer program that tests if any three points are on the same line. Let the coordinates of the points to be tested be (A, B), (C, D), and (E, F).

```
10  REM THIS PROGRAM DETERMINES WHETHER
        3 POINTS LIE ON THE SAME LINE
20  INPUT A, B, C, D, E, F
30  IF (D-B)/(C-A)=(F-D)/(E-C) THEN 60
40  PRINT A; "","";B,C;"","";D,E;"","";F,
        "POINTS ARE NOT ON THE SAME LINE"
50  GOTO 70
60  PRINT A;"","";B,C;"","";D,E;"","";F,
        "POINTS ARE ON THE SAME LINE"
70  GOTO 20
80  END
```

REM statements often tell how the program works.

Points on the same line are called <u>collinear</u> points.

When the ordered pairs $(1, 2)$, $(13, 23)$, and $(5, 9)$ are used, the output is as follows.

```
1, 2      13, 23      5, 9
POINTS ARE ON THE SAME LINE
```

When the ordered pairs $(-5, 2)$, $(-5, 7)$, $(3, 8)$ are used, a division-by-zero error occurs in line 30. Add the following lines to the program to eliminate the error.

```
25  IF C - A=0 OR E - C=0 THEN PRINT "AT LEAST ONE LINE HAS AN
        UNDEFINED SLOPE": GOTO 70
```

Exercises   Use the computer program to determine whether each set of points lies on one line.

1. $(1, 5)$, $(16, 14)$, $(-4, 2)$

2. $(-2, -3)$, $(2, 1)$, $(5, 6)$

3. $(-459, -80)$, $(865, 163)$, $(54, 1)$

4. $(7, 14)$, $(-5, 10)$, $(-5, 8)$

5. Modify the program to print the slope for each segment.

slope   (325)

point-slope form   (329)

standard form   (330)

x-intercept   (333)

y-intercept   (333)

slope-intercept form   (333)

slopes of parallel lines   (346)

slopes of perpendicular lines   (347)

midpoint of a line segment   (350)

Chapter Summary

1. The slope of a line is the ratio of the change in *y* to the corresponding change in *x*.
   $$\text{slope} = \frac{\text{change in } y}{\text{change in } x} \text{ or } m = \frac{\text{change in } y}{\text{change in } x} \quad (325)$$

2. Given the coordinates of two points on a line, (x_1, y_1) and (x_2, y_2), the slope, *m*, can be found as follows.
   $$m = \frac{y_2 - y_1}{x_2 - x_1} \text{ where } x_2 \neq x_1 \quad (326)$$

3. For a given point (x_1, y_1) on a nonvertical line, the point-slope form of a linear equation is
   $$y - y_1 = m(x - x_1). \quad (329)$$

4. The standard form of a linear equation is $Ax + By = C$ where A, B, and C are integers and A and B are not both zero.   (330)

5. The *x*-coordinate of the point where a line crosses the *x*-axis is called the *x*-intercept of the line.   (333)

6. The *y*-coordinate of the point where a line crosses the *y*-axis is called the *y*-intercept of the line.   (333)

7. Given the points (x, y) and $(0, b)$ and slope *m*, the slope-intercept form of a linear equation is $y = mx + b$.   (333)

8. If two lines have the same slope, then they are parallel. All vertical lines are parallel.   (346)

9. If the product of the slopes of two lines is -1, then the lines are perpendicular. In a plane, vertical lines are perpendicular to horizontal lines.   (347)

10. The midpoint of a line segment is the point on that segment that separates it into two segments of equal length.   (350)

11. The coordinate of the midpoint, *P*, of two points, x_1 and x_2, on a number line is
    $$P = \frac{x_1 + x_2}{2}. \quad (350)$$

12. The coordinates of the midpoint, (x, y), of a line segment whose endpoints are at (x_1, y_1) and (x_2, y_2) are
    $$(x, y) = \left(\frac{x_1 + x_2}{2}, \frac{y_1 + y_2}{2}\right). \quad (350)$$

10-1 Determine the slope of the line passing through each pair of points.

    1. $(-3, 5), (4, 5)$ **2.** $(8, 3), (2, 5)$

    3. $(-2, 5), (-2, 9)$ **4.** $(-3, -5), (9, -1)$

10-2 Write an equation in point-slope form and standard form of the line through each pair of points.

    5. $(8, 1), (-3, 5)$ **6.** $(-2, 5), (9, 5)$

    7. $(0, 5), (-2, 0)$ **8.** $(-3, 0), (0, -6)$

10-3 Determine the slope and y-intercept of the graph of each equation.

    9. $3x - 2y = 7$ **10.** $y = \frac{1}{4}x + 3$

   11. $8x + y = 4$ **12.** $x = 2y - 7$

    Determine the x- and y-intercepts of the graph of each equation.

   13. $2x + 5y = 10$ **14.** $3x + 4y = 15$

10-4 Graph each equation using the x- and y-intercepts.

   15. $3x - y = 9$ **16.** $5x + 2y = 12$

    Graph each equation using the slope and y-intercept.

   17. $y = \frac{2}{3}x + 4$ **18.** $y = -\frac{3}{2}x - 6$

10-5 Write an equation of the line satisfying the given conditions. Use slope-intercept form.

   19. slope $= 4$ and passes through $(6, -2)$

   20. passes through $(9, 5)$ and $(-3, -4)$

   21. passes through $(2, 2)$ and has y-intercept of 7

   22. slope $= -\frac{3}{5}$ and y-intercept $= 3$

10-6 Write an equation of the line that is parallel to the graph of each equation and that passes through the given point. Use slope-intercept form.

   23. $4x - y = 7; (2, -1)$ **24.** $3x + 9y = 1; (3, 0)$

    Write an equation of the line that is perpendicular to the graph of each equation and that passes through the given point. Use slope-intercept form.

   25. $2x - 7y = 1; (-4, 0)$ **26.** $8x - 3y = 7; (4, 5)$

10-7 Find the coordinates of the midpoint of line segment AB.

   27. $A(3, 5), B(9, -3)$ **28.** $A(2, 7), B(8, 4)$

Determine the slope of the line passing through each pair of points.

  1. $(9, 2), (3, -4)$ **2.** $(8, 3), (8, 1)$ **3.** $(4, -5), (-2, -5)$

Determine the slope and y-intercept of the graph of each equation.

  4. $x - 8y = 3$ **5.** $3x - 2y = 9$

  6. $\frac{1}{2}x + \frac{3}{4}y = 2$ **7.** $y = 7$

Write an equation in standard form of the line satisfying the given conditions.

  8. passes through $(2, 5)$ and $(8, -3)$

  9. passes through $(-2, -1)$ and $(6, -4)$

10. has slope of 2 and y-intercept $= 3$

11. has y-intercept $= -4$ and passes through $(5, -3)$

12. slope $= \frac{3}{4}$ and passes through $(6, -2)$

Write an equation in slope-intercept form of the line satisfying the given conditions.

13. passes through $(4, -2)$ and the origin

14. passes through $(-2, -5)$ and $(8, -3)$

15. passes through $(6, 4)$ with y-intercept $= -2$

16. slope $= -\frac{2}{3}$ and y-intercept $= 5$

17. slope $= 6$ and passes through $(-3, -4)$

18. parallel to $6x - y = 7$ and passes through $(-2, 8)$

19. parallel to $3x + 7y = 4$ and passes through $(5, -2)$

20. perpendicular to $5x - 3y = 9$ and passes through the origin

21. perpendicular to $x + 3y = 7$ and passes through $(5, 2)$

Find the coordinates of the midpoint of the segment whose endpoints are given.

22. $(9, 3), (3, 6)$ **23.** $(-2, -7), (6, -5)$

Graph each equation.

24. $4x - 3y = 24$ **25.** $2x + 7y = 16$

The test questions on this page deal with expressions and equations. The information at the right may help you with some of the questions.

Directions: Choose the best answer. Write A, B, C, or D.

1. If $(-9)(-9)(-9) = (-9)(-9)(-9)n$, then $n =$

    (A) 1   (B) 0   (C) -9 (D) 9

2. If $(1 - 2 + 3 - 4 + 5 - 6) = -3$, then $2(-1 + 2 - 3 + 4 - 5 + 6) =$

    (A) -6 (B) 6   (C) 3   (D) -3

3. If $ab + 6 = 5bc$, then $b =$

    (A) $\frac{a + 6}{5c}$ (B) $\frac{6}{a - 5c}$

    (C) $\frac{-6}{5c - a}$ (D) $\frac{6}{5c - a}$

4. If $23(66 + x) = 2300$, then $x =$

    (A) 1185   (B) 782
    (C) 34   (D) 44

5. If $3y + 2$ is an odd integer, what is the next consecutive odd integer?

    (A) $3y + 4$ (B) $5y + 2$
    (C) $5y + 4$ (D) $y + 2$

6. $\dfrac{3(1.8 - 2.6) - (1.8 - 2.6)}{2} =$

    (A) 0.8   (B) -0.8
    (C) -1.5 (D) -3.4

7. $-3(a - b) =$

    (A) $3(b - a)$ (B) $3(a + b)$
    (C) $3(-b) + (-a)$ (D) $3ab$

8. If $2m - n = 8$ and $m + p = 14$, what is the value of n in terms of p?

    (A) $2p - 20$ (B) $10 - p$
    (C) $20 - 2p$ (D) $p - 10$

9. If $\frac{x}{8} + 3 = 1$, the value of $\frac{x}{2}$ is

    (A) -32 (B) -16 (C) -8 (D) 4

---

1. Many problems can be solved without much calculating if the basic mathematical concepts are understood. Always look carefully at what is asked, and think of possible shortcuts for solving the problem.

2. Check your solutions by substituting values for the variables.

10. If $x = 1$, $y = -2$, and $z = 2$, then $\dfrac{x^2y}{(x - z)^2} =$

    (A) -2 (B) 2   (C) $-\frac{1}{3}$ (D) $\frac{1}{3}$

11. If $1 + \frac{c}{12} = 2\frac{3}{4}$, then $c =$

    (A) 33   (B) 32   (C) 21   (D) 12

12. A number added to one-third of itself results in a sum of 40. What is the number?

    (A) 32   (B) 30   (C) 10   (D) 27

13. How many dollars do you have if you have n nickels, d dimes, and k quarters?

    (A) $\dfrac{5n + 25k + 10d}{100}$

    (B) $\dfrac{25n + 10d + 4k}{100}$

    (C) $20n + 10d + 4k$

    (D) $\dfrac{20n + 10d + 4k}{100}$

14. Emily, who is 24 years old, is three times as old as Barb. Barb is four years younger than twice Pedro's age. How old is Pedro?

    (A) 2   (B) 12   (C) 10   (D) 6

15. If x, y, and z are three consecutive integers and $x > y > z$, then $(x - y)(x - z)(y - z) =$

    (A) 2   (B) -2 (C) -16 (D) 16

16. If $20x$ cartons fill $\frac{x}{5}$ trucks, how many trucks are needed to hold 400 cartons?

    (A) 20   (B) 8000   (C) 100   (D) 4

CHAPTER 11

Systems of Open Sentences

Adam and Sara travel in a small boat 36 miles downstream in the same time that it takes to travel 24 miles upstream. In still water, the motor on the boat has a rating of 12 mph. What is the rate of the current? In this chapter, you will learn how to solve problems such as this by using systems of equations.

11-1 Graphing Systems of Equations

The graphs of $y = 2x - 1$ and $y = -\frac{1}{2}x + 4$ are shown at the right.

Notice that the graphs intersect at the point $(2, 3)$. Since this point lies on the graph of each equation, it follows that the ordered pair $(2, 3)$ satisfies both equations.

This can be checked by substituting $(2, 3)$ into each equation.

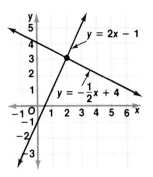

$$y = 2x - 1 \qquad\qquad y = -\frac{1}{2}x + 4$$
$$3 \overset{?}{=} 2(2) - 1 \qquad\qquad 3 \overset{?}{=} -\frac{1}{2}(2) + 4$$
$$3 = 3 \;\checkmark \qquad\qquad 3 = 3 \;\checkmark$$

The equations $y = 2x - 1$ and $y = -\frac{1}{2}x + 4$ together are called a **system of equations**. The solution of this system is $(2, 3)$.

Examples

1 Graph the equations $x + y = 6$ and $y = 2x$. Then find the solution of the system of equations.

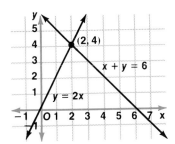

The graphs intersect at $(2, 4)$. Therefore, $(2, 4)$ is the solution of the system of equations $x + y = 6$ and $y = 2x$.

Check: $x + y = 6 \qquad y = 2x$
$$2 + 4 \overset{?}{=} 6 \qquad 4 \overset{?}{=} 2(2)$$
$$6 = 6 \;\checkmark \quad 4 = 4 \;\checkmark$$

2 Graph the equations $y = x - 4$ and $x + \frac{1}{2}y = \frac{5}{2}$. Then find the solution of the system of equations.

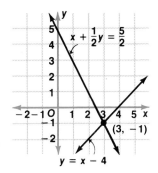

The graphs intersect at $(3, -1)$. Therefore, $(3, -1)$ is the solution of the system of equations $y = x - 4$ and $x + \frac{1}{2}y = \frac{5}{2}$.

Check: $y = x - 4 \qquad\quad x + \frac{1}{2}y = \frac{5}{2}$
$$-1 \overset{?}{=} 3 - 4 \qquad 3 + \frac{1}{2}(-1) \overset{?}{=} \frac{5}{2}$$
$$-1 = -1 \;\checkmark \qquad \frac{6}{2} - \frac{1}{2} \overset{?}{=} \frac{5}{2}$$
$$\frac{5}{2} = \frac{5}{2} \;\checkmark$$

Exploratory Exercises

State the ordered pair for the intersection of each pair of lines.

1. a and b
2. a and c
3. a and d
4. a and e
5. b and c
6. b and d
7. b and e
8. c and d
9. c and e
10. d and e

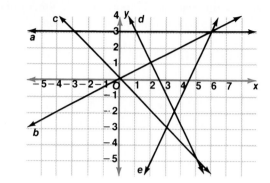

State which of the ordered pairs is a solution of the given equation.

11. $x + 2y = 7$ **a.** $(3, 2)$ **b.** $(6, 1)$ **c.** $(9, -1)$ **d.** $(-1, 4)$

12. $3x - 2y = 5$ **a.** $(1, 1)$ **b.** $(-1, 1)$ **c.** $(1, -1)$ **d.** $(5, 5)$

13. $1.5x + 0.3y = 3$ **a.** $(0, 10)$ **b.** $(2, 1)$ **c.** $(-3, 5)$ **d.** $(4, -10)$

14. $\frac{1}{2}x + \frac{2}{3}y = 3$ **a.** $(2, 3)$ **b.** $(2, 1)$ **c.** $(4, 5)$ **d.** $(3, 2)$

Written Exercises

Graph each pair of equations. Then state the solution of each system of equations.

15. $x - y = 2$
    $2x + 3y = 9$

16. $3x - 2y = 10$
    $x + y = 0$

17. $x + y = 8$
    $x - y = 2$

18. $5x - 3y = 12$
    $2x - 3y = 3$

19. $y = x$
    $x + y = 4$

20. $y = -3x$
    $4x + y = 2$

21. $x + y = 5$
    $x - 2y = -4$

22. $y = x - 1$
    $y + x = 11$

23. $y = -x$
    $y = 2x$

24. $x + y = 3$
    $y = x + 3$

25. $y = x + 3$
    $3y + x = 5$

26. $\frac{1}{2}x + y = -\frac{9}{2}$
    $x - y = 6$

27. $2x + 3y = -17$
    $y = x - 4$

28. $y = x - 3$
    $2x - y = 8$

29. $\frac{1}{3}x + 2y = 1$
    $\frac{3}{4}x + \frac{1}{4}y = -2$

30. $0.1x + 0.2y = 0.8$
    $0.5x - 0.5y = -0.5$

31. $0.4x + 0.3y = 2.4$
    $1.2x - 1.8y = -3.6$

32. $3x + y = -11$
    $\frac{1}{2}x + 2y = 0$

33. $\frac{1}{2}x + \frac{2}{3}y = -3$
    $4x - \frac{1}{3}y = -7$

34. $\frac{1}{3}x - \frac{1}{4}y = 2$
    $9x + \frac{3}{4}y = 24$

35. $2x = 3y$
    $2x - y = 4$

36. $3.4x + 6.3y = 4.4$
    $2.1x + 3.7y = 3.1$

37. $2x - 8y = 64$
    $\frac{1}{4}x + \frac{1}{7}y = 0$

38. $x + 2y = 0$
    $\frac{1}{3}x + \frac{1}{3}y = -1$

The Murphy Company, a leading gadget maker, wanted to know its break-even volume. This is the volume (number of gadgets) at which revenue (income) equals cost. Revenue is described by the equation $y = 5x$ and cost is described by the equation $y = 4x + 200$.

When using a graphing calculator, Jan Martinez used the numbers 0 to 700 along the x-axis with a scale of 100 and 0 to 1400 along the y-axis with a scale of 200. With the calculator in the COMP (computation) mode, she pressed the following keys to graph the equations $y = 5x$, which describes the revenue, and $y = 4x + 200$, which describes the cost.

ENTER: [GRAPH] 5 [ALPHA] [X] [EXE] [GRAPH] 4 [ALPHA] [X] [+] 200 [EXE]

Next, Jan used the [TRACE] key to find the point of intersection of the two lines. She moved the pointer along the line $y = 4x + 200$ by using the [→] and [←] keys until the pointer reached the intersection.

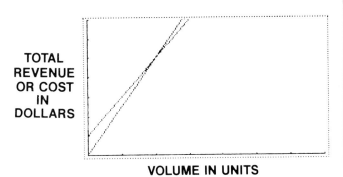

TOTAL REVENUE OR COST IN DOLLARS

VOLUME IN UNITS

Jan found the x-coordinate value to be 201.0638298 or 200 when rounded to the nearest ten. By pressing the keys [SHIFT] [X↔Y], she found the y-coordinate value to be 1004.255319 or 1000 when rounded to the nearest ten. This indicates that the Murphy Company must sell 200 gadgets to break even. At 200 units, the Murphy Company will incur $1000 of costs and receive $1000 in revenue. To make a profit, the Murphy Company must sell more than 200 gadgets.

Exercises

Make a break-even graph for each company. State the number of units necessary to break even and the revenue at that point. Use a graphing calculator if possible.

1. Zwik Company   revenue: $y = 4x$

cost: $y = 2x + 150$

2. Carilli Company   revenue: $y = 42x$

cost: $y = 3x + 780$

11-2 Systems of Equations

Not every system of equations has one ordered pair as its solution. Some systems have no solution. Others have infinitely many solutions.

A system of equations that has exactly one ordered pair as its solution is said to be consistent and independent.

Examples

1 Graph the equations $x + y = 3$ and $x + y = 4$. Then determine the number of solutions of the system of equations.

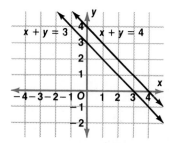

A system that has no solution is said to be inconsistent.

The graphs of the equations are parallel lines. Since they do not intersect, there is no solution to the system of equations. Note that the lines have the same slope but different x- and y-intercepts.

2 Graph the equations $2x + y = 3$ and $4x + 2y = 6$. Then determine the number of solutions of the system of equations.

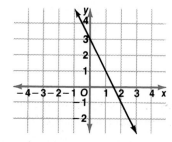

A system that has infinitely many solutions is said to be consistent and dependent.

The graphs of the equations have the same slope and intercepts. Any ordered pair on the graph satisfies both equations. So, there are infinitely many solutions of the system of equations.

Exploratory Exercises

State the slope and y-intercept of the graph of each equation. Then determine whether the system of equations has one solution, no solution, or infinitely many solutions.

1. $x + y = 4$
   $2x + 3y = 9$

2. $x + y = 6$
   $x - y = 2$

3. $x + y = 6$
   $3x + 3y = 3$

4. $x + y = 1$
   $2x - 2y = 8$

5. $x + 2y = 5$
   $3x - 15 = -6y$

6. $2x + 4y = 8$
   $x + 2y = 4$

7. $y = -3x$
   $6y - x = -38$

8. $x + y = 1$
   $3x + 5y = 7$

9. $3x + 6 = 7y$
   $x + 2y = 11$

10. $x + 5y = 10$
   $x + 5y = 15$

11. $3x - 8y = 4$
   $6x - 42 = 16y$

12. $2x + 3y = 5$
   $-6x - 9y = -15$

Written Exercises

Use the graphs at the right to determine whether each system has one solution, no solution, or infinitely many solutions. If the system has one solution, state it.

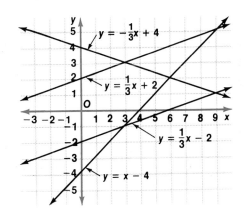

13. $y = -\frac{1}{3}x + 4$
   $y = \frac{1}{3}x + 2$

14. $y = \frac{1}{3}x - 2$
   $y = x - 4$

15. $y = \frac{1}{3}x - 2$
   $y = -\frac{1}{3}x + 4$

16. $y = \frac{1}{3}x + 2$
   $y = \frac{1}{3}x - 2$

17. $y = -\frac{1}{3}x + 4$
   $y = x - 4$

18. $y = x - 4$
   $y = \frac{1}{3}x + 2$

Graph each pair of equations. Then determine whether the system has one solution, no solution, or infinitely many solutions. If the system has one solution, state it.

19. $3x + y = 6$
   $y + 2 = x$

20. $3x - y = 4$
   $6x + 2y = -8$

21. $x + 2y = 5$
   $2x + 4y = 2$

22. $2x - y = -4$
   $-3x + y = -9$

23. $2x + 5 = 3y$
   $x = 2$

24. $x + y = 4$
   $x + y = 9$

25. $y = 5$
$y = 7$

26. $y = 6$
$x = 9$

27. $2x + 3y = 4$
$-4x - 6y = -8$

28. $9x - 2 = 4y$
$x + y = 6$

29. $x + y = 4$
$x - y = 10$

30. $x + 2y = 6$
$x + 2y = 8$

31. $\frac{1}{2}x + y = 3$
$\frac{1}{2}y + \frac{1}{4}x = 2$

32. $\frac{1}{2}x + \frac{1}{3}y = 6$
$y = \frac{1}{2}x + 2$

33. $\frac{2}{3}x + \frac{1}{4}y = 4$
$\frac{1}{3}x + \frac{1}{8}y = 2$

34. $\frac{1}{4}x - \frac{1}{3}y = -1$
$\frac{1}{3}x - \frac{1}{2}y = -2$

35. $0.3y = 0.4x - 1.2$
$1.2x + 0.9y = -3.6$

36. $0.2x + 0.1y = 1$
$x = 1 - 0.5y$

Challenge Exercises

Graph each pair of equations. Estimate the solution of each system of equations to the nearest half unit.

37. $y = \frac{1}{4}x + 2$
$y = -6x - 1$

38. $3x + y = 1$
$x - 2y = 6$

39. $5x + 3y = 18$
$4x - 18y = 45$

mini-review

Solve and check each equation.

1. $y^2 - 2y - 99 = 0$

2. $3x^2 + 15x + 18 = 0$

3. $2m^2 + 3m = 27$

Graph each inequality. Draw a separate set of axes for each.

4. $y \leq 4$

5. $3x > 2y$

6. $3x + 4y \geq 8$

Given $f(x) = -2x^2 - 3x + 7$, determine each value.

7. $f(-3)$

8. $f(2)$

9. $f\left(\frac{1}{3}\right)$

10. $f(4a)$

11. $f(-5) - f(3)$

Determine the slope, y-intercept, and x-intercept of the graph of each equation.

12. $4x - y = 11$

13. $2x - 13y = 52$

14. $-7x + 3y = 14$

Determine the slope of the line passing through each pair of points.

15. $(8, 7), (-3, 4)$

16. $(9, -2), (-3, 6)$

17. $\left(\frac{1}{3}, 4\right), \left(-\frac{1}{2}, 3\right)$

Write an equation in slope-intercept form of the line satisfying the given conditions.

18. $m = -\frac{2}{5}$ and passes through $(-4, 5)$

19. perpendicular to $3x + 2y = 7$ and passes through $(3, -1)$

20. parallel to $-5x + 6y = 12$ and passes through $(-2, -2)$

11-3 Substitution

Most systems of equations are easier to solve by algebraic methods than by graphing. One such algebraic method is called **substitution**.

Consider the system of equations $y = 2x$ and $3x + 4y = 11$. In the first equation, y is equal to $2x$. Since y must have the same value in the second equation, you can substitute $2x$ for y in $3x + 4y = 11$.

$$3x + 4y = 11$$
$$3x + 4(2x) = 11 \qquad \textit{Substitute 2x for y. Now the equation has only one variable, x.}$$
$$3x + 8x = 11 \qquad \textit{Solve to find the value of x.}$$
$$11x = 11$$
$$x = 1$$

Now find the value of y by substituting 1 for x in $y = 2x$.

$$y = 2x \qquad \textit{Could you also substitute 1 for x in 3x − 4y = 11?}$$
$$y = 2(1)$$
$$y = 2$$

Check:

$y = 2x$	$3x + 4y = 11$
$2 \stackrel{?}{=} 2(1)$	$3(1) + 4(2) \stackrel{?}{=} 11$
$2 = 2$ ✔	$3 + 8 \stackrel{?}{=} 11$
	$11 = 11$ ✔

The solution of the system of equations is $(1, 2)$.

Example

1 **Use substitution to solve the system of equations $x + 6y = 1$ and $3x - 10y = 17$.**

Solve the first equation for x.

Then substitute $1 - 6y$ for x in the second equation and find the value of y.

$$x + 6y = 1$$
$$x = 1 - 6y$$

$$3x - 10y = 17$$
$$3(1 - 6y) - 10y = 17$$
$$3 - 18y - 10y = 17$$
$$-28y = 14$$
$$y = -\frac{1}{2}$$

Substitute $-\frac{1}{2}$ for y in one of the equations to find the value of x.

$$x + 6y = 1$$
$$x + 6\left(-\frac{1}{2}\right) = 1$$
$$x - 3 = 1$$
$$x = 4$$

The solution of this system is $\left(4, -\frac{1}{2}\right)$. *Check this result.*

Examples

2 Use substitution to solve the system of equations $\frac{3}{2}x + y = 3$ and $3x + 2y = 12$.

Solve the first equation for y.

$$\frac{3}{2}x + y = 3$$
$$y = 3 - \frac{3}{2}x$$

Substitute $3 - \frac{3}{2}x$ for y in the second equation.

$$3x + 2y = 12$$
$$3x + 2\left(3 - \frac{3}{2}x\right) = 12$$
$$3x + 6 - 3x = 12$$
$$6 = 12$$

The statement, $6 = 12$, is false. This means that there are no values of x and y that are solutions to the equations. When you graph the equations, you find that they have the same slope but different intercepts. The graphs are parallel lines. Therefore, no solution exists.

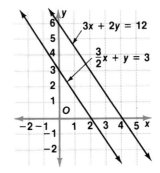

3 Define two variables. Then, use a system of equations and the substitution method to solve the problem.

The length of a rectangle is 13 inches more than its width. The perimeter is 86 inches. Find the dimensions of the rectangle.

explore Let ℓ = length of the rectangle.
Let w = width of the rectangle.

plan $\ell = w + 13$ *Write a system of equations.*
$2\ell + 2w = 86$

solve Substitute the value of ℓ into the equation $2\ell + 2w = 86$.

$$2\ell + 2w = 86$$
$$2(w + 13) + 2w = 86 \qquad \ell = w + 13$$
$$4w + 26 = 86$$
$$4w = 60$$
$$w = 15$$

Solve for ℓ.
$$\ell = w + 13$$
$$\ell = 15 + 13$$
$$\ell = 28$$

examine The length of the rectangle is 28 inches and the width of the rectangle is 15 inches.
Check these values in each equation.

Exploratory Exercises

Solve each equation for x. Then, solve each equation for y.

1. $y + 1 = x$

2. $y + 2x = 3$

3. $2x + 3y = 6$

4. $y + 13 = \frac{1}{2}x$

5. $3y - \frac{1}{2}x = 7$

6. $x + y = 5$

7. $2x + 2y = -10$

8. $3x + 5y = 8$

9. $\frac{2}{3}x - \frac{1}{2}y = 10$

10. $1.6x + 2.3y = 3.2$

11. $0.75x + 6 = -0.8y$

12. $\frac{2}{3}x - \frac{4}{5}y = 3$

For each system of equations, use the first equation to make a substitution in the second equation. Then solve for the remaining variable.

13. $y = 3 + 2x$

$x + y = 7$

14. $y = 7 - x$

$2x - y = 8$

15. $x = -6 + \frac{2}{3}y$

$x + 3y = 4$

16. $x = 5 - y$

$3y = 2x + 1$

17. $y = x$

$\frac{2}{3}x = 4y$

18. $x = \frac{3}{2} - \frac{1}{2}y$

$\frac{1}{4}x = \frac{1}{2}y + 3$

Written Exercises

Use substitution to solve each system of equations. State whether the system has no solution, one solution, or infinitely many solutions. If the system has one solution, state it.

19. $y = 2x$

$x + 2y = 8$

20. $y = 3x$

$x + 2y = -21$

21. $x = 2y$

$4x + 2y = 15$

22. $2x + 3y = 5$

$4x - 9y = 9$

23. $x = 3 - 2y$

$2x + 4y = 6$

24. $x - 3y = 3$

$2x + 9y = 11$

25. $3x + 2y = 14$

$x + \frac{2}{3}y = 4$

26. $3x + y = 2$

$2x - y = \frac{1}{2}$

27. $3x + 5y = 2x$

$x + 3y = y$

28. $x + \frac{1}{2}y = 4$

$y = -2x + 8$

29. $x - 2y = 5$

$3x - 5y = 8$

30. $2x - y = 7$

$\frac{3}{2}x = \frac{3}{4}y + 5$

31. $\frac{1}{4}x - 2y = -3$

$8x + 6y = 44$

32. $2x - y = 3x$

$2x + y = 3y$

33. $\frac{1}{2}x + \frac{1}{3}y = 3$

$-\frac{1}{4}x - \frac{2}{3}y = -3$

34. $-\frac{2}{5}x + \frac{2}{3}y = -8$

$\frac{1}{2}x + \frac{1}{2}y = 2$

35. $0.3x + 0.2y = 0.5$

$0.5x - 0.3y = 0.2$

36. $0.1x + 0.1y = 2$

$0.1x - 0.1y = 0$

For each problem, define two variables. Then, use a system of equations and the substitution method to solve the problem.

37. The length of a rectangle is 5 times the width. The perimeter is 24 meters. Find the dimensions of the rectangle.

38. Aaron has a total of 14 dimes and quarters. The total value of the coins is $2.15. How many dimes and how many quarters does he have?

39. The sum of two consecutive integers is 47. Find the numbers.

40. Lindsey is twice as old as Kyung. The sum of their ages is 51 years. Find their ages.

41. At a ski club meeting, there were 5 more males than females. There were 37 people present. How many males and how many females attended the meeting?

42. The difference between half of one number and twice a lesser number is 17. The sum of the lesser number and twice the greater number is 104. Find the numbers.

Challenge Exercises

Use substitution to solve each system of equations. Write each answer in the form (x, y, z).

43. $x + y + z = -54$
$x = -6y$
$z = 14y$

44. $2x + 3y - z = 17$
$x = \frac{1}{2}z + 1$
$y = -3z - 7$

45. $4x - \frac{1}{3}y + 5z = -95$
$x - 2z = -2$
$y + 3z = 9$

Excursions in Algebra   Checking Solutions

You can use a calculator to determine whether an ordered pair is a solution of a system of equations.

For example, follow the steps below to determine whether $(-2, -1)$ is the solution of the system $2x - 5y = 1$ and $3x - 4y = -2$.

ENTER: 2 $\boxed{\times}$ 2 $\boxed{+/-}$ $\boxed{-}$ 5 $\boxed{\times}$ 1 $\boxed{+/-}$ $\boxed{=}$

The display shows 1. Therefore $(-2, -1)$ is a solution of the first equation. Now check the second equation.

ENTER: 3 $\boxed{\times}$ 2 $\boxed{+/-}$ $\boxed{-}$ 4 $\boxed{\times}$ 1 $\boxed{+/-}$ $\boxed{=}$

The display shows -2. Therefore, $(-2, -1)$ is a solution of the second equation and a solution of the system of equations.

Exercises

Use a calculator to determine whether the given ordered pair is a solution of each system of equations. If the ordered pair is not a solution, find the solution.

1. $5m + 2n = -8$ $(-4, 6)$
$4m + 3n = 2$

2. $0.2a = 0.3b$ $(0.75, 0.5)$
$0.4a - 0.2b = 0.2$

3. $x + y = 40$ $(20, 20)$
$0.2x + 0.45y = 10.5$

4. $4x + 2y = 10.5$ $(1.25, 2.75)$
$2x + 3y = 10.75$

5. $108x + 537y = -1395$ $(2, 3)$
$-214x - 321y = 535$

6. $x + 5y = 20.25$ $(2.25, 3.6)$
$x + 3y = 13.05$

11-4 Elimination Using Addition

In some systems of equations the coefficients of either the x or y terms are additive inverses of each other. The simplest way to solve these systems is to add the equations. Because one of the variables is eliminated, this method is called *elimination*.

Examples

1 **Use elimination to solve the system of equations $x - 2y = 5$ and $2x + 2y = 7$.**

Add the two equations.

$x - 2y = 5$ *The y-coefficients, 2 and -2, are additive inverses.*
$2x + 2y = 7$ *Add like terms.*
$3x \quad\quad = 12$ *Notice the variable y is eliminated.*
$\quad\quad x = 4$

Substitute 4 for x in either equation and solve for y.

$x - 2y = 5$
$4 - 2y = 5$
$\quad -2y = 1$
$\quad\quad y = -\frac{1}{2}$

The solution of the system is $\left(4, -\frac{1}{2}\right)$.   *Check this result.*

2 **The sum of two numbers is 42. Their difference is 6. Find the numbers.**

explore	Let x = the greater number.
	Let y = the lesser number.

plan

$x + y = 42$ *The sum of two numbers is 42.*
$x - y = 6$ *Their difference is 6.*

solve

$x + y = 42$ *The y-coefficients, 1 and -1, are additive inverses.*
$x - y = 6$
$2x \quad\quad = 48$
$\quad\quad x = 24$

$x - y = 6$ *Substitute 24 for x in either equation and solve*
$24 - y = 6$ *for y.*
$\quad -y = -18$
$\quad\quad y = 18$

examine   The numbers are 18 and 24.   *Check this result.*

Exploratory Exercises

State whether addition can be used to solve each system.

1. $2x + y = 8$
   $x - y = 3$

2. $m + 3n = 5$
   $n = 3 - 2m$

3. $3a + b = 6$
   $4a + b = 7$

4. $y + 3x = 12$
   $3y - 3x = 6$

5. $y + 5x = 9$
   $y - 5x = 7$

6. $x = 5$
   $x + y = 7$

7. $y = -3$
   $2x + y = 6$

8. $\frac{1}{2}x - \frac{1}{5}y = 2$
   $3x + \frac{1}{5}y = -6$

9. $\frac{1}{2}x + \frac{1}{2}y = 3$
   $\frac{1}{4}x - \frac{1}{2}y = 3$

Written Exercises

Use elimination to solve each system of equations.

10. $x + y = 7$
    $x - y = 9$

11. $r - s = -5$
    $r + s = 25$

12. $-y + x = 6$
    $x + y = 5$

13. $m - n = 3$
    $n + m = 3$

14. $3x = 13 - y$
    $2x - y = 2$

15. $x + y = 8$
    $2x - y = 6$

16. $2x - 3y = -4$
    $x = 7 - 3y$

17. $5s + 4t = 12$
    $3s = 4 + 4t$

18. $0.6m - 0.2n = 0.9$
    $0.3m = 0.9 - 0.2n$

19. $x + y = 0$
    $\frac{1}{3}x = y$

20. $8x - 3y = 5$
    $\frac{1}{3}x + 3y = 20$

21. $9x + 2y = 8$
    $\frac{3}{2}x - 2y = 6$

22. $4x - \frac{1}{3}y = 8$
    $5x + \frac{1}{3}y = 6$

23. $\frac{1}{2}x - \frac{1}{2}y = 10$
    $\frac{3}{4}x + \frac{1}{2}y = 20$

24. $\frac{1}{3}x + \frac{1}{5}y = 5$
    $\frac{1}{3}x - \frac{1}{5}y = -5$

For each problem, define two variables. Then, use a system of equations and elimination to solve the problem.

25. The sum of two numbers is 45. Their difference is 7. Find the numbers.

26. The sum of two numbers is 48. Their difference is 24. Find the numbers.

27. The sum of two numbers is 64. Their difference is 42. Find the numbers.

28. The sum of two numbers is 101. Their difference is 25. Find the numbers.

29. A room is 12 feet longer than it is wide. Half the perimeter of the rectangular floor is 72 feet. Find the length and width.

30. Half the perimeter of a rectangular garden is 100 m. The garden is 10 m longer than it is wide. Find its length and width.

31. The difference between twice one number and a lesser number is 21. The sum of the lesser number and twice the greater number is 27. Find the numbers.

32. The difference between twice one number and a lesser number is 32. The sum of the lesser number and twice the greater number is 60. Find the numbers.

33. Sara McMenemy's farm is 82 acres larger than Jim Underwood's farm. Together the farms contain 276 acres. How large is each farm?

34. At a meeting for mathematics teachers, there were 7 more women than men. There were 43 teachers present. How many men and how many women were there?

Factor.

1. $18x^2 + 45xy + 18y^2$

2. $6m^2 - 9m - 60$

3. Simplify $\frac{x^2 - 9}{x^2 + 14x + 33}$.

Solve and check each equation.

4. $x^2 + 6x - 27 = 0$

5. $y + 2y^2 = 15$

Solve each equation for y in terms of x.

6. $5x - 3y = 12$

7. $-4x + 7y - 13 = 0$

State the domain, range, and inverse for each relation.

8. $\{(-1, 3), (-2, 3), (4, 2), (3, 2)\}$

9. $\{(3, -1), (3, -2), (4, 6), (6, 4)\}$

Graph each equation.

10. $x - 4y = 4$

11. $5x + 3y - 6 = 0$

Given $g(x) = 9x^2 - 13x + 5$, determine each value.

12. $g(-3)$

13. $g(1) - g(-2)$

14. Write an equation for the relationship between the variables in the following chart.

x	-1	1	3	5	7	9
y	1	5	9	13	17	21

Graph each inequality.

15. $2x + 3y < 12$

16. $x - 3y \le 6$

Determine the slope of the line passing through each pair of points named. Then write an equation in standard form of the line.

17. $(-2, 5), (5, 2)$

18. $(-2, -5), (3, -2)$

Determine the slope, y-intercept, and x-intercept of the graph of each equation.

19. $3x + 5y = 15$

20. $2x - 3y = 13$

Write an equation in slope-intercept form of the line satisfying the given conditions.

21. slope $= \frac{3}{2}$ and passes through $(4, 4)$

22. parallel to $x - 2y = 12$ and passes through $(4, 5)$

23. perpendicular to $y = 2x + 1$ and passes through $(-4, 5)$

Find the coordinates of the midpoint of the segment whose endpoints are named.

24. $(6, 4), (7, -12)$

25. $\left(-\frac{7}{2}, 4\right), \left(6\frac{3}{4}, -8\right)$

Graph the following pair of equations. Then state the solution of the system of equations.

26. $x - y = -1$
    $x + 4y = 14$

Use substitution or elimination to solve each system of equations.

27. $2x + y = 5$
    $-2x + 3y = 7$

28. $-3x + 2y = 16$
    $x - 2y = -12$

29. $4x + y = 5$
    $x + y = -4$

Solve each problem.

30. The square of a number added to 4 times the number equals 45. Find the number.

31. Find two integers whose sum is 16 and whose product is 48.

11-5 Elimination Using Multiplication

Some systems of equations cannot be solved directly by addition. Consider the following system.

$$6x + 12y = -5 \quad \textit{Would a variable be eliminated}$$
$$6x + 9y = -3 \quad \textit{by addition?}$$

If either equation is multiplied by -1, the system can be solved by addition.

$$6x + 12y = -5 \qquad \textbf{Multiply by } -1. \qquad -6x - 12y = 5 \qquad \textit{Then, add.}$$
$$6x + 9y = -3 \qquad\qquad\qquad\qquad\qquad \underline{6x + 9y = -3}$$
$$-3y = 2$$
$$y = -\frac{2}{3}$$

Substitute $-\frac{2}{3}$ for y in either of the original equations and solve for x.

$$6x + 9y = -3$$
$$6x + 9\left(-\frac{2}{3}\right) = -3$$
$$6x - 6 = -3$$
$$6x = 3$$
$$x = \frac{1}{2}$$

The solution of the system is $\left(\frac{1}{2}, -\frac{2}{3}\right)$. $\quad$ *Check this result.*

Example

1 **Use elimination to solve the system of equations $\frac{1}{3}x + 2y = 6$ and $x + 3y = -6$.**

Multiply the first equation by -3.

$$\frac{1}{3}x + 2y = 6 \qquad \textbf{Multiply by } -3. \qquad -x - 6y = -18 \qquad \textit{Then, add.}$$
$$x + 3y = -6 \qquad\qquad\qquad\qquad\qquad \underline{x + 3y = -6}$$
$$-3y = -24$$
$$y = 8$$

Substitute 8 for y in either of the original equations and solve for x.

$$x + 3y = -6$$
$$x + 3(8) = -6$$
$$x + 24 = -6$$
$$x = -30$$

The solution of the system of equations is $(-30, 8)$. $\quad$ *Check this result.*

In some cases it is necessary to multiply *each* equation by some number in order to solve the system by addition. Consider the following example.

Example

2 **Use elimination to solve the system of equations $3a + 4b = -25$ and $2a - 3b = 6$.**

Multiply the first equation by 2 and the second equation by -3 to eliminate a.

$$3a + 4b = -25 \quad \text{Multiply by 2.} \quad 6a + 8b = -50$$

$$2a - 3b = 6 \quad \text{Multiply by } -3. \quad \underline{-6a + 9b = -18} \quad \text{Add.}$$

$$17b = -68$$
$$b = -4$$

Substitute -4 for b in either of the original equations and solve for a.

$$2a - 3b = 6$$
$$2a - 3(-4) = 6$$
$$2a + 12 = 6$$
$$2a = -6$$
$$a = -3$$

The solution of the system of equations is $(-3, -4)$.   *Check this result.*

The above system of equations can also be solved by eliminating b. Multiply the first equation by 3 and the second equation by 4.

$$3a + 4b = -25 \quad \text{Multiply by 3.} \quad 9a + 12b = -75$$

$$2a - 3b = 6 \quad \text{Multiply by 4.} \quad \underline{8a - 12b = \quad 24}$$

$$17a \qquad = -51$$
$$a = -3$$

Substitute -3 for a in either of the original equations and solve for b.

$$2a - 3b = 6$$
$$2(-3) - 3b = 6$$
$$-6 - 3b = 6$$
$$-3b = 12$$
$$b = -4$$

The solution of the system of equations is $(-3, -4)$. Notice that the same result is obtained.

Exploratory Exercises

Explain how to eliminate the variable x in each system. Then, do the same for the variable y.

1. $x + 2y = 5$
  $3x + y = 7$

2. $4x + y = 8$
  $x - 7y = 2$

3. $x + 8y = 3$
  $4x - 2y = 7$

4. $y + x = 9$
  $2y - x = 1$

5. $4x - y = 4$
  $x + 2y = 3$

6. $y - 4x = 11$
  $2y + x = 6$

7. $3y - 8x = 9$
  $y - x = 2$

8. $0.6y + 0.9x = 1.1$
  $y + x = 2$

9. $\frac{3}{2}x + y = \frac{3}{2}$
  $\frac{3}{4}x + y = \frac{1}{4}$

10. $2x + y = 6$
  $3x - 7y = 9$

11. $1.2x + 1.4y = 3.6$
  $x - 2.8y = 7.3$

12. $3x + \frac{2}{3}y = 4$
  $4x + \frac{2}{3}y = 6$

Written Exercises

Use elimination to solve each system of equations.

13. $x - y = 6$
  $x + y = 5$

14. $x + y = 8$
  $2x - y = -6$

15. $y = 2x$
  $2x + y = 10$

16. $x + y = 1$
  $y = x$

17. $3x + 3y = 6$
  $2x - y = 1$

18. $3x + 4y = 7$
  $3x - 4y = 8$

19. $x + 2y = 8$
  $3x + 2y = 6$

20. $3x + 1 = -7y$
  $6x + 7y = 0$

21. $3x + 0.2y = 7$
  $3x = 0.4y + 6$

22. $9x + 8y = 7$
  $18x - 14 = 16y$

23. $x - 5y = 0$
  $2x - 3y = 7$

24. $5x + 3y = 12$
  $4x - 3y = 15$

25. $12x - 9y = 114$
  $12x + 7y = 82$

26. $2x + y = 3(x - 5)$
  $x + 5 = 4y + 2x$

27. $\frac{1}{3}x - y = -1$
  $\frac{1}{5}x - \frac{2}{5}y = -1$

28. $\frac{2}{5}x - \frac{1}{2}y = 1$
  $\frac{1}{5}x + \frac{1}{2}y = -1$

29. $\frac{1}{8}(x + y) = 1$
  $x - y = 4$

30. $2x - y = 36$
  $3x - 0.5y = 26$

31. $3x + \frac{1}{3}y = 10$
  $2x - 5 = \frac{1}{3}y$

32. $\frac{1}{2}x - \frac{2}{3}y = 2\frac{1}{3}$
  $\frac{3}{2}x + 2y = -25$

33. $\frac{2x + y}{3} = 15$
  $\frac{3x - y}{5} = 1$

34. $x + y = 6000$
  $0.06x + 0.08y = 460$

35. $x + y = 20$
  $0.4x + 0.15y = 4$

36. $1.6x + 0.4y = 1$
  $0.4x - 0.1y = 1$

37. $3.2y + 7.3x = 0$
  $y = -3.5x$

38. $0.25(x + 4y) = 3.5$
  $0.5x - 0.25y = 1$

39. $0.4x - 0.5y = 1$
  $0.4x + y = -2$

For each problem, define two variables. Then, use a system of equations and elimination to solve the problem.

40. The sum of two numbers is 57 and their difference is 7. Find the two numbers.

41. The difference between two numbers is 20 and their sum is 48. Find the two numbers.

42. The sum of two numbers is 25 and twice their difference is 14. Find the numbers.

43. Half the sum of two numbers is 26. Four times one of the numbers is the same as one-third of the other number. Find the numbers.

44. The sum of two numbers is 22. Five times one of the numbers equals six times the other number. Find the numbers.

45. The difference between two numbers is four. Twice the larger number equals three times the sum of the smaller number and 2. Find the numbers.

46. Mr. Agosto's farm is 32 acres smaller than Mr. Collins' farm. The two farms together contain 276 acres. How large is each farm?

47. A 140-meter rope is cut into two pieces. One piece is four times as long as the other. How long is each piece of rope?

48. The difference between the length and width of a rectangle is 7 cm. The perimeter of the rectangle is 50 cm. Find the length and width of the rectangle.

49. Layla is three times as old as Diana. In 10 years, Layla will be twice as old as Diana. Find their present ages.

50. A father is three times the age of his son. In six years, the father will be $2\frac{1}{2}$ times the age of the son. Find their present ages.

51. A rectangle has a perimeter of 40 cm. The length of the rectangle is 1 cm less than twice the width. Find the length and width of the rectangle.

52. The greater of two numbers is twice the lesser. If the greater is increased by 18, the result is 4 less than 4 times the lesser. Find the numbers.

53. The greater of two numbers is 1 less than 4 times the lesser. Three times the lesser number is 4 less than the greater. Find the numbers.

54. The perimeter of a rectangle is 86 cm. Twice the width exceeds the length by 2 cm. Find the dimensions of the rectangle.

55. The sum of Karl's age and his mother's age is 52. Karl's mother is 20 years older than Karl. How old is each?

Challenge Exercises

Use elimination to solve each system of equations.

56. $\dfrac{2}{x+7} - \dfrac{1}{y-3} = 0$

$\dfrac{1}{x-5} - \dfrac{3}{y+6} = 0$

57. $\dfrac{1}{x} + \dfrac{1}{y} = 7$

$\dfrac{2}{x} + \dfrac{3}{y} = 16$

58. $\dfrac{1}{x+y} = 2$

$\dfrac{1}{x-y} = \dfrac{1}{y}$

59. Explain how the elimination method shows that a system is inconsistent.

60. Explain how the elimination method shows that a system is dependent.

A system of three equations in three variables can be solved by elimination. Consider the following system of equations.

$$x + y + z = 6$$
$$x + 2y - z = 2$$
$$2x - 3y - z = -7$$

First, eliminate one of the variables. In this case, eliminate z by adding the first equation to each of the other equations.

$$x + y + z = 6 \qquad\qquad x + y + z = 6$$
$$x + 2y - z = 2 \qquad\qquad 2x - 3y - z = -7$$
$$\overline{2x + 3y \quad\; = 8} \qquad\qquad \overline{3x - 2y \quad\;\; = -1}$$

In this way, a system of two equations in two variables is obtained. Then solve the new system by elimination.

$$2x + 3y = 8 \qquad \text{Multiply by 2.} \qquad 4x + 6y = 16$$

$$3x - 2y = -1 \qquad \text{Multiply by 3.} \qquad \underline{9x - 6y = -3}$$
$$13x \quad\;\; = 13$$
$$x = 1$$

Substitute 1 for x in either of the two equations to solve for y.

$$2x + 3y = 8$$
$$2(1) + 3y = 8$$
$$3y = 6$$
$$y = 2$$

Finally, substitute 1 for x and 2 for y in any of the original equations to solve for z.

$$x + y + z = 6$$
$$1 + 2 + z = 6$$
$$z = 3$$

The solution of the system of equations is the ordered triple (1, 2, 3). This can be checked by substituting these values into the original system of equations.

Exercises

Solve each system of equations.

1. $x - y + z = 5$
$2x + y + z = 13$
$4x + y - 2z = 12$

2. $-3x + y - z = 6$
$3x - y + 2z = -7$
$x + y + z = 0$

3. $x + 2y - z = 12$
$x - 2y + z = -8$
$2x - y + z = -3$

4. $2x - 2y - z = -1$
$x + 2y + 2z = 9$
$x + y - z = 8$

5. $x + y + z = 5$
$2x + y - z = -1$
$3x - y + z = 1$

6. $5x + 3y + z = -4$
$3x + 2y - z = -4$
$x - 3y + 5z = 4$

7. $a - 2b + c = -9$
$2b + 3c = 16$
$4b = 8$

8. $a + b + c = 0$
$2a + b - c = 2$
$2a + 2b + c = 5$

9. $m + n + p = 3$
$m - p = 1$
$n - p = -4$

11-6 Graphing Systems of Inequalities

Consider the following system of inequalities.

$$y \geq x + 2$$
$$y \leq -2x - 1$$

The solution of this system is the set of all ordered pairs that satisfy both inequalities. To find the solution of this system, graph each inequality. The graph of each inequality is called a *half-plane*. The intersection of the half-planes represents the solution of the system.

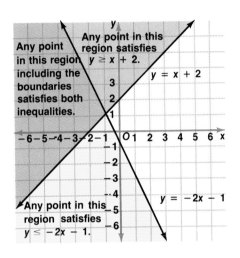

The graphs of $y = x + 2$ and $y = -2x - 1$ are the boundaries of the region. The solution of the system of inequalities is a region that contains an infinite number of ordered pairs.

Example

1 **Solve the system $y > x - 3$ and $y \leq -1$ by graphing.**

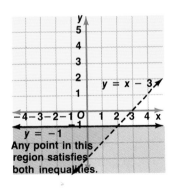

The graphs of $y = -1$ and $y = x - 3$ are boundaries of the region. The graph of $y = x - 3$ is a broken line and is not included in the solution of the system.

A system of inequalities may have no solution. This is illustrated by the next example.

2 **Solve the system** $y < x - 3$ **and** $x - y < -2$ **by graphing.**

Rewrite $x - y < -2$ as $y > x + 2$.
The graphs of the inequalities do not intersect. Thus, this system has no solutions.

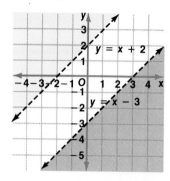

An inequality containing an absolute value expression can be graphed by graphing an equivalent system of inequalities.

3 **Solve the inequality** $|x| \leq y$ **by graphing.**

An equivalent system of inequalities is $x \leq y$ and $x \geq -y$.
The solution set is represented by the intersection of the shaded regions including the boundaries.

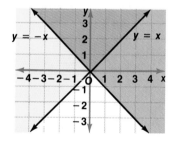

Exploratory Exercises

State whether each ordered pair is a solution of the system $x \leq 3$ **and** $y > 6$.

1. $(3, 7)$ **2.** $(6, 2)$ **3.** $(0, 8)$ **4.** $(-3, -3)$

State which region on the graph is the solution of the system of inequalities.

5. $y \geq 2x + 2$
$y \leq -x - 1$

6. $y \geq 2x + 2$
$y \geq -x - 1$

7. $y \leq 2x + 2$
$y \leq -x - 1$

8. $y \leq 2x + 2$
$y \geq -x - 1$

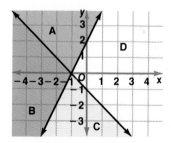

Written Exercises

Solve each system of inequalities by graphing.

9. $x > 3$
  $y < 6$

10. $y < -2$
  $x < -3$

11. $y > 0$
  $x \le 0$

12. $x > y$
  $y < 4$

13. $y > 2$
  $y > -x + 2$

14. $y \ge 2x + 1$
  $y \le -x + 1$

15. $y < -2$
  $y - x > 1$

16. $x \le 2$
  $y - 3 \ge 5$

17. $y \le x + 3$
  $y \ge x + 2$

18. $x \ge 1$
  $y + x \le 3$

19. $y \ge 3x$
  $3y \le 5x$

20. $y \ge x - 3$
  $y \ge -x - 1$

21. $y > x + 1$
  $y < x + 3$

22. $y - x < 1$
  $y - x > 3$

23. $2y + x < 4$
  $3x - y > 6$

24. $y + 2 < x$
  $2y - 3 > 2x$

Solve each inequality by graphing.

25. $|y| < x$

26. $|x| \ge y$

27. $|y| + 1 < x$

28. $|x - 1| < y$

29. $|y| > x + 3$

30. $|x| - 5 \ge y$

31. $|y - 4| > x$

32. $|2y + 4| \le x$

Write a system of inequalities for each graph.

33.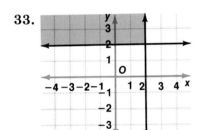

34.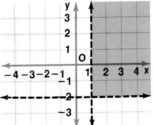

35.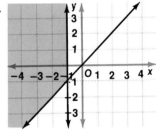

36.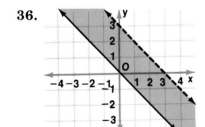

37.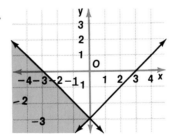

38.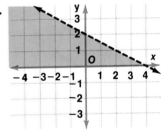

39.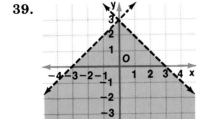

40.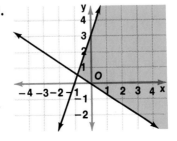

41.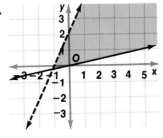

Determine whether the point of intersection of the two boundaries is part of the solution set of each system.

42. $y < -x$
$x \geq 6$

43. $y + 4 \leq 8$
$y \leq \frac{1}{2}x$

44. $x + 3y > 4$
$2x - y < 5$

45. $x + 1 \geq 0$
$3x - y > 7$

Challenge Exercises

Write a system of inequalities for each graph.

46.

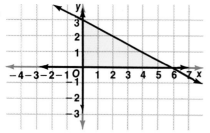

47.

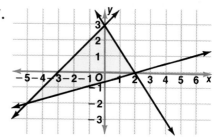

Solve each system of inequalities by graphing.

48. $x - y \leq -3$
$x + y \leq 1$
$y \geq 0$

49. $x + 4y < 4$
$5x - 8y < -8$
$3x - 2y \geq -16$

50. $x < 3$
$y > \frac{1}{5}x$
$x + 3y < 9$
$2x - y < -9$

51. $x + 4y \leq 13$
$x + y \leq 4$
$x - 5y \geq 16$
$x + 2y \geq -5$
$2x - y \leq -10$

mini-review

Use $h = vt - 16t^2$ to find the missing quantity.

1. Find v, if $t = 4$ seconds and $h = 272$ feet.

2. Find two values for t, if $v = 144$ feet per second and $h = 224$ feet.

Simplify each algebraic fraction. Assume that no denominator is equal to zero.

3. $\frac{21x^5y^4z^{21}}{35x^3y^8z^{12}}$

4. $\frac{2x + 3}{2x^2 - 7x - 15}$

Solve each equation if the domain is $\{-3, -1, 0, 1, 2, 4\}$.

5. $3x - y = -4$
6. $4a - b = 3$
7. $2x = 4 + 2y$
8. $3n = 9 - 6m$

9. Write the point-slope form of the equation of the line passing through $(-3, -4)$ and $(17, 12)$.

Write an equation in slope-intercept form of the line satisfying the given conditions.

10. passes through $(5, 3)$ and $(10, 0)$

11. parallel to $y = \frac{2}{3}x + 1$ and passes through $(-3, 3)$

Use substitution or elimination to solve each system of equations.

12. $5x + 4y = 4$
$2x + 3y = 10$

13. $3x - 2y = -1$
$y = 5x + 11$

14. $3x + y = -1$
$4x + 3y = -13$

Organizing information into a chart can be a helpful tool in problem solving. Consider this problem.

> **Mrs. Ruben, a grocer, mixes peanuts that sell for $1.65 a pound and almonds that sell for $2.10 a pound. She wants to make 30 pounds of the mixture that sells for $1.83 a pound. How many pounds of each should Mrs. Ruben include in the mixture?**

explore

First identify the variables.
Let x = the number of pounds of peanuts to use.
Let y = the number of pounds of almonds to use.

plan

Organize the variables and the information from the problem into a chart.

	Number of Pounds	×	Cost per Pound	=	Total Cost
Peanuts	x		1.65		$1.65x$
Almonds	y		2.10		$2.10y$
Mixture	30		1.83		$1.83(30)$

Note that the quantities in red were obtained by multiplying the number of pounds by the cost per pound.

Using the first and last columns of the chart, write two equations.

$$x + y = 30$$
$$1.65x + 2.10y = 1.83(30)$$

The mixture weighs 30 pounds. The cost of the peanuts plus the cost of the almonds equals the cost of the mixture.

Solve the system of equations to find the answer.

Exercises

Complete each chart and write two equations for each problem.

1. Carla invested $7000, part at 6% and the rest at 10%. The total yearly interest was $580. How much did she invest at each rate?

	Principal	×	Rate	=	Interest
First part					
Second part					
Total			✕		

2. A barge travels 32 miles up a river in 4 hours. The return trip takes 2 hours. Find the rate of the barge in still water and the rate of the current.

	Rate	×	Time	=	Distance
Up River					
Down River					

11-7 Problem Solving: Using Systems of Equations

Systems of equations can be used to solve many different problems.

Example

1 A barge travels 24 miles up a river in 3 hours. The return trip takes 2 hours. Find the rate of the barge in still water and the rate of the current.

> **explore**

Let b = the rate of the barge in still water.
Let c = the rate of the current.

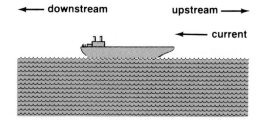

← downstream   upstream →

← current

Because of the current, the barge travels slower upstream than downstream.

Use the formula *rate × time = distance.*

	$r \times$	$t =$	d
Upstream	$b - c$	3	24
Downstream	$b + c$	2	24

Use a chart to organize the information.

> **plan**

$3(b - c) = 24 \;\rightarrow\; 3b - 3c = 24$
$2(b + c) = 24 \;\rightarrow\; 2b + 2c = 24$

> **solve**

$3b - 3c = 24$ Multiply by 2. $6b - 6c = 48$

$2b + 2c = 24$ Multiply by 3. $\dfrac{6b + 6c = 72}{12b \quad\;\; = 120}$ *Now add.*

$b = 10$

Solve for c.

$$2b + 2c = 24$$
$$2(10) + 2c = 24$$
$$20 + 2c = 24$$
$$2c = 4$$
$$c = 2$$

> **examine**

The rate of the barge in still water is 10 mph.
The rate of the current is 2 mph.

Example

2 A metal alloy is 25% copper. Another alloy is 50% copper. How much of each alloy should be used to make 1000 grams of an alloy that is 45% copper?

explore Let a = the number of grams of 25% copper alloy.
Let b = the number of grams of 50% copper alloy.

	25% Copper	50% Copper	45% Copper
Grams	a	b	1000
Percent	$0.25a$	$0.50b$	$0.45(1000)$

plan $a + b = 1000$
$0.25a + 0.50b = 0.45(1000)$

solve Use substitution.

$a + b = 1000$ $\qquad\qquad 0.25a + 0.50b = 0.45(1000)$
$\qquad a = 1000 - b \qquad 0.25(1000 - b) + 0.50b = 450$
$\qquad\qquad\qquad\qquad\qquad 250 - 0.25b + 0.50b = 450$
$\qquad\qquad\qquad\qquad\qquad\qquad\qquad 0.25b = 200$
$\qquad\qquad\qquad\qquad\qquad\qquad\qquad\qquad b = 800$

Find the value of a.

$a + b = 1000$
$a + 800 = 1000$
$\qquad a = 200$

examine 200 grams of the 25% copper alloy and 800 grams of the 50% copper alloy should be used. *Check this result.*

Written Exercises

Use a system of equations to solve each problem.

1. A boat is rowed 10 miles downstream in two hours, then rowed the same distance upstream in $3\frac{1}{3}$ hours. Find the rate of the boat in still water and the rate of the current.

2. While traveling with the wind, a plane flies 300 miles between Chicago and Columbus in 40 minutes. It returns against the wind in 45 minutes. Find the air speed of the plane and the rate of the wind.

3. A box contains two different types of candy, weighs 10 pounds, and costs $14.55. One type of candy costs $1.50 a pound. The other type costs $1.35 a pound. How many pounds of each kind are there?

4. How many pounds of candy that sells for 80 cents a pound should be mixed with candy that sells for $1.50 a pound to make 20 pounds of a mixture to sell at $1.01 a pound?

5. Joe sold 30 peaches from his tree for a total of $7.50. He sold the small ones for 20 cents each and the large ones for 35 cents each. How many of each kind did he sell?

6. Louise walks from her home to the city in 4 hours. She can travel the same distance on her bicycle in one hour. If she rides 6 mph faster than she walks, what is her speed on the bicycle?

7. An airplane travels 1800 miles in 3 hours flying with the wind. On the return trip, flying against the wind, it takes 4 hours to travel 2000 miles. Find the rate of the wind and the rate of the plane in still air.

8. A gas station attendant has some antifreeze that is 40% alcohol and another type of antifreeze that is 60% alcohol. He wishes to make 1000 gallons of antifreeze that is 48% alcohol. How much of each kind should he use?

9. A manufacturer has an order for 500 gallons of a 35% acid solution. The plant manager has on hand a 25% solution and a 50% solution. How many gallons of each type of solution should be mixed to fill the order?

10. A zinc alloy contains 2% zinc. A second alloy contains 8% zinc. How many pounds of each should be melted and blended to yield 1000 pounds of alloy containing 4% zinc?

11. A man invests $4000, part of it at 10% annual interest and the rest at 12% annual interest. If he receives $460 in interest at the end of one year, how much did he invest at each rate?

12. Koko opened her bank and found 83 coins in nickels and dimes. If she had $6.95 in all, how many coins of each did she have?

13. While driving to Northridge, Mrs. Winters averages 40 mph. On the return trip she averages 56 mph and saves two hours of traveling time. How far from Northridge does she live?

14. Two trains start toward each other on parallel tracks at the same time from towns 450 miles apart. One train travels 6 mph faster than the other train. What is the rate of each train if they meet in 5 hours?

15. A riverboat travels 48 miles downstream in the same time that it takes to travel 32 miles upstream. The speed of the boat in still water is 16 miles per hour greater than the speed of the current. Find the speed of the current.

16. In still water, a motorboat can travel five times as fast as the current in the river. A trip upriver and back totaling 96 miles can be done in 5 hours. Find the rate of the current.

17. When George was 2 miles upstream from the starting point on a canoe trip, he passed a log floating downstream with the current. He paddled upstream for one more hour and returned to the starting point just as the log arrived. What was the rate of the current?

18. The Rent-A-Car company rents a compact car for a fixed amount each day plus a fixed amount for each mile driven. Benito rented a car for 5 days, drove it 450 miles, and spent $134. Patti rented the same car for 4 days, drove it 350 miles, and spent $106. Find the daily charge and the charge per mile for the compact car.

11-8 Problem Solving: Digit Problems

Many problems relating to the digits of a number can be solved by using systems of equations.

Consider the two-digit number 86.

$$86 = 10(8) + 6$$

Let t = the tens digit.
Let u = the units digit.

Then a two-digit number can be represented as
$$10(t) + u.$$

Examples

1 **The sum of the digits of a two-digit number is 9. The number is 6 times the units digit. Find the number.**

> **explore**
>
> Let t = the tens digit.
> Let u = the units digit.
>
> Any two-digit number can be represented by $10t + u$.

> **plan**
>
> $t + u = 9$ *Write a system of equations.*
> $10t + u = 6u$

> **solve**
>
> Solve $t + u = 9$ for u and use substitution.
>
> $t + u = 9$ $\qquad\qquad$ $10t + u = 6u$
> $\qquad u = 9 - t$ $\qquad$ $10t + (9 - t) = 6(9 - t)$
> $\qquad\qquad\qquad\qquad\quad$ $10t + 9 - t = 54 - 6t$
> $\qquad\qquad\qquad\qquad\qquad\quad$ $15t = 45$
> $\qquad\qquad\qquad\qquad\qquad\qquad$ $t = 3$
>
> Solve for u.
>
> $t + u = 9$
> $3 + u = 9$
> $\quad u = 6$
>
> The number is $10(3) + 6$, or 36.

> **examine**
>
> Notice that the sum of the digits is 9 and that 36 is 6 times the units digit, 6.

2 **The tens digit of a two-digit number is twice the units digit. If the digits are reversed, the new number is 36 less than the original number. Find the original number.**

> **explore**
>
> Let t = the tens digit.
> Let u = the units digit.
> The original number can be represented by $10t + u$.
> The new number can be represented by $10u + t$.

$t = 2u$ *The tens digit is twice the units digit.*

$(10t + u) - 36 = 10u + t$ *The new number is 36 less than the original number.*

solve   Use substitution.

$$(10t + u) - 36 = 10u + t$$
$$10(2u) + u - 36 = 10u + 2u$$
$$20u + u - 36 = 12u$$
$$21u - 36 = 12u$$
$$9u = 36$$
$$u = 4$$

Solve for t.

$$t = 2u$$
$$t = 2(4)$$
$$t = 8$$

The original number is $10(8) + 4$, or 84.

examine   Notice that the tens digit is twice the units digit. If the digits are reversed the new number is 48, and 48 is 36 less than 84.

Written Exercises

Use a system of equations to solve each problem.

1. A two-digit number is 6 times its units digit. The sum of the digits is 6. Find the number.

2. The sum of the digits of a two-digit number is 13. Twice the tens digit is two less than 5 times the units digit. Find the number.

3. The tens digit of a two-digit number is 6 more than the units digit. The number is 2 more than 8 times the sum of the digits. Find the number.

4. The sum of the digits of a two-digit number is 12. If the digits are reversed, the new number is 18 less than the original number. Find the original number.

5. The sum of the digits of a two-digit number is 7. If the digits are reversed, the new number is 3 less than 4 times the original number. Find the original number.

6. A bank teller reversed the digits in the amount of a check and overpaid a customer by $9. The sum of the digits in the two-digit amount was 9. Find the amount of the check.

7. The sum of the digits of a two-digit number is 12. The units digit is twice the tens digit. Find the number.

8. The units digit of a two-digit number exceeds twice the tens digit by 1. The sum of the digits is 7. Find the number.

9. The sum of the digits of a two-digit number is 6. If the digits are reversed, the new number is 3 times the original tens digit. Find the original number.

10. The tens digit in a two-digit number exceeds twice its units digit by 1. If the digits are reversed, the number is 4 more than 3 times their sum. Find the original number.

11. A two-digit number is equal to 7 times the units digit. If 18 is added to the number, its digits are reversed. Find the number.

12. The tens digit of a two-digit number exceeds the units digit by 4. If the digits are reversed, the sum of the new number and the original number is 154. Find the original number.

13. The tens digit of a two-digit number exceeds twice the units digit by 1. If the digits are reversed, the sum of the new number and the original number is 143. Find the original number.

14. The ratio of the tens digit to the units digit of a two-digit number is 1 to 4. The number formed by reversing the digits is 2 less than 3 times the original number. Find the original number.

15. The sum of the digits of a two-digit number is 9. If 45 is subtracted from the number, the digits of the number are reversed. Find the number.

16. The sum of the digits of a three-digit number is 11. The tens digit is 3 times the hundreds digit and twice the units digit. Find the number.

17. The units digit of a three-digit number is 3. The sum of its digits is 9. If the units and hundreds digits are reversed, the sum of the new number and the original number is 909. Find the original number.

18. The units digit of a three-digit number is 5. The sum of its digits is 11. If the units and hundreds digits are reversed, the sum of the new number and the original number is 787. Find the original number.

19. The tens digit of a three-digit number is 7. The sum of its digits is 10. If the tens and hundreds digits are reversed, the sum of the new number and the original number is 1100. Find the original number.

20. The tens digit of a three-digit number is 3. The sum of its digits is 18. If the tens and hundreds digits are reversed, the sum of the new number and the original number is 1008. Find the original number.

21. The units digit of a two-digit number divided by the tens digit is 3. The sum of the digits is 12. Find the number.

22. The units digit of a two-digit number divides the tens digit twice with a remainder of 1. The sum of the digits is 10. Find the number.

23. Suppose that in a certain two-digit positive integer, the digits are reversed. The result is 36 less than the original integer. Find all integers for which this is true.

24. If the digits of a two-digit positive integer are reversed, the result is 6 less than twice the original. Find all integers for which this is true.

25. The numerator of a fraction is a positive two-digit integer. The denominator is found by reversing the digits of the numerator. If the value of the fraction is $\frac{7}{4}$, find all fractions that satisfy these conditions.

26. The sum of the digits of a three-digit number is 12. The tens digit is 3 less than the hundreds digit and the units digit is 3 times the hundreds digit. Find the number.

A square arrangement or *array* of numbers, called a **determinant**, has a number associated with it. For a 2×2 array, the value is found by multiplying numbers along the diagonals as shown.

$$\begin{vmatrix} 2 & 1 \\ 3 & 4 \end{vmatrix} = 2(4) - 3(1)$$

The determinant is denoted by vertical bars.

$$= 8 - 3$$
$$= 5$$

In general, the determinant of a 2×2 array is found as follows.

$$\begin{vmatrix} a & b \\ c & d \end{vmatrix} = ad - bc$$

Value of a Determinant

Determinants can be used to solve a system of linear equations. Consider solving the following system of equations.

$$ax + by = c$$
$$dx + ey = f$$

To solve for the variable x in this system, the variable y is eliminated.

$ax + by = c$	Multiply by e.	$(ae)x + (be)y = ce$	*Then, add.*
$dx + ey = f$	Multiply by $-b$.	$-(bd)x - (be)y = -bf$	

$$(ae - bd)x = ce - bf$$

$$x = \frac{ce - bf}{ae - bd}$$

To solve for y, eliminate the variable x.

$ax + by = c$	Multiply by d.	$(ad)x + (bd)y = cd$	*Then, add.*
$dx + ey = f$	Multiply by $-a$.	$-(ad)x - (ae)y = -af$	

$$(bd - ae)y = cd - af$$

$$y = \frac{cd - af}{bd - ae}$$

$$= \frac{af - cd}{ae - bd}$$

The general solution to this system of equations is $\left(\dfrac{ce - bf}{ae - bd}, \dfrac{af - cd}{ae - bd} \right)$.

Notice that the denominator of each fraction is the same. It can be written as a determinant.

$$ae - bd = \begin{vmatrix} a & b \\ d & e \end{vmatrix}$$

Each numerator also may be written as a determinant.

$$ce - bf = \begin{vmatrix} c & b \\ f & e \end{vmatrix} \quad \text{and} \quad af - cd = \begin{vmatrix} a & c \\ d & f \end{vmatrix}$$

Therefore, the solution to a system of equations in two variables can be found using determinants. This method is known as **Cramer's Rule**.

The solution to the system of equations $ax + by = c$
$dx + ey = f$
is (x, y)

where $x = \dfrac{\begin{vmatrix} c & b \\ f & e \end{vmatrix}}{\begin{vmatrix} a & b \\ d & e \end{vmatrix}}$ and $y = \dfrac{\begin{vmatrix} a & c \\ d & f \end{vmatrix}}{\begin{vmatrix} a & b \\ d & e \end{vmatrix}}$ and $\begin{vmatrix} a & b \\ d & e \end{vmatrix} \neq 0.$

Cramer's Rule

Example

Use Cramer's Rule to solve the system $3x - 5y = -7$ and $x + 2y = 16$.

$$x = \frac{\begin{vmatrix} -7 & -5 \\ 16 & 2 \end{vmatrix}}{\begin{vmatrix} 3 & -5 \\ 1 & 2 \end{vmatrix}}$$

$$y = \frac{\begin{vmatrix} 3 & -7 \\ 1 & 16 \end{vmatrix}}{\begin{vmatrix} 3 & -5 \\ 1 & 2 \end{vmatrix}}$$

$$= \frac{(-7)(2) - (16)(-5)}{(3)(2) - (1)(-5)}$$

$$= \frac{(3)(16) - (1)(-7)}{(3)(2) - (1)(-5)}$$

$$= \frac{66}{11}$$

$$= \frac{55}{11}$$

$$= 6$$

$$= 5$$

The solution is $(6, 5)$.

Exercises

Solve each system by using Cramer's Rule.

1. $2x + 2y = 6$
   $3x - 4y = 2$

2. $x - y = -3$
   $4x + 3y = -5$

3. $3x + 2y = 10$
   $x + y = 6$

4. $x - y = 4$
   $x - 2y = 8$

5. $x = 3 - 2y$
   $y = 2x - 3$

6. $3x + 2y = 7$
   $4x - y = 2$

Solving Systems of Equations

Computers are often used as tools for solving systems of equations. Consider the following system of equations.

$$ax + by = c$$
$$dx + ey = f$$

The general solution to the system of equations is $\dfrac{ce - bf}{ae - bd}$, $\dfrac{af - cd}{ae - bd}$.

This BASIC program will print the solution to a system of equations.

```
10   INPUT A,B,C,D,E,F
20   IF A * E − B * D = 0 THEN 70
30   LET X = (C * E − B * F) / (A * E − B * D)
40   LET Y = (A * F − C * D) / (A * E − B * D)
50   PRINT "(";X;",";Y;")" IS A SOLUTION"
60   GOTO 80
70   PRINT "NO UNIQUE SOLUTION"
80   END
```

Line 20 checks whether the denominator is zero.

Notice that line 70 makes no distinction between systems that have no solution and systems that have an infinite number of solutions. Study the following graphs.

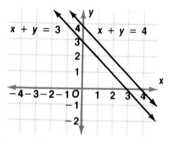

$$x + y = 4 \quad a = 1, b = 1, c = 4$$
$$x + y = 3 \quad d = 1, e = 1, f = 3$$
$$ae - bd = 1 \cdot 1 - 1 \cdot 1 = 0$$
$$ce - bf = 4 \cdot 1 - 1 \cdot 3 = 1$$
$$af - cd = 1 \cdot 3 - 1 \cdot 4 = -1$$

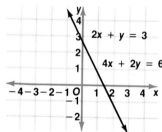

$$2x + y = 3 \quad a = 2, b = 1, c = 3$$
$$4x + 2y = 6 \quad d = 4, e = 2, f = 6$$
$$ae - bd = 2 \cdot 2 - 4 \cdot 1 = 0$$
$$ce - bf = 3 \cdot 2 - 1 \cdot 6 = 0$$
$$af - cd = 2 \cdot 6 - 3 \cdot 4 = 0$$

These and many other examples suggest that if only $ae - bd = 0$, the lines are parallel and the system has no solution. But if $ce - bf = 0$, the lines coincide and the system has an infinite number of solutions.

Exercises   Solve each system by using the computer program.

1. $5x + 5y = 16$
$2x + 2y = 5$

2. $7x - 3y = 5$
$14x - 6y = 10$

3. $x - 2y = 5$
$3x - 5y = 8$

4. Modify the computer program so that it will check whether the system has no solution or an infinite number of solutions.

Vocabulary

system of equations   (359)

substitution   (365)

elimination   (369, 372)

system of inequalities   (377)

Chapter Summary

1. A system of equations can be solved by graphing.   (359)
2. A system of equations may have either no solution, one solution, or infinitely many solutions.   (362)
3. Algebraic methods for solving systems of equations are as follows.
   substitution   (365)
   elimination   (369, 372)
4. A system of inequalities can be solved by graphing.   (377)
5. Systems of equations can be used to solve uniform motion problems   (382), mixture problems   (383), and digit problems   (385).

Chapter Review

11–1 **Graph each pair of equations. Then, state the solution of each system.**

1. $y = x$
$y = 2 - x$

2. $x + y = 6$
$x - y = 2$

3. $y = x - 1$
$x + y = 11$

11–2 **Graph each pair of equations. Then, determine whether the system has one solution, no solution, or infinitely many solutions.**

4. $3x + y = -4$
$6x + 2y = -8$

5. $y = 5$
$y = 7$

6. $x - y = 10$
$x + y = 11$

7. $9x + 2 = 4y$
$x + y = 6$

11–3 **Use substitution to solve each system of equations.**

8. $y = x - 1$
$4x - y = 19$

9. $2m + n = 1$
$m - n = 8$

10. $x = 2y$
$x + y = 6$

11. $5x = -10y$
$8x - 4y = 40$

12. $3m - 2n = -4$
$3m + n = 2$

13. $3x - y = 1$
$2x + 2y = 2$

Use elimination to solve each system of equations.

11-4 **14.** $x + y = 8$
$x - y = 8$

15. $2m - n = 4$
$m + n = 2$

16. $x + 2y = 6$
$-x + 3y = 4$

17. $2m - n = 5$
$2m + n = 3$

18. $3x - y = 11$
$x + y = 5$

19. $6r + 2s = 32$
$-6r + 9s = -21$

11-5 **20.** $3x + 3y = 6$
$2x - y = 1$

21. $x - 2y = 5$
$3x - 5y = 8$

22. $2x + 3y = 8$
$x - y = 2$

23. $6x + 8y = 4$
$2x - y = 5$

24. $2r + s = 9$
$r + 11s = -6$

25. $5m + 2n = -8$
$4m + 3n = 2$

Use a system of equations to solve each problem.

26. The sum of two numbers is 42. Their difference is 18. What are the numbers?

27. The difference between the length and width of a rectangle is 7 cm. The perimeter is 50 cm. Find the length and width.

11-6 **Solve each system of inequalities by graphing.**

28. $x \geq -3$
$y \leq x + 2$

29. $y > -x - 1$
$y \leq 2x + 1$

11-7 **Use a system of equations to solve each problem.**

30. Bill mixes candy costing 45¢ per pound with candy costing 65¢ per pound. A 7-pound box of the mix costs $3.65. How much of each kind of candy should Bill use?

31. A speedboat travels 60 miles with the current in $1\frac{1}{2}$ hours. The return trip takes 2 hours. What is the rate of the current?

11-8 **32.** A two-digit number is 6 times its units digit. The sum of its digits is 3. Find the number.

33. A two-digit number is 7 times its units digit. If 18 is added to the number, its digits are reversed. Find the original number.

Graph each pair of equations. Then, state the solution of each system.

1. $y = 3x$

$x + y = 4$

2. $y = x + 2$

$y = 2x + 7$

Graph each pair of equations. Then determine whether the system has one solution, no solution, or infinitely many solutions.

3. $3x + y = 5$

$6x + 2y = 10$

4. $x + y = 11$

$x + y = 14$

5. $x + y = 5$

$x - y = 1$

6. $x = 2y$

$x + y = 3$

Use substitution to solve each system of equations.

7. $y = 7 - x$

$y - x = 3$

8. $x + y = 8$

$x - y = 2$

9. $\frac{1}{8}(x + y) = 1$

$x - y = 4$

Use elimination to solve each system of equations.

10. $-3x - y = -10$

$3x - 2y = 16$

11. $5x - 3y = 12$

$2x - 3y = 3$

12. $2x + 5y = 12$

$x - 6y = -11$

13. $x + y = 6$

$x - y = 4\frac{1}{2}$

14. $3x + \frac{1}{3}y = 10$

$2x + \frac{1}{3}y = 5$

15. $8x - 6y = 14$

$-6x + 9y = -15$

Use a system of equations to solve each problem.

16. Jodi rode her bicycle against the wind for a distance of 15 km in 1 hour. The return trip took 36 minutes. What was the rate of the wind?

17. Mr. Salvatore mixes nuts costing $0.90 per pound and nuts costing $1.30 per pound. He wishes to make 50 pounds that cost $1.20 per pound. How many pounds of each should he use?

18. The units digit of a two-digit number exceeds twice the tens digit by 1. The sum of the digits is 10. Find the number.

19. The sum of the digits of a two-digit number is 10. If the digits are reversed, the new number is 18 less than the original number. Find the original number.

20. Solve the system of inequalities $y > -x + 2$ and $y \leq -3$ by graphing.

CHAPTER 12

Radical Expressions

When a sky diver jumps from an airplane, the time (t) it takes to reach the ground can be calculated by using the following formula.

$$t = \sqrt{\frac{2s_v}{g}}$$

In this formula, s_v is the vertical distance the sky diver falls and g is the acceleration due to gravity. To calculate the time, it is necessary to find the square root of $\frac{2s_v}{g}$. In this chapter, you will learn how to find and use square roots.

12-1 Square Roots

Squaring a number means using that number as a factor two times.

$$8^2 = 8 \cdot 8 = 64 \qquad \textit{8² is read "eight squared" and means 8 is used as a factor two times.}$$

$$(-8)^2 = (-8)(-8) = 64 \qquad \textit{-8 is used as a factor two times.}$$

The inverse of squaring is finding a **square root**. To find a square root of 64, you must find *two equal factors* whose product is 64.

$$x^2 = x \cdot x = 64$$

Since 8 times 8 is 64, one square root of 64 is 8. Since -8 times -8 is also 64, another square root of 64 is -8.

If $x^2 = y$, then x is a square root of y.	*Definition of Square Root*

An expression like $\sqrt{64}$ is called a **radical expression**. The symbol $\sqrt{}$ is a **radical sign**. It indicates the nonnegative or **principal square root** of the expression under the radical sign called the **radicand**.

The square root of a negative number is not defined for the sets of numbers covered thus far in this text.

$$\text{radical sign} \longrightarrow \sqrt{64} \longleftarrow \text{radicand}$$

$\sqrt{64} = 8 \qquad \sqrt{64}$ indicates the *principal* square root of 64.

$-\sqrt{64} = -8 \qquad -\sqrt{64}$ indicates the *negative* square root of 64.

$\pm\sqrt{64} = \pm 8 \qquad \pm\sqrt{64}$ indicates *both* square roots of 64.

± means positive or negative.

Examples

1 Find $\sqrt{81}$.

Since $9^2 = 81$, we know that $\sqrt{81} = 9$.

2 Find $-\sqrt{36}$.

Since $(6)^2 = 36$, we know that $-\sqrt{36} = -6$.

3 Find $\pm\sqrt{0.09}$.

Since $(0.3)^2 = 0.09$, we know that $\pm\sqrt{0.09} = \pm 0.3$.

To simplify a radical expression, find the square root of any factors of the radicand that are perfect squares. For example, to simplify $\sqrt{196}$, find any factors of 196 that are perfect squares. You can use the prime factorization of 196 as shown.

$$\begin{aligned}\sqrt{196} &= \sqrt{2 \cdot 2 \cdot 7 \cdot 7} \\ &= \sqrt{2^2 \cdot 7^2} \\ &= \sqrt{2^2} \cdot \sqrt{7^2} \\ &= 2 \cdot 7 \\ &= 14 \qquad \textbf{\textit{Check:}} \quad 14^2 = 196 \quad \text{✓}\end{aligned}$$

The property used to simplify $\sqrt{196}$ is stated below.

For any numbers a and b, where $a \geq 0$ and $b \geq 0$, $$\sqrt{ab} = \sqrt{a} \cdot \sqrt{b}.$$	***Product Property of Square Roots***

Example

4   **Simplify $\sqrt{576}$.**

$$\begin{aligned}\sqrt{576} &= \sqrt{2 \cdot 2 \cdot 2 \cdot 2 \cdot 2 \cdot 2 \cdot 3 \cdot 3} && \textit{Find the prime factorization of 576.} \\ &= \sqrt{2^6 \cdot 3^2} && \textit{Express the radicand using exponents.} \\ &= \sqrt{2^6} \cdot \sqrt{3^2} && \textit{Use the Product Property of Square Roots.} \\ &= 2^3 \cdot 3 && \textit{Simplify each radical.} \\ &= 24 && \textbf{\textit{Check:}} \quad 24^2 = 576 \quad \text{✓}\end{aligned}$$

A similar property for quotients can also be used to simplify square roots.

For any numbers a and b, where $a \geq 0$ and $b > 0$, $$\sqrt{\frac{a}{b}} = \frac{\sqrt{a}}{\sqrt{b}}.$$	***Quotient Property of Square Roots***

Examples

5   **Simplify $\sqrt{\frac{9}{16}}$.**

$$\begin{aligned}\sqrt{\frac{9}{16}} &= \frac{\sqrt{9}}{\sqrt{16}} && \textit{Use the Quotient Property} \\ && \textit{of Square Roots.} \\ &= \frac{3}{4}\end{aligned}$$

$\textbf{\textit{Check:}} \quad \left(\frac{3}{4}\right)^2 = \frac{9}{16} \quad \text{✓}$

6   **Simplify $-\sqrt{\frac{81}{121}}$.**

$$\begin{aligned}-\sqrt{\frac{81}{121}} &= -\frac{\sqrt{81}}{\sqrt{121}} \\ &= -\frac{9}{11}\end{aligned}$$

$\textbf{\textit{Check:}} \quad \left(-\frac{9}{11}\right)^2 = \frac{81}{121} \quad \text{✓}$

Exploratory Exercises

State the square of each number.

1. 10 **2.** 12 **3.** -7 **4.** -20 **5.** 0.3

6. 0.04 **7.** $\frac{1}{2}$ **8.** $\frac{4}{7}$ **9.** $-\frac{7}{8}$ **10.** $-\frac{11}{4}$

Simplify.

11. $\sqrt{121}$ **12.** $\pm\sqrt{36}$ **13.** $-\sqrt{81}$ **14.** $\sqrt{25}$ **15.** $\sqrt{\frac{4}{9}}$

16. $\sqrt{\frac{81}{64}}$ **17.** $\pm\sqrt{\frac{49}{121}}$ **18.** $-\sqrt{\frac{9}{100}}$ **19.** $\sqrt{0.0016}$ **20.** $\sqrt{0.09}$

Written Exercises

Find the principal square root of each number.

21. 49 **22.** 64 **23.** 16 **24.** 169

25. $\frac{25}{36}$ **26.** $\frac{81}{121}$ **27.** $\frac{36}{196}$ **28.** $\frac{25}{400}$

29. 0.0009 **30.** 0.0025 **31.** 0.0036 **32.** 0.16

Simplify.

33. $\sqrt{36}$ **34.** $\sqrt{9}$ **35.** $-\sqrt{100}$ **36.** $\pm\sqrt{144}$

37. $\pm\sqrt{25}$ **38.** $-\sqrt{169}$ **39.** $\sqrt{0.36}$ **40.** $\sqrt{0.0081}$

41. $\sqrt{529}$ **42.** $\sqrt{225}$ **43.** $\sqrt{676}$ **44.** $\sqrt{256}$

45. $-\sqrt{441}$ **46.** $-\sqrt{484}$ **47.** $\pm\sqrt{1024}$ **48.** $\pm\sqrt{289}$

49. $\sqrt{729}$ **50.** $\sqrt{961}$ **51.** $\sqrt{1764}$ **52.** $\sqrt{2025}$

53. $-\sqrt{\frac{289}{100}}$ **54.** $-\sqrt{\frac{169}{121}}$ **55.** $\sqrt{\frac{225}{25}}$ **56.** $\sqrt{\frac{144}{196}}$

57. $\sqrt{\frac{256}{361}}$ **58.** $\sqrt{\frac{196}{289}}$ **59.** $\pm\sqrt{\frac{961}{729}}$ **60.** $\pm\sqrt{\frac{484}{1024}}$

61. $\sqrt{\frac{576}{729}}$ **62.** $\sqrt{\frac{441}{529}}$ **63.** $\sqrt{\frac{0.09}{0.16}}$ **64.** $\sqrt{\frac{0.0025}{0.0036}}$

Excursions in Algebra   Radical Sign

Ancient mathematicians commonly wrote the word for root to indicate square roots. Late medieval Latin writers commonly used ℞, a contraction of radix (root), to indicate a square root. The symbol $\sqrt{}$ first appeared in print in Rudolff's *Coss* (1525). It is uncertain whether Rudolff used the symbol because it resembled a small *r* for radix or whether he invented a new symbol.

12-2 Irrational Numbers and Real Numbers

You have encountered natural numbers, whole numbers, integers, and rational numbers. The capital letters **N**, **W**, **Z**, and **Q** are often used to denote these sets of numbers.

Counting or Natural Numbers, **N**   {1, 2, 3, 4, . . .}

Whole Numbers, **W**   {0, 1, 2, 3, 4, . . .}

Integers, **Z**   {. . . , −3, −2, −1, 0, 1, 2, 3, . . .}

Rational Numbers, **Q** $\left\{ \begin{array}{l} \text{all numbers that can be expressed in the} \\ \text{form } \frac{a}{b}, \text{ where } a \text{ and } b \text{ are integers and } b \neq 0 \end{array} \right\}$

Recall that repeating or terminating decimals name rational numbers, since they can be represented as quotients of integers. The square roots of perfect squares also name rational numbers. For example, $\sqrt{0.16}$ is rational since it is equivalent to 0.4, a rational number. Consider the square roots of numbers that are *not* perfect squares.

$$\sqrt{2} = 1.414213 \ldots \quad \sqrt{3} = 1.732050 \ldots \quad \sqrt{7} = 2.645751 \ldots$$ *Notice that none of these decimals terminate or repeat.*

These numbers are *not* rational numbers. They are **irrational numbers**. The set of irrational numbers is often denoted by the capital letter **I**.

Irrational numbers are numbers that cannot be expressed in the form $\frac{a}{b}$, where a and b are integers and $b \neq 0$.	*Definition of Irrational Numbers*

Example

1 **Name the set or sets of numbers to which each number belongs.**

a. **−7** −7 is an integer. Since it can be expressed as $\frac{-7}{1}$, it is also a rational number.

b. **0.8888 . . .** This repeating decimal is equivalent to $\frac{8}{9}$. Thus, it names a rational number. This number can also be expressed as $0.\overline{8}$. The bar indicates repeating digits.

c. **3.141592 . . .** This decimal does not terminate or repeat. Thus, it names an irrational number. Often decimal approximations are used for such numbers. $\pi \approx 3.141592$

d. **$\sqrt{9}$** Notice $\sqrt{9} = 3$. The number 3 is a natural number, a whole number, an integer, and a rational number.

Irrational numbers together with rational numbers form the set of **real numbers, R.** You have graphed some rational numbers on the number line. Yet, if you graphed *all* rational numbers, the number line still would not be complete. Irrational numbers complete the number line. Therefore, the graph of all real numbers is the entire number line.

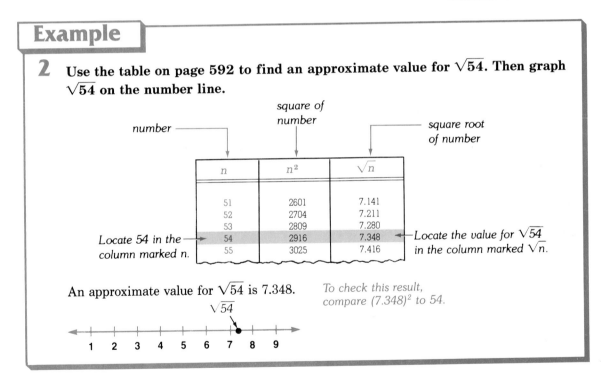

| Each real number corresponds to exactly one point on the number line. Each point on the number line corresponds to exactly one real number. | *Completeness Property for Points on the Number Line* |

You can approximate the graph of irrational numbers. The table of square roots on page 592 gives decimal approximations for square roots of integers from 1 to 100.

You can use a calculator, computer, or a table of square roots to approximate some irrational numbers.

Example

2 **Use the table on page 592 to find an approximate value for $\sqrt{54}$. Then graph $\sqrt{54}$ on the number line.**

number ──── square of number ──── square root of number

n	n^2	$\sqrt{n}$
51	2601	7.141
52	2704	7.211
53	2809	7.280
54	2916	7.348
55	3025	7.416

Locate 54 in the column marked n.   *Locate the value for $\sqrt{54}$ in the column marked $\sqrt{n}$.*

An approximate value for $\sqrt{54}$ is 7.348.   *To check this result, compare $(7.348)^2$ to 54.*

$\sqrt{54}$

Notice that the table on page 592 also has a column labeled n^2 that lists squares of integers from 1 to 100. You can use this column to determine whether an integer between 1 and 10,000 is a perfect square. Any integer that is *not* listed in the n^2 column is *not* a perfect square and has an irrational square root. You can also use the table to find the integers whose values are nearest to the irrational square root.

Example

3 Use the table on page 592 to determine whether each square root is rational or irrational. Then state the two integers between which the square root lies.

a. $\sqrt{9216}$

Look for 9216 in the column marked n^2. Since the number is listed, it is a perfect square. Thus, $\sqrt{9216}$ is a rational number. The corresponding value of n is 96. Thus, the value of $\sqrt{9216}$ is *exactly* 96.

n	n^2	$\sqrt{n}$
91	8281	9.539
92	8464	9.592
93	8649	9.644
94	8836	9.695
95	9025	9.747
96	9216	9.798

b. $\sqrt{8556}$

Notice that 8556 is *not* listed in the n^2 column. Thus, $\sqrt{8556}$ is irrational. Since 8556 is greater than 8464 and less than 8649, the value of $\sqrt{8556}$ is between 92 and 93.

Exploratory Exercises

Name the set or sets of numbers to which each number belongs. Use N for natural numbers, W for whole numbers, Z for integers, Q for rational numbers, and I for irrational numbers.

1. -5 **2.** $-\frac{1}{2}$ **3.** $\frac{6}{3}$ **4.** 16

5. $0.3333\ldots$ **6.** $\sqrt{11}$ **7.** $\sqrt{36}$ **8.** 0.125

9. $0.53694\ldots$ **10.** $0.\overline{35}$ **11.** 0.6125   **12.** $3.1416\ldots$

Use the table on page 592 to find an approximate value for each expression.

13. $\sqrt{2}$ **14.** $\sqrt{11}$ **15.** $\sqrt{20}$ **16.** $\sqrt{91}$ **17.** 16^2 **18.** 28^2

19. $\sqrt{31}$ **20.** $\sqrt{40}$ **21.** $\sqrt{89}$ **22.** $\sqrt{29}$ **23.** 66^2 **24.** $(-43)^2$

Written Exercises

Graph each number on the number line.

25. 3   **26.** -1.5 **27.** $1\frac{2}{3}$ **28.** $\sqrt{7}$ **29.** $-\sqrt{5}$ **30.** $\sqrt{20}$

State whether each decimal represents a rational number or an irrational number.

31. $0.10010001\ldots$ **32.** $1.23123412\ldots$ **33.** $0.\overline{571428}$ **34.** $0.\overline{37}$

35. $0.4444\ldots$ **36.** $4.34334333\ldots$ **37.** $7.6567876\ldots$ **38.** $0.777\ldots$

Use the table on page 592 to approximate each square root to the nearest hundredth.

39. $\sqrt{84}$ **40.** $-\sqrt{50}$ **41.** $\sqrt{66}$ **42.** $-\sqrt{19}$

Use the table on page 592 to approximate each square root to the nearest tenth.

43. $-\sqrt{20}$ **44.** $\sqrt{98}$ **45.** $-\sqrt{51}$ **46.** $\sqrt{31}$

The area of a square is given in each exercise. Find the length of a side of each square. Round answers to the nearest hundredth.

47. $A = 44$ m^2 **48.** $A = 84$ ft^2 **49.** $A = 13$ cm^2 **50.** $A = 24$ in^2

Use the table on page 592 to determine whether each square root is rational or irrational. Then state the two integers between which the square root lies if it is irrational.

51. $\sqrt{6724}$ **52.** $\sqrt{350}$ **53.** $\sqrt{3800}$ **54.** $\sqrt{1156}$

55. $\sqrt{888}$ **56.** $\sqrt{289}$ **57.** $\sqrt{9025}$ **58.** $\sqrt{3640}$

59. $\sqrt{7166}$ **60.** $\sqrt{7511}$ **61.** $\sqrt{1444}$ **62.** $\sqrt{3249}$

Solve each problem. Round answers to the nearest tenth.

63. The length of a rectangle is three times the width. If the area of the rectangle is 186 ft^2, what are the dimensions?

64. The area of a triangle is 38 m^2. If the base is twice the height, what is the base of the triangle? (Use the formula $A = \frac{1}{2}bh$.)

mini-review

Write the relation for each mapping as a set of ordered pairs.

1.

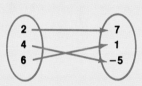

2.

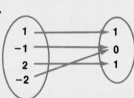

3.

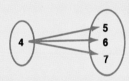

Find the missing terms for each sequence.

4. 5, 12, 19, 26, ——, ——, ——

5. $-4, 2, -1, \frac{1}{2},$ ——, ——, ——

Determine the slope of the line passing through each pair of points named.

6. $(3, 7), (9, 1)$ **7.** $(-2, 10), (4, -5)$ **8.** $\left(-\frac{1}{2}, -\frac{4}{5}\right), \left(1\frac{1}{4}, -1\right)$

Determine the slope and y-intercept of the graph of each equation. Then write each equation in slope-intercept form.

9. $6x - y = 5$ **10.** $\frac{1}{2}x + \frac{3}{4}y = -2$ **11.** $3(x - 1) = y + 6x - 5$

Graph each pair of equations. Then state whether the system has one solution, no solution, or infinitely many solutions. If there is one solution, state it.

12. $x + 3y = 12$
$x + 18 = -3y$

13. $x + 2y = 0$
$x - 4y = 12$

14. $\frac{3}{5}x - \frac{1}{2}y = -3$
$2x + 5y = -10$

Simplify.

15. $-\sqrt{121}$ **16.** $\sqrt{2.56}$ **17.** $\pm\sqrt{\frac{100}{25}}$ **18.** $\sqrt{\frac{144}{484}}$

12-3 Approximating Square Roots

Square roots can be approximated by using a calculator. The *square root key* on your calculator may be labeled $\sqrt{x}$ or $\sqrt{}$. When this key is pressed, the calculator replaces the number in the display with its principal square root. If a square root is irrational, the calculator display will show as many of the decimal places as it is capable. The result can be rounded to the nearest hundredth or thousandth.

Examples

Using Calculators

1 Find $\sqrt{196}$.

ENTER: 196 $\boxed{\sqrt{x}}$ DISPLAY: *196   14*

Therefore, $\sqrt{196} = 14$.

2 Find $-\sqrt{62.6}$. **Round to the nearest hundredth.**

ENTER: 62.6 $\boxed{\sqrt{x}}$ $\boxed{+/-}$ DISPLAY: *62.6   7.9120162   -7.9120162*

Therefore, $-\sqrt{62.6} \approx -7.91$.   *The symbol $\approx$ means is approximately equal to.*

Squaring and finding a square root are inverse operations. For example, since $4^2 = 16$, then $\sqrt{16} = 4$. Therefore, if your calculator does not have a square root key, it will have an inverse key, $\boxed{\text{INV}}$, and a square key, $\boxed{x^2}$. These keys can be used together to calculate square root.

Example

3 **Using Calculators**

Find $\sqrt{236}$. **Round to the nearest thousandth.**

ENTER: 236 $\boxed{\text{INV}}$ $\boxed{x^2}$ DISPLAY: *236   236   15.362292*

Therefore, $\sqrt{236} \approx 15.362$.

Exploratory Exercises

Answer each question.

1. Choose any positive number and enter it on your calculator. Then press the square root key, followed by the x^2 key. What is the result? Why?

2. Choose any negative number and enter it on your calculator. Then press the square root key. What happens? Why?

Written Exercises

Use a calculator to find each square root. Round decimal answers to the nearest hundredth.

3. $\sqrt{85}$ **4.** $\sqrt{71}$ **5.** $\sqrt{576}$ **6.** $\sqrt{1369}$ **7.** $\sqrt{60.3}$

8. $\sqrt{93.5}$ **9.** $-\sqrt{149}$ **10.** $-\sqrt{185}$ **11.** $\pm\sqrt{193}$ **12.** $\pm\sqrt{206}$

Use a calculator to find each square root. Round decimal answers to the nearest thousandth.

13. $\sqrt{131.4}$ **14.** $\sqrt{155.1}$ **15.** $\pm\sqrt{4624}$ **16.** $\pm\sqrt{6561}$ **17.** $\sqrt{175.6}$

18. $\sqrt{115.7}$ **19.** $-\sqrt{0.61}$ **20.** $-\sqrt{2.314}$ **21.** $\sqrt{0.00462}$ **22.** $\sqrt{0.00932}$

Excursions in Algebra   Divide-and-Average Method

One method for approximating square roots is called the *divide-and-average method*. This method is used below to approximate $\sqrt{88}$.

Step 1   Isolate 88 between two perfect squares. Choose 9 as an approximation for $\sqrt{88}$ because 88 is closer to 81 than 100.

$$81 < 88 < 100$$
$$9^2 < 88 < 10^2$$
$$9 < \sqrt{88} < 10$$

Step 2   Divide 88 by 9. Carry the quotient to one more decimal place than in the divisor. Do not round.

$$\begin{array}{r} 9.7 \leftarrow \text{one more decimal place than divisor} \\ 9)\overline{88.0} \end{array}$$

Step 3   Average the quotient, 9.7, and the divisor, 9. Carry the average to one more decimal place than in the quotient.

$$\frac{9.7 + 9}{2} = \frac{18.7}{2}$$
$$\approx 9.35$$

Step 4   Use the average, 9.35, as the new divisor. Repeat Steps 2 and 3.

Divide. $\begin{array}{r} 9.411 \\ 9.35)\overline{88.000} \end{array}$

Average. $\dfrac{9.35 + 9.411}{2} = \dfrac{18.761}{2}$

$$\approx 9.3805$$

An approximation for $\sqrt{88}$ is 9.3805. More accurate approximations may be made by continuing the divide-and-average method.

Exercises

1–10. Use the divide-and-average method twice to approximate each square root in Written Exercises 3–12 above.

12-4 Simplifying Square Roots

The square root of a positive integer is in *simplest form* if the radicand has no perfect square factor other than one.

The Product Property of Square Roots and prime factorization can be used to simplify radical expressions in which the radicand is *not* a perfect square.

Examples

1 Simplify $\sqrt{72}$.

$$\sqrt{72} = \sqrt{2 \cdot 2 \cdot 2 \cdot 3 \cdot 3} \quad \textit{prime factorization}$$
$$= \sqrt{2} \cdot \sqrt{2^2} \cdot \sqrt{3^2} \quad \textit{Product Property}$$
$$= \sqrt{2} \cdot 2 \cdot 3$$
$$= 6\sqrt{2}$$

2 Simplify $\sqrt{150}$.

$$\sqrt{150} = \sqrt{2 \cdot 3 \cdot 5 \cdot 5}$$
$$= \sqrt{2} \cdot \sqrt{3} \cdot \sqrt{5^2}$$
$$= \sqrt{2} \cdot 3 \cdot 5$$
$$= 5\sqrt{6}$$

When finding the positive square root of an expression containing variables, you must be sure that the result is *not* negative. Consider that $5^2 = 25$ and $(-5)^2 = 25$. When you find $\sqrt{25}$, however, you want *only* the principal square root. Therefore, absolute values are used as needed to ensure nonnegative results.

$$\sqrt{x^2} = |x| \qquad \sqrt{x^3} = x\sqrt{x} \qquad \sqrt{x^4} = x^2 \qquad \sqrt{x^6} = |x^3|$$

Examples

3 Simplify $\sqrt{81y^2}$.

$$\sqrt{81y^2} = \sqrt{81} \cdot \sqrt{y^2} \qquad \textit{Use the Product Property of Square Roots.}$$
$$= 9|y| \qquad \textit{The absolute value of y ensures a nonnegative result.}$$

4 Simplify $\sqrt{200a^4b^3}$.

$$\sqrt{200a^4b^3} = \sqrt{2 \cdot 2 \cdot 2 \cdot 5 \cdot 5 \cdot a \cdot a \cdot a \cdot a \cdot b \cdot b \cdot b}$$
$$= \sqrt{2} \cdot \sqrt{2^2} \cdot \sqrt{5^2} \cdot \sqrt{a^4} \cdot \sqrt{b^2} \cdot \sqrt{b} \qquad \textit{Product Property of Square Roots}$$
$$= \sqrt{2} \cdot 2 \cdot 5 \cdot a^2 \cdot b \cdot \sqrt{b}$$
$$= 10a^2b\sqrt{2b} \qquad \textit{Note that absolute value is not used because } a^2 \textit{ is always nonnegative}$$

and b is nonnegative in this case. If b were negative, then b^3 would be negative and $\sqrt{200a^4b^3}$ would not be defined.

The Product Property can be used to multiply square roots.

5 Simplify $\sqrt{10} \cdot \sqrt{20}$.

$$\sqrt{10} \cdot \sqrt{20} = \sqrt{10 \cdot 20} \qquad \text{Notice that } \sqrt{10 \cdot 20} = \sqrt{2} \cdot \sqrt{2^2} \cdot \sqrt{5^2}.$$
$$= \sqrt{10 \cdot 10 \cdot 2}$$
$$= 10\sqrt{2}$$

6 Simplify $\sqrt{7}(\sqrt{7} + \sqrt{5})$.

$$\sqrt{7}(\sqrt{7} + \sqrt{5}) = \sqrt{7} \cdot \sqrt{7} + \sqrt{7} \cdot \sqrt{5} \qquad \text{Use the Distributive Property.}$$
$$= \sqrt{49} + \sqrt{35} \qquad \text{Notice that 35 contains no perfect}$$
$$= 7 + \sqrt{35} \qquad \text{square factor other than 1 and there-fore cannot be simplified.}$$

Exploratory Exercises

Simplify. Use absolute value symbols when necessary.

1. $\sqrt{8}$ 2. $\sqrt{12}$ 3. $\sqrt{20}$ 4. $\sqrt{18}$ 5. $\sqrt{24}$ 6. $\sqrt{32}$

7. $\sqrt{48}$ 8. $\sqrt{19}$ 9. $\sqrt{m^2}$ 10. $\sqrt{y^6}$ 11. $\sqrt{x^5}$ 12. $\sqrt{a^3}$

13. $\sqrt{8a^3}$ 14. $\sqrt{9a^4}$ 15. $\sqrt{a^3b^3}$ 16. $\sqrt{x^4y^4}$ 17. $\sqrt{4} \cdot \sqrt{9}$ 18. $\sqrt{8} \cdot \sqrt{3}$

19. $\sqrt{5} \cdot \sqrt{10}$ 20. $\sqrt{6} \cdot \sqrt{9}$ 21. $\sqrt{11} \cdot \sqrt{11}$ 22. $\sqrt{10} \cdot \sqrt{10}$

23. $\sqrt{3}(\sqrt{3} + \sqrt{2})$ 24. $\sqrt{5}(\sqrt{3} + \sqrt{5})$ 25. $\sqrt{7}(\sqrt{5} - \sqrt{7})$ 26. $\sqrt{10}(\sqrt{10} - \sqrt{3})$

Written Exercises

Simplify. Use absolute value symbols when necessary.

27. $\sqrt{27}$ 28. $\sqrt{75}$ 29. $\sqrt{45}$ 30. $\sqrt{58}$

31. $\sqrt{72}$ 32. $\sqrt{80}$ 33. $\sqrt{90}$ 34. $\sqrt{98}$

35. $\sqrt{128}$ 36. $\sqrt{280}$ 37. $\sqrt{500}$ 38. $\sqrt{1000}$

39. $\sqrt{720}$ 40. $\sqrt{784}$ 41. $\sqrt{5184}$ 42. $\sqrt{2916}$

43. $\sqrt{32x^2}$ 44. $\sqrt{20a^2}$ 45. $\sqrt{40b^4}$ 46. $\sqrt{48m^4}$

47. $\sqrt{36a^2b^2}$ 48. $\sqrt{80a^2b^3}$ 49. $\sqrt{120a^3b}$ 50. $\sqrt{44m^4n}$

51. $\sqrt{60x^2y^4}$ 52. $\sqrt{54a^2b^2}$ 53. $\sqrt{20m^2n^7}$ 54. $\sqrt{147x^5y^4}$

55. $\sqrt{320m^4n^6}$ 56. $\sqrt{88x^{10}y^{10}}$ 57. $\sqrt{21x^2y}$ 58. $\sqrt{42xy}$

59. $\sqrt{6} \cdot \sqrt{8}$ 60. $\sqrt{10} \cdot \sqrt{30}$ 61. $2\sqrt{5} \cdot \sqrt{5}$ 62. $4\sqrt{2} \cdot \sqrt{2}$

63. $2\sqrt{3} \cdot 7\sqrt{5}$ 64. $3\sqrt{7} \cdot 6\sqrt{2}$ 65. $5\sqrt{10} \cdot 3\sqrt{10}$ 66. $6\sqrt{8} \cdot 7\sqrt{8}$

67. $4\sqrt{5} \cdot 3\sqrt{15}$ 68. $7\sqrt{30} \cdot 2\sqrt{6}$ 69. $\sqrt{3}(\sqrt{3} + \sqrt{6})$ 70. $\sqrt{7}(\sqrt{14} + \sqrt{7})$

71. $\sqrt{6}(\sqrt{7} - \sqrt{3})$ 72. $\sqrt{2}(\sqrt{8} - \sqrt{4})$ 73. $\sqrt{5}(\sqrt{10} - \sqrt{2})$ 74. $\sqrt{10}(\sqrt{10} - \sqrt{2})$

75. $\sqrt{3}(\sqrt{5} + \sqrt{27})$ 76. $\sqrt{5}(\sqrt{3} + \sqrt{125})$ 77. $\sqrt{3}(2\sqrt{12} + 4\sqrt{7})$ 78. $\sqrt{8}(2\sqrt{3} + 5\sqrt{6})$

79. Find the area of a rectangle whose length is $\sqrt{343}$ cm and whose width is $\sqrt{7}$ cm.

80. Find the area of a triangle whose base is $\sqrt{6}$ in. and whose altitude is $2\sqrt{6}$ in.

Many expressions can be written in several different ways. Consider the expression $\frac{1}{xy}$.

$$\frac{1}{xy} = \frac{1}{x \cdot y} = 1/(xy) = 1 \div (xy)$$

All four of these expressions are equivalent because they name the same number.

Notice the importance of the parentheses in the expressions $1/(xy)$ and $1 \div (xy)$. Without the parentheses, the order of operations dictates that the division would come *before* the multiplication. As a result, $1/xy$ and $1 \div xy$ would *not* be equivalent to $\frac{1}{xy}$.

Sometimes parentheses are *not* important when writing equivalent expressions. For example, both $\sqrt{3}^2$ and $(\sqrt{3})^2$ indicate that $\sqrt{3}$ is to be squared.

But sometimes the placement of the parentheses in an expression is *very* important, as shown in the examples below.

$3\sqrt{5^3}$ is equivalent to $3(\sqrt{5^3})$.

$3\sqrt{5^3}$ is *not* equivalent to $(3\sqrt{5})^3$.

$\sqrt{7} \cdot \sqrt{3} + \sqrt{2}$ is equivalent to $(\sqrt{7} \cdot \sqrt{3}) + \sqrt{2}$.

$\sqrt{7} \cdot \sqrt{3} + \sqrt{2}$ is *not* equivalent to $\sqrt{7}(\sqrt{3} + \sqrt{2})$.

Exercises

Determine whether each pair of expressions is equivalent. Write *yes* or *no*.

1. $\frac{2}{3a}$ and $\frac{2}{3 \cdot a}$

2. $\frac{4}{a+b}$ and $4/a + b$

3. ab^2 and $(ab)^2$

4. $(7 + \sqrt{2}) \div 8$ and $\frac{7 + \sqrt{2}}{8}$

5. $a + b^2$ and $(a + b)^2$

6. $2 + \sqrt{3} \cdot \sqrt{4}$ and $(2 + \sqrt{3})\sqrt{4}$

Choose the expressions that are equivalent to the first expression. There may be more than one correct answer.

7. $\frac{m + 2n}{4}$
        a. $(m + 2n)/4$
        b. $(m + 2n) \div 4$
        c. $m + (2n/4)$

8. $(2\sqrt{4})^6$
        a. $2(\sqrt{4})^6$
        b. $(2 \cdot \sqrt{4})^6$
        c. $2\sqrt{4}^6$

9. $2(\sqrt{6} + \sqrt{3})$
        a. $2 \cdot \sqrt{6} + \sqrt{3}$
        b. $2 \cdot (\sqrt{6} + \sqrt{3})$
        c. $2\sqrt{6} + \sqrt{3}$

10. $6\sqrt{3} \cdot 4\sqrt{7}$
        a. $6\sqrt{3}(4\sqrt{7})$
        b. $(6\sqrt{3})(4\sqrt{7})$
        c. $(6\sqrt{3} \cdot 4\sqrt{7})$

12-5 Simplifying Radical Expressions Involving Division

The Quotient Property of Square Roots can be used to simplify expressions and to divide radicals. In simplifying fractions, a radical is not left in the denominator.

Examples

1 Simplify $\dfrac{\sqrt{48}}{\sqrt{3}}$.

$\dfrac{\sqrt{48}}{\sqrt{3}} = \sqrt{\dfrac{48}{3}}$ *Quotient Property of Square Roots*

$= \sqrt{16}$ *Notice that 16 is a perfect square.*

$= 4$

2 Simplify $\dfrac{\sqrt{32}}{\sqrt{3}}$.

$\dfrac{\sqrt{32}}{\sqrt{3}} = \dfrac{\sqrt{32}}{\sqrt{3}} \cdot \dfrac{\sqrt{3}}{\sqrt{3}}$ *Notice that $\dfrac{\sqrt{3}}{\sqrt{3}} = 1$.*

$= \dfrac{\sqrt{32 \cdot 3}}{\sqrt{3 \cdot 3}}$

$= \dfrac{\sqrt{96}}{\sqrt{3^2}}$ $\sqrt{96} = \sqrt{2 \cdot 2 \cdot 2 \cdot 2 \cdot 2 \cdot 3}$

$= \dfrac{\sqrt{2^4} \cdot \sqrt{2} \cdot \sqrt{3}}{3}$ *Do you see why $\dfrac{\sqrt{3}}{\sqrt{3}}$ was used?*

$= \dfrac{4\sqrt{6}}{3}$

The method used to simplify $\dfrac{\sqrt{32}}{\sqrt{3}}$ in the above example is called **rationalizing the denominator**.

Notice that the denominator becomes a rational number.

Example

3 Simplify $\sqrt{\dfrac{25}{18}}$.

$\sqrt{\dfrac{25}{18}} = \dfrac{\sqrt{25}}{\sqrt{18}}$ *Quotient Property of Square Roots*

$= \dfrac{5}{\sqrt{3^2} \cdot \sqrt{2}}$ *Product Property of Square Roots*

$= \dfrac{5}{3\sqrt{2}} \cdot \dfrac{\sqrt{2}}{\sqrt{2}}$ *Rationalize the denominator.*

$= \dfrac{5\sqrt{2}}{6}$ $3\sqrt{2} \cdot \sqrt{2} = 3(2)$ or 6

Some quotients may have binomial expressions, such as $8 + \sqrt{2}$, as denominators. Multiplying $8 + \sqrt{2}$ by $8 - \sqrt{2}$ results in a rational number.

$$(8 + \sqrt{2})(8 - \sqrt{2}) = 8^2 - (\sqrt{2})^2$$

Use the pattern
$(a + b)(a - b) = a^2 - b^2$
to simplify.

$$= 64 - 2$$
$$= 62$$

Binomials of the form $a\sqrt{b} + c\sqrt{d}$ and $a\sqrt{b} - c\sqrt{d}$ are **conjugates** of each other.

Examples

4 **Simplify** $\dfrac{3}{3 - \sqrt{5}}$.

To rationalize the denominator, multiply both numerator and denominator by the conjugate of $3 - \sqrt{5}$, which is $3 + \sqrt{5}$.

$$\frac{3}{3 - \sqrt{5}} = \frac{3}{3 - \sqrt{5}} \cdot \frac{3 + \sqrt{5}}{3 + \sqrt{5}} \qquad \textit{Notice that } \frac{3 + \sqrt{5}}{3 + \sqrt{5}} = 1.$$

$$= \frac{3^2 + 3\sqrt{5}}{3^2 - (\sqrt{5})^2} \qquad \textit{Use the Distributive Property to multiply the numerators.}$$
$$\qquad\qquad\qquad \textit{Use the pattern } (a + b)(a - b) = a^2 - b^2 \textit{ to multiply the}$$
$$= \frac{9 + 3\sqrt{5}}{9 - 5} \qquad \textit{denominators.}$$

$$= \frac{9 + 3\sqrt{5}}{4}$$

5 **Simplify** $\dfrac{2\sqrt{3}}{3\sqrt{6} + 4\sqrt{2}}$.

$$\frac{2\sqrt{3}}{3\sqrt{6} + 4\sqrt{2}} = \frac{2\sqrt{3}}{3\sqrt{6} + 4\sqrt{2}} \cdot \frac{3\sqrt{6} - 4\sqrt{2}}{3\sqrt{6} - 4\sqrt{2}} \quad \textit{The conjugate of } 3\sqrt{6} + 4\sqrt{2} \textit{ is } 3\sqrt{6} - 4\sqrt{2}.$$

$$= \frac{2\sqrt{3} \cdot 3\sqrt{6} - 2\sqrt{3} \cdot 4\sqrt{2}}{(3\sqrt{6})^2 - (4\sqrt{2})^2}$$

$$= \frac{6\sqrt{18} - 8\sqrt{6}}{54 - 32} \qquad (3\sqrt{6})^2 = 3^2 \cdot \sqrt{6}^2 \textit{ or 54, and } (4\sqrt{2})^2 = 4^2 \cdot \sqrt{2}^2 \textit{ or 32}$$

$$= \frac{18\sqrt{2} - 8\sqrt{6}}{22} \qquad 6\sqrt{18} = 6\sqrt{2 \cdot 3^2} = 18\sqrt{2}$$

$$= \frac{2(9\sqrt{2} - 4\sqrt{6})}{22} \qquad \textit{The GCF of the numerator is 2.}$$

$$= \frac{9\sqrt{2} - 4\sqrt{6}}{11}$$

A radical expression is in simplest form when the following conditions are met.

1. **No radicands have perfect square factors other than one.**

2. **No radicands contain fractions.**

3. **No radical appears in the denominator of a fraction.**

Simplified Form
for Radicals

Exploratory Exercises

Simplify.

1. $(3 - \sqrt{7})(3 + \sqrt{7})$

2. $(\sqrt{6} - 5)(\sqrt{6} + 5)$

3. $(4 + 2\sqrt{2})(4 - 2\sqrt{2})$

State a conjugate of each expression. Then multiply the expression by this conjugate.

4. $3 + \sqrt{2}$

5. $\sqrt{3} + 4$

6. $7 - \sqrt{5}$

7. $6 + \sqrt{8}$

8. $9 - \sqrt{3}$

9. $\sqrt{2} + \sqrt{5}$

10. $\sqrt{3} - \sqrt{7}$

11. $2\sqrt{5} - \sqrt{6}$

12. $2\sqrt{8} + 3\sqrt{5}$

State the fraction by which each expression should be multiplied to rationalize the denominator.

13. $\frac{3}{\sqrt{5}}$

14. $\frac{7}{\sqrt{11}}$

15. $\frac{3\sqrt{2}}{\sqrt{6}}$

16. $\frac{2\sqrt{3}}{\sqrt{8}}$

17. $\sqrt{\frac{3}{5}}$

18. $\sqrt{\frac{8}{7}}$

19. $\frac{1}{3 + \sqrt{7}}$

20. $\frac{2\sqrt{5}}{4 - \sqrt{3}}$

Written Exercises

Simplify.

21. $\frac{\sqrt{42}}{\sqrt{6}}$

22. $\frac{\sqrt{20}}{\sqrt{5}}$

23. $\frac{\sqrt{10}}{\sqrt{7}}$

24. $\frac{\sqrt{7}}{\sqrt{3}}$

25. $\frac{\sqrt{6}}{\sqrt{18}}$

26. $\frac{\sqrt{5}}{\sqrt{10}}$

27. $\sqrt{\frac{3}{7}}$

28. $\sqrt{\frac{5}{2}}$

29. $\sqrt{\frac{7}{20}}$

30. $\sqrt{\frac{11}{32}}$

31. $\sqrt{\frac{2}{3}} \cdot \sqrt{\frac{5}{2}}$

32. $\sqrt{\frac{7}{11}} \cdot \sqrt{\frac{10}{7}}$

33. $\sqrt{\frac{4}{7}} \cdot \sqrt{\frac{3}{4}}$

34. $\sqrt{\frac{1}{6}} \cdot \sqrt{\frac{6}{11}}$

35. $\sqrt{\frac{a}{3}}$

36. $\sqrt{\frac{b}{6}}$

37. $\sqrt{\frac{a^2}{5}}$

38. $\sqrt{\frac{m^4}{11}}$

39. $\sqrt{\frac{27}{b^2}}$

40. $\sqrt{\frac{54}{r^2}}$

41. $\sqrt{\frac{5n^5}{4m^5}}$

42. $\sqrt{\frac{11a^3}{10b^3}}$

43. $\frac{\sqrt{3a^3b^4}}{\sqrt{8ab^6}}$

44. $\frac{\sqrt{9x^5y}}{\sqrt{12x^2y^6}}$

45. $\frac{1}{7 - \sqrt{3}}$

46. $\frac{1}{6 + \sqrt{3}}$

47. $\frac{11}{\sqrt{2} + 5}$

48. $\frac{10}{\sqrt{5} - 9}$

49. $\frac{6}{\sqrt{3} + \sqrt{2}}$

50. $\frac{12}{\sqrt{6} - \sqrt{5}}$

51. $\frac{10a}{2 - \sqrt{a}}$

52. $\frac{9b}{6 + \sqrt{b}}$

53. $\frac{2\sqrt{5}}{-3 + \sqrt{6}}$

54. $\frac{-3\sqrt{5}}{-2 - \sqrt{6}}$

55. $\frac{-9\sqrt{2}}{-4 + \sqrt{8}}$

56. $\frac{-10\sqrt{6}}{3 + \sqrt{6}}$

57. $\frac{2\sqrt{7}}{3\sqrt{5} + 5\sqrt{3}}$

58. $\frac{3\sqrt{11}}{7\sqrt{2} - 6\sqrt{5}}$

59. $\frac{6\sqrt{5}}{4\sqrt{8} - 2\sqrt{7}}$

60. $\frac{4\sqrt{19}}{3\sqrt{7} + 4\sqrt{12}}$

Challenge Exercises

Simplify. Assume that the value of each variable is positive.

61. $\frac{5 + 3\sqrt{2}}{4 - 6\sqrt{2}}$

62. $\frac{3\sqrt{2} - \sqrt{7}}{2\sqrt{3} - 5\sqrt{2}}$

63. $\frac{\sqrt{a} - \sqrt{b}}{\sqrt{a} + \sqrt{b}}$

64. $\frac{\sqrt{x} + \sqrt{y}}{\sqrt{x} - \sqrt{y}}$

Name the quadrant for the graph of (x, y) given each condition.

1. $x < 0, y > 0$

2. $x < 0, y < 0$

3. $x > 0, y = 0$

State whether each equation is a linear equation.

4. $2xy = 5$

5. $x = 3y$

6. $x^2 + y^2 = 4$

Determine whether each relation is a function.

7. $\{(5, -2), (4, -2), (3, 6), (1, -3)\}$

8. $\{(1, 1), (2, 2), (3, 3)\}$

9. $\{(-2, 0), (-1, 4), (0, 1), (2, 3), (-1, 5)\}$

Given $f(x) = x - x^2$, determine each value.

10. $f(-3)$

11. $f(4)$

12. $f(a - 2)$

13. $3f\left(\frac{2}{3}\right)$

14. Graph $3x - 2y = 12$ using the x- and y-intercepts.

15. Graph $3x - y = 2$ using the slope-intercept method.

16. Solve this system of inequalities by graphing.

    $x - y > -1$
    $2x + 3y < 8$

Write an equation in slope-intercept form of the line satisfying the given conditions.

17. $m = -\frac{3}{5}$ and passes through $(4, -1)$

18. x-intercept $= 3$ and y-intercept $= 7$

19. passes through $(-6, 1)$ and $(2, 7)$

20. passes through $(-2, 3)$ and $(-2, 8)$

21. perpendicular to $3x - 5y = 2$ and passes through $(4, -1)$

22. Find the coordinates of the midpoint of the segment whose endpoints are $(-2, 1)$ and $(6, -7)$.

Solve each system of equations.

23. $6x + y = 7$
    $x - 2y = 12$

24. $3x + 4y = 4$
    $9x - 2y = 5$

25. $3x - 2y = 9$
    $6x = 4y + 1$

Simplify. Use absolute value symbols when necessary.

26. $-\sqrt{49}$

27. $\sqrt{40}$

28. $\sqrt{18a^2b^3}$

29. $\sqrt{6}(\sqrt{2} - 3\sqrt{3})$

30. $\sqrt{\frac{6}{11}}$

31. $\frac{\sqrt{2a^3}}{\sqrt{5a^4b^6}}$

32. $\frac{4}{\sqrt{3} + 2\sqrt{2}}$

Use a system of equations to solve each problem.

33. A riverboat travels 42 miles upstream in $5\frac{1}{4}$ hours and returns in 3 hours. Find the rate of the current and the rate of the boat in still water.

34. The sum of the digits of a two-digit number is 11. If the digits are reversed, the new number is 63 more than the original number. Find the original number.

12-6 Adding and Subtracting Radical Expressions

Radical expressions in which the radicands are alike can be added or subtracted in the same way that monomials are added or subtracted.

$$3x + 2x = (3 + 2)x = 5x$$
$$3\sqrt{2} + 2\sqrt{2} = (3 + 2)\sqrt{2} = 5\sqrt{2}$$

$$7y - 4y = (7 - 4)y = 3y$$
$$7\sqrt{5} - 4\sqrt{5} = (7 - 4)\sqrt{5} = 3\sqrt{5}$$

Notice that the Distributive Property was used to simplify each radical expression.

Examples

1 **Simplify $3\sqrt{11} + 6\sqrt{11} - 2\sqrt{11}$.**

$3\sqrt{11} + 6\sqrt{11} - 2\sqrt{11} = (3 + 6 - 2)\sqrt{11}$ *Distributive Property*
$$= 7\sqrt{11}$$

2 **Simplify $9\sqrt{7} - 4\sqrt{2} + 3\sqrt{2} - 5\sqrt{7}$.**

$9\sqrt{7} - 4\sqrt{2} + 3\sqrt{2} - 5\sqrt{7} = 9\sqrt{7} - 5\sqrt{7} - 4\sqrt{2} + 3\sqrt{2}$ *Commutative Property*
$$= (9 - 5)\sqrt{7} + (-4 + 3)\sqrt{2}$$ *Distributive Property*
$$= 4\sqrt{7} - \sqrt{2}$$

If the radical expressions are *not* in simplest form, first simplify. Then, if possible, use the Distributive Property to further simplify the expression. Some expressions, such as $3\sqrt{5} - 2\sqrt{7}$ and $4\sqrt{2} + \sqrt{17}$, *cannot* be simplified further because the radicands are different and there are no common factors.

Example

3 **Simplify $6\sqrt{8} + 11\sqrt{18}$.**

$6\sqrt{8} + 11\sqrt{18} = 6\sqrt{2^2 \cdot 2} + 11\sqrt{3^2 \cdot 2}$ *Simplify each term.*
$$= 6(2\sqrt{2}) + 11(3\sqrt{2})$$
$$= 12\sqrt{2} + 33\sqrt{2}$$
$$= 45\sqrt{2}$$ *Distributive Property*

4 Simplify $7\sqrt{98} + 5\sqrt{32} - 2\sqrt{75}$.

$$
\begin{aligned}
7\sqrt{98} + 5\sqrt{32} - 2\sqrt{75} &= 7\sqrt{7^2 \cdot 2} + 5\sqrt{2^4 \cdot 2} - 2\sqrt{5^2 \cdot 3} \\
&= 7(7\sqrt{2}) + 5(4\sqrt{2}) - 2(5\sqrt{3}) \\
&= 49\sqrt{2} + 20\sqrt{2} - 10\sqrt{3} \\
&= 69\sqrt{2} - 10\sqrt{3}
\end{aligned}
$$

5 Simplify $4\sqrt{7} - 5\sqrt{28} + 5\sqrt{\frac{2}{7}}$.

$$
\begin{aligned}
4\sqrt{7} - 5\sqrt{28} + 5\sqrt{\tfrac{2}{7}} &= 4\sqrt{7} - 5\sqrt{2^2 \cdot 7} + 5\left(\frac{\sqrt{2}}{\sqrt{7}} \cdot \frac{\sqrt{7}}{\sqrt{7}}\right) \quad \text{\textit{Rationalize the}} \\
&\qquad\qquad\qquad\qquad\qquad\qquad\quad \text{\textit{denominator of } } 5\sqrt{\tfrac{2}{7}}. \\
&= 4\sqrt{7} - 5(2\sqrt{7}) + 5\left(\frac{\sqrt{14}}{7}\right) \\
&= 4\sqrt{7} - 10\sqrt{7} + \frac{5\sqrt{14}}{7} \\
&= \frac{28\sqrt{7} - 70\sqrt{7} + 5\sqrt{14}}{7} \quad \text{\textit{The least common}} \\
&\qquad\qquad\qquad\qquad\qquad \text{\textit{denominator is 7.}} \\
&= \frac{5\sqrt{14} - 42\sqrt{7}}{7}
\end{aligned}
$$

A calculator can be used to find approximations of expressions containing radicals.

6 **Using Calculators**

Simplify $4\sqrt{7} - 5\sqrt{28} + 5\sqrt{\frac{2}{7}}$.

ENTER: 4 [×] 7 [√] [−] 5 [×] 28 [√] [+] 5 [×] [(] 2 [÷] 7 [)] [√] [=]

The display shows -13.20189545. Therefore, $4\sqrt{7} - 5\sqrt{28} + 5\sqrt{\frac{2}{7}} \approx -13.20$.

You can also use a calculator to verify the answer to Example 5, $\frac{5\sqrt{14} - 42\sqrt{7}}{7}$.

ENTER: [(] 5 [×] 14 [√] [−] 42 [×] 7 [√] [)] [÷] 7 [=]

The display again shows -13.20189545 as in Example 6.

The answer in Example 5, $\frac{5\sqrt{14} - 42\sqrt{7}}{7}$ is the exact value of $4\sqrt{7} - 5\sqrt{28} + 5\sqrt{\frac{2}{7}}$, while -13.20 is an approximation of $4\sqrt{7} - 5\sqrt{28} + 5\sqrt{\frac{2}{7}}$.

Therefore, $\frac{5\sqrt{14} - 42\sqrt{7}}{7} \approx -13.20$, and the answer to Example 5 has been verified.

Exploratory Exercises

Name the expressions in each group that have the same radicand.

1. $5\sqrt{3}, 4\sqrt{6}, 3\sqrt{3}$
2. $5\sqrt{14}, -3\sqrt{7}, 4\sqrt{7}$
3. $3\sqrt{12}, 2\sqrt{6}, 5\sqrt{12}$
4. $4\sqrt{2}, 3\sqrt{2}, 2\sqrt{3}$
5. $-3\sqrt{3}, 2\sqrt{6}, 12\sqrt{3}$
6. $3\sqrt{3}, 2\sqrt{3}, 3\sqrt{6}$
7. $2\sqrt{10}, -6\sqrt{10}, 4\sqrt{2}, 7\sqrt{10}$
8. $3\sqrt{2}, 5\sqrt{6}, 2\sqrt{3}, 4\sqrt{7}$
9. $2\sqrt{5}, -\sqrt{10}, 3\sqrt{5}, -5\sqrt{5}$

Simplify.

10. $3\sqrt{5} + 2\sqrt{5}$
11. $4\sqrt{3} - 7\sqrt{3}$
12. $4\sqrt{7} + 11\sqrt{7}$
13. $8\sqrt{6} + 3\sqrt{6}$
14. $4\sqrt{x} - 5\sqrt{x}$
15. $5\sqrt{y} + 6\sqrt{y}$
16. $2\sqrt{11} + 3\sqrt{5}$
17. $3\sqrt{15} - 2\sqrt{5}$
18. $10\sqrt{6} - 3\sqrt{6}$
19. $25\sqrt{13} + \sqrt{13}$
20. $15\sqrt{6} - \sqrt{6}$
21. $21\sqrt{19} + 19\sqrt{19}$
22. $18\sqrt{2x} + 3\sqrt{2x}$
23. $2\sqrt{3a} - 7\sqrt{3a}$
24. $3\sqrt{5m} - 5\sqrt{5m}$

Written Exercises

Simplify.

25. $4\sqrt{11} + 7\sqrt{11}$
26. $10\sqrt{21} - 8\sqrt{21}$
27. $3\sqrt{13} - 7\sqrt{13}$
28. $-7\sqrt{7} + 6\sqrt{7}$
29. $5\sqrt{7} + 12\sqrt{7}$
30. $2\sqrt{11} + 4\sqrt{11}$
31. $4\sqrt{3} + 7\sqrt{3} - 2\sqrt{3}$
32. $5\sqrt{2} + 4\sqrt{2} + 7\sqrt{2}$
33. $8\sqrt{7} - 10\sqrt{7} + 3\sqrt{7}$
34. $2\sqrt{11} - 6\sqrt{11} - 3\sqrt{11}$
35. $5\sqrt{5} + 3\sqrt{5} - 18\sqrt{5}$
36. $2\sqrt{10} + 2\sqrt{10} + 2\sqrt{10}$
37. $8\sqrt{3} - 2\sqrt{2} + 3\sqrt{2} + 5\sqrt{3}$
38. $-3\sqrt{5} + 2\sqrt{5} + 5\sqrt{2} + 3\sqrt{2}$
39. $4\sqrt{6} + \sqrt{7} - 6\sqrt{2} + 4\sqrt{7}$
40. $10\sqrt{10} + 4\sqrt{10} - \sqrt{3} + 6\sqrt{10}$
41. $\sqrt{3} + \sqrt{5}$
42. $4\sqrt{6} + 2\sqrt{2}$
43. $2\sqrt{3} + \sqrt{12}$
44. $3\sqrt{8} + 7\sqrt{2}$
45. $3\sqrt{7} - 2\sqrt{28}$
46. $3\sqrt{50} - 4\sqrt{5}$
47. $4\sqrt{20} + 9\sqrt{45}$
48. $3\sqrt{27} + 5\sqrt{48}$
49. $3\sqrt{28} - 8\sqrt{63}$
50. $2\sqrt{50} - 3\sqrt{32}$
51. $\sqrt{18} + \sqrt{108} + \sqrt{50}$
52. $\sqrt{48} - \sqrt{12} + \sqrt{300}$
53. $8\sqrt{72} + 2\sqrt{20} - 3\sqrt{5}$
54. $2\sqrt{20} - 3\sqrt{24} - \sqrt{180}$
55. $2\sqrt{108} - \sqrt{27} + \sqrt{363}$
56. $8\sqrt{50} + 5\sqrt{72} - 2\sqrt{98}$
57. $\sqrt{7} + \sqrt{\frac{1}{7}}$
58. $\sqrt{6} + \sqrt{\frac{2}{3}}$
59. $\sqrt{3} - \sqrt{\frac{1}{3}}$
60. $\sqrt{10} - \sqrt{\frac{2}{5}}$
61. $3\sqrt{3} - \sqrt{45} + 3\sqrt{\frac{1}{3}}$
62. $14\sqrt{\frac{3}{2}} + 9\sqrt{\frac{2}{3}} + \sqrt{24}$
63. $3\sqrt{\frac{7}{4}} - 10\sqrt{\frac{1}{7}} + 3\sqrt{28}$
64. $6\sqrt{\frac{3}{5}} - 3\sqrt{60} + 2\sqrt{\frac{3}{5}}$

The ability to use and compute square roots is often a necessary tool in physics. For example, square roots are used to compute the minimum velocity a spacecraft must have to escape the gravitational force of a planet. The escape velocity, v, can be calculated using the following formula.

$$v = \sqrt{\frac{2GM}{r}}$$

In this formula, G is a gravitational constant, M is the mass of the planet, and r is the radius of the planet.

Example

Compute the escape velocity for Earth if its mass is 5.98×10^{24} kg, its radius is 6.37×10^6 m, and $G = 6.67 \times 10^{-11}$ N-m^2/kg^2. The abbreviation N represents newton, a unit of force in the metric system. Using these constants the answers will be in m/s.

$$1 \text{ N} = \frac{1 \text{ kg} \cdot m}{s^2}$$

$$v = \sqrt{\frac{2GM}{r}} \qquad G = 6.67 \times 10^{-11}, M = 5.98 \times 10^{24}, r = 6.37 \times 10^6$$

$$= \sqrt{\frac{2(6.67 \times 10^{-11})(5.98 \times 10^{24})}{6.37 \times 10^6}}$$

$$= \sqrt{\frac{7.97732 \times 10^{14}}{6.37 \times 10^6}}$$

$$= \sqrt{1.2523265 \times 10^8} \approx 1.12 \times 10^4$$

The escape velocity for Earth is approximately 1.12×10^4 m/s.

Exercises

Compute the escape velocity for the indicated planet using the formula above. Use $G = 6.67 \times 10^{-11}$ N-m^2/kg^2.

1. Mars: mass 6.46×10^{23} kg; radius 3.39×10^6 m

2. Jupiter: mass 1.90×10^{27} kg; radius 7.15×10^7 m

3. Mercury: mass 3.35×10^{23} kg; radius 2.44×10^6 m

4. Neptune: mass 1.03×10^{26} kg; radius 2.25×10^7 m

5. Venus: mass 4.90×10^{24} kg; radius 6.06×10^6 m

6. Uranus: mass 8.73×10^{25} kg; radius 2.35×10^7 m

12-7 Radical Equations

Equations containing radicals with variables in the radicand are called **radical equations**. To solve such equations, first isolate the radical on one side of the equation. Then square each side of the equation to eliminate the radical.

Examples

1 **Solve $\sqrt{y} = 11$ and check.**

$\sqrt{y} = 11$

$(\sqrt{y})^2 = (11)^2$ *Square each side to eliminate the radical.*

$y = 121$

Check: $\sqrt{y} = 11$

$\sqrt{121} \overset{?}{=} 11$

$11 = 11$ ✔

The solution is 121.

2 **Solve $\sqrt{3y - 5} - 4 = 0$ and check.**

$\sqrt{3y - 5} - 4 = 0$

$\sqrt{3y - 5} = 4$ *Isolate the radical by adding 4 to each side.*

$3y - 5 = 16$ *Square each side.*

$3y = 21$

$y = 7$

Check: $\sqrt{3y - 5} - 4 = 0$

$\sqrt{3 \cdot 7 - 5} - 4 \overset{?}{=} 0$

$\sqrt{16} - 4 \overset{?}{=} 0$

$4 - 4 \overset{?}{=} 0$

$0 = 0$ ✔

The solution is 7.

Consider this equation.

$x = 2$ *The solution is 2.*

Now square each side.

$x^2 = 4$ *The solutions are 2 and -2.*

Notice that squaring each side of an equation does not necessarily produce results that satisfy the *original* equation. Therefore you must check *all* solutions when solving radical equations.

Example

3 Solve $\sqrt{3x - 14} = 6 - x$ and check.

$\sqrt{3x - 14} = 6 - x$

$3x - 14 = (6 - x)^2$ *Square each side.*

$3x - 14 = 36 - 12x + x^2$ *Recall that $(a - b)^2 = a^2 - 2ab + b^2$.*

$0 = x^2 - 15x + 50$

$0 = (x - 5)(x - 10)$ *Factor.*

$x - 5 = 0$ or $x - 10 = 0$ *Use the Zero Product Property and set*

$x = 5$ or $x = 10$ *each factor equal to zero.*

Check: $\sqrt{3x - 14} = 6 - x$

$\sqrt{3 \cdot 5 - 14} \overset{?}{=} 6 - 5$ or $\sqrt{3 \cdot 10 - 14} \overset{?}{=} 6 - 10$

$\sqrt{15 - 14} \overset{?}{=} 1$ $\sqrt{30 - 14} \overset{?}{=} -4$

$\sqrt{1} \overset{?}{=} 1$ $\sqrt{16} \overset{?}{=} -4$

$1 = 1$ ✔ $4 \neq -4$

Notice that 10 satisfies $3x - 14 = (6 - x)^2$, but does not satisfy the *original* equation. Therefore, 5 is the *only* solution of $\sqrt{3x - 14} = 6 - x$.

Exploratory Exercises

Solve and check.

1. $\sqrt{y} = 3$ **2.** $\sqrt{a} = 4$ **3.** $\sqrt{m} = -2$ **4.** $\sqrt{s} = -5$

5. $-\sqrt{r} = -2$ **6.** $-\sqrt{z} = -3$ **7.** $-\sqrt{y} = 4$ **8.** $\sqrt{x - 3} = 6$

Written Exercises

Solve and check.

9. $\sqrt{r} = 5$ **10.** $\sqrt{s} = 9$ **11.** $\sqrt{3x} = 3$ **12.** $\sqrt{2m} = 4$

13. $\sqrt{3m} = -9$ **14.** $\sqrt{4a} = -1$ **15.** $\sqrt{b} - 5 = 0$ **16.** $\sqrt{s} + 3 = 0$

17. $\sqrt{2d} + 1 = 0$ **18.** $4 - \sqrt{3a} = 0$ **19.** $5 + \sqrt{2x} = 8$ **20.** $2 + \sqrt{m} = 13$

21. $\sqrt{4x + 1} = 3$ **22.** $\sqrt{2x + 7} = 5$ **23.** $\sqrt{8s + 1} - 5 = 0$ **24.** $\sqrt{3b - 5} = -4$

25. $\sqrt{\frac{x}{4}} = 6$ **26.** $\sqrt{\frac{x}{4}} = 10$ **27.** $\sqrt{\frac{4a}{3}} - 2 = 0$ **28.** $\sqrt{\frac{9s}{2}} - 6 = 0$

29. $\sqrt{\frac{5x}{4}} - 8 = 2$ **30.** $\sqrt{\frac{k}{7}} - 10 = -3$ **31.** $\sqrt{r} = 3\sqrt{5}$ **32.** $3\sqrt{7} = \sqrt{m}$

33. $5\sqrt{2n^2 - 28} = 20$ **34.** $4\sqrt{3m^2 - 15} = 4$ **35.** $\sqrt{2z^2 - 121} = z$ **36.** $\sqrt{5x^2 - 7} = 2x$

37. $\sqrt{x + 2} = x - 4$ **38.** $\sqrt{1 - 2x} = 1 + x$ **39.** $4 + \sqrt{x - 2} = x$ **40.** $\sqrt{x^2 + 3} = 2x$

41. The square root of the product of 4 and a number is 26. Find the number.

42. The square root of the product of 9 and a number is 27. Find the number.

43. The square root of 6 less than a number is 13. Find the number.

44. Eight more than the square root of 3 times a number is 23. Find the number.

45. The time, t, in seconds that it takes an object, initially at rest, to fall a vertical distance of s_v meters is given by the formula $t = \sqrt{\dfrac{2s_v}{g}}$. ($g$ is the acceleration due to gravity.) Find s_v when $t = 4$ and the acceleration due to gravity is 9.8 m/s^2.

Challenge Exercises

Solve and check.

46. $\sqrt{x + 16} = \sqrt{x} + 4$

47. $6 - \sqrt{x} = \sqrt{x - 12}$

48. $\sqrt{x + 5} = 5 + \sqrt{x}$

Solve each system of equations.

49. $3\sqrt{x} - 5\sqrt{y} = 9$
$2\sqrt{x} + 5\sqrt{y} = 6$

50. $-4\sqrt{a} + 6\sqrt{b} = 3$
$-3\sqrt{a} + 3\sqrt{b} = 1$

mini-review

Write an equation for the relationship between the variables in each chart. Then copy and complete each chart.

1.

x	1	2	3	4	5	6
y	3	5	7	9		

2.

a	-3	-2	-1	0	1	2	3
b	21	18	15	12			

Write the standard form of the equation of the line passing through the given point and having the given slope.

3. $(3, 8)$, $\dfrac{1}{6}$

4. $(-4, -10)$, undefined

5. $\left(-\dfrac{1}{2}, 8\right)$, 0

Use substitution or elimination to solve each system of equations. State whether the system has no solution, one solution, or infinitely many solutions. If the system has one solution, state it.

6. $2x + y = 14$
$x = 3y$

7. $2x + 5y = 8$
$\dfrac{1}{2}x + \dfrac{5}{4}y = 2$

8. $\dfrac{4x - y}{2} = 11$
$\dfrac{2x + y}{3} = -8$

Solve.

9. While driving from home to Fort Worth, Mr. Sharp averages 60 mph. On the return trip, he averages 45 mph and travels one hour longer. How far from Fort Worth does he live?

Use the table on page 592 to determine whether each square root is rational or irrational. Then state the two integers between which the square root lies if it is irrational.

10. $\sqrt{6065}$

11. $\sqrt{9025}$

12. $\sqrt{1089}$

13. $\sqrt{2816}$

Simplify.

14. $\sqrt{\dfrac{4ab^2}{7c^3}}$

15. $\dfrac{6}{\sqrt{2} - 4\sqrt{3}}$

16. $\sqrt{20} - 4\sqrt{3} + \sqrt{125}$

12-8 The Pythagorean Theorem

On a baseball diamond the distance from one base to the next is 90 feet. How far from home plate is second base? To answer this question, notice that a baseball diamond can be separated into two right triangles.

The side opposite the right angle in a right triangle is called the **hypotenuse**. This side is *always* the longest side of a right triangle. The other two sides are called the **legs** of the triangle.

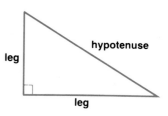

To find the length of the hypotenuse, given the lengths of the legs, you can use a formula proposed by the Greek mathematician Pythagoras.

In a right triangle, if a and b are the measures of the legs and c is the measure of the hypotenuse, then $$c^2 = a^2 + b^2.$$	*The Pythagorean Theorem*

If c is the measure of the longest side of a triangle and $c^2 \neq a^2 + b^2$, then the triangle is not a right triangle.

For the baseball diamond, $a = 90$ and $b = 90$. Find the distance from home plate to second base, that is, the length of the hypotenuse.

$$c^2 = a^2 + b^2$$
$$c^2 = 90^2 + 90^2$$
$$c^2 = 8100 + 8100$$
$$c^2 = 2 \cdot 8100$$
$$c = \sqrt{8100 \cdot 2}$$
$$c = 90\sqrt{2}$$
$$\approx 90(1.414) \qquad \sqrt{2} \approx 1.414$$
$$\approx 127$$

The distance from home plate to second base is approximately 127 feet.

The Pythagorean Theorem can be illustrated geometrically as shown below.

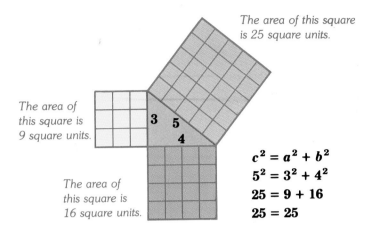

The area of this square is 25 square units.

The area of this square is 9 square units.

3 5

4

The area of this square is 16 square units.

$$c^2 = a^2 + b^2$$
$$5^2 = 3^2 + 4^2$$
$$25 = 9 + 16$$
$$25 = 25$$

The Pythagorean Theorem can be used to find the length of a side of a right triangle when the lengths of the other two sides are known.

Examples

1 **Find the length of the hypotenuse of a right triangle if $a = 15$ and $b = 8$.**

$$c^2 = 15^2 + 8^2$$
$$c^2 = 225 + 64$$
$$c^2 = 289$$
$$c = 17$$

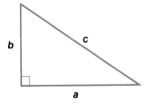

The length of the hypotenuse is 17 units.

2 **Find the length of a leg of a right triangle if $a = 6$ and $c = 14$.**

$$c^2 = a^2 + b^2$$
$$14^2 = 6^2 + b^2$$
$$196 = 36 + b^2$$
$$160 = b^2$$
$$\sqrt{160} = b$$
$$b = 4\sqrt{10}$$
$$\approx 4(3.162) \qquad \sqrt{10} \approx 3.162$$
$$\approx 12.648$$

The length of the leg is $4\sqrt{10}$ or approximately 12.648 units.

Example

3 The measures of three sides of a triangle are 5, 7, and 9. Determine whether this triangle is a right triangle.

Since the measure of the longest side is 9, let $c = 9$, $a = 5$, and $b = 7$. Then determine whether $c^2 = a^2 + b^2$.

$$c^2 = a^2 + b^2$$
$$9^2 \stackrel{?}{=} 5^2 + 7^2$$
$$81 \stackrel{?}{=} 25 + 49$$
$$81 \neq 74$$

The numbers do not satisfy the Pythagorean Theorem. Therefore, the triangle is *not* a right triangle.

Exploratory Exercises

State whether each sentence is *true* or *false*.

1. $3^2 + 4^2 = 5^2$
2. $9^2 + 10^2 = 11^2$
3. $6^2 + 8^2 = 9^2$
4. $5^2 + 12^2 = 13^2$

Solve each equation.

5. $3^2 + 4^2 = c^2$
6. $6^2 + 8^2 = c^2$
7. $5^2 + 12^2 = c^2$
8. $a^2 + 12^2 = 13^2$
9. $a^2 + 15^2 = 17^2$
10. $a^2 + 24^2 = 25^2$
11. $6^2 + b^2 = 10^2$
12. $3^2 + b^2 = 5^2$

Written Exercises

If *c* is the measure of the hypotenuse of a right triangle, find each missing measure. Use the table on page 592.

13. $a = 5$, $b = 12$, $c = ?$
14. $a = 6$, $b = 3$, $c = ?$
15. $a = 4$, $b = \sqrt{11}$, $c = ?$
16. $a = \sqrt{7}$, $b = 9$, $c = ?$
17. $b = 12$, $c = 15$, $a = ?$
18. $b = 10$, $c = 11$, $a = ?$
19. $b = 30$, $c = 34$, $a = ?$
20. $a = 11$, $c = 61$, $b = ?$
21. $a = 20$, $c = 29$, $b = ?$
22. $a = \sqrt{5}$, $c = \sqrt{30}$, $b = ?$
23. $a = \sqrt{7}$, $b = \sqrt{9}$, $c = ?$
24. $a = \sqrt{11}$, $c = \sqrt{47}$, $b = ?$

The measures of three sides of a triangle are given. Determine whether each triangle is a right triangle.

25. 9, 16, 20
26. 9, 40, 41
27. 45, 60, 75
28. 11, 12, 16

For each problem, make a drawing. Then use an equation to solve the problem. Use the table on page 592.

29. Find the length of the diagonal of a rectangle whose length is 8 meters and whose width is 5 meters.

30. Judy hikes 7 miles due east and then 3 miles due north. How far is she from the starting point?

31. The diagonal of a rectangular wall measures 14 meters. One side of the wall is 10 meters long. What is the length of the other side?

32. A rope from the top of a mast on a sailboat is attached to a point 2 meters from the base of the mast. The rope is 8 meters long. How high is the mast?

12-9 The Distance Formula

The Pythagorean Theorem can be used to find the distance between any two points in the coordinate plane. For example, find the distance between $(-5, 2)$ and $(4, 5)$.

In the diagram at the right, the distance to be found is the length of segment d. Notice that a right triangle can be formed by drawing lines parallel to the axes from $(-5, 2)$ and $(4, 5)$. The coordinates of the point where these lines meet are $(4, 2)$.

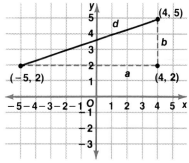

The length of side a of the triangle is the difference of the x-coordinates, $4 - (-5)$, or 9 units. The length of side b of the triangle is the difference of the y-coordinates, $5 - 2$, or 3 units.

Now the Pythagorean Theorem can be used to find the length of d, that is, the distance between $(-5, 2)$ and $(4, 5)$.

$$d^2 = 9^2 + 3^2$$
$$d^2 = 81 + 9$$
$$d^2 = 90$$
$$d = \sqrt{90}$$
$$d = 3\sqrt{10}$$

The distance between $(-5, 2)$ and $(4, 5)$ is $3\sqrt{10}$ units. The method for finding the distance between any two points can be generalized as the following formula.

> The distance between any two points with coordinates (x_1, y_1) and (x_2, y_2) is given by the following formula.
>
> $$d = \sqrt{(x_2 - x_1)^2 + (y_2 - y_1)^2}$$

The Distance Formula

Example

1 Find the distance between the points with coordinates $(8, -2)$ and $(5, 3)$.

$d = \sqrt{(x_2 - x_1)^2 + (y_2 - y_1)^2}$ *Let $x_2 = 5$, $x_1 = 8$, $y_2 = 3$, and $y_1 = -2$.*

$d = \sqrt{(5 - 8)^2 + [3 - (-2)]^2}$

$d = \sqrt{(-3)^2 + 5^2}$

$d = \sqrt{9 + 25}$

$d = \sqrt{34}$ *This is approximately 5.831 units.*

Example

2 Find a if the distance between the points with coordinates $(5, a)$ and $(7, -3)$ is $\sqrt{85}$ units.

$$d = \sqrt{(x_2 - x_1)^2 + (y_2 - y_1)^2} \quad \text{Let } x_2 = 7, x_1 = 5, y_2 = -3, \text{ and } y_1 = a.$$
$$\sqrt{85} = \sqrt{(7 - 5)^2 + (-3 - a)^2}$$
$$\sqrt{85} = \sqrt{2^2 + (-3 - a)^2}$$
$$\sqrt{85} = \sqrt{4 + 9 + 6a + a^2}$$
$$\sqrt{85} = \sqrt{13 + 6a + a^2}$$
$$85 = 13 + 6a + a^2 \quad \text{Square each side.}$$
$$0 = a^2 + 6a - 72$$
$$0 = (a + 12)(a - 6) \quad \text{Factor.}$$
$$a + 12 = 0 \quad \text{or} \quad a - 6 = 0$$
$$a = -12 \qquad a = 6$$

The value of a is -12 or 6. *These answers can be checked by substituting -12 and 6 for a in the equation $\sqrt{85} = \sqrt{(7 - 5)^2 + (-3 - a)^2}$.*

Exploratory Exercises

State the values of x_1, x_2, y_1, and y_2 for each pair of points.

1. $(3, 4), (6, 8)$ **2.** $(5, -1), (11, 7)$ **3.** $(-4, 2), (4, 17)$

4. $(-2, 8), (3, 20)$ **5.** $(-3, 5), (2, 7)$ **6.** $(5, 4), (-3, 8)$

7. $(-8, -4), (-3, 8)$ **8.** $(2, 7), (10, -4)$ **9.** $(3, 7), (-2, -5)$

10. $(3, 2), (0, 5)$ **11.** $(2, 2), (5, -1)$ **12.** $(-8, -7), (0, 8)$

Written Exercises

Find the distance between each pair of points whose coordinates are given.

13. $(-4, 2), (4, 17)$ **14.** $(5, -1), (11, 7)$ **15.** $(-3, 5), (2, 7)$

16. $(5, 4), (-3, 8)$ **17.** $(-8, -4), (-3, 8)$ **18.** $(2, 7), (10, -4)$

19. $(7, -9), (4, -3)$ **20.** $(9, -2), (3, -6)$ **21.** $(10, 8), (2, -3)$

22. $(11, -2), (-4, 5)$ **23.** $(-2, 5), \left(-\frac{1}{2}, 3\right)$ **24.** $(4, 2), \left(6, -\frac{2}{3}\right)$

25. $\left(\frac{2}{3}, -4\right), (3, -2)$ **26.** $\left(6, -\frac{2}{7}\right), (5, -1)$ **27.** $\left(\frac{4}{5}, -1\right), \left(2, -\frac{1}{2}\right)$

The coordinates of a pair of points are given in each exercise. Find two possible values for a if the points are the given distance apart.

28. $(4, 7), (a, 3); d = 5$ **29.** $(3, a), (-4, 2); d = \sqrt{170}$

30. $(8, 1), (5, a); d = 5$ **31.** $(-3, a), (5, 2); d = 17$

32. $(a, 5), (-7, 3); d = \sqrt{29}$ **33.** $(5, 9), (a, -3); d = 13$

34. $(a, -4), (2, -3); d = \sqrt{65}$ **35.** $(-6, -5), (-3, a); d = \sqrt{13}$

36. $(4, -7), (7, a); d = \sqrt{34}$ **37.** $(-5, a), (4, -2); d = \sqrt{130}$

Challenge Exercises

Use the Distance Formula to solve each problem.

38. Show that the points with the coordinates (0, 0), (7, 0), (7, 4), and (0, 4) are the vertices of a rectangle.

39. Show that the point with coordinates (2, 6) is the midpoint of the segment joining the points with coordinates $(-1, 2)$ and $(5, 10)$.

40. Find the distance between $M(2\sqrt{5}, \sqrt{5})$ and $N(4\sqrt{5}, -3\sqrt{5})$.

41. Find the distance between $A(\sqrt{8}, \sqrt{3})$ and $B(\sqrt{3}, -\sqrt{8})$.

42. The coordinates of the vertices of a triangle are $(2, -1)$, $(-2, 2)$, and $(-6, 14)$. Find the perimeter of the triangle.

Excursions in Algebra — Equations of Circles

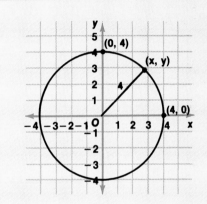

A circle is the set of all points in the coordinate plane that are the same distance from a given point, called the *center*. All points of the circle shown at the right are 4 units from the point with coordinates (0, 0), the center. Choose any point of the circle. The distance from a point (x, y) to the center is 4 units. Now apply the Distance Formula.

$$d = \sqrt{(x_2 - x_1)^2 + (y_2 - y_1)^2}$$
$$4 = \sqrt{(x - 0)^2 + (y - 0)^2}$$
$$4 = \sqrt{x^2 + y^2}$$
$$16 = x^2 + y^2 \quad \text{\textit{Square each side.}}$$

Thus, $16 = x^2 + y^2$ is the equation of a circle with center at (0, 0) and radius of 4 units.

> **The equation of a circle with center at (0, 0) and a radius of r units is $x^2 + y^2 = r^2$.**
>
> *Equation of a Circle*

Example Find the equation of a circle with center at (0, 0) and a radius of 7 units.

$$x^2 + y^2 = r^2$$
$$x^2 + y^2 = 7^2 \quad \text{\textit{Substitute 7 for r.}}$$
$$x^2 + y^2 = 49$$

Exercises State the radius of each circle whose equation is given below.

1. $x^2 + y^2 = 100$ **2.** $x^2 + y^2 = 81$ **3.** $x^2 + y^2 = 8$

4. $x^2 + y^2 = 10$ **5.** $x^2 + y^2 = \frac{1}{4}$ **6.** $x^2 + y^2 = \frac{4}{9}$

7. $x^2 + y^2 = 256$ **8.** $x^2 + y^2 = 576$ **9.** $x^2 + y^2 = 0.09$

10. $x^2 + y^2 = 0.16$ **11.** $x^2 + y^2 = \frac{100}{121}$ **12.** $x^2 + y^2 = \frac{196}{625}$

Computing Square Roots

One method for computing a square root, called the **divide-and-average** method, involves first choosing any approximation for the square root and then calculating more accurate approximations.

Another method for computing $\sqrt{N}$ is to substitute any approximation, A, into the expression $\frac{A^2 + N}{2A}$. Then repeat this process until successive values of A are close together. This method is demonstrated in the following computer program.

```
10   PRINT "FOR WHAT NUMBER DO YOU WANT
          TO FIND THE SQUARE ROOT?"
20   INPUT N
30   PRINT "CHOOSE ANY APPROXIMATION FOR THE SQUARE ROOT"
40   INPUT A
50   LET L = A
60   PRINT A
70   LET A = (A ↑ 2 + N) / (2 * A)
80   IF ABS (A − L) > .000001 THEN 50      The ABS function finds the absolute value.
90   PRINT "AN APPROXIMATION OF THE SQUARE ROOT OF ";
          N;" IS ";A
100  END
```

Notice that the variable L is used to store the old value of A. Then, in line 70, the next value of A is computed. Finally, in line 80, the values of A and L are compared. If the values are close together (in this case, if their difference is less than 0.000001), then the program ends. If the difference is not less than 0.000001, the loop is executed again to obtain a closer approximation.

Exercises

Use the program above to approximate each square root.

1. $\sqrt{10}$ 2. $\sqrt{19}$ 3. $\sqrt{23}$ 4. $\sqrt{92}$ 5. $\sqrt{198}$

6. $\sqrt{147}$ 7. $\sqrt{323}$ 8. $\sqrt{582}$ 9. $\sqrt{1901}$ 10. $\sqrt{5460}$

11. In the BASIC language, SQR(A) can be used to find an approximation of the square root of A. Write an additional line for the program above to compute and print an approximation using the SQR function. Run the program to compare the two approximations for Exercises 1–10.

square root   (395)

radical expression   (395)

radical sign   (395)

radicand   (395)

principal square root   (395)

irrational numbers   (398)

real numbers   (399)

square root key   (402)

rationalizing the denominator   (407)

conjugate   (408)

radical equations   (415)

hypotenuse   (418)

legs   (418)

Pythagorean Theorem   (418)

Chapter Summary

1. If $x^2 = y$, then x is a square root of y.   (395)

2. The symbol $\sqrt{}$, called a radical sign, indicates a nonnegative square root. The expression under the radical sign is called the radicand.   (395)

3. Product Property of Square Roots: For any numbers a and b, if $a \geq 0$ and $b \geq 0$, then $\sqrt{ab} = \sqrt{a} \cdot \sqrt{b}$.   (396)

4. Quotient Property of Square Roots: For any numbers a and b, if $a \geq 0$ and $b > 0$, then $\sqrt{\frac{a}{b}} = \frac{\sqrt{a}}{\sqrt{b}}$.   (396)

5. The square roots of numbers such as 2, 3, 5, and 7 are irrational numbers. Irrational numbers are numbers that cannot be expressed in the form $\frac{a}{b}$, where a and b are integers, $b \neq 0$.   (398)

6. Irrational numbers together with the rational numbers form the set of real numbers. Each real number corresponds to exactly one point on the number line. Each point on the number line corresponds to exactly one real number. (399)

7. A calculator can be used to approximate square roots.   (402)

8. Rationalizing the denominator changes the denominator to a rational number.   (407)

9. Binomials of the form $a\sqrt{b} + c\sqrt{d}$ and $a\sqrt{b} - c\sqrt{d}$ are called conjugates of each other. They can be used to rationalize some denominators.   (408)

10. A radical expression is in simplest form when the following conditions are met.

    1. No radicands have perfect square factors other than one.
    2. No radicands contain fractions.
    3. No radical appears in the denominator of a fraction.   (408)

11. Square roots having like radicands can be added or subtracted.   (411)

12. To solve radical equations, first isolate the radical on one side of the equation. Then square each side of the equation to eliminate the radical.   (415)

13. The Pythagorean Theorem: In a right triangle, if a and b are the measures of the legs and c is the measure of the hypotenuse, then $c^2 = a^2 + b^2$.   (418)

14. The Distance Formula: The distance between any two points with coordinates (x_1, y_1) and (x_2, y_2) is given by the following formula.

$$d = \sqrt{(x_2 - x_1)^2 + (y_2 - y_1)^2} \quad (421)$$

Chapter Review

12–1 **Simplify.**

    1. $\sqrt{121}$ **2.** $-\sqrt{64}$ **3.** $\pm\sqrt{\frac{4}{81}}$ **4.** $-\sqrt{\frac{100}{225}}$

12–2 **Graph each number on the number line.**

    5. $-\sqrt{5}$ **6.** $\sqrt{8}$ **7.** $\sqrt{15}$ **8.** $-\sqrt{2}$

    State whether each decimal represents a rational number or an irrational number.

    9. $0.4\overline{6}$ **10.** $4.2302300230002\ldots$ **11.** 7.634

    Use the table on page 592 to determine the two integers between which the square root lies.

    12. $\sqrt{250}$ **13.** $\sqrt{490}$ **14.** $-\sqrt{172}$ **15.** $-\sqrt{365}$

12–3 **Use a calculator to find each square root. Round decimal answers to the nearest thousandth.**

    16. $-\sqrt{19}$ **17.** $\pm\sqrt{2304}$ **18.** $\sqrt{61.7}$

Simplify. Use absolute value symbols when necessary.

12–4 **19.** $\sqrt{18}$ **20.** $\sqrt{108}$ **21.** $\sqrt{720}$ **22.** $\sqrt{2916}$

23. $\sqrt{4b^2}$ **24.** $\sqrt{\dfrac{60}{x^2}}$ **25.** $\sqrt{44a^2b}$ **26.** $\sqrt{6} \cdot \sqrt{8}$

27. $5\sqrt{3} \cdot \sqrt{3}$ **28.** $\sqrt{3}(\sqrt{3} + \sqrt{6})$ **29.** $\sqrt{5}(\sqrt{10} - \sqrt{3})$

12–5 **30.** $\dfrac{\sqrt{35}}{\sqrt{5}}$ **31.** $\dfrac{\sqrt{20}}{\sqrt{7}}$ **32.** $\dfrac{7}{2 - \sqrt{3}}$

33. $\dfrac{9}{3 + \sqrt{2}}$ **34.** $\dfrac{5\sqrt{2}}{7\sqrt{3} + 6\sqrt{5}}$ **35.** $\dfrac{3\sqrt{7}}{4\sqrt{6} - 7\sqrt{7}}$

12–6 **36.** $2\sqrt{13} + 3\sqrt{13}$ **37.** $8\sqrt{15} - 3\sqrt{15}$ **38.** $3\sqrt{7} - 5\sqrt{28}$

39. $4\sqrt{27} + 6\sqrt{48}$ **40.** $\sqrt{8} + \sqrt{\dfrac{1}{8}}$ **41.** $\sqrt{3} - \sqrt{\dfrac{1}{3}}$

12–7 **Solve and check.**

42. $\sqrt{3x} = 6$ **43.** $\sqrt{3x + 5} = 4$ **44.** $\sqrt{7x - 1} - 5 = 0$

45. $\sqrt{w} = 5\sqrt{6}$ **46.** $\sqrt{x + 4} = x - 8$ **47.** $\sqrt{\dfrac{4a}{3}} - 2 = 0$

12–8 **Use the Pythagorean Theorem to find each missing measure.**

48. $a = 6$, $b = 10$, $c = ?$

49. $a = 10$, $c = 15$, $b = ?$

50. $b = 6$, $c = 12$, $a = ?$

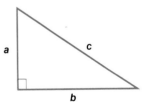

51. Find the length of the diagonal of a rectangle whose length is 15 meters and whose width is 8 meters.

12–9 **Find the distance between each pair of points whose coordinates are given.**

52. $(4, 2)$, $(7, -9)$ **53.** $(9, -2)$, $(1, 13)$ **54.** $(5, -2)$, $(-8, -3)$

Chapter Test

Simplify.

1. $\sqrt{3} \cdot \sqrt{15}$

2. $\sqrt{40}$

3. $\dfrac{5}{\sqrt{2} + \sqrt{3}}$

4. $\sqrt{\dfrac{a^2}{7}}$

5. $\sqrt{54x^4 y}$

6. $(4 + \sqrt{3})(4 - \sqrt{3})$

7. $\dfrac{7}{7 + \sqrt{5}}$

8. $2\sqrt{27} + \sqrt{48} - 3\sqrt{3}$

9. $3\sqrt{50} - 2\sqrt{8}$

10. $\sqrt{\dfrac{3x^2}{4n^3}}$

11. $\sqrt{3xy^3}$

12. $15\sqrt{6} - \sqrt{72}$

13. $\sqrt{\dfrac{32}{9}}$

14. $\sqrt{\dfrac{64x^4}{8x^2}}$

15. $\dfrac{6\sqrt{5}}{2\sqrt{2} + 5\sqrt{3}}$

16. $\sqrt{2}(\sqrt{2} + 4\sqrt{3})$

17. $\sqrt{6} + \sqrt{\dfrac{2}{3}}$

18. $\sqrt{96} \cdot \sqrt{48}$

Name the set or sets of numbers to which each of the following numbers belongs. Use N for natural numbers, W for whole numbers, Z for integers, Q for rational numbers, and I for irrational numbers.

19. $\sqrt{16}$

20. $4.\overline{56}$

21. $\dfrac{3}{8}$

22. $-\sqrt{18}$

Solve and check.

23. $\sqrt{t} + 5 = 3$

24. $\sqrt{5x^2 - 9} = 2x$

25. $\sqrt{4x + 1} = 5$

26. $\sqrt{4x - 3} = 6 - x$

Use the Pythagorean Theorem to find each missing measure.

27. Find c if $a = 8$ and $b = 10$.

28. Find b if $a = 2$ and $c = 8$.

29. Find a if $a = b$ and $c = 12$.

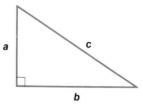

Find the distance between each pair of points whose coordinates are given.

30. $(4, 3), (6, -3)$

31. $(-8, 2), (4, -3)$

32. $(4, 7), (4, -2)$

The test questions on this page deal with coordinates and geometry. The information at the right may help you with some of the questions.

Directions: Choose the best answer. Write A, B, C, or D.

1. The midpoint of $\overline{AB}$ is M. If the coordinates of A are $(-3, 2)$ and the coordinates of M are $(-1, 5)$, what are the coordinates of B?

   (A) $(1,10)$ **(B)** $(1, 8)$
   (C) $(0, 7)$ **(D)** $(-5, 8)$

2.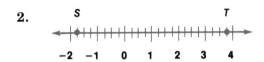

   The length of $\overline{ST}$ is

   (A) 5 **(B)** $4\frac{1}{2}$ **(C)** $5\frac{1}{2}$ **(D)** 6

3. What is the area of the shaded triangle in square units?

   (A) 12 **(B)** $12\sqrt{3}$ **(C)** $24\sqrt{3}$ **(D)** 24

4. Seven squares of the same size form a rectangle when placed side-by-side. The perimeter of the rectangle is 496. What is the area of each square?

   (A) 72 ft^2 **(B)** 324 ft^2
   (C) 900 ft^2 **(D)** 961 ft^2

5. What is the total length of fencing needed to enclose a rectangular area 46 feet by 34 feet?

   (A) 26 yards 1 foot **(B)** $26\frac{2}{3}$ yards

   (C) 52 yards 2 feet **(D)** $53\frac{1}{3}$ yards

---

1. The diameter of a circle is twice the length of the radius.
2. The area of a circle with radius r is πr^2.
3. The area of a triangle is the product of $\frac{1}{2}$ the base and the height.

---

6. The number of degrees in the smaller angle formed by the hands of a clock at 12:15 is

   (A) 120 **(B)** $82\frac{1}{2}$ **(C)** $92\frac{1}{2}$ **(D)** 90

7. The length of each side of a square is $\frac{3x}{5} + 1$. The perimeter of the square is

   (A) $\frac{12x + 20}{5}$ **(B)** $\frac{12x + 4}{5}$

   (C) $\frac{3x + 4}{5}$ **(D)** $\frac{3x}{5} + 16$

8. If a line passes through point $(0, 2)$ and has a slope of 4, what is the equation of the line?

   (A) $x = 2y + 4$ **(B)** $x = 4y + 2$
   (C) $y = 4x + 2$ **(D)** $y = 2x + 4$

9. The larger circle has a diameter of b. The area of the shaded ring in square units is

   (A) $b^2 - c^2$ **(B)** $\pi b^2 - \pi c^2$

   (C) $\frac{1}{4}\pi (b^2 - c^2)$ **(D)** $\frac{1}{2}\pi (b^2 - c^2)$

10. If a circle of radius 10 meters has its radius decreased by 2 meters, by what percent is its area decreased?

    (A) 20% **(B)** 40% **(C)** 80% **(D)** 36%

11. What is the value of x if the area of the triangle is $\frac{1}{4}$ the area of the square?

    (A) $\sqrt{2}$ **(B)** $2\sqrt{2}$ **(C)** 4 **(D)** 8

CHAPTER 13

Quadratics

Mars Station is the site of a tracking antenna for spacecraft. It is capable of tracking spacecraft to the edge of the solar system.

The reflector of this antenna is 64 meters in diameter and has a shape whose cross section may be described by a quadratic equation. In this chapter, you will learn about quadratic equations and quadratic functions.

13-1 Graphing Quadratic Functions

An equation such as $y = x^2 - 4x + 1$ describes a type of function known as a **quadratic function**.

A quadratic function is a function that can be described by an equation of the form $y = ax^2 + bx + c$, where $a \neq 0$.	*Definition of Quadratic Function*

Graphs of quadratic functions have certain common characteristics. For instance, they all have a general shape called a **parabola**.

The table and graph below illustrate some other common characteristics of quadratic functions.

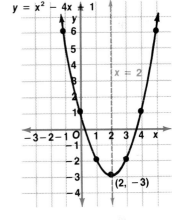

x	$x^2 - 4x + 1$	y
-1	$(-1)^2 - 4(-1) + 1$	6
0	$0^2 - 4(0) + 1$	1
1	$1^2 - 4(1) + 1$	-2
2	$2^2 - 4(2) + 1$	-3
3	$3^2 - 4(3) + 1$	-2
4	$4^2 - 4(4) + 1$	1
5	$5^2 - 4(5) + 1$	6

Notice the matching values in the y-column.

Notice that in the y-column of the table, -3 does *not* have a matching value. Also, -3 is the y-coordinate of the lowest point on the graph of $y = x^2 - 4x + 1$. For the graph of $y = x^2 - 4x + 1$, the *lowest point*, called the **minimum point**, has coordinates $(2, -3)$.

The vertical line containing the minimum point is the **axis of symmetry**. The equation of the axis of symmetry for the graph above is $x = 2$.

If the graph of any quadratic function is folded along the axis of symmetry, the two halves coincide. In other words, the two halves of the parabola are *symmetric*.

Example

1 Graph $y = -x^2 + 2x + 3$.

x	$-x^2 + 2x + 3$	y
-2	$-(-2)^2 + 2(-2) + 3$	-5
-1	$-(-1)^2 + 2(-1) + 3$	0
0	$-0^2 + 2(0) + 3$	3
1	$-(1)^2 + 2(1) + 3$	4
2	$-(2)^2 + 2(2) + 3$	3
3	$-(3)^2 + 2(3) + 3$	0
4	$-(4)^2 + 2(4) + 3$	-5

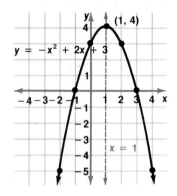

The graph of $y = -x^2 + 2x + 3$ opens *downward*. The equation of the axis of symmetry is $x = 1$. The graph has a *highest point*, or **maximum point**, at (1, 4).

In general, a parabola will open upward and have a minimum point when the coefficients of y and x^2 have the same sign. It will open downward and have a maximum point when the coefficient of y and x^2 have the opposite sign. The maximum or minimum point of the graph *always* lies on the axis of symmetry.

Notice that in Example 1 the axis of symmetry is halfway between any two points having the same y-coordinate. Consider the points on the graph whose coordinates are $(-1, 0)$ and $(3, 0)$.

From these coordinates, the equation of the axis of symmetry may be found as shown below.

$x = \dfrac{-1 + 3}{2}$ *Add the x-coordinates and divide by 2.*

$\quad = 1$

The equation of the axis of symmetry is $x = 1$.

In general, the equation of the axis of symmetry for the graph of a quadratic function can be found by using the following rule.

The equation of the axis of symmetry for $y = ax^2 + bx + c$**, where** $a \neq 0$**, is** $$x = -\frac{b}{2a}.$$	*Equation of Axis of Symmetry*

Example

2 Find the equation of the axis of symmetry and the coordinates of the maximum or minimum point of the graph of $y = x^2 - x - 6$. Then use the information to draw the graph.

First, find the equation of the axis of symmetry.

$$x = -\frac{b}{2a}$$

$$= -\frac{-1}{2 \cdot 1} \qquad a = 1, b = -1$$

$$= \frac{1}{2}$$

The equation of the axis of symmetry is $x = \frac{1}{2}$.

Next, find the coordinates of the maximum or minimum point. Since the coefficients of y and x^2 have the same sign, the graph of the function has a minimum point. The minimum point lies on the axis of symmetry. Since the axis of symmetry is $x = \frac{1}{2}$, the minimum point will have an x-coordinate of $\frac{1}{2}$. Find the y-coordinate by substituting $\frac{1}{2}$ for x in $y = x^2 - x - 6$.

$$y = x^2 - x - 6$$

$$= \left(\frac{1}{2}\right)^2 - \frac{1}{2} - 6$$

$$= \frac{1}{4} - \frac{1}{2} - 6$$

$$= -\frac{25}{4}$$

The coordinates of the minimum point are $\left(\frac{1}{2}, -\frac{25}{4}\right)$.

Then, construct a table. For the values of x, choose some integers greater than $\frac{1}{2}$, and some less than $\frac{1}{2}$. This insures that points on both sides of the axis of symmetry are plotted. Use this information to draw the graph.

x	$x^2 - x - 6$	y
-2	$(-2)^2 - (-2) - 6$	0
-1	$(-1)^2 - (-1) - 6$	-4
0	$0^2 - 0 - 6$	-6
1	$1^2 - 1 - 6$	-6
2	$2^2 - 2 - 6$	-4
3	$3^2 - 3 - 6$	0

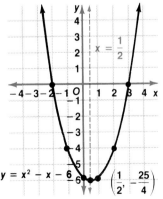

Example

3 **Find the coordinates of the maximum point for the graph of $y = -2x^2 - 8x + 9$.**

*Since the coefficients of y and x^2 have different signs,
the graph of the function has a maximum point.*

$x = -\frac{b}{2a}$ First, find the equation of the axis of symmetry.

$= -\frac{-8}{2(-2)}$ *$a = -2, b = -8$*

$= -2$ The equation of the axis of symmetry is $x = -2$.

$y = -2(-2)^2 - 8(-2) + 9$ *Since the maximum point lies on
the axis of symmetry, substitute
-2 for x in $y = -2x^2 - 8x + 9$.*

$= -8 + 16 + 9$

$= 17$

The coordinates of the maximum point are $(-2, 17)$.

Exploratory Exercises

State whether the graph of each quadratic function opens upward or downward.

1. $y = x^2 - 1$ **2.** $y = -x^2 + 3x$ **3.** $y = -x^2 + x + 1$

4. $y = x^2 + x + 3$ **5.** $y = -x^2 + 4x + 5$ **6.** $y = 2x^2 - 5x + 2$

7. $y = 3x^2 + 9x - 1$ **8.** $y = 2x^2 - 8x + 1$ **9.** $y = -5x^2 + 3x + 2$

Find the equation of the axis of symmetry of the graph of each quadratic function.

10. $y = x^2 + x + 3$ **11.** $y = -x^2 + 4x + 5$ **12.** $y = -x^2 + 3x$

13. $y = x^2 + 6x + 8$ **14.** $y = 3x^2 + 6x + 16$ **15.** $y = 5x^2 + 20x + 37$

16. $y = 8x + 3 + x^2$ **17.** $y = \frac{3}{2} + 7x + 5x^2$ **18.** $y = -4x^2 + 4x + \frac{5}{2}$

Written Exercises

Find the equation of the axis of symmetry and the coordinates of the maximum or minimum point of the graph of each quadratic function.

19. $y = -x^2 + 5x + 6$ **20.** $y = -x^2 + 5x - 14$ **21.** $y = x^2 - 4x + 13$

22. $y = x^2 + 2x$ **23.** $y = -5x^2 + 15x + 23$ **24.** $y = -3x^2 + 4$

25. $y = 3x^2 + 6x - 17$ **26.** $y = 3x^2 + 24x + 80$ **27.** $y = -2x^2 - 9$

28. $y = -3x^2 - 6x + 5$ **29.** $y = 5x^2 + 10x + 6$ **30.** $y = 7x^2 + 14x - 9$

31. $y = -7x^2 + 14x + 15$ **32.** $y = -4x^2 + 8x + 13$ **33.** $y = 3x^2 + 4$

34. $y = 2x^2 + 12x - 17$ **35.** $y = -5x^2 + 10x + 37$ **36.** $y = 2x^2 - 6x + 19$

Find the equation of the axis of symmetry and the coordinates of the maximum or minimum point of the graph of each quadratic function. Then draw the graph.

37. $y = x^2 - 4x - 5$ **38.** $y = -x^2 + 4x + 5$ **39.** $y = -x^2 + 6x + 5$

40. $y = x^2 - x - 6$ **41.** $y = x^2 - 3$ **42.** $y = -x^2 + 7$

43. $y = 2x^2 + 3$ **44.** $y = x^2 - 2x - 8$ **45.** $y = x^2 - x - 12$

46. $y = \frac{1}{4}x^2 - 4x + 3\frac{3}{4}$ **47.** $y = \frac{1}{2}x^2 + 3x + \frac{9}{2}$ **48.** $y = -3x^2 - 6x + 4$

Many times businesses will raise the prices of their goods or services to increase their profit. However, when they raise their prices, they usually lose some customers. In such situations, the price at which the maximum profit occurs needs to be found.

Example

An auditorium has seats for 1200 people. For the past several days, the auditorium has been filled to capacity for each show. Tickets currently cost $5.00 and the owner wants to increase the ticket prices. He estimates that for each $0.50 increase in price, 100 fewer people will attend. What ticket price will maximize the profit?

Let x = number of $0.50 price increases. Thus $5.00 + 0.50x$ represents the single ticket price and $1200 - 100x$ represents the number of tickets sold.

$$\begin{aligned} \text{Income} &= \text{number of tickets sold} \cdot \text{ticket price} \\ &= (1200 - 100x) \cdot (5.00 + 0.50x) \\ &= 6000 + 100x - 50x^2 \end{aligned}$$

Notice that the result is a quadratic equation. The graph of the related function, $y = -50x^2 + 100x + 6000$, opens downward and thus has a maximum point. Since this is a maximum point, the x-coordinate gives the number of price increases needed to maximize the profit.

Recall that the x-coordinate of the maximum point is given by the equation of the axis of symmetry.

$$x = -\frac{b}{2a}$$

$$= -\frac{100}{2(-50)} \qquad a = -50, b = 100$$

$$= 1$$

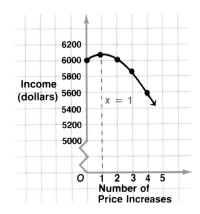

The equation of the axis of symmetry is $x = 1$.

Thus, the profit is maximized when the owner makes one $0.50 price increase. So, the price of one ticket should be $5.50.

Exercises   Solve each problem.

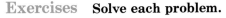

1. A grocer sells 50 loaves of bread a day. The cost is $0.65 a loaf. The grocer estimates that for each $0.05 price increase, 2 fewer loaves of bread will be sold. What cost will maximize the profit?

2. A bus company transports 500 people a day between Morse Rd. and High St. A one-way fare is $0.50. The owners estimate that for each $0.10 price increase 50 passengers will be lost. What price will maximize their profit?

Graphing Calculators: Graphing Quadratic Functions

A graphing calculator is a powerful tool for studying graphs of functions and relations. The graphs of certain functions, such as the quadratic function $y = x^2$, are built into the graphing mode of the calculator. Functions that are not built in will be called user-generated functions.

The calculator will produce the graphical representation of a particular built-in function by pressing the single key on the calculator that corresponds to that function.

Example

1   **Graph $y = x^2$.**

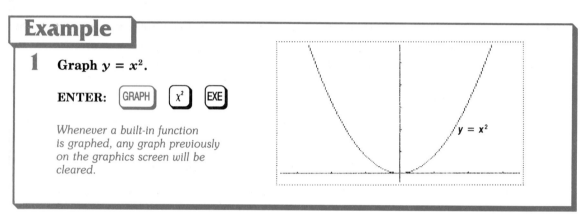

ENTER: GRAPH x^2 EXE

Whenever a built-in function is graphed, any graph previously on the graphics screen will be cleared.

$y = x^2$

When the calculator draws the graph of a built-in function, the x-axis scale, y-axis scale, and position of the x-axis and y-axis on the graphics screen are automatically set. These values were chosen so as to give an appropriate graphical representation of the function on the graphics screen.

When the calculator draws the graph of a user-generated function, the range parameter values are not automatically set. In order to get an appropriate graphical representation of a user-generated function, it is sometimes necessary to manually set the range parameter values.

Example

2   Set the range parameters so the Xmin is -8, Xmax is 6, Xscl is 2, Ymin is -7, Ymax is 28, and Yscl is 4. Then graph $y = x^2$, $y = x^2 - 4$, and $y = (x + 2)^2 + 5$ on the same set of axes.

Enter $y = x^2$ as a user-generated function by entering [X] before pressing the [x^2] key.

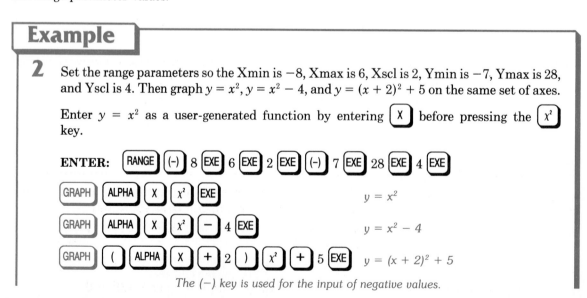

ENTER: RANGE (-) 8 EXE 6 EXE 2 EXE (-) 7 EXE 28 EXE 4 EXE

GRAPH ALPHA X x^2 EXE $y = x^2$

GRAPH ALPHA X x^2 − 4 EXE $y = x^2 - 4$

GRAPH (ALPHA X + 2) x^2 + 5 EXE $y = (x + 2)^2 + 5$

The (−) key is used for the input of negative values.

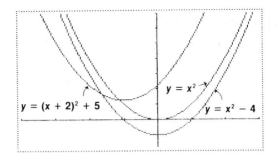

Notice that the graphs of $y = x^2 - 4$ and $y = (x + 2)^2 + 5$ look like the graph of $y = x^2$. However, the graph of $y = x^2 - 4$ is translated downward 4 units and has $(0, -4)$ as its lowest, or minimum, point. The graph of $y = (x + 2)^2 + 5$ is translated 2 units to the left and 5 units upward. *What are the coordinates of its minimum point?*

Example

3 **Graph $y = x^2$, $y = 3x^2$, and $y = 0.5x^2$ on the same set of axes.**

The range parameter values for the built-in function $y = x^2$ will produce appropriate graphical representations of the other two functions. Therefore, graph $y = x^2$ as a built-in function. Then graph the other functions.

ENTER: GRAPH x^2 EXE GRAPH 3

ALPHA X x^2 EXE

GRAPH 0.5 ALPHA X x^2

EXE

The graphs of $y = 3x^2$ and $y = 0.5x^2$ look like the graph of $y = x^2$. However, the graph of $y = 3x^2$ is narrower and the graph of $y = 0.5x^2$ is wider than the graph of $y = x^2$.

Written Exercises

Graph each pair of equations on the same set of axes on a graphing calculator. Then compare the graph of the second equation to the graph of the first equation.

1. $y = x^2$; $y = x^2 - 5$

2. $y = x^2$; $y = x^2 + 3$

3. $y = x^2$; $y = (x + 8)^2$

4. $y = x^2$; $y = (x - 10)^2$

5. $y = x^2$; $y = 4x^2$

6. $y = x^2$; $y = 0.25x^2$

7. $y = x^2$; $y = (x - 3)^2 + 6$

8. $y = x^2$; $y = (x + 4)^2 - 8$

9. $y = 2x^2$; $y = -2(x + 1)^2 + 4$

10. $y = x^2$; $y = -\frac{1}{3}(x - 12)^2 - 6$

13-2 Solving Quadratic Equations by Graphing

Recall that many quadratic equations can be solved by factoring. For example, solve $x^2 + x - 6 = 0$.

$$x^2 + x - 6 = 0 \qquad \textit{Find the factors of } x^2 + x - 6.$$
$$(x + 3)(x - 2) = 0$$
$$x + 3 = 0 \quad \text{or} \quad x - 2 = 0 \qquad \textit{Zero Product Property}$$
$$x = -3 \qquad\qquad x = 2$$

The solutions of an equation are called the **roots** of the equation. The roots of $x^2 + x - 6 = 0$ are -3 and 2. Notice that the roots of $x^2 + x - 6 = 0$ are the x-intercepts of the graph of the related function, $y = x^2 + x - 6$.

x	$x^2 + x - 6$	y
-3	$(-3)^2 + (-3) - 6$	0
-2	$(-2)^2 + (-2) - 6$	-4
-1	$(-1)^2 + (-1) - 6$	-6
0	$0^2 + 0 - 6$	-6
1	$1^2 + 1 - 6$	-4
2	$2^2 + 2 - 6$	0

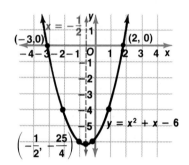

In general, the roots of any quadratic equation of the form $ax^2 + bx + c = 0$ are the x-intercepts of the graph of the related function $y = ax^2 + bx + c$.

Many times, only approximations of roots can be found by graphing. In the following example, the consecutive integers between which the roots lie are found.

Examples

1 **Locate the roots of $x^2 - 6x + 6 = 0$ by graphing the related function.**

The graph of the related function, $y = x^2 - 6x + 6$, is a parabola that opens upward.

axis of symmetry

$$x = -\frac{b}{2a}$$

$$= -\frac{-6}{2 \cdot 1} \text{ or } 3$$

minimum point

Substitute 3 for x in $y = x^2 - 6x + 6$.

$$y = 3^2 - 6(3) + 6$$

$$= -3$$

The equation of the axis of symmetry is $x = 3$ and the coordinates of the minimum point are $(3, -3)$.

438 *Quadratics*

x	$x^2 - 6x + 6$	y
1	$1^2 - 6(1) + 6$	1
2	$2^2 - 6(2) + 6$	-2
3	$3^2 - 6(3) + 6$	-3
4	$4^2 - 6(4) + 6$	-2
5	$5^2 - 6(5) + 6$	1

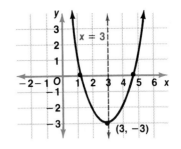

From the graph it can be seen that one root lies between 1 and 2. The other root is between 4 and 5.

2 **Locate the roots of $x^2 - 4x + 5 = 0$ by graphing the related function.**

The graph of the related function $y = x^2 - 4x + 5$ opens upward and has as its axis of symmetry $x = 2$. The coordinates of its minimum point are $(2, 1)$.

x	$x^2 - 4x + 5$	y
0	$0^2 - 4(0) + 5$	5
1	$1^2 - 4(1) + 5$	2
2	$2^2 - 4(2) + 5$	1
3	$3^2 - 4(3) + 5$	2
4	$4^2 - 4(4) + 5$	5

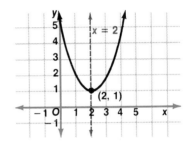

The graph has no x-intercepts since it does not cross the x-axis. Therefore, the equation $x^2 - 4x + 5 = 0$ has no roots that are real numbers.

Exploratory Exercises

State the roots of each quadratic equation whose related function is graphed below.

1.

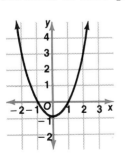

2.

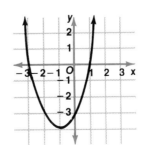

3.

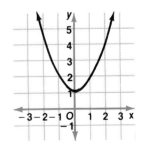

State the roots of each quadratic equation whose related function is graphed below.

4.

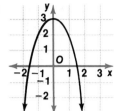

5.

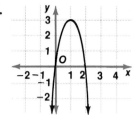

6.

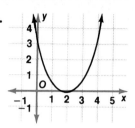

Written Exercises

The maximum or minimum point and the roots of a quadratic function are given. Graph the related function.

7. minimum point: $(4, -2)$
roots: 2, 6

8. maximum point: $(-3, 4)$
roots: 0, -6

9. maximum point: $(-5, 0)$
root: -5

10. minimum point: $(2, 0)$
root: 2

11. minimum point: $(-5, 1)$
roots: no real roots

12. maximum point: $(3, -2)$
roots: no real roots

13. maximum point: $(-1, 6)$
roots: $-4 < x < -3$
$1 < x < 2$

14. minimum point: $(-5, -1)$
roots: $-6 < x < -5$
$-5 < x < -4$

Locate the roots of each equation by graphing the related function.

15. $x^2 - x - 12 = 0$

16. $x^2 + 7x + 12 = 0$

17. $x^2 - 5x - 6 = 0$

18. $x^2 + 3x - 18 = 0$

19. $x^2 - 4 = 0$

20. $x^2 - 9 = 0$

21. $x^2 - 10x = -21$

22. $x^2 + 4x = 12$

23. $x^2 - 2x + 2 = 0$

24. $x^2 + 6x + 10 = 0$

25. $x^2 - 4x + 1 = 0$

26. $x^2 - 6x + 4 = 0$

27. $x^2 - 8x + 16 = 0$

28. $x^2 + 6x + 9 = 0$

29. $3x^2 + 4x - 1 = 0$

30. $6x^2 - 13x - 15 = 0$

31. $x^2 - 8x + 18 = 0$

32. $x^2 + 2x + 4 = 0$

33. $x^2 - 2x - 1 = 0$

34. $x^2 + 6x + 7 = 0$

35. $x^2 + 5x + 9 = 0$

36. $x^2 - 3x + 7 = 0$

37. $4x^2 - 12x + 3 = 0$

38. $4x^2 + 4x - 35 = 0$

Challenge Exercises

Locate the y-intercepts of each quadratic relation by graphing. Use appropriate values of y.

39. $x = 2y^2 - 8y + 7$

40. $x = -y^2 - 4y - 1$

41. $x = y^2 - 2y + 3$

42. $x = -3y^2 + 6y - 4$

43. $x = \frac{1}{9}y^2 - \frac{2}{3}y$

44. $x = -\frac{3}{4}y^2 - 6y - 9$

Write in standard form the equation of a line that passes through each pair of points.

1. $(-4, 7), (4, 3)$

2. $\left(-\frac{1}{2}, 3\right), (-4, -5)$

3. $\left(\frac{1}{3}, 2\frac{1}{2}\right), \left(-\frac{2}{3}, 1\frac{1}{2}\right)$

Graph each equation using the slope-intercept method.

4. $y = \frac{4}{3}x + 2$

5. $2x - 4y = 1$

6. $4x + \frac{2}{3}y = 6$

Solve each system of equations.

7. $9m + n = 1$
$\quad m - 3n = 4$

8. $4x + 5y = 3$
$\quad -2x - 3y = 3$

9. $3a + 5b = 1$
$\quad 2a + 3b = -2$

Simplify.

10. $\sqrt{28} - \sqrt{63} + 4\sqrt{7}$

11. $\sqrt{\frac{50x^3}{3}}$

12. $\frac{\sqrt{5}}{4\sqrt{3} + \sqrt{5}}$

Solve each equation and check your solution.

13. $\sqrt{4x} = 12$

14. $\sqrt{7x - 3} - 2 = 3$

15. $\sqrt{\frac{3a}{2}} + 1 = 0$

16. Solve the system of inequalities $y < 3x - 1$ and $2x + 3y \geq 6$ by graphing.

17. Find the distance between the points whose coordinates are $(-3, 4)$ and $(4, 7)$.

Excursions in Algebra   Parabolic Surfaces

Suppose a parabola is rotated about its axis of symmetry. The surface described by this rotation is known as a paraboloid of revolution, or a parabolic surface.

Parabolic surfaces are used in automobile headlamps, radar antennas, solar collectors, and other devices. This is because a parabolic surface can focus incoming energy at a single point, or reflect the energy from a point and send it in one direction.

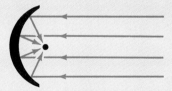

This point is known as the focus. The focus lies on the axis of symmetry.

The focus for a parabola with equation $y = ax^2 + bx + c$ has the following coordinates.

$$\left(-\frac{b}{2a}, \frac{4ac - b^2 + 1}{4a}\right)$$

Exercises

Find the coordinates of the focus of each parabola whose equation is given.

1. $y = x^2 + 6x + 9$

2. $y = 2x^2 + 4x + 7$

3. $y = \frac{1}{4}x^2 - 2x + 3$

13-3 Completing the Square

An equation like $x^2 - 36 = 0$ can be solved in the following way.

$$x^2 - 36 = 0$$

$$x^2 = 36 \qquad \textit{Add 36 to each side.}$$

$$\sqrt{x^2} = \sqrt{36} \qquad \textit{Find the square root of each side.}$$

$$|x| = 6 \qquad \textit{Recall that } |x| = \sqrt{x^2}.$$

$$x = \pm 6 \qquad \textit{x} = \pm 6 \textit{ since } |+6| = |-6| = 6.$$

This same method can be used to solve the equation $x^2 - 4x + 4 = 3$.

Example

1 **Solve $x^2 - 4x + 4 = 3$.**

$$x^2 - 4x + 4 = 3$$

$$(x - 2)^2 = 3 \qquad \textit{Factor } x^2 - 4x + 4.$$

$$\sqrt{(x - 2)^2} = \sqrt{3} \qquad \textit{Find the square root of each side.}$$

$$|x - 2| = \sqrt{3}$$

$$x - 2 = \pm\sqrt{3} \qquad \textit{Why is this so?}$$

$$x = 2 \pm \sqrt{3} \qquad \textit{Add 2 to each side.}$$

The solution set is $\{2 + \sqrt{3}, 2 - \sqrt{3}\}$.

The quadratic expression must be a perfect square in order to use the above method. If it is not a perfect square, then a method called **completing the square** may be used.

Consider the pattern for squaring a binomial.

$$(x + 6)^2 = x^2 + 2(6)(x) + 6^2$$

$$= x^2 + 12x \quad + 36$$

$$\left(\frac{12}{2}\right) \;\rightarrow\; 6^2 \qquad \textit{Notice that 36 is } 6^2 \textit{ and 6 is one-half of 12.}$$

To complete the square for an expression of the form $x^2 + bx$, follow the steps listed below.

Step 1 Find one-half of b, the coefficient of x.

Step 2 Square the result of **Step 1**.

Step 3 Add the result of **Step 2** to $x^2 + bx$.

2 **Find the value of c that makes $x^2 + 14x + c$ a perfect square.**

    Step 1   Find one-half of 14. $\frac{14}{2} = 7$

    Step 2   Square the result of **Step 1**. $7^2 = 49$

    Step 3   Add the result of **Step 2** to $x^2 + 14x$. $x^2 + 14x + 49$

Thus, $c = 49$. Notice that $x^2 + 14x + 49$ is equal to $(x + 7)^2$, which is a perfect square.

3 **Solve $x^2 + 6x - 16 = 0$ by completing the square.**

$x^2 + 6x - 16 = 0$	*Notice that $x^2 + 6x - 16$ is not a perfect square.*
$x^2 + 6x = 16$	*Add 16 to each side. Then complete the square.*
$x^2 + 6x + 9 = 16 + 9$	$\left(\frac{6}{2}\right)^2 = 9$, *so add 9 to each side of the equation.*
$(x + 3)^2 = 25$	*Factor $x^2 + 6x + 9$.*
$x + 3 = \pm 5$	*Find the square root of each side.*
$x = \pm 5 - 3$	*Subtract 3 from each side.*

    $x = 5 - 3$ or $x = -5 - 3$

    $x = 2$ $x = -8$

The roots of $x^2 + 6x - 16 = 0$ are 2 and -8.   *Check this result.*

Notice that the roots can be used to factor $x^2 + 6x - 16$.

$$x^2 + 6x - 16 = (x - 2)(x + 8)$$

4 **Solve $2x^2 - 9x + 8 = 0$ by completing the square.**

$2x^2 - 9x + 8 = 0$	*To complete the square, the coefficient of x^2 must be 1. Divide each side of the equation by 2.*
$x^2 - \frac{9}{2}x + 4 = 0$	
$x^2 - \frac{9}{2}x = -4$	*Subtract 4 from each side. Then complete the square.*
$x^2 - \frac{9}{2}x + \frac{81}{16} = -4 + \frac{81}{16}$	*Why was $\frac{81}{16}$ added to each side?*
$\left(x - \frac{9}{4}\right)^2 = \frac{17}{16}$	
$x - \frac{9}{4} = \pm\frac{\sqrt{17}}{4}$	
$x = \frac{9 \pm \sqrt{17}}{4}$	

The roots of $2x^2 - 9x + 8 = 0$ are $\frac{9 + \sqrt{17}}{4}$ and $\frac{9 - \sqrt{17}}{4}$.

The decimal form of irrational roots can be found by using a calculator.

5 **Using Calculators**

Use a calculator to express $\frac{9 \pm \sqrt{17}}{4}$ in decimal form to the nearest hundredth.

ENTER: (9 + 17 $\sqrt{x}$) ÷ 4 =   DISPLAY: 3.2807762

ENTER: (9 − 17 $\sqrt{x}$) ÷ 4 =   DISPLAY: 1.2192236

In decimal form, $\frac{9 \pm \sqrt{17}}{4}$ is approximately 3.28 and 1.22.

Exploratory Exercises

State whether each trinomial is a perfect square.

1. $y^2 + 8y + 7$
2. $b^2 + 4b + 3$
3. $m^2 - 10m + 25$

4. $r^2 - 8r + 16$
5. $d^2 + 12d + 27$
6. $h^2 - 13h + \frac{169}{4}$

Find the value of c that makes each trinomial a perfect square.

7. $x^2 + 8x + c$
8. $x^2 + 4x + c$
9. $m^2 - 10m + c$

10. $x^2 - 6x + c$
11. $z^2 + 2z + c$
12. $x^2 + 7x + c$

13. $x^2 - 7x + c$
14. $a^2 + 5a + c$
15. $x^2 - 13x + c$

Written Exercises

Solve each equation by completing the square.

16. $y^2 + 4y + 3 = 0$
17. $n^2 + 8n + 7 = 0$
18. $t^2 - 4t = 21$

19. $z^2 - 4z = 2$
20. $p^2 + 2p - 3 = 0$
21. $y^2 - 8y = 4$

22. $r^2 + 14r - 10 = 5$
23. $y^2 + 7y + 10 = -2$
24. $n^2 + 5n + 6 = 0$

25. $x^2 - 5x + 2 = -2$
26. $n^2 - 11n - 12 = 0$
27. $x^2 - 4x + 1 = 0$

28. $x^2 - 6x + 7 = 0$
29. $b^2 - 6b + 4 = 0$
30. $\frac{1}{2}t^2 - 2t - \frac{3}{2} = 0$

31. $4x^2 - 20x + 25 = 0$
32. $2b^2 - b - 14 = 7$
33. $2d^2 + 3d - 20 = 0$

34. $x^2 - \frac{7}{2}x + \frac{3}{2} = 0$
35. $a^2 - \frac{1}{2}a - \frac{3}{2} = 0$
36. $3y^2 - y - 10 = 0$

37. $0.3x^2 + 0.1x - 0.2 = 0$
38. $0.3n^2 - 0.2n = 0.1$
39. $\frac{1}{2}q^2 - \frac{5}{4}q - 3 = 0$

40. $x^2 + 2x - 10 = 0$
41. $r^2 + 0.25r - 0.5 = 0$
42. $2m^2 - 5m + 1 = 0$

43. $a^2 - 6a + 6 = 0$
44. $x^2 - 10x = 23$
45. $y^2 - 8y = 13$

46. $2x^2 - 6x - 5 = 0$
47. $4y^2 - 2y = 1$
48. $3x^2 - 7x - 3 = 0$

Challenge Exercises

Solve by completing the square. Leave irrational roots in simplest radical form.

49. $x^2 + 4x + c = 0$
50. $x^2 - bx + 8 = 0$
51. $x^2 + bx + c = 0$

52. $ax^2 + bx + c = 0$
53. $x^2 + xy - 2y^2 = 0$
54. $x^2 + 4bx + b^2 = 0$

13-4   The Quadratic Formula

The method of completing the square can be used to develop a general formula for solving any quadratic equation. Begin with the general form of a quadratic equation, $ax^2 + bx + c = 0$, where $a \neq 0$.

$$ax^2 + bx + c = 0$$

$$x^2 + \frac{b}{a}x + \frac{c}{a} = 0 \qquad \textit{Divide by a so the coefficient of } x^2 \textit{ becomes 1.}$$

$$x^2 + \frac{b}{a}x = -\frac{c}{a} \qquad \textit{Subtract } \frac{c}{a} \textit{ from each side.}$$

Now complete the square.

$$x^2 + \frac{b}{a}x + \left(\frac{b}{2a}\right)^2 = -\frac{c}{a} + \left(\frac{b}{2a}\right)^2$$

$$\left(x + \frac{b}{2a}\right)^2 = -\frac{c}{a} + \frac{b^2}{4a^2} \qquad \textit{Factor the left side.}$$

$$\left(x + \frac{b}{2a}\right)^2 = \frac{b^2 - 4ac}{4a^2} \qquad \textit{Simplify the right side.}$$

Find the square root of each side.

$$x + \frac{b}{2a} = \pm \sqrt{\frac{b^2 - 4ac}{4a^2}}$$

$$x + \frac{b}{2a} = \frac{\pm\sqrt{b^2 - 4ac}}{2a} \qquad \textit{Simplify the square root on the right side.}$$

$$x = \frac{\pm\sqrt{b^2 - 4ac}}{2a} - \frac{b}{2a} \qquad \textit{Subtract } \frac{b}{2a} \textit{ from each side.}$$

$$x = \frac{-b \pm \sqrt{b^2 - 4ac}}{2a} \qquad \textit{The result is an expression for x.}$$

This result is called the **Quadratic Formula** and can be used to solve any quadratic equation.

> **The roots of a quadratic equation of the form $ax^2 + bx + c = 0$, where $a \neq 0$, are given by:**
>
> $$x = \frac{-b \pm \sqrt{b^2 - 4ac}}{2a}$$
>
> *The Quadratic Formula*

In order to find a real value for $\sqrt{b^2 - 4ac}$, the value of $b^2 - 4ac$ must be nonnegative. If $b^2 - 4ac$ is negative, the equation has no real roots.

Examples

1 **Use the Quadratic Formula to solve $x^2 - 6x - 3 = 0$.**

$$x = \frac{-b \pm \sqrt{b^2 - 4ac}}{2a}$$

$$= \frac{-(-6) \pm \sqrt{(-6)^2 - 4(1)(-3)}}{2(1)} \qquad a = 1, b = -6, c = -3$$

$$= \frac{6 \pm \sqrt{36 + 12}}{2}$$

$$= \frac{6 \pm \sqrt{48}}{2} \qquad\qquad \sqrt{48} = \sqrt{16 \cdot 3} \text{ or } 4\sqrt{3}$$

$$x = \frac{6 + 4\sqrt{3}}{2} \quad \text{or} \quad x = \frac{6 - 4\sqrt{3}}{2}$$

$$x = 3 + 2\sqrt{3} \qquad x = 3 - 2\sqrt{3} \qquad \textit{Check this result.}$$

The solution set is $\{3 + 2\sqrt{3}, 3 - 2\sqrt{3}\}$.

2 **Using Calculators**

Use the Quadratic Formula to find the decimal approximations of the roots of $\frac{4}{3}x^2 - 2x = -\frac{1}{3}$. Round to the nearest hundredth.

First, change the equation to the general form, $ax^2 + bx + c = 0$.

$$\frac{4}{3}x^2 - 2x = -\frac{1}{3}$$

$$\frac{4}{3}x^2 - 2x + \frac{1}{3} = 0 \qquad \textit{Add } \frac{1}{3} \textit{ to each side.}$$

$$4x^2 - 6x + 1 = 0 \qquad \textit{Multiply each side by 3. Now } a = 4, b = -6, \text{ and } c = 1.$$

Second, simplify $\sqrt{b^2 - 4ac}$ or $\sqrt{(-6)^2 - 4(4)(1)}$ and store the value.

ENTER: 6 [+/−] [x^2] [−] 4 [×] 4 [×] 1 [=]   [$\sqrt{x}$] [STO]

$\qquad\qquad$ *b* *squared* *minus 4ac* *equals square root*

DISPLAY: *4.472136*

Third, find the value of $\frac{-b + \sqrt{b^2 - 4ac}}{2a}$.   *Remember, you have stored the value of $\sqrt{b^2 - 4ac}$.*

ENTER: 6 [+] [RCL] [=] [÷] [(] 2 [×] 4 [)] [=]   DISPLAY: *1.309017*

$\qquad\qquad$ *−b + √b² − 4ac*   *÷ 2a*   *Why does −b = 6?*

Finally, find the value of $\frac{-b - \sqrt{b^2 - 4ac}}{2a}$.

ENTER: 6 [−] [RCL] [=] [÷] [(] 2 [×] 4 [)] [=]   DISPLAY: *0.190983*

The decimal approximations of the roots are 1.31 and 0.19.

Exploratory Exercises

State the values of a, b, and c for each quadratic equation.

1. $x^2 + 7x + 6 = 0$

2. $y^2 + 8y + 15 = 0$

3. $m^2 + 4m + 3 = 0$

4. $2t^2 - t = 15$

5. $4x^2 + 8x = -3$

6. $2y^2 + 3 = -7y$

7. $y^2 - 25 = 0$

8. $2y^2 = 98$

9. $2x^2 + 8x = 0$

10. $3n^2 - 18n = 0$

11. $3m^2 + 2 = -5m$

12. $3k^2 + 11k = 4$

State the value of $b^2 - 4ac$ for each quadratic equation.

13. $x^2 + 5x - 6 = 0$

14. $r^2 + 10r + 9 = 0$

15. $y^2 - 7y - 8 = 0$

16. $z^2 - 13z + 36 = 0$

17. $m^2 - 2m = 8$

18. $y^2 - 2y = 35$

19. $y^2 + y = 12$

20. $4n^2 - 20n = 0$

21. $5t^2 = 125$

22. $2x^2 - x = 3$

23. $3x^2 + 14x = 5$

24. $3x^2 + 23x + 14 = 0$

Written Exercises

Solve each equation.

25. $x^2 + 7x + 6 = 0$

26. $y^2 + 8y + 15 = 0$

27. $m^2 + 4m + 2 = 0$

28. $p^2 + 5p + 3 = 0$

29. $2r^2 + r - 15 = 0$

30. $8t^2 + 10t + 3 = 0$

31. $2t^2 - t = 4$

32. $-4x^2 + 8x = -3$

33. $2y^2 + 3 = -7y$

34. $y^2 - 25 = 0$

35. $2y^2 = 98$

36. $-2x^2 + 8x + 3 = 3$

37. $3n^2 - 5n = -1$

38. $3m^2 + 2 = -8m$

39. $3k^2 + 11k = 4$

40. $-x^2 + 5x - 6 = 0$

41. $r^2 + 10r + 9 = 0$

42. $y^2 - 7y - 8 = 0$

43. $z^2 - 13z + 32 = 0$

44. $m^2 - 2m - 4 = -3$

45. $y^2 - 2y = 35$

46. $y^2 + y = 12$

47. $4n^2 - 20n = 0$

48. $5t^2 = 125$

49. $3n^2 - 2n = 1$

50. $y^2 - \frac{3}{5}y + \frac{2}{25} = 0$

51. $3x^2 + 23x + 10 = -4$

52. $3x^2 - \frac{5}{4}x - \frac{1}{2} = 0$

53. $24x^2 - 2x - 15 = 0$

54. $21x^2 + 5x - 6 = 0$

55. $35x^2 - 11x - 6 = 0$

56. $2x^2 - 0.7x - 0.3 = 0$

57. $x^2 - 1.1x - 0.6 = 0$

58. $-r^2 - 6r + 3 = 0$

59. $2x^2 - x = 2$

60. $3x^2 + 8x = -3$

61. $k^2 - 6k + 1 = 0$

62. $4x^2 - 8x + 3 = 6$

63. $a^2 + 3a + 1 = 0$

64. $4b^2 + 20b + 23 = 0$

65. $4x^2 + 8x - 1 = 0$

66. $-4y^2 + 16y + 13 = 0$

Challenge Exercises

Solve each equation by factoring, if possible. Then, solve each equation using the Quadratic Formula.

67. $2x^2 + 4x - 15 = 0$

68. $3x^2 + 2x - 21 = 0$

69. $x^2 + 2x - 2 = 0$

70. If a quadratic equation can be factored, what can you say about the value of $b^2 - 4ac$?

Cumulative Review

Graph each equation using the x- and y-intercepts.

1. $2x + 5y = -10$ **2.** $\frac{1}{3}x - \frac{1}{2}y = 4$

Write the equation of the line satisfying the given conditions. Express answers in slope-intercept form and standard form.

3. passes through $(2, -1)$ and $(4, 1)$

4. slope $= \frac{2}{3}$ and passes through $(5, 1)$

5. parallel to $x + 2y = -1$ and passes through $(1, 4)$

6. perpendicular to $x + 2y = -1$ and passes through $(1, 4)$

7. Find the coordinates of endpoint A if the midpoint of $\overline{AB}$ is at $(7, 3)$ and the endpoint B is at $(2, -5)$.

Solve each system of equations.

8. $2x - 3y = 5$
$x - y = 2$

9. $2x + 3y = 1$
$3x + 24 = 4y$

10. $4m - 1 = -3n$
$8m + 6n = 2$

11. $3x + y = 1$
$-5 - 6x = 2y$

Solve each system by graphing.

12. $y \geq x + 1$
$2x + 3y \leq 6$

13. $y > x + 2$
$3x - 2y < 4$

Use a calculator to find each square root. Round decimal answers to the nearest hundredth.

14. $\sqrt{86}$ **15.** $-\sqrt{41}$

16. $\pm\sqrt{21.1}$ **17.** $\sqrt{0.651}$

Simplify.

18. $\frac{\sqrt{42}}{\sqrt{6}}$ **19.** $\frac{\sqrt{15}}{\sqrt{7}}$

20. $\sqrt{28m^3}$ **21.** $\sqrt{162x^4}$

22. $\sqrt{\frac{x^3}{8}}$ **23.** $\sqrt{\frac{6a^4}{5b^3}}$

24. $\sqrt{3}(\sqrt{6} + \sqrt{15})$ **25.** $\frac{7\sqrt{2}}{3 - \sqrt{2}}$

26. $3\sqrt{8} - 4\sqrt{2} + \sqrt{12}$

27. Solve $\sqrt{3x - 5} + x = 11$ and check.

The measures of three sides of a triangle are given. Determine whether each triangle is a right triangle.

28. 7, 11, 14   **29.** 10, 24, 26

Find the equation of the axis of symmetry and the coordinates of the maximum or minimum point for the graph of each quadratic function. Then draw the graph.

30. $y = x^2 - 6x + 8$

31. $y = -x^2 + 4x - 2$

Solve each equation by completing the square. Use a calculator to express irrational roots in decimal form to the nearest hundredth.

32. $x^2 - 8x + 3 = 0$

33. $x^2 + 12x + 24 = 0$

Use a system of equations to solve each problem.

34. An airplane travels 620 miles in 2 hours flying with the wind. Flying against the wind, it takes 4 hours to travel 1160 miles. Find the rate of the wind and the rate of the plane in still air.

35. A gas station attendant has some antifreeze that is 50% glycol and another type of antifreeze that is 30% glycol. He wishes to make 100 gallons of antifreeze that is 45% glycol. How much of each kind should he use?

36. A two-digit number is 5 times the sum of its digits. The units digit is 1 more than the tens digit. Find the number.

Problem Solving

Another strategy for solving problems is to list possibilities. When making a list, use a systematic approach so you do not omit important items. This strategy is often helpful when you need to find the *number* of solutions to a problem. Study the following examples.

Example 1 **Which whole numbers less than 30 are divisible by both 3 and 4?**

List the numbers divisible by 3.   3, 6, 9, 12, 15, 18, 21, 24, 27
List the numbers divisible by 4.   4, 8, 12, 16, 20, 24, 28

Which numbers are in both lists? Since 12 and 24 are in both lists, they are divisible by both 3 and 4.

Example 2 **How many ways can you receive change for a quarter if at least one coin is a dime?**

List the possibilities. List ways that use the fewest number of coins first.

1. dime, dime, nickel   *This is the same as dime, nickel, dime.*
2. dime, dime, 5 pennies
3. dime, nickel, nickel, nickel
4. dime, nickel, nickel, 5 pennies
5. dime, nickel, 10 pennies
6. dime, 15 pennies   *Are there any other solutions?*

Thus, there are 6 ways to receive change for a quarter if at least one coin is a dime.

Exercises

Solve each problem.

1. Which whole numbers less than 50 are divisible by both 8 and 12?

2. Which whole numbers less than 100 are divisible by both 3 and 11?

3. How many ways can you receive change for a half-dollar if you receive at least one dime and one quarter?

4. How many different whole numbers can be written using the digits 3, 6, 6, and 7? Each number must use all four digits.

5. An ice cream shop makes chocolate, butterscotch, or strawberry sundaes. Any sundae can be served with whipped cream, nuts, neither, or both. How many ways can a sundae be served?

6. The president, vice president, secretary, and treasurer of a club are to be seated in four chairs in the front of a meeting room. How many seating arrangements are possible?

7. In how many ways can you write 45 as the sum of positive consecutive integers?

8. How many positive whole numbers less than 50 can be written as the sum of 2 squares?

13-5 The Discriminant

In the Quadratic Formula, the expression $b^2 - 4ac$ is called the **discriminant**. The discriminant can give information about the nature of the roots of a quadratic equation.

In particular, the discriminant is used to determine how many real roots there are. Real roots are roots that are real numbers, that is, *either* rational numbers *or* irrational numbers. Recall that the real roots of an equation are represented by the intersection of the x-axis and the graph of the related function.

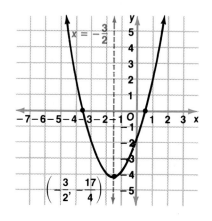

Solve $x^2 + 3x - 2 = 0$.

$$x = \frac{-b \pm \sqrt{b^2 - 4ac}}{2a}$$

$$= \frac{-3 \pm \sqrt{3^2 - 4(1)(-2)}}{2(1)} \qquad \begin{array}{l} a = 1, b = 3, \\ c = -2 \end{array}$$

$$= \frac{-3 \pm \sqrt{9 + 8}}{2}$$

$$= \frac{-3 \pm \sqrt{17}}{2}$$

The roots are approximately −3.56 and 0.56.

The graph of the corresponding function intersects the x-axis at two different points.

Notice that $b^2 - 4ac > 0$ and there are two real distinct roots.

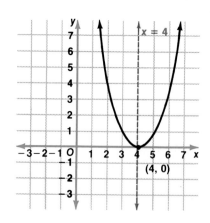

Solve $x^2 - 8x + 16 = 0$.

$$x = \frac{-b \pm \sqrt{b^2 - 4ac}}{2a}$$

$$= \frac{-(-8) \pm \sqrt{(-8)^2 - 4(1)(16)}}{2(1)}$$

$$= \frac{8 \pm \sqrt{64 - 64}}{2}$$

$$= \frac{8 \pm 0}{2} \quad \text{or} \quad 4$$

The graph of the corresponding function intersects the x-axis at one point.

Notice that $b^2 - 4ac = 0$ and there is one real distinct root.

Solve $x^2 + 6x + 10 = 0$.

$$x = \frac{-b \pm \sqrt{b^2 - 4ac}}{2a}$$

$$= \frac{-6 \pm \sqrt{6^2 - 4(1)(10)}}{2(1)}$$

$$= \frac{-6 \pm \sqrt{36 - 40}}{2}$$

$$= \frac{-6 \pm \sqrt{-4}}{2}$$

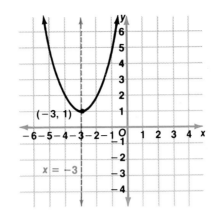

(−3, 1)

$x = -3$

The graph of the corresponding function does not intersect the x-axis.

Notice that $b^2 - 4ac < 0$ and there are no real roots since *no* real number can be the square root of a negative number.

The relationship between the value of the discriminant and the nature of the roots of an equation can be summarized as follows.

Discriminant	Nature of Roots
$b^2 - 4ac > 0$	two real distinct roots
$b^2 - 4ac = 0$	one real distinct root
$b^2 - 4ac < 0$	no real roots

Nature of Roots of a Quadratic Equation

Examples

1 **State the value of the discriminant of $2x^2 - 10x + 11 = 0$. Then determine the nature of the roots of the equation.**

$b^2 - 4ac = (-10)^2 - 4(2)(11)$ *a = 2, b = −10, c = 11*

$\qquad\qquad = 100 - 88$

$\qquad\qquad = 12$

Since $b^2 - 4ac > 0$, then $2x^2 - 10x + 11 = 0$ has two real distinct roots.

2 **State the value of the discriminant of $4x^2 - 12x + 9 = 0$. Then determine the nature of the roots of the equation.**

$b^2 - 4ac = (-12)^2 - 4(4)(9)$ *a = 4, b = −12, c = 9*

$\qquad\qquad = 144 - 144$

$\qquad\qquad = 0$

Since $b^2 - 4ac = 0$, then $4x^2 - 12x + 9 = 0$ has one real distinct root.

Example

3 State the value of the discriminant of $3x^2 + 4x + 2 = 0$. Then determine the nature of the roots of the equation.

$$b^2 - 4ac = 4^2 - 4(3)(2) \qquad a = 3, b = 4, c = 2$$
$$= 16 - 24$$
$$= -8$$

Since $b^2 - 4ac < 0$, then $3x^2 + 4x + 2 = 0$ has no real roots.

Exploratory Exercises

State the value of the discriminant of each equation. Then determine the nature of the roots of the equation.

1. $x^2 + 3x - 4 = 0$
2. $y^2 + 3y + 1 = 0$
3. $m^2 + 5m - 6 = 0$
4. $s^2 + 8s + 16 = 0$
5. $2k^2 - 7k + 6 = 0$
6. $2p^2 - p - 3 = 0$
7. $x^2 - 1.2x = 0$
8. $2z^2 + 7z + 50 = 0$
9. $y^2 - 4y - 8 = 0$
10. $3x^2 + x + 1 = 0$
11. $3x^2 + 7x - 2 = 0$
12. $2x^2 - 2x - 1 = 0$
13. $\frac{4}{3}x^2 + 4x + 3 = 0$
14. $\frac{3}{2}m^2 + m = -\frac{7}{2}$
15. $4a^2 + 10a + 6.25 = 0$

Written Exercises

Determine the nature of the roots of each equation by using the discriminant. Find all real roots. Leave irrational roots in simplest radical form.

16. $x^2 + 5x + 3 = 0$
17. $y^2 - 4y + 1 = 0$
18. $r^2 + 4r - 12 = 0$
19. $z^2 + 8z - 5 = 0$
20. $m^2 + 7m + 6 = 0$
21. $p^2 - 7p + 6 = 0$
22. $k^2 + 6k + 10 = 0$
23. $a^2 + 8a - 12 = 0$
24. $d^2 + 4d + 7 = 0$
25. $2x^2 + 3x + 1 = 0$
26. $2x^2 - 3x + 1 = 0$
27. $3y^2 + y - 1 = 0$
28. $m^2 - 14m + 49 = 0$
29. $3n^2 - n - 5 = 0$
30. $15a^2 + 2a + 16 = 0$
31. $h^2 - 16h + 64 = 0$
32. $y^2 - 4y - 32 = 0$
33. $11z^2 - z - 3 = 0$
34. $3p^2 - 4p - 1 = 0$
35. $9y^2 - 6y + 1 = 0$
36. $3g^2 - 4g + 1 = 0$

Determine the nature of the roots of each equation by using the discriminant. Find all real roots. Use a calculator to express irrational roots in decimal form to the nearest hundredth.

37. $2x^2 - x - 2 = 0$
38. $6r^2 - 5r = 7$
39. $-9m + m^2 = -14$
40. $2a^2 + a = 5$
41. $0.3a^2 + 0.8a + 0.4 = 0$
42. $3y^2 - 2y = 1$
43. $3z^2 + 10z + 5 = 0$
44. $2b^2 + 7b + 2 = 0$
45. $7r^2 - 3r - 1 = 0$
46. $8p^2 + 7p + 1 = 0$
47. $3a^2 + 2 = 5a$
48. $0.6v^2 + v = 1.8$
49. $2z^2 + 5 = 8z$
50. $5b^2 = 1 + 6b$
51. $2x^2 - 8x = 7$
52. $x^2 - \frac{5}{3}x = -\frac{2}{3}$
53. $\frac{1}{3}x^2 - 4x + 13\frac{1}{3} = 0$
54. $5c^2 - 7c = 1$

The equation $x^2 = -1$ has *no* solution among the real numbers. This is because the square of a real number is always greater than or equal to zero.

We define a new number to be a solution to $x^2 = -1$. This number is designated by the letter i and is called the ***imaginary unit***. The imaginary unit i is *not* a real number.

$$i^2 = -1 \quad \text{or} \quad i = \sqrt{-1}$$

Using i as you would any constant, you can rewrite square roots of negative numbers.

$$\sqrt{-4} = \sqrt{-1} \cdot \sqrt{4} \quad \text{or} \quad i \cdot 2 = 2i$$
$$\sqrt{-3} = \sqrt{-1} \cdot \sqrt{3} \quad \text{or} \quad i\sqrt{3}$$
$$\sqrt{-18} = \sqrt{-1} \cdot \sqrt{18} \quad \text{or} \quad i \cdot 3\sqrt{2} = 3i\sqrt{2}$$

If the value of the discriminant of an equation is less than zero, the roots of the equation are *not* real numbers. However, the roots can be expressed using the imaginary unit.

Consider the equation $x^2 + 6x + 10 = 0$. The discriminant of this equation is a negative number.

$$b^2 - 4ac = 6^2 - 4(1)(10)$$
$$= 36 - 40$$
$$= -4$$

Even though the discriminant is less than zero, the Quadratic Formula can still be used to find the roots of the equation.

$$x = \frac{-b \pm \sqrt{b^2 - 4ac}}{2a}$$

$$= \frac{-6 \pm \sqrt{-4}}{2}$$

$$= \frac{-6 \pm 2i}{2} \quad \text{or} \quad -3 \pm i$$

The roots of $x^2 + 6x + 10 = 0$ are $-3 + i$ and $-3 - i$. Roots that contain the imaginary unit are known as **imaginary roots**.

Imaginary roots always appear in pairs.

Exercises

Simplify.

1. $\sqrt{-36}$
2. $\sqrt{-144}$
3. $\sqrt{-49}$
4. $\sqrt{-64}$
5. $\sqrt{-12}$
6. $\sqrt{-50}$
7. $\sqrt{-75}$
8. $\sqrt{-200}$

Find the roots of each equation.

9. $x^2 + 5x + 7 = 0$
10. $2x^2 - 3x + 4 = 0$
11. $3x^2 + x + 1 = 0$
12. $r^2 - 4r + 10 = 0$
13. $5x^2 - 2x + 8 = 0$
14. $3k^2 + 3k + 2 = 0$

13-6 Solving Quadratic Equations

You have studied a variety of methods for solving quadratic equations. The table summarizes these methods.

Method	Can be Used	Comments
graphing	always	Not always exact; use only when a picture of function is needed.
factoring	sometimes	Use if constant term is 0 or factors are easily determined.
completing the square	always	Useful for equations of form $x^2 + bx + c = 0$ where b is even.
Quadratic Formula	always	Other methods may be easier, but this method *always* works.

Use the information in the above table to help you determine how to solve a quadratic equation.

To solve:

$$x^2 + 9x + 20 = 0$$
$$2m^2 + 7m + 5 = 0$$
$$x^2 - 2x - 3 = 0$$

$$x^2 + 6x - 315 = 0$$

$$6n^2 - 5n = 0$$

Try:

factoring; Quadratic Formula
Quadratic Formula
factoring; completing the
   square; Quadratic Formula
completing the square;
   Quadratic Formula
factoring

To solve an equation with fractional or decimal coefficients, it is sometimes easier to first change the coefficients to integers by multiplication.

Example

1 Solve $y^2 - \frac{5}{12}y - \frac{1}{4} = 0$.

$$y^2 - \frac{5}{12}y - \frac{1}{4} = 0 \qquad \textit{The LCM of 12 and 4 is 12.}$$

$$12\left(y^2 - \frac{5}{12}y - \frac{1}{4}\right) = 12(0) \qquad \textit{Multiply each side by 12.}$$

$$12y^2 - 5y - 3 = 0$$

$$(3y + 1)(4y - 3) = 0 \qquad \textit{Factor or use the Quadratic Formula.}$$

$$3y + 1 = 0 \quad \text{or} \quad 4y - 3 = 0$$

$$y = -\frac{1}{3} \quad \text{or} \qquad y = \frac{3}{4} \qquad \textit{Check this result.}$$

The solutions are $-\frac{1}{3}$ and $\frac{3}{4}$.

2 Solve $x^2 + 0.4x - 3.2 = 0$.

$x^2 + 0.4x - 3.2 = 0$

$10(x^2 + 0.4x - 3.2) = 10(0)$ *Multiply each side by 10.*

$10x^2 + 4x - 32 = 0$

$x = \dfrac{-4 \pm \sqrt{4^2 - 4(10)(-32)}}{2(10)}$ *Try factoring or the Quadratic Formula.*

$= \dfrac{-4 \pm \sqrt{16 + 1280}}{20}$

$= \dfrac{-4 \pm 36}{20}$

$x = 1.6$ or $x = -2$ *Check this result.*

The solutions are 1.6 and -2.

Exploratory Exercises

State the method that seems easiest for solving each equation.

1. $x^2 - 12x + 27 = 0$ **2.** $y^2 - 19y = -84$ **3.** $2m^2 + 19m + 9 = 0$

4. $a^2 - 12a - 4 = 0$ **5.** $3r^2 - 7r - 5 = 0$ **6.** $z^2 - 2z - 120 = 0$

7. $3x^2 - 2x - 5 = 0$ **8.** $2b^2 + 1 = 6b$ **9.** $y^2 + 4y = 9$

State the number by which to multiply each side of each equation before solving.

10. $x^2 - \frac{1}{2}x - \frac{1}{9} = 0$ **11.** $x^2 - \frac{7}{6}x - \frac{1}{2} = 0$ **12.** $3x^2 - \frac{5}{4}x - \frac{1}{2} = 0$

13. $x^2 - 1.3x - 0.3 = 0$ **14.** $2x^2 - 0.7x - 0.3 = 0$ **15.** $0.2x^2 - 0.33x - 0.35 = 0$

Written Exercises

Solve each quadratic equation by an appropriate method. Express irrational roots in simplest radical form and in decimal form to the nearest hundredth.

16. $x^2 - 9x + 20 = 0$ **17.** $y^2 + 10y - 2 = 0$ **18.** $2x^2 + 4x + 1 = 0$

19. $3z^2 - 7z - 3 = 0$ **20.** $x^2 + 3x = -2$ **21.** $r^2 + 13r = -42$

22. $2y^2 - 5y + 2 = 0$ **23.** $3x^2 - 7x - 6 = 0$ **24.** $4x^2 - 7x - 2 = 0$

25. $2k^2 + k - 5 = 0$ **26.** $3h^2 - 5h - 2 = 0$ **27.** $x^2 - 5x - 7 = 0$

28. $r^2 + 4r + 1 = 0$ **29.** $m^2 + 2m + 8 = 0$ **30.** $y^2 - 3y + 3 = 0$

31. $3z^2 = 5z - 1$ **32.** $9b = -5b^2 - 3$ **33.** $a^2 - 15a = -52$

34. $2z^2 + 4z = 5$ **35.** $x^2 = 4x + 2$ **36.** $-2x - 2 = -x^2$

37. $x^2 - x + \frac{3}{16} = 0$ **38.** $2a^2 - 8a + \frac{15}{2} = 0$ **39.** $r^2 + r + \frac{2}{9} = 0$

40. $x^2 - \frac{17}{20}x + \frac{3}{20} = 0$ **41.** $y^2 - \frac{3}{5}y + \frac{2}{25} = 0$ **42.** $x^2 - 1.1x - 0.6 = 0$

43. $2x^2 - 0.7x - 0.6 = 0$ **44.** $5x^2 - 0.5x - 0.3 = 0$ **45.** $0.7a^2 - 2.8a = 7$

13-7 Problem Solving: Quadratic Equations

You can use the methods for solving quadratic equations to solve some types of verbal problems.

Examples

1   **A rectangle has a perimeter of 19 centimeters. Its area is 21 square centimeters. Find its dimensions.**

explore   Let ℓ = the measure of the length.
Let w = the measure of the width.

plan   Use the formula $P = 2\ell + 2w$.

$19 = 2\ell + 2w$ *Substitute 19 for P.*

$9.5 = \ell + w$ *Solve for ℓ.*

$\ell = 9.5 - w$

Now write an equation using the formula $A = \ell w$.

$21 = (9.5 - w)w$ *Substitute 21 for A and $9.5 - w$ for ℓ.*

solve $21 = 9.5w - w^2$

$w^2 - 9.5w + 21 = 0$ *Subtract $9.5w - w^2$ from each side.*

$2w^2 - 19w + 42 = 0$ *Multiply each side by 2.*

$(2w - 7)(w - 6) = 0$ *Factor.*

$w = 3.5$ or $w = 6$

If $w = 3.5$ then $\ell = 9.5 - 3.5$ or 6.

If $w = 6$, then $\ell = 9.5 - 6$ or 3.5.

The dimensions are 3.5 cm and 6 cm.

examine   The perimeter of a rectangle with dimensions 3.5 cm and 6 cm is $2(3.5) + 2(6)$ or 19 cm. The area of this rectangle is $(3.5)(6)$ or 21 cm².

2   **A pan is to be formed by cutting squares measuring 2 cm on a side from a square piece of sheet metal and then folding the sides. If the volume of the pan is to be 392 cm³, what is the original size of the sheet metal?**

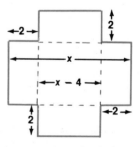

Let x = the measure of each side.

Use the formula $V = \ell w h$ to write an equation.

$392 = (x - 4)(x - 4)(2)$ $V = 392, \ell = x - 4, w = x - 4, h = 2$

$$392 = 2x^2 - 16x + 32$$
$$2x^2 - 16x - 360 = 0 \qquad \textit{Subtract 392 from each side.}$$
$$x^2 - 8x - 180 = 0 \qquad \textit{Divide each side by 2.}$$
$$(x - 18)(x + 10) = 0 \qquad \textit{Factor.}$$
$$x = 18 \quad \text{or} \quad x = -10$$

The length of each side of the sheet metal must be 18 cm since -10 cm is not an appropriate answer.   *Examine this solution.*

Written Exercises

Find the dimensions of each rectangle described below.

1. The perimeter is 30 m. The area is 56 m^2.

2. The perimeter is 34 in. The area is 72 in^2.

3. The perimeter is 46 yd. The area is 130 yd^2.

4. The perimeter is 70 cm. The area is 294 cm^2.

5. The length is 2 m more than 3 times the width. The area is 56 m^2.

6. The width is 7 cm less than the length. The area is 78 cm^2.

7. The perimeter is 60 in. The area is 200 in^2.

8. The perimeter is 37 in. The area is 78 in^2.

9. The length is 4 in. more than the width. The area is 45 in^2.

10. The width is 3 ft less than the length. The area is 54 ft^2.

11. The length is $\frac{8}{5}$ times the width. The area is 160 m^2.

12. The width is 0.5 m less than one-half the length. The area is 21 m^2.

Solve each problem. Approximate solutions that are irrational roots.

13. Find two integers whose sum is 14 and whose product is 48.

14. Find two integers whose sum is 13 and whose product is 42.

15. Find two integers whose difference is 6 and whose product is 135.

16. Find two integers whose sum is 12 and whose squares differ by 24.

17. A rectangular piece of sheet metal is 3 times as long as it is wide. Squares measuring 2 cm on a side are cut from each corner and the sides are folded to form a pan. If the volume of the pan is 512 cm^3, what are the dimensions of the sheet metal?

18. A rectangular piece of sheet metal is twice as long as it is wide. Squares measuring 5 inches on a side are cut from each corner and the sides are folded to form a box. If the volume of the box is 1,760 in^3, what are the dimensions of the sheet metal?

19. A rectangular piece of glass is twice as long as it is wide. If the length and width are both reduced by 1 cm, the area of the glass becomes 10 cm^2. What are the original dimensions of the glass?

20. A rectangular piece of sheet metal is 3 in. longer than it is wide. If the length and width are both increased by 2 in. the area increases by 34 in^2. What are the original dimensions of the sheet metal?

21. The perimeter of a rectangle is 8 meters and its area is 3.84 square meters. Find its dimensions.

22. A rectangle has a perimeter of 15.4 cm and an area of 14.4 cm^2. Find the dimensions of the rectangle.

23. The length of a rectangle is $\frac{8}{5}$ times its width. The area is 56 square meters. Find the dimensions of the rectangle.

24. The width of a rectangle is one-half meter less than one-half its length. The area is 21 m^2. Find its dimensions.

25. Dan Kurtz has a rectangular flower garden that measures 15 m by 20 m. He wishes to place a concrete walk of uniform width around the garden. His budget allows him to cover 74 m^2. How wide can the walk be?

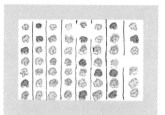

26. A picture has a square frame that is 5 cm wide. The area of the picture is two-thirds of the total area of the picture and the frame. What are the dimensions of the frame?

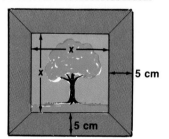

---

mini-review

Find the coordinates of the midpoint of the line segment whose endpoints are given.

1. (6, 8), (4, 10)

2. (−2, 7), (−3, −9)

3. $\left(\frac{1}{4}, -4\right)$, $\left(2\frac{1}{2}, -1\frac{1}{3}\right)$

Graph each pair of equations. Then state the solution of each system of equations.

4. $x + y = 3$
    $x - 3y = 3$

5. $2x - 3y = -14$
    $y = 2x + 10$

6. $\frac{1}{2}x - y = 4$
    $x + \frac{1}{3}y = 1$

Simplify.

7. $-\sqrt{\frac{1.69}{0.49}}$

8. $\sqrt{\frac{12a^2b}{5c^4}}$

9. $\sqrt{2}(\sqrt{10} - \sqrt{11})$

State whether each decimal represents a rational number or an irrational number.

10. 2.3333 . . .

11. 0.2020020002 . . .

12. $4.\overline{65}$

Find the distance between each pair of points whose coordinates are given. Give answers in simplest radical form and in decimal form to the nearest hundredth.

13. (2, −3), (5, 7)

14. (−8, −1), (−11, 3)

15. $\left(\frac{1}{4}, 2\right)$, $\left(-1, \frac{1}{2}\right)$

Locate the roots of each equation by graphing the related function.

16. $x^2 - 6x + 5 = 0$

17. $x^2 + 16 = 8x$

18. $x^2 - 2x + 2 = 0$

State the value of the discriminant of each equation. Then determine the nature of the roots of the equation.

19. $t^2 + 5t - 9 = 0$

20. $2y^2 - 3y + 6 = 0$

21. $p^2 + \frac{1}{4} = p$

13-8 The Sum and Product of Roots

An engineer or scientist often must find an equation to describe a certain situation. For example, suppose the roots of a quadratic equation are known to be -3 and 8.

> If $x = -3$, then $x + 3 = 0$.   If $x = 8$, then $x - 8 = 0$.
>
> $(x + 3)(x - 8) = 0$ *Why is this so?*
>
> $x^2 - 5x - 24 = 0$

The quadratic equation $x^2 - 5x - 24 = 0$ has roots -3 and 8.

Suppose you find the sum and the product of the roots.

$$\text{sum} = -3 + 8 = 5$$
$$\text{product} = -3 \cdot 8 = -24$$

Now look at the equation.

product of roots   *The product of the roots is the constant term.*

$x^2 - 5x - 24 = 0$

opposite of sum of roots   *The opposite of the sum of the roots is the coefficient of x.*

This *always* works. Suppose the two roots are r_1 and r_2. Then $(x - r_1)(x - r_2) = 0$. Multiplying the binomials results in the equation $x^2 - (r_1 + r_2)x + r_1 r_2 = 0$. Notice that the coefficient of x, $-(r_1 + r_2)$, is the opposite of the sum of the roots. The constant term, $r_1 r_2$, is the product of the roots.

> **To find a quadratic equation of the general form**
> $x^2 + bx + c = 0$ **given its roots:**
> 1. **The coefficient of x is the opposite of the sum of the roots.**
> 2. **The constant term is the product of the roots.**
>
> *Finding Quadratic Equations*

Example

1 **Find a quadratic equation whose roots are -4 and 7.**

opposite of sum of roots $= -(-4 + 7) = -3$

product of roots $= -4 \cdot 7 = -28$

The opposite of the sum of the roots, -3, is the coefficient of x. The product of the roots, -28, is the constant term.

Thus, a quadratic equation whose roots are -4 and 7 is $x^2 - 3x - 28 = 0$.

The method of the sum and the product of roots for finding quadratic equations is especially useful when the roots involve radicals.

> ## Example
>
> **2** Find a quadratic equation whose roots are $1 + \sqrt{5}$ and $1 - \sqrt{5}$.
>
> $$\text{opposite of sum of roots} = -[(1 + \sqrt{5}) + (1 - \sqrt{5})]$$
> $$= -(1 + \sqrt{5} + 1 - \sqrt{5})$$
> $$= -(2) \text{ or } -2$$
> $$\text{product of roots} = (1 + \sqrt{5})(1 - \sqrt{5})$$
> $$= 1 - 5 \text{ or } -4$$
>
> The opposite of the sum of the roots, -2, is the coefficient of x. The product of the roots, -4, is the constant term. Thus, a quadratic equation whose roots are $1 + \sqrt{5}$ and $1 - \sqrt{5}$ is $x^2 - 2x - 4 = 0$.

By studying Example 2, the following rule may be discovered.

> Given the quadratic equation $ax^2 + bx + c = 0$, where $a \neq 0$, the sum of the roots of the equation is $-\frac{b}{a}$ and the product of the roots of the equation is $\frac{c}{a}$.

Sum and Product of Roots

> ## Example
>
> **3** Are $\frac{3}{2}$ and $-\frac{4}{3}$ roots of $6x^2 - x - 12 = 0$?
>
> *Sum of Roots*
>
> $$\frac{3}{2} + \left(-\frac{4}{3}\right) = \frac{9}{6} + \left(-\frac{8}{6}\right)$$
> $$= \frac{1}{6} \qquad \textit{Is this equal to } -\frac{b}{a}?$$
>
> *Product of Roots*
>
> $$\frac{3}{2}\left(-\frac{4}{3}\right) = -2 \qquad \textit{Is this equal to } \frac{c}{a}?$$
>
> Since $\frac{-b}{a} = \frac{1}{6}$ and $\frac{c}{a} = -2$, $\frac{3}{2}$ and $-\frac{4}{3}$ are roots of the equation.

Exploratory Exercises

State the sum and product of the roots of each equation.

1. $x^2 - 5x + 6 = 0$

2. $y^2 - 8y - 20 = 0$

3. $z^2 + 12z - 28 = 0$

4. $3m^2 + 6m - 3 = 0$

5. $4k^2 + 20k - 16 = 0$

6. $4t^2 + 8t + 3 = 0$

7. $4x^2 + 4x = 35$

8. $6a^2 - 13a = 15$

9. $6b^2 - 5b = 21$

State whether the given numbers are roots of each equation.

10. $-4, 6; x^2 - 2x - 24 = 0$

11. $-6, 3; y^2 + 3y - 18 = 0$

12. $3, 7; a^2 - 10a + 21 = 0$

13. $2, 3; n^2 + 5n - 6 = 0$

14. $-1, 7; b^2 - 8b + 7 = 0$

15. $-2, 4; r^2 + 2r + 8 = 0$

16. $4 + \sqrt{3}, 4 - \sqrt{3}; p^2 + 8p + 13 = 0$

17. $1 + \sqrt{7}, 1 - \sqrt{7}; x^2 - 2x - 6 = 0$

18. $-\frac{1}{3}, \frac{1}{2}; x^2 - \frac{1}{6}x - \frac{1}{6} = 0$

19. $-\frac{1}{4}, \frac{1}{6}; x^2 + \frac{5}{12}x + \frac{1}{24} = 0$

Find a quadratic equation having the given roots.

20. $4, 1$

21. $5, 2$

22. $-4, 5$

23. $3, 8$

24. $0, -5$

25. $1, -6$

26. $3 + \sqrt{2}, 3 - \sqrt{2}$

27. $7 - \sqrt{5}, 7 + \sqrt{5}$

28. $\frac{2}{3}, 7$

29. $\frac{1}{2}, \frac{-3}{2}$

30. $1.4, 2.2$

31. $0.3, -0.6$

Written Exercises

Find the sum and product of the roots of each equation.

32. $y^2 + 15y + 54 = 0$

33. $a^2 - 5a - 24 = 0$

34. $m^2 - m - 6 = 0$

35. $b^2 + 12b - 28 = 0$

36. $6x^2 + 31x + 35 = 0$

37. $21r^2 + 2r - 8 = 0$

38. $k^2 + 6k - 1 = 0$

39. $4y^2 + 4y + 1 = 0$

40. $7h^2 - 33h - 10 = 0$

41. $n^2 - 10n + 23 = 0$

42. $\frac{1}{2}m^2 - \frac{3}{2}m + 4 = 0$

43. $z^2 + \frac{13}{2}z - \frac{9}{4} = 0$

44. $m^2 - 1.2m + 0.27 = 0$

45. $6k^2 - 0.4k - 0.02 = 0$

46. $2c^2 - \frac{2}{3}c = \frac{1}{6}$

47. $12a^2 - \frac{7}{2}a + \frac{1}{4} = 0$

48. $a^2 + 4a\sqrt{3} + 9 = 0$

49. $2y^2 + y\sqrt{2} - 6 = 0$

Find a quadratic equation having the given roots.

50. $5, 1$

51. $4, 7$

52. $6, -5$

53. $7, -8$

54. $1, -10$

55. $8, -3$

56. $5, -9$

57. $-2, -17$

58. $16, -5$

59. $\frac{5}{2}, 2$

60. $\frac{7}{3}, -3$

61. $-\frac{3}{4}, 8$

62. $\frac{2}{3}, \frac{-3}{2}$

63. $\sqrt{2}, -\sqrt{2}$

64. $\sqrt{3}, \sqrt{3}$

65. $2 + \sqrt{3}, 2 - \sqrt{3}$

66. $5 - \sqrt{2}, 5 + \sqrt{2}$

67. $4 + \sqrt{10}, 4 - \sqrt{10}$

68. $3 - \sqrt{5}, 3 + \sqrt{5}$

69. $\frac{1 + \sqrt{7}}{2}, \frac{1 - \sqrt{7}}{2}$

70. $\frac{2 - \sqrt{11}}{3}, \frac{2 + \sqrt{11}}{3}$

71. Find the general form of the quadratic equation with roots q and r.

Excursions in Algebra   Decimal Coefficients

Use the Quadratic Formula and a calculator to solve each equation. Approximate the roots to the nearest hundredth.

1. $1.4x^2 + 0.2x - 4.1 = 0$

2. $0.2x^2 - 1.3x + 0.6 = 0$

3. $2.1y^2 + 4.6y - 1.3 = 0$

4. $3.8m^2 - 2.1m - 1.7 = 0$

5. $0.4r^2 + 5.2r - 2.3 = 0$

6. $7.9z^2 - 5.1z + 0.6 = 0$

Graphing Quadratic Functions

The following BASIC program can be used as an aid for graphing quadratic functions.

```
10   PRINT "FOR THE QUADRATIC FUNCTION Y=A*X↑2+B*X+C,
        ENTER A,B, AND C"
20   INPUT A,B,C
30   LET S = −B / (2 * A)        Line 30 determines the axis of symmetry.
40   PRINT "THE EQUATION OF THE AXIS OF SYMMETRY IS X = ";S
50   LET Y = A * S↑2 + B * S + C
60   IF A > 0 THEN PRINT "THE MINIMUM POINT IS (";S;",";Y;")"
70   IF A < 0 THEN PRINT "THE MAXIMUM POINT IS (";S;",";Y;")"
80   IF A = 0 THEN PRINT "A MUST NOT BE ZERO"
90   FOR X = S − 4 TO S + 4
95   LET Y = A * X↑2 + B * X + C
100   PRINT "(";X;",";Y;")"
110   NEXT X
120   END
```

The zeros of a quadratic function are the roots of the related quadratic equation. The program above can be extended to compute the zeros of the function. To do this, delete line 120 and add the lines shown at the right. Notice that the discriminant is used in lines 130 and 140 to determine the number of real zeros. If there are one or two real zeros, the Quadratic Formula is used to compute the zeros.

```
120   LET D = B↑2 − 4 * A * C
130   IF D > 0 THEN 170
140   IF D = 0 THEN 200
150   PRINT "NO REAL ZEROS"
160   GOTO 220
170   PRINT "THE ZEROS ARE"
180   PRINT ( − B + SQR (D)) /
        (2 * A);" AND ";( − B −
        SQR (D)) / (2 * A)
190   GOTO 220
200   PRINT "THE ZERO IS"
210   PRINT − B / (2 * A)
220   END
```

Exercises

Use the program above to compute the coordinates of several points of the graphs of the following quadratic functions. Also use lines 120–220 above to compute the zeros for each function.

1. $y = x^2 - 2x - 6$
2. $y = x^2 + x - 3$
3. $y = 2x^2 - 4x + 5$
4. $y = 3x^2 + 6x - 2$
5. $y = -x^2 + 4x - 4$
6. $y = -2x^2 + 8x - 6$

7. Graph the following equations on one coordinate axis:

$y = 2x^2 + x - 6$, $y = x^2 + x - 6$, and $y = \frac{1}{2}x^2 + x - 6$

8. What effect does the coefficient of x^2 have on the shape of the graph?

quadratic function (431)

parabola (431)

minimum point (431)

axis of symmetry (431)

maximum point (432)

roots (438)

completing the square (442)

Quadratic Formula (445)

discriminant (450)

Chapter Summary

1. A quadratic function is a function described by an equation of the form $y = ax^2 + bx + c$, where $a \neq 0$. (431)

2. The graph of a quadratic function has a general shape called a parabola. (431)

3. The minimum point is the lowest point on the graph of a parabola that opens upward. The maximum point is the highest point on the graph of a parabola that opens downward. (431–432)

4. If the graph of a parabola is folded along the axis of symmetry, the two halves of the graph coincide. (431)

5. Equation of Axis of Symmetry: The equation of the axis of symmetry for the graph of $y = ax^2 + bx + c$, where $a \neq 0$, is given by $x = -\frac{b}{2a}$. (432)

6. The roots of a quadratic equation are the x-coordinates of the points where the graph of the corresponding quadratic function crosses the x-axis. (438)

7. To complete the square for $x^2 + bx$, follow these steps.
   Step 1   Find one-half of b, the coefficient of x.
   Step 2   Square the result of **Step 1**.
   Step 3   Add the result of **Step 2** to $x^2 + bx$. (438)

8. The Quadratic Formula: The roots of a quadratic equation of the form $ax^2 + bx + c = 0$, where $a \neq 0$, are given by
$$x = \frac{-b \pm \sqrt{b^2 - 4ac}}{2a}.$$ (445)

9. Nature of Roots of a Quadratic Equation: The discriminant, $b^2 - 4ac$, gives the following information about the roots of a quadratic equation.
   1. $b^2 - 4ac > 0$, two real distinct roots
   2. $b^2 - 4ac = 0$, one real distinct root
   3. $b^2 - 4ac < 0$, no real roots (451)

10. Methods for solving quadratic equations include graphing, factoring, completing the square, and applying the Quadratic Formula. (454)

11. To find a quadratic equation of the form $x^2 + bx + c = 0$ given its roots:
   1. The coefficient of x is the opposite of the sum of the roots.
   2. The constant term is the product of the roots. (459)

12. Sum and Product of Roots: Given the quadratic equation $ax^2 + bx + c = 0$, where $a \neq 0$, the sum of the roots of the equation is $-\frac{b}{a}$ and the product of the roots of the equation is $\frac{c}{a}$. (460)

Chapter Review

13–1 Find the equation of the axis of symmetry and the coordinates of the maximum or minimum point of the graph of each quadratic function.

1. $y = x^2 - 3x - 4$ **2.** $y = -x^2 + 6x + 16$ **3.** $y = 2x^2 + 9x + 9$

13–2 Locate the roots of each equation by graphing the related function.

4. $x^2 - x - 12 = 0$ **5.** $x^2 + 6x + 9 = 0$ **6.** $x^2 - 8x + 12 = 0$

13–3 Find the value of c that makes each trinomial a perfect square.

7. $h^2 + 12h + c$ **8.** $m^2 + 8m + c$ **9.** $r^2 - 5r + c$

Solve each equation by completing the square.

10. $x^2 - 16x + 32 = 0$ **11.** $y^2 + 6y + 4 = 0$

12. $x^2 - 7x - 5 = 0$ **13.** $4a^2 + 16a + 15 = 0$

13–4 Use the Quadratic Formula to solve each equation.

14. $x^2 - 8x = 20$ **15.** $2x^2 + 7x - 15 = 0$

16. $2m^2 + 3 = 7m$ **17.** $9k^2 - 12k - 1 = 0$

18. $3s^2 - 7s - 2 = 0$ **19.** $5b^2 + 9b + 3 = 0$

13–5 Use the discriminant to determine the nature of the roots of each equation.

20. $3m^2 - 8m - 40 = 0$ **21.** $7x^2 - 6x + 5 = 0$

22. $4p^2 + 4p = 15$ **23.** $9k^2 - 13k + 4 = 0$

13–6 State the method that seems easiest for solving each equation. Then use it to solve the equation.

24. $r^2 - 12r = -27$ **25.** $m^2 + 10m - 7 = 0$

26. $3a^2 - 11a + 10 = 0$ **27.** $9x^2 + 11 = 20x$

28. $2x^2 - \frac{17}{6}x + 1 = 0$ **29.** $x^2 - \frac{27}{20}x + \frac{3}{5} = 0$

30. $x^2 - x + 0.21 = 0$ **31.** $x^2 - 2.3x + 0.6 = 0$

13–7 Solve each problem.

32. A rectangle has a perimeter of 38 inches. Its area is 84 square inches. Find its dimensions.

33. Find two integers whose sum is 21 and whose product is 90.

13–8 State the sum and product of the roots of each equation.

34. $y^2 + 8y - 14 = 0$ **35.** $4a^2 - 6a + 11 = 0$

Find a quadratic equation having the given roots.

36. $1, -8$ **37.** $\frac{3}{2}, -4$ **38.** $3 + \sqrt{5}, 3 - \sqrt{5}$

Find the equation of the axis of symmetry and the coordinates of the maximum or minimum point of the graph of each quadratic function.

1. $y = 4x^2 - 8x - 17$

2. $y = -3x^2 + 12x + 34$

Locate the roots of each equation by graphing the related function.

3. $x^2 + x - 2 = 0$

4. $x^2 - 8x + 15 = 0$

Find the value of c that makes each trinomial a perfect square.

5. $x^2 + 14x + c$

6. $x^2 - 21x + c$

Solve by completing the square.

7. $k^2 - 8k - 4 = 0$

8. $m^2 - 6m + 6 = 0$

Use the Quadratic Formula to solve each equation.

9. $2x^2 - 5x - 12 = 0$

10. $2m^2 - 9m + 8 = 0$

11. $3y^2 - 2y - 4 = 0$

12. $2y^2 + 3y - 20 = 0$

Use the discriminant to determine the nature of the roots of each equation.

13. $3y^2 - y - 10 = 0$

14. $4y^2 + 12y + 9 = 0$

15. $y^2 + \sqrt{3}y - 5 = 0$

16. $x^2 + 2x + 2 = 0$

Solve each equation.

17. $4x^2 - 5x + 1 = 0$

18. $m^2 + 18m + 75 = 0$

19. $3x^2 + 2x = 5$

20. $2x^2 - 3x = 10$

21. $7x^2 - \frac{23}{3}x + 2 = 0$

22. $x^2 - 4.4x + 4.2 = 0$

Solve each problem.

23. A rectangle has a perimeter of 44 centimeters. Its area is 105 square centimeters. Find its dimensions.

24. Find two integers whose sum is 22 and whose product is 72.

State the sum and product of the roots of each equation.

25. $m^2 + 15m + 41 = 0$

26. $2x^2 - x - 6 = 0$

Find a quadratic equation having the given roots.

27. $-2, 5$

28. $6 + \sqrt{3}, 6 - \sqrt{3}$

CHAPTER 14

Rational Expressions

Many hot air balloons can be seen in the photograph above. Although each balloon is different in design or size, the same method of inflating and piloting apply to each. Similarly, the properties and operations you have applied to common fractions can also be applied to rational expressions.

14-1 Simplifying Rational Expressions

Rational expressions are algebraic expressions whose numerator and denominator are polynomials. The expressions $\frac{5x + 3}{y}$, $\frac{2}{x}$, and $\frac{a - 2}{a^2 + 4}$ are examples of rational expressions or algebraic fractions.

A fraction indicates division. Zero cannot be used as a denominator because division by zero is undefined. Therefore, any value assigned to a variable that results in a denominator of zero must be excluded from the domain of the variable.

For $\frac{5}{x}$, exclude $x = 0$.   For $\frac{3x + 7}{x + 4}$, exclude $x = -4$.

For $\frac{y^2 - 5}{x^2 - 5x + 6}$, exclude $x = 2$ and $x = 3$.   *Factor $x^2 - 5x + 6$ to see why.*

Example

1 **For each fraction, state the values of the variable that must be excluded.**

a. $\frac{5x}{x + 7}$

Exclude the values for which $x + 7 = 0$.

$$x + 7 = 0$$
$$x = -7$$

Therefore, x cannot equal -7.

b. $\frac{2a - 3}{a^2 - a - 12}$

Exclude the values for which $a^2 - a - 12 = 0$.

$$a^2 - a - 12 = 0$$
$$(a - 4)(a + 3) = 0 \qquad \textit{Factor } a^2 - a - 12 = 0.$$
$$a = 4 \text{ or } a = -3 \qquad \textit{Use the Zero Product Property.}$$

Therefore, a cannot equal 4 or -3.

Recall that to simplify an algebraic fraction such as $\frac{14a^2bc}{42abc^2}$, first factor the numerator and denominator. Then simplify common factors.

$$\frac{14a^2bc}{42abc^2} = \frac{2 \cdot 7 \cdot a \cdot a \cdot b \cdot c}{2 \cdot 3 \cdot 7 \cdot a \cdot b \cdot c \cdot c} \qquad \textit{Notice that } a \neq 0, b \neq 0, \text{ and } c \neq 0.$$

$$= \frac{\overset{1}{\cancel{2}} \cdot \overset{1}{\cancel{7}} \cdot \overset{1}{\cancel{a}} \cdot a \cdot \overset{1}{\cancel{b}} \cdot \overset{1}{\cancel{c}}}{\underset{1}{\cancel{2}} \cdot 3 \cdot \underset{1}{\cancel{7}} \cdot \underset{1}{\cancel{a}} \cdot \underset{1}{\cancel{b}} \cdot \underset{1}{\cancel{c}} \cdot c} \quad \text{or} \quad \frac{a}{3c} \qquad \textit{The GCF is 14abc.}$$

The same procedure can be used to simplify rational expressions having polynomials in the numerator and denominator.

Examples

2 Simplify $\dfrac{3a^2 + a - 2}{a^2 + 7a + 6}$. **State the excluded values of a.**

$$\dfrac{3a^2 + a - 2}{a^2 + 7a + 6} = \dfrac{(a + 1)(3a - 2)}{(a + 6)(a + 1)} \qquad \text{Factor } 3a^2 + a - 2.$$
$$\text{Factor } a^2 + 7a + 6.$$

$$= \dfrac{\overset{1}{\cancel{(a + 1)}}(3a - 2)}{(a + 6)\underset{1}{\cancel{(a + 1)}}} \qquad \text{The GCF is } (a + 1).$$

$$= \dfrac{3a - 2}{a + 6}$$

The excluded values of a are any values for which $a^2 + 7a + 6 = 0$.

$$a^2 + 7a + 6 = 0$$
$$(a + 6)(a + 1) = 0$$
$$a = -6 \quad \text{or} \quad a = -1 \qquad \textit{Zero Product Property}$$

Therefore, a cannot equal -6 or -1.

3 Simplify $\dfrac{2x - 2y}{y^2 - x^2}$. **State the excluded values of x and y.**

$$\dfrac{2x - 2y}{y^2 - x^2} = \dfrac{2(x - y)}{(y - x)(y + x)} \qquad \textit{Distributive Property}$$
$$\textit{Factor } y^2 - x^2.$$

$$= \dfrac{2(-1)(y - x)}{(y - x)(y + x)} \qquad \textit{Notice that } x - y = -1(y - x).$$

$$= \dfrac{-2\overset{1}{\cancel{(y - x)}}}{\underset{1}{\cancel{(y - x)}}(y + x)} \qquad \textit{Simplify.}$$

$$= \dfrac{-2}{y + x} \quad \text{or} \quad -\dfrac{2}{y + x}$$

The excluded values of x and y are any values for which $y^2 - x^2 = 0$.

$$y^2 - x^2 = 0$$
$$(y - x)(y + x) = 0$$
$$y = x \quad \text{or} \quad y = -x \qquad \textit{Zero Product Property}$$

Therefore, y cannot equal x or $-x$.

Exploratory Exercises

Simplify each rational expression. State the excluded values of the variables.

1. $\dfrac{(y + 4)}{(y - 4)(y + 4)}$

2. $\dfrac{(a - 7)}{(a + 1)(a - 7)}$

3. $\dfrac{r(r + 3)}{(r + 3)}$

4. $\dfrac{m(m - 1)}{m - 1}$

5. $\dfrac{2x^3}{2x^2(x^2 - 4)}$

6. $\dfrac{-3z^2}{z(z^2 - 5)}$

7. $\dfrac{(a - 4)(a + 4)}{(a - 2)(a - 4)}$

8. $\dfrac{(c + 6)(c - 2)}{(c - 2)(c - 2)}$

9. $\dfrac{(t + 2)(t - 2)(t + 3)(t - 3)}{(t + 2)(t - 3)}$

10. $\dfrac{(x + 5)(x - 6)}{(x + 5)(x - 5)(x^2 - 6)}$

11. $\dfrac{-1(3w - 2)}{(3w - 2)(w + 4)}$

12. $\dfrac{a(b + 2)(b - 7)}{(b + 1)(b + 2)(b + 3)}$

Written Exercises

Simplify each rational expression. State the excluded values of the variables.

13. $\dfrac{y - 3}{y^2 - 9}$

14. $\dfrac{s + 6}{s^2 - 36}$

15. $\dfrac{a^2 - 25}{a^2 + 3a - 10}$

16. $\dfrac{x^2 - 49}{x^2 - 2x - 35}$

17. $\dfrac{r^3 - r^2}{r - 1}$

18. $\dfrac{z^2 - 3z}{z - 3}$

19. $\dfrac{4n^2 - 8}{4n - 4}$

20. $\dfrac{6y^3 - 12y^2}{12y^2 - 18}$

21. $\dfrac{3m^3}{6m^2 - 3m}$

22. $\dfrac{7a^3b^2}{21a^2b + 49ab^3}$

23. $\dfrac{x + 3}{x^2 + 6x + 9}$

24. $\dfrac{c - 6}{c^2 - 12c + 36}$

25. $\dfrac{g^2 + g - 2}{g^2 - 3g + 2}$

26. $\dfrac{r^2 - r - 20}{r^2 + 9r + 20}$

27. $\dfrac{m^2 - 36}{m^2 + 5m - 6}$

28. $\dfrac{a^2 - 9}{a^2 + 6a - 27}$

29. $\dfrac{2y - 4}{y^2 + 3y - 10}$

30. $\dfrac{4x + 8}{x^2 + 6x + 8}$

31. $\dfrac{k^2 - 1}{k^2 + 2k + 1}$

32. $\dfrac{b^2 - 9}{b^2 + 6b + 9}$

33. $\dfrac{9 - a^2}{a^2 - a - 6}$

34. $\dfrac{25 - x^2}{x^2 + x - 30}$

35. $\dfrac{-x^2 + 6x - 9}{x^2 - 6x + 9}$

36. $\dfrac{a^2 - 2a + 1}{-a^2 + 2a - 1}$

37. $\dfrac{4y^2 + 7y - 2}{8y^2 + 15y - 2}$

38. $\dfrac{4x^2 - 6x - 4}{2x^2 - 8x + 8}$

39. $\dfrac{3m^2 + 9m + 6}{4m^2 + 12m + 8}$

40. $\dfrac{6r^2 + 12r - 48}{5r^2 - 5r - 10}$

41. $\dfrac{2t^2 - t - 21}{28 - 15t + 2t^2}$

42. $\dfrac{3z^2 + 5z - 2}{4 - 13z + 3z^2}$

43. $\dfrac{b^2 + 2b - 8}{b^4 - 20b^2 + 64}$

44. $\dfrac{a^2 + 2a - 3}{a^4 - 10a^2 + 9}$

45. $\dfrac{2x^2 + 7x - 4}{4x^2 - 4x + 1}$

46. $\dfrac{6y^2 + 7y + 2}{6y^2 + 5y + 1}$

47. $\dfrac{x^4 - 1}{x^4 - 5x^2 + 4}$

48. $\dfrac{x^4 - 16}{x^4 - 8x^2 + 16}$

49. $\dfrac{6s^2 + 17s - 14}{3s^2 - 20s + 12}$

50. $\dfrac{8t^2 - 14t - 15}{12t^2 - 19t - 21}$

51. $\dfrac{c^2 - c - 20}{c^3 + 10c^2 + 24c}$

52. $\dfrac{n^2 - 8n + 12}{n^3 - 12n^2 + 36n}$

53. $\dfrac{a^4 - 5a^2 + 4}{a^2 - a - 2}$

54. $\dfrac{y^4 - 13y^2 + 36}{y^2 + 5y + 6}$

55. $\dfrac{12x^3 + 12x^2 - 9x}{12x^3 + 18x^2 - 12x}$

56. $\dfrac{16a^3 - 24a^2 - 160a}{8a^4 - 36a^3 + 16a^2}$

Sometimes the solution to a problem involves several steps. An important strategy for solving such problems is to identify subgoals. This involves taking steps that will either produce part of the solution or will make the problem easier to solve.

Example 1 **Find all whole numbers less than 100 whose digits have a sum of 10.**

Choose a subgoal. Find all pairs of digits that have a sum of 10.

$$1, 9 \quad 2, 8 \quad 3, 7 \quad 4, 6 \quad 5, 5$$

Then use each pair of digits to write whole numbers. The solutions are 19, 91, 28, 82, 37, 73, 46, 64, and 55.

Example 2 **How many pairs of unit fractions have a sum of $\frac{1}{2}$?**

The unit fractions are $\left\{\frac{1}{2}, \frac{1}{3}, \frac{1}{4}, \frac{1}{5}, \ldots\right\}$

Suppose one fraction is $\frac{1}{3}$. Use subtraction to find the other fraction.

$$\frac{1}{2} - \frac{1}{3} = \frac{1}{6}$$

Thus, $\frac{1}{3} + \frac{1}{6} = \frac{1}{2}$. This is one solution to the problem.

Suppose one fraction is $\frac{1}{4}$. $\quad \frac{1}{2} - \frac{1}{4} = \frac{1}{4}$

Thus, $\frac{1}{4} + \frac{1}{4} = \frac{1}{2}$. This is a second solution to the problem.

Suppose one fraction is $\frac{1}{5}$. $\quad \frac{1}{2} - \frac{1}{5} = \frac{3}{10}$ $\quad \frac{3}{10}$ is *not* a unit fraction.

There is no unit fraction that can be added to $\frac{1}{5}$ to get $\frac{1}{2}$.

Are there any other pairs of unit fractions whose sum is $\frac{1}{2}$? If so, one would have to be greater than $\frac{1}{4}$ and the other would have to be less than $\frac{1}{4}$. Why?

Since $\frac{1}{3}$ and $\frac{1}{2}$ are the *only* unit fractions greater than $\frac{1}{4}$, there are only two pairs of unit fractions that have a sum of $\frac{1}{2}$.

Exercises

Solve each problem.

1. How many pairs of unit fractions have a sum of $\frac{1}{6}$?

2. How many whole numbers less than 1000 have digits whose sum is 10?

3. Find all whole numbers between 10 and 1000 that stay the same when the digits are written in reverse order. For example, 686 has this property.

4. Suppose the scoring in football is simplified to 7 points for a touchdown and 3 points for a field goal. What scores are impossible to achieve?

14-2 Multiplying Fractions

To multiply fractions, you multiply the numerators and multiply the denominators.

$$\frac{3}{5} \cdot \frac{4}{7} = \frac{3 \cdot 4}{5 \cdot 7}$$

$$= \frac{12}{35}$$

This method can be generalized as follows.

For all rational numbers $\frac{a}{b}$ and $\frac{c}{d}$, where $b \neq 0$ and $d \neq 0$,

$$\frac{a}{b} \cdot \frac{c}{d} = \frac{ac}{bd}.$$

Multiplying Fractions

The same method can be used to multiply rational expressions.

Examples

1 Find $\frac{5}{a} \cdot \frac{b}{7}$ and simplify. State any excluded values.

$$\frac{5}{a} \cdot \frac{b}{7} = \frac{5 \cdot b}{a \cdot 7}$$ *Multiply the numerators.*
Multiply the denominators.

$$= \frac{5b}{7a}$$

Since $7a$ cannot equal 0, $a \neq 0$.

2 Find $\frac{2a^2d}{3bc} \cdot \frac{9b^2c}{16ad^2}$ and simplify. State any excluded values.

$$\frac{2a^2d}{3bc} \cdot \frac{9b^2c}{16ad^2} = \frac{18a^2b^2cd}{48abcd^2}$$ *The GCF is $6abcd$.*

$$= \frac{3ab}{8d}$$ *Change the fraction to simplest form.*

Since bc and ad^2 cannot equal 0, $a \neq 0$, $b \neq 0$, $c \neq 0$, and $d \neq 0$.

From this point on, it will be assumed that all replacements for variables in rational expressions that result in denominators equal to zero will be excluded.

You may have used the shortcut shown below for simplifying and finding products at the same time.

$$\frac{3}{4} \cdot \frac{16}{21} = \frac{\overset{1}{\cancel{3}}}{\underset{1}{\cancel{4}}} \cdot \frac{\overset{4}{\cancel{16}}}{\underset{7}{\cancel{21}}} = \frac{4}{7}$$

The same method can be used with rational expressions.

Examples

3 **Find** $\dfrac{x + 5}{3x} \cdot \dfrac{12x^2}{x^2 + 7x + 10}$ **and simplify.**

$$\frac{x + 5}{3x} \cdot \frac{12x^2}{x^2 + 7x + 10} = \frac{x + 5}{3 \cdot x} \cdot \frac{2 \cdot 2 \cdot 3 \cdot x \cdot x}{(x + 5)(x + 2)} \qquad x \neq 0, \, x \neq -5, \text{ or } x \neq -2$$

$$= \frac{\overset{1}{\cancel{x + 5}}}{3 \cdot \cancel{x}} \cdot \frac{2 \cdot 2 \cdot \cancel{3} \cdot \cancel{x} \cdot x}{\cancel{(x + 5)}(x + 2)} \qquad \textit{Simplify.}$$

$$= \frac{4x}{x + 2}$$

4 **Find** $\dfrac{4a + 8}{a^2 - 25} \cdot \dfrac{a - 5}{5a + 10}$ **and simplify.**

$$\frac{4a + 8}{a^2 - 25} \cdot \frac{a - 5}{5a + 10} = \frac{4(a + 2)}{(a - 5)(a + 5)} \cdot \frac{a - 5}{5(a + 2)} \qquad a \neq 5, \, a \neq -5, \text{ or } a \neq -2$$

$$= \frac{4\overset{1}{\cancel{(a + 2)}}}{(a \cancel{- 5})(a + 5)} \cdot \frac{\overset{1}{a \cancel{- 5}}}{5\cancel{(a + 2)}} \qquad \textit{Simplify.}$$

$$= \frac{4}{5(a + 5)}$$

$$= \frac{4}{5a + 25}$$

5 **Find** $\dfrac{x^2 - x - 6}{9 - x^2} \cdot \dfrac{x^2 + 7x + 12}{x^2 + 4x + 4}$ **and simplify.**

$$\frac{x^2 - x - 6}{9 - x^2} \cdot \frac{x^2 + 7x + 12}{x^2 + 4x + 4} = \frac{(x - 3)(x + 2)}{(3 - x)(3 + x)} \cdot \frac{(x + 3)(x + 4)}{(x + 2)(x + 2)} \qquad \begin{array}{l} x \neq 3, \, x \neq -3, \\ \text{or } x \neq -2 \end{array}$$

$$= \frac{(x - 3)(x + 2)}{-1(x - 3)(x + 3)} \cdot \frac{(x + 3)(x + 4)}{(x + 2)(x + 2)} \qquad \begin{array}{l} \textit{Notice that} \\ 3 - x = -1(x - 3). \end{array}$$

$$= \frac{\overset{1}{\cancel{(x - 3)}}\overset{1}{\cancel{(x + 2)}}}{-1\cancel{(x - 3)}\cancel{(x + 3)}} \cdot \frac{\overset{1}{\cancel{(x + 3)}}(x + 4)}{\cancel{(x + 2)}(x + 2)} \qquad \textit{Simplify.}$$

$$= \frac{x + 4}{-1(x + 2)}$$

$$= -\frac{x + 4}{x + 2}$$

Exploratory Exercises

Find each product and simplify.

1. $\frac{1}{3} \cdot \frac{5}{8}$

2. $\frac{3}{4} \cdot \frac{5}{7}$

3. $-\frac{5}{6} \cdot \frac{7}{8}$

4. $\frac{2}{3}\left(-\frac{5}{9}\right)$

5. $-\frac{7}{8}\left(-\frac{5}{9}\right)$

6. $\frac{3}{a} \cdot \frac{b}{4}$

7. $\frac{a}{3} \cdot \frac{a}{5}$

8. $\frac{3a}{5} \cdot \frac{2x}{y}$

9. $\frac{2}{3}\left(\frac{1}{2}\right)$

10. $\frac{4}{5} \cdot \frac{5}{8}$

11. $\left(-\frac{5}{9}\right)\left(\frac{3}{10}\right)$

12. $\left(-\frac{4}{5}\right)\left(-\frac{3}{8}\right)$

13. $\left(-\frac{4}{9}\right)\left(\frac{3}{8}\right)$

14. $\frac{5}{12} \cdot \frac{4}{9}$

15. $\left(\frac{4}{7}\right)\left(\frac{11}{16}\right)$

Written Exercises

Find each product and simplify.

16. $\frac{2}{9} \cdot \frac{3}{5}$

17. $\frac{4}{9} \cdot \frac{1}{3}$

18. $\frac{16}{75} \cdot \frac{5}{8}$

19. $\frac{32}{7} \cdot \frac{35}{8}$

20. $\left(\frac{1}{4}\right)^2$

21. $\left(\frac{2}{3}\right)^2$

22. $\left(\frac{4}{9}\right)^3$

23. $\left(\frac{8}{5}\right)^3$

24. $\frac{ab}{ac} \cdot \frac{c}{d}$

25. $\frac{a^2b}{b^2c} \cdot \frac{c}{d}$

26. $\frac{6a^2n}{8n^2} \cdot \frac{12n}{9a}$

27. $\frac{10n^3}{6x^3} \cdot \frac{12n^2x^4}{25n^2x^2}$

28. $\frac{8}{m^2}\left(\frac{m^2}{2c}\right)^2$

29. $\left(\frac{2a}{b}\right)^2 \frac{5c}{6a}$

30. $\frac{6m^3n}{10a^2} \cdot \frac{4a^2m}{9n^3}$

31. $\frac{7xy^3}{11z^2} \cdot \frac{44z^3}{21x^2y}$

32. $\frac{y-3}{7} \cdot \frac{14}{y-3}$

33. $\frac{5n-5}{3} \cdot \frac{9}{n-1}$

34. $\frac{3a-3b}{a} \cdot \frac{a^2}{a-b}$

35. $\frac{-(2a+7c)}{6} \cdot \frac{36}{-7c-2a}$

36. $\frac{2a+4b}{5} \cdot \frac{25}{6a+8b}$

37. $\frac{3x+30}{2x} \cdot \frac{4x}{4x+40}$

38. $\frac{3}{x-y} \cdot \frac{(x-y)^2}{6}$

39. $\frac{a^2-b^2}{4} \cdot \frac{16}{a+b}$

40. $\frac{m^2-4}{2} \cdot \frac{4}{m-2}$

41. $\frac{9}{m-3} \cdot \frac{m^2-9}{12}$

42. $\frac{r^2}{r-s} \cdot \frac{r^2-s^2}{s^2}$

43. $\frac{a^2-b^2}{a-b} \cdot \frac{7}{a+b}$

44. $\frac{x^2-16}{9} \cdot \frac{x+4}{x-4}$

45. $\frac{y^2-4}{y^2-1} \cdot \frac{y+1}{y+2}$

46. $\frac{x^2-y^2}{x^2-1} \cdot \frac{x-1}{x-y}$

47. $\frac{r^2+s^2}{r^2-s^2} \cdot \frac{r-s}{r+s}$

48. $\frac{m^2+16}{m^2-16} \cdot \frac{m-4}{m+4}$

49. $\frac{3k+9}{k} \cdot \frac{k^2}{k^2-9}$

50. $\frac{3a-6}{a^2-9} \cdot \frac{a+3}{a^2-2a}$

51. $\frac{y^2-x^2}{y} \cdot \frac{x}{x-y}$

52. $\frac{b+a}{b-a} \cdot \frac{a^2-b^2}{a}$

53. $\frac{3mn^2-3m}{n} \cdot \frac{3m}{n^2-1}$

54. $\frac{x+3}{x+4} \cdot \frac{x}{x^2+7x+12}$

55. $\frac{x}{x^2+8x+15} \cdot \frac{2x+10}{x^2}$

56. $\dfrac{1}{x^2 + x - 12} \cdot \dfrac{x - 3}{x + 5}$

57. $\dfrac{x - 5}{x^2 - 7x + 10} \cdot \dfrac{x - 2}{3}$

58. $\dfrac{b^2 + 20b + 99}{b + 9} \cdot \dfrac{b + 7}{b^2 + 12b + 11}$

59. $\dfrac{z^2 - 51z + 50}{z^2 - 9z + 20} \cdot \dfrac{z^2 - 11z + 24}{z^2 - 18z + 80}$

60. $\dfrac{b^2 + 19b + 84}{b - 3} \cdot \dfrac{b^2 - 9}{b^2 + 15b + 36}$

61. $\dfrac{z^2 + 16z + 39}{z^2 + 9z + 18} \cdot \dfrac{z + 5}{z^2 + 18z + 65}$

62. $\dfrac{y^2 + 3y^3}{y^2 - 4} \cdot \dfrac{2y + y^2}{y + 4y^2 + 3y^3}$

63. $\dfrac{3t^3 - 14t^2 + 8t}{2t^2 - 3t - 20} \cdot \dfrac{16t^2 + 34t - 15}{24t^2 - 25t + 6}$

64. $\dfrac{6y^2 - 5y - 6}{3y^2 - 20y - 7} \cdot \dfrac{y^2 - 49}{12y^3 + 23y^2 + 10y}$

65. $\dfrac{2m^2 - 9m + 9}{3m^2 + 19m - 14} \cdot \dfrac{m^2 + 14m + 49}{9 - 6m + m^2}$

Use the formula $A = \ell w$ to represent the area of each rectangle. Write each expression in simplest form.

66.

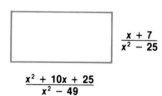

$\dfrac{x + 7}{x^2 - 25}$

$\dfrac{x^2 + 10x + 25}{x^2 - 49}$

67.

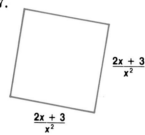

$\dfrac{2x + 3}{x^2}$

$\dfrac{2x + 3}{x^2}$

68.

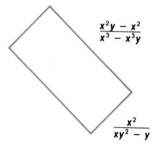

$\dfrac{x^2y - x^2}{x^3 - x^3y}$

$\dfrac{x^2}{xy^2 - y}$

69.

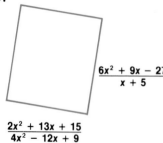

$\dfrac{6x^2 + 9x - 27}{x + 5}$

$\dfrac{2x^2 + 13x + 15}{4x^2 - 12x + 9}$

70.

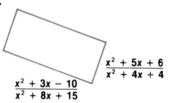

$\dfrac{x^2 + 5x + 6}{x^2 + 4x + 4}$

$\dfrac{x^2 + 3x - 10}{x^2 + 8x + 15}$

71.

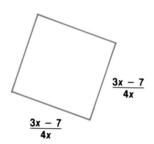

$\dfrac{3x - 7}{4x}$

$\dfrac{3x - 7}{4x}$

Challenge Exercises

Find each product and simplify.

72. $\dfrac{a}{a^2 + 6a + 9} \cdot \dfrac{a + 3}{a - 5} \cdot \dfrac{a^2 - 25}{5a + 25}$

73. $\dfrac{a^2x - b^2x}{y} \cdot \dfrac{y^2 + y}{a - 2} \cdot \dfrac{4 - 2a}{axy - bxy}$

74. $\dfrac{x^2y}{x^2 + 4xy + 4y^2} \cdot \dfrac{x^2 + 2xy}{xy} \cdot \dfrac{y}{x^4 - 9x^2}$

75. $\dfrac{2x^2 + x - 15}{4x^2 + 2x - 30} \cdot \dfrac{6x^2 - 8x + 2}{3x^2 + 8x - 3} \cdot \dfrac{x^2 + x - 2}{x^2 - x}$

14-3 Dividing Fractions

Consider the following products. What is the pattern of the factors in each example?

$$\frac{5}{1} \cdot \frac{1}{5} = 1 \qquad \frac{1}{4} \cdot \frac{4}{1} = 1 \qquad \frac{2}{3} \cdot \frac{3}{2} = 1 \qquad \frac{x}{y} \cdot \frac{y}{x} = 1$$

Remember that all replacements for variables that yield a denominator of zero are excluded.

You should recall that two numbers whose product is 1 are called **multiplicative inverses** or **reciprocals**.

To find the quotient of two fractions, you multiply by the reciprocal of the second fraction.

$$\frac{2}{3} \div \frac{3}{4} = \frac{2}{3} \cdot \frac{4}{3} \qquad \text{\textit{The reciprocal of} } \frac{3}{4} \text{ \textit{is} } \frac{4}{3}.$$

$$= \frac{8}{9}$$

This method can be generalized as follows.

> **For all rational numbers $\frac{a}{b}$ and $\frac{c}{d}$, where $b \neq 0$, $c \neq 0$, and $d \neq 0$,**
>
> $$\frac{a}{b} \div \frac{c}{d} = \frac{a}{b} \cdot \frac{d}{c}.$$

Dividing Fractions

The same method is used to divide rational expressions.

Examples

1 Simplify $\frac{5}{x} \div \frac{y}{z}$.

$$\frac{5}{x} \div \frac{y}{z} = \frac{5}{x} \cdot \frac{z}{y} \qquad \text{\textit{The reciprocal of} } \frac{y}{z} \text{ \textit{is} } \frac{z}{y}.$$

$$= \frac{5z}{xy}$$

2 Simplify $\frac{2x}{x+1} \div (x-1)$.

$$\frac{2x}{x+1} \div (x-1) = \frac{2x}{x+1} \cdot \frac{1}{x-1} \qquad \text{\textit{The reciprocal of} } (x-1) \text{ \textit{is} } \frac{1}{x-1}.$$

$$= \frac{2x}{(x+1)(x-1)}$$

$$= \frac{2x}{x^2 - 1}$$

Example

3 Simplify $\dfrac{x-y}{x^2-y^2} \div \dfrac{x+y}{x^2+2xy+y^2}$.

$$\frac{x-y}{x^2-y^2} \div \frac{x+y}{x^2+2xy+y^2} = \frac{x-y}{x^2-y^2} \cdot \frac{x^2+2xy+y^2}{x+y}$$

$$= \frac{x-y}{(x+y)(x-y)} \cdot \frac{(x+y)(x+y)}{x+y}$$

$$= \frac{\overset{1}{\cancel{x-y}}}{(x+y)\cancel{(x-y)}} \cdot \frac{\overset{1}{\cancel{(x+y)}}\overset{1}{\cancel{(x+y)}}}{\cancel{x+y}}$$

$$= 1$$

Exploratory Exercises

State the reciprocal of each expression.

1. $\dfrac{3}{4}$

2. $\dfrac{-5}{8}$

3. $\dfrac{8}{-3}$

4. $-\dfrac{9}{10}$

5. $\dfrac{m}{2}$

6. $\dfrac{x^2}{4}$

7. $\dfrac{5}{2p}$

8. $\dfrac{-8}{3n}$

9. 6

10. x

11. a^2

12. $\dfrac{1}{3}a$

13. $2bc$

14. $\dfrac{2}{5}m^2$

15. $\dfrac{x+y}{x-y}$

16. $\dfrac{a^2+b}{a-b}$

17. $\dfrac{3}{5}a^2b^2$

18. $x-2$

State a multiplication expression for each division expression.

19. $\dfrac{3}{8} \div \dfrac{1}{4}$

20. $\dfrac{1}{7} \div \dfrac{-1}{49}$

21. $\dfrac{1}{3} \div -6$

22. $\dfrac{x}{y} \div \dfrac{y^2}{x}$

23. $\dfrac{2x}{4-2a} \div \dfrac{a^2}{b-2}$

24. $\dfrac{b-a}{a-b} \div (a+b)$

Written Exercises

Find each quotient and simplify.

25. $\dfrac{5}{6} \div \dfrac{2}{3}$

26. $\dfrac{3}{8} \div \dfrac{-1}{2}$

27. $-\dfrac{5}{6} \div \dfrac{1}{3}$

28. $\dfrac{5}{8} \div 5$

29. $\dfrac{a^2}{b^2} \div \dfrac{b^2}{a^2}$

30. $\dfrac{a}{b} \div \dfrac{b^3}{a^4}$

31. $\dfrac{y^2}{x^2} \div \dfrac{a^2}{x^2}$

32. $\dfrac{a^2}{b} \div \dfrac{a^2}{b^2}$

33. $\dfrac{(-a)^2}{b} \div \dfrac{a}{b}$

34. $\dfrac{p^3}{2q} \div \dfrac{-(p^2)}{4q}$

35. $\dfrac{3m}{m+1} \div (m-2)$

36. $\dfrac{n^2}{n-3} \div (n+4)$

37. $\dfrac{b^2-9}{4b} \div (b-3)$

38. $\dfrac{y^2+8y+16}{y^2} \div (y+4)$

39. $\dfrac{y^2}{x+2} \div \dfrac{y}{x+2}$

40. $\dfrac{2a^3}{a+1} \div \dfrac{a^2}{a+1}$

41. $\dfrac{p^2}{y^2-4} \div \dfrac{p}{2-y}$

42. $\dfrac{q}{y^2-4} \div \dfrac{q^2}{y+2}$

43. $\dfrac{x^2-16}{16-x^2} \div \dfrac{7}{x}$

44. $\dfrac{y}{5} \div \dfrac{y^2-25}{5-y}$

45. $\dfrac{m^2+2m+1}{2} \div \dfrac{m+1}{m-1}$

46. $\dfrac{x^2-4x+4}{3} \div \dfrac{x^2-4}{2x}$

47. $\dfrac{m^2+2mn+n^2}{3m} \div \dfrac{m^2-n^2}{2}$

48. $\dfrac{a^2+2ab+b^2}{2x} \div \dfrac{a+b}{x^2}$

49. $\dfrac{y^2-16}{y^2-64} \div \dfrac{y+4}{y-8}$

50. $\dfrac{k^2-81}{k^2-36} \div \dfrac{k-9}{k+6}$

51. $\dfrac{t^2+8t+16}{w^2-6w+9} \div \dfrac{2t+8}{3w-9}$

52. $\dfrac{k+2}{m^2+4m+4} \div \dfrac{4k+8}{m+4}$

53. $\dfrac{x}{x+2} \div \dfrac{x^2}{x^2+5x+6}$

54. $\dfrac{x^2+x-2}{x^2+5x+6} \div \dfrac{x^2+2x-3}{x^2+7x+12}$

55. $\dfrac{x^2+2x-15}{x^2-x-30} \div \dfrac{x^2-3x-18}{x^2-2x-24}$

56. $\dfrac{2m^2+7m-15}{m+5} \div \dfrac{9m^2-4}{3m+2}$

57. $\dfrac{a^2+3a-10}{a^2+8a+15} \div \dfrac{a^2-6a+8}{12+a-a^2}$

58. $\dfrac{2x^2-x-15}{x^2-2x-3} \div \dfrac{2x^2+3x-5}{1-x^2}$

Challenge Exercises

Perform the indicated operations and simplify.

59. $\dfrac{x^2+5x+6}{x^2-x-12} \cdot \dfrac{x-4}{x^2+11x+18} \div \dfrac{x+7}{x^2+14x+45}$

60. $\dfrac{2x-3}{2x^2-7x+6} \cdot \dfrac{x^2+3x-10}{5x+1} \div \dfrac{3x^2+14x-5}{3x^2+2x-1}$

61. $\dfrac{x^2+2x-3}{6x^2-5x+1} \cdot \dfrac{2x^2+9x+4}{2x^2+7x+3} \div \dfrac{x^2-5x+4}{6x^2-x-1}$

62. $\dfrac{4x^2-6x-4}{8x^2-2} \div \dfrac{8x^2-8x+2}{4x^2-10x-6} \cdot \dfrac{4x^2-4x+1}{x^2-5x+6}$

63. $\dfrac{x^2-1}{2x^2+14x+12} \div \dfrac{2x^2-3x+1}{8x^2+36x-72} \cdot \dfrac{x^2+3x-4}{x^2+5x-6}$

1. Write a quadratic equation whose roots are -2 and 7.

2. Solve the equation $2x^2 - 4x - 13 = 0$ by completing the square.

Simplify.

3. $7\sqrt{11} - 4\sqrt{11}$

4. $3\sqrt{5} + 2\sqrt{20}$

5. $\dfrac{5\sqrt{2}}{3 - \sqrt{2}}$

Use a calculator to simplify. Round answers to the nearest hundredth.

6. $\dfrac{2 - \sqrt{3}}{5 + 3\sqrt{3}}$

7. $\dfrac{6\sqrt{3}}{\sqrt{3} - 2}$

8. $\sqrt{\dfrac{2}{5}} + \sqrt{40} + \sqrt{10}$

Solve each system of equations by graphing.

9. $y \geq \frac{2}{3}x - 2$
   $x + 3y \leq 6$

10. $3x + 2y = 0$
    $x - 2y = 8$

11. $y = \frac{x}{2}$
    $2y = x + 4$

Solve each system of equations.

12. $x - 4y = 3$
    $-2x + 7y = -8$

13. $4x + y = 4$
    $7x + 3y = -3$

14. $5x + 2y = -8$
    $4x + 3y = 2$

Solve each equation and check your solution.

15. $\sqrt{3x + 15} - x = 5$

16. $x^2 - 6x + 1 = 0$

17. $3x^2 + 19x - 14 = 0$

18. An ocean freighter sailed 60 miles up the St. Lawrence River in five hours. The return trip downstream took only three hours for the same distance. Find the rate of the freighter in still water and the rate of the current.

Excursions in Algebra
Does 2 = 1?

Find the fallacy in the following "proof" of "2 = 1."

$$a = b$$
$$a \cdot a = b \cdot a \qquad \text{Multiply each side by } a.$$
$$a^2 = ab$$
$$a^2 - b^2 = ab - b^2 \qquad \text{Subtract } b^2 \text{ from each side.}$$
$$(a - b)(a + b) = b(a - b) \qquad \text{Factor.}$$
$$\frac{(a - b)(a + b)}{(a - b)} = \frac{b(a - b)}{(a - b)} \qquad \text{Divide each side by } (a - b).$$
$$a + b = b$$
$$b + b = b \qquad \text{Substitute } b \text{ for } a.$$
$$2b = b$$
$$\frac{2b}{b} = \frac{b}{b} \qquad \text{Divide each side by } b.$$
$$2 = 1$$

14-4 Dividing Polynomials

To divide a polynomial by a polynomial, you can use a long division process similar to that used in arithmetic. For example, you can divide $x^2 + 8x + 15$ by $x + 5$ as shown below.

Step 1

To find the first term of the quotient, divide the first term of the dividend (x^2) by the first term of the divisor (x).

$$\begin{array}{r} x \\ x + 5 \overline{)\, x^2 + 8x + 15} \\ \underline{x^2 + 5x} \\ 3x \end{array}$$

When dividing x^2 by x, the result is x.

Multiply $x(x + 5)$.

Subtract.

Step 2

To find the next term of the quotient, divide the first term of the partial dividend ($3x$) by the first term of the divisor (x).

$$\begin{array}{r} x \; + 3 \\ x + 5 \overline{)\, x^2 + 8x + 15} \\ \underline{x^2 + 5x} \\ 3x + 15 \\ \underline{3x + 15} \\ 0 \end{array}$$

When dividing $3x$ by x, the result is 3.

Multiply $3(x + 5)$.

Subtract.

Therefore, $x^2 + 8x + 15$ divided by $x + 5$ is $x + 3$. Since the remainder is 0, the divisor is a factor of the dividend. This means that $(x + 5)(x + 3) = x^2 + 8x + 15$.

If the divisor is **not** a factor of the dividend, there will be a nonzero remainder. When there is a nonzero remainder, the quotient can be expressed as follows.

$$\text{quotient} = \text{partial quotient} + \frac{\text{remainder}}{\text{divisor}}$$

Example

1 Find the quotient of $(2x^2 - 11x - 20) \div (2x + 3)$.

$$\begin{array}{r} x \; - \; 7 \\ 2x + 3 \overline{)\, 2x^2 - 11x - 20} \\ \underline{2x^2 + \; 3x} \\ -14x - 20 \\ \underline{-14x - 21} \\ 1 \end{array}$$

⟵ *Multiply $x(2x + 3)$.*
⟵ *Subtract, then bring down -20.*
⟵ *Multiply $-7(2x + 3)$.*
⟵ *Subtract. The remainder is 1.*

Therefore, the quotient is $x - 7$ with remainder 1.

Thus, $(2x^2 - 11x - 20) \div (2x + 3) = x - 7 + \frac{1}{2x + 3}$.   ⟵ *remainder* ⟵ *divisor*

Notice that in an expression such as $s^3 + 9$ there is no s^2 term and no s term. In such situations, the expression can be renamed using 0 as the coefficient of these terms.

$$s^3 + 9 = s^3 + 0s^2 + 0s + 9$$

Example

2 Find the quotient of $\frac{s^3 + 9}{s - 3}$.

$$
\begin{array}{r}
s^2 + 3s\ + 9 \\
s - 3\overline{)s^3 + 0s^2 + 0s +\ \ 9} \\
\underline{s^3 - 3s^2} \\
3s^2 + 0s \\
\underline{3s^2 - 9s} \\
9s +\ \ 9 \\
\underline{9s - 27} \\
36
\end{array}
$$

Insert $0s^2$ and $0s$. Why is this permissible?

←*remainder*

Therefore, $\frac{s^3 + 9}{s - 3} = s^2 + 3s + 9 + \frac{36}{s - 3}$.

Exploratory Exercises

State the first term of each quotient.

1. $\frac{a^2 + 3a + 2}{a + 1}$

2. $\frac{b^2 + 8b - 20}{b - 2}$

3. $\frac{8m^3 + 27}{2m + 3}$

4. $\frac{2x^2 + 3x - 2}{2x - 1}$

5. $\frac{x^3 + 2x^2 - 5x + 12}{x + 4}$

6. $\frac{2x^3 - 5x^2 + 22x + 51}{2x + 3}$

Written Exercises

Find each quotient.

7. $(x^2 + 7x + 12) \div (x + 3)$

8. $(x^2 + 9x + 20) \div (x + 5)$

9. $(a^2 - 2a - 35) \div (a - 7)$

10. $(x^2 + 6x - 16) \div (x - 2)$

11. $(c^2 + 12c + 36) \div (c + 9)$

12. $(y^2 - 2y - 30) \div (y + 7)$

13. $(2r^2 - 3r - 35) \div (2r + 7)$

14. $(3t^2 - 14t - 24) \div (3t + 4)$

15. $\frac{10x^2 + 29x + 21}{5x + 7}$

16. $\frac{12n^2 + 36n + 15}{6n + 3}$

17. $\frac{x^3 - 7x + 6}{x - 2}$

18. $\frac{4m^3 + 5m - 21}{2m - 3}$

19. $\frac{4t^3 + 17t^2 - 1}{4t + 1}$

20. $\frac{2a^3 + 9a^2 + 5a - 12}{a + 3}$

21. $\frac{27c^2 - 24c + 8}{9c - 2}$

22. $\frac{48b^2 + 8b + 7}{12b - 1}$

23. $\frac{6n^3 + 5n^2 + 12}{2n + 3}$

24. $\frac{t^3 - 19t + 9}{t - 4}$

25. $\frac{3s^3 + 8s^2 + s - 7}{s + 2}$

26. $\frac{9d^3 + 5d - 8}{3d - 2}$

27. $\frac{20t^3 - 27t^2 + t + 6}{4t - 3}$

28. $\frac{6x^3 - 9x^2 - 4x + 6}{2x - 3}$

29. $\frac{56x^3 + 32x^2 - 63x - 36}{7x + 4}$

30. Find the value of k if $(x + 2)$ is a factor of $x^3 + 7x^2 + 7x + k$.

31. Find the value of k if $(2m - 5)$ is a factor of $2m^3 - 5m^2 + 8m + k$.

32. When $x^3 - 7x^2 + 4x + k$ is divided by $(x - 2)$, the remainder is 15. Find k.

Excursions in Algebra   Finding Factors

Is $(x - 2)$ a factor of $x^3 + 3x^2 - 6x - 8$? One way to determine this is to divide $x^3 + 3x^2 - 6x - 8$ by $(x - 2)$ using the long division process. If the remainder is 0, then $(x - 2)$ is a factor of $x^3 + 3x^2 - 6x - 8$.

Another method can be used to determine whether $(x - 2)$ is a factor. If $(x - 2)$ is a factor, then when $x^3 + 3x^2 - 6x - 8 = 0$, $x - 2 = 0$ and $x = 2$. Substitute 2 for x and evaluate $x^3 + 3x^2 - 6x - 8$. If the result is 0, then $(x - 2)$ is a factor.
    You can use your calculator to evaluate $x^3 + 3x^2 - 6x - 8$ when $x = 2$.

The [STO] and [RCL] keys are used to store and recall the value for x that is used to evaluate the polynomial.

Example

Evaluate $x^3 + 3x^2 - 6x - 8$ when $x = 2$.

ENTER: 2 [STO] [y^x] 3 [+] 3 [×] [RCL] [x^2] [−] 6 [×] [RCL] [−] 8 [=]

DISPLAY: 2 2 2 3 8 3 3 2 4 20 6 6 2 8 8 0

Since the result is 0, $(x - 2)$ is a factor of $x^3 + 3x^2 - 6x - 8$.

Exercises

Use your calculator to determine whether the given binomial is a factor.

1. $x^3 + 2x^2 - 22x + 21$; $(x - 3)$

2. $x^3 - 6x^2 + 14x - 12$; $(x - 2)$

3. $x^3 - 8x^2 + 13x - 16$; $(x - 8)$

4. $x^4 + 3x^3 + 2x^2 - x + 6$; $(x - 2)$

14-5 Adding and Subtracting Fractions with Like Denominators

To add or subtract fractions with *like* denominators, you simply add or subtract the numerators. Then write this sum or difference over the common denominator.

$$\frac{3}{7} + \frac{2}{7} = \frac{5}{7} \qquad\qquad \frac{7}{9} - \frac{2}{9} = \frac{5}{9}$$

Sometimes the result needs to be changed to simplest form.

$$\frac{3}{8} + \frac{1}{8} = \frac{4}{8} \qquad \text{\textit{The GCF of}} \qquad \frac{9}{16} - \frac{3}{16} = \frac{6}{16} \qquad \text{\textit{The GCF of}}$$
$$\text{\textit{4 and 8 is 4.}} \qquad\qquad\qquad\qquad\qquad \text{\textit{6 and 16 is 2.}}$$
$$= \frac{1}{2} \qquad\qquad\qquad\qquad\qquad\qquad = \frac{3}{8}$$

These methods can be generalized as follows.

For all rational numbers $\frac{a}{c}$ and $\frac{b}{c}$, where $c \neq 0$, $$\frac{a}{c} + \frac{b}{c} = \frac{a + b}{c} \text{ and } \frac{a}{c} - \frac{b}{c} = \frac{a - b}{c}.$$

Adding and Subtracting Fractions with Like Denominators

These methods can be used to add or subtract rational expressions.

Examples

1 Find $\dfrac{3}{x + 2} + \dfrac{1}{x + 2}$ and simplify.

$$\frac{3}{x + 2} + \frac{1}{x + 2} = \frac{3 + 1}{x + 2} \qquad \text{\textit{Since the denominators are both x + 2, add the numerators.}}$$

$$= \frac{4}{x + 2}$$

2 Find $\dfrac{3a + 2}{a - 7} - \dfrac{a - 3}{a - 7}$ and simplify.

$$\frac{3a + 2}{a - 7} - \frac{a - 3}{a - 7} = \frac{(3a + 2) - (a - 3)}{a - 7}$$

$$= \frac{3a + 2 - a + 3}{a - 7}$$

$$= \frac{2a + 5}{a - 7}$$

The parentheses indicate that the quantity (a − 3) is subtracted from the quantity (3a + 2). Remember to change the signs of both terms of the second polynomial.

Example

3 Find $\dfrac{8n + 3}{3n + 4} - \dfrac{2n - 5}{3n + 4}$ and simplify.

$$\dfrac{8n + 3}{3n + 4} - \dfrac{2n - 5}{3n + 4} = \dfrac{(8n + 3) - (2n - 5)}{3n + 4}$$ *Since the denominators are both $3n + 4$, subtract the numerators.*

$$= \dfrac{8n + 3 - 2n + 5}{3n + 4}$$ *Why?*

$$= \dfrac{6n + 8}{3n + 4}$$

$$= \dfrac{2(\overset{1}{\cancel{3n + 4}})}{(\underset{1}{\cancel{3n + 4}})}$$ *Factor and simplify.*

$$= 2$$

Sometimes the denominators are additive inverses.

Example

4 Find $\dfrac{x}{x - 2} - \dfrac{x + 1}{2 - x}$ and simplify.

$$\dfrac{x}{x - 2} - \dfrac{x + 1}{2 - x} = \dfrac{x}{x - 2} - \dfrac{x + 1}{-(x - 2)}$$ *Notice that $2 - x = -(x - 2)$.*

$$= \dfrac{x}{x - 2} - \left(-\dfrac{x + 1}{x - 2}\right)$$

$$= \dfrac{x}{x - 2} + \dfrac{x + 1}{x - 2}$$

$$= \dfrac{2x + 1}{x - 2}$$

Exploratory Exercises

Find each sum or difference. Simplify.

1. $\dfrac{5}{8} + \dfrac{2}{8}$

2. $\dfrac{9}{16} + \dfrac{5}{16}$

3. $\dfrac{4}{a} + \dfrac{3}{a}$

4. $\dfrac{6}{b} + \dfrac{7}{b}$

5. $\dfrac{b}{x} + \dfrac{2}{x}$

6. $\dfrac{5}{2z} + \dfrac{-7}{2z}$

7. $\dfrac{-4k}{t} + \dfrac{6k}{t}$

8. $\dfrac{2n}{m^2} + \dfrac{3n}{m^2}$

9. $\dfrac{3}{11} - \dfrac{2}{11}$

10. $\dfrac{14}{16} - \dfrac{15}{16}$

11. $\dfrac{a}{5} - \dfrac{b}{5}$

12. $\dfrac{7}{a} - \dfrac{4}{a}$

13. $\dfrac{4}{x} - \dfrac{6}{x}$

14. $\dfrac{7}{a} - \dfrac{c}{a}$

15. $\dfrac{8k}{5m} - \dfrac{3k}{5m}$

16. $\dfrac{r}{y^2} - \dfrac{s}{y^2}$

Written Exercises

Find each sum or difference. Simplify.

17. $\frac{y}{2} + \frac{y}{2}$

18. $\frac{a}{12} + \frac{2a}{12}$

19. $\frac{5x}{24} - \frac{3x}{24}$

20. $\frac{7t}{t} - \frac{8t}{t}$

21. $\frac{3}{a} + \frac{6}{a}$

22. $\frac{2y}{b} - \frac{y}{b}$

23. $\frac{3}{x} - \frac{7}{x}$

24. $\frac{8}{x} + \frac{2}{x}$

25. $\frac{12a}{7} - \frac{3a}{7}$

26. $\frac{y}{14} - \frac{3y}{14}$

27. $\frac{y}{2} + \frac{y-6}{2}$

28. $\frac{m+4}{5} + \frac{m-1}{5}$

29. $\frac{a+2}{6} - \frac{a+3}{6}$

30. $\frac{b+3}{7} - \frac{b+1}{7}$

31. $\frac{x}{x-1} + \frac{1}{x-1}$

32. $\frac{x}{x+1} + \frac{1}{x+1}$

33. $\frac{8}{y-2} - \frac{6}{y-2}$

34. $\frac{y}{b+6} - \frac{2y}{b+6}$

35. $\frac{2n}{2n-5} + \frac{5}{5-2n}$

36. $\frac{x+y}{y-2} + \frac{x-y}{2-y}$

37. $\frac{y}{a+1} - \frac{y}{a+1}$

38. $\frac{y}{y-1} - \frac{1}{y-1}$

39. $\frac{a+b}{x-3} + \frac{a+b}{3-x}$

40. $\frac{a+b}{x-3} - \frac{a+b}{3-x}$

41. $\frac{r^2}{r-s} + \frac{s^2}{r-s}$

42. $\frac{x^2}{x-y} - \frac{y^2}{x-y}$

43. $\frac{m^2}{m+n} + \frac{2mn+n^2}{m+n}$

44. $\frac{x^2}{x-y} - \frac{2xy+y^2}{x-y}$

45. $\frac{12n}{3n+2} + \frac{8}{3n+2}$

46. $\frac{6x}{x+y} + \frac{6y}{x+y}$

47. $\frac{12n}{3n-2} - \frac{8}{3n-2}$

48. $\frac{a}{2a-2b} - \frac{b}{2a-2b}$

49. $\frac{a^2}{a-b} + \frac{-(b^2)}{a-b}$

50. $\frac{a^2}{a^2-b^2} + \frac{-(b^2)}{a^2-b^2}$

51. $\frac{r^2}{r-3} + \frac{9}{3-r}$

52. $\frac{m^2}{4+m} - \frac{16}{m+4}$

53. $\frac{x^2}{x^2-1} + \frac{2x+1}{x^2-1}$

54. $\frac{2x+1}{(x+1)^2} + \frac{x^2}{(x+1)^2}$

55. $\frac{x-1}{(x+1)^2} - \frac{x-1}{(x+1)^2}$

56. $\frac{25}{k+5} - \frac{k^2}{k+5}$

57. $\frac{x}{x^2+2x+1} + \frac{1}{x^2+2x+1}$

58. $\frac{x}{x^2-6x+9} - \frac{3}{x^2-6x+9}$

59. $\frac{3m}{m^2+2m+1} + \frac{3}{m^2+2m+1}$

60. $\frac{8y}{4y^2+12y+9} + \frac{12}{4y^2+12y+9}$

61. $\frac{2}{t^2-t-2} - \frac{t}{t^2-t-2}$

Using the formula $P = 2\ell + 2w$, find the perimeter of each rectangle.

62.
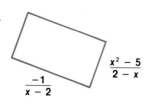
$\frac{t^2-2t}{t^2+t-6}$

$\frac{3t-6}{t^2+t-6}$

63.
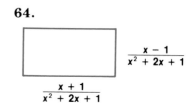
$\frac{x^2-5}{2-x}$

$\frac{-1}{x-2}$

64.
$\frac{x-1}{x^2+2x+1}$

$\frac{x+1}{x^2+2x+1}$

Find each missing measure if c is the measure of the hypotenuse of a right triangle and a and b are the measures of the legs.

1. Find a if $b = 15$ and $c = 17$.
2. Find c if $a = 4$ and $b = 5$.

Simplify.

3. $\sqrt{96}$

4. $\sqrt{200}$

5. $\sqrt{2}(3 + \sqrt{18})$

6. $\sqrt{3}(\sqrt{6} + \sqrt{15})$

7. $\dfrac{4}{4 + \sqrt{3}}$

8. $\dfrac{5}{3 - \sqrt{5}}$

9. $7\sqrt{11} - \sqrt{44}$

10. $4\sqrt{3} - \sqrt{48}$

Solve each system of equations.

11. $3x + 2y = 6$
    $x - 3y = 13$

12. $4x + 3y = 8$
    $-3x + y = 7$

13. $2x - 5y + 14 = 0$
    $5x - 3y = 3$

14. $2x + 7y = 7$
    $3x + 8y = 13$

15. $3x - 2y = -4$
    $3x + 6y = 20$

16. $3x + 4y = 3$
    $12x + 18y = 13$

Solve each equation and check your solution.

17. $\sqrt{3x + 7} - x = 1$

18. $\sqrt{5 - 10x} - 3 = 2$

Graph each system of inequalities.

19. $x \le 4$
    $y \ge -x + 2$

20. $y \le -x + 3$
    $2x - 3y \ge 3$

Solve each problem.

21. The sum of two numbers is 34. Their difference is 8. Find the numbers.

22. A grain barge moves 56 miles up river on the Mississippi in seven hours. The return trip downstream takes four hours. Find the rate of the barge in still water and the rate of the current.

Find a quadratic equation having the given roots.

23. $\frac{2}{3}$ and -3

24. $2 + \sqrt{3}$ and $2 - \sqrt{3}$

Solve each equation.

25. $2x^2 + 5x - 3 = 0$

26. $3x^2 - 4x - 4 = 0$

27. $x^2 - 10x + 22 = 0$

28. $x^2 - 6x + 7 = 0$

State the axis of symmetry and the coordinates for the maximum or minimum point of the graph.

29. $y = x^2 - 8x + 12$

30. $y = x^2 - 4x - 5$

Solve each equation by completing the square.

31. $x^2 - 4x + 1 = 0$

32. $x^2 - 6x + 2 = 0$

33. Simplify $\dfrac{x - 2}{3x^2 + x - 14}$.

Find each quotient.

34. $\dfrac{2x^2 + 11x + 15}{2x^2 - 5x - 3} \div \dfrac{x^2 + 7x + 12}{x^2 + x - 12}$

35. $m - 4 \overline{) m^3 - 6m^2 - m + 32}$

14-6 Adding and Subtracting Fractions with Unlike Denominators

When you add or subtract fractions with unlike denominators, you must first rename the fractions so the denominators are alike. Any common denominator could be used. However, the computation is easier if the **least common denominator** (LCD) is used. Recall that the least common denominator is the **least common multiple** (LCM) of the denominators. Consider the following example.

$$\frac{1}{6} + \frac{3}{8}$$ *To add these fractions, first find the LCM of 6 and 8.*

The set of multiples of 6 and 8 can be found by multiplying 6 and 8 by each positive integer.

multiples of 6: $1 \cdot 6, 2 \cdot 6, 3 \cdot 6, 4 \cdot 6, 5 \cdot 6, 6 \cdot 6, \ldots$
6,   12,   18,   24,   30,   36, ...

multiples of 8: $1 \cdot 8, 2 \cdot 8, 3 \cdot 8, 4 \cdot 8, 5 \cdot 8, 6 \cdot 8, \ldots$
8,   16,   24,   32,   40,   48, ...

Compare these multiples. The least number that is common to both sets of multiples is 24. Thus, 24 is the LCM of 6 and 8.

Prime factorization can also be used to find the LCM.

1. Find the prime factorization of each number.

   6 = 2 · 3
   8 = 2 · 2 · 2

2. Use each prime factor the greater number of times it appears in either factorization.

   2 appears as a factor three times in 8.
   3 appears as a factor one time in 6.

Therefore, the LCM of 6 and 8 is $2 \cdot 2 \cdot 2 \cdot 3$, or 24.

Thus, the LCD of $\frac{1}{6}$ and $\frac{3}{8}$ is 24. Rename $\frac{1}{6}$ and $\frac{3}{8}$ so that they have denominators of 24.

$$\frac{1}{6} + \frac{3}{8} = \frac{1}{6} \cdot \frac{4}{4} + \frac{3}{8} \cdot \frac{3}{3} \qquad \frac{1}{6} = \frac{?}{24} \text{ and } \frac{3}{8} = \frac{?}{24}$$

$$= \frac{4}{24} + \frac{9}{24}$$

$$= \frac{13}{24}$$

Similarly, to add or subtract rational expressions with different denominators, find the LCM of the denominators.

1 Find the LCM of $12x^2y$ and $15x^2y^2$.

$12x^2y = 2 \cdot 2 \cdot 3 \cdot x \cdot x \cdot y$ *Factor each expression.*

$15x^2y^2 = 3 \cdot 5 \cdot x \cdot x \cdot y \cdot y$

$\text{LCM} = 2 \cdot 2 \cdot 3 \cdot 5 \cdot x \cdot x \cdot y \cdot y$ *Use each factor the greater number of times it appears*

       $= 60x^2y^2$ *in either factorization.*

2 Find the LCM of $x^2 + x - 2$ and $x^2 + 5x - 6$.

$x^2 + x - 2 = (x - 1)(x + 2)$ *Factor each expression.*

$x^2 + 5x - 6 = (x + 6)(x - 1)$ *Use each factor the greater number of times*

      $\text{LCM} = (x - 1)(x + 2)(x + 6)$ *it appears in either factorization.*

After finding the LCM, rename the fractions so that the denominators are alike. Then, add or subtract the numerators.

3 Find $\dfrac{6}{5x} + \dfrac{7}{10x^2}$ and simplify.

$5x = 5 \cdot x$ *Use each factor the greater*

$10x^2 = 2 \cdot 5 \cdot x \cdot x$ *number of times it appears.*

The LCD for $\dfrac{6}{5x}$ and $\dfrac{7}{10x^2}$ is $2 \cdot 5 \cdot x \cdot x$ or $10x^2$. Since the denominator of $\dfrac{7}{10x^2}$ is already $10x^2$,

only $\dfrac{6}{5x}$ needs to be renamed.

$\dfrac{6}{5x} + \dfrac{7}{10x^2} = \dfrac{6}{5x} \cdot \dfrac{2x}{2x} + \dfrac{7}{10x^2}$ *Why do you multiply $\dfrac{6}{5x}$ by $\dfrac{2x}{2x}$?*

            $= \dfrac{12x}{10x^2} + \dfrac{7}{10x^2}$

            $= \dfrac{12x + 7}{10x^2}$

4 Find $\dfrac{a}{a^2 - 4} - \dfrac{4}{a + 2}$ and simplify.

Since $a^2 - 4 = (a - 2)(a + 2)$ and $a + 2 = (a + 2)$, the LCD for $\dfrac{a}{a^2 - 4}$ and $\dfrac{4}{a + 2}$ is $(a - 2)(a + 2)$, or $a^2 - 4$.

$\dfrac{a}{a^2 - 4} - \dfrac{4}{a + 2} = \dfrac{a}{(a - 2)(a + 2)} - \dfrac{4}{(a + 2)} \cdot \dfrac{(a - 2)}{(a - 2)}$

                $= \dfrac{a - 4(a - 2)}{(a - 2)(a + 2)}$ $a - 4(a - 2) = a - 4a + 8$

                $= \dfrac{-3a + 8}{a^2 - 4}$

5 Find $\dfrac{x+4}{(2-x)(x+3)} + \dfrac{x-5}{(x-2)^2}$ and simplify.

Multiply the first fraction by $\dfrac{-1}{-1}$ to change $(2-x)$ to $(x-2)$.

$$\dfrac{x+4}{(2-x)(x+3)} + \dfrac{x-5}{(x-2)^2} = \dfrac{-1}{-1} \cdot \dfrac{(x+4)}{(2-x)(x+3)} + \dfrac{x-5}{(x-2)^2}$$

$$= \dfrac{-(x+4)}{(x-2)(x+3)} + \dfrac{x-5}{(x-2)^2}$$

The LCD for $(x-2)(x+3)$ and $(x-2)^2$ is $(x+3)(x-2)(x-2)$. *Why?*

$$\dfrac{-(x+4)}{(x-2)(x+3)} + \dfrac{x-5}{(x-2)^2} = \dfrac{-(x+4)}{(x-2)(x+3)} \cdot \dfrac{(x-2)}{(x-2)} + \dfrac{(x-5)}{(x-2)^2} \cdot \dfrac{(x+3)}{(x+3)}$$

$$= \dfrac{-x^2 - 2x + 8 + x^2 - 2x - 15}{(x+3)(x-2)(x-2)}$$

$$= \dfrac{-4x - 7}{(x+3)(x-2)^2}$$

Exploratory Exercises

State the LCD for each pair of fractions.

1. $\dfrac{3}{8}, \dfrac{5}{12}$

2. $\dfrac{6}{a}, \dfrac{7}{13}$

3. $\dfrac{6}{11a}, \dfrac{2}{5}$

4. $\dfrac{2}{a^2}, \dfrac{5}{7}$

5. $\dfrac{2}{a}, \dfrac{3}{b}$

6. $\dfrac{4}{a^2}, \dfrac{5}{a}$

7. $\dfrac{6}{b^3}, \dfrac{7}{ab}$

8. $\dfrac{4}{b^4}, \dfrac{5}{b^5}$

9. $\dfrac{3}{20a^2}, \dfrac{1}{24ab^3}$

10. $\dfrac{7}{60x^2y^2}, \dfrac{6}{35xz^3}$

11. $\dfrac{1}{12an^2}, \dfrac{3}{40a^4}$

12. $\dfrac{11}{56x^3y}, \dfrac{10}{49ax^2}$

13. $\dfrac{7}{a+5}, \dfrac{a}{a-3}$

14. $\dfrac{m}{m+n}, \dfrac{6}{n}$

15. $\dfrac{x+5}{3x-6}, \dfrac{x-3}{x-2}$

Written Exercises

Find each sum or difference. Simplify.

16. $\dfrac{3}{4} + \dfrac{1}{8}$

17. $\dfrac{5}{7} + \dfrac{3}{14}$

18. $\dfrac{3}{4} + \dfrac{2}{9}$

19. $\dfrac{2}{3} + \dfrac{-7}{8}$

20. $\dfrac{-1}{4} - \dfrac{1}{8}$

21. $\dfrac{7}{10} - \dfrac{3}{20}$

22. $\dfrac{6}{7} - \dfrac{2}{11}$

23. $\dfrac{-3}{12} - \dfrac{-5}{18}$

24. $\dfrac{8}{11} + \dfrac{y}{4}$

25. $\dfrac{t}{3} + \dfrac{2t}{7}$

26. $\dfrac{x}{7} - \dfrac{4}{5}$

27. $\dfrac{2n}{5} - \dfrac{3m}{4}$

28. $\dfrac{5}{2a} + \dfrac{-3}{6a}$

29. $\dfrac{6}{5} + \dfrac{a}{5b}$

30. $\dfrac{7}{3a} - \dfrac{3}{6a^2}$

31. $\dfrac{7}{a} - \dfrac{x+1}{2a}$

32. $\dfrac{5b}{7x} + \dfrac{3a}{21x^2}$

33. $\dfrac{5}{xy} + \dfrac{6}{yz}$

34. $\dfrac{3z}{7w^2} - \dfrac{2z}{w}$

35. $\dfrac{6}{x} - \dfrac{5}{x^2}$

36. $\dfrac{2}{t} + \dfrac{t+3}{s}$

37. $\dfrac{7}{a} - \dfrac{x-1}{b}$

38. $\dfrac{a}{1(a+b)} + \dfrac{6}{b}$

39. $\dfrac{m}{1(m-n)} - \dfrac{5}{m}$

40. $\dfrac{4a}{2a+6} + \dfrac{3}{a+3}$

41. $\dfrac{5}{3x-9} + \dfrac{3}{x-3}$

42. $\dfrac{2x}{x^2-5x} - \dfrac{-3x}{x-5}$

43. $\dfrac{-3}{a-5} + \dfrac{-6}{a^2-5a}$

44. $\dfrac{b+8}{b^2-64} + \dfrac{1}{b-8}$

45. $\dfrac{2y}{y^2-25} + \dfrac{y+5}{y-5}$

46. $\dfrac{m-n}{m+n} - \dfrac{1}{m^2-n^2}$

47. $\dfrac{3a+2}{3a-6} - \dfrac{a+2}{a^2-4}$

48. $\dfrac{a^2}{a^2-b^2} + \dfrac{a}{(a-b)^2}$

49. $\dfrac{x^2}{4x^2-9} + \dfrac{x}{(2x+3)^2}$

50. $\dfrac{x}{x^2+2x+1} + \dfrac{1}{x+1}$

51. $\dfrac{y}{y^2-2y+1} - \dfrac{1}{y-1}$

52. $\dfrac{x^2-1}{x+1} + \dfrac{x^2+1}{x-1}$

53. $\dfrac{a}{a-b} + \dfrac{b}{2b+3a}$

54. $\dfrac{k}{2k+1} - \dfrac{2}{k+2}$

55. $\dfrac{m-1}{m+1} + \dfrac{4}{2m+5}$

56. $\dfrac{-18}{y^2-9} + \dfrac{7}{3-y}$

57. $\dfrac{-3}{5-a} + \dfrac{5}{a^2-25}$

58. $\dfrac{a-1}{4ab} + \dfrac{a^2-a}{16b^2}$

59. $\dfrac{x-1}{3xy} - \dfrac{x^2-x}{9y^2}$

60. $\dfrac{6}{a^2-2ab+b^2} - \dfrac{6}{a-b}$

61. $\dfrac{x^2+4x-5}{x^2-2x-3} + \dfrac{2}{x+1}$

62. $\dfrac{3}{m-2} + \dfrac{2}{2-m}$

63. $\dfrac{4}{5-p} - \dfrac{3}{p-5}$

64. $\dfrac{a+2}{a^2-9} - \dfrac{2a}{6a^2-17a-3}$

65. $\dfrac{a-2}{a^2+4a+4} + \dfrac{a+2}{a-2}$

66. $\dfrac{3m}{m^2+3m+2} - \dfrac{3m-6}{m^2+4m+4}$

67. $\dfrac{4a}{6a^2-a-2} - \dfrac{5a+1}{2-3a}$

68. $\dfrac{2x+1}{(x-1)^2} + \dfrac{x-2}{(1-x)(x+4)}$

69. $\dfrac{a+3}{3a^2-10a-8} + \dfrac{2a}{a^2-8a+16}$

70. $\dfrac{m-n}{m^2+2mn+n^2} - \dfrac{m+n}{m-n}$

Excursions in Algebra Evaluating Sums and Differences

Use a calculator to evaluate each sum or difference for the values given.

1. $\dfrac{3}{x-2} + \dfrac{4}{x^2-4}$; $x = 2.1$

2. $\dfrac{5y}{4-w^2} - \dfrac{3y}{w}$; $y = 0.2$, $w = -0.1$

3. $\dfrac{m^2-1}{m+1} + \dfrac{m^2+1}{m-1}$; $m = 0.5$

4. $\dfrac{r-s}{r^2+4rs+4} - \dfrac{r+2s}{r-s}$; $r = -0.2$, $s = -0.1$

14-7 Mixed Expressions and Complex Fractions

Algebraic expressions such as $a + \frac{b}{c}$, and $5 + \frac{x-y}{x+3}$ are called **mixed expressions**. Changing mixed expressions to rational expressions is similar to changing mixed numbers to improper fractions.

Mixed expressions contain monomials and algebraic fractions.

Mixed Number to Improper Fraction

$$3\tfrac{2}{5} \text{ or } 3 + \frac{2}{5} = \frac{3(5) + 2}{5}$$

$$= \frac{15 + 2}{5}$$

$$= \frac{17}{5}$$

Mixed Expression to Rational Expression

$$a + \frac{a^2 + b}{a - b} = \frac{a(a - b) + (a^2 + b)}{a - b}$$

$$= \frac{a^2 - ab + a^2 + b}{a - b}$$

$$= \frac{2a^2 - ab + b}{a - b}$$

Example

1 Find $8 + \frac{x^2 - y^2}{x^2 + y^2}$ and simplify.

Notice that $8 + \frac{x^2 - y^2}{x^2 + y^2}$ is a mixed expression. Therefore, use the method shown above to find the sum.

$$8 + \frac{x^2 - y^2}{x^2 + y^2} = \frac{8(x^2 + y^2) + (x^2 - y^2)}{x^2 + y^2}$$

$$= \frac{8x^2 + 8y^2 + x^2 - y^2}{x^2 + y^2}$$

$$= \frac{9x^2 + 7y^2}{x^2 + y^2}$$

If a fraction has one or more fractions in the numerator *or* denominator, it is called a **complex fraction**. Some complex fractions are shown below.

$$\frac{3\tfrac{1}{2}}{5\tfrac{2}{3}} \qquad \frac{8}{\frac{a}{b}} \qquad \frac{\frac{a+b}{a}}{\frac{a-b}{b}} \qquad \frac{\frac{1}{x} - \frac{1}{y}}{\frac{1}{x} + \frac{1}{y}}$$

Consider the complex fraction $\dfrac{\frac{3}{5}}{\frac{7}{8}}$.

To simplify this fraction, rewrite it as $\frac{3}{5} \div \frac{7}{8}$ and proceed as follows.

$$\frac{3}{5} \div \frac{7}{8} = \frac{3}{5} \cdot \frac{8}{7} \text{ or } \frac{24}{35}$$

Recall that to find the quotient, you multiply by $\frac{8}{7}$, the reciprocal of $\frac{7}{8}$.

Similarly, to simplify $\dfrac{\frac{a}{b}}{\frac{c}{d}}$, rewrite it as $\frac{a}{b} \div \frac{c}{d}$ and proceed as follows.

$$\frac{a}{b} \div \frac{c}{d} = \frac{a}{b} \cdot \frac{d}{c} \text{ or } \frac{ad}{bc}$$

The reciprocal of $\frac{c}{d}$ is $\frac{d}{c}$.

Any complex fraction $\dfrac{\frac{a}{b}}{\frac{c}{d}}$, where $b \neq 0$, $c \neq 0$, and $d \neq 0$, may be expressed as $\frac{ad}{bc}$.

Simplifying Complex Fractions

Example

2 Simplify $\dfrac{\frac{1}{x} + \frac{1}{y}}{\frac{1}{x} - \frac{1}{y}}$.

$$\dfrac{\frac{1}{x} + \frac{1}{y}}{\frac{1}{x} - \frac{1}{y}} = \dfrac{\frac{y}{xy} + \frac{x}{xy}}{\frac{y}{xy} - \frac{x}{xy}}$$

The LCD of the numerator $\frac{1}{x} + \frac{1}{y}$ and the denominator $\frac{1}{x} - \frac{1}{y}$ is xy.

$$= \dfrac{\frac{y + x}{xy}}{\frac{y - x}{xy}}$$

Add to simplify the numerator.

Subtract to simplify the denominator.

$$= \frac{y + x}{xy} \cdot \frac{xy}{y - x}$$

The reciprocal of $\frac{y - x}{xy}$ is $\frac{xy}{y - x}$.

$$= \frac{y + x}{\cancel{xy}} \cdot \frac{\cancel{xy}}{y - x}$$

Eliminate common factors.

$$= \frac{y + x}{y - x}$$

3 Simplify $\dfrac{x + 4 - \dfrac{1}{x + 4}}{x + 11 + \dfrac{48}{x - 3}}$.

4 Simplify $\dfrac{x - \dfrac{x + 4}{x + 1}}{x - 2}$.

$\dfrac{x + 4 - \dfrac{1}{x + 4}}{x + 11 + \dfrac{48}{x - 3}}$

$= \dfrac{\dfrac{(x + 4)(x + 4) - 1}{x + 4}}{\dfrac{(x + 11)(x - 3) + 48}{x - 3}}$

$= \dfrac{\dfrac{x^2 + 8x + 16 - 1}{x + 4}}{\dfrac{x^2 + 8x - 33 + 48}{x - 3}}$

$= \dfrac{\dfrac{x^2 + 8x + 15}{x + 4}}{\dfrac{x^2 + 8x + 15}{x - 3}}$

$= \dfrac{x^2 + 8x + 15}{x + 4} \cdot \dfrac{x - 3}{x^2 + 8x + 15}$

$= \dfrac{x - 3}{x + 4}$

$\dfrac{x - \dfrac{x + 4}{x + 1}}{x - 2}$

$= \dfrac{\dfrac{x(x + 1) - (x + 4)}{x + 1}}{x - 2}$

$= \dfrac{\dfrac{x^2 + x - x - 4}{x + 1}}{x - 2}$

$= \dfrac{\dfrac{x^2 - 4}{x + 1}}{\dfrac{x - 2}{1}}$

$= \dfrac{x^2 - 4}{x + 1} \cdot \dfrac{1}{x - 2}$

$= \dfrac{(x + 2)(x - 2)}{(x + 1)(x - 2)}$

$= \dfrac{x + 2}{x + 1}$

Exploratory Exercises

Find each sum and simplify.

1. $4 + \dfrac{2}{x}$

2. $8 + \dfrac{5}{3y}$

3. $x + \dfrac{x}{y}$

4. $z + \dfrac{2z}{w}$

5. $2m + \dfrac{4 + m}{m}$

6. $3a + \dfrac{a + 1}{2a}$

7. $b^2 + \dfrac{2}{b - 2}$

8. $3r^2 + \dfrac{4}{2r + 1}$

Written Exercises

Simplify.

9. $\dfrac{3\frac{1}{2}}{4\frac{3}{4}}$

10. $\dfrac{\dfrac{x^2}{y}}{\dfrac{y}{x^3}}$

11. $\dfrac{\dfrac{x + 4}{y - 2}}{\dfrac{x^2}{y^3}}$

12. $\dfrac{\dfrac{x^3}{y^2}}{\dfrac{x + y}{x - y}}$

13. $\dfrac{\dfrac{x+y}{a+b}}{\dfrac{x^2-y^2}{a^2-b^2}}$

14. $\dfrac{\dfrac{x-y}{x+y}}{\dfrac{x+y}{x-y}}$

15. $\dfrac{\dfrac{1}{x}+\dfrac{1}{y}}{\dfrac{1}{y}-\dfrac{1}{x}}$

16. $\dfrac{\dfrac{a+b}{x}}{\dfrac{a-b}{y}}$

17. $\dfrac{\dfrac{x^2+8x+15}{x^2+x-6}}{\dfrac{x^2+2x-15}{x^2-2x-3}}$

18. $\dfrac{\dfrac{a^2-6a+5}{a^2+13a+42}}{\dfrac{a^2-4a+3}{a^2+3a-18}}$

19. $\dfrac{\dfrac{y^2-1}{y^2+3y-4}}{y+1}$

20. $\dfrac{\dfrac{a^2-2a-3}{a^2-1}}{a-3}$

21. $\dfrac{\dfrac{a^2+2a}{a^2+9a+18}}{\dfrac{a^2-5a}{a^2+a-30}}$

22. $\dfrac{\dfrac{x^2+4x-21}{x^2-9x+18}}{\dfrac{x^2+3x-28}{x^2-10x+24}}$

23. $\dfrac{x-\dfrac{15}{x-2}}{x-\dfrac{20}{x-1}}$

24. $\dfrac{m+\dfrac{35}{m+12}}{m-\dfrac{63}{m-2}}$

25. $7+\dfrac{x^2+y^2}{x^2-4y^2}$

26. $5+\dfrac{a^2+11}{a^2-1}$

27. $\dfrac{x+2+\dfrac{2}{x+5}}{x+6+\dfrac{6}{x+1}}$

28. $\dfrac{x+5+\dfrac{3}{x+1}}{x-1-\dfrac{3}{x+1}}$

mini-review

Solve each system of equations.

1. $3x + 2y = 5$
$7x + 3y = 0$

2. $3x - 5y = -18$
$9x - 2y = -2$

Simplify.

3. $12\sqrt{3} - 3\sqrt{12}$

4. $\dfrac{6}{3-\sqrt{3}}$

5. $\sqrt{3}(\sqrt{21} - \sqrt{6})$

6. $\sqrt{147}$

Solve each equation.

7. $4x^2 - 8x + 3 = 0$

8. $x^2 - 6x + 3 = 0$

9. Use substitution to solve the following system of equations.
$$2x + 5y = -1$$
$$x + 2y = 2$$

10. Solve $2x^2 + 12x + 13 = 0$ by completing the square.

11. The sum of two numbers is 116. Their difference is 18. Find the numbers.

12. The difference of two integers is 6. Their product is 135. Find the numbers.

Find each sum or difference. Simplify.

13. $\dfrac{a}{a^2-b^2} + \dfrac{b}{a^2-b^2}$

14. $\dfrac{x}{x+3} - \dfrac{1}{x-1}$

14-8 Solving Rational Equations

Recall that you can solve equations containing fractions by using the least common denominator of all the fractions in the equation. The fractions are eliminated by multiplying each side of the equation by the common denominator. This method can also be used with **rational equations**. Rational equations are equations containing rational expressions.

Examples

1 Solve $\frac{x-4}{4} + \frac{x}{3} = 6$.

$$\frac{x-4}{4} + \frac{x}{3} = 6 \qquad \text{\textit{The LCD of the fractions is 12.}}$$

$$12\left(\frac{x-4}{4} + \frac{x}{3}\right) = 12(6) \qquad \textit{Multiply each side of the equation by 12.}$$

$$3(x-4) + 4(x) = 72 \qquad \textit{The fractions are eliminated.}$$

$$3x - 12 + 4x = 72$$

$$7x = 84$$

$$x = 12$$

Check: $\quad \frac{x-4}{4} + \frac{x}{3} = 6$

$$\frac{12-4}{4} + \frac{12}{3} \overset{?}{=} 6$$

$$2 + 4 \overset{?}{=} 6$$

$$6 = 6 \quad \checkmark$$

The solution is 12.

2 Solve $\frac{3}{2x} - \frac{2x}{x+1} = -2$.

$$\frac{3}{2x} - \frac{2x}{x+1} = -2 \qquad \textit{Note that } x \neq -1 \textit{ and } x \neq 0. \\ \textit{The LCD of the fractions is } 2x(x+1).$$

$$2x(x+1)\left(\frac{3}{2x} - \frac{2x}{x+1}\right) = 2x(x+1)(-2) \qquad \textit{Multiply each side of the equation by } 2x(x+1).$$

$$3(x+1) - 2x(2x) = -4x^2 - 4x$$

$$3x + 3 - 4x^2 = -4x^2 - 4x$$

$$7x = -3$$

$$x = -\frac{3}{7} \qquad \textit{Check this result.}$$

The solution is $-\frac{3}{7}$.

3 Solve $x - \dfrac{2}{x-3} = \dfrac{x-1}{3-x}$.

$$x - \frac{2}{x-3} = \frac{x-1}{3-x} \qquad \begin{array}{l} x \neq 3 \\ \text{Note that } 3 - x = -(x - 3). \end{array}$$

$$x - \frac{2}{x-3} = -\frac{x-1}{(x-3)} \qquad \text{The LCD is } (x - 3).$$

$$(x-3)\left(x - \frac{2}{x-3}\right) = (x-3)\left(-\frac{x-1}{x-3}\right) \qquad \begin{array}{l} \text{Multiply each side of} \\ \text{the equation by } (x - 3). \end{array}$$

$$x(x-3) - 2 = -(x-1)$$

$$x^2 - 3x - 2 = -x + 1$$

$$x^2 - 2x - 3 = 0$$

$$(x-3)(x+1) = 0 \qquad \textit{Factor.}$$

$$x - 3 = 0 \quad \text{or} \quad x + 1 = 0 \qquad \textit{Zero Product Property}$$

$$x = 3 \quad \text{or} \quad x = -1$$

Since x cannot equal 3, the only solution is -1.

4 Solve $\dfrac{2m}{m-1} + \dfrac{m-5}{m^2-1} = 1$.

$$\frac{2m}{m-1} + \frac{m-5}{m^2-1} = 1 \qquad \textit{What values for m are excluded?}$$

$$\frac{2m}{m-1} + \frac{m-5}{(m+1)(m-1)} = 1 \qquad \begin{array}{l} \textit{Factor } m^2 - 1. \\ \textit{The LCD is } (m+1)(m-1). \end{array}$$

$$(m+1)(m-1)\left(\frac{2m}{m-1} + \frac{m-5}{(m+1)(m-1)}\right) = (m+1)(m-1) \cdot 1$$

$$2m(m+1) + (m-5) = m^2 - 1$$

$$2m^2 + 2m + m - 5 = m^2 - 1$$

$$m^2 + 3m - 4 = 0$$

$$(m+4)(m-1) = 0 \qquad \textit{Factor.}$$

$$m + 4 = 0 \quad \text{or} \quad m - 1 = 0 \qquad \textit{Zero Product Property}$$

$$m = -4 \quad \text{or} \quad m = 1$$

The solution is -4. Why is 1 not a solution?

Exploratory Exercises

State the LCD for each pair of fractions.

1. $\dfrac{m}{2}, \dfrac{m}{3}$

2. $\dfrac{x}{5}, \dfrac{2x}{3}$

3. $\dfrac{3b}{4}, \dfrac{5b}{8}$

4. $\dfrac{1}{x}, \dfrac{5x}{x+1}$

5. $\dfrac{4}{r^2-1}, \dfrac{5}{r-1}$

6. $\dfrac{m}{2m^2+3m-35}, \dfrac{8}{2m-7}$

State the LCD for each set of fractions.

7. $\frac{6}{x}, \frac{7}{x-1}, \frac{1}{4}$

8. $\frac{5}{k-1}, \frac{7}{k}, \frac{1}{k+1}$

9. $\frac{5k}{k+5}, \frac{k^2}{k+3}, \frac{1}{k+3}$

10. $\frac{7}{h+1}, \frac{1}{2}, \frac{2h+5}{h-1}$

11. $\frac{3x}{x^2-1}, \frac{1}{x+1}, \frac{1}{x-1}$

12. $\frac{2x+1}{4x^2-1}, \frac{1}{3}, \frac{1}{2x+1}$

Written Exercises

Solve each equation.

13. $\frac{2a-3}{6} = \frac{2a}{3} + \frac{1}{2}$

14. $\frac{3x}{5} + \frac{3}{2} = \frac{7x}{10}$

15. $\frac{3a}{2} + \frac{5}{4} = \frac{5a}{2}$

16. $\frac{2b-3}{7} - \frac{b}{2} = \frac{b+3}{14}$

17. $\frac{x+1}{x} + \frac{x+4}{x} = 6$

18. $\frac{18}{b} = \frac{3}{b} + 3$

19. $\frac{3}{5x} + \frac{7}{2x} = 1$

20. $\frac{11}{2x} - \frac{2}{3x} = \frac{1}{6}$

21. $\frac{5x}{x+1} + \frac{1}{x} = 5$

22. $\frac{5k}{k+2} + \frac{2}{k} = 5$

23. $\frac{2}{3r} - \frac{3r}{r-2} = -3$

24. $\frac{3n}{n+2} - \frac{5}{7} = 4$

25. $\frac{m}{m+1} + \frac{5}{m-1} = 1$

26. $\frac{r-1}{r+1} - \frac{2r}{r-1} = -1$

27. $\frac{4x}{2x+3} - \frac{2x}{2x-3} = 1$

28. $\frac{4x}{3x-2} + \frac{2x}{3x+2} = 2$

29. $\frac{c}{c-4} - \frac{6}{4-c} = c$

30. $\frac{5}{z-3} - \frac{z}{3-z} = z$

31. $\frac{5}{5-p} - \frac{p^2}{5-p} = -2$

32. $\frac{r^2}{r-7} + \frac{50}{7-r} = 14$

33. $\frac{14}{b-6} = \frac{1}{2} + \frac{6}{b-8}$

34. $\frac{2a-3}{a-3} - 2 = \frac{12}{a+3}$

35. $\frac{r}{3r+6} - \frac{r}{5r+10} = \frac{2}{5}$

36. $\frac{x-2}{x} - \frac{x-3}{x-6} = \frac{1}{x}$

37. $\frac{2b-5}{b-2} - 2 = \frac{3}{b+2}$

38. $\frac{z+3}{z-1} + \frac{z+1}{z-3} = 2$

39. $\frac{7}{k-3} - \frac{1}{2} = \frac{3}{k-4}$

40. $\frac{x+2}{x-2} - \frac{2}{x+2} = \frac{-7}{3}$

41. $\frac{3w}{w^2-5w+4} = \frac{2}{w-4} + \frac{3}{w-1}$

42. $\frac{7}{x^2-5x} + \frac{3}{5-x} = \frac{4}{x}$

43. $\frac{9}{b^2-7b+12} = \frac{5}{b-3} + \frac{2}{b-4}$

44. $\frac{6}{z+2} + \frac{3}{z^2-4} = \frac{2z-7}{z-2}$

45. $\frac{m+3}{m+5} + \frac{2}{m-9} = \frac{-20}{m^2-4m-45}$

46. $\frac{4}{k^2-8k+12} = \frac{k}{k-2} + \frac{1}{k-6}$

47. $\frac{h^2-7h-8}{3h^2+2h-8} + \frac{1}{h+2} = 0$

Challenge Exercise

48. The measures of the sides of a triangle are $\frac{x}{x+1}, \frac{x+2}{x+3}$, and $\frac{3}{3x-1}$. If the measure of its perimeter is $\frac{2x^2+7x+8}{x^2+4x+3}$, find the value of x.

14-9 Problem Solving: Work and Uniform Motion

Katina Marsh can wash and wax cars at the rate of 1 every 3 hours. Working at this rate, in 1 hour she can complete $\frac{1}{3}$ of the job. In 2 hours she can complete $\frac{1}{3} \cdot 2$ or $\frac{2}{3}$ of the job. In t hours, she can complete $\frac{1}{3} \cdot t$ or $\frac{t}{3}$ of the job.

This and many other similar examples suggest the following formula.

$$(\text{rate of working}) \cdot (\text{time}) = (\text{work done})$$
$$r \qquad \cdot \qquad t \quad = \qquad w$$

This formula is applied in the example below.

Example

1 **Katina Marsh washes and waxes 1 car every 3 hours. Toshio Meko washes and waxes 1 car in 4 hours. If Katina and Toshio work together, find how long it will take them to wash and wax one car.**

explore Let t = time in hours for Katina and Toshio to wash and wax one car.

plan In t hours, Katina can do $\frac{1}{3} \cdot t$ or $\frac{t}{3}$ of the job.

In t hours, Toshio can do $\frac{1}{4} \cdot t$ or $\frac{t}{4}$ of the job.

	r	$\cdot$ t	$=$ w
Katina	$\frac{1}{3}$	t	$\frac{t}{3}$
Toshio	$\frac{1}{4}$	t	$\frac{t}{4}$

$\frac{t}{3} + \frac{t}{4} = 1$ *Together they complete 1 job.*

solve
$4t + 3t = 12$ *Multiply each side of the equation by the LCD, 12.*
$7t = 12$
$t = \frac{12}{7}$

Katina and Toshio can do the job in $\frac{12}{7}$, or $1\frac{5}{7}$ hours.   *This is about 1 hour and 43 minutes.*

examine Katina does $\frac{1}{3}t$ or $\frac{1}{3} \cdot \frac{12}{7}$ of the job. $\frac{1}{3} \cdot \frac{12}{7} = \frac{4}{7}$

Toshio does $\frac{1}{4}t$ or $\frac{1}{4} \cdot \frac{12}{7}$ of the job. $\frac{1}{4} \cdot \frac{12}{7} = \frac{3}{7}$

Does $\frac{4}{7} + \frac{3}{7}$ equal 1?

Recall that uniform motion problems can be solved by using a formula similar to the one used to solve work problems.

(rate) $\cdot$ (time) = (distance)
r $\cdot$ t = d

Example

2 While on a fishing trip, Sally and her brother rented a boat. The maximum speed of the boat in still water was 3 miles per hour. At this rate, a 9-mile trip downstream with the current took the same amount of time as a 3-mile trip upstream against the current. What was the rate of the current?

explore

Let c = the rate of the current.
The rate of the boat when traveling downstream, or *with the current*, is 3 mph *plus* the rate of the current. That is, $3 + c$. The rate when traveling upstream, or *against the current*, is 3 mph *minus* the rate of the current. That is, $3 - c$.

plan

	r	$\cdot$ t	$=$ d
downstream	$3 + c$	$\dfrac{9}{3 + c}$	9
upstream	$3 - c$	$\dfrac{3}{3 - c}$	3

To represent the time t, solve $rt = d$ for t.

Thus, $t = \dfrac{d}{r}$.

solve

$$\frac{9}{3 + c} = \frac{3}{3 - c}$$ *time downstream = time upstream*

$$9(3 - c) = 3(3 + c)$$ *Multiply each side of the equation by the LCD, $(3 + c)(3 - c)$.*

$$27 - 9c = 9 + 3c$$

$$-12c = -18$$

$$c = \frac{3}{2}$$

The rate of the current is $\frac{3}{2}$ or $1\frac{1}{2}$ miles per hour.   *Examine this solution.*

Exploratory Exercises

Answer each question.

1. Luisa can paint her house in 8 days. What part of it will she paint: **a.** in 1 day?   **b.** in 3 days?   **c.** in x days?

2. Drew can build a garage in n days. What part of it will he build: **a.** in 1 day?   **b.** in 4 days?   **c.** in x days?

3. Dimas can do a job alone in 8 days. Stephen can do the same job alone in 10 days. What part of the job can Dimas do: **a.** in 1 day?   **b.** in x days?   What part of the job can Stephen do: **c.** in 1 day?   **d.** in x days? What part of the job can they do together: **e.** in 1 day? **f.** in x days?

4. Pat can do a job alone in 4 days. Mini can do the same job alone in 12 days. What part of the job can Pat do: **a.** in 1 day?   **b.** in x days?   What part of the job can Mini do: **c.** in 1 day? **d.** in x days?   What part of the job can they do together: **e.** in 1 day? **f.** in x days?

5. The top flying speed of an open cockpit biplane is 120 mph. At this speed, a 420-mile trip flying with the wind takes the same amount of time as a 300-mile trip flying against the wind. What is the speed of the wind?

a. Let s = the speed of the wind. Copy and fill in each entry in the chart at the right.

b. Write an equation to solve this problem.

	r	$\cdot$	t	$=$	d
with the wind					
against the wind					

Written Exercises

Solve each problem. Use $rt = w$ or $rt = d$ and a chart.

6. Jane can wash the windows of a building in 4 hours. Jim can do the same job in 6 hours. If they work together, how long will it take them to wash the windows?

7. Mark can clean the garage in 6 hours. Rosetta can do the same job in 8 hours. If they work together, how long will it take them to clean the garage?

8. Helena can do a job in 5 days. Jefferson can do the same job in 8 days. If they work together, how long will it take them to complete the job?

9. Frank can do a job in 10 hours. Keith can do the same job in 15 hours. If they work together, how long will it take them to complete the job?

10. A swimming pool can be filled by one pipe in 12 hours and by another pipe in 4 hours. How long will it take to fill the pool if the water flows through both pipes?

11. A swimming pool can be filled by one pipe in 10 hours. The drain pipe can empty the pool in 15 hours. If both pipes are open, how long will it take to fill the pool?

12. Hugo and Denise together can mow a lawn in 12 minutes. It takes Hugo 20 minutes to do the job alone. How long would it take Denise to do the job alone?

13. Bernice and Erica together can do a job in $3\frac{3}{5}$ hours. Bernice can do the job alone in 6 hours. How many hours will it take Erica to do the job alone?

14. A long distance cyclist pedaling at a steady rate travels 30 miles with the wind. He can travel only 18 miles against the wind in the same amount of time. If the rate of the wind is 3 mph, what is the cyclist's rate without the wind?

15. A tugboat pushing a barge up the Ohio River takes 1 hour longer to travel 36 miles up the river than to travel the same distance down the river. If the rate of the current is 3 mph, find the speed of the tugboat and barge in still water.

16. An airplane can fly at a rate of 600 mph in calm air. It can fly 2520 miles with the wind in the same time it can fly 2280 miles against the wind. Find the speed of the wind.

17. A motorboat takes $\frac{2}{3}$ as much time to travel 10 miles downstream as it does to travel the same distance upstream. If the rate of the current is 5 mph, find the speed of the motorboat in still water.

Challenge Exercise

18. An aircraft is flying from Tokyo to Honolulu, a distance of 3860 miles. In still air, the aircraft flies at 600 mph. It has a 75 mph tailwind. At what point would it be quicker to go on to Honolulu than return to Tokyo in case of emergency? Give your answer in terms of distance from Tokyo.

To make good pictures, photographers must make sure the correct amount of light enters the camera. The light enters the camera through an adjustable opening. The sizes of the opening are called *f*-stops. The *f*-stops are determined by dividing the focal length of the lens by the diameter of the opening. For example, suppose a camera lens has a focal length of 50 mm. The diameter of the opening is 6.25 mm.

$$f\text{-stop} = \frac{\text{focal length}}{\text{diameter of opening}}$$

$$= \frac{50 \text{ mm}}{6.25 \text{ mm}}$$

$$= 8 \quad \text{The } f\text{-stop is 8.}$$

The dial that shows *f*-stops has the numbers 1.4, 2, 2.8, 4, 5.6, 8, 11, and 16. Each *f*-stop lets twice as much light into the camera as the next *f*-stop. For example, *f*/4 lets in *twice* as much light as *f*/5.6.

Another dial marked 1, 2, 4, 8, 15, 30, 60, 125, 250, 500, and 1000 shows the shutter speed. The shutter speed, in seconds, is the multiplicative inverse of the number on the dial. For example, a shutter speed marked 30 means light enters the camera for $\frac{1}{30}$ of a second. Each shutter speed lets light into the camera for twice as long as the next setting.

Suppose you want to photograph a parked sports car on a sunny day. The setting *f*/11 at $\frac{1}{60}$ would allow the correct amount of light into the camera. For a moving car, you need a faster shutter speed, such as $\frac{1}{125}$. At $\frac{1}{125}$, only half the amount of light is allowed into the camera. For proper exposure, the *f*-stop must be increased to *f*/8.

Exercises

A lens has a focal length of 200 mm. Find the *f*-stop for each opening whose diameter is given.

1. 50 mm　　　　**2.** 25 mm　　　　**3.** 12.5 mm　　　　**4.** 36 mm

For each exposure combination, state two others that give the same exposure.

5. *f*/8 at $\frac{1}{60}$　　　**6.** *f*/5.6 at $\frac{1}{60}$　　　**7.** *f*/2.8 at $\frac{1}{125}$　　　**8.** *f*/4 at $\frac{1}{30}$

14-10 Using Formulas

Expressions and equations involving rational expressions often contain more than one variable. Sometimes it is useful to solve the equations or formulas for one of the variables. This is often true with formulas. Study the following example.

> ### Example
>
> **1** **Solve for n in $S = \frac{n}{2}(A + t)$.**
>
> $S = \frac{n}{2}(A + t)$ *The LCD is 2.*
>
> $2S = 2\left[\frac{n}{2}(A + t)\right]$
>
> $2S = n(A + t)$
>
> $\frac{2S}{A + t} = n$

The formula below applies to camera and lens systems. In the formula, f is the focal length of the lens, a is the distance from the object to the lens, and b is the distance from the image to the lens.

$$\frac{1}{f} = \frac{1}{a} + \frac{1}{b}$$

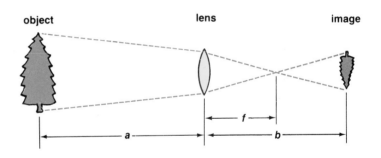

> ### Example
>
> **2** **Solve for f in $\frac{1}{f} = \frac{1}{a} + \frac{1}{b}$.**
>
> $\frac{1}{f} = \frac{1}{a} + \frac{1}{b}$
>
> $abf\left(\frac{1}{f}\right) = abf\left(\frac{1}{a} + \frac{1}{b}\right)$ *The LCD is abf.*
>
> $ab = bf + af$
> $ab = (b + a)f$ *Factor bf + af.*
>
> $\frac{ab}{b + a} = f$

Exploratory Exercises

Answers each question.

1. In Example 1, what steps would you take to solve for A?

2. In Example 1, what steps would you take to solve for t?

3. In Example 2, what steps would you take to solve for a?

4. In Example 2, what steps would you take to solve for b?

Written Exercises

Solve each motion formula for the variable indicated.

5. $a = \frac{v}{t}$, for t

6. $v = r + at$, for a

7. $s = vt + \frac{1}{2}at^2$, for v

8. $s = vt + \frac{1}{2}at^2$, for a

9. $F = G\left(\frac{Mm}{d^2}\right)$, for M

10. $F = G\left(\frac{Mm}{d^2}\right)$, for d

11. $f = \frac{W}{g} \cdot \frac{V^2}{R}$, for V

12. $f = \frac{W}{g} \cdot \frac{V^2}{R}$, for R

Solve each business formula for the variable indicated.

13. $A = P + Prt$, for P

14. $Prt = I$, for r

15. $I = \left(\frac{100 - P}{P}\right)\frac{365}{R}$, for P

16. $I = \frac{365d}{360 - dr}$, for d

17. $a = \frac{r}{2y} - 0.25$, for y

18. $c = \frac{P - 100}{P}$, for P

Solve each electronics formula for the variable indicated.

19. $H = (0.24)I^2Rt$, for R

20. $P = \frac{E^2}{R}$, for E

21. $\frac{1}{R_T} = \frac{1}{R_1} + \frac{1}{R_2}$, for R_1

22. $I = \frac{E}{r + R}$, for R

23. $I = \frac{nE}{nr + R}$, for n

24. $I = \frac{E}{\frac{r}{n} + R}$, for r

Solve each mathematical formula for the variable indicated.

25. $y = mx + b$, for m

26. $A = \frac{1}{2}h(a + b)$, for h

27. $m = \frac{y_2 - y_1}{x_2 - x_1}$, for y_2

28. $m = \frac{y_2 - y_1}{x_2 - x_1}$, for x_1

29. $\frac{P}{D} = Q + \frac{R}{D}$, for R

30. $\frac{P}{D} = Q + \frac{R}{D}$, for D

Solve each equation for n.

31. $\frac{1}{2}n + b = n$

32. $\frac{n}{x} = \frac{y}{r}$

33. $\frac{n}{a} + \frac{b}{c} = d$

34. $\frac{a}{c} = n + bn$

35. $\frac{n}{a + b} = c - n$

36. $\frac{1}{n} = \frac{1}{a} + \frac{1}{b}$

37. $\frac{n + 2}{b} = \frac{n + b}{c}$

38. $r = \frac{16M}{a(n + 1)}$

39. $\frac{a}{n} = \frac{n}{b}$

40. $\frac{a}{n} + \frac{a}{b} = \frac{b}{n}$

41. $\frac{r}{n} - \frac{2}{k} = \frac{k}{n}$

42. $\frac{1}{2n} = \frac{n}{2x}$

1. Write a quadratic equation whose roots are 13 and -4.

2. For the equation $y = 3x^2 - 6x + 4$, state the axis of symmetry and the coordinates of the maximum or minimum point of its graph.

3. Solve the following system of inequalities by graphing.
$$y \leq 3$$
$$x - 2y \leq 2$$

Simplify.

4. $\sqrt{135}$

5. $3\sqrt{80} - 8\sqrt{5}$

6. $\dfrac{4}{5 - \sqrt{5}}$

Solve each equation and check your solution.

7. $\sqrt{3x - 11} + 4 = 7$

8. $2x^2 - 3x - 9 = 0$

9. $4x^2 - 12x + 7 = 0$

10. A 24-ft pleasure boat moves 36 miles up the Hudson River in 4 hours. The return trip covering the same distance takes $2\frac{1}{4}$ hours. Find the rate of the boat in still water and the rate of the current.

Excursions in Algebra   Formulas

On the way to the amusement park, the Palmer family took the scenic country route but returned home that night on the expressway. If their car averaged 40 mph to the park and 60 mph home, find a formula for their average speed.

Recall that distance = (rate)(time), or d = rt.

Likewise, $r = \frac{d}{t}$ and $t = \frac{d}{r}$.

Take the formula average rate $= \dfrac{\text{total distance}}{\text{total time}}$ and apply it to this situation.

$$\text{average rate} = \frac{\text{distance to park} + \text{distance home}}{\text{time to park} + \text{time home}}$$

$$\text{average rate} = \frac{d + d}{\dfrac{d}{40} + \dfrac{d}{60}} \qquad \text{Solve } d = 40t \text{ and } d = 60t \text{ for } t.$$

Mathematicians would create a formula to find the average rate that would apply to all cases where distance does not change but rates do.

$$\text{average rate} = \frac{d + d}{\dfrac{d}{r_1} + \dfrac{d}{r_2}} \quad \text{or} \quad \frac{2r_1 \cdot r_2}{r_1 + r_2} \qquad \textit{Use this formula to determine the Palmers' average rate above.}$$

14-11 Rational Expressions in Science

Many scientific formulas, such as the ones for electrical resistance, contain rational expressions.

Electricity can be described as the flow of electrons through a conductor, such as a copper wire. Electricity flows more freely through some conductors than others. The force opposing the flow is called *resistance*. The unit of resistance commonly used is the *ohm*.

Resistances can occur one after another, in *series*. They can also occur in branches of the conductor going in the same direction, in *parallel*. Study the diagrams below. Formulas for the total resistance, R_T, are given under the diagrams.

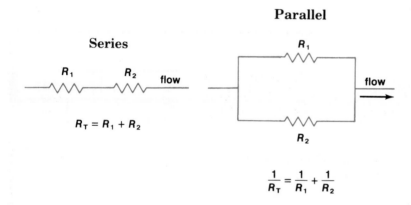

Series

$$R_T = R_1 + R_2$$

Parallel

$$\frac{1}{R_T} = \frac{1}{R_1} + \frac{1}{R_2}$$

Example

1 **Assume that R_1 = 4 ohms and R_2 = 3 ohms. Compute the total resistance of the conductor when the resistances are in series.**

$R_T = R_1 + R_2$

$R_T = 4 + 3$

$R_T = 7$

Thus, the total resistance is 7 ohms.

2 Assume that R_1 = 5 ohms and R_2 = 6 ohms. Compute the total resistance of the conductor when the resistances are in parallel.

$$\frac{1}{R_T} = \frac{1}{R_1} + \frac{1}{R_2}$$

$$\frac{1}{R_T} = \frac{1}{5} + \frac{1}{6}$$

$$\frac{1}{R_T} = \frac{11}{30}$$

$1 \cdot 30 = R_T \cdot 11$ *Find the cross products.*

$$\frac{30}{11} = R_T$$

Thus, the total resistance is $\frac{30}{11}$, or $2\frac{8}{11}$ ohms.

A *circuit*, or path for the flow of electrons, often has some resistances connected in series and others in parallel.

3 A parallel circuit has one branch in series as shown at the right. Given that the total resistance is $2\frac{1}{4}$ ohms, R_1 = 3 ohms, and R_2 = 4 ohms, find R_3.

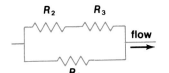

$$\frac{1}{R_T} = \frac{1}{R_1} + \frac{1}{R_2 + R_3}$$ *The total resistance of the branch in series is $R_2 + R_3$.*

$$\frac{1}{2\frac{1}{4}} = \frac{1}{3} + \frac{1}{4 + R_3}$$

$$\frac{4}{9} = \frac{1}{3} + \frac{1}{4 + R_3}$$ *Note that $\frac{1}{2\frac{1}{4}} = \frac{1}{\frac{9}{4}} = 1 \div \frac{9}{4}$ or $1 \cdot \frac{4}{9}$.*

$$\frac{4}{9} - \frac{1}{3} = \frac{1}{r + R_3}$$

$$\frac{1}{9} = \frac{1}{4 + R_3}$$ $\frac{4}{9} - \frac{3}{9} = \frac{1}{9}$

$$4 + R_3 = 9$$ *Find the cross products.*

$$R_3 = 5$$

Thus, the third resistance is 5 ohms.

Written Exercises

Problems 1-6 relate to the diagram below. Solve each problem.

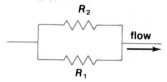

1. Find the total resistance, R_T, given that $R_1 = 8$ ohms and $R_2 = 6$ ohms.

2. Find the total resistance, R_T, given that $R_1 = 4.5$ ohms and $R_2 = 3.5$ ohms.

3. Find R_1, given that total resistance is $2\frac{2}{9}$ ohms and $R_2 = 5$ ohms.

4. Find R_1, given that total resistance is $3\frac{3}{7}$ ohms and $R_2 = 8$ ohms.

5. Find R_1 and R_2, given that total resistance is $2\frac{2}{3}$ ohms and R_1 is two times as great as R_2.

6. Find R_1 and R_2, given that total resistance is $2\frac{1}{4}$ ohms and R_1 is three times as great as R_2.

Solve each problem.

7. Resistances of 3 ohms, 6 ohms, and 9 ohms are connected in series. What is the total resistance?

8. Eight lights on a decorated tree are connected in series. Each has a resistance of 12 ohms. What is the total resistance?

9. Three coils with resistances of 3 ohms, 4 ohms, and 6 ohms are connected in parallel. What is the total resistance?

10. Three coils with resistances of 4 ohms, 6 ohms, and 15 ohms are connected in parallel. What is the total resistance?

11. Three appliances are connected in parallel: a lamp of resistance 120 ohms, a toaster of resistance 20 ohms, and an iron of resistance 12 ohms. Find the total resistance.

12. Three appliances are connected in parallel: a lamp of resistance 60 ohms, an iron of resistance 20 ohms, and a heating coil of resistance 80 ohms. Find the total resistance.

Problems 13-15 relate to the diagram below. Solve each problem.

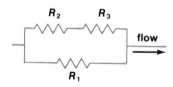

13. Find the total resistance, given that $R_1 = 5$ ohms, $R_2 = 4$ ohms, and $R_3 = 3$ ohms.

14. Find R_1, given that the total resistance is $2\frac{10}{13}$ ohms, $R_2 = 3$ ohms, and $R_3 = 6$ ohms.

15. Find R_2, given that the total resistance is $3\frac{1}{2}$ ohms, $R_1 = 5$ ohms, and $R_3 = 4$ ohms.

Challenge Exercises

Solve each problem.

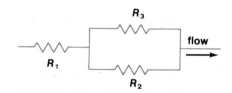

16. Write an expression for the total resistance for the diagram at the right.

17. Find the total resistance, given that $R_1 = 5$ ohms, $R_2 = 4$ ohms, and $R_3 = 6$ ohms.

Adding Fractions

You do not have to write special programs to tell a computer how to add integers and decimals. However, this is not the case with addition of fractions. The following program tells the computer how to add two fractions. The program uses the definition $\frac{a}{b} + \frac{c}{d} = \frac{ad + bc}{bd}$.

```
10   PRINT "ENTER THE NUMERATOR AND THE
        DENOMINATOR OF THE FIRST FRACTION."
20   INPUT A,B
30   PRINT "ENTER THE NUMERATOR AND THE
        DENOMINATOR OF THE SECOND FRACTION."
40   INPUT C,D
50   LET N = A * D + B * C
60   LET X = B * D
70   PRINT "THE SUM IS ";N;"/";X
80   END
```

However, the sum given by this program may not always be in simplest form. The following lines can be added to the program so that the sum is given in simplest form.

```
70    IF N < X THEN 100
80    LET Y = X
90    GOTO 110
100   LET Y = N
110   FOR I = Y TO 1 STEP - 1
120   IF INT (N / I) < > N / I THEN 140
130   IF INT (X / I) = X / I THEN 150
140   NEXT I
150   LET N = N / I
160   LET X = X / I
170   PRINT "THE SUM IS ";N;"/";X
180   END
```

Exercises

Use the program to find each sum in simplest form.

1. $\frac{2}{3} + \frac{4}{9}$ **2.** $\frac{8}{11} + \frac{1}{22}$ **3.** $\frac{4}{6} + \frac{1}{4}$ **4.** $\frac{7}{16} + \frac{9}{10}$ **5.** $\frac{5}{12} + \frac{7}{18}$

6. Write a program similar to the one above to subtract two fractions.

rational expressions   (467)

multiplicative inverse   (475)

reciprocal   (475)

least common denominator   (486)

least common multiple   (486)

mixed expression   (490)

complex fraction   (490)

rational equation   (494)

Chapter Summary

1. Rational expressions are algebraic expressions whose numerators and denominators are polynomials.   (467)

2. Zero cannot be used as a denominator because division by zero is undefined. (467)

3. To simplify an algebraic fraction, first factor the numerator and denominator. Then eliminate common factors.   (467)

4. Multiplying Fractions: For all rational numbers $\frac{a}{b}$ and $\frac{c}{d}$, where $b \neq 0$ and $d \neq 0$, $\frac{a}{b} \cdot \frac{c}{d} = \frac{ac}{bd}$.   (471).

5. Dividing Fractions: For all rational numbers $\frac{a}{b}$ and $\frac{c}{d}$, where $b \neq 0$, $c \neq 0$, and $d \neq 0$, $\frac{a}{b} \div \frac{c}{d} = \frac{a}{b} \cdot \frac{d}{c}$.   (475)

6. Adding or Subtracting Fractions with Like Denominators: For all rational numbers $\frac{a}{c}$ and $\frac{b}{c}$, where $c \neq 0$,

$$\frac{a}{c} + \frac{b}{c} = \frac{a+b}{c} \text{ and } \frac{a}{c} - \frac{b}{c} = \frac{a-b}{c}.   (482)$$

7. To add or subtract fractions that have unlike denominators, rename the fractions by using the least common denominator.   (486)

8. Simplifying Complex Fractions: Any complex fraction $\dfrac{\frac{a}{b}}{\frac{c}{d}}$, where $b \neq 0$, $c \neq 0$,

and $d \neq 0$, may be expressed as $\frac{ad}{bc}$.   (491)

9. Rational equations can be solved by first multiplying every term of the equation by the least common denominator.   (494)

Chapter Review

14–1 Simplify each algebraic fraction. State the excluded values of the variables.

1. $\dfrac{3x^2y}{12xy^3z}$

2. $\dfrac{x+y}{x^2 + 3xy + 2y^2}$

3. $\dfrac{x^2 + 10x + 21}{x^3 + x^2 - 42x}$

Find each product and simplify.

14–2 **4.** $\dfrac{7}{9} \cdot \dfrac{a^2}{b}$ **5.** $\dfrac{5x^2y}{8ab} \cdot \dfrac{12a^2b}{25x}$ **6.** $\dfrac{x^2 + x - 12}{x + 2} \cdot \dfrac{x + 4}{x^2 - x - 6}$

Find each quotient.

14–3 **7.** $\dfrac{7a^2b}{x^2 + x - 30} \div \dfrac{3a}{x^2 + 15x + 54}$ **8.** $\dfrac{m^2 + 4m - 21}{m^2 + 8m + 15} \div \dfrac{m^2 - 9}{m^2 + 12m + 35}$

14–4 **9.** $x + 3\overline{)x^3 + 7x^2 + 10x - 6}$ **10.** $2a - 5\overline{)6a^3 - 19a^2 + 2a + 15}$

Find each sum or difference and simplify.

14–5 **11.** $\dfrac{x}{x^2 - 1} + \dfrac{1}{x^2 - 1}$ **12.** $\dfrac{7}{x^2} + \dfrac{a}{x^2}$ **13.** $\dfrac{2x}{x - 3} - \dfrac{6}{x - 3}$ **14.** $\dfrac{2}{x - y} + \dfrac{x}{y - x}$

14–6 **15.** $\dfrac{5a}{3x} - \dfrac{2}{4x^2y}$ **16.** $\dfrac{x}{x + 3} - \dfrac{5}{x - 2}$ **17.** $\dfrac{2x + 3}{x^2 - 4} + \dfrac{6}{x + 2}$

14–7 **Simplify.**

18. $\dfrac{\frac{x^2}{y^3}}{\frac{3x}{9y^2}}$ **19.** $\dfrac{\frac{x - 3}{x + 5}}{\frac{x + 5}{x}}$ **20.** $\dfrac{\frac{a^2 - 13a + 40}{a^2 - 4a - 32}}{\frac{a - 5}{a + 7}}$ **21.** $\dfrac{x - \frac{35}{x + 2}}{x + \frac{42}{x + 13}}$

14–8 **Solve each equation.**

22. $\dfrac{4x}{3} + \dfrac{7}{2} = \dfrac{7x}{12}$ **23.** $\dfrac{3}{x} + \dfrac{1}{x - 5} = \dfrac{1}{2x}$ **24.** $\dfrac{1}{h + 1} + \dfrac{2}{3} = \dfrac{2h + 5}{h - 1}$

25. $\dfrac{3x + 2}{x^2 + 7x + 6} = \dfrac{1}{x + 6} + \dfrac{4}{x + 1}$ **26.** $\dfrac{3m - 2}{2m^2 - 5m - 3} - \dfrac{2}{2m + 1} = \dfrac{4}{m - 3}$

14–9 **Solve each problem.**

27. Bill can paint the outside of a house in 40 hours. Roberta can do the job in 48 hours. If they work together, how long will it take them to complete the job?

28. An airplane can fly at a rate of 525 mph in calm air. It can fly 1605 miles with the wind in the same amount of time it can fly 1545 miles against the wind. Find the speed of the wind.

14–10 **Solve each formula for the variable indicated.**

29. $\dfrac{1}{f} = \dfrac{1}{a} + \dfrac{1}{b}$, for a **30.** $S = \dfrac{a}{1 - r}$, for r

14–11 **Solve each problem.**

31. Assume that $R_1 = 4$ ohms and $R_2 = 6$ ohms. What is the total resistance of the conductor if R_1 and R_2 are: **a.** connected in series? **b.** connected in parallel?

32. A parallel circuit has one branch in series. The total resistance is expressed by the formula: $\dfrac{1}{R_T} = \dfrac{1}{R_1} + \dfrac{1}{R_2 + R_3}$. Find R_1 if $R_T = 3$ ohms, $R_2 = 3$ ohms, and $R_3 = 9$ ohms.

Simplify.

1. $\dfrac{\frac{5}{9}}{\frac{2}{3}}$

2. $\dfrac{21x^2y}{28ax}$

3. $\dfrac{x^2 + 7x - 18}{x^2 + 12x + 27}$

4. $\dfrac{x^2 - x - 56}{x^2 + x - 42}$

5. $\dfrac{7x^2 - 28}{5x^3 - 20x}$

6. $\dfrac{2x^2 - 5x - 3}{x^2 + 2x - 15}$

Perform the indicated operations and simplify.

7. $\dfrac{3x}{x + 3} + \dfrac{5x}{x + 3}$

8. $\dfrac{2x}{x - 7} - \dfrac{14}{x - 7}$

9. $\dfrac{2x}{x + 7} + \dfrac{4}{x + 4}$

10. $\dfrac{2a + 1}{2a - 3} + \dfrac{a - 3}{3a + 2}$

11. $\dfrac{x + 5}{x + 2} + 6$

12. $\dfrac{x - 2}{x - 8} + x + 5$

13. $\dfrac{3x + 2}{4x + 1} + \dfrac{7}{x}$

14. $\dfrac{3x - 8}{x + 4} + \dfrac{9}{x + 1}$

15. $\dfrac{x^2 + 4x - 32}{x + 5} \cdot \dfrac{x - 3}{x^2 - 7x + 12}$

16. $\dfrac{3x^2 + 2x - 8}{x^2 - 4} \div \dfrac{6x^2 + 13x - 28}{2x^2 - 3x - 35}$

17. $\dfrac{4x^2 + 11x + 6}{x^2 - x - 6} \div \dfrac{x^2 + 8x + 16}{x^2 + x - 12}$

18. $\dfrac{3x^2 + 5x - 28}{x^2 - 3x - 28} \cdot \dfrac{x^2 - 8x + 7}{3x - 7}$

19. $\dfrac{x - \dfrac{24}{x + 5}}{x - \dfrac{72}{x - 1}}$

20. $\dfrac{\dfrac{x^2 - x - 6}{x^2 + 2x - 15}}{\dfrac{x^2 - 2x - 8}{x^2 + x - 20}}$

21. $\dfrac{\dfrac{2}{3m} + \dfrac{3}{m^2}}{\dfrac{2}{5m} + \dfrac{5}{m}}$

Solve each equation.

22. $\dfrac{y + 3}{6} = \dfrac{y + 2}{12} - \dfrac{2}{5}$

23. $\dfrac{x + 1}{x} + \dfrac{6}{x} = x + 7$

24. $\dfrac{4m}{m - 3} + \dfrac{6}{3 - m} = m$

25. $\dfrac{-2b - 9}{b^2 + 7b + 12} = \dfrac{b}{b + 3} + \dfrac{2}{b + 4}$

26. $\dfrac{1}{y - 4} - \dfrac{2}{y - 8} = \dfrac{-1}{y + 6}$

27. $\dfrac{m + 3}{m - 1} + \dfrac{m + 1}{m - 3} = \dfrac{22}{3}$

Solve each formula for the variable indicated.

28. $F = G\left(\dfrac{Mm}{d^2}\right)$, for G

29. $\dfrac{1}{R_T} = \dfrac{1}{R_1} + \dfrac{1}{R_2}$, for R_2

Solve each problem.

30. Willie can do a job in 6 days. Myra can do the same job in $4\frac{1}{2}$ days. If they work together how long will it take to complete the job?

31. The top speed of a boat in still water is 5 mph. At this speed, a 21-mile trip downstream took the same amount of time as a 9-mile trip upstream. Find the rate of the current.

The test questions on this page deal with averages and rational expressions. The information at the right may help you with some of the questions.

Directions: Choose the best answer. Write A, B, C, or D.

1. $\dfrac{68 + 68 + 68 + 68}{4} =$

   (A) 17   (B) 68   (C) 136   (D) 272

2. Carrie's bowling scores for four games are $b + 2$, $b + 3$, $b - 2$, and $b - 1$. What must her score be on her fifth game to average $b + 2$?

   (A) $b + 8$ (B) b
   (C) $b - 2$ (D) $b + 5$

3. Which of the following *cannot* be the average of 10, 7, 13, 2, and x, if $x > 4$?

   (A) 7   (B) 9   (C) 13   (C) 258

4. The average of eight integers can be
   (A) 127.6   (B) 130.8
   (C) 131.3   (D) 135.5

5. The sum of six integers is what percent of the average of six integers?

   (A) 0.001% (B) 2% (C) 10% (D) 600%

6. What is the average of $70 - c$, $70 + 2c$, and $45 - c$?

   (A) 75   (B) $61\frac{2}{3} + \frac{2}{3}c$(C) $61\frac{2}{3}$ (D) 75

7. What is the average of $2b - 4$, $b + 5$, and $3b + 8$?

   (A) $b + 3$ (B) $2b$
   (C) $6b + 9$ (D) $2b + 3$

8. If $\frac{x}{2}$, $\frac{x}{5}$, and $\frac{x}{7}$ are whole numbers, x may be

   (A) 20   (B) 35   (C) 50   (D) 70

---

1. To find the average of a group of numbers, first add the numbers. Then divide by the quantity of numbers added.

2. The average of a group of numbers cannot be greater than the greatest number in the group, nor less than the least number.

3. Most questions involving rational expressions can be solved by writing the expression in a different form.

---

9. If $abc = 4$ and $b = c$, then $a =$
   (A) c^2 (B) $\frac{1}{4c}$ (C) $\frac{4}{c^2}$ (D) $\frac{1}{c^2}$

10. If $\frac{1}{b - d} = 4$, then $d =$

    (A) $b - \frac{1}{4}$ (B) $4b - 1$

    (C) $\frac{b + 1}{4}$ (D) $b + 4$

11. If $\frac{2a}{5b} = 6$, then $\frac{2a - 5b}{5b} =$

    (A) 6   (B) $\frac{2}{5}$ (C) 15   (D) 5

12. If $3\frac{1}{5}c = 2\frac{1}{2}b$ and $c \neq 0$, then $\frac{b}{c} =$

    (A) $\frac{25}{32}$ (B) $\frac{32}{25}$ (C) $\frac{7}{8}$ (D) $\frac{11}{10}$

13. If the product of a number and b is increased by y, the result is t. Find the number in terms of b, y, and t.

    (A) $\frac{t - y}{b}$ (B) $\frac{y - t}{b}$

    (C) $\frac{b + y}{t}$ (D) $t - by$

14. Mark can type 60 words per minute and there is an average of 360 words per page. At this rate, how many *hours* would it take him to type k pages?

    (A) $\frac{k}{6}$ (B) $\frac{k}{60}$ (C) $\frac{k}{10}$ (D) $\frac{10}{k}$

CHAPTER 15

Statistics and Probability

The Stryker Panthers are playing the Defiance Bulldogs in a tournament game for the Buckeye Border League championship. In order to keep an accurate account of team scores and player performance, each school has students who serve as statisticians. They record data such as number of serves in-bounds, the number of spikes attempted, and the number of blocks missed. The collection, organizing, and interpreting of these numerical facts is known as *statistics*.

15-1 What Is Statistics?

Numerical information called **data** can be useful in our daily lives. A branch of mathematics called **statistics** provides methods for collecting, organizing, and interpreting data. One way to organize data is by using tables.

Mark and Debbie Hutsko are going to buy a new car. They have narrowed their choices down to three cars. A consumer magazine containing information on Model A, Model B, and Model C gives them additional information to consider. The table below contains some facts and figures for each model.

Dimensions (in.)	Model A	Model B	Model C
Front shoulder room	55.5	56.0	56.0
Maximum front leg room	42.0	41.0	43.0
Front head room	3.4	3.4	3.4
Rear shoulder room	55.0	55.5	55.5
Rear head room	0.5	1.0	2.0

Examples

Use the information in the table above to answer each question.

1 **Which car would give the Hutskos minimum leg room in the front seat?**

Leg room in Model A is 42.0 inches, in Model B is 41.0 inches, and in Model C is 43.0 inches.

Therefore, Model B would provide the least amount of leg room in the front seat.

2 **Which car would give the Hutskos maximum back-seat room for their children?**

Comparing the rear shoulder room, both Model B and Model C provide the most room, 55.5 inches, while Model A has 55.0 inches. Comparing the rear head room, Model A has 0.5 inch, Model B has 1.0 inch, and Model C has 2.0 inches. Model C provides the most rear head room.

Therefore, Model C provides the most room in the back seat.

Statistical data in the table on page 513 were collected by actually measuring cars. Data may be collected when they occur, such as keeping score at a volleyball game. Data may also be collected by conducting surveys or polls. When you study and analyze data to draw conclusions, it is important to know how the data were obtained.

Written Exercises

The following table lists the nutritional value found in food items purchased at some fast food restaurants.

ITEM	CALORIES	PROTEIN (g)	CARBOHYDRATES (g)	FATS (g)	SODIUM (mg)
McDonald's Big Mac	541	26	39	31	962
Burger King Whopper	606	29	51	32	909
Burger Chef Hamburger	258	11	24	13	393
Pizza Hut Thin 'N Crispy Cheese Pizza (1/2 of 10-in. pie)	450	25	54	15	N.A.*
Pizza Hut Thick 'N Chewy Pepperoni Pizza (1/2 of 10-in. pie)	560	31	68	18	N.A.*
Arthur Treacher's Fish Sandwich	440	16	68	18	836
Burger King Whaler	486	18	64	46	735
McDonald's Filet-O-Fish	402	15	34	23	709
Long John Silver's Fish (2 pieces)	318	19	19	19	N.A.*
Kentucky Fried Chicken Original Dinner (3 pieces chicken)	830	52	56	46	2285
Kentucky Fried Chicken Extra Crispy Dinner (3 pieces chicken)	950	52	63	54	1915

*Not Available

1. Which hamburger is highest in protein?

2. Which item is lowest in fat content?

3. If you were on a 1000-calorie-a-day diet and had already eaten 560 calories, what options do you have for one more item?

4. If you were on a low sodium (salt) diet, what food should you avoid?

5. According to the table, is it true that items higher in calories are higher in protein?

6. According to the table, is it true that foods that have a high carbohydrate content are also high in fat content?

15-2 Line Plots

Statistical data can be organized and presented on a number line. Numerical information displayed on a number line is called a **line plot**.

The hourly wages earned by the principal wage earner in ten families is shown in the chart at the right.

The data range from $8.00 per hour to $20.25 per hour. In order to represent each on a number line, the scale shown must include these values. A "W" is used to represent each hourly wage. If more than one "W" has the same location on the number line, additional W's will be placed one above the other. A line plot on hourly wages is shown below.

Family	Hourly Wage
A	$ 8.00
B	$10.50
C	$20.25
D	$ 9.40
E	$11.00
F	$13.75
G	$ 8.50
H	$10.50
I	$ 9.00
J	$11.00

Note that some data values are approximated between integer values on the number line.

Example

1 **The number of takeoffs and landings of the busiest airports in the United States for a one-year period are listed below. Make a line plot of the data.**

Since the numbers are too large to represent on a number line, change each number to represent the number of 100,000 takeoffs and landings. Round each number to the nearest tenth.
Since 768,079 = 7.68079 × 100,000, you would plot 7.7 for Chicago O'Hare.

Airport	Takeoffs/ Landings	Plot
Chicago O'Hare	768,079	7.7
Atlanta	749,909	7.5
Dallas/Fort Worth	547,901	5.5
Los Angeles	545,973	5.5
Santa Ana	521,360	5.2

Airport	Takeoffs/ Landings	Plot
Van Nuys	503,488	5.0
Denver	502,897	5.0
St. Louis	411,288	4.1
Boston	402,695	4.0
Newark	400,204	4.0

The line plot shown in Example 1 should enable you to analyze the given data. For example, there are five airports with approximately 500,000 takeoffs and landings in one year.

How many airports have a much greater number of takeoffs and landings than other airports?

Exploratory Exercises

State the scale you would use to make a line plot for the following data.

1. 4.2, 5.3, 7.8, 9.1, 7.3, 6.9

2. 123, 234, 734, 456, 111, 482, 379

3. 30, 30, 30, 40, 50, 10, 20

4. 7890, 3875, 9879, 1267, 4444, 6754, 5555, 3791, 7956, 9347

Written Exercises

Use the table to complete Exercises 5–10.

5. Make a line plot of the data.

6. Identify any points that seem to stand out.

7. What fish was the largest?

8. What fish was the smallest?

9. Are the weights of any fish clustered? If so, which ones?

10. Find similar data on a subject of interest to you and make a line plot.

| Record Weights for Freshwater Fish ||
Fish	Weight
Bass, largemouth	22 lb 4 oz
Bluegill	4 lb 12 oz
Carp	57 lb 13 oz
Catfish, blue	97 lb
Catfish, channel	58 lb
Muskellunge	69 lb 15 oz
Perch, white	4 lb 12 oz
Pike, northern	46 lb 2 oz
Salmon, Atlantic	79 lb 2 oz
Salmon, coho	31 lb
Salmon, pink	12 lb 9 oz
Trout, brown	35 lb 15 oz
Trout, rainbow	42 lb 2 oz
Walleye	25 lb

The players with the most runs batted in (RBI) for the National League (1967–1988) are listed below. Use the data to complete Exercises 11–15.

Year	Name	RBI	Year	Name	RBI
1967	Orlando Cepeda	111	1979	Dave Winfield	118
1968	Willie McCovey	105	1980	Mike Schmidt	121
1969	Willie McCovey	126	1981	Mike Schmidt	91
1970	Johnny Bench	148	1982	Dale Murphy	109
1971	Joe Torre	137		Al Oliver	
1972	Johnny Bench	125	1983	Dale Murphy	121
1973	Willie Stargell	119	1984	Mike Schmidt	106
1974	Johnny Bench	129		Gary Carter	
1975	Greg Luzinski	120	1985	Dave Parker	125
1976	George Foster	121	1986	Mike Schmidt	119
1977	George Foster	149	1987	Andre Dawson	137
1978	George Foster	120	1988	Will Clark	109

11. Make a line plot of the data.

12. What is the greatest number of RBI's during a season?

13. What is the least number of RBI's during a season?

14. Identify the heaviest cluster of data for any 10-unit span on the scale.

15. What is the most common number of RBI's per season since 1967?

15-3 Stem and Leaf Plots

A **stem and leaf plot** is another method used to organize statistical data. The greatest common place value of the data is used to form the *stem*. The next greatest common place value is used to form the *leaves*.

stem   leaf

Example

1   **Make a stem and leaf plot of the algebra test scores given below. Then complete each question.**

56, 65, 98, 82, 64, 71, 78, 77, 86, 95, 91, 59,   *The greatest common place*
69, 70, 80, 92, 76, 82, 85, 91, 92, 99, 73   *value is tens.*

Since the data range from 56 to 99, the stems range from 5 to 9. To plot the data, make a vertical list of the stems. Each number is assigned to the graph by pairing the units digit, or leaf, with the correct stem. The score 56 is plotted by placing the units digit, 6, to the right of stem 5.

Stem	Leaf
5	6 9
6	5 4 9
7	1 8 7 0 6 3
8	2 6 0 2 5
9	8 5 1 2 1 2 9

A second stem and leaf plot can be made to arrange the leaves in numerical order from least to greatest.

Stem	Leaf
5	6 9
6	4 5 9
7	0 1 3 6 7 8
8	0 2 2 5 6
9	1 1 2 2 5 8 9

8|2 represents the number 82.   *The leaves are in numerical order.*

a. What was the lowest score on the algebra test? *56*
b. What was the highest score on the algebra test? *99*
c. In which interval did most students score? *91 to 99*

Data with more than two digits can be rounded to two digits before plotting or can be truncated to two digits. To truncate means to cut off. For a stem and leaf plot, you would truncate everything after the second digit.

The number 355 would round to 36.   *355 → 36*

The number 355 would truncate to 35.   *355 → 35*

To what does 389 round?   **To what does 389 truncate?**

A *back-to-back* stem and leaf plot is sometimes used to compare two sets of data or rounded and truncated values of the same data. In a back-to-back plot, the same stem is used for the leaves of both plots.

Example

2 Estimated populations of counties in California are listed below. Make a back-to-back stem and leaf plot of the populations comparing rounded values and truncated values.

County	Pop. (thousands)	County	Pop. (thousands)
Butte	149	San Bernardina	893
Contra Costa	657	San Francisco	679
Fresno	515	San Mateo	588
Kern	403	Santa Barbara	299
Marin	223	Santa Cruz	188
Sacramento	783	Sonoma	300

Rounded		Truncated
9 5	1	4 8
2	2	2 9
0 0	3	0
0	4	0
9 2	5	1 8
8 6	6	5 7
8	7	8
9	8	9

Using rounded data,
2|2 represents
215,000–224,999 people.

Using truncated data,
2|2 represents
220,000–229,999 people.

Does there appear to be much difference between the two plots? *no*

Exploratory Exercises

Truncate each number to two digits.

1. 456,876   **2.** 34,591   **3.** 1234   **4.** 1,234,567

Write the stems that would be used to plot each set of data.

5. 23, 45, 56, 12, 27, 56, 37   **6.** 8, 11, 23, 37, 31, 42, 59

7. 230, 456, 784, 245, 745, 357   **8.** 4.5, 6.1, 5.8, 9.8, 4.1, 3.2

Written Exercises

Use the stem and leaf plot to answer these questions.

History Test Scores

Stem	Leaf
6	1 1 4 6 7 8
7	2 3 5 7 9
8	1 3 5 6 6 7 7 8 9
9	0 0 3 4 6 8 9 9
10	0 0

9|3 = 93

9. What is the best test score?

10. How many students took the test?

11. How many students scored 90?

12. What is the lowest score?

13. Find the difference between the high and low scores.

Use the ages of the people who attended a gymnastics meet to complete Exercises 14–17.

AGES: 12, 17, 15, 14, 19, 17, 13,
16, 15, 16, 17, 18, 24, 23,
28, 45, 48, 36, 12, 23, 15,
14, 13, 15, 17, 18, 19, 15,
15, 16, 16, 16, 16, 17

14. Make a stem and leaf plot of the data.

15. How many people attended the meet?

16. What are the ages of the youngest and oldest persons attending?

17. Which age group was more widely represented?

The stem and leaf plot below gives the average weekly incomes of several families.

Stem	Leaves
1	9 9
2	1 5
3	1 1 3 4 9 9
4	0 2 9
5	5
6	1 2

2|5 represents $250–$259.

18. What was the highest weekly income?

19. What was the lowest weekly income?

20. Find the difference between the highest and lowest weekly income.

21. How many families earn more than $500 per week?

22. What range does 4|2 represent?

23. Have the leaf values been rounded or truncated?

mini-review

Use a calculator to simplify. Round answers to the nearest hundredth.

1. $2\sqrt{14} - 3\sqrt{6} + 4\sqrt{5}$

2. $7\sqrt{41} + 6\sqrt{13} - 5\sqrt{3}$

3. $2\sqrt{3} - 4\sqrt{3} + 4\sqrt{5}$

4. In a right triangle, the measures of the legs are x and $x + 7$. If the measure of the hypotenuse is $x + 9$, find x.

5. Find two integers whose sum is 15 and whose product is 54.

6. Find the distance between two points whose coordinates are (2, 3) and (5, 7).

7. Find the coordinates of the midpoint of a segment whose endpoints have coordinates (0, 3) and (5, 7).

8. Solve $x^2 - 12x = 22$ by completing the square.

9. Use the Quadratic Formula to solve $m^2 - 3m = 2$.

10. Simplify $\frac{2x}{x-3} - \frac{x+3}{x-2}$.

11. Solve $3x + \frac{x}{2} = \frac{1}{x-1}$.

12. Make a line plot of the following data. 123, 235, 164, 188, 193, 222, 156, 178, 162, 179, 153

13. Make a stem and leaf plot of the following data. Truncate entries. 4267, 5679, 3623, 6791, 3471, 3124, 5629, 4444, 3812, 5814

15-4 Measures of Central Tendency: Mean, Median, and Mode

In analyzing statistical data, it is often useful to have numbers describe the complete set of data. *Measures of central tendency* are used because they represent centralized or middle values of the data. These measures of central tendency are called the **mean**, **median**, and **mode**.

The *mean* is a number that represents an *average* of a set of data. It is found by adding the elements in the set and then dividing that sum by the number of elements in the set.

The mean of a set of data is the sum of the elements in the set divided by the number of elements in the set.	*Definition of Mean*

Example

1 The high temperatures for a 7-day week during December in Chicago were 29°, 31°, 28°, 32°, 29°, 27°, and 55°. Find the mean high temperature for the week.

$$\text{mean} = \frac{29 + 31 + 28 + 32 + 29 + 27 + 55}{7} \qquad \text{\textit{The mean is the sum of 7 numbers divided by 7.}}$$

$$= \frac{231}{7}$$

$$= 33$$

The mean, or average, high temperature for the week was 33°.

Notice in Example 1 that the mean temperature, 33°, is greater than all of the daily temperatures except one, 55°. Thus, 33° is not a very good representation of the average of the set of data. Extremely high or low values, such as 55°, affect the mean.

Another measure of central tendency is the *median*. The same number of values are above the median as below the median.

The median is the middle number of a set of data when the numbers are arranged in numerical order.	*Definition of Median*

To find the median temperature for the temperatures in Chicago listed in Example 1, arrange them in order from least to greatest.

$$27° \quad 28° \quad 29° \quad 29° \quad 31° \quad 32° \quad 55°$$

Since there are an odd number of temperatures, 7, the middle one is the fourth value, which is 29°. The median temperature for the 7-day period is 29°.

The median is not affected by the extremely high temperature.

If a set of data contains an even number of elements, the median is the value halfway between the two middle elements.

Example

2 The batting averages for 10 members of a baseball team are 0.234, 0.256, 0.321, 0.333, 0.290, 0.240, 0.198, 0.222, 0.300, and 0.276. Find the median batting average.

Arrange the batting averages in order.

0.198 0.222 0.234 0.240 0.256 0.276 0.290 0.300 0.321 0.333

Since there are an even number of batting averages, the median is halfway between the two middle elements, 0.256 and 0.276.

$$\frac{0.256 + 0.276}{2} = 0.266 \qquad \textit{Find the mean of the two middle elements.}$$

The median batting average is 0.266.
There are five averages above the median and five below the median.

Another measure of central tendency is called the *mode*.

The mode is the number that occurs most often in a set of data.	*Definition of Mode*

If no number occurs more often than the other numbers, then a set of data has no mode. A set of data may have more than one mode. For example, in {2, 3, 3, 4, 6, 6}, 3 and 6 are both modes for the set of data.

Example

3 The stem and leaf plot represents the scores on the Chapter 5 test in Mrs. Theesfeld's geometry class. Find the median and mode scores.

There are 31 scores shown. The median will be the middle score, or the 16th value. You can count the leaves from the bottom up to the 16th score or from the top down.

Stem	Leaf
5	6 8 9
6	1 6 9
7	4 5 7 7 9 9
8	2 4 6 7 7 8 8 9
9	1 3 3 4 4 5 5 5 7
10	0 0

$9|3 = 93$

The median score is 87.

The mode is the score that appears most often. Note that for stem 9, there are three leaves with a value of 5.

The mode is 95.

Exploratory Exercises

Find the mean, median, and mode for each set of data.

1. 4, 6, 9, 12, 5   **2.** 7, 13, 4, 7   **3.** 10, 3, 8, 15

4. 9, 9, 9, 9, 8   **5.** 300, 34, 40, 50, 60   **6.** 23, 23, 12, 12

Written Exercises

Find the median and mode of the data represented in each stem and leaf plot.

7.

Stem	Leaf
7	3 5
8	2 2 4
9	0 4 7 9
10	5 8
11	4 6

$9|4 = 94$

8.

Stem	Leaf
5	3
6	5 8
7	3 7 7
8	4 8 8 9

$6|8 = 68$

9.

Stem	Leaf
9	3 5
10	2 5 8
11	5 8 9 9
12	7 8 9

$11|5 = 115$

Solve.

10. The price list for computers shown in a magazine advertisement was $899, $1295, $1075, $1597, and $1800. Find the median price.

11. The prices of six different models of printers in a computer store are $299, $349, $495, $329, $198, and $375. Find the median price.

12. In a basketball game between Kennedy High School and Grant High School, the Grant players' individual points were 23, 4, 6, 11, 4, 7, 8, 12, 3, and 5. Find the mean, median, and mode of the individual points.

13. Olivia swims the 50-yard freestyle for the Darien High School swim team. Her times in the last six meets were 26.89 seconds, 26.27 seconds, 25.18 seconds, 25.63 seconds, 27.16 seconds, and 27.18 seconds. Find the mean and median of her swimming times.

14. The prices of slacks in five different stores are $29.95, $31.50, $25.45, $33.49, and $28.49. Find the mean price of slacks.

15. Shane bowled 6 games, and his scores were 147, 134, 132, 157, 123, and 140. Find his mean and median bowling scores.

16. One of the events in the Winter Olympics is the Men's 500-meter Speed Skating. The times for this event are shown to the right. Find the mean, median, and mode times.

Year	Time (s)	Year	Time (s)
1928	43.4	1964	40.1
1932	43.4	1968	40.3
1936	43.4	1972	39.44
1948	43.1	1976	39.17
1952	43.2	1980	38.03
1956	40.2	1984	38.19
1960	40.2	1988	36.45

A department store sold 95 jean jackets at a regular price of $82. For a spring sale, the jackets were reduced to $70 and an additional 110 jackets sold. The final clearance price was $45 and the remaining 54 jackets sold.

Can the mean selling price of the jackets be determined by the expression $\frac{\$82 + \$70 + \$45}{3}$?

The above expression would be true only if there are the same number of jackets sold at each price.

Since there were 95 jackets sold at $82, 110 jackets sold at $70, and 54 jackets sold at $45, a **weighted mean** must be found. To find the weighted mean, each price must be multiplied by the number of jackets sold at that price.

$\overline{X}$ is often used to represent the mean value of a set of data. $\overline{X}$ *is read "X-bar."*
Let $\overline{X}_W$ represent the weighted mean.

$$\overline{X}_W = \frac{(95)(\$82) + (110)(\$70) + (54)(\$45)}{95 + 110 + 54}$$

$$= \frac{\$7790 + \$7700 + \$2430}{259}$$

$$= \$69.189189$$

The weighted mean cost of a jean jacket is $69.19.

Exercises

1. A company employs 200 workers at hourly rates. One hundred workers earn $6.50 per hour, ten earn $7.25 per hour, ten earn $8.00 per hour, twenty earn $10.25 per hour, and sixty earn $9.75 per hour. Find the weighted mean hourly wage of the workers.

2. Anthony's sold 92 men's suits for a regular price of $425. For the spring sale, the suits were reduced to $250 and 126 were sold. At the final clearance, the remaining 79 suits sold for $125. Find the weighted mean price for the suits.

15-5 Measures of Variation: Range, Quartiles, and Interquartile Ranges

A *measure of variation*, called the **range**, describes the spread of numbers in a set of data. To find the range, determine the difference between the greatest value and the least value in the set.

The difference between the greatest and the least values of a set of data is called the range.	*Definition of Range*

From the table, determine the greatest and least amounts of precipitation for each city to find the range. In Albany, the range of the monthly precipitation is 3.3 − 2.3 or 1.0 inches. In Phoenix, the range is 1.0 − 0.1 or 0.9 inches.

Normal Monthly Precipitation		
	Albany, NY	**Phoenix, AZ**
January	2.4	0.7
February	2.3 ← *least*	0.6
March	3.0	0.8
April	2.9	0.3
May	3.3	*least* → 0.1
June	3.3	0.2
July	3.0	0.7
August	3.3 ← *greatest* → 1.0	
September	3.2	0.6
October	2.9	0.6
November	3.0	0.5
December	3.0	0.8

Quartiles are values that divide the data into four equal parts. The median divides the data in half. The **upper quartile** (UQ) divides the top half into two equal parts. The **lower quartile** (LQ) divides the bottom half into two equal parts.

Another measure of variation uses the upper quartile and lower quartile values to determine the **interquartile range**.

The difference between the upper and lower quartiles of a set of data is called the interquartile range. It represents the middle half of the data.	*Definition of Interquartile Range*

Example

1 Mr. Alvarez gave a quiz in his computer programming class. The scores were 23, 12, 22, 14, 15, 27, 30, 18, 26, 27, 28, 15, 16, and 18. Find the median, upper and lower quartiles, and interquartile range.

First, order the scores. Then, find the median.

$$12 \quad 14 \quad 15 \quad \boxed{15} \quad 16 \quad 18 \quad 18 \mid 22 \quad 23 \quad 26 \quad \boxed{27} \quad 27 \quad 28 \quad 30$$

The lower quartile is the median of the lower half.   median = 20   *The upper quartile is the median of the upper half.*

The LQ is 15.   The UQ is 27.

The interquartile range is 27 − 15 or 12. Therefore, the middle 50% or middle half of the quiz scores varies 12 points.

2 Find the interquartile range for the data in the stem and leaf plot.

Stem	Leaf
4	2 5 8
5	1 3 6 9
6	0 1 5 7 9
7	5 6 8
8	3 9 9

The bottom half begins at 42 and ends at 61.

The upper half begins at 65 and ends at 89.

$LQ = 53$

$median = \frac{61 + 65}{2}$

$= 63$

$UQ = 76$

The interquartile range is $76 - 53$ or 23. This indicates that the middle half, or 50%, of the data varies by 23 points.

3 The Burch Manufacturing Company held its annual golf tournament for its employees. The scores for 18 holes were 89, 90, 102, 105, 88, 115, 99, 100, 103, 167, 111, 99, 95, 95, 80, 102, 103, 106, 94, 112, and 85. Find the median, upper and lower quartiles, and interquartile range.

Stem	Leaf
8	0 5 8 9
9	0 4 5 5 9 9
10	0 2 2 3 3 5 6
11	1 2 5
12	
13	
14	
15	
16	7

The brackets group the values in the bottom half and the values in the upper half.

What do the boxes contain?

$LQ = \frac{90 + 94}{2}$

$= 92$

$median = 100$

$UQ = \frac{105 + 106}{2}$

$= 105.5$

$interquartile\ range = 105.5 - 92$

$= 13.5$

In the example above, one score, 167, seems to be very different from the others. A score that is much higher or much lower than the rest of the data is called an **outlier**. An outlier is any element of the set of data that is at least 1.5 interquartile ranges above the upper quartile or below the lower quartile.

An outlier will not affect the median, LQ, or UQ, but it will affect the mean.

To determine if 167 is an outlier, multiply 1.5 times the interquartile range, 13.5. Next, add this product to the UQ. Then, subtract this product from the LQ. These values give boundaries for determining outliers.

(1.5)(13.5) = 20.25
105.5 + 20.25 = 125.75
92 − 20.25 = 71.75

Since $167 > 125.75$, the value is an outlier. There are no elements that are more than 1.5 interquartile ranges below the lower quartile.

No values are below 71.75.

Exploratory Exercises

Find the median and upper and lower quartiles for each set of data.

1. 12, 23, 17, 18, 16

2. 3, 6, 7, 2, 4, 5, 8, 1, 5

3. 34, 45, 56, 25, 37, 43

4. 78, 96, 65, 84, 99, 68, 77

5. 100, 90, 80, 70, 60, 50, 40, 30

6. 3.2, 4.5, 2.6, 3.4, 5.3, 7.8

Written Exercises

Find the median and upper and lower quartiles for each set of data.

7.

Stem	Leaf
7	3 5 7 8
8	0 1 3 5 7
9	4 6 8
10	0 1 6
11	8 9

$9|4 = 94$

8.

Stem	Leaf
3	0 2
4	5 7 9
5	0 2 4 6
6	3 4 4 5 5
7	7 8 8

$5|2 = 52$

9.

Stem	Leaf
5	5 8 9
6	2 7 9 9
7	3 5 6 7 7
8	2 5
9	2

$6|7 = 67$

Solve.

10. The points scored by individuals on the Atlanta Hawks professional basketball team in the 1986–1987 season are 2294, 1304, 1053, 900, 788, 657, 223, 381, 463, 405, and 342. Find the median, upper and lower quartiles, and interquartile range. Identify any outlier values.

11. The Goodwill Games were held in Moscow in July 1986. The total number of medals won by the top eleven countries are 241, 28, 31, 18, 6, 10, 6, 142, 9, 11, and 6. Find the interquartile range. Identify any outliers.

12. From 1960 to 1968, the leading bowlers in the PBA had annual earnings of $22,525, $26,280, $49,972, $46,333, $33,592, $47,674, $54,720, $54,165, and $67,377. Find the range of earnings and the interquartile range.

13. During the 1986 National Football League season, the number of passes attempted by the leading quarterbacks was 372, 307, 305, 468, 421, 541, 294, 245, 378, 492, 342, and 363. Find the range and the interquartile range.

14. The top earnings of the Professional Rodeo Cowboy champions from 1966 to 1975 were $40,358, $51,996, $49,129, $57,726, $41,493, $49,245, $60,852, $64,447, $66,929, and $50,300. Find the interquartile range.

15. The number of yards gained by the top seven rushers in the National Football League in 1986 was 1203, 1327, 1353, 1333, 1516, 1821, and 903. Find the range and the interquartile range.

State whether the graph of each quadratic function opens upward or downward.

1. $y = x^2 - 1$ **2.** $f(x) = -x^2 - 2x + 3$

Find the equation of the axis of symmetry and the coordinates of the maximum or minimum point of the graph of each quadratic function.

3. $y = 3x^2 + 6x - 17$ **4.** $y = -4x^2 + 8x + 13$

Use the Quadratic Formula to solve each equation. Leave irrational roots in simplest form.

5. $x^2 + 8x + 15 = 0$ **6.** $4x^2 + 20x = -23$

7. $24w^2 - 2w - 15 = 0$ **8.** $-2x - 2 = -x^2$

Use the discriminant to find the nature of the roots of each equation.

9. $x^2 - 3x + 8 = 0$ **10.** $x^2 + 10x + 25 = 0$

11. $x^2 - 2x - 8 = 0$ **12.** $x^2 - 3x - 10 = 0$

13. Find two integers whose difference is 6 and whose product is 135.

14. The perimeter of a rectangle is 48 cm and its area is 143 sq cm. Find its length and width.

State the sum and product of the roots of each equation.

15. $x^2 - 3x + 5 = 0$ **16.** $3x^2 - 4x - 7 = 0$

Write a quadratic equation that has the given roots.

17. $9, -11$ **18.** $\frac{3}{4}, -\frac{2}{3}$

Simplify.

19. $\frac{r^3 - r^2}{r - 1}$ **20.** $\frac{25 - x^2}{x^2 + x - 30}$

21. $\frac{\frac{x + y}{a + b}}{\frac{x^2 - y^2}{a^2 - b^2}}$ **22.** $\frac{\frac{x - y}{x + y}}{\frac{x + y}{x - y}}$

Perform the indicated operation and simplify.

23. $\frac{x - 3}{7} \cdot \frac{21}{x - 3}$ **24.** $\frac{2a + 4b}{5} \cdot \frac{25}{6a + 8b}$

25. $\frac{x + 4}{5} + \frac{x - 1}{5}$ **26.** $\frac{8}{x - 1} - \frac{6}{x - 1}$

27. $\frac{5}{6} + \frac{3}{14}$ **28.** $\frac{8}{3} - \frac{10}{8}$

29. $\frac{m - 1}{m + 1} + \frac{4}{2m + 5}$ **30.** $\frac{-18}{y^2 - 9} - \frac{7}{3 - y}$

Divide.

31. $\frac{5}{6} \div \frac{3}{4}$

32. $\frac{k + 2}{m^2 + 4m + 4} \div \frac{4k + 8}{m + 4}$

Solve each equation.

33. $\frac{2x - 3}{6} = \frac{2x}{3} + \frac{1}{2}$

34. $\frac{2}{3x} - \frac{3x}{x - 2} = -3$

35. Make a stem and leaf plot of the given data.
34, 27, 46, 18, 36, 31, 48, 23, 36, 31, 30, 29, 33, 33

36. Find the mean, median, and mode of the given data.
45, 56, 48, 27, 100, 38, 48, 27, 39, 15, 24, 14, 14, 14, 26, 37, 39, 20, 22

Find the range, upper and lower quartiles, and interquartile range of each set of data.

37.

Stem	Leaf
3	4 5 7
4	0 2 5 7 8
5	2 3 6 6
6	1 4 4
7	1 2

$5|3 = 53$

38.

Stem	Leaf
2	1 6
3	0 3 6 7
4	2 5 7
5	0 7 8
6	4 5

$s|6 = 36$

15-6 Box and Whisker Plots

The quartiles of a set of data can be displayed in graphical form using a **box plot**. Data for a box plot must be in numerical order or shown in a stem and leaf plot.

Example

1 **Organize the given information in a stem and leaf plot. Then, use the stem and leaf plot to draw a box plot.**

39, 23, 46, 67, 55, 38, 46, 61, 27, 52, 49

First, make a stem and leaf plot.

Next, draw a vertical line beside the stem and leaf plot. Assign a scale to the line to include the *extreme* values of the data. Plot dots to represent the extremes, the upper and lower quartiles and the median.

Then, draw a box around the interquartile range. Mark the median by a line through its point in the box. Draw a segment from the lower quartile to the lower extreme and from the upper quartile to the upper extreme.

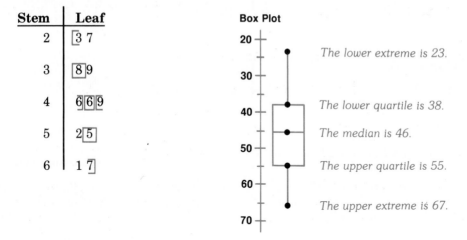

Stem	Leaf
2	3 7
3	8 9
4	6 6 9
5	2 5
6	1 7

Box Plot

The lower extreme is 23.

The lower quartile is 38.

The median is 46.

The upper quartile is 55.

The upper extreme is 67.

The segments from the box to the extremes are called *whiskers*, thus the name **box and whisker plot**.

In a box and whisker plot, the extreme values, upper and lower quartiles, and median are shown as dots. The rest of the data is included in either the box or a whisker. Each whisker contains one-fourth of the data while the box contains one-half of the data.

A set of data with extremes 34 and 99, upper and lower quartiles of 93 and 81, and median of 88.5 is graphed on a *horizontal* box plot. Note that one value is not included in the box and whisker graph, but appears as a dot by itself. This indicates that the value 34 is an *outlier*.

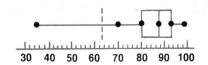

Exploratory Exercises

Complete each question for the indicated box plot.

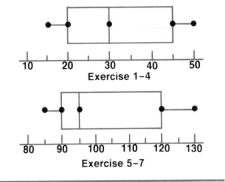

1. What percent of the data is between 45 and 50?
2. What is the interquartile range?
3. What part of the data is between 20 and 45?
4. What percent of the data is less than 20?

Exercise 1–4

5. What is the median?
6. What is the least element in this set of data?
7. What percent of the data is between 95 and 120?

Exercise 5–7

Written Exercises

Complete each question.

8. What part of the data is between 60 and 90?
9. What is the lower quartile?
10. Name a set of data that would give a box plot with no whiskers.

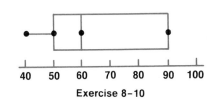

Exercise 8–10

Compare box plots X and Y to complete Exercises 11–14.

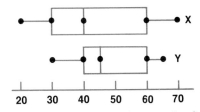

11. Which plot has the lesser median?
12. Which plot has the greater range?
13. Which plot has the lesser interquartile range?
14. Which plot has the widest spread of data?

15. Speeds of the fastest train runs in the U.S. and Canada are given below in miles per hour. Make a box and whisker plot of the data.
97.8, 95.6, 83.7, 91.2, 88.5, 90.2, 93.3, 85.5, 88.4, 87.8, 84.7, 83.2, 86.2, 95.4, 83.7, 86.3

16. The average number of points scored per game by each player on the Utah Jazz basketball team during the 1986–1987 season are 14.6, 11.7, 8.7, 8.5, 29.8, 14.9, 7.7, 4.7, 4.6, 4.5, and 4.4. Make a box and whisker plot of the data. Are there any outliers?

17. The number of completed passes for the leading receivers for the teams in the National Football Conference in 1986 was 63, 81, 80, 76, 66, 73, 74, 64, 86, and 61. Make a box and whisker plot of this data.

18. The number of completed passes for leading receivers for each team in the American Football Conference in 1986 was 77, 65, 84, 70, 95, 71, 80, 69, 67, 85, and 65. Make a box and whisker plot of this data. How does this box plot compare with the one in Exercise 17?

15-7 Scatter Plots

Jenelle's assignment in her sociology class was to conduct a survey and interpret the results. She wanted to determine whether there was a relationship between grades earned on a test and time spent studying for that test. She obtained the necessary information from ten classmates and organized it in the table shown.

The information in the table can be displayed in a graph called a **scatter plot**.

A scatter plot is a graph of two variables or paired data.

Student	Study Time	Test Score
Rebeca	15 min	68
Bradley	1 hr 15 min	87
Justine	1 hr	92
Allison	45 min	73
Tami	1 hr 30 min	95
Mick	1 hr	83
Montega	30 min	77
Christy	2 hr	98
Doug	10 min	65
Shelley	2 hr	94

Example

1 **Using the information in the table below, construct a scatter plot of the data. Refer to the scatter plot to answer the questions below.**

First, express individual study times in minutes.

Plot the data on vertical and horizontal axes. Place the data for time studied on the horizontal axis and the data for test scores on the vertical axis.

Student	Study Time	Test Score
Rebeca	15 min	68
Bradley	75 min	87
Justine	60 min	92
Allison	45 min	73
Tami	90 min	95
Mick	60 min	83
Montega	30 min	77
Christy	120 min	98
Doug	10 min	65
Shelley	120 min	94

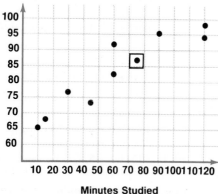

Test Score

Minutes Studied

a. Who studied and still did poorly? *Allison*

b. Describe the point with the box around it. *The point represents a study time of around 75 minutes with a resulting test score of 87.*

c. Does the graph show a relationship between time studied and test scores? *In general, the plot seems to indicate the greater amount of time spent studying, the higher the grade.*

There is a pattern, or association, between the variables on a scatter plot. This association can be negative or positive. Note in Example 1 how the points *slant* upward, resembling a line with a positive slope. Thus, this association is said to be *positive*.

In what direction does a line with a negative slope slant?

The scatter plot at the right compares hourly wages with the number of hours worked. Since the points are so scattered, there appears to be no association between the variables in this scatter plot. It is difficult to make predictions from a scatter plot when there is no association between the variables.

The scatter plot can still be used to answer questions. How many people worked less than 20 hours? How many people earned $16 per hour?

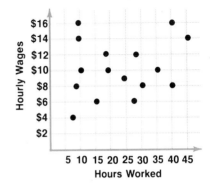

Example

2 The scatter plot shown represents the number of hours per week individuals watch television paired with the number of hours per week they spend in physical activity. Is there any association between time spent watching television and physical activity? *yes*

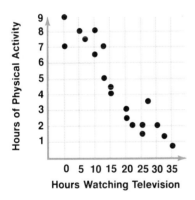

There appears to be a negative association between watching television and physical activity. The more time spent watching television, the less time spent in physical activity.

Exploratory Exercises

1. Collect data from the class of the number of hours per week students watch television paired with the number of hours per week they spend studying. Construct a scatter plot and compare it to the scatter plot in the example.

Determine whether a scatter plot of the data for the following would show positive, negative, or no association between the variables.

2. heights of sons and fathers

3. speed of a car and miles per gallon

4. amount of money earned and spent

5. your height and month of birth

6. age of a car and its value

7. playing time and points scored

Written Exercises

The defensive statistics for the number of tackles for the 1985 football season of the Chicago Bears is listed below. Use this data to complete Exercises 8–12.

Player	Solo Tackles	Assists
Fencik	98	20
Singletary	89	24
Marshall	60	18
Wilson	54	15
Duerson	55	9
Frazier	45	7
McMichael	39	5
Richardson	37	4
Dent	33	5
Hampton	26	6
Perry	26	5
Hartenstine	15	4
Gayle	16	2
Rivera	12	2

8. Construct a scatter plot of the data.

9. Does the scatter plot show a positive or negative association?

10. Is it possible to predict the number of assists a defensive player would have if you were given the number of solo tackles for that player?

11. Does the position played affect the number of tackles made by a player? How?

12. Does the amount of time played affect the number of tackles made by a player?

The maximum and minimum monthly temperatures in July of ten cities in eastern states are listed in the table at the right.

City	Maximum	Minimum
Hartford, CT	85	62
Baltimore, MD	87	67
Boston, MA	82	65
Portland, ME	79	57
Concord, NH	83	56
Albany, NY	83	60
New York, NY	84	69
Burlington, VT	81	59
Norfolk, VA	90	70
Huntington, WV	86	65

13. Make a scatter plot of the data.

14. Which city has the highest maximum and minimum temperature?

15. Which city has the lowest minimum and maximum temperature?

16. Is there a cluster of cities? If so, explain why.

mini-review

Simplify.

1. $\dfrac{\frac{4 - x^2}{2}}{\frac{2 - x}{5}}$

2. $\dfrac{2a^2 - 3a - 9}{a^3 - 9a}$

3. $\dfrac{t^2 - 3t - 28}{2t^2 - 128} \div \dfrac{t^2 - t - 42}{t^3 + 6t^2 - 16t}$

Solve.

4. $\dfrac{p - 8}{2} + \dfrac{p}{3} = 10$

5. $\sqrt{7q + 1} - 6 = 2$

6. $9x^2 - 1 = 12x$

7. Make a box and whisker plot of the set of data given.

8, 12, 15, 23, 16, 7, 11, 9, 24, 26, 45, 55

8. Make a scatter plot of the set of ordered pairs given.

{(23, 12), (5, 8), (13, 22), (2, 27), (35, 34), (23, 4), (12, 34), (20, 30), (40, 15), (30, 25), (15, 8), (8, 15)}

Is the number of police officers related to the number of crimes? Is the amount of time spent studying related to test scores? The relationship between two measures made on a population can be examined through statistics.

The relationships described above involve two variables. Therefore, they can be denoted with ordered pairs (x, y).

The ordered pairs of data can be graphed on a coordinate system as points. All such points may lie on the same line or lie close to a line. If so, they are said to be *linearly related*. A **regression line** is a line that best fits data that are linearly related.

x	y
Number of police officers	Number of crimes
Time spent studying	Test Scores

A graphing calculator can be used to graph the ordered pairs (x, y) for a particular relationship and the regression line that best fits the data.

Example **Graph the ordered pairs of data given in the table at the right and draw a regression line.**

x	y
-8	-4
-6	3
2	-1
4	5
-1	9

A. Set the LR2 mode. `SHIFT` `MODE` `÷`

B. Set the range values as desired.

C. Clear the statistical memories. `SHIFT` `SCL` `EXE`

D. Enter the data.

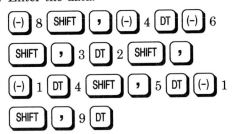

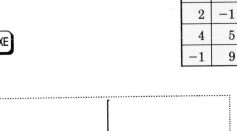

E. Draw the graph.

`GRAPH` `SHIFT` `LINE` `1` `EXE`

Exercises

Use the above procedure to graph each set of data and draw a regression line.

1.

x	-2	-1	2	3	4	7
y	1	-3	-2	5	2	6

2.

x	-3	-3	-2	-1	1	2
y	6	3	5	1	-1	-1

3. Do the data in Exercise 1 or 2 appear to have a stronger relationship? Why?

15-8 Probability and Odds

Will the Cubs or the Giants be more likely to win the game? What is the chance of drawing an ace from a deck of cards? What are the possibilities of rain today? What are the chances of getting *heads* in one toss of the coin? When we are uncertain about the occurrence of an event, we can attempt to measure the chances of it happening with *probability*.

The **probability** of an event is a ratio that tells how likely it is that an event will take place. The numerator is the number of favorable outcomes and the denominator is the number of possible outcomes. For example, when you toss a die, there are six ways it can fall. The probability of getting a "2" on one roll of a die is one chance out of six, or $\frac{1}{6}$. $P(2)$ means *the probability of getting a 2 on one toss of a die*.

$$P(\text{event}) = \frac{\text{number of favorable outcomes}}{\text{number of possible outcomes}}$$

Definition of Probability

Examples

1   Diego has a collection of tapes that he plays regularly. He has six rock tapes, three country tapes, and four movie sound track tapes. If Diego chooses a tape at random, what is the probability that he will pick a country tape?

$P(\text{country tape}) = \dfrac{\text{number of country tapes}}{\text{total number of tapes}}$ *number of favorable outcomes*
number of possible outcomes

$\phantom{P(\text{country tape})} = \dfrac{3}{13}$

The probability of choosing a country tape is 3 out of 13.

2   Kendra has a collection of various cereals on a shelf in the cabinet. Five of the cereals contain corn, two contain rice, and four contain oats. Without looking, she selects a box of cereal for breakfast. What is the probability that the cereal she selects will contain oats?

$P(\text{oats}) = \dfrac{\text{number of oat cereals}}{\text{total boxes of cereal}}$ *number of favorable outcomes*
number of possible outcomes

$\phantom{P(\text{oats})} = \dfrac{4}{11}$

The probability of choosing an oat cereal is 4 out of 11.

The probability of any event is always a value from 0 to 1, inclusive. In algebra this is written $0 \le P(\text{event}) \le 1$. If the probability of an event is 0, it is impossible for that event to occur. An event that is certain to occur has a probability of 1.

The probability of the occurrence of an event may be written as a percent, a fraction, or a decimal. In this way, probabilities can be compared. For example, suppose the probability of event A is 0.6 and the probability of event B is 0.5. Since $0.6 > 0.5$, event A is more likely to occur than event B.

Suppose a weather forecaster states that the probability of rain today is 0.25 or $\frac{1}{4}$. This means that the probability that it will not rain is 0.75 or $\frac{3}{4}$. The odds that it will rain today are $1:3$. The odds that it will not rain today are $3:1$.

The odds of an event occurring is the ratio of the number of ways the event can occur (successes) to the number of ways the event cannot occur (failures). *Odds = successes : failures*	*Definition of Odds*

Examples

3   **Eleven poker chips are numbered consecutively 1 through 10 with two of them labeled with a 6 and placed in a jar. A chip is drawn at random. Find the probability of drawing a 6.**

$P(6) = \dfrac{\text{number of chips labeled 6}}{\text{number of chips in the jar}}$

$= \dfrac{2}{11}$ The probability of drawing a 6 is 2 out of 11.

4   **Find the odds of drawing a 6 from the jar in Example 3. Find the odds of not drawing a 6.**

Odds of drawing 6 = $\dfrac{\text{number of chances}}{\text{to draw 6}} : \dfrac{\text{number of chances to}}{\text{draw other numbers}}$

$= 2:9$ *Read "2 to 9."*

Odds of not drawing a 6 = $\dfrac{\text{number of chances to}}{\text{draw other numbers}} : \dfrac{\text{number of chances}}{\text{to draw 6}}$

$= 9:2$ *Read "9 to 2."*

Exploratory Exercises

The probability of occurrence of any event can be shown on the number line below. Copy the number line and locate the probability of each event described.

```
0                              1/2                         1
|------------------------------|---------------------------|
Impossible    Unlikely    Equally Likely    Likely    Certain
```

1. It will rain today.
2. Today is Saturday.
3. You are in Algebra class.
4. A coin will land tails up.
5. You will pass the next test.
6. You will go skiing tomorrow.

Find each probability if a die is rolled.

7. $P(3)$
8. $P(\text{even number})$
9. $P(\text{number less than } 1)$
10. $P(\text{a number divisible by } 4)$
11. $P(\text{a number greater than } 1)$

Find the odds in favor of each outcome if a die is rolled.

12. a number greater than 3
13. a multiple of 2
14. not a 4
15. a number divisible by 3

Written Exercises

16. If the probability of an event occurring is $\frac{2}{3}$, what are the odds of the event occurring?

17. If the probability of an event occurring is $\frac{3}{7}$, what are the odds that it will not occur?

18. If the odds in favor of an event occurring are 7:5, what is the probability of the event occurring?

19. If the odds against an event occurring are 9:14, what is the probability of the event occurring?

The number of males and females enrolled in Darien High School are listed per class in the table below.

Darien High School		
Grade	Male	Female
9	120	150
10	100	100
11	130	110
12	150	175

20. If a student is chosen at random, what is the probability that the student is a female?

21. If a student is chosen at random, what is the probability that the student is a male in Grade 11?

22. If one student is chosen to represent the student body, what are the odds in favor of selecting a female?

23. If one student is chosen from Grade 12, which is more likely, selecting a male or selecting a female?

A card is selected at random from a deck of 52 cards.

24. What are the odds in favor of selecting a heart?

25. What is the probability of selecting an ace?

15-9 Drawing Conclusions from Experiments

Mike conducted an experiment by using two bags of candy-coated chocolate candies. The number of each color of candy was counted and recorded in the table below.

Color	Bag 1	Bag 2	Total
Red	12	12	24
Brown	16	16	32
Orange	10	8	18
Yellow	9	11	20
Green	8	7	15
Tan	5	5	10
	60	60	120

Mike found the following probabilities as a result of his experiment.

$$P(\text{yellow}) = \frac{\text{number of yellows}}{\text{total number of trials}} = \frac{20}{120} = 0.167$$

$$P(\text{tan}) = \frac{10}{120} = 0.083 \qquad P(\text{brown}) = \frac{32}{120} = 0.267$$

$$P(\text{red}) = \frac{24}{120} = 0.2 \qquad P(\text{green}) = \frac{15}{120} = 0.125$$

$$P(\text{orange}) = \frac{18}{120} = 0.15$$

Based on these results, which color is most likely to occur in a bag of candy-coated chocolate candies? Which color is least likely to occur?

Probability calculated by performing experiments is called **experimental** or **empirical probability**. Experimental probabilities are not exact; they are only approximations. The results will vary for each trial of an experiment. However, if a large number of trials are performed, the results will be more accurate.

If an experiment is repeated many times and the probabilities for each event are the same, then the events are **equally likely**. Notice that the events above are not equally likely.

Example

1 When a die is tossed, each number has an equal chance of appearing on top. The probability of any number is $\frac{1}{6}$ or 0.167. The results of 100 tosses of a die are shown in the following table.

Experimental Trial						
Number on die	1	2	3	4	5	6
Number of times each occurred	15	15	19	17	14	20

$P(1) = \frac{15}{100} = 0.15 \qquad P(2) = \frac{15}{100} = 0.15 \qquad P(3) = \frac{19}{100} = 0.19$

$P(4) = \frac{17}{100} = 0.17 \qquad P(5) = \frac{14}{100} = 0.14 \qquad P(6) = \frac{20}{100} = 0.20$

Notice that all of these results are close to $\frac{1}{6}$ or 0.167.

Exploratory Exercises

Determine whether each event described below has equally likely outcomes.

1. tossing a fair coin

2. passing a test

3. winning a golf game

4. tossing a die

5. winning a lottery

6. choosing a soft drink at random from a machine

Written Exercises

Toss two dice at least 100 times. Record the sums of the faces. Use the results to answer each question.

7. What is the probability of a sum of 2?

8. What is the probability of a sum of 7?

9. What is the probability of a sum of 11?

10. What is the probability of a sum of 10?

11. What is the probability of a sum less than 5?

12. What sum is most likely to occur?

Toss three coins at least 100 times. Record the results and use them to answer each question.

13. What is the probability of tossing three heads?

14. What is the probability of tossing two heads?

15. What is the probability of tossing one head?

16. What is the probability of tossing zero heads?

17. What is the probability of tossing at least two heads?

18. What is the probability of tossing at most two heads?

Complete the following experiment. Then answer each question.

19. From a deck of cards, remove the ace to 10 of one suit. Shuffle these cards, then deal out one card. Record the result and replace the card. Repeat the experiment 100 times.

20. From the results in Exercise 19, determine the probability of getting the same card twice in a row. Then, determine the same for three times in a row.

21. From the results in Exercise 19, determine the longest streak of consecutively numbered cards, using the ace as a *one*. What is the probability of that streak?

15-10 Compound Events

A **compound event** consists of two or more *simple events*. Tossing a die is a simple event. Tossing two dice is a compound event. The probability of a compound event can be calculated if its outcomes are equally likely.

Examples

1   **If three coins are tossed, what is the probability of getting exactly two heads?**

To calculate the probability, you need to know how many outcomes are possible. This may be done by using a tree diagram.

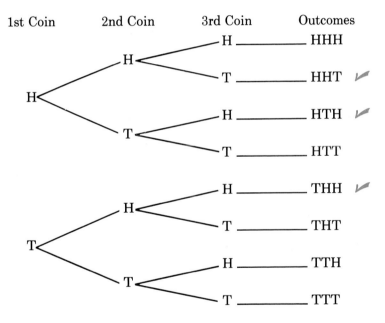

There are eight possible outcomes and three of them have exactly two heads. Therefore, the probability of getting exactly two heads in one toss of three coins is $\frac{3}{8}$ or 0.375.

2   **Jody has four bottles of soft drink—one bottle of cola, one of root beer, one of ginger ale, and one of orange. She chooses three of these bottles to take to a party. If she chooses the ginger ale, what is the probability she also chooses the root beer?**

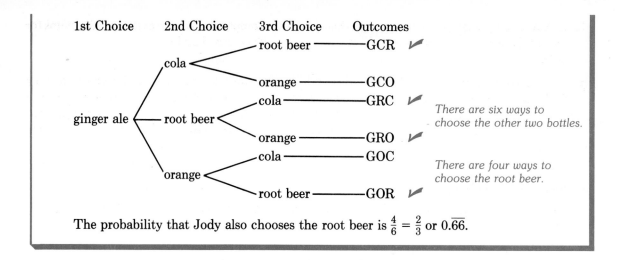

There are six ways to choose the other two bottles.

There are four ways to choose the root beer.

The probability that Jody also chooses the root beer is $\frac{4}{6} = \frac{2}{3}$ or $0.\overline{66}$.

Exploratory Exercises

An automobile dealer has cars available with the combinations of colors, engines, and transmissions indicated in the following tree diagram. A selection is made at random.

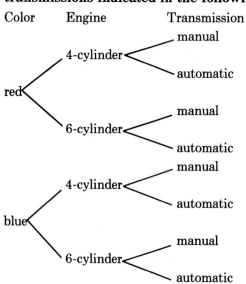

1. What is the probability of selecting a car with manual transmission?

2. What is the probability of selecting a blue car with manual transmission?

3. What is the probability of selecting a car with a 4-cylinder engine and a manual transmission?

4. What is the probability of selecting a blue car with a 6-cylinder engine and an automatic transmission?

Written Exercises

Draw a tree diagram for each exercise. Use the results to answer each question.

5. Find the probability of getting exactly three tails when four coins are tossed.

6. Find the probability that a family with four children has exactly four girls. Assume that the probability a girl is born is the same as the probability a boy is born.

7. In Exercise 6, what is the probability that the family has two boys and two girls in any order?

8. Compare and contrast the tree diagrams for Exercises 5 and 6.

For each shrimp, lobster, or chicken dinner in a restaurant, you have a choice of soup or salad. With shrimp you may have hash browns or a baked potato. With lobster you may have rice or hash browns. With chicken you may have rice, hash browns, or a baked potato. If all combinations are equally likely to be ordered, find each probability of an order containing each of the following.

9. shrimp

10. rice

11. shrimp and rice

12. soup and hash browns

13. chicken, salad, and rice

Bill, Raul, and Joe are in a bicycle race. If each boy has an equal chance of winning, find each probability.

14. Joe wins the race.

15. Raul finishes last.

16. Joe, Raul, and Bill finish first, second, and third, respectively.

Adam's class set up a lottery with two-digit numbers. The first digit is a number from 1 to 4. The second digit is a number from 3 to 8.

17. What is the probability that 44 was the winning number?

18. What is the probability that a number with a 2 in it wins?

Box A contains one blue marble and one green marble. Box B contains one green marble and one red marble. Box C contains one white marble and one green marble.

19. If a marble is drawn at random from each box, what is the probability that all three marbles are green?

20. What is the probability that exactly two marbles are green?

mini-review

Find the equation of the axis of symmetry and the coordinates of the maximum or minimum point of the graph of each quadratic function.

1. $y = x^2 + 3x - 12$

2. $y = -x^2 - 5x - 1$

3. $y = 3x^2 - 8x + 2$

Simplify.

4. $\frac{n^2 - 4}{2n} \div (n + 2)$

5. $\frac{16}{b + 4} - \frac{b^2}{b + 4}$

6. $\frac{3a}{15a + 9} + \frac{2}{5a + 3}$

Solve.

7. $\frac{n + 4}{6} = \frac{3n - 1}{2} + 1$

8. $\frac{a + 1}{3} = \frac{4a}{3} - \frac{1}{2}$

9. $\frac{y + 2}{y - 3} + \frac{y + 3}{y - 2} = 1$

Solve each problem.

10. The perimeter of a rectangle is 18 yards and its area is 19.44 square yards. Find its dimensions.

11. Seki bowled six games, and her scores were 124, 132, 141, 123, 136, and 145. Find her mean and median bowling scores.

Bar Graphs

Graphs are often used to present data and show relationships. The following program uses the data given at the right to construct a **bar graph**. Specific quantities can then be easily compared. The program uses low resolution graphics to draw the graph.

```
10   REM  ENTER NUMBER OF MEDALS, YEAR
20   FOR N = 1 TO 6
30   READ A(N),Y(N)
40   NEXT N
50   REM  ENTER GRAPHICS MODE
60   GR
70   REM  SET COLOR,MAKE BORDER
80   COLOR = 15
90   HLIN 0,39 AT 0
100  HLIN 0,39 AT 39
110  VLIN 0,39 AT 0
120  VLIN 0,39 AT 39
130  REM  CONSTRUCT GRAPH
140  FOR N = 1 TO 6
150  VLIN (38 − A(N)),38 AT N * 5
160  NEXT N
170  FOR N = 1 TO 6
180  PRINT  TAB( N * 5 + 1);A(N);
190  NEXT N
200  PRINT
210  FOR N = 1 TO 6
220  PRINT  TAB(N * 5);Y(N);
230  NEXT N
240  DATA 7,1968,8,1972,10,1976
250  DATA 12,1980,8,1984,6,1988
260  END
```

Number of Medals Won by the U.S. Winter Olympic Games Teams					
1968	1972	1976	1980	1984	1988
7	8	10	12	8	6

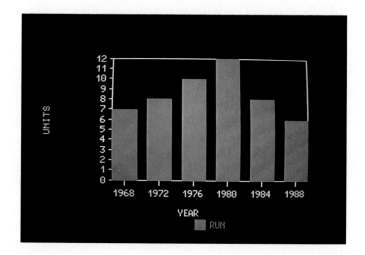

Exercises

Change lines 240 and 250 to make bar graphs for each table.

1.

Median Age of U.S. Population					
1940	1950	1960	1970	1980	1986
29.0	30.2	29.5	28.0	30.0	31.8

2.

Number of Unemployed Persons in the U.S. (in millions)					
1981	1982	1983	1984	1985	1986
8.3	10.7	10.7	8.5	8.3	8.2

3. What do you think line 180 does?

4. How would you change the program to construct a bar graph with five bars instead of six?

Vocabulary

data (513)
statistics (513)
line plot (515)
stem and leaf plot (517)
mean (520)
median (520)
mode (521)

range (524)
upper quartile (524)
lower quartile (524)
interquartile range (524)
outlier (525)
box and whisker plot (528)
scatter plot (530)

probability (534)
odds (535)
experimental or empirical
 probability (537)
equally likely events (537)
compound event (539)

Chapter Summary

1. Statistics is a branch of mathematics that provides methods for collecting, organizing, and interpreting data. (513)

2. The mean of a set of data is the sum of the elements in the set divided by the number of elements in the set. (520)

3. The median is the middle number of a set of data when the numbers are arranged in numerical order. (520)

4. The mode is the number that occurs most often in a set of data. (521)

5. The difference between the greatest and the least values of a set of data is called the range. (524)

6. The upper quartile (UQ) divides the top half of a set of data into two equal parts. The lower quartile (LQ) divides the bottom half of a set of data into two equal parts. The difference between the UQ and LQ is called the interquartile range. (524)

7. Ways to display statistical data include line plots, stem and leaf plots, box plots, box and whisker plots, and scatter plots. (515, 517, 528, 530)

8. The probability of an event is a ratio that tells how likely it is that an event will take place.
$$P(\text{event}) = \frac{\text{number of favorable outcomes}}{\text{number of possible outcomes}} \quad (534)$$

9. The odds of an event occurring is the ratio of the number of ways the event can occur (successes) to the number of ways the event cannot occur (failures).
$$\text{Odds} = \text{successes} : \text{failures} \quad (535)$$

10. Probability calculated by performing experiments is called experimental or empirical probability. (537)

11. If an experiment is repeated many times and the probabilities for each event are the same, then the events are equally likely. (537)

12. A compound event consists of two or more simple events. (539)

Chapter Review

15–1 The following table lists the cost of selected goods and services from 1980 to 1985. Use the table to complete the exercises below.

1. Which item had the greatest percentage of increase over a 1-year change?

2. Which item had the greatest percentage of decrease over a 5-year change?

3. The cost of which item changed the least from 1980 to 1985?

Item	1-Year Change	5-Year Change
New cars	+3.5%	+18.9%
Gasoline	+3.0%	+ 0.9%
Used cars	−1.9%	+60.2%
Stereos	−3.1%	− 6.6%
Long-distance calls	−3.8%	+ 9.8%
Television sets	−8.0%	−19.1%
Driver's-license fees	+6.5%	+65.5%

15–2
15–3
4. Make a line plot and a stem and leaf plot of the following data.
123, 167, 133, 138, 125, 134, 145, 155, 152, 159, 164, 135, 144, 156

15–4 **5.** From the data in Exercise 4, determine the mean, median, and mode.

15–5 Find the median, the upper and lower quartiles, and the interquartile range for each set of data.

6.

Stem	Leaf
6	2 4
7	0 2 5 8
8	3 6 6 6
9	0 2 6 7 7
10	2 5 7

$8|6 = 86$

7.

Stem	Leaf
23	4 6
24	3 6 8
25	6 7 8 8
26	0 3 5
27	8

$23|4 = 234$

15–6 Complete each question for the indicated box plot.

8. What is the upper extreme?

9. What is the median?

10. What is the upper quartile?

11. What percent of the data is between 50 and 85?

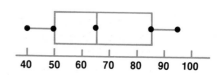

15–7 **12.** Make a scatter plot of the set of ordered pairs given. Then determine whether there is a positive, negative, or no association between the variables. {(40, 2), (10, 10), (5, 12), (35, 3), (25, 5)}

15–8 **13.** If the odds in favor of an event are 9:4, what is the probability of the event occurring?

15–9 **14.** Toss two coins 100 times. What is the probability of tossing one head and one tail?

15–10 **15.** Gary has one blue coat, one gray coat, one pair of gray slacks, one pair of blue slacks, one white shirt, one gray shirt, and one red shirt. If Gary chooses the gray slacks, what is the probability he also chooses the blue coat?

1. Construct a line plot of the following data.

78, 74, 86, 88, 99, 103, 85, 85, 85

2. Make a stem and leaf plot of the following data.

42, 44, 76, 56, 78, 62, 65, 69, 41, 40, 55, 66, 42

3. Find the median, quartiles, interquartile range, and range for the following data.

Stem	Leaf
6	3 4
7	0 1 5
8	7 8 8 9
9	1 6
10	2

$9|6 = 96$

4. Construct a box plot of the following data. State the upper and lower quartiles, the upper and lower extremes and the median.

8, 3, 6, 13, 14, 11, 10, 9, 7, 8, 8, 5, 11, 14, 15

5. Construct a scatter plot of the given data. Let the horizontal axis be *Shots Attempted*. State if there is a positive, negative, or no association shown.

Basketball Player	A	B	C	D	E	F	G	H	I	J
Shots Attempted	60	25	35	12	80	4	15	42	11	22
Shots Completed	25	10	15	4	36	1	6	16	4	8

6. A radio station played 7 rock records, 5 country records, and 8 easy listening records in one hour. If you just tuned in to this station, what is the probability you will hear a rock record?

7. A restaurant menu offers chicken, steak, and turkey as entrees. You may order soup or salad. For dessert, you may have pie, cake, or cheesecake. You order these items at random. Draw a tree diagram to show all possible dinners you might choose.

A table is shown below giving the ages of students in a class. Use the data to complete Exercises 8–10.

Age	Female	Male
15	5	6
16	6	3
17	8	2

8. If a student is chosen at random from this class, what is the probability that the student is a girl?

9. If a student is chosen at random, what is the probability that the student is a 16-year-old male?

10. If a student is chosen at random, what are the odds in favor of a female?

11. A weather forecaster states that the probability of rain is 0.35. What are the odds that it will not rain?

CHAPTER 16

Trigonometry

In playing pool, similar triangles are used when banking a shot. Pool is one of many games that involve triangles. Trigonometry is a branch of mathematics that involves triangles.

16-1 Angles

A protractor can be used to measure angles as shown below.

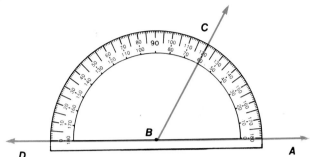

The symbol used for angle is ∠

∠ABC measures 60°. However, where ray *BC* intersects the curve of the protractor, there are two readings—60° and 120°. 120° is the measure of ∠DBC. What is the sum of the degree measures of ∠ABC and ∠DBC?

Two angles are **supplementary** if the sum of their degree measures is 180.	*Supplementary Angles*

Example

1 The measure of an angle is three times its supplement. Find the measure of each angle.

$x + 3x = 180$ *Let x be the measure of the lesser angle*
  $4x = 180$ *and 3x the measure of the greater angle.*
   $x = 45$

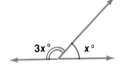

The measures are 45° and 3 · 45°, or 135°.   *Check this result.*

Two angles are **complementary** if the sum of their degree measures is 90.	*Complementary Angles*

Example

2 An angle is 16° greater than its complement. Find the measure of each angle.

$x + (x + 16) = 90$ *Let x be the measure of the lesser angle*
  $2x + 16 = 90$ *and x + 16 the measure of the greater angle.*
     $2x = 74$
      $x = 37$

The measures are 37° and 37° + 16°, or 53°.   *Check this result.*

What is the sum of the degree measures of the three angles of a triangle? Use a protractor to measure the angles of each triangle below. Then find the sum of the angle measures of each triangle.

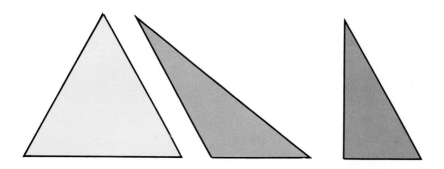

What did you discover? In each case, your sum should be approximately 180°.

The sum of the degree measures of the angles in any triangle is 180.	*Sum of the Angles of a Triangle*

Examples

3   **What are the measures of the angles of an equilateral triangle?**

In an equilateral triangle, the sides are congruent and the angles are congruent.

$x + x + x = 180$ *Let x be the measure of each angle.*

$3x = 180$

$x = 60$

Each angle measures 60°.

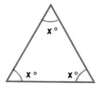

4   **What are the measures of the angles of an isosceles right triangle?**

An isosceles right triangle contains a right angle and two congruent angles.

$x + x + 90 = 180$ *What does x represent?*

$2x + 90 = 180$ *How do you know one angle is 90°?*

$2x = 90$

$x = 45$

The measures are 45°, 45°, and 90°.

Exploratory Exercises

Find the complement of each angle measure.

1. 85°
2. 42°
3. 13°
4. 45°
5. 55°
6. 24°
7. 11°
8. 76°
9. $x°$
10. $3x°$
11. $(2x + 40)°$
12. $(x - 7)°$

Find the supplement of each angle measure.

13. 130°
14. 65°
15. 127°
16. 87°
17. 90°
18. 32°
19. 108°
20. 156°
21. $y°$
22. $6m°$
23. $(3x + 5)°$
24. $(x - 20)°$

Written Exercises

Find both the complement and the supplement of each angle measure.

25. 42°
26. 87°
27. 125°
28. 160°
29. 90°
30. 68°
31. 21°
32. 174°
33. 99°
34. $x°$
35. $3y°$
36. $(x + 30)°$
37. $(x - 38)°$
38. $5x°$
39. $(90 - x)°$
40. $(180 - y)°$

Find the measure of the third angle of each triangle in which the measures of two angles of the triangle are given.

41. 16°, 42°
42. 40°, 70°
43. 50°, 45°
44. 90°, 30°
45. 89°, 90°
46. 63°, 12°
47. 43°, 118°
48. 4°, 38°
49. $x°$, $y°$
50. $x°$, $(x + 20)°$
51. $y°$, $(y - 10)°$
52. $m°$, $(2m + 1)°$

Solve.

53. One of the congruent angles of an isosceles triangle measures 37°. Find the measures of the other angles.

54. The measures of the angles of a certain triangle are consecutive even integers. Find their measures.

55. An angle measures 38° less than its complement. Find the measures of the two angles.

56. One angle of a triangle measures 53°. Another angle has measure 37°. What is the measure of the third angle?

57. One angle of a triangle measures 10° more than the second. The measure of the third angle is twice the sum of the first two angles. Find the measure of each angle.

58. One of two complementary angles measures 30° more than three times the other. Find the measure of each angle.

59. Find the measure of an angle that is 10° more than its complement.

60. Find the measure of an angle that is 30° less than its supplement.

61. Find the measure of an angle that is one-half the measure of its complement.

62. Find the measure of an angle that is one-half the measure of its supplement.

16-2 30°–60° Right Triangles

Suppose the measure of one acute angle of a right triangle is 30°. The measure of the other acute angle is 90° − 30° or 60°. Such triangles are called **30°–60° right triangles**.

An acute angle has a measure less than 90°.

Examine the equilateral triangle shown at the right. What is the measure of ∠B?

Line segment CD (denoted $\overline{CD}$) can be drawn perpendicular to $\overline{AB}$. What kind of right triangle is triangle CDB (denoted $\triangle CDB$)?

$\overline{CD}$ bisects $\overline{AB}$. Since the measure of $\overline{DB}$ is one-half the measure of $\overline{AB}$, it is also one-half the measure of the hypotenuse, $\overline{BC}$. Why?

The hypotenuse of a right triangle is the side opposite the right angle.

In a 30°–60° right triangle, the measure of the side opposite the 30° angle is one-half the measure of the hypotenuse.

This can also be stated as: In a 30°–60° right triangle, the measure of the hypotenuse is twice the measure of the side opposite the 30° angle.

In calculations, lower case letters are used to designate the measures of the sides of a triangle. For example, the measure of the side opposite angle R is r.

Example

1 Find the length of $\overline{BC}$ in $\triangle ABC$.

$a = \frac{1}{2}c$ *The measure of the hypotenuse is c.*

$a = \frac{1}{2}(8)$

$a = 4$

Thus, $\overline{BC}$ is 4 units long.

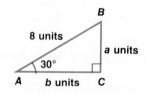

Example

2 **Find the length of $\overline{PQ}$ in $\triangle PQR$.**

$r = 2q$ *The measure of the side opposite*
$ = 2(6)$ *the 30° angle is q.*
$ = 12$

Thus, $\overline{PQ}$ is 12 units long.

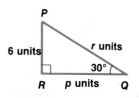

Suppose a represents the measure of the side opposite the 30° angle. The Pythagorean Theorem is used to find the length of the side opposite the 60° angle in a 30°–60° right triangle.

$a^2 + b^2 = c^2$ *Pythagorean Theorem*
$a^2 + b^2 = (2a)^2$ *The measure of the hypotenuse is*
$a^2 + b^2 = 4a^2$ *twice the measure of the side*
$ b^2 = 4a^2 - a^2$ *opposite the 30° angle.*
$ b^2 = 3a^2$
$ b = a\sqrt{3}$ *How is this obtained?*

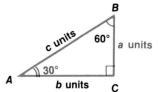

In a **30°–60° right triangle**, if a is the measure of the side opposite the **30°** angle, then $2a$ is the measure of the hypotenuse, and $a\sqrt{3}$ is the measure of the side opposite the **60°** angle.	*30°–60° Right Triangles*

Examples

3 **Find the length of $\overline{AC}$ in $\triangle ABC$.**

$b = a\sqrt{3}$ *The measure of the side opposite the 30° angle is a.*
$ = 5\sqrt{3}$ *The calculator entry is*
$ \approx 5(1.732)$ $5 \boxed{\times} 3 \boxed{\sqrt{x}} \boxed{=}$
$ \approx 8.660$

Thus, $\overline{AC}$ is approximately 8.660 units long.

4 **Find the lengths of $\overline{XY}$ and $\overline{XZ}$ in $\triangle XYZ$.**

$x = z\sqrt{3}$
$8\sqrt{3} = z\sqrt{3}$
$\phantom{8\sqrt{3}} 8 = z$ *Divide each side by $\sqrt{3}$.*

Since $y = 2z$, then $y = 2(8)$, or 16.

Thus, $\overline{XY}$ is 8 units long and $\overline{XZ}$ is 16 units long.

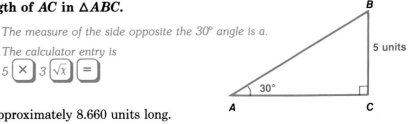

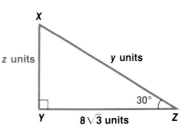

Exploratory Exercises

The length of the hypotenuse of a 30°–60° right triangle is given. Find the length of the side opposite the 30° angle in each triangle.

  1. 8 m
  2. 16 cm
  3. 13 mm
  4. 9 mi
  5. $4\frac{1}{2}$ in.
  6. $3\frac{3}{8}$ in.
  7. 16.36 m
  8. 4.63 cm

The length of the side opposite the 30° angle in a 30°–60° right triangle is given. Find the length of the hypotenuse in each triangle.

  9. 7 m
  10. 6.2 cm
  11. 4.35 mm
  12. $4\frac{1}{2}$ mi
  13. $6\frac{3}{8}$ in.
  14. 13 m
  15. 3.86 cm
  16. $7\frac{3}{4}$ in.

The length of the side opposite the 60° angle in a 30°–60° right triangle is given. Find the length of the other two sides in each triangle.

  17. $4\sqrt{3}$ ft
  18. $2\sqrt{3}$ cm
  19. $8\sqrt{3}$ m
  20. $\sqrt{3}$ mi
  21. $7\sqrt{3}$ yd
  22. $9\sqrt{3}$ mm

Written Exercises

Find the length of the third side in each triangle.

23.

24.
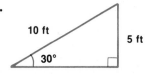

25.
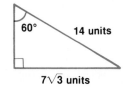

26.
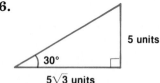

27.
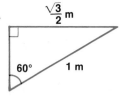

28.
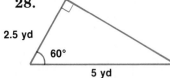

Find the lengths of the missing sides of each 30°–60° right triangle described below.

	Hypotenuse	Side Opposite 30° Angle	Side Opposite 60° Angle
29.	6 m	____	____
30.	4.75 mm	____	____
31.	$3\frac{1}{2}$ in.	____	____
32.	____	8 cm	
33.	____	6.5 m	____
34.	____	$3\frac{1}{4}$ in.	____
35.	____	____	$\sqrt{12}$ m
36.	____	____	$3.5\sqrt{3}$ cm

Challenge Exercises

Solve each problem.

37. In △ABC below, $\overline{AC}$ is 10 meters long, $\overline{AB}$ is $3\sqrt{3}$ meters long, and ∠A measures 30°. Find the length of $\overline{BC}$.

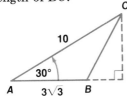

38. In △PQR below, $\overline{PS}$ is $8\sqrt{3}$ yards long. Find the perimeter of △PQR.

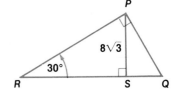

39. In rectangle JKLM shown at the right, the measure of ∠NJK is 30° and $\overline{KN}$ is 5 meters long. Find the perimeter of rectangle JKLM.

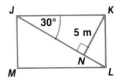

---

mini-review

Approximate the roots of each equation by graphing the related function.

1. $x^2 - 2x - 8 = 0$

2. $4x^2 - 8x + 3 = 0$

3. $x^2 + 2x + 4 = 0$

Use the Quadratic Formula to find the roots of each equation. Round each answer to the nearest hundredth.

4. $3x^2 + x - 5 = 0$

5. $m^2 - 6m = 10$

6. $-2y^2 + 7y - 1 = 0$

Simplify.

7. $\frac{6}{a^2}\left(\frac{a^2}{3b}\right)^2$

8. $\frac{4m+8}{2m} \cdot \frac{m^2}{m^2-4}$

9. $\frac{x}{y} \div \frac{x^7}{y^4}$

10. $\frac{k^2+2k-3}{k^2+6k+5} \div \frac{3k+9}{k^2-1}$

11. $\frac{3\frac{1}{4}}{6\frac{3}{4}}$

12. $\frac{\frac{n^2-4}{n^2-4n-12}}{n-2}$

State the scale you would use to plot each set of data.

13. 6.1, 6.7, 7, 7.5, 8.2, 8.8

14. 95%, 72%, 50%, 87%, 46%, 75%

Find the median from each stem and leaf plot.

15.
2	1 3 4
3	0 2 6
4	7 9
5	5

$4|7 = 47$

16.
1	0 5
3	1 2 3 7
4	6 8
9	0 1 5

$9|0 = 90$

17.
9	1 5 6
10	0 1 4 8
11	6 9
12	3 7 9

$12|3 = 123$

Solve each problem.

18. The perimeter of a rectangle is 16 yards and its area is 8.4 square yards. Find its dimensions.

19. Gary can paint the outside of a house in 40 hours. Sue can do the job in 36 hours. If they work together, how long will it take them to complete the job?

16-3 Similar Triangles

Are the triangles below the same size? Do they have the same shape?

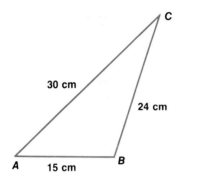

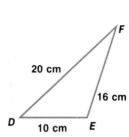

Measure the angles of $\triangle ABC$. Compare these with the measures of $\triangle DEF$. What do you discover?

Two triangles that have congruent angles are called **similar triangles**. The two triangles above are similar triangles. Similar triangles have the same shape, but not necessarily the same size.

If two triangles are similar, the congruent angles are called **corresponding angles**. Also, the sides opposite congruent angles are called **corresponding sides**.

Compare the measures of the corresponding sides. Note that AB means the measure of $\overline{AB}$, BC means the measure of $\overline{BC}$, and so on.

$$\frac{AB}{DE} = \frac{15}{10} \qquad \frac{BC}{EF} = \frac{24}{16} \qquad \frac{AC}{DF} = \frac{30}{20}$$

$$= \frac{3}{2} \qquad\qquad = \frac{3}{2} \qquad\qquad = \frac{3}{2}$$

What do you discover?

If two triangles are similar, the measures of their corresponding angles are equal and the measures of their corresponding sides are proportional.	*Similar Triangles*

This means that for the triangles above you can write the following proportions.

$$\frac{AB}{DE} = \frac{BC}{EF} \qquad \frac{AB}{DE} = \frac{AC}{DF} \qquad \frac{BC}{EF} = \frac{AC}{DF}$$

Proportions can be used to find the missing measures of similar triangles.

Example

1 **If a tree 6 feet tall casts a shadow 4 feet long, how high is a flagpole that casts a shadow 18 feet long?**

$\triangle JKL$ is similar to $\triangle PQR$.

$\dfrac{JK}{PQ} = \dfrac{KL}{QR}$ *Corresponding sides of similar triangles are proportional.*

$\dfrac{6}{x} = \dfrac{4}{18}$

$4x = 6(18)$ *Cross multiply.*

$4x = 108$

$x = 27$

The flagpole is 27 feet high.

Tell how you could use the method shown above to find the height of a building.

Example

2 **Find the distance, *UV*, across the pond shown below.**

$\triangle STU$ is similar to $\triangle WVU$.

$\dfrac{TU}{VU} = \dfrac{ST}{WV}$ *Corresponding sides of similar triangles are proportional.*

$\dfrac{70}{x} = \dfrac{50}{75}$

$50x = 70(75)$

$50x = 5250$

$x = 105$

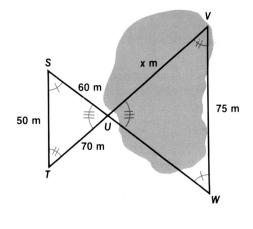

The pond is 105 meters across.

Name some other instances where you can use similar triangles as above to find distances.

Exploratory Exercises

For each pair of similar triangles, list the corresponding angles and the corresponding sides.

1.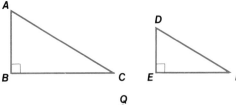

2.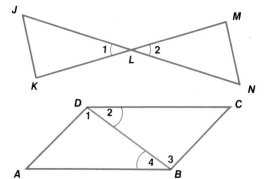

3.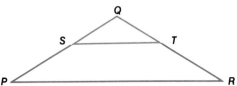

4.

Written Exercises

$\triangle ABC$ and $\triangle DEF$ are similar. For each set of measures, find the measure of the missing sides.

5. $a = 5, d = 7, f = 6, e = 5$

6. $c = 11, f = 6, d = 5, e = 4$

7. $b = 4.5, d = 2.1, e = 3.4, f = 3.2$

8. $a = 16, c = 12, b = 13, e = 7$

9. $a = 17, b = 15, c = 10, f = 6$

10. $c = 18, f = 12, d = 18, e = 16$

11. $a = 4\frac{1}{4}, b = 5\frac{1}{2}, e = 2\frac{3}{4}, f = 1\frac{3}{4}$

12. $c = 7\frac{1}{2}, f = 5, a = 10\frac{1}{2}, b = 15$

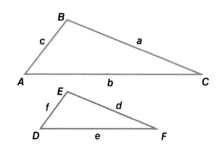

Use similar triangles to solve each problem.

13. In building a roof, a 5-foot support is to be placed at point B as shown on the diagram. Find the length of the support that is to be placed at point A.

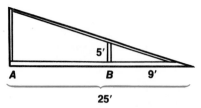

14. $\triangle ABC$ is similar to $\triangle EDC$. Find the distance across the lake from point A to point B.

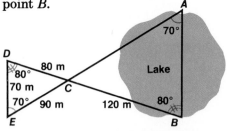

15. A triangle has sides 9 cm, 15 cm, and 18 cm. The longest side of a similar triangle is 22 cm in length. Find the length of the shortest side of that triangle.

16. A fence post one meter high casts a shadow 170 cm long. Find the height of a flagpole whose shadow is 8000 cm long.

16-4 Trigonometric Ratios

For similar triangles, ratios of the measures of corresponding sides can be written. If enough of these measures are known, these ratios can be used to find the measures of the remaining sides.

For every right triangle, certain ratios can be set up. These ratios, called **trigonometric ratios**, involve not only the measures of the sides but the measures of the acute angles as well. Again, if enough is known, these ratios can be used to find the measures of the remaining parts of the triangle.

A typical right triangle is shown at the right.

$\overline{BC}$ is *opposite* $\angle A$.

$\overline{AC}$ is *adjacent* to $\angle A$.

$\overline{AB}$ is the *hypotenuse* and is opposite right angle C.

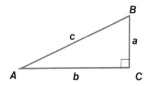

Notice that in this triangle

a denotes the measure of the side opposite $\angle A$.

b denotes the measure of the side opposite $\angle B$.

c denotes the measure of the side opposite $\angle C$.

Three common trigonometric ratios are defined as follows.

The sine, cosine, and tangent ratios are abbreviated as sin, cos, and tan, respectively.

$$\textbf{sine of } \angle A = \frac{\text{measure of side opposite } \angle A}{\text{measure of hypotenuse}}$$ $$\sin A = \frac{a}{c}$$ $$\textbf{cosine of } \angle A = \frac{\text{measure of side adjacent to } \angle A}{\text{measure of hypotenuse}}$$ $$\cos A = \frac{b}{c}$$ $$\textbf{tangent of } \angle A = \frac{\text{measure of side opposite } \angle A}{\text{measure of side adjacent to } \angle A}$$ $$\tan A = \frac{a}{b}$$	*Definition of Trigonometric Ratios*

1 **Find the sine of ∠D in the 30°–60° right triangle below.**

$$\sin D = \frac{EF}{DE}$$

Since $\overline{DE}$ is the hypotenuse, its measure is twice the measure of $\overline{EF}$, the side opposite the 30° angle.

Let $x = EF$. Then $2x = DE$.

$$\sin D = \frac{EF}{DE} \qquad sin = \frac{opposite}{hypotenuse}$$

$$= \frac{x}{2x}$$

$$= \frac{1}{2} \qquad \text{Thus, } \sin D = \frac{1}{2} \text{ or } \sin 30° = \frac{1}{2}.$$

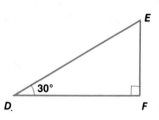

Are all 30°–60° right triangles the same size?

Regardless of the size, the measure of the hypotenuse is twice the measure of the side opposite the 30° angle. Therefore, we see that a trigonometric ratio like sin 30° will always be the same regardless of the triangle's size.

2 **Using Calculators**

For △ABC, use a calculator to express sin A, cos A, tan A, sin B, cos B, and tan B. Round answers to the nearest thousandth.

Notice that a = 7, b = 24, and c = 25.

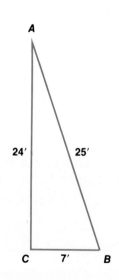

DISPLAY:

$$\sin A = \frac{a}{c} = \frac{7}{25} \qquad\qquad 0.28$$

$$\cos A = \frac{b}{c} = \frac{24}{25} \qquad\qquad 0.96$$

$$\tan A = \frac{a}{b} = \frac{7}{24} \qquad\qquad 0.291667 \approx 0.292$$

$$\sin B = \frac{b}{c} = \frac{24}{25} \qquad\qquad 0.96$$

$$\cos B = \frac{a}{c} = \frac{7}{25} \qquad\qquad 0.28$$

$$\tan B = \frac{b}{a} = \frac{24}{7} \qquad\qquad 3.4285714 \approx 3.429$$

Exploratory Exercises

For the three triangles below, express each trigonometric ratio as a fraction in simplest form.

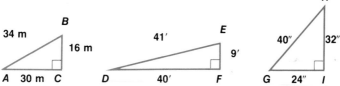

1. sin A	**2.** cos A	**3.** tan A
4. cos B	**5.** sin B	**6.** tan B
7. sin D	**8.** cos D	**9.** tan D
10. cos E	**11.** sin E	**12.** tan E
13. sin G	**14.** cos G	**15.** tan G
16. cos H	**17.** sin H	**18.** tan H

19. Are angles A and B complementary angles? Is sin A = cos B?

20. Are angles D and E complementary angles? Is cos D = sin E?

21. Based on your answers for problems 19–20, what is the relationship of the sine and cosine of complementary angles?

Written Exercises

Express the sine, cosine, and tangent of each acute angle in each triangle to the nearest thousandth.

22.

23.

24.

25.

26.

27.

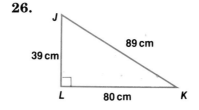

28.

29.

30.

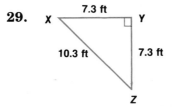

Simplify each algebraic fraction. State the excluded values of the variables.

1. $\dfrac{b^2 + 5b + 6}{b^2 - 9}$

2. $\dfrac{6m^2}{6m^2 - 2m}$

Simplify.

3. $\dfrac{-18x^6y^2}{24x^4y^6}$

4. $\dfrac{5mn^3 - 5n^2}{m} \cdot \dfrac{2m^2}{m^2n^2 - 1}$

5. $\dfrac{9}{x - 3} - \dfrac{4}{3 - x}$

6. $2 + \dfrac{a^2 + 1}{a - 1}$

7. $\dfrac{\frac{a + b}{x - y}}{\frac{a^2 - b^2}{x^2 - y^2}}$

Find each quotient.

8. $\dfrac{2m^2 + 5m - 3}{m^2 - 9} \div \dfrac{4m^2 - 1}{2m^2 - 5m - 3}$

9. $(3t^3 - 11t^2 - 31t + 7) \div (3t + 1)$

10. Solve $\dfrac{4}{t - 2} - \dfrac{2t - 3}{t^2 - 4} = \dfrac{5}{t - 2}$

Solve each equation for n.

11. $\dfrac{2}{3}n + 4b = n$

12. $\dfrac{a}{b} - \dfrac{a}{n} = \dfrac{b}{n}$

13. Find the lower quartile, median, and upper quartile for the following set of data.
23, 24, 24, 28, 30, 39, 40, 42, 45

A card is selected at random from a deck of 52 playing cards

14. What are the odds in favor of selecting a club?

15. What is the probability of selecting a jack?

16. What are the odds against selecting a jack?

Box X contains one red marble and one blue marble. Box Y contains one yellow marble and one red marble. Box Z contains one green marble and one red marble. A marble is drawn at random from each box.

17. What is the probability that all three marbles are not red?

18. What is the probability that two marbles are red and one marble is yellow?

Find both the complement and the supplement of each angle measure.

19. $56°$

20. $(2x + 15)°$

21. The length of the side opposite the 60° angle in a 30°–60° right triangle is $5\sqrt{3}$ cm. Find the length of the other two sides.

22. Express the sine, cosine, and tangent of each acute angle in $\triangle ABC$ to the nearest thousandth.

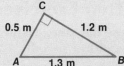

Solve each problem.

23. Carla can do a job in 5 days. Cameron can do the same job in 3 days. If they work together, how long will it take them to complete the job?

24. Find the measure of an angle that is 22° less than its complement.

25. The prices for some running shoes are $46, $42, $38, $57, $70, $64, $40, $75, $68, and $54. Find the mean and median price of these shoes.

26. Pam's times in the 200-meter run during competitions last season were 28.60 s, 29.21 s, 31.00 s, 27.43 s, 27.25 s, 28.40 s, 30.74 s, 31.30 s, 29.08 s, and 28.48 s. Make a box plot of the data.

16-5 Using Tables and Calculators

If you know the measures of the sides of a right triangle, you can calculate the sine, cosine, and tangent of any of its angles.

The table on page 593 lists approximations of these ratios in decimal form.

Examples

1 **Find tan 28°.**

Find 28° in the angle column.

Angle	sin	cos	tan
26°	0.4384	0.8988	0.4877
27°	0.4540	0.8910	0.5095
28°	0.4695	0.8829	0.5317
29°	0.4848	0.8746	0.5543
30°	0.5000	0.8660	0.5774

Find the corresponding reading in the tan column.

$$\tan 28° \approx 0.5317$$

2 **If cos $A = \frac{1}{2}$, find the measure of $\angle A$.**

First express the ratio as a decimal.

$$\frac{1}{2} = 0.5000$$

Find this decimal in the cos column.

Angle	sin	cos	tan
56°	0.8290	0.5592	1.4826
57°	0.8387	0.5446	1.5399
58°	0.8480	0.5299	1.6003
59°	0.8572	0.5150	1.6643
60°	0.8660	0.5000	1.7321

The corresponding angle is 60°, so the measure of $\angle A$ *is* 60°.

3 **Find the measure of $\angle A$ if sin $A = 0.9711$.**

The measure of $\angle A$ must be between 76° and 77° since 0.9711 is between 0.9703 and 0.9744.

Angle	sin	cos	tan
76°	0.9703	0.2419	4.0108
77°	0.9744	0.2250	4.3315
78°	0.9781	0.2079	4.7046
79°	0.9816	0.1908	5.1446

Because 0.9711 is closer to 0.9703 than it is to 0.9744, the measure of $\angle A$ is closer to 76° than 77°. Hence, the measure of $\angle A$ is approximately 76°.

A calculator can also be used to find trigonometric ratios. It will give the same approximate values you find in the table on page 593.

Examples

Using Calculators

4 Find cos 25°.

ENTER: 25 [COS]   DISPLAY: $0.9630\overline{1}8$

The calculator must be in degree mode.

Rounded to ten-thousandths, cos 25° ≈ 0.9631.

5 Find the measure of ∠A if sin A = 0.4384.

ENTER: 0.4384 [INV] [SIN]   DISPLAY: 26.001839

The inverse key "undoes" the original function.

To the nearest degree, the measure of ∠A is 26°.

Exploratory Exercises

Use the table on page 593 to answer each question.

1. At what angle is the sine at its maximum (greatest value)? At its minimum (least value)?
2. At what angle is the cosine at its minimum? At its maximum?
3. For what angle are the sine and cosine equal?
4. What is the tangent of the angle at which the sine and cosine are equal?
5. What are the minimum and maximum values of the tangent?

Written Exercises

Use the table on page 593 or a calculator to find the value of each trigonometric ratio.

6. cos 25°	**7.** tan 31°	**8.** sin 89°	**9.** tan 14°
10. cos 76°	**11.** sin 22°	**12.** cos 83°	**13.** tan 24°
14. sin 68°	**15.** tan 9°	**16.** sin 27°	**17.** cos 18°
18. cos 42°	**19.** tan 50°	**20.** sin 45°	**21.** cos 30°
22. sin 30°	**23.** tan 30°	**24.** sin 60°	**25.** cos 60°

Use the table on page 593 or a calculator to find the measure of each angle to the nearest degree.

26. sin A = 0.4384	**27.** cos B = 0.4848	**28.** tan D = 1.3250
29. cos ∠1 = 0.9781	**30.** tan ∠ABC = 5.1446	**31.** sin A = 0.9620
32. cos B = 0.3900	**33.** tan ∠2 = 1.7321	**34.** sin D = 0.2756
35. cos A = 0.3746	**36.** cos B = 0.7660	**37.** sin ∠DEF = 0.9848
38. tan C = 57.2900	**39.** sin X = 0.0520	**40.** cos Z = 0.9986
41. tan A = 0.0524	**42.** sin B = 0.5000	**43.** cos B = 0.8660

Find the value of each expression.

44. sin 30° − cos 60°	**45.** tan 45° + sin 0°	**46.** sin 90° + cos 0°
47. tan 0° + sin 90°	**48.** sin 0° − cos 60°	**49.** sin 90° − tan 0°
50. sin 30° − cos 30°	**51.** cos 60° − sin 30°	**52.** tan 45° − tan 45°

As shown below, letters from the English alphabet can be used to convey mathematical concepts.

$$x^2 - 7 = 0 \qquad A = \tfrac{1}{2}bh \qquad \overline{AB} \qquad \sin A$$

In higher mathematics, letters from the Greek alphabet are often used in the same way. The table below shows uppercase and lowercase letters of the Greek alphabet.

GREEK ALPHABET

Greek Letter		Greek Name	English Equivalent	Greek Letter		Greek Name	English Equivalent
A	α	Alpha	a	N	ν	Nu	n
B	β	Beta	b	Ξ	ξ	Xi	x
Γ	γ	Gamma	g	O	o	Omicron	ŏ
Δ	δ	Delta	d	Π	π	Pi	p
E	ϵ	Epsilon	ĕ	P	ρ	Rho	r
Z	ζ	Zeta	z	Σ	σ	Sigma	s
H	η	Eta	ē	T	τ	Tau	t
Θ	θ	Theta	th	Y	υ	Upsilon	u
I	ι	Iota	i	Φ	φ	Phi	ph
K	κ	Kappa	k	X	χ	Chi	ch
Λ	λ	Lambda	l	Ψ	ψ	Psi	ps
M	μ	Mu	m	Ω	ω	Omega	ō

A commonly used Greek letter in mathematics is π. This letter is used to represent the ratio of the circumference of a circle to its diameter. An approximate value for π is 3.14.

Other Greek letters are commonly used in higher mathematics. For example, Δ is commonly used to represent the difference between two measures, θ is often used to represent the measure of an angle, and Σ stands for the sum of a set of numbers.

Exercises

State how each expression, equation, or inequality is read.

1. $\sin \theta$
2. $\cos \phi$
3. $\tan \psi$
4. Δx

5. $r = 0.7 \, \Omega$
6. $\alpha - \lambda$
7. $\sin (\alpha + \beta)$
8. $|x - y| < \gamma$

16-6 Solving Right Triangles

To solve a triangle means to find all the missing measures of the triangle. The trigonometric ratios can be used to solve a triangle. The ratio used depends upon what measures are given and what measures are missing. Sometimes, more than one ratio can be used.

Example

1 **Find the measures of $\angle A$ and $\angle C$ in $\triangle ABC$.**

Because the lengths of all sides are given, the sine, cosine, or tangent ratio can be used. Suppose you use the sine ratio.

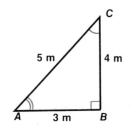

$$\sin A = \frac{4}{5}$$

$$= 0.8000$$

Using the table or a calculator, we see that the measure of $\angle A$ is approximately 53°. Thus, the measure of $\angle C$ is about 90° − 53°, or 37°.

How would you use the cosine ratio to find the measure of $\angle A$? How would you use the tangent ratio to find the measure of $\angle A$?

Example

2 **Find the length of $\overline{DE}$ in $\triangle DEF$.**

With respect to $\angle E$, only the lengths of the adjacent side and the hypotenuse are given. In this case, use the cosine ratio.

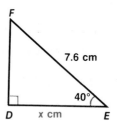

$$\cos 40° = \frac{x}{7.6}$$

$$0.7660 \approx \frac{x}{7.6}$$

$$(0.7660)7.6 \approx x$$

$$5.8 \approx x$$

Thus, $\overline{DE}$ is approximately 5.8 cm long.

Use $\sin 50° = \frac{x}{7.6}$ to solve for the measure of $\overline{DE}$. Do you get the same result as in Example 2?

Examples

3 **Find the length of $\overline{JK}$ in △JKL.**

In this case, use the tangent ratio.

$\tan 48° = \frac{5.8}{x}$ *Let x be the measure of $\overline{JK}$.*

$1.1106 \approx \frac{5.8}{x}$

$1.1106x \approx 5.8$

$x \approx 5.2$

Thus, $\overline{JK}$ is about 5.2 mm long.

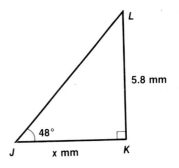

4 **Find the missing measures in the triangle below.**

The measure of $\angle R$ is $90° - 36°$, or $54°$.

$\sin 36° = \frac{x}{18}$

$0.5878 \approx \frac{x}{18}$

$10.6 \approx x$

Thus, $\overline{QR}$ is about 10.6 inches long.

$\cos 36° = \frac{y}{18}$

$0.8090 \approx \frac{y}{18}$

$14.56 \approx y$

Thus, $\overline{PQ}$ is about 14.56 inches long.

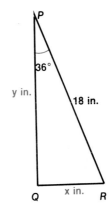

Exploratory Exercises

State which trigonometric ratios you would use to find the missing measures in each triangle.

1.

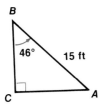

2.

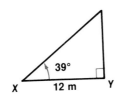

3.

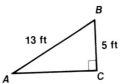

4.

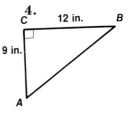

Written Exercises

Solve each triangle.

5.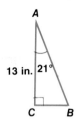

6.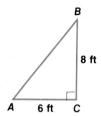

7.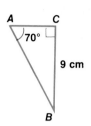

8.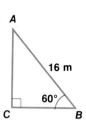

9.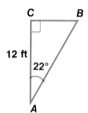

10.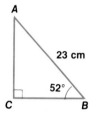

11.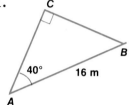

12.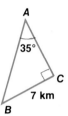

Solve each right triangle (∠C is the right angle).

13. angle $A = 31°$, $a = 6$ m

14. $a = 6$ in., $c = 10$ in.

15. angle $B = 42°$, $c = 10$ in.

16. $b = 5$ ft, $a = 4$ ft

17. $c = 14$ ft, $b = 11$ ft

18. $c = 11$ m, $b = 6$ m

19. angle $B = 40°$, $b = 6$ cm

20. angle $B = 28°$, $a = 16$ cm

21. angle $A = 45°$, $c = \sqrt{2}$ ft

22. angle $A = 75°$, $b = 3$ km

mini-review

Solve each quadratic equation. Express irrational roots in simplest radical form and in decimal form to the nearest hundredth.

1. $x^2 + 4x + 10 = 0$

2. $3y^2 - 13y = -14$

3. $2m^2 + 1 = 8m$

Simplify.

4. $\frac{6}{a^3} \cdot \left(\frac{a^2}{2}\right)^2$

5. $\frac{x^2 - y^2}{x} \cdot \frac{y}{y - x}$

6. $\frac{n + 1}{n - 1} + \frac{3}{3n - 2}$

7. $\frac{6\frac{2}{3}}{\frac{4}{5}}$

Four bicyclists, Alfonso, Jana, Teresa, and Hakeem, have an equal chance of winning a race. Draw a tree diagram of the possible outcomes. Then answer each question.

8. What is the probability that Hakeem wins the race?

9. What is the probability that Jana finishes first and Alfonso finishes second?

10. What is the probability that Teresa finishes first, Hakeem second, Alfonso third, and Jana fourth?

16-7 Problem Solving: Using Trigonometry

In order to use the trigonometric ratios to solve problems, it is helpful to understand the meaning of angles of **elevation** and **depression**.

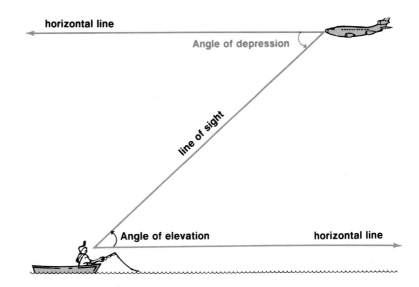

The angle of elevation or depression is the angle formed by the line of sight to an object and a horizontal line.

These concepts are used with trigonometric ratios to find missing measurements of right triangles. Sometimes, as shown in the examples, these measures are difficult to obtain directly.

Example

1   A radio tower casts a shadow 120 meters long when the angle of elevation of the sun is 41°. How tall is the tower?

> **explore**   It would be helpful to first draw a picture for the problem.

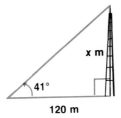

Note that the angle of elevation opens upward.

> **plan** $\tan 41° = \frac{x}{120}$

> **solve** $0.8693 \approx \frac{x}{120}$
>
> $(0.8693)120 \approx x$
>
> $104.3 \approx x$

The height of the tower is approximately 104.3 meters.   *Examine the solution.*

2 From the top of an observation tower 50 meters high, a forest ranger spotted a deer at an angle of depression of 28°. How far was the deer from the base of the tower?

First make a drawing.

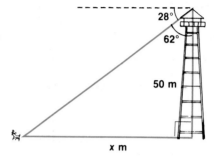

Notice that the angle of depression opens downward from the horizontal.

Since the angle of depression is 28°, its complement, which is an angle of the triangle shown, is 62°.

$$\tan 62° = \frac{x}{50}$$

$$1.8807 \approx \frac{x}{50}$$

$$(1.8807)50 \approx x$$

$$94 \approx x$$

The deer is about 94 meters from the tower.

Exploratory Exercises

Name the angles of elevation and depression in each drawing.

1.

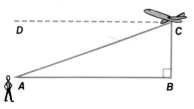

2.

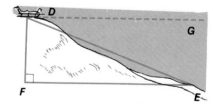

3.

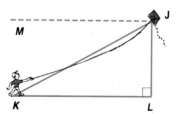

4.

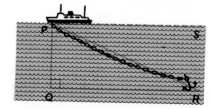

Written Exercises

Use trigonometric ratios to solve each problem.

5. A road rises 38 feet vertically over a horizontal distance of 540 feet. What is the angle of elevation of the road?

6. At a point 200 feet from the base of a flagpole, the angle of elevation is 62°. Find the height of the flagpole and the distance from the point to the top.

7. A chimney casts a shadow 75 feet long when the angle of elevation of the sun is 41°. How tall is the chimney?

8. A train in the mountains rises 8 feet for every 200 feet it moves along the track. Find the angle of elevation of the tracks.

9. At a point 210 feet from the base of a building, the angle of elevation to the top of the building is 55°. How tall is the building?

10. How far will a submarine travel when going to a depth of 300 feet if its course has an angle of depression of 25°?

11. A guy wire is fastened to a TV tower 40 feet above the ground and forms an angle of 52° with the tower. How long is the wire?

12. A roof is constructed as shown in the diagram below. Find the pitch (angle of elevation) of the roof.

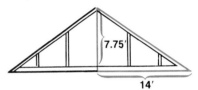

13. Find the area of a right triangle in which one acute angle measures 25° and the leg opposite that angle is 40 cm long.

14. From the top of a 70-meter lighthouse, an airplane was observed that was directly over a ship. The angle of elevation of the plane was 18°, while the angle of depression of the ship was 25°. Find the distance from the ship to the foot of the lighthouse and the height of the plane.

15. An airplane, at an altitude of 2000 feet, is directly over a power plant. The navigator finds the angle of depression of the airport to be 19°. How far is the plane from the airport? How far is the power plant from the airport?

16. In a parking garage, each level is 20 feet apart. Each ramp to a level is 130 feet long. Find the measure of the angle of elevation for each ramp.

17. A train in the mountains rises 15 feet for every 250 feet it moves along the track. Find the angle of elevation of the track.

Challenge Exercises

Solve each triangle.

18.

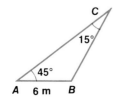

19.

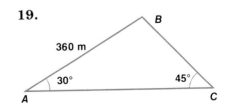

Glenn Hatfield is a licensed surveyor. In addition to measuring land, he uses his surveying instruments to check building construction. Glenn frequently uses trigonometry in his work.

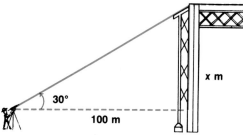

As shown above, Glenn has used his instrument to measure the angle of elevation and find the height from the top of the instrument to the top of the bridge.

$$\tan 30° = \frac{x}{100}$$

$$0.5774 \approx \frac{x}{100}$$

$$57.74 \approx x$$

If the instrument height is 1.54 m, what is the height of the bridge?

Exercises

Use trigonometry to find the heights of various objects for the following measurements.

	Angle of Elevation	Distance from Instrument	Instrument Height
1.	35°	100 m	1.45 m
2.	26°	62.56 m	1.73m
3.	11°	131.37 m	1.68 m
4.	3°	895 m	1.54 m

Trigonometric Functions

Trigonometric functions are used in many applications of mathematics. A computer can calculate values of trigonometric functions very rapidly. However, the values are found for angle measures in *radians*, not for angle measures in degrees. For an angle whose measure is D degrees, the measure in radians, R, is $D \cdot \pi/180$.

Trigonometric functions usually available in BASIC include SIN (sine), COS (cosine), and TAN (tangent). The following program uses the SIN function to print a table of sines.

```
5   PRINT "ANGLE" , "SINE"
10  FOR D = 0 TO 360 STEP 30
20  LET R = D * 3.1416 / 180
25  LET R1 = INT ( SIN (R) * 1000 + .5)/1000
30  PRINT D, R1
40  NEXT D
50  END
```

ANGLE	SINE
0	0
30	.5
60	.866
90	1
120	.866
150	.5
180	0
210	−.5
240	−.866
270	−1
300	−.866
330	−.5
360	0

The sine function generates ordered pairs. To graph the sine function, use the horizontal axis for the degree values. Use the vertical axis for the sine. After plotting the points, complete the graph by connecting the plotted points with a smooth continuous curve as shown.

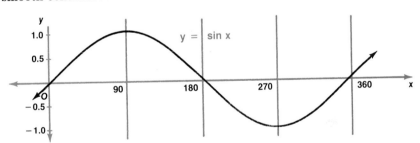

Exercises

1. Modify the program to print a table of cosines and then graph the cosine function.

State whether the value of each trigonometric ratio is positive or negative.

2. sin 30°   **3.** cos 150°   **4.** sin 330°   **5.** cos 60°

State which is greater.

6. cos 30° or cos 90°   **7.** sin 0° or sin 90°

Find the values of x for which each equation is true.

8. cos x = 1   **9.** sin x = 1   **10.** sin x = 0   **11.** cos x = −1

12. How are the graphs of the sine function and cosine function related?

Vocabulary

supplementary angles (547)
complementary angles (547)
30°–60° right triangles (550)
similar triangles (554)
corresponding angles (554)
corresponding sides (554)
trigonometric ratios (557)

sine (557)
cosine (557)
tangent (557)
solving triangles (564)
angle of elevation (567)
angle of depression (567)

Chapter Summary

1. Two angles are supplementary if the sum of their measures is 180°. (547)
2. Two angles are complementary if the sum of their measures is 90°. (547)
3. The sum of the measures of the angles in any triangle is 180°. (548)
4. In a 30°–60° right triangle, if a is the measure of the side opposite the 30° angle, then $2a$ is the measure of the hypotenuse, and $a\sqrt{3}$ is the measure of the side opposite the 60° angle. (551)
5. Two triangles are similar if the measures of their corresponding angles are equal. (554)
6. If two triangles are similar, the measures of their corresponding sides are proportional. (554)
7. Sine of an angle:

$$\sin A = \frac{\text{measure of side opposite } \angle A}{\text{measure of hypotenuse}} \quad (557)$$

8. Cosine of an angle:

$$\cos A = \frac{\text{measure of side adjacent to } \angle A}{\text{measure of hypotenuse}} \quad (557)$$

9. Tangent of an angle:

$$\tan A = \frac{\text{measure of side opposite } \angle A}{\text{measure of side adjacent to } \angle A} \quad (557)$$

10. The trigonometric ratios can be used to find the missing measures of a triangle. (564)
11. The angle of elevation or depression is the angle formed by the line of sight to an object and a horizontal line. (567)

Chapter Review

16–1 Find both the complement and the supplement of each angle measure.

    1. 66°   **2.** 62°   **3.** 148°   **4.** $y°$

Find the measure of the third angle of the triangle in which the measures of two angles of the triangle are given.

    5. 16°, 72°   **6.** 42°, 121°   **7.** 37°, 90°   **8.** $y°$, $x°$

16–2 Find the lengths of the missing sides of each 30°–60° right triangle described below. Express irrational answers in simplest radical form and in decimal form to the nearest thousandth.

	Hypotenuse	Side Opposite 30° Angle	Side Opposite 60° Angle
9.	8 cm	_____	_____
10.	4.25 cm	_____	_____
11.	_____	$3\frac{1}{2}$ in.	_____
12.	_____	_____	$2\sqrt{3}$ in.

16–3 For similar triangles *ABC* and *DEF*, find the measures of the missing sides.

    13. $a = 5$, $d = 11$, $f = 6$, $e = 14$

    14. $c = 16$, $b = 12$, $a = 10$, $f = 9$

    15. $a = 8$, $c = 10$, $b = 6$, $f = 12$

    16. $c = 12$, $f = 9$, $a = 8$, $e = 11$

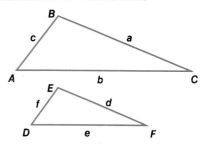

16–4 For $\triangle ABC$, express each trigonometric ratio as a fraction in simplest form.

    17. $\sin A$ **18.** $\cos B$

    19. $\cos A$ **20.** $\sin B$

    21. $\tan A$ **22.** $\tan B$

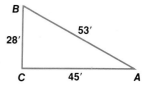

16–5 Use the table on page 593 or a calculator to find the value of each trigonometric ratio.

    23. $\cos 61°$ **24.** $\sin 42°$ **25.** $\tan 13°$ **26.** $\sin 84°$

16–6 Solve each right triangle ($\angle C$ is the right angle).

    27. angle $A = 62°$, $a = 7$ m   **28.** $a = 9$ in., $c = 13$ in.

    29. angle $B = 36°$, $c = 8$ yd   **30.** $a = 10$ cm, $b = 15$ cm

16–7 Solve each problem.

    31. The diagonal of a rectangle is 16 cm long and makes an angle of 50° with a side of the rectangle. Find the length and width of the rectangle.

    32. A weather balloon is directly above a tree. The angle of elevation is 60° when you are 100 meters from the tree. How high is the balloon?

Find both the complement and the supplement of each angle measure.

1. 28°

2. 69°

3. $(y + 20)°$

Find the measure of the third angle of a triangle in which the measures of two angles of the triangle are given.

4. 16°, 47°

5. 89°, 66°

6. 45°, 120°

Find the lengths of the missing sides of each 30°–60° right triangle described below. Express irrational answers in simplest radical form and in decimal form to the nearest thousandth.

	Hypotenuse	Side Opposite 30° Angle	Side Opposite 60° Angle
7.	17 in.	————	————
8.	————	8 ft	————
9.	————	————	$9\sqrt{3}$ m

Triangles *ABC* and *JKL* are similar. For each set of measures, find the measures of the missing sides.

10. $c = 20,\ l = 15,\ k = 16,\ j = 12$

11. $c = 12,\ b = 13,\ a = 6,\ l = 10$

12. $k = 5,\ c = 6.5,\ b = 7.5,\ a = 4.5$

13. $l = 1\frac{1}{2},\ c = 4\frac{1}{2},\ k = 2\frac{1}{4},\ a = 3$

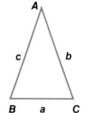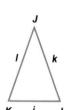

Solve each right triangle ($\angle C$ is the right angle).

14. angle $A = 56°,\ a = 17$

15. $a = 12,\ b = 16$

16. angle $B = 42°,\ c = 10$

17. $b = 21,\ c = 29$

Solve each problem.

18. A kite is flying at the end of a 300-foot string. Assuming the string is straight and forms an angle of 58° with the ground, how high is the kite?

19. A plane is 1000 feet above the ground. The angle of depression of the landing strip is 20°. How far is the plane from the landing strip?

20. A 6-foot pole casts a 4-foot shadow. How tall is a tree that casts a 50-foot shadow?

The questions on this page involve comparing two quantities, one in Column A and one in Column B. In certain questions, information related to one or both quantities is centered above them. All variables used stand for real numbers.

Directions:
Write A if the quantity in Column A is greater.
Write B if the quantity in Column B is greater.
Write C if the quantities are equal.
Write D if there is not enough information to determine the relationship.

Column A	Column B
1. $(0.64)^2$	$\sqrt{64}$
2. $\dfrac{1}{\sqrt{2}}$	$\dfrac{1}{2}\sqrt{2}$
3. $-(9^2)$	$\sqrt{900}$

$$x^2 = 25$$
$$y^2 = 36$$

Column A	Column B
4. x	y

$$x > 1$$

Column A	Column B
5. $\dfrac{x^3 + x^2}{x}$	$\dfrac{x^3 + x^2}{x^2}$
6. a given chord in a given circle	the radius of the same circle

7.

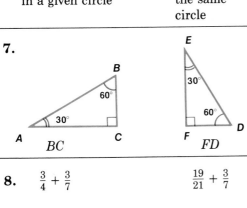

BC	FD

Column A	Column B
8. $\dfrac{3}{4} + \dfrac{3}{7}$	$\dfrac{19}{21} + \dfrac{3}{7}$

Examples

Column A	Column B

$$\frac{1}{k} < 0$$

1. $-\dfrac{1}{k}$ k

The answer is A because $-\dfrac{1}{k}$ is positive, while k is negative.

2. $\dfrac{3^4 + 3^5}{3^4}$ $\dfrac{3^2 + 3^3}{3^2}$

The answer is C because the simplest form of each quantity is $1 + 3$ or 4.

Hint: $\dfrac{3^4 + 3^5}{3^4} = \dfrac{3^4}{3^4} + \dfrac{3^5}{3^4}$

Column A	Column B
9. $(5 + 7)^2$	$5^2 + 7^2$
10. The distance from City A to City B is 50 miles.	
The time required to travel from A to B at 55 mph.	60 minutes

11. $a > b > c$
$a \neq 0, b \neq 0, c \neq 0$

ab	bc

12. $a > b > c > 0$

ab	bc

Column A	Column B
13. $(x + 2)(x + 3)$	$(x + 1)(x + 4)$

14. $r > 7, 0 < t < 6$

$\dfrac{1}{r}$	$\dfrac{1}{t}$

$$3x + 2 = y$$

15. $30x + 15$	$10y$

Appendix: BASIC

The BASIC Language

BASIC (Beginner's All-Purpose Symbolic Instruction Code) is a computer language. The symbols used in BASIC are similar to those used in algebra. Compare the symbols in the following lists.

Algebra	BASIC	Algebra	BASIC
$+$	$+$	$>$	$>$
$-$	$-$	$\geq$	$> =$
$\times$ or $\cdot$	$*$	$<$	$<$
$\div$	$/$	$\leq$	$< =$
$=$	$=$	$\neq$	$< >$
4^3	$4\uparrow3$ or $4 \wedge 3$	x	X

A numeric variable in BASIC is represented by a capital letter or a capital letter followed by a numeral. Examples are A, B, X, N1, and M2.

In BASIC, the multiplication symbol may never be omitted. To write A times B, write A $*$ B, not AB as in algebra.

Examples of algebraic and BASIC expressions, equations, and inequalities are shown below.

Algebra	BASIC
$6 + 3$	$6 + 3$
$a - 4$	A $-$ 4
$7t$, $7 \times t$, or $7 \cdot t$	$7 * T$
$24 \div x$ or $\frac{24}{x}$	24/X
$S \geq n^2$	S $> =$ N$\uparrow$2
$y = 8\frac{1}{2}$	Y $= 8.5$
$y_2 \neq x_1 + 4$	Y2 $< >$ X1 $+ 4$

BASIC uses parentheses as they are used in algebra for grouping. Also, the same order of operations is used as in algebra.

> 1. **Do all operations in parentheses from the innermost parentheses outward.**
> 2. **Evaluate all powers from left to right.**
> 3. **Then do all multiplications and divisions from left to right.**
> 4. **Then do all additions and subtractions from left to right.**

Order of Operations in BASIC

1 **Evaluate $3 + 9 - 2 * (6 - 1)\uparrow 2/10$.**

$$
\begin{aligned}
3 + 9 - 2 * (6 - 1)\uparrow 2/10 &= 3 + 9 - 2 * (5)\uparrow 2/10 \\
&= 3 + 9 - 2 * 25/10 \\
&= 3 + 9 - 50/10 \\
&= 3 + 9 - 5 \\
&= 12 - 5 \\
&= 7
\end{aligned}
$$

Do all operations in parentheses.
Evaluate powers.
Do all multiplications and divisions from left to right. Then, do all additions and subtractions from left to right.

A computer can be used in calculator mode to evaluate expressions. In calculator mode, there are certain BASIC words that can be used to give a computer instructions, or *commands*, that it will carry out at once. To use the computer in this way, enter a command by typing it on the keyboard and then pressing the RETURN or ENTER key. The monitor shows the command as it is typed. When RETURN is pressed, the output is printed on the line below the command.

PRINT 6 * 4/3 + ((5 + 3)/4)   *Press RETURN.*
10   *The value of the expression is 10.*

▪ A **computer program** is a series of statements that gives directions to the computer. The purpose of a program is to get information into the computer (**input**), do the calculations, and then get the results out of the computer (**output**).

A sample program is shown below.

10   PRINT 11.2 + 9.3 ⎫
20   END   ⎬ *statements*

line numbers ────↑

In a computer program, each statement has a **line number**. Numbering by tens permits statements to be inserted later. The computer follows the instructions in numerical order. A program should end with an END statement.

One way to get information into the computer is to use an INPUT statement. When the computer executes an INPUT statement it prints a question mark and waits until the user types in a number.

The programmer chooses the line numbers, usually integers from 1 to 9999.

2 **Write a program to compute the sum of two numbers.**

10   INPUT A   *Type a number and press RETURN.*
20   INPUT B   *Type a number and press RETURN.*
30   PRINT A + B   *The computer adds the two numbers and prints the sum.*
40   END

The program may also be written this way.

```
10   INPUT A, B        Type a number, a comma, and a number.
20   PRINT A + B
30   END
```

Another way to get information into the computer is to use a READ statement and a DATA statement. Each READ statement needs a DATA statement. The DATA statement may appear anywhere in the program but is usually placed after the READ statement. To modify the program in Example 2 to include READ and DATA statements, we would replace lines 10 and 20 with

```
10   READ A, B
20   DATA 22, 37
```

After the program is entered, the command RUN is entered into the computer. This command instructs the computer to execute, or run, the program.

Commands, such as RUN, do not have line numbers.

If the program in Example 2 is entered into the computer followed by RUN, the value of A + B is computed and printed.

Written Exercises

Change each algebraic expression to a BASIC expression.

1. $a + b + c$

2. $6m - 7n + 8$

3. $a + b \cdot 9$

4. $4(y + 3)$

5. $38 \div y$

6. $3x^2 + 4x + 9$

7. $a + \dfrac{5}{3 + a}$

8. $\dfrac{a}{b} + n$

9. $\dfrac{5x + 3}{2x - 1}$

Use the computer in calculator mode to evaluate each expression.

10. $4 + 15 - 3$

11. $3 * 9 * 4 - 2$

12. $3\uparrow 2 + 5$

13. $(3 + 4)\uparrow(2 + 1)$

14. $2 * ((4 + 9) * 3)$

15. $8 * (4 * (-6 + 9)/2)/16$

If A = 1, B = 2, C = 3, X = 12, Y = 0, Z = 0.5, A1 = 0.25, and A2 = 6, determine the output when each statement is executed by the computer.

16. 60   PRINT B * C

17. 20   PRINT C * X

18. 40   PRINT A1 * X + A2

19. 110   PRINT A * C − Y * Z

20. 70   PRINT (B * B) * C↑2

21. 80   PRINT B * X/A2

22. 100   PRINT X * A1 * Z

23. 130   PRINT (X − 2 * C)/B

24. 50   PRINT (X − 2 * C) * (2 * C − B)

25. 90   PRINT 4 * (B * C↑2 − X * Z)/B

Write a program for each exercise using either an INPUT statement or READ and DATA statements.

26. Find the sum of 31, 64, and 82.

27. Find the product of 31, 64, and 82.

28. Find the quotient of 382 and 94.

29. Find the difference of 382 and 94.

30. Find the perimeter of a rectangle with a width of 3.9 cm and a length of 8.2 cm.

Assignment Statements

In BASIC, the equals sign, =, is used in a slightly different way than in algebra. In algebra, each side of an equation may have many terms and variables. In BASIC only the left side of an equation may have a variable.

Algebra	BASIC
$x + 16 = 7$	X = 7 − 16
$4x + 7y = 5x + 16$	X = 7 * Y − 16
$y − 4 = 3(x − 2)^2$	Y = 3 * (X − 2)↑2 + 4

The LET statement is used to assign values to variables. The general form of a LET statement is shown below.

line number LET *variable* = *expression*

In a LET statement, the equals sign tells the computer to assign the value of the expression on the right to the variable on the left. The LET statement is another way to provide a program with data.

```
10   LET X = 6
20   LET Y = 12
30   LET Z = 20
40   LET W = X + Y + Z
50   PRINT W
60   END
```

In this program 6 is assigned to X, 12 to Y, and 20 to Z. The value of X + Y + Z is computed and assigned to W. The output is 38, the value of W.

READ-DATA statements also assign values to variables.

```
10   READ R
20   DATA 0.05
30   LET I = 1000 * R * 10
40   PRINT R, I
50   END
```

Lines 10 and 20 assign 0.05 to R.

A comma in a PRINT statement causes the output to be printed in columns.

Sometimes a programmer wishes to repeat the same operation in a program. This can be done using a GO TO statement.

The general form of the GO TO statement follows.

line number GO TO *line number*

Study the use of the GO TO statement in the following program.

```
10   READ R
20   DATA 0.05, 0.055, 0.06, 0.065
30   I = 1000 * R * 10
40   PRINT R, I
50   GO TO 10
60   END
```

The computer assigns the first value in line 20 to R.

Line 50 instructs the computer to return to line 10. Then the second value in line 20 is assigned to R.

The process of returning to line 10 is continued until all the values in line 20 are used. Then the computer prints an error message such as OUT OF DATA.

Example

1   Write and run a program that finds the areas of four rectangles with dimensions 8 cm and 5 cm, 7 cm and 2 cm, 9 cm and 6.2 cm, and 4.5 cm and 3 cm.

```
10   READ L, W
20   DATA 8, 5, 7, 2, 9, 6.2, 4.5, 3
30   LET A = L * W
40   PRINT A,
50   GO TO 10
60   END
RUN
40          14          55.8
13.5
OUT OF DATA IN 10
```

Use A = lw, the formula for the area of a rectangle.

The result of assigning 8 to L and 5 to W is 40. The results from the other values are 14, 55.8, and 13.5.

Compare the two programs below. Notice that the program on the right uses READ-DATA statements in place of LET statements. What advantage do you see in using READ-DATA statements rather than LET statements?

```
10   LET A = 3
11   LET B = 4
12   LET C = 5
13   LET D = 6
14   LET X = 2
20   LET Y = (A + B + C)/(D * X)
30   PRINT Y
40   END
```

```
10   READ A, B, C, D, X
15   DATA 3, 4, 5, 6, 2
20   LET Y = (A + B + C)/(D * X)
30   PRINT Y
40   END
```

Written Exercises

If A = 3, B = 4, and M1 = 16, find the value assigned to X when each statement is executed by the computer.

1. 190 LET X = 6 * A

2. 25 LET X = A * B + 5

3. 20 LET X = B + M1 * B

4. 30 LET X = M1/B

5. 20 LET X = A↑4

6. 40 LET X = M1 − B + 3 * A

Correct the error in each BASIC statement.

7. 30 LET Y = 2A

8. 60 LET 4 * A=B+3+C

9. 90 LET A + B = C

10. 20 LET X = (3 + Y/5

11. 40 LET L * W = A

12. 60 LET 30 = X * 5

Change each algebraic equation to a BASIC equation solved for y.

13. $Y - 10 = 35$

14. $2x + 6y = 3x - 1$

15. $4 + y = 2(x - 5)^2$

16. $\sqrt{y} + 1 = 7x - 42$

Given the values of A, B, and C, write a program to find the value of X.

17. A = 2, B = 4, C = −1,
X = A + B * (C − 1)

18. A = 5, B = 2, C = 10,
X = A/B + C↑2 − A

Determine the output for each program.

19. 10 LET A = 4
20 LET B = 5
30 PRINT A * B
40 END

20. 10 READ P, R, T
20 DATA 300, 0.06, 5
30 LET I = P * R * T
40 PRINT P, R, T, I
50 END

21. 10 READ R
20 LET P = 3.14159
30 LET C = 2 * P * R
40 DATA 2, 4, 7, 9
50 PRINT C
60 GO TO 10
70 END

22. 10 READ A, B, C
20 LET S = A + B + C
30 LET A1 = S/B
40 PRINT A, B, C, A1
50 DATA 4, 5, 6
60 END

Write a BASIC program to solve each problem.

23. Find the area of a triangle with a base of 15 units and an altitude of 8 units.

24. Find the volume of a sphere with a radius of 14 cm if $V = \frac{4}{3}\pi r^3$.

25. Find the total surface area of a box with a length of 17 cm, a width of 9 cm, and a height of 4.5 cm. Use $A = 2lw + 2lh + 2hw$.

IF-THEN Statements

The IF-THEN statement is used to compare two numbers. It instructs the computer what to do based on the results of the comparison. The general form of the IF-THEN statement is as follows.

line number IF *sentence* THEN *line number*

The sentence uses one of the following symbols.

Symbol	Meaning	Example
=	is equal to	X = 3
>	is greater than	A > B
<	is less than	X + 10 < 21
> =	is greater than or equal to	A > = 16
< =	is less than or equal to	4 * R < = 10
< >	is not equal to	6 < > 10

The IF-THEN statement is used to compare two numbers. If the sentence containing the two numbers is true, then the program execution goes to the line whose number follows THEN. If the sentence is false, then the program execution simply goes to the next line of the program.

Study the use of IF-THEN statements in the programs below.

```
10   LET X = 5
20   IF X > 10 THEN 40
30   PRINT X
40   END
```

Line 10 instructs the computer to assign the value 5 to X.

In line 20, the sentence X > 10 is false since 5 > 10 is false. Therefore, the program execution goes to line 30.

The output for this program is 5, the value of X.

```
10   LET X = 15
20   IF X > 10 THEN 40
30   GO TO 50
40   PRINT X
50   END
```

Since X = 15, the sentence in line 20 is true and the program execution goes to line 40.

To what line would the program execution go if the sentence in line 20 were X < 10?

Suppose you wish to find the areas of ten different squares with sides of 1 unit, 2 units, 3 units, and so on up to 10 units. You can use a LET statement to assign the variables. Consider the following.

```
10   LET S = 1
20   LET S = S + 1
```

In algebra, the statement S = S + 1 is nonsense. But in BASIC, it means that S should be assigned a new value equal to 1 more than its previous value.

The program continues as follows.

```
10  LET S = 1
12  LET A = S↑2        Use A = s², the formula for the area of a square.
14  PRINT A
20  LET S = S + 1      In line 20, the value of S is increased by 1.
30  GO TO 12           Line 30 repeats the program for each new value of S.
40  END
```

As the program is written now, it will continue indefinitely. A line is needed to stop the program when S is greater than 10. An IF-THEN statement can be used to do this.

$$25 \quad \text{IF } S > 10 \text{ THEN } 40$$

Study the use of the IF-THEN statement in the program below.

```
10  LET S = 1
12  LET A = S↑2
14  PRINT A
20  LET S = S + 1
25  IF S > 10 THEN 40      If S is less than or equal to 10, the
        ↓ no       yes     computer is sent back to line 12. If S
30  GO TO 12               is greater than 10, the program ends.
40  END
```

Example

1 Write a program that determines whether a number is positive or negative. Use −3, 6, and −11 as data.

```
10  READ N
20  DATA −3, 6, −11, 1000       Line 30 stops the program when 1000 is read in
30  IF N = 1000 THEN 90         the DATA line.
40  IF N > 0 THEN 70            Line 40 asks "Is N positive?" If yes, the
50  PRINT N; " IS NEGATIVE"     computer goes to line 70. If no, the computer
60  GO TO 10                    goes to the next statement, line 50.
70  PRINT N; " IS POSITIVE"
80  GO TO 10                    The computer prints characters enclosed in
90  END                         quotation marks. A semicolon causes output to
RUN                             be printed close together.
−3 IS NEGATIVE
6 IS POSITIVE
−11 IS NEGATIVE                 No OUT OF DATA line is printed.
```

Written Exercises

State the line number to which the program execution goes after line 10. Let A = 10, B = 6, and X = 20.

1. 10   IF A < 20 THEN 75
   20   PRINT A

2. 10   IF A > = 20 THEN 80
   20   PRINT 3 * A

3. 10   IF A <> B THEN 60
   20   PRINT "HELLO"

4. 10   IF A + X < B THEN 40
   20   PRINT A + X

5. 10   IF X↑3 > A * B THEN 60
   20   X = X + 2

6. 10   IF A − X = B − 16 THEN 40
   20   PRINT A, B, X

Use the program at the right to determine whether the values of X, Y, or both X and Y will be printed. Give the values to be printed.

7. X = 12, Y = 12
8. X = 14, Y = 19
9. X = 9, Y = 10
10. X = 21, Y = 5

```
10  IF X > Y THEN 40
20  LET X = X + 3
30  LET Y = Y + 2
40  IF X > = Y THEN 60
50  PRINT X
60  PRINT Y
70  END
```

Using an IF-THEN statement, write a BASIC program to solve each problem.

11. Print the integers from 1 to 10 in descending order.

12. Print the squares of the integers from 20 to 40 inclusive.

13. Print the cubes of the odd integers from 7 to 21 inclusive.

14. Read a real number. Print the number, its additive inverse, and its multiplicative inverse if it exists.

Compound statements may be used in IF-THEN statements as shown in the programs in problems 15 and 16. Determine the output when each statement is executed by the computer.

15.
```
10  READ A, B
20  DATA 3, 4, 2, 4
30  IF A = 3 AND B = 4 THEN 60
40  PRINT "HOW ARE YOU?"
50  GO TO 80
60  PRINT "HELLO"
70  GO TO 10
80  END
```

16.
```
10  READ A, B
20  DATA 3, 5, 5, 6
30  IF A = 3 OR B = 4 THEN 60
40  PRINT "HOW ARE YOU?"
50  GO TO 80
60  PRINT "HELLO"
70  GO TO 10
80  END
```

584   *BASIC*

FOR-NEXT Statements

FOR-NEXT statements can increase the efficiency of a program. Compare the following programs.

```
10   LET S = 1
12   LET A = S↑2
14   PRINT A
20   LET S = S + 1
25   IF S > 10 THEN 40
30   GO TO 12
40   END
```

```
100   FOR S = 1 TO 10 STEP 1
120   LET A = S↑2
140   PRINT A
300   NEXT S
400   END
```

In a single line in the program at the right, line 100, the value of S is stated at 1 and is increased in steps of 1 until it reaches 10.

100 FOR S = 1 ...	TO 10 ...	STEP 1
LET S = 1	IF S > 10 THEN 40	S = S + 1

Line 100 in the program at the right above accomplishes the same thing as lines 10, 20, and 25 do in the program on the left above.

When the computer reads line 100 the first time, S is assigned the value 1. When line 300, NEXT S, is encountered, the computer returns to line 100 and increases the value of S by 1. This process continues until the value of S is greater than 10. The computer then goes to the line following NEXT S.

The general form for FOR-NEXT statements is shown below.

line number FOR *variable* = *number* TO *number* STEP *number*

line number NEXT *variable*

A FOR statement must always be paired with a NEXT statement. The number following STEP may be positive or negative.

Example

1 **Write a program to print the integers from 1 to 10. Use FOR-NEXT statements.**

```
10   FOR I = 1 TO 10     STEP is not needed when increasing by 1.
20   PRINT I
30   NEXT I
40   END
```

2 **Write a program that prints the even integers from 2 to 100.**

```
10   FOR I = 2 TO 100 STEP 2
20   PRINT I
30   NEXT I
40   END
```

3 **Write a program that prints a table of the integers from 1 to 3 and their squares.**

```
10   PRINT "NUMBER", "SQUARE"    Line 10 prints headings for a table.
20   FOR I = 1 TO 3
30   PRINT I, I↑2
40   NEXT I
50   END
```

NUMBER	SQUARE	*This is how the table appears.*
1	1	
2	4	
3	9	

FOR-NEXT statements, like GO TO statements and IF-THEN statements, can be used in programs to create loops so that parts of the programs are repeated. There are only two ways these loops can appear.

Nested Loops

```
┌─ FOR X
│ ┌─ FOR Y
│ └─ NEXT Y
└── NEXT X
```

The loops do not cross.

Independent Loops

```
┌─ FOR X
└─ NEXT X
┌─ FOR Y
└─ NEXT Y
```

The loops do not cross. They are not nested.

Not Acceptable

```
┌─ FOR X
├─ FOR Y
└─ NEXT X
└── NEXT Y
```

These loops cross.

The program below utilizes a nested loop to generate all ordered pairs with first element from set A and second element from set B where A = {1, 2, 3, 4} and B = {6, 7, 8}.

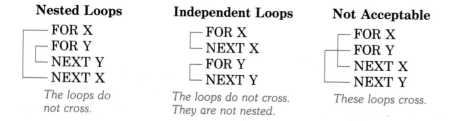

```
10   FOR A = 1 TO 4
20   FOR B = 6 TO 8
30   PRINT A; ","; B,
40   NEXT B
50   NEXT A
60   END
```

Note that the loop for B is nested within the loop for A.

The output for the program on the preceding page follows.

1, 6	1, 7	1, 8
2, 6	2, 7	2, 8
3, 6	3, 7	3, 8
4, 6	4, 7	4, 8

Notice that A equals the same value while B goes through the loop.

Written Exercises

Find the error or errors in each program. Then enter and run the corrected program on your computer.

1.
```
10   IF I = 1 TO 5
20   LET Y = 3 * I + 4
30   PRINT I, Y
40   NEXT I
50   END
```

2.
```
10   FOR I = 1 TO 10 STEP 2
20   LET Y = 3 * I + 10
30   PRINT I, Y
40   NEXT Y
50   END
```

3.
```
 5   LET S = 0
10   READ M, N
20   FOR J = M + N STEP 3
30   LET S = S + J
40   NEXT J
50   PRINT S
60   END
```

4.
```
10   FOR X = 1 TO 5
20   FOR Y = 2 TO 10
30   PRINT X * Y
40   NEXT X
50   NEXT Y
60   END
```

Using FOR-NEXT statements, write a BASIC program to solve each problem.

5. Print the odd numbers from 1 to 25.

6. Find the sum of any five numbers. Use READ-DATA statements.

7. Print a table of squares and cubes from 1 to 25.

8. The formula for converting Fahrenheit to Celsius temperature is $C = \frac{5}{9}(F - 32)$. Generate Celsius temperatures for all even Fahrenheit values from 0 to 100.

9. Use the formula in problem 8 to generate Fahrenheit values for all integer Celsius values from -20 to 40.

10. Print an addition table showing all addition facts from $1 + 1 = 2$ to $4 + 4 = 8$.

11. Print a multiplication table showing all multiplication facts from $1 \cdot 1 = 1$ to $10 \cdot 10 = 100$.

12. Print a table of all positive integers and their squares such that none of the squares is greater than 1,000.

13. Find the pairs of integers that satisfy the equation $x^2 + y^2 = 100$.

14. Find the pairs of integers that satisfy the equation $x^2 \cdot y^2 = 400$.

Special Features of BASIC

When the results of computations exceed six significant digits, the computer will use E notation. This is the computer equivalent to scientific notation. The E means "times 10 to the given power."

Result of computation	E Notation	Meaning
150,000,000,000	1.5E + 11	1.5×10^{11}
37,867,275	3.7867275E + 07	3.7867275×10^7
0.003629	3.629E − 03	3.629×10^{-3}

Sometimes it is necessary to save the values of variables for use later in a program. Consider the following program in which values of variables are *not* saved.

```
10   READ N
20   DATA 3, 4, 7, 11, −2, 14
30   PRINT N
40   GO TO 10
50   END
```

Each time the computer executes line 10, the previous value of N is erased.

The use of subscripted variables allows you to save values for future use. A subscripted variable may be any valid numeric variable followed by a nonnegative integer, or a variable or expression whose value is a nonnegative integer. $X(1)$, $A(I)$, and $F(N+1)$ are valid subscripted variables.

```
10   DIM A(12)
20   FOR I = 1 TO 12
30   READ A(I)
35   PRINT A(I)
40   NEXT I
50   DATA 2, −6, 7.3, 10, 13, −8.6, 5, 9, −12, 1.5, 4, −1
60   END
```

A(1) is a subscripted variable where I represents an integer from 1 to 12.

Line 10 uses a DIMENSION statement that tells the computer to reserve 12 spaces for the values of the subscripted variables $A(I)$. The number of spaces reserved must be greater than or equal to the number of spaces needed. Line 30 reads and stores numbers from the data line in subscripted variables. For example, 2 is stored in position $A(1)$, 7.3 is stored in position $A(3)$, 13 is stored in position $A(5)$, and so on. Line 35 will cause the computer to print all stored values of the subscripted variables $A(I)$.

BASIC contains useful internal functions. These are functions that are built into the computer to perform special operations. Some of them are shown below. Examples of their use follow.

ABS(X)   This function finds the absolute value of X.

SQR(X)   This function finds the square root of X.

INT(X)   This function finds the greatest integer less than or equal to X.

Examples

1 **Write a program to read a number and print it and its absolute value.**

```
10   READ N
20   DATA 6, −3, 0, 4, 0.31, −9, 10000
30   IF N = 10000 THEN 60
40   PRINT N, ABS(N)
50   GO TO 10
60   END
```

The programmer does not need the absolute value of 10000. 10000 is a flag that signals the end of the data.

2 **Write a program to print the square and square root of the first ten positive integers.**

```
 5   PRINT "N", "N SQUARED", "SQRT of N"
10   FOR N = 1 TO 10
20   PRINT N, N↑2, SQR(N)
30   NEXT N
40   END
```

Some examples of the greatest integer function are given as follows.

$$INT(5.72) = 5 \qquad INT(−8.9) = −9$$
$$INT(−13) = −13 \qquad INT(44.99) = 44$$

Suppose you wish to determine if 2 is a factor of 7. If INT(7/2) = 7/2, then 7/2 must be an integer. Of course, you know that 7/2 = 3.5 and INT(7/2) = 3. Therefore, 2 is *not* a factor of 7.

In general, if INT(N/X) = N/X, then X is a factor of N. Example 3 illustrates this principle.

3 Write a program to find all the factors of 48.

```
10   FOR I = 1 TO 48
20   IF INT(48/I) = 48/I THEN 40
30   GO TO 50
40   PRINT I
50   NEXT I
60   END
```

Written Exercises

Write each number using scientific notation and in E notation.

1. 8,200,000

2. 0.00000005108

3. 27,372,800,000

Write the value of each expression in decimal notation.

4. 8.091E + 08

5. 2.2771E − 04

6. 1.461E + 11

If A = −5.1, B = 9, C = 5.1, and D = 1, determine the output of each PRINT statement.

7. 30   PRINT ABS(A)

8. 90   PRINT ABS(A * D)

9. 200   PRINT SQR(B) − SQR(D)

10. 70   PRINT INT(C)

11. 40   PRINT INT(ABS(A))

12. 110   PRINT ABS(INT(A))

Write a BASIC program to solve each problem.

13. Store fifteen given numbers in a subscripted variable. Print the 3rd, 7th, and 14th numbers.

14. Store twenty given numbers in a subscripted variable. Print the numbers in reverse order.

15. Read a given number and test if it is negative. If it is, then print its absolute value.

16. Subtract two given numbers and print the absolute value of their difference.

17. Test each of the following to see if it is negative: 7, −3, 0, −9, 42, 36. If not, print the number and its square root.

18. Print the square roots of all integers from 93 to 121 inclusive.

19. Find the greatest integer less than or equal to each of these numbers: 3.7, 0, 6, 0.31, −2.56, −4.01.

20. Use the greatest integer function to determine if nine given numbers are even or odd.

21. Use the greatest integer function to round a given decimal to the nearest tenth.

22. Find all integers between 1 and 100 that are multiples of 6.

Symbols

$=$	is equal to	π	pi
$\neq$	is not equal to	$\{\ \}$	set
$>$	is greater than	$\%$	percent
$<$	is less than	$^\circ$	degrees
$\geq$	is greater than or equal to	$a{:}b$	ratio of a to b
$\leq$	is less than or equal to	$f(x)$	f of x, the value of f at x
$\approx$	is approximately equal to	(a, b)	ordered pair a, b
$\cdot$	times	$\overline{AB}$	line segment AB
$-$	negative	AB	measure of $\overline{AB}$
$+$	positive	$\sqrt{\ }$	principal square root
$\pm$	positive or negative	$\cos A$	cosine of A
$-a$	opposite or additive inverse of a	$\sin A$	sine of A
$\lvert a \rvert$	absolute value of a	$\tan A$	tangent of A
$a \stackrel{?}{=} b$	Does a equal b?		

Metric System

mm	millimeter	h	hour
cm	centimeter	min	minute
m	meter	s	second
km	kilometer	km/h	kilometer per hour
g	gram	m/s	meters per second
kg	kilogram	°C	degrees Celsius
mL	milliliter		
L	liter		

SQUARES AND APPROXIMATE SQUARE ROOTS

n	n^2	$\sqrt{n}$	n	n^2	$\sqrt{n}$
1	1	1.000	51	2601	7.141
2	4	1.414	52	2704	7.211
3	9	1.732	53	2809	7.280
4	16	2.000	54	2916	7.348
5	25	2.236	55	3025	7.416
6	36	2.449	56	3136	7.483
7	49	2.646	57	3249	7.550
8	64	2.828	58	3364	7.616
9	81	3.000	59	3481	7.681
10	100	3.162	60	3600	7.746
11	121	3.317	61	3721	7.810
12	144	3.464	62	3844	7.874
13	169	3.606	63	3969	7.937
14	196	3.742	64	4096	8.000
15	225	3.873	65	4225	8.062
16	256	4.000	66	4356	8.124
17	289	4.123	67	4489	8.185
18	324	4.243	68	4624	8.246
19	361	4.359	69	4761	8.307
20	400	4.472	70	4900	8.367
21	441	4.583	71	5041	8.426
22	484	4.690	72	5184	8.485
23	529	4.796	73	5329	8.544
24	576	4.899	74	5476	8.602
25	625	5.000	75	5625	8.660
26	676	5.099	76	5776	8.718
27	729	5.196	77	5929	8.775
28	784	5.292	78	6084	8.832
29	841	5.385	79	6241	8.888
30	900	5.477	80	6400	8.944
31	961	5.568	81	6561	9.000
32	1024	5.657	82	6724	9.055
33	1089	5.745	83	6889	9.110
34	1156	5.831	84	7056	9.165
35	1225	5.916	85	7225	9.220
36	1296	6.000	86	7396	9.274
37	1369	6.083	87	7569	9.327
38	1444	6.164	88	7744	9.381
39	1521	6.245	89	7921	9.434
40	1600	6.325	90	8100	9.487
41	1681	6.403	91	8281	9.539
42	1764	6.481	92	8464	9.592
43	1849	6.557	93	8649	9.644
44	1936	6.633	94	8836	9.695
45	2025	6.708	95	9025	9.747
46	2116	6.782	96	9216	9.798
47	2209	6.856	97	9409	9.849
48	2304	6.928	98	9604	9.899
49	2401	7.000	99	9801	9.950
50	2500	7.071	100	10000	10.000

TRIGONOMETRIC RATIOS

Angle	sin	cos	tan	Angle	sin	cos	tan
0°	0.0000	1.0000	0.0000	45°	0.7071	0.7071	1.0000
1°	0.0175	0.9998	0.0175	46°	0.7193	0.6947	1.0355
2°	0.0349	0.9994	0.0349	47°	0.7314	0.6820	1.0724
3°	0.0523	0.9986	0.0524	48°	0.7431	0.6691	1.1106
4°	0.0698	0.9976	0.0699	49°	0.7547	0.6561	1.1504
5°	0.0872	0.9962	0.0875	50°	0.7660	0.6428	1.1918
6°	0.1045	0.9945	0.1051	51°	0.7771	0.6293	1.2349
7°	0.1219	0.9925	0.1228	52°	0.7880	0.6157	1.2799
8°	0.1392	0.9903	0.1405	53°	0.7986	0.6018	1.3270
9°	0.1564	0.9877	0.1584	54°	0.8090	0.5878	1.3764
10°	0.1736	0.9848	0.1763	55°	0.8192	0.5736	1.4281
11°	0.1908	0.9816	0.1944	56°	0.8290	0.5592	1.4826
12°	0.2079	0.9781	0.2126	57°	0.8387	0.5446	1.5399
13°	0.2250	0.9744	0.2309	58°	0.8480	0.5299	1.6003
14°	0.2419	0.9703	0.2493	59°	0.8572	0.5150	1.6643
15°	0.2588	0.9659	0.2679	60°	0.8660	0.5000	1.7321
16°	0.2756	0.9613	0.2867	61°	0.8746	0.4848	1.8040
17°	0.2924	0.9563	0.3057	62°	0.8829	0.4695	1.8807
18°	0.3090	0.9511	0.3249	63°	0.8910	0.4540	1.9626
19°	0.3256	0.9455	0.3443	64°	0.8988	0.4384	2.0503
20°	0.3420	0.9397	0.3640	65°	0.9063	0.4226	2.1445
21°	0.3584	0.9336	0.3839	66°	0.9135	0.4067	2.2460
22°	0.3746	0.9272	0.4040	67°	0.9205	0.3907	2.3559
23°	0.3907	0.9205	0.4245	68°	0.9272	0.3746	2.4751
24°	0.4067	0.9135	0.4452	69°	0.9336	0.3584	2.6051
25°	0.4226	0.9063	0.4663	70°	0.9397	0.3420	2.7475
26°	0.4384	0.8988	0.4877	71°	0.9455	0.3256	2.9042
27°	0.4540	0.8910	0.5095	72°	0.9511	0.3090	3.0777
28°	0.4695	0.8829	0.5317	73°	0.9563	0.2924	3.2709
29°	0.4848	0.8746	0.5543	74°	0.9613	0.2756	3.4874
30°	0.5000	0.8660	0.5774	75°	0.9659	0.2588	3.7321
31°	0.5150	0.8572	0.6009	76°	0.9703	0.2419	4.0108
32°	0.5299	0.8480	0.6249	77°	0.9744	0.2250	4.3315
33°	0.5446	0.8387	0.6494	78°	0.9781	0.2079	4.7046
34°	0.5592	0.8290	0.6745	79°	0.9816	0.1908	5.1446
35°	0.5736	0.8192	0.7002	80°	0.9848	0.1736	5.6713
36°	0.5878	0.8090	0.7265	81°	0.9877	0.1564	6.3138
37°	0.6018	0.7986	0.7536	82°	0.9903	0.1392	7.1154
38°	0.6157	0.7880	0.7813	83°	0.9925	0.1219	8.1443
39°	0.6293	0.7771	0.8098	84°	0.9945	0.1045	9.5144
40°	0.6428	0.7660	0.8391	85°	0.9962	0.0872	11.4301
41°	0.6561	0.7547	0.8693	86°	0.9976	0.0698	14.3007
42°	0.6691	0.7431	0.9004	87°	0.9986	0.0523	19.0811
43°	0.6820	0.7314	0.9325	88°	0.9994	0.0349	28.6363
44°	0.6947	0.7193	0.9657	89°	0.9998	0.0175	57.2900
45°	0.7071	0.7071	1.0000	90°	1.0000	0.0000	∞

Algebraic Skills Review

Integer Equations: Addition and Subtraction

Solve each equation.

1. $-5 + (-8) = x$
2. $-7 + 4 = y$
3. $-4 + 8 = t$
4. $9 + (-2) = a$
5. $6 + 6 = b$
6. $3 + (-8) = m$
7. $-7 + (-9) = v$
8. $-5 + 5 = z$
9. $-19 + 43 = c$
10. $51 + (-26) = w$
11. $-37 + (-48) = d$
12. $-93 + 44 = e$
13. $67 + (-82) = n$
14. $28 + 46 = f$
15. $29 + (-37) = s$
16. $-94 + (-58) = g$
17. $-18 + 63 = p$
18. $28 + (-52) = j$
19. $77 + 57 = r$
20. $47 + (-29) = x$
21. $-18 + 26 = a$
22. $-65 + (-75) = k$
23. $21 + (-47) = h$
24. $-15 + 52 = q$
25. $y = -13 + (-98)$
26. $u = -5 + 82$
27. $s = -47 + 26 + (-18)$
28. $a = -71 + (-85) + (-16)$
29. $41 + 57 + (-32) = m$
30. $82 + (-14) + (-35) = c$

31. $-4 - (-2) = t$
32. $5 - (-6) = p$
33. $-9 - 3 = x$
34. $5 - (-5) = k$
35. $-3 - (-8) = m$
36. $3 - 9 = w$
37. $-6 - 8 = w$
38. $0 - 6 = v$
39. $j = -10 - (-4)$
40. $-23 - 45 = a$
41. $28 - (-14) = z$
42. $-53 - (-61) = f$
43. $c = -16 - 47$
44. $90 - 43 = g$
45. $71 - (-47) = q$
46. $-99 - (-26) = s$
47. $38 - (-19) = t$
48. $-20 - (-92) = j$
49. $18 - 47 = y$
50. $h = -15 - (-81)$
51. $-42 - 63 = b$
52. $-84 - 47 = r$
53. $42 - (-47) = d$
54. $y = -19 - (-63)$
55. $16 - (-84) = n$
56. $42 - (-26) = k$
57. $-52 - (-33) = x$
58. $-35 - 86 = a$
59. $v = -8 - (-47)$
60. $33 - 51 = t$

61. $-2 + g = 7$
62. $9 + s = -5$
63. $-7 + k = -2$
64. $-4 + y = -9$
65. $m + 6 = 2$
66. $t + (-4) = 10$
67. $h - (-2) = 6$
68. $v - 7 = -4$
69. $a - (-6) = -5$
70. $r - (-3) = -8$
71. $j - (-8) = 5$
72. $x - 8 = -9$
73. $-2 - x = -8$
74. $14 = -48 + b$
75. $c + (-26) = 45$
76. $z - (-57) = -39$
77. $d + (-44) = -61$
78. $n - 38 = -19$
79. $-77 = w + 23$
80. $e - (-26) = 41$
81. $p - 47 = 22$
82. $-63 - f = -82$
83. $87 = t + (-14)$
84. $q + (-53) = 27$

Integer Equations: Multiplication and Division

Solve each equation.

1. $x = (-8)(-4)$

2. $(-3)5 = t$

3. $(-7)(-2) = a$

4. $(-9)8 = b$

5. $6(-5) = v$

6. $k = 8(6)$

7. $14(-26) = s$

8. $(-46)(-25) = g$

9. $(-71)(-20) = y$

10. $(-42)66 = h$

11. $(-97)47 = w$

12. $53(-32) = c$

13. $19(-46) = x$

14. $(-82)0 = e$

15. $72(43) = m$

16. $(-18)(-18) = d$

17. $24(-29) = u$

18. $f = (-39)45$

19. $(-76)(-34) = s$

20. $(-81)(-18) = q$

21. $(-65)28 = t$

22. $71(-38) = p$

23. $j = 49(-92)$

24. $36(24) = a$

25. $(-42)78 = z$

26. $(-54)(-77) = r$

27. $n = (-6)(-127)(-4)$

28. $(13)(-12)(95) = w$

29. $(-1)(45)(-45) = v$

30. $(-3)(61)(99) = y$

31. $72 \div (-8) = g$

32. $-64 \div 8 = b$

33. $-45 \div (-9) = y$

34. $56 \div (-7) = z$

35. $42 \div 6 = e$

36. $-24 \div (-6) = m$

37. $992 \div (-32) = a$

38. $-4428 \div 54 = k$

39. $x = -600 \div (-24)$

40. $1472 \div (-64) = p$

41. $-564 \div (-47) = h$

42. $-504 \div 14 = j$

43. $-2201 \div 71 = r$

44. $1512 \div (-28) = n$

45. $765 \div (-85) = q$

46. $-1591 \div (-37) = f$

47. $s = -1080 \div 36$

48. $3432 \div (-52) = v$

49. $2730 \div 78 = k$

50. $-3936 \div 96 = c$

51. $-1476 \div 41 = z$

52. $1496 \div (-22) = a$

53. $2646 \div (-63) = t$

54. $w = -4730 \div (-55)$

55. $-1092 \div (-26) = x$

56. $-2700 \div (-75) = e$

57. $1127 \div 49 = y$

58. $d = 1900 \div (-38)$

59. $-845 \div 13 = w$

60. $-1596 \div (-42) = a$

61. $-5p = 35$

62. $7g = -49$

63. $-3x = -24$

64. $a \div (-6) = -2$

65. $m \div (-8) = 8$

66. $q \div 9 = -3$

67. $41j = 1476$

68. $62y = -2356$

69. $b \div (-21) = 13$

70. $-33n = -1815$

71. $k \div 46 = -41$

72. $w \div 17 = 24$

73. $c \div (-59) = -7$

74. $-56h = 1792$

75. $-42z = 1512$

76. $j \div (-27) = 27$

77. $89s = -712$

78. $-18v = -1044$

79. $d \div (-34) = -43$

80. $f \div 14 = -63$

81. $45t = 810$

82. $-74w = 1554$

83. $-49e = -2058$

84. $r \div (-16) = -77$

85. $x \div (-26) = 47$

86. $-23 = t \div 44$

87. $-962 = -37g$

88. $-3040 = 95k$

89. $84 = x \div 97$

90. $-108 = m \div (-12)$

Fraction Equations: Addition and Subtraction

Solve each equation and express answers in simplest form.

1. $\frac{3}{11} + \frac{6}{11} = x$

2. $\frac{4}{7} + \frac{5}{7} = a$

3. $\frac{5}{9} - \frac{2}{9} = t$

4. $\frac{17}{18} - \frac{5}{18} = w$

5. $\frac{1}{3} + \frac{2}{9} = b$

6. $\frac{1}{2} - \frac{1}{3} = v$

7. $\frac{3}{4} - \frac{9}{16} = s$

8. $\frac{2}{3} + \frac{8}{15} = r$

9. $\frac{5}{6} - \frac{3}{4} = d$

10. $\frac{4}{9} + \frac{1}{6} = c$

11. $m = \frac{7}{9} + \frac{3}{8}$

12. $\frac{11}{12} - \frac{7}{10} = j$

13. $\frac{5}{6} - \frac{5}{12} = p$

14. $4\frac{2}{3} + 1\frac{8}{15} = k$

15. $5\frac{1}{2} - 2\frac{1}{3} = w$

16. $8\frac{1}{12} - 5\frac{5}{12} = e$

17. $7 - 1\frac{4}{9} = h$

18. $n = \frac{3}{16} + \frac{7}{12}$

19. $4\frac{1}{2} - 2\frac{2}{3} = q$

20. $7\frac{1}{12} - 4\frac{5}{8} = x$

21. $11\frac{5}{6} + 9\frac{7}{15} = f$

22. $y = 9\frac{2}{7} - 5\frac{5}{6}$

23. $\frac{1}{4} + \frac{5}{6} + \frac{7}{12} = c$

24. $\frac{5}{6} + \frac{2}{9} + \frac{3}{4} = z$

25. $-\frac{2}{13} + \left(-\frac{3}{13}\right) = t$

26. $-\frac{11}{18} + \frac{17}{18} = f$

27. $-\frac{9}{10} - \frac{7}{10} = n$

28. $-\frac{7}{11} - \left(-\frac{3}{11}\right) = a$

29. $\frac{1}{12} - \left(-\frac{7}{12}\right) = w$

30. $\frac{17}{21} + \left(-\frac{10}{21}\right) = g$

31. $\frac{1}{4} + \left(-\frac{2}{3}\right) = p$

32. $b = -\frac{1}{6} - \frac{8}{9}$

33. $\frac{1}{3} - \frac{5}{6} = m$

34. $-\frac{1}{2} + \left(-\frac{3}{5}\right) = a$

35. $\frac{3}{7} + \left(-5\right) = s$

36. $-\frac{5}{9} - 2 = k$

37. $t = 1\frac{1}{2} - \left(-\frac{3}{4}\right)$

38. $-\frac{3}{8} + \frac{4}{7} = c$

39. $\frac{3}{5} - \left(-3\frac{1}{4}\right) = v$

40. $-8\frac{7}{8} - \left(-4\frac{5}{12}\right) = r$

41. $-3\frac{1}{6} + 5\frac{1}{15} = d$

42. $-1\frac{8}{9} + \left(-5\frac{7}{12}\right) = h$

43. $7\frac{5}{6} + \left(-8\frac{7}{8}\right) = j$

44. $-3\frac{1}{2} - 4\frac{5}{9} = e$

45. $q = \frac{11}{16} - 12$

46. $-5\frac{11}{20} + 4\frac{7}{12} = z$

47. $-1\frac{1}{12} - \left(-\frac{2}{3}\right) = w$

48. $-4\frac{16}{21} + \left(-7\frac{5}{9}\right) = y$

49. $\frac{3}{13} + p = \frac{10}{13}$

50. $e + \frac{4}{15} = \frac{13}{15}$

51. $\frac{2}{5} + n = \frac{2}{3}$

52. $j - \frac{5}{18} = \frac{17}{18}$

53. $r - \frac{1}{4} = \frac{5}{16}$

54. $b - \frac{1}{2} = \frac{2}{5}$

55. $s + \frac{2}{7} = 2$

56. $\frac{7}{10} - a = \frac{1}{2}$

57. $1\frac{5}{6} + x = 2\frac{1}{4}$

58. $4\frac{1}{4} = w + 2\frac{1}{3}$

59. $d - 1\frac{5}{7} = 6\frac{1}{4}$

60. $h - \frac{3}{4} = 2\frac{5}{8}$

61. $t - \frac{2}{3} = 1\frac{5}{8}$

62. $g + \frac{5}{6} = \frac{4}{9}$

63. $q - \frac{7}{10} = -\frac{11}{15}$

64. $-\frac{3}{7} + c = \frac{1}{2}$

65. $-\frac{3}{4} = v + \left(-\frac{1}{8}\right)$

66. $f - \left(-\frac{1}{8}\right) = \frac{3}{10}$

67. $m - \left(-1\frac{3}{8}\right) = -2\frac{1}{2}$

68. $-6\frac{5}{6} + y = 7\frac{7}{15}$

69. $7\frac{1}{6} - z = -5\frac{2}{3}$

70. $-2\frac{1}{3} + w = -5\frac{5}{6}$

71. $-6\frac{1}{7} + k = -\frac{4}{21}$

72. $-4\frac{5}{12} = t - \left(-10\frac{1}{36}\right)$

Fraction Equations: Multiplication and Division

Solve each equation and express answers in simplest form.

1. $\frac{1}{7}\left(\frac{1}{3}\right) = x$

2. $\frac{2}{3}\left(\frac{1}{5}\right) = v$

3. $\frac{5}{6}\left(\frac{3}{10}\right) = y$

4. $2 \div \frac{1}{3} = j$

5. $\frac{5}{6} \div \frac{1}{6} = f$

6. $\frac{1}{4} \div \frac{5}{8} = c$

7. $\frac{2}{3}(9) = b$

8. $\frac{5}{18}\left(\frac{3}{10}\right) = r$

9. $\frac{8}{15} \div \frac{1}{10} = k$

10. $g = \frac{1}{2} \div 8$

11. $\frac{3}{14} \div \frac{2}{7} = y$

12. $\frac{7}{12}\left(\frac{4}{5}\right) = w$

13. $\frac{6}{13} \div \frac{5}{7} = z$

14. $\frac{24}{25}\left(\frac{15}{32}\right) = a$

15. $\frac{7}{10}\left(\frac{5}{28}\right) = w$

16. $2\frac{2}{3}\left(\frac{4}{5}\right) = n$

17. $t = \frac{7}{8}\left(4\frac{1}{4}\right)$

18. $1\frac{3}{4} \div \frac{7}{12} = e$

19. $1 \div 2\frac{3}{5} = d$

20. $2\frac{1}{10}\left(4\frac{2}{7}\right) = q$

21. $1\frac{3}{5} \div 11\frac{1}{5} = s$

22. $3\frac{1}{8}\left(2\frac{4}{5}\right)\left(\frac{5}{7}\right) = p$

23. $3\frac{2}{3}\left(\frac{1}{8}\right)\left(1\frac{1}{11}\right) = h$

24. $m = 3\frac{1}{21} \div 1\frac{21}{35}$

25. $\frac{1}{5}\left(-\frac{1}{8}\right) = p$

26. $-\frac{2}{9}\left(\frac{1}{3}\right) = w$

27. $-\frac{5}{8}\left(-\frac{4}{5}\right) = c$

28. $-3 \div \frac{1}{2} = y$

29. $-\frac{7}{9} \div \left(-\frac{1}{9}\right) = r$

30. $\frac{2}{3} \div \left(-\frac{4}{9}\right) = t$

31. $-\frac{2}{5}(-10) = m$

32. $\frac{2}{3} \div \left(-\frac{7}{9}\right) = h$

33. $-\frac{9}{15}\left(\frac{5}{9}\right) = s$

34. $-\frac{9}{14} \div \left(-\frac{3}{7}\right) = a$

35. $\frac{7}{16} \div \left(-\frac{7}{11}\right) = z$

36. $j = -\frac{4}{5}(30)$

37. $-7 \div 4 = q$

38. $-4\frac{9}{10}\left(-1\frac{5}{21}\right) = b$

39. $-5\frac{3}{5} \div 4\frac{1}{5} = g$

40. $-5\frac{3}{5} \div \left(-4\frac{1}{5}\right) = e$

41. $v = 6\frac{1}{4}\left(-1\frac{7}{15}\right)$

42. $3\frac{1}{3}\left(-4\frac{1}{2}\right) = w$

43. $-2\left(1\frac{5}{18}\right) = k$

44. $4\frac{2}{5} \div \left(-\frac{11}{15}\right) = p$

45. $-2\frac{5}{8} \div 7\frac{1}{2} = x$

46. $5 \div (-11) = n$

47. $d = -2\frac{3}{10}\left(-\frac{5}{12}\right)$

48. $-9\frac{1}{3}\left(-3\frac{3}{4}\right) = f$

49. $\frac{1}{3}a = 5$

50. $\frac{4}{7}k = 4$

51. $\frac{2}{5}x = \frac{4}{7}$

52. $w \div 5 = 3$

53. $w \div \frac{1}{4} = \frac{3}{8}$

54. $c \div \frac{3}{10} = \frac{1}{2}$

55. $\frac{7}{11}t = \frac{4}{5}$

56. $h \div \frac{1}{8} = \frac{4}{11}$

57. $z \div 6 = \frac{5}{12}$

58. $1\frac{1}{2}d = \frac{6}{7}$

59. $\frac{10}{33} = b \div 4\frac{2}{5}$

60. $2\frac{1}{6}j = 5\frac{1}{5}$

61. $s \div 2\frac{1}{6} = 2\frac{2}{5}$

62. $1\frac{3}{24}g = 3\frac{1}{8}$

63. $3 = 1\frac{7}{11}q$

64. $n \div \frac{2}{3} = -\frac{4}{9}$

65. $-1\frac{3}{4}p = -\frac{5}{8}$

66. $v \div \left(-\frac{7}{11}\right) = 1\frac{2}{7}$

67. $-1\frac{3}{5} = e \div \left(-3\frac{1}{5}\right)$

68. $-\frac{5}{9}r = 7\frac{1}{2}$

69. $3\frac{4}{7}x = -3\frac{3}{4}$

70. $a \div 3\frac{2}{7} = -8\frac{3}{4}$

71. $-2\frac{4}{7}m = -3\frac{3}{8}$

72. $f \div \left(-3\frac{1}{8}\right) = -3\frac{2}{5}$

Decimal Equations: Addition and Subtraction

Solve each equation.

1. $0.53 + 0.26 = x$
2. $14.756 + 0.185 = k$
3. $0.711 - 0.158 = z$
4. $12.01 - 0.83 = s$
5. $0.4 + 0.86 = n$
6. $1.4 - 0.12 = a$
7. $57.5 + 7.94 = m$
8. $10.04 - 0.18 = f$
9. $5 - 1.63 = r$
10. $5.92 + 7.3 = b$
11. $12 + 9.6 = y$
12. $28.05 - 9.95 = c$
13. $0.2 + 6.51 + 2.03 = y$
14. $4.4 + 30.6 + 11.2 = z$
15. $0.007 + 3 + 10.02 = h$
16. $w = 20.13 - 12.5$
17. $2.3 - 0.846 = t$
18. $11 - 1.1 = p$
19. $6.2 + 5.54 + 13.66 = g$
20. $a = 412 - 0.007$
21. $101.12 + 9.099 = s$
22. $66.4 - 5.288 = d$
23. $84.083 - 17 = m$
24. $q = 0.046 + 5.8 + 11.37$
25. $8 - 3.49 = n$
26. $8.77 + 0.3 + 52.9 = x$
27. $14.7 - 5.8364 = e$
28. $66.68421 - 18.465 = v$
29. $y = 0.0013 + 2.881$
30. $127.11 + 48 + 0.143 = u$

31. $-0.47 + 0.62 = h$
32. $-4.5 + (-12.8) = x$
33. $-1.7 + 0.24 = p$
34. $-6.831 - (-2.648) = c$
35. $-4.23 - 2.47 = b$
36. $2.64 - (-5.9) = k$
37. $10 + (-0.43) = r$
38. $6.7 - (-0.64) = v$
39. $-6.71 - (-8) = e$
40. $14.14 + (-1.4) = a$
41. $1.2 - 6.73 = j$
42. $-9.7 + (-0.86) = d$
43. $-7 - 4.63 = w$
44. $-0.17 - (-14.6) = g$
45. $m = 1.8 + (-14.14)$
46. $5.003 + (-0.47) = f$
47. $0.88 - 42 = s$
48. $-6.2 + (-27.47) = j$
49. $n = -1.4962 + 2.118$
50. $2.4 - (-1.736) = q$
51. $4.16 + (-5.909) = t$
52. $17 + (-0.45) = w$
53. $10 - 13.463 = a$
54. $f = -82.007 - 3.218$
55. $-11.264 + (-8.2) = z$
56. $-56 + 2.783 = s$
57. $-0.682 - (-0.81) = y$
58. $r = -23 + 4.093$
59. $2.08 - (-0.094) = t$
60. $-51.34 + (-5.1346) = x$

61. $2.2 + a = 11.4$
62. $h + 1.83 = 8.42$
63. $c + 5.4 = -11.33$
64. $m - 0.41 = 0.85$
65. $p - 1.1 = 14.9$
66. $r - 0.76 = -3.2$
67. $t + (-6.47) = -22.3$
68. $-6.11 + b = 14.321$
69. $k - (-4) = 7.9$
70. $k - 99.7 = -46.88$
71. $w + (-17.8) = -5.63$
72. $-5 = y - 22.7$
73. $13.475 + d = 4.09$
74. $-5 - q = 1.19$
75. $-3.214 + f = -16.04$
76. $-88.9 = s - 6.21$
77. $2 + e = 1.008$
78. $n + (-4.361) = 59.78$
79. $4.8 - j = -5.834$
80. $w - 0.73 = -1.8$
81. $-8 = g - (-4.821)$
82. $-2.315 + x = -15$
83. $m + (-1.4) = 0.07$
84. $v - 5.234 = -1.051$
85. $7.1 = v - (-0.62)$
86. $s + 6.4 = -0.11$
87. $t - (-46.1) = -3.673$
88. $k + (-1.604) = -0.45$
89. $81.6 + p = -6.73$
90. $-0.1448 - z = -2.6$

Decimal Equations: Multiplication and Division

Solve each equation.

1. $46(0.5) = e$
2. $108(0.9) = b$
3. $g = 6.47(39)$
4. $0.04(197) = f$
5. $r = 67(5.892)$
6. $2.8(4.27) = d$
7. $0.061(5.5) = m$
8. $0.62(0.13) = c$
9. $4.007(1.95) = q$
10. $6.25 \div 5 = w$
11. $t = 91.8 \div 27$
12. $7.31 \div 43 = h$
13. $5.91 \div 0.3 = a$
14. $167.5 \div 2.5 = k$
15. $4.7208 \div 0.84 = v$
16. $p = 278.1 \div 6.18$
17. $30{,}176 \div 9.43 = n$
18. $0.1001 \div 0.77 = j$
19. $2.11(0.059) = w$
20. $s = 0.4484 \div 1.18$
21. $0.0062(84.7) = x$
22. $0.03912 \div 1.63 = z$
23. $230.4 \div 0.072 = m$
24. $w = 59.8(100.23)$
25. $v = 432 \div 9.6$
26. $0.008(0.0045) = x$
27. $1.21(0.47)(9.3) = s$
28. $0.0418 \div 0.19 = x$
29. $0.032(13)(2.6) = t$
30. $0.0001926 \div 0.00321 = y$

31. $-5(0.2) = x$
32. $-1.7(-44) = f$
33. $72(1.01) = c$
34. $627(-0.14) = a$
35. $-2.3(7.81) = n$
36. $r = -1.02(-4.4)$
37. $57.6 \div (-12) = b$
38. $160.8 \div 24 = h$
39. $-16.38 \div (-0.7) = t$
40. $m = -15.54 \div 2.1$
41. $-0.405 \div (-0.27) = a$
42. $-598 \div 0.13 = p$
43. $0.45(-0.0016) = k$
44. $y = -0.002052 \div 0.054$
45. $6.7284 \div 1.08 = d$
46. $-0.0066(-91.8) = w$
47. $455 \div (-1.82) = q$
48. $-0.905(0.208) = g$
49. $-2.4827 \div (-6.71) = e$
50. $0.153 \div (-0.017) = z$
51. $j = -462.1(0.0094)$
52. $56.1(2.3) = y$
53. $0.07553 \div 0.0083 = v$
54. $-1.7(-0.121) = s$
55. $t = -0.6612 \div (-0.114)$
56. $-0.026(45.1) = x$
57. $59(-0.00042) = w$
58. $7.93(-5.036) = c$
59. $9.2397 \div 1.9 = t$
60. $-0.000101 \div 0.001 = m$

61. $7c = 4.2$
62. $37p = 81.4$
63. $57k = 0.1824$
64. $1.5m = 9.9$
65. $1.296 = 0.48d$
66. $0.0022b = 0.1958$
67. $t \div 110 = 2.8$
68. $x \div 71 = 0.33$
69. $r \div 0.85 = 10$
70. $h \div 1.98 = 6.7$
71. $a \div 0.002 = 0.109$
72. $n \div 40.6 = 0.021$
73. $100.8x = 9374.4$
74. $2.61 = f \div 9.5$
75. $1.7118 = 0.317e$
76. $0.0603g = 0.0043416$
77. $w \div 0.0412 = 60$
78. $q \div 1.07 = 0.088$
79. $5j = -32.15$
80. $-1.2v = 112.8$
81. $-0.013s = -0.00923$
82. $w \div (-2) = -2.48$
83. $z \div 2.8 = -6.2$
84. $a \div (-0.53) = -0.034$
85. $k \div (-0.013) = -0.7$
86. $-4.63\,t = -125.473$
87. $7.9y = 1583.16$
88. $6.05p = -1573$
89. $g \div 9.9 = 12$
90. $x \div (-0.063) = 0.015$

Forms of Real Numbers

Write each fraction in simplest form.

1. $\frac{13}{26}$ 2. $\frac{9}{12}$ 3. $-\frac{36}{42}$ 4. $\frac{5}{60}$

5. $-\frac{24}{32}$ 6. $-\frac{10}{35}$ 7. $\frac{54}{63}$ 8. $-\frac{45}{60}$

9. $\frac{48}{84}$ 10. $-\frac{28}{42}$ 11. $-\frac{72}{96}$ 12. $\frac{75}{105}$

13. $-\frac{16}{100}$ 14. $\frac{24}{60}$ 15. $\frac{15}{27}$ 16. $-\frac{99}{111}$

17. $\frac{126}{700}$ 18. $-\frac{198}{462}$ 19. $-\frac{84}{1080}$ 20. $-\frac{525}{1155}$

Change each fraction to a decimal.

21. $\frac{1}{4}$ 22. $-\frac{3}{10}$ 23. $\frac{1}{50}$ 24. $\frac{2}{3}$

25. $-\frac{1}{9}$ 26. $-\frac{16}{25}$ 27. $-\frac{9}{20}$ 28. $\frac{1}{11}$

29. $\frac{5}{9}$ 30. $-\frac{5}{8}$ 31. $\frac{43}{100}$ 32. $-\frac{5}{6}$

33. $-\frac{7}{11}$ 34. $\frac{3}{7}$ 35. $\frac{4}{5}$ 36. $-\frac{7}{12}$

37. $-\frac{15}{16}$ 38. $-\frac{8}{15}$ 39. $\frac{1}{6}$ 40. $-\frac{11}{32}$

41. $\frac{9}{11}$ 42. $\frac{11}{16}$ 43. $-\frac{11}{15}$ 44. $\frac{124}{125}$

Change each mixed numeral to a decimal.

45. $-5\frac{1}{2}$ 46. $14\frac{17}{100}$ 47. $6\frac{3}{25}$ 48. $-7\frac{1}{3}$

49. $4\frac{3}{25}$ 50. $-20\frac{2}{9}$ 51. $12\frac{3}{4}$ 52. $-10\frac{5}{6}$

53. $-1\frac{4}{9}$ 54. $-9\frac{16}{50}$ 55. $-2\frac{2}{11}$ 56. $13\frac{13}{40}$

57. $3\frac{5}{12}$ 58. $-8\frac{5}{7}$ 59. $2\frac{3}{5}$ 60. $11\frac{1}{12}$

61. $-44\frac{3}{8}$ 62. $19\frac{8}{15}$ 63. $-67\frac{7}{10}$ 64. $5\frac{3}{16}$

65. $78\frac{2}{9}$ 66. $-108\frac{1}{20}$ 67. $-51\frac{6}{7}$ 68. $8\frac{1}{15}$

Change each decimal to a fraction in simplest form.

69. 0.3 70. 0.14 71. 0.013 72. -1.25

73. 4.2 74. -20.05 75. $0.\overline{3}$ 76. -14.50

77. $-12.\overline{7}$ 78. 6.125 79. -8.6 80. $-8.\overline{6}$

81. 23.15 82. -0.37 83. -33.85 84. 1.16

85. -2.27 86. 16.75 87. -5.375 88. 4.26

89. 7.1875 90. -9.45 91. $5.2\overline{6}$ 92. -0.324

Percents

Write each decimal as a percent.

1. 0.71   2. 0.4   3. 0.835   4. 1.05

5. 0.009   6. 0.27   7. 2.5   8. 0.706

Write each fraction as a percent.

9. $\frac{31}{100}$ 10. $\frac{1}{2}$ 11. $\frac{4}{5}$ 12. $\frac{3}{10}$

13. $\frac{1}{8}$ 14. $\frac{5}{4}$ 15. $\frac{2}{3}$ 16. $\frac{4}{11}$

Write each percent as a decimal.

17. 14%   18. 10%   19. 450%   20. 6%

21. 27.5%   22. 4.2%   23. 190.5%   24. 0.3%

Write each percent as a fraction in simplest form.

25. 17%   26. 40%   27. 8%   28. 75%

29. 0.9%   30. 2.5%   31. 45.6%   32. 1.05%

Solve each of the following.

33. 10% of 70 is __.   34. 20% of 35 is __.   35. 4% of 250 is __.

36. 255% of 160 is __.   37. 115% of 24 is __.   38. 130% of 60 is __.

39. __ is 3.7% of 300.   40. __ is 22.5% of 260.   41. __ is 52.6% of 150.

42. 5 is __% of 20.   43. 3 is __% of 10.   44. 17 is __% of 68.

45. __% of 500 is 55.   46. __% of 96 is 12.   47. __% of 81 is 27.

48. 40% of __ is 12.   49. 10% of __ is 16.   50. 65% of __ is 26.

51. 25 is $33\frac{1}{3}$% of __.   52. 54 is 108% of __.   53. 1.28 is 16% of __.

54. 6.3% of 400 is __.   55. __% of 40 is 25.   56. 68% of __ is 85.

57. 44% of __ is 37.4.   58. 53% of 62 is __.   59. __% of 16 is 56.

60. __ is 235% of 270.   61. __ is 5.8% of 45.   62. 28 is __% of 21.

63. 2.5% of __ is 1.   64. __% of 20 is 26.2.   65. 420% of __ is 336.

66. __% of 45 is 30.   67. $87\frac{1}{2}$% of __ is 14.   68. $66\frac{2}{3}$% of 81 is __.

69. __ is 12.4% of 15.   70. 135 is 675% of __.   71. 45 is __% of 36.

72. 14.5% of 18 is __.   73. 180.5% of 200 is __.   74. 98.1 is __% of 90.

75. __% of 85 is 102.   76. 3% of __ is 18.   77. 44% of __ is 37.4.

78. __% of 170 is 153.   79. 738 is 72% of __.   80. $266\frac{2}{3}$% of 561 is __.

Evaluating Expressions

Evaluate each expression if $a = 3$, $b = 5$, $c = 12$, and $d = 9$.

1. $8 + c$

2. $d - 4$

3. $b \cdot d$

4. $c \div a$

5. $a + c + d$

6. $c - d$

7. $a \cdot b \cdot c$

8. $d - a$

9. $a \cdot c$

10. $\dfrac{d}{a}$

11. $\dfrac{13 + c}{b}$

12. $\dfrac{a + d}{4}$

Evaluate each expression if $e = 2$, $f = 5$, $g = 6$, and $h = 10$.

13. $8g$

14. fh

15. g^2

16. e^5

17. $7f^2$

18. $g^2 h^3$

19. $\dfrac{h^4}{f^2}$

20. $3e^2 g$

21. $8g^2 f^2$

22. $e^4 f^2 h^3$

23. $20e^3 f^3 g$

24. $\dfrac{3e^2 f^3}{g}$

Evaluate each expression if $x = 3$, $j = 4$, $k = 9$, and $m = 20$.

25. $k^2 - 4k + 6$

26. $(m + j) \div 3$

27. $x^3 j^2 - 4m$

28. $(xj + k) \div x$

29. $5j^2 \div m + k^2$

30. $j^3 + mk + 4x^4$

31. $(j^3 + m)k - 4x$

32. $(xj)^2 + km^2$

33. $k^3 + m \div j - 5j^2$

34. $(5 + j)^2 \div k + m^2$

35. $(m - k)^3 \div (2j + 3)$

36. $(x^4 + k)m^2 - kx$

Evaluate each expression if $n = -1$, $p = 6$, $q = -8$, $r = 15$, and $s = -24$.

37. $pr + 2q$

38. $pn^4 + s$

39. $r^2 - q + 5s$

40. $pq^2 \div ns$

41. $(p + 2q)n - r$

42. $(p + q)^5 n^5 + s$

43. $pr^2 + ns - 6q$

44. $p(r^2 + ns) - 6q$

45. $(q + r + s)p + n$

46. $\dfrac{4(p^2 + q^2)}{2q} - s$

47. $(p + n)^3 + \left(\dfrac{s}{p}\right)q$

48. $[(r + s)q]p$

The formula for the total surface area of a rectangular solid is $T = 2lw + 2wh + 2lh$ where T is the total surface area of the solid, l is its length, w is its width, and h is its height. Find the total surface area of each rectangular solid.

49. $l = 8$, $w = 5$, $h = 14$.

50. $l = 4$, $w = 2.5$, $h = 3$

51. $l = 7$, $w = 7$, $h = 16$

52. $l = 14$, $w = 17$, $h = 11$

53. $l = 21$, $w = 18$, $h = 6$

54. $l = 3.7$, $w = 1.2$, $h = 3.5$

The formula to change Fahrenheit degrees to Celsius degrees is $C = \frac{5}{9}(F - 32)$ where C is the temperature in Celsius degrees and F is the temperature in Fahrenheit degrees. Change each temperature in Fahrenheit degrees to Celsius degrees.

55. $86°$ F

56. $5°$ F

57. $-13°$ F

58. $41°$ F

59. $-40°$ F

60. $23°$ F

61. $-58°$ F

62. $374°$ F

Inequalities

Replace each ▪ with >, <, or = to make each sentence true.

1. $9 \ \blacksquare \ 12$
2. $14 \ \blacksquare \ 7$
3. $-3 \ \blacksquare \ 0$
4. $-7 \ \blacksquare \ -3$
5. $-5 \ \blacksquare \ 3$
6. $7 + 8 \ \blacksquare \ 15$
7. $-8 + 5 \ \blacksquare \ -6$
8. $-24 \div (-8) \ \blacksquare \ -3$
9. $-3 - (-9) \ \blacksquare \ -12$
10. $-9 \ \blacksquare \ -3 + (-4)$
11. $-6 \cdot 3 \ \blacksquare \ -18$
12. $10 \ \blacksquare \ 36 \div 4$
13. $8 \ \blacksquare \ -40 \div 5$
14. $5 - (-4) \ \blacksquare \ 9$
15. $27 \ \blacksquare \ 4 \cdot 7$
16. $4 - 7 \ \blacksquare \ 3$

Solve each inequality.

17. $x + 4 < 10$
18. $a + 7 \geq 15$
19. $g + 5 > -8$
20. $c + 9 \leq 3$
21. $z - 4 > 20$
22. $h - (-7) > -2$
23. $m - 14 \leq -9$
24. $d - (-3) < 13$
25. $\frac{g}{-8} < 4$
26. $\frac{w}{3} > -12$
27. $\frac{p}{5} < 8$
28. $\frac{t}{-4} \geq -10$
29. $7b \geq -49$
30. $-5j < -60$
31. $-8f < 48$
32. $-2 + 9n \leq 10n$
33. $-5e + 9 > 24$
34. $3y - 4 > -37$
35. $7s - 12 < 13$
36. $-6v - 3 \geq -33$
37. $-2k + 12 < 30$
38. $-2x + 1 < 16 - x$
39. $15t - 4 > 11t - 16$
40. $13 - y \leq 29 + 2y$
41. $5q + 7 \leq 3(q + 1)$
42. $2(w + 4) \geq 7(w - 1)$
43. $-4t - 5 > 2t + 13$
44. $9m + 7 < 2(4m - 1)$
45. $3\left(a + \frac{2}{3}\right) \geq a - 1$
46. $3(3y + 1) < 13y - 8$

47. $2 + x < -5$ or $2 + x > 5$
48. $-4 + t > -5$ or $-4 + t < 7$
49. $3 \leq 2g + 7$ and $2g + 7 \leq 15$
50. $7 - 3s < 13$ and $7s < 3s + 12$
51. $2x + 1 < -3$ or $3x - 2 > 4$
52. $2v - 2 \leq 3v$ and $4v - 1 \geq 3v$
53. $3b - 4 \leq 7b + 12$ and $8b - 7 \leq 25$
54. $-9 < 2z + 7 < 10$
55. $5m - 8 \geq 10 - m$ or $5m + 11 < -9$
56. $12c - 4 \leq 5c + 10$ or $-4c - 1 \leq c + 24$
57. $2h - 2 \leq 3h \leq 4h - 1$
58. $3p + 6 < 8 - p$ and $5p + 8 \geq p + 6$
59. $4a + 3 < 3 - 5a$ or $a - 1 \geq -a$
60. $d - 4 < 5d + 14 < 3d + 26$
61. $2r + 8 > 16 - 2r$ and $7r + 21 < r - 9$
62. $-4j + 3 < j + 22$ and $j - 3 < 2j - 15$
63. $3n \neq 9$ and $6n - 5 \leq 2n + 7$
64. $7e \neq -21$ and $5e + 8 \geq e + 6$
65. $2(q - 4) \leq 3(q + 2)$ or $q - 8 \leq 4 - q$
66. $\frac{1}{2}w + 5 \geq w + 2 \geq \frac{1}{2}w + 9$

67. $|g + 6| > 8$
68. $|t - 5| \leq 3$
69. $|a + 5| \geq 0$
70. $|y - 9| < 19$
71. $|2m - 5| > 13$
72. $|14 - w| \geq 20$
73. $|3p + 5| \leq 23$
74. $|6b - 12| \leq 36$
75. $|25 - 3x| < 5$
76. $|7 + 8x| > 39$
77. $|4c + 5| \geq 25$
78. $|4 - 5s| > 46$
79. $|8 - 2h| \leq 25$
80. $|-8 - 10q| \geq 17$
81. $|7r - 5| < 3$
82. $|-3n - 12| < 15$

Polynomials: Addition and Subtraction

Find each sum.

1. $(3x - 4y) + (8x + 6y)$

2. $(12b + 2a) + (7b - 13a)$

3. $(7m - 8n) + (4m - 5n)$

4. $(5x^2 + 3x) + (4x^2 + 2x)$

5. $(-6s - 11t) + (5s - 6t)$

6. $(-14g - h) + (-8g + 5h)$

7. $(4p - 7q) + (5q - 8p)$

8. $(5y^2 - 7y) + (7y - 3y^2)$

9. $(9b^3 - 3b^2) + (12b^2 + 4b)$

10. $(2r + 8s) + (-3s - 9t)$

11. $(2a^2 + 4a + 5) + (2a^2 - 10a + 6)$

12. $(7x - 2y - 5z) + (x + 7y - 8z)$

13. $(-3m + 9mn - 5n) + (14m - 2n - 5mn)$

14. $(5x + 8y + 3z) + (-6z + 6y)$

15. $(6 - 4g - 9h) + (12g - 4h - 6j)$

16. $(x^2 - 4x + 8) + (12 + 7x - 4x^2)$

17. $(-7t^2 + 4ts - 6s^2) + (3s^2 - 12ts - 5t^2)$ 18. $(7g + 8h - 9) + (-g - 3h - 6k)$

19. $(8a^2 - 4ab - 3b^2 + a - 4b) + (3a^2 + 6ab - 9b^2 + 7a + 9b)$

20. $(-3v + 14w - 12x - 13y + 6z - 8) + (16 - 6v + 2x - 5y - 2z)$

21. $(3y^2 - 7y + 6) + (3 - 2y^2 - 5y) + (y^2 - 8y - 12)$

22. $(4a^2 - 10b^2 + 7c^2) + (2c^2 - 5a^2 + 2b) + (7b^2 - 7c^2 + 7a)$

23. $(5x^2 + 3) + (4 - 7x - 9x^2) + (2x - 3x^2 - 5) + (2x - 6)$

24. $(9p - 13p^2) + (7p^2 + 5q^2) + (-6p - 12q) + (3q - 8q^2)$

Find each difference.

25. $(5g + 3h) - (2g + 7h)$

26. $(2e - 5f) - (7e - f)$

27. $(-3m + 8n) - (6m - 4n)$

28. $(6a^2 - 9a) - (4a^2 + 2a)$

29. $(-11k + 6) - (-6k - 8)$

30. $(9y^2 - 4y) - (-6y^2 - 8y)$

31. $(-r - 3s) - (2s - 5r)$

32. $(13c^2 - 4c) - (5c - 12c^2)$

33. $(g^3 - 2g^2) - (5g^2 - 7)$

34. $(7a + 4b) - (7b - 6c)$

35. $(z^2 + 6z - 8) - (4z^2 - 7z - 5)$

36. $(6v - 12w - 2x) - (2v + 8w - 10x)$

37. $(6a^2 - 7ab - 4b^2) - (6b^2 + 2a^2 + 5ab)$

38. $(3r - 7t) - (2t + 2s + 9r)$

39. $(7ax^2 + 2ax - 4a) - (5ax - 2ax^2 + 8a)$

40. $(h^3 + 4h^2 - 7h) - (3h^2 - 7h - 8)$

41. $(4d + 3e - 8f) - (-3d + 10e - 5f + 6)$

42. $(-3z^2 + 4x^2 - 8y^2) - (7x^2 - 14z^2 - 12)$

43. $(2b^2 + 7b - 2) - (2b^2 + 3b - 16)$

44. $(15j^4k^2 - 7j^2k + 8) - (8j^2k + 11)$

45. $(9x^2 - 11xy - 3y^2) - (12y^2 + x^2 - 16xy)$

46. $(17z^4 - 5z^2 + 3z) - (4z^4 + 2z^3 + 3z)$

47. $(-4p - 7q - 3t) - (-8t - 5q - 8p)$

48. $(-14h + 16j - 7k) - (-3j + 5h - 6k - 3)$

49. $(14a + 9b - 2x - 11y + 4z) - (8a + 8b + 6x - 5y - 7z)$

50. $(7m^2 - 3mn + 4n^2 - 2m - 8n) - (4m^2 + 3m - 4n^2 - 13n + 4mn)$

Polynomials: Multiplication and Division

Find each product.

1. $t^3 \cdot t^6$

2. $g^5 \cdot g^9$

3. $(3x^2y)(-5x^3y^8)$

4. $(7p^4q^7r^2)(4p^5r^7)$

5. $(2a^2b)(-b^2c^3)(-8ab^2c^4)$

6. $(e^4f^6g)^5$

7. $(-2m^6n^2)^6$

8. $(-3h^2k^3)^3(5hj^6k^8)^2$

9. $(v^4w)^6(-1v^3w^2)^8$

10. $5y(y^2 - 3y + 6)$

11. $-ab(3b^2 + 4ab - 6a^2)$

12. $4st^2(-4s^2t^3 + 7s^5 - 3st^3)$

13. $(d + 2)(d + 3)$

14. $(z + 7)(z - 4)$

15. $(m - 5)(m - 8)$

16. $(2x - 5)(x + 6)$

17. $(7a - 4)(2a - 5)$

18. $(t + 7)^2$

19. $(q - 4h)^2$

20. $(w - 12)(w + 12)$

21. $(2b + 4d)(2b - 4d)$

22. $(4x + y)(2x - 3y)$

23. $(7v + 3)(v + 4)$

24. $(4e + 3)(4e + 3)$

25. $(7s - 8)(3s - 2)$

26. $(5b - 6)(5b + 6)$

27. $(4g + 3h)(2g - 5h)$

28. $(5c - 2d)^2$

29. $(10x + 11y)(10x - 11y)$

30. $(12r - 4s)(5r + 8s)$

Divide.

31. $\dfrac{24x^5}{8x^2}$

32. $\dfrac{s^7t^4}{s^5}$

33. $\dfrac{-9h^2k^4}{18h^5j^3k^4}$

34. $\dfrac{3m^7n^2p^4}{9m^2np^3}$

35. $\dfrac{9a^2b^7c^3}{12a^5b^4c}$

36. $\dfrac{-15xy^5z^7}{-10x^4y^6z^4}$

37. $\dfrac{-5w^4v^2 - 3w^3v}{w^3}$

38. $\dfrac{8g^4h^4 + 4g^3h^3}{2gh^2}$

39. $\dfrac{x^2 - 2x - 15}{x - 3}$

40. $\dfrac{q^2 - 10q + 24}{q - 4}$

41. $\dfrac{2j^2 + 10j + 12}{j + 3}$

42. $\dfrac{12d^2 + d - 6}{3d - 2}$

43. $\dfrac{4s^3 + 4s^2 - 9s - 18}{2s + 3}$

44. $\dfrac{z^3 - 27}{z - 3}$

45. $\dfrac{6n^3 + 7n^2 - 29n + 12}{3n - 4}$

46. $\dfrac{3e^2 + 3e - 80}{e - 5}$

47. $\dfrac{-6k^2 + 3k + 12}{2k + 3}$

48. $\dfrac{t^3 + 3t}{t - 1}$

49. $\dfrac{4g^3 - 4g - 20}{2g - 4}$

50. $\dfrac{9a^3 - 3a^2 + 7a + 2}{3a + 1}$

51. $\dfrac{8c^3 - 2c^2 + 2c - 4}{2c - 2}$

52. $\dfrac{8y^3 - 1}{2y + 1}$

53. $\dfrac{12m^3 + m^2 - 20}{4m - 5}$

54. $\dfrac{5b^3 - 2b^2 + 7b + 4}{5b + 3}$

Factoring

Find the prime factorization of each integer. Write each negative integer as the product of -1 and its prime factors.

1. 35
2. 12
3. 72
4. 64
5. -75
6. 70
7. 85
8. -92
9. -117
10. -114
11. 243
12. -360
13. 405
14. 605
15. -5292
16. 5076

Factor.

17. $10g + 35h$
18. $t^3s^2 - t^2$
19. $15a^2b - 24a^5b^2$
20. $18c^4d - 30c^3e$
21. $36m^4n^2p + 12m^5n^3p^2$
22. $6g^5h^3 - 12g^4h^6k - 18g^6h^5$
23. $p^2 - q^2$
24. $144x^2 - 49y^2$
25. $75r^2 - 48$
26. $64v^2 - 100w^4$
27. $g^2 + 4g + 4$
28. $t^2 - 22t + 121$
29. $9n^2 - 36nm + 36m^2$
30. $2a^2b^2 + 4ab^2c^2 + 2b^2c^4$
31. $g^2 - 14g + 48$
32. $z^2 + 15z + 36$
33. $12 - 13b + b^2$
34. $x^2 + 17xy + 16y^2$
35. $g^2 - 4g - 32$
36. $h^2 + 12h - 28$
37. $s^2 - 13st - 30t^2$
38. $3a^2 + 11a + 10$
39. $6y^2 + 2y - 20$
40. $12j^2 - 34j - 20$
41. $24a^2 - 57ax + 18x^2$
42. $2sx - 4tx + 2sy - 4ty$
43. $8ac - 2ad + 4bc - bd$
44. $2e^2g + 2fg + 4e^2h + 4fh$
45. $5x^3 - 2x^2y - 5xy^2 + 2y^3$
46. $4p^2 + 12pr + 9r^2$
47. $169 - 16t^2$
48. $b^2 - 11b - 42$
49. $30g^2h - 15g^3$
50. $3b^2 - 13bd + 4d^2$
51. $a^2x - 2a^2y - 5x + 10y$
52. $s^2 + 30s + 225$
53. $18v^2 + 42v + 12$
54. $4k^2 + 2k - 12$
55. $5z^3 - 8z^2 - 21z$
56. $5g^2 - 20h^2$
57. $30x^2 - 125x + 70$
58. $a^2b^2 - b^2 + a^2 - 1$
59. $8t^4 + 56t^3 + 98t^2$
60. $3p^3q^2 + 27pq^2$
61. $a^2c^2 + b^2c^2 - 4a^2d^2 - 4b^2d^2$
62. $36m^3n - 90m^2n^2 + 36mn^2$
63. $4x^2z^2 + 7xyz^2 - 36y^2z^2$
64. $4g^2j^2 - 25h^2j^2 - 4g^2 + 25h^2$

Algebraic Fractions

Simplify.

1. $\dfrac{48a^2b^5c}{32a^7b^2c^3}$

2. $\dfrac{-28x^3y^4z^5}{42xyz^2}$

3. $\dfrac{k+3}{4k^2+7k-15}$

4. $\dfrac{t^2-s^2}{5t^2-2st-3s^2}$

5. $\dfrac{6g^2-19g+15}{12g^2-6g-18}$

6. $\dfrac{2d^2+4d-6}{d^4-10d^2+9}$

Find each product or quotient in simplest form.

7. $\dfrac{5m^2n}{12a^2} \cdot \dfrac{18an}{30m^4}$

8. $\dfrac{25g^7h}{28t^3} \cdot \dfrac{42s^2t^3}{5g^5h^2}$

9. $\dfrac{6a+4b}{36} \cdot \dfrac{45}{3a+2b}$

10. $\dfrac{x^2y}{18z} \div \dfrac{2yz}{3x^2}$

11. $\dfrac{p^2}{14qr^3} \div \dfrac{2r^2p}{7q}$

12. $\dfrac{3d}{2d^2-3d} \div \dfrac{9}{2d-3}$

13. $\dfrac{t^2-2t-15}{t-5} \cdot \dfrac{t+5}{t+3}$

14. $\dfrac{5e-f}{5e+f} \div (25e^2-f^2)$

15. $\dfrac{8}{6c^2+17c+10} \div \dfrac{6}{c+2}$

16. $\dfrac{3v^2-27}{15v} \cdot \dfrac{v^2}{v+3}$

17. $\dfrac{3k^2-10k+3}{5} \div \dfrac{3k-1}{15k}$

18. $\dfrac{3g^2+15g}{4} \cdot \dfrac{g^2}{g+5}$

19. $\dfrac{x^2-2x-15}{2x^2-7x-15} \cdot \dfrac{4x^2-4x-15}{2x^2+x-15}$

20. $\dfrac{a^2-16}{3a^2-13a+4} \div \dfrac{3a^2+11a-4}{a^3}$

Find each sum or difference in simplest form.

21. $\dfrac{j+4}{3} + \dfrac{j-7}{3}$

22. $\dfrac{15n}{5n+3} + \dfrac{9}{5n+3}$

23. $\dfrac{2a-3}{b} - \dfrac{a-5}{b}$

24. $\dfrac{25}{5-g} - \dfrac{g^2}{5-g}$

25. $\dfrac{s}{t^2} - \dfrac{r}{3t}$

26. $\dfrac{7}{ab} + \dfrac{4}{bc}$

27. $\dfrac{2}{2p+3} + \dfrac{p}{3p+2}$

28. $\dfrac{x}{x+y} - \dfrac{5}{y}$

29. $\dfrac{c}{c^2-4c} - \dfrac{5c}{c-4}$

30. $\dfrac{t+10}{t^2-100} + \dfrac{1}{t-10}$

31. $\dfrac{1}{g^2-6gh+9h^2} - \dfrac{3}{g-3h}$

32. $\dfrac{x}{x+2} + \dfrac{x^2+3x}{x^2+5x+6}$

33. $\dfrac{3d}{d^2-3d-10} + \dfrac{d+1}{d^2-8d+15}$

34. $\dfrac{k+2}{k^2-8k+16} - \dfrac{k+3}{k^2+k-20}$

Solving Equations

Solve each equation.

1. $2x - 5 = 3$

2. $4t + 5 = 37$

3. $7a + 6 = -36$

4. $47 = -8g + 7$

5. $-3c - 9 = -24$

6. $5k - 7 = -52$

7. $5s + 4s = -72$

8. $6(y - 5) = 18$

9. $-21 = 7(p - 10)$

10. $2m + 5 - 6m = 25$

11. $3z - 1 = 23 - 3z$

12. $5b + 12 = 3b - 6$

13. $\frac{e}{5} + 6 = -2$

14. $\frac{d}{4} - 8 = -5$

15. $\frac{p + 10}{3} = 4$

16. $\frac{h - 7}{6} = 1$

17. $\frac{5f + 1}{8} = -3$

18. $\frac{4n - 8}{-2} = 12$

19. $\frac{2a}{7} + 9 = 3$

20. $\frac{-3t - 4}{2} = 8$

21. $\frac{6v - 9}{3} = v$

22. $|s - 4| = 7$

23. $|5g + 8| = 33$

24. $|16 - 3b| = 22$

25. $t(t - 4) = 0$

26. $4p(p + 7) = 0$

27. $7m(2m - 12) = 0$

28. $(x - 5)(x + 8) = 0$

29. $(3s + 6)(2s - 7) = 0$

30. $(4g + 5)(2g + 10) = 0$

31. $h^2 + 4h = 0$

32. $3b^2 - 3b = 0$

33. $m^2 - 16 = 0$

34. $w^2 + w - 30 = 0$

35. $2c^2 - 14c + 24 = 0$

36. $6p^2 + 10p - 24 = 0$

37. $5n^2 = 15n$

38. $4x^2 + 20x + 25 = 0$

39. $3y^2 = 75$

40. $\frac{a}{3} - \frac{a}{4} = 3$

41. $\frac{k}{6} + \frac{2k}{3} = -\frac{5}{2}$

42. $\frac{r + 3}{r} + \frac{r - 12}{r} = 5$

43. $\frac{2y}{y - 4} - \frac{3}{5} = 3$

44. $\frac{2t}{t + 3} + \frac{3}{t} = 2$

45. $\frac{2g}{2g + 1} - \frac{1}{2g - 1} = 1$

46. $\frac{8n^2}{2n^2 - 5n - 3} = 4$

47. $5 - \frac{5x}{x - 4} = \frac{2}{x^2 - 4x}$

48. $\frac{2}{e + 1} - \frac{3}{e + 2} = 0$

49. $\frac{1}{2d - 1} - \frac{12d}{6d^2 + d - 2} = \frac{-4}{3d + 2}$

50. $\frac{4z}{2z + 1} - \frac{6 - z}{2z^2 - 5z - 3} = 2$

51. $\frac{b + 1}{b - 2} - \frac{3b}{3b + 2} = \frac{20}{3b^2 - 4b - 4}$

52. $\frac{2s}{9s^2 - 3s - 2} + \frac{2}{3s + 1} = \frac{4}{3s - 2}$

53. $\frac{m}{20} = \frac{9}{15}$

54. $\frac{12}{21} = \frac{20}{f}$

55. $\frac{4}{9} = \frac{16}{t + 8}$

56. $\frac{4}{14} = \frac{2h - 1}{21}$

57. $\frac{2 + c}{c - 5} = \frac{8}{9}$

58. $\frac{2y + 4}{y - 3} = \frac{2}{3}$

Algebraic Fractions

Simplify.

1. $\dfrac{48a^2b^5c}{32a^7b^2c^3}$

2. $\dfrac{-28x^3y^4z^5}{42xyz^2}$

3. $\dfrac{k+3}{4k^2+7k-15}$

4. $\dfrac{t^2-s^2}{5t^2-2st-3s^2}$

5. $\dfrac{6g^2-19g+15}{12g^2-6g-18}$

6. $\dfrac{2d^2+4d-6}{d^4-10d^2+9}$

Find each product or quotient in simplest form.

7. $\dfrac{5m^2n}{12a^2}\cdot\dfrac{18an}{30m^4}$

8. $\dfrac{25g^7h}{28t^3}\cdot\dfrac{42s^2t^3}{5g^5h^2}$

9. $\dfrac{6a+4b}{36}\cdot\dfrac{45}{3a+2b}$

10. $\dfrac{x^2y}{18z}\div\dfrac{2yz}{3x^2}$

11. $\dfrac{p^2}{14qr^3}\div\dfrac{2r^2p}{7q}$

12. $\dfrac{3d}{2d^2-3d}\div\dfrac{9}{2d-3}$

13. $\dfrac{t^2-2t-15}{t-5}\cdot\dfrac{t+5}{t+3}$

14. $\dfrac{5e-f}{5e+f}\div(25e^2-f^2)$

15. $\dfrac{8}{6c^2+17c+10}\div\dfrac{6}{c+2}$

16. $\dfrac{3v^2-27}{15v}\cdot\dfrac{v^2}{v+3}$

17. $\dfrac{3k^2-10k+3}{5}\div\dfrac{3k-1}{15k}$

18. $\dfrac{3g^2+15g}{4}\cdot\dfrac{g^2}{g+5}$

19. $\dfrac{x^2-2x-15}{2x^2-7x-15}\cdot\dfrac{4x^2-4x-15}{2x^2+x-15}$

20. $\dfrac{a^2-16}{3a^2-13a+4}\div\dfrac{3a^2+11a-4}{a^3}$

Find each sum or difference in simplest form.

21. $\dfrac{j+4}{3}+\dfrac{j-7}{3}$

22. $\dfrac{15n}{5n+3}+\dfrac{9}{5n+3}$

23. $\dfrac{2a-3}{b}-\dfrac{a-5}{b}$

24. $\dfrac{25}{5-g}-\dfrac{g^2}{5-g}$

25. $\dfrac{s}{t^2}-\dfrac{r}{3t}$

26. $\dfrac{7}{ab}+\dfrac{4}{bc}$

27. $\dfrac{2}{2p+3}+\dfrac{p}{3p+2}$

28. $\dfrac{x}{x+y}-\dfrac{5}{y}$

29. $\dfrac{c}{c^2-4c}-\dfrac{5c}{c-4}$

30. $\dfrac{t+10}{t^2-100}+\dfrac{1}{t-10}$

31. $\dfrac{1}{g^2-6gh+9h^2}-\dfrac{3}{g-3h}$

32. $\dfrac{x}{x+2}+\dfrac{x^2+3x}{x^2+5x+6}$

33. $\dfrac{3d}{d^2-3d-10}+\dfrac{d+1}{d^2-8d+15}$

34. $\dfrac{k+2}{k^2-8k+16}-\dfrac{k+3}{k^2+k-20}$

Solving Equations

Solve each equation.

1. $2x - 5 = 3$

2. $4t + 5 = 37$

3. $7a + 6 = -36$

4. $47 = -8g + 7$

5. $-3c - 9 = -24$

6. $5k - 7 = -52$

7. $5s + 4s = -72$

8. $6(y - 5) = 18$

9. $-21 = 7(p - 10)$

10. $2m + 5 - 6m = 25$

11. $3z - 1 = 23 - 3z$

12. $5b + 12 = 3b - 6$

13. $\frac{e}{5} + 6 = -2$

14. $\frac{d}{4} - 8 = -5$

15. $\frac{p + 10}{3} = 4$

16. $\frac{h - 7}{6} = 1$

17. $\frac{5f + 1}{8} = -3$

18. $\frac{4n - 8}{-2} = 12$

19. $\frac{2a}{7} + 9 = 3$

20. $\frac{-3t - 4}{2} = 8$

21. $\frac{6v - 9}{3} = v$

22. $|s - 4| = 7$

23. $|5g + 8| = 33$

24. $|16 - 3b| = 22$

25. $t(t - 4) = 0$

26. $4p(p + 7) = 0$

27. $7m(2m - 12) = 0$

28. $(x - 5)(x + 8) = 0$

29. $(3s + 6)(2s - 7) = 0$

30. $(4g + 5)(2g + 10) = 0$

31. $h^2 + 4h = 0$

32. $3b^2 - 3b = 0$

33. $m^2 - 16 = 0$

34. $w^2 + w - 30 = 0$

35. $2c^2 - 14c + 24 = 0$

36. $6p^2 + 10p - 24 = 0$

37. $5n^2 = 15n$

38. $4x^2 + 20x + 25 = 0$

39. $3y^2 = 75$

40. $\frac{a}{3} - \frac{a}{4} = 3$

41. $\frac{k}{6} + \frac{2k}{3} = -\frac{5}{2}$

42. $\frac{r + 3}{r} + \frac{r - 12}{r} = 5$

43. $\frac{2y}{y - 4} - \frac{3}{5} = 3$

44. $\frac{2t}{t + 3} + \frac{3}{t} = 2$

45. $\frac{2g}{2g + 1} - \frac{1}{2g - 1} = 1$

46. $\frac{8n^2}{2n^2 - 5n - 3} = 4$

47. $5 - \frac{5x}{x - 4} = \frac{2}{x^2 - 4x}$

48. $\frac{2}{e + 1} - \frac{3}{e + 2} = 0$

49. $\frac{1}{2d - 1} - \frac{12d}{6d^2 + d - 2} = \frac{-4}{3d + 2}$

50. $\frac{4z}{2z + 1} - \frac{6 - z}{2z^2 - 5z - 3} = 2$

51. $\frac{b + 1}{b - 2} - \frac{3b}{3b + 2} = \frac{20}{3b^2 - 4b - 4}$

52. $\frac{2s}{9s^2 - 3s - 2} + \frac{2}{3s + 1} = \frac{4}{3s - 2}$

53. $\frac{m}{20} = \frac{9}{15}$

54. $\frac{12}{21} = \frac{20}{f}$

55. $\frac{4}{9} = \frac{16}{t + 8}$

56. $\frac{4}{14} = \frac{2h - 1}{21}$

57. $\frac{2 + c}{c - 5} = \frac{8}{9}$

58. $\frac{2y + 4}{y - 3} = \frac{2}{3}$

Radicals

Simplify.

1. $\sqrt{25}$
2. $-\sqrt{64}$
3. $\pm\sqrt{576}$
4. $\sqrt{900}$

5. $\pm\sqrt{0.01}$
6. $\sqrt{1.44}$
7. $-\sqrt{0.0016}$
8. $\sqrt{4.84}$

9. $\sqrt{\dfrac{36}{121}}$
10. $-\sqrt{\dfrac{16}{36}}$
11. $\sqrt{\dfrac{81}{49}}$
12. $\pm\sqrt{\dfrac{64}{100}}$

13. $\sqrt{75}$
14. $\sqrt{20}$
15. $\sqrt{162}$
16. $\sqrt{700}$

17. $\sqrt{4x^4y^3}$
18. $\sqrt{12ts^3}$
19. $\sqrt{175m^4n^6}$
20. $\sqrt{99a^3b^7}$

21. $\sqrt{\dfrac{54}{g^2}}$
22. $\sqrt{\dfrac{32c^5}{9d^2}}$
23. $\sqrt{\dfrac{27p^4}{3p^2}}$
24. $\sqrt{\dfrac{243y^7}{3y^4}}$

25. $\sqrt{7}\,\sqrt{3}$
26. $\sqrt{3}\,\sqrt{15}$
27. $6\sqrt{2}\,\sqrt{3}$
28. $5\sqrt{6}\cdot 2\sqrt{3}$

29. $\sqrt{5}(\sqrt{3}+\sqrt{10})$
30. $\sqrt{2}(\sqrt{6}+\sqrt{32})$
31. $\sqrt{5}\,\sqrt{t}$
32. $\sqrt{18}\,\sqrt{g^3}$

33. $\sqrt{12k}\,\sqrt{3k^5}$
34. $\sqrt{15m^2}\,\sqrt{6n^3}$
35. $\dfrac{\sqrt{3}}{\sqrt{5}}$
36. $\sqrt{\dfrac{2}{7}}$

37. $\dfrac{3\sqrt{6}}{\sqrt{2}}$
38. $\sqrt{\dfrac{x}{8}}$
39. $\sqrt{\dfrac{t^2}{3}}$
40. $\sqrt{\dfrac{20}{a}}$

41. $(5+\sqrt{3})(5-\sqrt{3})$
42. $(\sqrt{17}+\sqrt{11})(\sqrt{17}-\sqrt{11})$
43. $(2\sqrt{5}+\sqrt{7})(2\sqrt{5}-\sqrt{7})$

44. $\dfrac{1}{3+\sqrt{5}}$
45. $\dfrac{2}{\sqrt{3}-5}$
46. $\dfrac{12}{\sqrt{8}-\sqrt{6}}$
47. $\dfrac{10}{\sqrt{13}+\sqrt{7}}$

48. $\dfrac{14}{3\sqrt{2}+\sqrt{5}}$
49. $\dfrac{\sqrt{3}}{\sqrt{3}-5}$
50. $\dfrac{\sqrt{6}}{7-2\sqrt{3}}$
51. $\dfrac{5\sqrt{10}}{3\sqrt{3}+2\sqrt{5}}$

52. $6\sqrt{13}+7\sqrt{13}$
53. $9\sqrt{15}-4\sqrt{15}$
54. $2\sqrt{11}-8\sqrt{11}$
55. $2\sqrt{12}+5\sqrt{3}$

56. $2\sqrt{27}-4\sqrt{12}$
57. $4\sqrt{8}-3\sqrt{5}$
58. $8\sqrt{32}+4\sqrt{50}$
59. $6\sqrt{20}+\sqrt{45}$

60. $2\sqrt{63}+8\sqrt{45}-6\sqrt{28}$
61. $10\sqrt{\dfrac{1}{5}}-\sqrt{45}-12\sqrt{\dfrac{5}{9}}$
62. $3\sqrt{\dfrac{1}{3}}-9\sqrt{\dfrac{1}{12}}+\sqrt{243}$

Solve and check.

63. $\sqrt{t}=10$
64. $\sqrt{3g}=6$
65. $\sqrt{y}-2=0$
66. $5+\sqrt{a}=9$

67. $\sqrt{2k}-4=8$
68. $\sqrt{5y+4}=7$
69. $\sqrt{10x^2-5}=3x$
70. $\sqrt{2a^2-144}=a$

71. $\sqrt{b^2+16}+2b=5b$
72. $\sqrt{m+2}+m=4$
73. $\sqrt{3-2c}+3=2c$

Use the quadratic formula to solve each equation.

74. $s^2+8s+7=0$
75. $d^2-14d+24=0$
76. $3h^2=27$

77. $n^2-3n+1=0$
78. $2z^2+5z-1=0$
79. $3w^2-8w+2=0$

80. $3f^2+2f=6$
81. $2r^2-r-3=0$
82. $x^2-9x=5$

Equations in Two Variables

Write an equation in slope-intercept form for each given slope and y-intercept.

1. $m = 2, b = 5$
2. $m = -4, b = 1$
3. $m = \frac{1}{2}, b = -3$
4. $m = -1, b = -6$

5. $m = \frac{3}{2}, b = 1$
6. $m = 5, b = \frac{1}{4}$
7. $m = \frac{2}{5}, b = -4$
8. $m = -\frac{3}{4}, b = \frac{1}{2}$

Write each equation in slope-intercept form.

9. $-3x + y = 2$
10. $6x + y = -5$
11. $2x - y = -3$
12. $-x - y = 4$

13. $2x + 5y = 10$
14. $x - 4y = -2$
15. $-9x + 3y = -18$
16. $4x + 7y = 3$

Write each equation in standard form.

17. $y = 3x + 6$
18. $y = -4x + 1$
19. $y = \frac{2}{3}x - 7$
20. $y = \frac{1}{4}x + \frac{1}{2}$

21. $y = -\frac{5}{3}x - \frac{1}{3}$
22. $y = 2x - \frac{1}{2}$
23. $\frac{5}{6}y = \frac{1}{4}x + \frac{2}{3}$
24. $\frac{1}{5}x = \frac{7}{10}y - \frac{3}{4}$

Write the equation of the line with each x-intercept and y-intercept. Use slope-intercept form.

25. x-intercept = 2; y-intercept = -1

26. x-intercept = -3; y-intercept = -2

27. x-intercept = 1; y-intercept = 5

28. x-intercept = $-\frac{1}{2}$; y-intercept = 3

Write the equation of the line that passes through each pair of points. Use slope-intercept form.

29. $(1, 3), (2, 7)$
30. $(4, 8), (2, 4)$
31. $(-2, 3), (3, 1)$
32. $(0, 0), (-2, -3)$

33. $(5, 1), (3, -2)$
34. $(2, -1), (5, -4)$
35. $(8, 6), (10, 3)$
36. $(-4, -1), (-1, -7)$

Solve each system of equations.

37. $y = 3x$
$4x + 2y = 30$

38. $a = -2b$
$3a + 5b = 21$

39. $n = m + 4$
$3m + 2n = 19$

40. $h = k - 7$
$2h - 5k = -2$

41. $s + 2t = 6$
$3s - 2t = 2$

42. $c + 2d = 10$
$-c + d = 2$

43. $3v + 5w = -16$
$3v - 2w = -2$

44. $e - 5f = 12$
$3e - 5f = 6$

45. $-3p + 2q = 10$
$-2p - q = -5$

46. $2a + 5b = 13$
$4a - 3b = -13$

47. $5s + 3t = 4$
$-4s + 5t = -18$

48. $2g - 7h = 9$
$-3g + 4h = 6$

49. $2c - 6d = -16$
$5c + 7d = -18$

50. $6m - 3n = -9$
$-8m + 2n = 4$

51. $3x - 5y = 8$
$4x - 7y = 10$

52. $9a - 3b = 5$
$a + b = 1$

Find the equation of the axis of symmetry and the maximum or minimum point for the graph of each quadratic function.

53. $y = -x^2 + 2x - 3$
54. $y = x^2 - 4x - 4$
55. $y = 3x^2 + 6x + 3$

56. $y = 2x^2 + 12x$
57. $y = x^2 - 6x + 5$
58. $y = 4x^2 - 1$

59. $y = -2x^2 - 2x + 4$
60. $y = \frac{1}{2}x^2 + 4x + \frac{1}{4}$
61. $y = 6x^2 - 12x - 4$

Glossary

abscissa (289)   The first component of an ordered pair; the x-coordinate.

absolute value (46)   The absolute value of a number is the number of units that it is from zero on the number line.

addend (46)   In an addition expression of the form $a + b = c$, the addends are a and b.

Addition Property of Equality (60)   For any numbers a, b, and c, if $a = b$, then $a + c = b + c$.

additive identity (12)   The number 0 is the additive identity since the sum of any number and 0 is equal to the number.

additive inverse (55)   Two numbers are additive inverses if their sum is zero. The additive inverse, or opposite, of a is $-a$.

algebraic expression (3)   An expression consisting of one or more numbers and variables along with one or more arithmetic operations.

algebraic fractions (282)   Fractions that contain variables are algebraic fractions.

area of a polygon (212)   The measurement of the region bounded by the polygon.

Associative Property of Addition (19)   For any numbers a, b, and c,
$$(a + b) + c = a + (b + c).$$

Associative Property of Multiplication (19)   For any numbers a, b, and c,
$$(ab)c = a(bc).$$

axis of symmetry (431)   A straight line with respect to which a figure is symmetric.

base (4)   In an expression of the form x, the base is x.
(110)   In the proportion $\frac{17}{25} = \frac{r}{100}$, the base is 25.

binomial (191)   A polynomial with exactly two terms.

boundary (312)   A line that separates a graph into half-planes.

coefficient (16)   *See* numerical coefficient.

Commutative Property of Addition (19)   For any numbers a and b,
$$a + b = b + a.$$

Commutative Property of Multiplication (19)   For any numbers a and b, $ab = ba$.

comparison property (123)   For any two numbers a and b, exactly one of the following sentences is true.
$$a < b \qquad a = b \qquad a > b$$

complementary angles (547)   Two angles are complementary if the sum of their measures is 90°.

completeness property for points in the plane (290)   When plotting points, the following is true.
1. Exactly one point in the plane is located by a given ordered pair of numbers.
2. Exactly one ordered pair of numbers locates a given point in the plane.

complex fraction (490)   If a fraction has one or more fractions in the numerator or denominator, it is called a complex fraction.

composite number (229)   Any positive integer, except 1, that is not prime.

compound sentence (140)   Two sentences connected by *and* or *or*.

conjugates (408)   Two binomials of the form $a\sqrt{b} + c\sqrt{d}$ and $a\sqrt{b} - c\sqrt{d}$.

consecutive even integers (101)   Numbers given when beginning with an even integer and counting by two's.

consecutive numbers (101)   Numbers in counting order.

consecutive odd integers (101)   Numbers given when beginning with an odd integer and counting by two's.

consistent (362)   A system of equations is consistent and independent if it has one ordered pair as its solution. A system of equations is consistent and dependent if it has infinitely many ordered pairs as its solution.

constant (157)   A monomial that does not contain variables.

coordinate (42)   The coordinate of a point is the number that corresponds to it on the number line.

coordinate plane (289)   The number plane formed by two perpendicular number lines that intersect at their zero points.

cosine (557)   In a right triangle, the cosine of angle $A =$

$$\frac{\text{measure of side adjacent to angle } A}{\text{measure of hypotenuse}}.$$

degree (191)   The degree of a monomial is the sum of the exponents of its variables. The degree of a nonzero constant is 0. The degree of a polynomial is the greatest of the degrees of its terms.

Density Property (137)   Between every pair of distinct rational numbers, there is another rational number.

direct variation (178)   A direct variation is described by an equation of the form $y = kx$, where k is not zero.

discriminant (450)   In the quadratic formula, the expression $b^2 - 4ac$ is called the discriminant.

distance formula (421)   The distance between any two points (x_1, y_1) and (x_2, y_2) is given by the formula
$$d = \sqrt{(x_2 - x_1)^2 + (y_2 - y_1)^2}.$$

Distributive Property (15)   For any numbers a, b, and c:
1. $a(b + c) = ab + ac$ and $(b + c)a = ba + ca$.
2. $a(b - c) = ab - ac$ and $(b - c)a = ba - ca$.

dividend (3)   In the division expression $a \div b = c$, the dividend is a.

Division Property of Equality (87)   For any numbers a, b, and c, with $c \neq 0$, if $a = b$, then $\frac{a}{c} = \frac{b}{c}$.

divisor (3)   In the division expression $a \div b = c$, the divisor is b.

domain (296)   The domain of a relation is the set of all first components from each ordered pair.

element (11)   One of the members of a set.

elimination method (369)   A method for solving systems of equations in which the equations are added or subtracted to eliminate one of the variables. Multiplication of one or both equations may occur before the equations are added or subtracted.

empty set (11)   A set with no elements.

equals sign (9)   The equals sign, $=$, between two expressions indicates that if the sentence is true, the expressions name the same number.

equation (9)   A mathematical sentence that contains the equals sign.

equivalent equations (60)   Equations that have the same solution.

evaluate (6)   To find the value of an expression when the values of the variables are known.

exponent (4)   A number used to tell how many times a number is used as a factor. In an expression of the form x^n, the exponent is n.

extremes (105)   *See* proportion.

factor (3)   In a multiplication expression, the quantities being multiplied are called factors.

FOIL method for multiplying binomials (201)   To multiply two binomials, find the sum of the products of
F   the first terms,
O   the outer terms,
I   the inner terms, and
L   the last terms.

proportion (105) An equation of the form $\frac{a}{b} = \frac{c}{d}$ which states that two ratios are equal. The first and fourth terms (a and d) are called the extremes. The second and third terms (b and c) are called the means.

Pythagorean Theorem (418) In a right triangle if a and b are the measures of the legs, and c is the measure of the hypotenuse, then $c^2 = a^2 + b^2$.

quadrant (290) One of the four regions into which two perpendicular number lines separate the plane.

quadratic formula (445) The solutions of a quadratic equation of the form $ax^2 + bx + c = 0$, where $a \neq 0$, are given by
$$x = \frac{-b \pm \sqrt{b^2 - 4ac}}{2a}.$$

quadratic function (431) A quadratic function is a function described by an equation of the form $y = ax^2 + bx + c$, where $a \neq 0$.

quotient (3) The result of division.

quotient property of square roots (396) For any numbers a and b, where $a \geq 0$ and $b > 0$,
$$\sqrt{\frac{a}{b}} = \frac{\sqrt{a}}{\sqrt{b}}.$$

radical equations (415) Equations containing radicals with variables in the radicand.

radical expression (395) An expression of the form $\sqrt{a}$.

radical sign (395) The symbol $\sqrt{\ }$ indicating the principal or nonnegative square root.

radicand (395) The expression under the radical sign.

range (296) The range of a relation is the set of all second components from each ordered pair.
(524) The difference between the greatest and the least values of a set of data.

ratio (105) A comparison of two numbers by division. The ratio of a to b is $\frac{a}{b}$.

rational numbers (Q) (49) Numbers that can be expressed in the form $\frac{a}{b}$, where a and b are integers, $b \neq 0$.

real numbers (R) (399) Irrational numbers together with rational numbers form the set of real numbers.

reciprocal (84) The reciprocal of a number is its multiplicative inverse.

Reflexive Property of Equality (12) For any number a, $a = a$.

relation (296) A set of ordered pairs.

repeating decimal (398) A decimal numeral in which a digit or a group of digits repeats is called a repeating decimal.

right triangle (418) A triangle that has a 90° angle.

root of an equation (438) A solution of the equation.

scientific notation (165) A number is expressed in scientific notation when it is in the form $a \times 10^n$, where $1 \leq a < 10$ and n is an integer.

set (11) A collection of objects or numbers.

similar triangles (554) If two triangles are similar, the measures of their corresponding angles are equal and the measures of their corresponding sides are proportional.

simplest form of an expression (16) An expression in simplest form has no like terms and no parentheses.

sine (557) In a right triangle, the sine of angle $A = $
$$\frac{\text{measure of side opposite angle } A}{\text{measure of hypotenuse}}.$$

slope (325) The slope of a line is the ratio of the change in y to the corresponding change in x.
$$\text{slope} = \frac{\text{change in } y}{\text{change in } x}$$

slope-intercept form (333) The slope-intercept form of the equation of a line is $y = mx + b$. The slope of the line is m, and the y-intercept is b.

solution (9)   A replacement for the variable in an open sentence which results in a true sentence.

solution set (125)   The set of all replacements for the variable in an open sentence which make the sentence true.

solve (9)   To solve an open sentence means to find all the solutions.

square root (395)   If $x^2 = y$, then x is a square root of y.

standard form of linear equation (330)   A linear equation in standard form is $Ax + By = C$ where A, B, and C are integers, and A and B are not both zero.

statistics (513)   A branch of mathematics which provides methods for collecting, organizing, and interpreting data.

subset (11)   A set that is made from the elements of another set.

substitution method (365)   A method for solving systems of equations. One variable is expressed in terms of the other variable in one equation. Then the expression is substituted into the other equation.

Substitution Property of Equality (13)   For any numbers a and b, if $a = b$ then a may be replaced by b.

Subtraction Property of Equality (61)   For any numbers a, b, and c, if $a = b$, then $a - c = b - c$.

supplementary angles (547)   Two angles are supplementary if the sum of their measures is 180°.

Symmetric Property of Equality (12)   For any numbers a and b, if $a = b$ then $b = a$.

system of equations (359)   A set of equations with the same variables.

tangent (557)   In a right triangle the tangent of angle $A = \dfrac{\text{measure of side opposite angle } A}{\text{measure of side adjacent to angle } A}$.

term (15)   A number, a variable, or a *product* or *quotient* of numbers and variables. The terms of an expression are separated by the symbols $+$ and $-$.

terminating decimal (398)   A specific type of repeating decimal in which only a zero repeats.

Transitive Property of Equality (12)   For any numbers a, b, and c, if $a = b$ and $b = c$, then $a = c$.

trigonometric ratios (557)   Ratios in a right triangle that involve the measures of the sides and the measures of the angles.

trinomial (191)   A polynomial having exactly three terms.

uniform motion (220)   When an object moves at a constant speed, or rate, it is said to be in uniform motion.

union (141)   The union of two sets consists of all the points that belong to at least one of the sets.

variable (3)   In a mathematical sentence, a variable is a symbol used to represent an unspecified number.

vertical line test (306)   If any vertical line drawn on the graph of a relation passes through no more than one point of its graph, then the relation is a function.

whole numbers (W) (41)   The set of numbers $\{0, 1, 2, 3, \ldots\}$.

x-axis (289)   The horizontal number line which helps to form the coordinate plane.

x-coordinate (289)   The first component of an ordered pair.

x-intercept (333)   The value of x when y is 0.

y-axis (289)   The vertical number line which helps to form the coordinate plane.

y-coordinate (289)   The second component of an ordered pair.

y-intercept (333)   The value of y when x is 0.

zero product property (263)   For all numbers a and b, if $ab = 0$, then $a = 0$ or $b = 0$.

Selected Answers

CHAPTER 1 THE LANGUAGE OF ALGEBRA

Page 5 Lesson 1-1

1. $7x$ **3.** $a + 19$ **5.** b^3 **7.** 25^2 **9.** x squared or x to the second power **11.** 5 cubed or 5 to the third power
13. n to the first power **15.** 5^3 **17.** $7a^4$ **19.** $\frac{1}{2}a^2b^3$
21. 5^3x^2y **23.** $x + 17$ **25.** x^3 **27.** $2x^2$ **29.** $6x - 17$
31. $94 + 2x$ **33.** $\frac{3}{4}x^2$ **35.** 16 **37.** 125 **39.** 2548.040
41. $a + b - ab$

Page 8 Lesson 1-2

1. Multiply 4 by 2. Then add 3. **3.** Divide 8 by 4. Then multiply by 2. **5.** Multiply 2 by 6. Then subtract from 12. **7.** Subtract 3 from 9. Then square the result.
9. Square 9. Square 3. Then subtract. **11.** Square 3. Then multiply by 7. **13.** Subtract 3 from 5. Square the result. Then multiply by 4. **15.** Add 8 and 6. Divide by 2. Then add 2. **17.** 2 **19.** 48 **21.** 48 **23.** 14
25. 6 **27.** 8 **29.** 316 **31.** 1 **33.** 0.846 **35.** $\frac{26}{3}$; $8.\overline{6}$
37. 15 **39.** 1.32 **41.** 31 **43.** 75 **45.** 3 **47.** 2
49. 13.2 **51.** 3.04 **53.** 413 **55.** $\frac{11}{18}$; $0.6\overline{1}$ **57.** 1920
59. $2(a + b)$; 7 **61.** $b^2 + c$; $\frac{1}{4}$; 0.25 **63.** 53

Page 10 Lesson 1-3

1. false **3.** false **5.** false **7.** true **9.** false **11.** true
13. true **15.** false **17.** (name of current president)
19. 2 **21.** Foster, Rath, or Winters **23.** 11
25. 11.97 **27.** 9 **29.** 1.45 **31.** 5 **33.** 2 **35.** 11.05
37. 6 **39.** $\frac{7}{13}$ **41.** $\frac{7}{8}$ **43.** $\frac{7}{4}$ **45.** $5\frac{5}{6}$ **47.** {6, 7, 8}
49. {4, 5, 6} **51.** {5} **53.** {4, 5, 6, 7, 8} **55.** {8}

Page 13 Lesson 1-4

1. Reflexive (=) **3.** Substitution (=) **5.** Transitive (=)
7. 5 **9.** 0 **11.** Multiplicative Property of Zero
13. Substitution (=) **15.** Additive Identity Property
17. Multiplicative Identity Property **19.** Symmetric (=)

Page 17 Lesson 1-5

1. 5 **3.** 0.2 **5.** $\frac{1}{5}$ **7.** 0.5 **9.** $6bc$, bc **11.** $4xy$, $5xy$
13. $\frac{m^2n}{2}$, $5m^2n$ **15.** $6a + 6b$ **17.** $24x - 56$ **19.** $10a$
21. $18a$ **23.** $15am - 12$ **25.** $38mn$ **27.** $22x^2$
29. $22y^2 + 3$ **31.** in simplified form **33.** $37a + 23b$
35. $3x + 4y$ **37.** $23a + 42$ **39.** $10a + 2b$
41. $8.827xy^3 - 0.012y^3$ **43.** $1.042a^2 + 13.3529a$
45. $\frac{3y}{4} + \frac{13x}{4}$ **47.** $1.8rs^3 + 9.61r^3s$

Page 20 Lesson 1-6

1. Associative Property of Addition **3.** Distributive Property **5.** Commutative Property of Multiplication
7. Associative Property of Multiplication
9. Commutative Property of Addition **11.** no, no, no
13. Commutative Property of Addition; Associative Property of Addition; Distributive Property; Substitution Property of Equality **15.** Commutative Property of Addition **17.** Associative Property of Addition
19. Commutative Property of Multiplication
21. Additive Identity Property **23.** Commutative Property of Addition **25.** Distributive Property
27. Multiplicative Identity Property **29.** $12a + 6b$
31. $5x + 10y$ **33.** $\frac{5}{3}x^2 + 5x$ **35.** $3a + 13b + 2c$
37. $5 + 9ac + 14b$ **39.** $15x + 10y$ **41.** $\frac{3}{4} + \frac{5}{3}x + \frac{4}{3}y$
43. $\frac{23}{10}x + \frac{6}{5}y$ **45.** $12 + 30x + 45y$

Page 24 Lesson 1-7

1. $A = s^2$ **3.** $P = 4s$ **5.** 236 square units **7.** 499 square units **9.** 503.68 square units **11.** 129 square units **13.** 384 square units **15.** $\frac{25}{64}$ square units
17. 63.2 square units **19.** 21 square units **21.** 3740 square units **23.** 8 square units **25.** 191 square units
27. $2x + y^2 = z$ **29.** $x + a^2 = n$ **31.** $r = (a - b)^3$
33. $(abc)^2 = k$ **35.** $2m + n^2 = y$ **37.** $29 - xy = z$
39. 150 mi **41.** 3300 m **43.** $A = a^2 - b^2$
45. $A = \frac{1}{2}\pi a^2$

Page 26 Lesson 1-8

1. a. $1 bills **b.** 7 **c.** $267 **d.** none **3. a.** does not say **b.** 7¢ **c.** $7.18 **d.** more **e.** $(n - 7)$¢ or $(359 - n)$¢
5. a. nickels **b.** pennies **c.** 3 **d.** 2 **e.** does not say **f.** dimes **7. a.** $\frac{3}{4}$ **b.** $\frac{1}{3}$ **c.** $\frac{n}{4}$ **d.** yes **9. a.** no **b.** $x - 8$ **c.** juniors **d.** juniors **11. a.** 48 **b.** 72 **c.** no

Page 31 Lesson 1-9

1. $49 - n$ **3.** $w + 4$ **5.** $2t + 8$ **7.** $1.11 \div 3$ or $\frac{\$1.11}{3}$
9. $5.65n$ **11.** Let n = the number; $n + 24 = 89$
13. Let n = the number; $n - 19 = 83$ **15.** Let a = Tyrone's age; $2a + 17 = 53$ **17.** Let a = Sonia's age now; $(a + 4) + (a - 3 + 4) = 59$ **19.** Let x = Bob's height in inches; $x + (x + 5) = 137$ **21.** Let x = Ling's age now; $(x - 5) + (x + 27 - 5) = 45$ **23.** Let y = number of yards gained in both games; $134 + (134 - 17) = y$ **25.** Let x = number of years for a tree to become $33\frac{1}{3}$ feet tall; $17 + 1\frac{1}{2}x = 33\frac{1}{2}$ **27.** Let q = number of quarters; $q + (q + 4) + (q + 4 - 7) = 28$

29. Let p = number of pounds lost each week; $145 - 6p = 125$ **31.** In 7 years, Olivia will be 29 years old. How old is she now? **33.** The sum of two numbers is 33. One number is 7 less than the other. Find the numbers. **35.** The sum of Irene's age and her mother's age is 58. Irene's mother is 26 years older than Irene. How old is Irene? **37.** Ramon's car weighs 250 pounds more than Seth's car. The sum of the weights of both cars is 7140 pounds. How much does each car weigh? **39.** There are 33 students enrolled in Mr. Wyatt's class. If 5 are absent, how many are present? **41.** Reggie is 31 cm shorter than Soto. The sum of Soto's height and twice Reggie's height is 502 cm. How tall is Soto? **43.** Bonnie drove 80 miles further west than Cheryl drove east. Together they drove 520 miles. How far did each drive? **45.** Alex won 12 fewer games than Greg. Greg won twice as many games as Alex. How many games did Greg win? **47.** Emilio has 29 bills. He has 7 fewer $10 bills than $5 bills, and 3 more $1 bills than $5 bills. How many $5 bills does he have? **49.** Teresa has 37 coins. She has 4 more nickels than dimes and 9 fewer quarters than dimes. How many dimes does she have?

Page 38 Chapter Review

1. a^4 **3.** $2x - 17$ **5.** $8x$ **7.** 320 **9.** 4 **11.** 0.45 **13.** 4 **15.** 2 **17.** 2.2 **19.** Additive Identity Property **21.** Multiplicative Property of Zero **23.** Reflexive Property of Equality **25.** $11x$ **27.** $8b + 6$ **29.** Commutative Property of Addition **31.** Associative Property of Addition **33.** $14a + 9b$ **35.** $18 - d^2 = f$ **37.** less **39.** Let x = Minal's weight; $x + (x + 8) = 182$

CHAPTER 2 ADDING AND SUBTRACTING RATIONAL NUMBERS

Page 42 Lesson 2-1

1. 3 **3.** −10 **5.** 6 **7.** −7 **9.** −9 **11.** −2 **13.** {−3, 2, 1} **15.** {−3, −2, −1} **17.** {..., −4, −3, −2, −1, 0, 1} **19.** −3 **21.** +650 **23.** +12 **25.** −450 **27.** +37 **29.** {−1, 0, 1, 2} **31.** {−5, −3, −1} **33.** {−3, −2, −1, 0} **35.** {−6, −4, −2, 0, 2} **51.** 9, 1, −7 **53.** −6, −1, 4 **55.** 81, 243, 729 **57.** 25, 36, 49 **59.** −6, 7, −8

Page 45 Lesson 2-2

1. $-3 + 5 = 2$ **3.** $4 + (-5) = -1$ **5.** $-1 + (-4) = -5$ **7.** $-4 + 3 = -1$ **9.** $4 + (-6) = -2$ **11.** The parentheses in $4 + (-6)$ shows the sign of the number. The parentheses in $4 + (5 - 3)$ indicate that $5 - 3$ should be evaluated first. **13.** 16 **15.** −19 **17.** −9 **19.** −5 **21.** 0 **23.** 6 **25.** −1 **27.** −8 **29.** 11 **31.** −6 **33.** 0 **35.** −13 **37.** −22 **39.** −42 **41.** $400 + (-300) = x$; 100 m **43.** $-75 + 35 = x$; −45 m

Page 47 Lesson 2-3

1. 8 **3.** 6 **5.** 17 **7.** 21 **9.** 0 **11.** − **13.** + **15.** + **17.** − **19.** + **21.** − **23.** 16 **25.** −19 **27.** 13 **29.** 26 **31.** −29 **33.** 4 **35.** −9 **37.** 16 **39.** 90 **41.** −54 **43.** 8 **45.** 9 **47.** −15 **49.** 0 **51.** 6 **53.** 2 **55.** 5 **57.** 3 **59.** −2 **61.** −30 **63.** 613 **65.** −539 **67.** −203 **69.** 500 **71.** 926 **73.** 587 **75.** 286 **77.** n **79.** n

Page 50 Lesson 2-4

1. $\frac{3}{4}$ **3.** 1.76 **5.** $\frac{3}{11}$ **7.** 1.82 **9.** $\frac{3}{82}$ **11.** + **13.** − **15.** − **17.** + **19.** − **21.** + **23.** −2 **25.** $-\frac{1}{11}$ **27.** $-\frac{1}{6}$ **29.** 0.88 **31.** −5.6 **33.** −14.7 **35.** $\frac{1}{6}$ **37.** $-\frac{1}{6}$ **39.** $-\frac{21}{8}$ **41.** 0.622 **43.** −0.665 **45.** −0.3005 **47.** $\frac{7}{30}$ **49.** $\frac{29}{60}$ **51.** $\frac{3}{20}$ **53.** $\$20 + (-\$12.87) = x$; $\$7.13 **55.** −4.6 **57.** 0.5 **59.** 5.18 **61.** −4.2125

Page 54 Lesson 2-5

1. 0 **3.** −0.1 **5.** 1 **7.** $4m$ **9.** $4r$ **11.** $3b - 4y$ **13.** −26 **15.** 6 **17.** 80 **19.** $-5a$ **21.** $30b$ **23.** $-38z$ **25.** $\frac{-19}{12}$ **27.** $\frac{5}{14}$ **29.** $\frac{7}{48}$ **31.** −14 **33.** −12 **35.** −8.7 **37.** −7.54 **39.** −17.4 **41.** $-97a + 48k$ **43.** $89mp - 24ps$ **45.** $3a$ **47.** $24w$ **49.** $2.2k$ **51.** $-0.73x$ **53.** −6 **55.** −$4.66 (decrease)

Page 57 Lesson 2-6

1. −6 **3.** −5 **5.** 13 **7.** a **9.** 0 **11.** −3.7 **13.** $\frac{-3}{7}$ **15.** $\frac{8}{17}$ **17.** −7 **19.** 13 **21.** 7 **23.** 16 **25.** −56 **27.** −7.1 **29.** $\frac{3}{4}$ **31.** $8 + (-13)$ **33.** $-17 + (-8)$ **35.** $-1.7 + 1.5$ **37.** $9y + (-3y)$ **39.** 8 **41.** 15 **43.** −18 **45.** 40 **47.** 52 **49.** −11 **51.** $\frac{4}{5}$ **53.** $\frac{1}{4}$ **55.** −21 **57.** 11.5 **59.** −28.9 **61.** −153.8 **63.** −17 **65.** −1.4 **67.** $7m$ **69.** $-22p$ **71.** $33b$ **73.** $-35z$ **75.** −2 **77.** 58 **79.** 8.9 **81.** −16.7 **83.** −5.5 **85.** 8 **87.** $-\frac{5}{8}$ **89.** $-\frac{52}{21}$ **91.** $-\frac{67}{34}$

Page 62 Lesson 2-7

1. −21 **3.** 5 **5.** 10 **7.** 16 **9.** 5 **11.** −3 **13.** −3 **15.** −19 **17.** −32 **19.** −19 **21.** −19 **23.** −16 **25.** 15 **27.** 52 **29.** 28 **31.** −10 **33.** 8 **35.** −6 **37.** −3.6 **39.** −0.8 **41.** −1.3 **43.** $-\frac{17}{16}$ **45.** 2 **47.** $-\frac{52}{45}$ **49.** 29 **51.** 122

Page 65 Lesson 2-8

1. $m - 8$ **3.** $y + 11$ **5.** $z - 31$ **7.** $p + 47$ **9.** −17 **11.** −25 **13.** −12 **15.** 20 **17.** −14 **19.** −15 **21.** −19 **23.** 58 **25.** 25 **27.** −28 **29.** −53 **31.** −9 **33.** 7 **35.** −60 **37.** 12.4 **39.** −4.7 **41.** 0.0706 **43.** $\frac{2}{9}$; $0.\overline{2}$ **45.** $\frac{34}{15}$; $2.2\overline{6}$ **47.** $-\frac{23}{72}$; 0.319 **49.** −2.366

51. -3537.53 **53.** 12   **55.** -101 **57.** 76

1. a. How far downstream was she from her starting point? **b.** 110 km **c.** 320 km **d.** upstream-positive, downstream-negative **e.** Let d = distance from starting point; $110 + (-320) = d$ **3. a.** What is the second integer in the sum of the two integers? **b.** 9 **c.** -23 **d.** addition **e.** Let x = the missing integer; $9 + x = -23$ **5.** Let f = floor; $1 + 14 + (-9) = f$ **7.** Let n = the number; $n - 13 = -5$ **9.** Let r = runs; $41 - 17 = r$ **11.** -210 (210 km downstream) **13.** -32 **15.** 6th floor **17.** 8 **19.** 24 runs **21.** 63 cars **23.** $54.44 **25.** 153 seconds **27.** $-19°$C **29.** 77 **31.** -38 **33.** -43 m (43 m below sea level) **35.** -74 m (74 m below sea level) **37.** -85 m (85 m below cave entrance) **39.** $35\frac{5}{8}$; 35.625 points

1. $\{-3, 1, 2, 5\}$ **3.** $\{3, 4, 5, \ldots\}$ **9.** -1 **11.** -6 **13.** 8 **15.** -4 **17.** -53 **19.** -1 **21.** 2.7 **23.** 3.649 **25.** 66 **27.** $-23b$ **29.** $\frac{23}{12}$ **31.** -3.9 **33.** $-51pq + 53k$ **35.** -22 **37.** 4 **39.** 12.37 **41.** $-6x$ **43.** 17 **45.** -6.7 **47.** $-\frac{11}{14}$ **49.** -8 **51.** -24 **53.** 38 **55.** 53 **57.** -13 **59.** 19 **61.** 55 **63.** $17°$F

CHAPTER 3   MULTIPLYING AND DIVIDING RATIONAL NUMBERS

1. $-$ **3.** $-$ **5.** $-$ **7.** $+$ **9.** $+$ **11.** $-$ **13.** a and b are both positive or both negative **15.** a, b, and c are all positive; a is positive and b and c are negative; b is positive and a and c are both negative; c is positive and a and b are both negative **17.** a is positive **19.** positive **21.** -6 **23.** -15 **25.** -63 **27.** $\frac{49}{9}$ **29.** 24 **31.** $-\frac{6}{5}$ **33.** 60 **35.** $-\frac{12}{35}$ **37.** $-\frac{7}{2}$ **39.** 3 **41.** $-\frac{1}{5}$ **43.** 0 **45.** -7.805 **47.** 114.148 **49.** 0 **51.** -112 **53.** 8 **55.** 54 **57.** $-\frac{3}{4}$ **59.** $\frac{10}{7}$ **61.** $98xy$ **63.** -68.706 **65.** $-8x$ **67.** $88x$ **69.** $14x - 22a$ **71.** $-40a + 44b$ **73.** $-\frac{11}{18}a - \frac{4}{9}b$ **75.** $5.1x - 7.6y$ **77.** $3(-10) = -30$ **79.** $(-3)(-10) = 30$ **81.** 3 **83.** 8244.768

1. $\frac{1}{3}$ **3.** none **5.** $\frac{3}{2}$ **7.** $\frac{3}{4}$ **9.** $-\frac{11}{3}$ **11.** $\frac{7}{10}$ **13.** $-\frac{5}{3}$ **15.** $\frac{4}{13}$ **17.** $-\frac{7}{17}$ **19.** 6 **21.** -5 **23.** 14 **25.** -5 **27.** 6 **29.** 4 **31.** -19 **33.** $-4x$ **35.** $6a$ **37.** $-7a$ **39.** 7 **41.** -5 **43.** $-\frac{5}{48}$ **45.** $-\frac{7}{80}$ **47.** $\frac{1}{12}$ **49.** $-\frac{35}{2}$

51. $\frac{21}{2}$ **53.** $\frac{27}{2}$ **55.** $a + 3$ **57.** $-a + 5$ **59.** $-10a - 15b$ **61.** $-10a + 5b$

1. 3 **3.** $\frac{4}{3}$ **5.** $\frac{9}{4}$ **7.** $-\frac{1}{8}$ **9.** 9 **11.** 4 **13.** 4 **15.** -7 **17.** -8 **19.** -6 **21.** 7 **23.** -7 **25.** -7 **27.** 17 **29.** $\frac{40}{9}$ **31.** $-\frac{11}{3}$ **33.** -14 **35.** 0.188 **37.** 0.456 **39.** 48 **41.** 70 **43.** -275 **45.** -90 **47.** -1885 **49.** 112 **51.** -25 **53.** 35 **55.** $-\frac{81}{4}$ **57.** $\frac{250}{3}$ **59.** -9 **61.** 15 **63.** -10 **65.** $\frac{14}{9}$ **67.** 6 **69.** $\frac{1}{3}$ **71.** 45 **73.** 23 **75.** -30 **77.** $-\frac{1}{3}$ **79.** 12 **81.** 5 **83.** 13 **85.** -65.3 **87.** $9.50 **89.** 22 cans

1. Add 7 to each side. Then divide each side by 3. **3.** Subtract 5. Then divide by 2. **5.** Multiply by 5. Then subtract 2. **7.** Subtract 2. Then multiply by $\frac{11}{3}$. **9.** 3 **11.** 5 **13.** $\frac{6}{5}$ **15.** -9 **17.** $-\frac{25}{3}$ **19.** 4 **21.** -153 **23.** -69 **25.** -81 **27.** -32 **29.** 25 **31.** -104 **33.** 11 **35.** $\frac{65}{7}$ **37.** 28 **39.** 8 **41.** 6.236 **43.** 16 **45.** -12 **47.** -3 **49.** 3 **51.** -6

1. Add 1 to each side. Subtract $3x$ from each side. **3.** Add 3 to each side. Add $5y$ to each side. Divide each side by 9. **5.** Add 10 to each side. Add $3y$ to each side. Divide each side by 11. **7.** Multiply $x + 1$ by 3. Subtract 3 from each side. Divide each side by 3. **9.** Multiply $3 + 5y$ by 4. Subtract 12 from each side. Divide each side by 20. **11.** Multiply $2x - 3$ by 6. Add 18 to each side. Divide each side by 12. **13.** 10 **15.** $-\frac{1}{2}$ **17.** no solution **19.** all numbers **21.** $\frac{7}{8}$ **23.** all numbers **25.** 2 **27.** 3.453 **29.** all numbers **31.** $\frac{13}{5}$ **33.** no solution **35.** -5.077 **37.** -2 **39.** no solution **41.** $\frac{106}{23}$ **43.** 2 **45.** all numbers **47.** 15 **49.** 10 **51.** -30 **53.** 10 **55.** -9 **57.** 4.2 **59.** $\frac{45}{26}$ **61.** $\frac{1}{3}$ **63.** $\frac{7}{5}$ **65.** -1.278 **67.** $\frac{5}{3}$ **69.** all numbers

1. 12; $9x - 84 = 96 + 8x$ **3.** 2; $8t - 14 = 5t - 6$ **5.** 10; $52z = 30 + 17z$ **7.** 16 **9.** $\frac{11}{6}$ **11.** 49 **13.** 44 **15.** $\frac{8}{5}$ **17.** 18 **19.** $\frac{3}{8}$ **21.** -2 **23.** 8.860 **25.** $\frac{21}{2}$ **27.** -1 **29.** $2d - r$ **31.** $ef - d$ **33.** $\frac{3z + 2y}{e}$ **35.** $\frac{c - d}{2}$ **37.** $\frac{b}{a} - 1$ **39.** $\frac{5}{3}(b - a)$ **41.** 14 **43.** 10 **45.** 18 and 30 or 150 and 162

1. 2, 3, 4   **3.** $-4, -2, 0$ **5.** 13, 15, 17, 19   **7.** Let $x = $ least integer; $x + (x + 1) = 17$ **9.** Let $x = $ least odd integer; $x - (x - 2) = -36$ **11.** Let $x = $ a number; $3x + 4 = -11$ **13.** 28, 29, 30   **15.** 31, 32, 33, 34   **17.** no solution   **19.** 31, 33   **21.** 31, 33, 35   **23.** 21.41   **25.** 46 customers   **27.** 38 vans   **29.** 260 m   **31.** 8th, 9th, 10th, 11th   **33.** $64   **35.** 7 m, 9 m, 11 m   **37.** 16   **39.** 140.195   **41.** 24   **43.** 30, 32, 34, 36   **45.** 17, 19   **47.** 21, 22   **49.** 50 m by 60 m   **51.** 11 ft, 15 ft

1. $\frac{3}{11}$ **3.** $\frac{21}{16}$ **5.** $\frac{2}{1}$ **7.** $\frac{24}{7}$ **9.** $\frac{4}{1}$ **11.** $\frac{7}{24}$ **13.** 6   **15.** 5   **17.** $\frac{28}{3}$; 9.$\overline{3}$ **19.** 6   **21.** 2   **23.** 2.28   **25.** 1.251   **27.** 63.368   **29.** 5   **31.** $\frac{8}{5}$; 1.6   **33.** $\frac{77}{4}$; 19.25   **35.** $\frac{3}{5}$; 0.6   **37.** 20 days   **39.** 33 gallons   **41.** 267.9 km   **43.** 27 teeth   **45.** $\frac{5}{2}$ **47.** $\frac{20}{7}$ **49.** $-\frac{20}{67}$

1. 31%   **3.** 30%   **5.** 80%   **7.** $37\frac{1}{2}$%; 37.5%   **9.** 175%   **11.** 40%   **13.** 70%   **15.** $12\frac{1}{2}$%; 12.5%   **17.** 32   **19.** 70.6   **21.** 25%   **23.** 24   **25.** 60   **27.** 65%   **29.** 30.375   **31.** 140   **33.** 25%   **35.** $242.80   **37.** $62\frac{1}{2}$%; 62.5%   **39.** $6540   **41.** $560   **43.** 160.045%   **45.** 40   **47.** $6.24   **49.** 540 seats   **51.** 40 questions   **53.** 480 people   **55.** 22.434%   **57.** $1271.50

1. 10   **3.** 8   **5.** 45   **7.** 36   **9.** 40   **11.** 10   **13.** 39   **15.** 28   **17.** 50.130   **19.** $3.00   **21.** D, $6, 6%   **23.** $172, $28   **25.** $47.89, 25.496%   **27.** 39   **29.** 56.220%   **31.** 20%   **33.** $27   **35.** $303.60   **37.** $78.81   **39.** $41.18   **41.** $88.50   **43.** $17.97   **45.** 4%

1. -99 **3.** $-\frac{3}{7}$ **5.** $-5a - 12b$ **7.** $-9b$ **9.** 18   **11.** 8   **13.** -16 **15.** 10   **17.** -18 **19.** 69   **21.** 7   **23.** 3   **25.** -6 **27.** no solution   **29.** $cd - y$ **31.** $\frac{7a + 9b}{8}$ **33.** 25, 27, 29   **35.** 18   **37.** $\frac{16}{3}$; 5.$\overline{3}$ **39.** $87\frac{1}{2}$%; 87.5%   **41.** 60%   **43.** 60   **45.** 35

CHAPTER 4   INEQUALITIES

1. false   **3.** true   **5.** true   **7.** false   **9.** false   **11.** {all numbers greater than 3}   **13.** {all numbers less than 6}   **15.** {-4 and all numbers greater than -4}   **17.** yes

19. no   **21.** yes   **23.** $x > 3$ **25.** $x \neq 3$ **27.** $x \geq 0$ **29.** $x < -3$ **31.** $<$ **33.** $=$ **35.** $<$ **37.** $>$ **39.** $=$ **41.** $<$ **43.** $<$ **45.** $>$ **47.** $<$ **49.** $<$ **63.** none

1. -7 **3.** 13   **5.** $-3y$ **7.** $-5z$ **9.** -3 **11.** 7   **13.** $a < 8$ **15.** $r \geq 32$ **17.** $b < 8$ **19.** $m > -19$ **21.** $y \leq -8$ **23.** $n > 11$ **25.** $n > 34$ **27.** $x > 8$ **29.** $n < -11$ **31.** $-15 < s$ **33.** $t \leq 34.6$ **35.** $m > 6.38$ **37.** $r > 1$ **39.** $x < \frac{5}{6}$ **41.** $-6 < f$ **43.** $n > 6$ **45.** $14 > x$ **47.** $-4 \leq a$ **49.** $r < -6.6$ **51.** $z > 168.93$ **53.** $t \leq -1.753$ **55.** $t \geq \frac{3}{20}$ **57.** $\left\{y | y \geq \frac{1}{12}\right\}$; $\{y | y \geq 0.08\overline{3}\}$ **59.** $\left\{y | y < \frac{17}{18}\right\}$; $\{y | y < 0.9\overline{4}\}$

1. 3, no   **3.** -8, yes   **5.** $\frac{1}{3}$, no   **7.** $-\frac{1}{4}$, yes   **9.** 6, no   **11.** -11, yes   **13.** $\frac{3}{4}$, no   **15.** $-\frac{3}{5}$, yes   **17.** $\{p | p < 6\}$ **19.** $\{s | s < -7\}$ **21.** $\left\{r | r \leq -\frac{35}{2}\right\}$ **23.** $\left\{s | s > \frac{17}{4}\right\}$ **25.** $\{h | h < 450\}$ **27.** $\{b | -128 \leq b\}$ **29.** $\{r | r < -0.7\}$ **31.** $\left\{y | y > \frac{10}{3}\right\}$ **33.** $\{s | s \leq -6\}$ **35.** $\left\{k | k < -\frac{8}{7}\right\}$ **37.** $\{m | m > 0.7\}$ **39.** $\{t | t \geq -1.396\}$ **41.** $\{z | z < 0.08\}$ **43.** $\left\{x | x > -\frac{4}{5}\right\}$ **45.** $\left\{w | \frac{13}{12} \leq w\right\}$ **47.** $\{a | -9 > a\}$ **49.** $\{x | x < 4.598\}$ **51.** $\{y | y \geq -5.427\}$ **53.** $\{a | a \geq 0.030\}$ **55.** $\{b | b \leq 0.033\}$ **57.** $x > 0, x > |y|$

1. Add 1 to each side. Then divide each side by 3.   **3.** Add 7, then divide by 4.   **5.** Subtract 32, then divide by 14.   **7.** Add 12, then divide by 11.   **9.** Subtract 7, then multiply by 4.   **11.** Add 5, then multiply by -4 and reverse the inequality symbol.   **13.** Subtract 13, then divide by -2 and reverse the inequality symbol.   **15.** Subtract 7 and subtract k.   **17.** $r > 8$ **19.** $p < -\frac{3}{2}$ **21.** $y \geq -1$ **23.** $x > 80$ **25.** $x < -5$ **27.** $n < \frac{21}{5}$ **29.** $y \leq 0$ **31.** $c < 2$ **33.** $x < 37.097$ **35.** $x > -\frac{1}{2}$ **37.** $x \geq -34.100$ **39.** $y < 4$ **41.** $x < -1$ **43.** $n \geq \frac{23}{3}$

1. $\frac{4}{5}$ **3.** $\frac{10}{11}$ **5.** $\frac{6}{5}$ **7.** $-\frac{1}{4}$ **9.** $-\frac{9}{7}$ **11.** $-\frac{15}{13}$ **13.** $\frac{459}{10}$ **15.** $<$ **17.** $>$ **19.** $<$ **21.** $=$ **23.** $<$ **25.** $<$ **27.** $\frac{19}{28}$ **29.** $-\frac{47}{99}$ **31.** 0   **33.** 0   **35.** $\frac{79}{21}, \frac{97}{28}$; 28-ounce can for 97¢   **37.** $\frac{91}{184}, \frac{189}{340}$; 184-gram can of peanuts for 91¢   **39.** $\frac{179}{2.1}, \frac{169}{1.9}$; six-pack of cola containing 2.1 liters for $1.79   **41.** $\frac{93}{27}, \frac{79}{20}$; 27-ounce loaf of bread for 93¢   **43.** Dudley's Market

1. false   **3.** true   **5.** false   **7.** true   **9.** true   **11.** true
13. $0 \le m < 9$ **15.** $\frac{3}{4} < p \le \frac{11}{9}$ **17.** $-\frac{4}{5} < z < \frac{2}{3}$
19. $-\frac{13}{7} < m < -\frac{6}{5}$ **21.** $-4.9 \le a \le -2.4$
39. {all numbers}   **41.** $m < -1$ **43.** $q < -2$ or $q > -1$
45. $x \le 4$ **47.** $x < \frac{3}{2}$ **49.** $x \le -3$ or $x \ge 1$
51. $x < -3$ or $x > 3$ **53.** $x \ge -3$ and $x < 5$ **55.** no
solutions   **57.** $x > 0$ or $x < -\frac{3}{2}$ **59.** $x > -1$ and
$x < 3$ **61.** $x > -4$ and $x < 1$

1. $x = 4$ or $x = -4$ **3.** $y > 3$ or $y < -3$ **5.** $y < \frac{5}{2}$ and
$y > -\frac{5}{2}$ **7.** $x + 2 > 3$ or $x + 2 < -3$ **9.** $x + 2 < 3$ and
$x + 2 > -3$ **11.** $2x - 5 \ge 3$ or $2x - 5 \le -3$
13. $7 - x = 4$ or $7 - x = -4$ **15.** $3x + 1 > 6$ or
$3x + 1 < -6$ **17.** $\{y | y > 3$ or $y < -5\}$
19. $\{y | y > -3$ and $y < 5\}$ **21.** $\{y | y \ge 1$ and $y \le 3\}$
23. {all numbers}   **25.** $\{y | y \ge 8$ or $y \le 2\}$
27. $\{x | x = 1\}$ **29.** $\left\{x | x \le \frac{5}{2}$ and $x \ge -\frac{9}{2}\right\}$
31. $\{x | x \ge 8$ or $x \le -10\}$ **33.** $\left\{y | y \ge \frac{9}{2}$ or $y \le \frac{1}{2}\right\}$
35. $\left\{x | x = \frac{4}{3}\right\}$ **37.** $\left\{t | t < \frac{4}{3}$ and $t > -1\right\}$
39. $\{y | y = 2$ or $y = -2\}$
41. $\{a | a \ge -1.37$ and $a \le 1.43\}$
43. $\{x | x < -10.44$ or $x > 5.36\}$
45. $\{t | t \ge -0.088$ and $t \le 0.064\}$
47. $\{x | x \ge -3.972$ and $x \le 0.383\}$
49. $\{b | b > 0.018$ or $b < -0.029\}$ **51.** $|x| < 8$
53. $|x - 1.5| \le 0.005$ **55.** $|x| = 1$ **57.** $|x| < 3$
59. $|x| \ge 2$ **61.** $-2, -1, 0, 1, 2$ **63.** $2a + 1$ **65.** when
a is any real number $\ne 0$ **67.** no, $|x| = |y|$

1. The number is greater than or equal to 5.   **3.** 7 and 9;
5 and 7; 3 and 5; or 1 and 3   **5.** He must deliver 70 or
more papers per day.   **7.** 145 kg or greater   **9.** His sales
were greater than $150,000 and less than $250,000.
11. $12.10 or less   **13.** 3, 4, or 5   **15.** Her score must be
9.7 or greater.   **17.** His score must be greater than 9.4.
19. 38 and 40

1. false   **3.** true   **5.** >   **7.** <   **13.** $\{y | y \ge 6\}$
15. $\{n | n < -11\}$ **17.** $\left\{y | y < -\frac{1}{6}\right\}$ **19.** $\{a | a \le -4\}$
21. $\{r | r < -24\}$ **23.** $\{t | t < -1.4\}$ **25.** $\{x | x > -15\}$
27. $\{m | m > 6\}$ **29.** $\{a | a > 6\}$ **31.** $\{m | m > -4\}$
33. >   **35.** >   **37.** $-\frac{71}{120}$ **39.** 1.25 liters of soda for
$1.31   **41.** $-3 \le x \le 17$ **47.** $\{m | m \le 6$ and $m \ge -4\}$
49. The number is between $\frac{7}{8}$ and $\frac{17}{8}$; the number is
between 0.875 and 2.125.

CHAPTER 5   POWERS

1. $3 \cdot 9 = 27$ **3.** $16 \cdot 16 = 256$ **5.** $25 \cdot 5 = 125$ **7.** x^8
9. a^8 **11.** m^4 **13.** a^7 **15.** m^6 **17.** t^6 **19.** a^{13}
21. a^3b^5 **23.** m^4n^3 **25.** $12a^5$ **27.** $-20x^5y$ **29.** y^5z^4
31. $6a^3b^6$ **33.** $-6x^8y^3$ **35.** $12x^6y^3$ **37.** $-6x^9y$
39. $6x^4y^4z^4$ **41.** $-35a^5b^2c$ **43.** $6a^3m^3n$ **45.** $a^2b^2c^2$
47. $9ab^2$ **49.** $3a^3b^2$ **51.** abc **53.** $\frac{1}{8}a^2b^2c$ **55.** y^{2+b}
57. a^{x+3} **59.** x^{6a} **61.** y^{4a-5} **63.** $(x + 3)^{a+b}$

1. 64; 16   **3.** yes ($b \ne 0$)   **5.** 5^9 or $1,953,125$ **7.** $(-5)^6$ or
$15,625$ **9.** $65,353.43$ **11.** $16,777,216$ **13.** x^{12}
15. $100y^2$ **17.** $\frac{1}{4}c^2$ **19.** $0.16d^2$ **21.** a^3b^6 **23.** $4a^4b^2$
25. $4a^6b^9$ **27.** $\frac{1}{8}x^3y^6$ **29.** $0.01x^4$ **31.** $0.008a^6$
33. $-48x^3y^2$ **35.** $-24a^3b$ **37.** x^4y^4 **39.** $100,000$
41. $152a^6$ **43.** $262,144$

1. 1   **3.** $\frac{1}{4}$ **5.** $\frac{1}{100}$ **7.** $\frac{1}{25}$ **9.** $\frac{1}{64}$ **11.** 18   **13.** $\frac{1}{m^5}$
15. $\frac{1}{d^2e}$ **17.** k^5 **19.** $\frac{b^7}{a^2}$ **21.** x **23.** $\frac{b^3}{a^4}$ **25.** n^3 **27.** $\frac{1}{x}$
29. a^2 **31.** $\frac{1}{k^6}$ **33.** an **35.** $\frac{a}{n^2}$ **37.** b^3c^3 **39.** $-\frac{1}{y^3}$
41. $3b$ **43.** $\frac{b^3}{5}$ **45.** y^3 **47.** ab^2 **49.** $\frac{w^2}{t^5}$ **51.** $-\frac{5y^5}{x^9}$
53. $-\frac{4b^3}{c^2}$ **55.** $-2ab^4c^5$ **57.** $-\frac{9yz^5}{x^3}$ **59.** $\frac{7x^3}{4z^{10}}$
61. $a^{10}b^6$ **63.** $\frac{125}{r^3s^3}$ **65.** 1   **67.** $\frac{m}{7nr^2}$ **69.** x^5 **71.** a^{10}
73. 2   **75.** 3   **77.** -1

1. 5000; 5.79×10^7 **3.** $12,760$; 1.4959×10^8 **5.** $142,700$;
7.7812×10^8 **7.** 4.293×10^3 **9.** 2.4×10^5
11. 3.19×10^{-4} **13.** 9.2×10^{-8} **15.** 3.2×10^6
17. 7.6×10^6 **19.** 3×10^2; 300   **21.** 6×10^2; 600
23. 5.5×10^{-9}; 0.0000000055 **25.** 6×10^{-3}; 0.006
27. 6.51×10^3; 6510   **29.** 7.8×10^8; 780,000,000
31. 2.1×10^{-1}; 0.21   **33.** 4×10^8; 400,000,000

1. a. $n - 5$ **b.** $n + 11$ **c.** $n - x$ **3. a.** $n + 14$
b. $n + 3$ **c.** 7   **5. a.** $2m$ **b.** $\frac{m}{3}$ **c.** $m - 5$ **d.** $m + 8$
7. Jack, 44; Charlie, 30   **9.** not enough information
11. 6   **13.** 22   **15.** 22

1. $0.60x + 0.75(16 - x) = 10.95$
3. $6x + 36 = 4.20(12 + x)$ **5.** 12 nickels   **7.** 32 pennies,
21 nickels   **9.** 45 pounds   **11.** 6 nickels, 12 quarters,
9 dimes   **13.** 110 at 95¢; 154 at $1.25   **15.** 5 pounds
17. adult, $15.95; child, $9.95

Page 175   Lesson 5-7

1. $0.10(5) + 1.00n = 0.40(5 + n)$
3. $0.10n + 0.25(30 - n) = 0.20(30)$ **5.** 1 liter
7. 8 grams   **9.** 25.038 mL   **11.** 36 pounds
13. \$4500 at 6.2%, \$8000 at 8.6%   **15.** 20 liters

Page 180   Lesson 5-8

1. yes, 3   **3.** no   **5.** yes, 7   **7.** no   **9.** 28   **11.** -5
13. 6   **15.** $26\frac{1}{4}$ **17.** 0.608   **19.** 4, $y = 4x$
21. $\frac{1}{5}$, $y = \frac{1}{5}x$ **23.** $-\frac{2}{3}$, $y = -\frac{2}{3}x$ **25.** $\frac{4}{5}$, $y = \frac{4}{5}x$
27. $\frac{11,723}{2060}$, $y = \frac{11,723}{2060}x$ **29.** 9.75 kg   **31.** \$467.50
33. 6.993 gallons   **35.** $3\frac{1}{3}$ ft^3 **37.** $\frac{567}{8}$

Page 182   Lesson 5-9

1. yes, 6   **3.** yes, 50   **5.** yes, -13 **7.** yes, 1   **9.** yes, 40
11. no   **13.** 48   **15.** 99   **17.** 6.075   **19.** 3.61 h or about
3 h 37 min   **21.** 30 m^3 **23.** 640 cycles per second

Page 185   Lesson 5-10

1. Emilio   **3.** same distance   **5.** Betty   **7.** 8 feet
9. $8\frac{1}{43}$ feet; about 8 ft $\frac{1}{4}$ in.   **11.** 12 feet

Page 188   Chapter Review

1. b^9 **3.** a^4b^3 **5.** $-12a^3b^4$ **7.** $49a^2$ **9.** $\frac{1}{9}b^4$
11. $576x^5y^2$ **13.** y^4 **15.** $3b^3$ **17.** $\frac{a^4}{2b}$ **19.** 2.4×10^5
21. 6×10^{11} **23.** 6×10^{-7} **25.** Jim, 27; Joe, 37
27. 11 dimes, 21 nickels   **29.** 90   **31.** $4\frac{1}{2}$ ft or 4 ft 6 in.
from the end where Lee is seated

CHAPTER 6   POLYNOMIALS

Page 192   Lesson 6-1

1. trinomial   **3.** monomial   **5.** monomial   **7.** 1   **9.** 3
11. 3   **13.** no degree   **15.** 5   **17.** 1   **19.** 1   **21.** 12
23. 6   **25.** 3   **27.** 7   **29.** 1   **31.** 2   **33.** 9   **35.** 4
37. 5   **39.** 2   **41.** 3   **43.** 7   **45.** 7
47. $1 + x^2 + x^3 + x^5$ **49.** $3xy^3 - 2x^2y + x^3$
51. $p^4 + 21p^2x + 3px^3$ **53.** $\frac{1}{4}x - \frac{2}{5}s^4x^2 + \frac{1}{3}s^2x^3 + 4x^4$
55. $-3x^3 + 5x^2 + 2x + 7$ **57.** $7ax^3 + 11x^2 - 3x + 2a$
59. $\frac{1}{5}x^5 - 8a^3x^3 + \frac{2}{3}x^2 + 7a^3x$
61. $2.4tx^5 + 5.1x^3 - 0.3tx^2 + 1.7t^3x$

Page 196   Lesson 6-2

1. $-5a - 9b$ **3.** $-4x^2 + 5$ **5.** $-2y^2 + 7y - 12$
7. $10r^2 - 4rs + 6s$ **9.** $x^3 + x + 1$ **11.** $15x^2y$
13. $8m - 7n$ **15.** $-x^2 - 3x - 7$
17. $3ab^2 - 5a^2b + b^3$ **19.** $5m$ and $-3m$; $4mn$ and $-mn$;
$2n$ and $8n$ **21.** $8a^2b$ and $16a^2b$; $11b^2$ and $-2b^2$ **23.** $6x^2$
and $-8x^2$; $7y^3$ and $3y^3$ **25.** $7x + 14y$ **27.** $7m + 5n$
29. $7x - 4y$ **31.** $3a - 11m$ **33.** $13m + 3n$

35. $13x - 2y$ **37.** $3n^2 + 13n + 11$ **39.** $-2 - 6a$
41. $x^2 + \frac{4}{3}x + \frac{1}{4}$ **43.** $7ax^2 + 3a^2x - 5ax + 2x$
45. $-3x^2y + 6xy^2 - 2y^2$ **47.** $-2mn^2 + 3mn - n - 3n^3$
49. $9a - 3b - 4c + 16d$ **51.** $7ax^2 - 5a^2x - 7a^3 + 4$
53. $x^2y^2 - 5xy - 10$ **55.** $6m^2n^2 + 8mn - 28$
57. $6m^2n^2 + 10mn - 23$ **59.** $-3y^2$ **61.** $2x + 6y$
63. 5   **65.** $92 - x$ **67.** 135   **69.** $190 - 8x$ **71.** 30
73. $176 - 2x$ **75.** $168 - 6x$ **77.** $-7x^2 - 2x + 184$

Page 199   Lesson 6-3

1. $-60a^3$ **3.** -6 **5.** $35a^2 + 56a$ **7.** $15a^2b - 9ab$
9. $-24a^4 - 56a^3$ **11.** $9x^2y^2 + 6x^2y$ **13.** $15a + 35$
15. $-24x - 15$ **17.** $4x^2 + 3x$ **19.** $15b^2 + 24b$
21. $-10x^2 - 22x$ **23.** $2.2a^2 + 7.7a$ **25.** $21a^3 - 14a^2$
27. $15s^3t + 6s^2t^2$ **29.** $35x^3y - 7xy^3$
31. $10a^4 - 14a^3 + 4a$ **33.** $35x^4y - 21x^3y^2 + 7x^2y^2$
35. $40y^4 + 35y^3 - 15y^2$ **37.** $-28x^3 + 16x^2 - 12x$
39. $15x^4y - 35x^3y^2 + 5x^2y^3$
41. $36m^4n + 4m^3n - 20m^2n^2$
43. $-32x^2y^2 - 56x^2y + 112xy^3$ **45.** $-3x^3 - \frac{1}{3}x^2 + \frac{5}{3}x$
47. $-16m^3n + 6m^2n^2 - 2mn^3$
49. $-\frac{1}{4}ab^4 + \frac{1}{3}ab^3 - \frac{3}{4}ab^2$ **51.** $7a^4 + 21a^3 - a + 25$
53. $-3m^3 + 41m^2 - 14m - 16$
55. $21a^3 - 6a^2 - 46a + 28$ **57.** $50.6t^2 + 21t - 102$
59. $6m^3 + 21m^2 - \frac{33}{2}m$
61. $0.55a^3 + 0.3565a^2 - 0.14125a$

Page 203   Lesson 6-4

1. $8x$ **3.** $11a$ **5.** $22x$ **7.** $13x$ **9.** $-b$ **11.** $14m$
13. $a^2 + 10a + 21$ **15.** $m^2 - 16m + 55$
17. $x^2 + 7y - 44$ **19.** $2x^2 + 17x + 8$
21. $8a^2 + 2a - 3$ **23.** $10a^2 + 11ab - 6b^2$
25. $14y^2 - 23y + 3$ **27.** $20r^2 - 13rs - 21s^2$
29. $12x^2 - \frac{3}{2}x - \frac{3}{8}$ **31.** $\frac{3}{16}y^2 + \frac{1}{12}xy - \frac{2}{3}x^2$
33. $6a^2 + 3.7a + 0.56$ **35.** $0.63x^2 + 3.9xy + 6y^2$
37. $5a^3 - 13a^2 + 49a + 22$ **39.** $12s^3 + 47s^2 + 4s - 45$
41. $30x^3 - 37x^2y + 55xy^2 - 18y^3$
43. $0.63t^3 - 0.40t^2 + 1.41t + 0.34$
45. $\frac{4}{9}x^3 - \frac{5}{6}x^2y + \frac{13}{24}xy^2 - \frac{1}{8}y^3$
47. $a^4 - a^3 - 8a^2 - 29a - 35$
49. $-6x^4 - 5x^3 + 7x^2 - 71x + 40$
51. $-25x^3 + 20x^2 + 31x - 14$
53. $-35b^5 + 14b^4 - 18b^3 - 19b^2 + 14b - 12$
55. $-\frac{10}{3}x^5 + \frac{7}{3}x^4 - 6x^3 + \frac{4}{3}x^2 - x$
57. $8x^{3y} - 20x^{6y} + 12x^{4y}$ **59.** $-8m^{7x} - 24m^{5x}$

Page 207   Lesson 6-5

1. $a^2 + 4ab + 4b^2$ **3.** $4x^2 + 4xy + y^2$
5. $9m^4 + 12m^2n + 4n^2$ **7.** $a^4 + 2a^2b + b^2$
9. $9a^2 - 48ab + 64b^2$ **11.** $25a^2 - 30ab + 9b^2$
13. $9x^2 - 6xy + y^2$ **15.** $4x^4 - 12x^2y^3 + 9y^6$
17. $25x^2 + 10xy + y^2$ **19.** $25a^2 - 10ab + b^2$

21. $49x^2 + 42xy + 9y^2$ **23.** $25r^2 - 70rs + 49s^2$
25. $36a^2 - 60ab + 25b^2$ **27.** $121m^2 + 154mn + 49n^2$
29. $\frac{1}{9}x^2 + \frac{2}{3}xy + y^2$ **31.** $\frac{9}{16}a^2 - 3ab + 4b^2$
33. $1 + 2x + x^2$ **35.** $9.61 + 12.4y + 4y^2$
37. $x^6 - 10x^3y^2 + 25y^4$ **39.** $64a^2 - 4b^2$
41. $3.24a^4 - 8.28a^2y^2 + 5.29y^4$ **43.** $\frac{16}{9}x^4 - y^2$
45. $64x^4 - 48x^2y + 9y^2$ **47.** $16x^6 - 24x^3y^2 + 9y^4$
49. $x^4 - 25x^2 + 144$ **51.** $16x^3 - 64x^2 - x + 4$
53. $a^4 + 4a^3b + 6a^2b^2 + 4ab^3 + b^4$
55. $81m^4 + 216m^3n + 216m^2n^2 + 96mn^3 + 16n^4$
57. $a^5 - 5a^4b + 10a^3b^2 - 10a^2b^3 + 5ab^4 + b^5$
59. $16x^4 - 96x^3y + 216x^2y^2 - 216xy^3 + 81y^4$
61. $0.0081m^4 - 0.108m^3n + 0.54m^2n^2 - 1.2mn^3 + n^4$
63. $x^{4n} - y^{2n}$ **65.** $x^{6n} + 2x^{3n}y^{5n} + y^{10n}$

Page 211   Lesson 6-6

1. $7x - 11$ **3.** $8y - 42$ **5.** $-36y + 34$ **7.** $10q + 48$
9. $2x - 79$ **11.** $-\frac{1}{5}$ **13.** $-\frac{13}{9}$ **15.** $\frac{15}{7}$ **17.** $-\frac{7}{2}$
19. 8 **21.** $\frac{21}{2}$ **23.** -3 **25.** -83 **27.** 2 **29.** $\frac{58}{33}$
31. $\frac{23}{4}$ **33.** $\frac{1}{3}$ **35.** 2 **37.** 0 **39.** $-\frac{3}{2}$ **41.** 17
43. $-\frac{15}{8}$ **45.** $2ac$ **47.** $2a$ **49.** $\frac{b-3}{a}$ **51.** $\frac{3-b}{7}$
53. mn **55.** $10a$ **57.** $\frac{a}{5}$

Page 213   Lesson 6-7

1. a. $n + 5$ **b.** $2t + 5$ **c.** $2n + 8$ **d.** $t + 4$ **3.** 15 m
5. 18 ft by 40 ft **7.** 19 cm **9.** 11 in. by 11 in.
11. 8 cm, 16 cm, 13 cm **13.** 7 m **15.** 60 yd by 80 yd
17. 4.5 ft

Page 217   Lesson 6-8

1. \$480 **3.** $3\frac{1}{2}$ years **5.** 12% **7.** \$2400 **9.** $15\frac{1}{2}\%$
11. owner–\$148,800; real estate company–\$11,200
13. \$240 **15.** \$6400 at 8%, \$3600 at 12% **17.** \$3800
19. \$3000 **21.** $11\frac{1}{2}\%$ **23.** \$18,000

Page 221   Lesson 6-9

1. a. 160 km **b.** 480 km **c.** $80h$ km **3. a.** 40 mph
b. $\frac{240}{x}$ mph **5. a.** 8 hours **b.** 4 hours **7.** $3\frac{1}{2}$ hours
9. 44 mph **11.** $2\frac{1}{2}$ hours **13.** 11:00 A.M. **15.** 240 km
17. 450 mph, 530 mph **19.** 320 mph **21.** 225 miles

Page 225   Chapter Review

1. $3x^4 + x^2 - x - 5$ **3.** $-3x^3 + x^2 - 5x + 5$
5. $x^2 - 8x + 7$ **7.** $7x^3y + 28x^2y^2 - 56xy^3$
9. $x^3 + 56x^2 - 30x + 3$ **11.** $4x^2 + 13x - 12$
13. $\frac{3}{8}a^3 - \frac{1}{4}a^2 + 4$ **15.** $25x^2 - 9y^2$
17. $64x^2 - 80x + 25$ **19.** 4 **21.** 18 in. **23.** \$5240
25. 2 hours

CHAPTER 7   FACTORING

Page 230   Lesson 7-1

1. yes **3.** no **5.** yes **7.** composite; $3 \cdot 13$
9. composite; $7 \cdot 13$ **11.** 4 **13.** 9 **15.** 11 **17.** 10
19. 36 **21.** $3 \cdot 7$ **23.** $2^2 \cdot 3 \cdot 5$ **25.** $3 \cdot 13$
27. $2 \cdot 17$ **29.** $2^4 \cdot 7$ **31.** $2^4 \cdot 19$ **33.** $2^2 \cdot 3 \cdot 5^2$
35. $2^2 \cdot 5 \cdot 7 \cdot 11$ **37.** $-1 \cdot 2 \cdot 13$
39. $-1 \cdot 2 \cdot 2 \cdot 5 \cdot 5 \cdot 5$ **41.** $2 \cdot 7 \cdot 7 \cdot a \cdot a \cdot b$
43. $2 \cdot 2 \cdot 7 \cdot 7 \cdot b \cdot b$ **45.** $-1 \cdot 2 \cdot 3 \cdot 17 \cdot x \cdot x \cdot x \cdot y$
47. 5 **49.** 29 **51.** 126 **53.** 19 **55.** $17a$ **57.** $2x$
59. 5 **61.** 6 **63.** 2 **65.** $4a$ **67.** $4a$ **69.** mn **71.** 1
73. $2ab$ **75.** $4a$

Page 233   Lesson 7-2

1. 3 **3.** y **5.** 1 **7.** $3y$ **9.** $4x$ **11.** $7a$ **13.** xy **15.** 1
17. $12(2x^2 + y^2)$ **19.** $5ab(a + 2)$ **21.** $8x(2x + 1)$
23. $12ab(3ab - 1)$ **25.** $9b(3a^2 + b^2)$ **27.** $6xy(3y - 4x)$
29. $y^3(15x + y)$ **31.** $x(29y - 3)$ **33.** $a(1 + a^3b^3)$
35. $5(a^2 + 2ab - 3b^2)$ **37.** $3xy(x^2 + 3y + 12)$
39. $7abc(4abc + 3ac - 2)$ **41.** $a(1 + ab + a^2b^3)$
43. $a^3(14x + 19y + 11z)$ **45.** $3a(14bc - 4ab^2 + ac^2)$
47. $x(x^4 + 5x^3 + 3x + 2)$ **49.** $\frac{1}{3}(2x + y)$
51. $\frac{2}{5}(a - b + 2c)$

Page 236   Lesson 7-3

1. yes **3.** yes **5.** no **7.** no **9.** yes **11.** yes
13. $(a + 3)(a - 3)$ **15.** $(2x + 3y)(2x - 3y)$
17. $(a + 2b)(a - 2b)$ **19.** $(4a + 3b)(4a - 3b)$
21. $(4a + 5)(4a - 5)$ **23.** $2a(a + 9)$ **25.** $4(2x^2 - 3y^2)$
27. $12(a + 2)(a - 2)$ **29.** $5(3x + 2z)(3x - 2z)$
31. $(3x^2 + 5y^2)(3x^2 - 5y^2)$
33. $(0.1n - 1.3r)(0.1n + 1.3r)$
35. $(1.3p - 1.6q)(1.3p + 1.6q)$ **37.** $4(x + 4y)(x - 4y)$
39. $(3x^2 + 4y)(3x^2 - 4y)$ **41.** $7(2x + 1)(2x - 1)$
43. $(3x + 4)(3x - 4)$ **45.** $15(x + 2y)(x - 2y)$
47. $(a^2 + b)(a^2 - b)$ **49.** $\left(\frac{1}{2}x + 4\right)\left(\frac{1}{2}x - 4\right)$
51. $\frac{1}{2}(3x + 7y)(3x - 7y)$
53. $(x^4 + 1)(x^2 + 1)(x + 1)(x - 1)$ **55.** $(x - 2y)x$
57. $(a - b + c)(a + b - c)$
59. $(x - y - 0.6z)(x - y + 0.6z)$
61. $(x^2 + 4)(x + 2)(x - 2)$
63. $2a(9a^2 + 4b^2)(3a + 2b)(3a - 2b)$
65. $3x(4x^2 + y^2)(2x + y)(2x - y)$ **67.** $m = 6, n = 5$
69. $a = 23, b = 22; a = 9, b = 6; a = 7, b = 2$

Page 240   Lesson 7-4

1. yes, $(a + 2)^2$ **3.** no **5.** no **7.** no **9.** yes,
$(2x - 1)^2$ **11.** yes, $(3b - 1)^2$ **13.** $(a + 6)^2$
15. $(x + 8)^2$ **17.** $(n - 4)^2$ **19.** $(2a + 1)^2$
21. $(1 - 5a)^2$ **23.** not factorable **25.** $2(5x + 2)^2$
27. $(11y + 1)^2$ **29.** $(3x + 7)^2$ **31.** $(5b - 3)^2$ **33.** not
factorable **35.** $(m + 8n)^2$ **37.** $(3x + 4y)^2$
39. $9(4p + 1)^2$ **41.** $(8x - 5)^2$ **43.** $3(x + 6)^2$

45. $6(x^2 + 3x + 4)$ **47.** $(2x + z^2)^2$ **49.** $\left(\frac{1}{2}a + 3\right)^2$
51. $(m^2 + 6n^2)^2$ **53.** $\left(3a + \frac{4}{5}\right)^2$ **55.** not factorable
57. 42 or -42 **59.** 4 **61.** 9 **63.** y^2 **65.** $4y^2$
67. $x - 10$ **69.** $4a - 18b^2$

Page 245 Lesson 7-5

1. 1, 14 **3.** $-2, -6$ **5.** 9, -4 **7.** 3, -10 **9.** 3, 15
11. 3, -20 **13.** 10 **15.** 7 **17.** $+8$ **19.** $+7m$
21. $(y + 3)(y + 9)$ **23.** $(c + 3)(c - 1)$
25. $(y - 3)(y - 5)$ **27.** $(m - 5)(m + 4)$ **29.** prime
31. $(z - 13)(z + 3)$ **33.** $(r - 12)(r + 2)$
35. $(y + 10)(y - 2)$ **37.** $(p + 5)^2$ **39.** $(b - 4)(b - 7)$
41. $(15 + m)(1 + m)$ **43.** $(y + 7)(y + 4)$
45. $(a - 20)(a + 2)$ **47.** $(x - 6)(x + 3)$ **49.** prime
51. $(m - 11)(m + 5)$ **53.** $(11 - j)(6 - j)$
55. $(21 - a)(2 - a)$ **57.** $(c - 4d)(c + 2d)$
59. $(a + 3b)(a - b)$ **61.** $(a + b - 6)(a + b + 1)$
63. 7, -7, 11, -11 **65.** 12, -12 **67.** 8, -8
69. 4, -4 **71.** 10, -10, 22, -22 **73.** 5, -5, 13, -13

Page 248 Lesson 7-6

1. $(k - m)(r + s)$ **3.** $(4b + y)(x - y)$
5. $(7rp - 4q)(r - 3p)$ **7.** $(8m + 1)(x + y)$
9. $(5mp - 3bc)(2m + 3p)$ **11.** $(4m - 3p)(y - 5)$
13. $(a + b)(7x + 3m - 4p)$ **15.** $(a + b - c)(8r - 3y)$
17. $(x + y)$ **19.** $(a + 3b)$ **21.** $(a - 4c)$
23. $(4m - 3p)$ **25.** $(6x + 7y)$ **27.** $(3 + 2b^2)$
29. $(2x + b)(a + 3c)$ **31.** $(a + p)(y + 4)$
33. $(m + x)(2y + 7)$ **35.** $(a - c)(y - b)$
37. $(a + 1)(a - 2b)$ **39.** $(4x + 3y)(a + b)$
41. $(m^2 + p^2)(3 - 5p)$ **43.** $(a - 3b^2)(5a - 4b)$
45. $3(2a - c)(a - b)$ **47.** $a(x + ax - 1 - 2a)$
49. $(7m + 2n)(p + q)(p - q)$
51. $(x + 1)(x - 1)(x + 2)$ **53.** $(a^2 + b^2)(a - b)$
55. $(7m - 5y)(a + b)(a - b)$
57. $(m - 2)(a + 2)(a + 3)$ **59.** $(2x + 3y)(a - 3)(a - 4)$

Page 253 Lesson 7-7

1. 2, 9 **3.** 3, 8 **5.** $-2, -6$ **7.** $-4, -5$ **9.** $-4, 10$
11. $-15, 2$ **13.** $-4, 15$ **15.** $(3y + 5)(y + 1)$
17. $(3x + 2)(x + 2)$ **19.** $(4m - 3)(2m - 1)$
21. $(2h - 3)(h + 1)$ **23.** $(3k - 2)(k + 3)$
25 $(3p - 2)(a - 2)$ **27.** $(2a + 7)(a - 2)$
29. $(3t - 2)(2t + 3)$ **31.** $(2y - 3)(y - 1)$
33. $(2q + 3)(q - 6)$ **35.** $(3m + 2)(2m + 5)$
37. $(3x - 11)(2x + 1)$ **39.** prime
41. $(4p - 3)(2p - 3)$ **43.** $(5p - 2)(3p + 4)$
45. $(2y - 3)(3y - 5)$ **47.** $(2s + 5)(3s - 4)$
49. $(9x + 5)(2x + 5)$ **51.** $(9c - 2)(2c + 5)$
53. $(2r - 3)(9r + 5)$ **55.** $(4m - n)(2m - 3n)$
57. $(4x - 5y)(4x + y)$ **59.** $(4p - q)(5p + 4q)$
61. $(3k + 5m)^2$ **63.** $(5s + 8t)(4s - 3t)$
65. $(4x - 3q)(2x - 9q)$ **67.** $(2c + 5d)(7c + 3d)$
69. $(10a - 11)(3a + 8)$ **71.** prime
73. $(1 - 9y)(2 - 3y)$ **75.** $(6a - b)(8a - 3b)$
77. prime **79.** $4x^2(5x - 7)(2x - 3)$

81. $5t(4t - 7)(3t + 2)$ **83.** $a^2b(4a - 7b)(5a - 6b)$
85. $4x + 14$

Page 257 Lesson 7-8

1. $3(x^2 + 5)$ **3.** $a(5x + 6y)$ **5.** $6a(2x^2 + 3y^2)$
7. $5x^2(1 - 2y)$ **9.** $(a + 3b)(a - 3b)$
11. $3ab(a + 2 + 3b)$ **13.** $2(a + 6)(a - 6)$
15. $m(m + 3)^2$ **17.** $4a(a + 3)(a - 3)$
19. $(m^2 + p)(m^2 - p)$ **21.** $(2k + 1)(k + 1)$
23. $6(y + 2x)(y - 2x)$ **25.** $3(y + 8)(y - 1)$
27. $2(5y + 1)(2y + 3)$ **29.** $2(b^2 + 3b + 1)$
31. $(m + 4n)^2$ **33.** $3(2p - 7)(p + 5)$
35. $a(3x + 7)(x + 3)$ **37.** $3a(3a - 2)(a + 8)$
39. $3(x - 5y)(x + 2y)$ **41.** $m(mn + 7)(mn - 7)$
43. $3(x + 11y)(x - 3y)$ **45.** $4(x^2 + 9)(x + 3)(x - 3)$
47. $(3y + 1)(3y - 1)(y^2 + 1)$ **49.** $6r(5r - 2s)(2r - s)$
51. $(y^2 + z^2)(x + 1)(x - 1)$ **53.** $0.7(y + 3)(y - 3)$
55. $\frac{1}{3}(b + 3)^2$ **57.** $\frac{1}{4}(r + 4)(r + 2)$
59. $0.7(y + 2)(y + 3)$

Page 260 Chapter Review

1. $2 \cdot 3 \cdot 7$ **3.** $2 \cdot 2 \cdot 2 \cdot 2 \cdot 5$ or $2^4 \cdot 5$
5. $11 \cdot 11 \cdot a \cdot a \cdot a \cdot a \cdot b \cdot b \cdot b \cdot b$ **7.** 5 **9.** $2ab$
11. $6(x^2y + 2xy + 1)$ **13.** $2a(12b + 9c + 16a)$
15. $\frac{1}{4}(3x + y)$ **17.** $(5 + 3y)(5 - 3y)$
19. $2(y + 8)(y - 8)$ **21.** $(x^2 + 1)(x + 1)(x - 1)$
23. $(4x - 1)^2$ **25.** $2(4x - 5)^2$ **27.** $\left(y - \frac{3}{4}\right)^2$
29. $(b - 3)(b - 5)$ **31.** $(b - 1)(b + 6)$
33. $(a - 7)(a + 4)$ **35.** $(s + 13)(s - 3)$
37. $(m - 7n)(m + 6n)$ **39.** $(r - 13s)(r + 5s)$
41. $(m - 1)(4m - 3p)$ **43.** $(m + 7)(b + r)$
45. $(4k - p)(4k - 7p)$ **47.** $(4m + 3)(m + 2)$
49. $(2r + 5)(r - 4)$ **51.** $(2m - 1)(2m - 3)$
53. $(3x - 4)(x + 5)$ **55.** $3(x + 2)(x - 2)$
57. $a(3y + 5)(5y + 4)$ **59.** $(x + 7)(x - 7)(m + b)$

CHAPTER 8   APPLICATIONS OF FACTORING

Page 264 Lesson 8-1

1. $x = 0$ or $x + 3 = 0$ **3.** $3r = 0$ or $r - 4 = 0$
5. $3t = 0$ or $4t - 32 = 0$ **7.** $x - 6 = 0$ or $x + 4 = 0$
9. $a + 3 = 0$ or $3a - 12 = 0$
11. $2y + 8 = 0$ or $3y + 24 = 0$ **13.** $x - 3 = 0$
15. $3x + 2 = 0$ or $x - 7 = 0$
17. $4x - 7 = 0$ or $3x + 5 = 0$
19. $2x + 3 = 0$ or $x + 7 = 0$ **21.** $\{0, -3\}$ **23.** $\{0, 4\}$
25. $\{0, 8\}$ **27.** $\{6, -4\}$ **29.** $\{-3, 4\}$ **31.** $\{-4, -8\}$
33. $\{3\}$ **35.** $\left\{-\frac{2}{3}, 7\right\}$ **37.** $\left\{\frac{7}{4}, -\frac{5}{3}\right\}$ **39.** $\left\{-\frac{3}{2}, -7\right\}$
41. 16 years **43.** -2 **45.** $(x - 3)(x - 5) = 0$
47. $(x + 9)(x + 11) = 0$ **49.** $\left(x + \frac{3}{4}\right)\left(x + \frac{5}{6}\right) = 0$

Page 267 Lesson 8-2

1. greatest common factor **3.** perfect square trinomial

5. trinomial that has two binomial factors   **7.** greatest common factor   **9.** difference of squares   **11.** trinomial that has two binomial factors   **13.** greatest common factor   **15.** perfect square trinomial   **17.** greatest common factor   **19.** pairs of terms that have a common monomial factor   **21.** $x^2(x - 5)$ **23.** $(3x - 2y)^2$
25. $(5z + 4)(7z - 3)$ **27.** $2(2x - 9)^2$
29. $(1 - 7k)(1 + 7k)$ **31.** $(x - 3)(x - 2)$
33. $x(x^2 - 3)(4x - 3y)$ **35.** $(3a - 5)^2$ **37.** $4(m + 5)^2$
39. $(x + 2y)(2x - 1)$ **41.** $2x(3x + 5y)$
43. $(2a - 3b)(2a + 3b)$ **45.** $(x + 6)^2$
47. $(4y - 3)(3y + 7)$ **49.** $(x - 3)(x + 3)(x + 5)$
51. $m^2(m - n)(m + n)$ **53.** $5(x + 9)^2$
55. $5a^2(2c + 3)^2$ **57.** $3x^2y(5x - 2y^2 + z)$
59. $3(2x + y)(2x - y)$ **61.** $8x(x - 2y^2)(x + 2y^2)$
63. $(x^2 + 8)(x + 2)$ **65.** $12(x - 3y)(x + 2y)$
67. $(x + y - a + b)(x + y + a - b)$
69. $2(3x - 2y)(x + y)$ **71.** $\left(\frac{x^2}{3} - 7y\right)\left(\frac{x^2}{3} + 7y\right)$
73. $\left(\frac{m}{2} - \frac{3}{5}\right)\left(\frac{m}{2} + \frac{3}{5}\right)$ **75.** $(3a + 5b)(2x - 7y)$
77. prime   **79.** $(x + 3 - y)(x + 3 + y)$
81. $0.01(x - 3y)^2$ **83.** $x(x - 1)$
85. $0.001m^2n(120mn - 1)$ **87.** $\left(\frac{1}{2}x - \frac{3}{5}y\right)^2$
89. $x(x + 3)(x^2 - 2y)$

Page 271   Lesson 8-3

1. $n(n - 3) = 0$ **3.** $8c(c + 4) = 0$ **5.** $3x\left(x - \frac{1}{4}\right) = 0$
7. $7y(y - 2) = 0$ **9.** $13y(2y - 1) = 0$
11. $(y + 4)(y - 4) = 0$ **13.** $(y - 8)(y - 8) = 0$
15. $(y - 5)(y - 5) = 0$ **17.** $\{0, -36\}$ **19.** $\left\{\frac{3}{2}, -\frac{3}{2}\right\}$
21. $\{2, -2\}$ **23.** $\{6, -6\}$ **25.** $\{0, -9\}$ **27.** $\{0, -8\}$
29. $\{6, -6\}$ **31.** $\{0, -5\}$ **33.** $\left\{\frac{2}{3}, -\frac{2}{3}\right\}$ **35.** $\{3, -3\}$
37. $\{2, -2\}$ **39.** $\{-5\}$ **41.** $\left\{-\frac{2}{9}\right\}$ **43.** $\left\{\frac{4}{9}, -\frac{4}{9}\right\}$
45. $\left\{0, \frac{1}{6}\right\}$ **47.** 0 or -6 **49.** 5   **51.** 15 years
53. -10 or 10   **55.** 25 or -25 **57.** 11

Page 275   Lesson 8-4

1. $\{-7, 3\}$ **3.** $\{8, 2\}$ **5.** $\left\{-\frac{3}{2}, -1\right\}$ **7.** $\left\{\frac{4}{3}, \frac{1}{2}\right\}$
9. $\{5, -3\}$ **11.** $\left\{0, -\frac{7}{2}, -\frac{4}{3}\right\}$ **13.** $\{-4, -9\}$
15. $\{3, -7\}$ **17.** $\{-9, 7\}$ **19.** $\{-7, 7\}$ **21.** $\{0, 12\}$
23. $\{8, -3\}$ **25.** $\left\{\frac{3}{2}, -5\right\}$ **27.** $\left\{-\frac{1}{3}, -\frac{5}{2}\right\}$ **29.** $\left\{\frac{3}{2}, -8\right\}$
31. $\{0, 9, -9\}$ **33.** $\{0, 2, 4\}$ **35.** $\left\{0, \frac{1}{5}, -7\right\}$
37. $\left\{0, -\frac{5}{2}, 3\right\}$ **39.** $\{12, -4\}$ **41.** $\left\{\frac{10}{3}, 30\right\}$
43. $\{-4, -5\}$ **45.** $\left\{\frac{2}{3}, -4\right\}$ **47.** $\{-2, 2, -1\}$
49. $\left\{\frac{1}{3}, -\frac{1}{3}, 2\right\}$ **51.** $\{2, -2\}$
53. $\{(x, y)|x = 3 \text{ or } y = -4\}$
55. $\left\{(r, s)|r = -2 \text{ or } s = \frac{4}{3}\right\}$ **57.** $x^2 - 5x + 6 = 0$
59. $3x^2 + x - 2 = 0$

Page 277   Lesson 8-5

1. Let x = the smaller integer; $x(x + 1) = 110$ **3.** Let x = the smaller integer; $x(x + 2) = 168$ **5.** Let x = one integer; $x(15 - x) = 44$ **7.** Let x = the smaller integer; $x^2 + (x + 1)^2 = 181$ **9.** Let x = the smaller integer; $(x + 1)^2 - x^2 = 17$ or $x^2 - (x + 1)^2 = 17$
11. 10, 11; $-10, -11$ **13.** 12, 14; $-12, -14$ **15.** 4, 11
17. 9, 10; $-9, -10$ **19.** 8, 9; $-8, -9$
21. 10, 12; $-10, -12$ **23.** 13, 15   **25.** 13, 14   **27.** 3, 8
29. 8, 11; $-8, -11$ **31.** 9, 11   **33.** 4, 9
35. 5, 6; $-8, -7$ **37.** 6, 8; $-12, -10$ **39.** 13, 15, 17

Page 279   Lesson 8-6

1. Let x = width. $x(x + 3) = 40$; length, 8 m; width, 5 m
3. Let x = amount that length and width are increased. $(4 + x)(7 + x) = 54$; length, 9 in.; width, 6 in.   **5.** Let x = width of the strip. $(15 + 2x)(10 + 2x) = 300$ The width of the strip = $2\frac{1}{2}$ feet.   **7.** 176 feet per second
9. 11,116 ft   **11.** 3 seconds; 7 seconds   **13.** 12 seconds
15. 20 seconds; 125 seconds   **17.** $\frac{1}{2}$ second; 102 seconds

Page 283   Lesson 8-7

1. $24, \frac{1}{3}$ **3.** $6y, \frac{7}{3x}$ **5.** $3x^2y^2, \frac{-y^3}{6x^3}$ **7.** $x, \frac{y + 1}{y - 2}$
9. $m + 5, \frac{1}{2}$ **11.** $z + 3, \frac{1}{z - 3}$ **13.** $\frac{19a}{21b}$ **15.** $\frac{1}{a - b}$
17. $\frac{c - 2}{c + 2}$ **19.** a **21.** $\frac{1}{x^2 - 4}$ **23.** $\frac{x^2 + 3}{x + 3}$ **25.** $-\frac{2}{1 - 2y}$
27. $-\frac{4a}{a^2 + 6ab + 9b}$ **29.** $\frac{1}{x + y}$ **31.** $\frac{1}{x + 4}$ **33.** $\frac{m - 4}{m + 4}$
35. $\frac{y - 3}{y + 3}$ **37.** $x + 5$ **39.** $\frac{6x}{x + 4}$ **41.** $\frac{2k + 5}{2k - 5}$ **43.** $\frac{1}{2 - 5x}$
45. $\frac{5}{3}$ **47.** $\frac{m - 4}{m - 3}$ **49.** $\frac{p - 2}{p + 6}$ **51.** $\frac{2x - 3}{3x - 2}$ **53.** $y + 3$
55. $\frac{2x + 8}{x - 11}$ **57.** $\frac{1}{x}$

Page 285   Chapter Review

1. $\{0, -11\}$ **3.** $\{0, 5\}$ **5.** $\left\{\frac{2}{3}, -\frac{7}{4}\right\}$ **7.** 6
9. $m(3m + 5)(2m - 3)$ **11.** $(3a + 5b)(8m - 3n)$
13. $(7m + 5n)^2$ **15.** $8(a^2 + 4b^2)$ **17.** $\{0, -7\}$
19. $\{0, 3\}$
21. $\left\{-\frac{2}{5}\right\}$ **23.** $\{-5, -8\}$ **25.** $\left\{\frac{3}{2}, -8\right\}$ **27.** $\left\{\frac{4}{3}, \frac{1}{2}\right\}$
29. 11, 13; $-11, -13$
31. 8 inches by 10 inches   **33.** $\frac{x}{4y^2z}$ **35.** $\frac{x + 3}{x - 3}$

CHAPTER 9   FUNCTIONS AND GRAPHS

Page 291   Lesson 9-1

1. $(1, 4)$ **3.** $(-3, 3)$ **5.** $(-1, -2)$ **7.** $(4, 0)$ **9.** $(1, 1)$
11. $(-1, -1)$ **13.** $(-2, 2)$ **15.** $(-3, -4)$ **17.** $(2, 2)$
19. $(3, -1)$ **21.** I   **23.** II   **25.** none   **27.** IV
29. IV   **31.** II   **33.** II   **35.** III   **37.** IV   **39.** none
41. I   **43.** III

Page 293   Lesson 9-2

1. $D = \{0, 1, 2\}$ $R = \{-2, 2, 4\}$
3. $D = \{-3, -2, -1, 0\}$ $R = \{1, 0, 2\}$
5. $D = \{7, -2, 4, 5, -9\}$ $R = \{5, -3, 0, -7, 2\}$
7. $D = \{-3, -2, 3, 0\}$ $R = \{0, -5, 6, 7, -17\}$
9. $D = \{3.1, -4.7, 2.4, -9\}$ $R = \{-1, 3.9, -3.6, 12.12\}$
11. $D = \{\frac{1}{2}, 1\frac{1}{2}, -3, -5\frac{1}{4}\}$ $R = \{\frac{1}{4}, -\frac{2}{3}, \frac{2}{5}, -7\frac{2}{7}\}$
13. $\{(1, 5), (2, 7), (3, 9), (4, 11)\}$ $D = \{1, 2, 3, 4\}$
$R = \{5, 7, 9, 11\}$
15. $\{(-4, 1), (-2, 3), (0, 1), (2, 3), (4, 1)\}$
$D = \{-4, -2, 0, 2, 4\}$ $R = \{1, 3\}$
17. $\{(1, 5), (2, 3), (3, -2), (4, 9)\}$
19. $\{(1, 7), (-2, 7), (3, 7)\}$
21. $\{(4, 2), (-3, 2), (8, 2), (8, 9), (7, 5)\}$
23. $\{(4, 2), (1, 3), (3, 3), (6, 4)\}$ $\{(2, 4), (3, 1), (3, 3), (4, 6)\}$
25. $\{(8, 1), (7, 3), (6, 5), (2, -2)\}$
$\{(1, 8), (3, 7), (5, 6), (-2, 2)\}$
27. $\{(1, 3), (2, 5), (1, -7), (2, 9)\}$
$\{(3, 1), (5, 2), (-7, 1), (9, 2)\}$
29. $\{(-2, 2), (-1, 1), (0, 1), (1, 1), (2, -1), (3, 1)\}$
$D = \{-2, -1, 0, 1, 2, 3\}$ $R = \{2, 1, -1\}$
31. $\{(-3, 0), (-2, 2), (-1, 3), (0, 1), (1, -1),$
$(1, -2), (1, -3), (3, -2)\}$ $D = \{-3, -2, -1, 0, 1, 3\}$
$R = \{0, 2, 3, 1, -1, -2, -3\}$
33. $\{(-3, 3), (-1, 2), (1, 1), (1, 3), (2, 0), (2, -1), (3, -1)\}$
$D = \{-3, -1, 1, 2, 3\}$ $R = \{3, 2, 1, 0, -1\}$

Page 298   Lesson 9-3

1. $\{(-4, -12), (-2, -6), (0, 0), (1, 3), (2, 6), (3, 9)\}$
3. $\{(-4, -1), (-2, \frac{1}{3}), (0, \frac{5}{3}), (1, \frac{7}{3}), (3, \frac{11}{3})\}$
5. $y = 7 - 3x$ **7.** $n = 7 - 4m$ **9.** $y = 4 - 2x$
11. $s = \frac{2 - 6r}{5}$ **13.** $b = \frac{3a - 8}{7}$ **15.** $q = \frac{4 + 5m}{7}$
17. b, c **19.** a, b, c **21.** a
23. $\{(-2, -13), (-1, -8), (0, -3), (2, 7), (5, 22)\}$
25. $\{(-2, -6), (-1, -5), (0, -4), (2, -2), (5, -1)\}$
27. $\{(-2, 14), (-1, 9), (0, 4), (2, -6), (5, -21)\}$
29. $\{(-2, 8), (-1, \frac{20}{3}), (0, \frac{16}{3}), (2, \frac{8}{3}), (5, -\frac{4}{3})\}$
31. $\{(-2, -\frac{1}{4}), (-1, \frac{1}{4}), (0, \frac{3}{4}), (2, \frac{7}{4}), (5, \frac{13}{4})\}$
33. $\{(-2, \frac{13}{2}), (-1, 4), (0, \frac{3}{2}), (2, -\frac{7}{2}), (5, -11)\}$
35. $\{(-2, -7), (-1, -2), (0, 3), (2, 13), (5, 28)\}$
37. $\{(-2, 6), (-1, \frac{14}{3}), (0, \frac{10}{3}), (2, \frac{2}{3}), (5, -\frac{10}{3})\}$

Page 303   Lesson 9-4

1. yes   **3.** no   **5.** yes   **7.** no   **9.** no   **11.** no   **13.** no
15. yes   **17.** yes   **19.** $y = \frac{16 - x}{5}$ **21.** $y = \frac{-6x + 7}{14}$
23. $y = \frac{32x - 8}{3}$ **25.** no   **27.** no   **29.** yes   **31.** no
33. yes   **35.** no

Page 308   Lesson 9-5

1. no   **3.** yes   **5.** yes   **7.** yes   **9.** yes   **11.** no   **13.** -9
15. -1 **17.** 4 **19.** no; $\{(3, 1), (5, 1), (7, 1)\}$; yes
21. yes; $\{(3, -3), (2, -2), (1, -1), (0, 0)\}$; yes

23. yes; $\{(4, 5), (1, 6), (3, -2), (3, 0)\}$; no
25. no; $\{(-2, 3), (-1, 4), (5, -2), (5, 4)\}$; no   **27.** yes
29. yes   **31.** no   **33.** yes   **35.** no   **37.** no   **39.** -14
41. 20   **43.** -3 **45.** $-\frac{2}{9}$ **47.** $4b^2 - 2b$ **49.** 12
51. -21 **53.** $b^2 - 7b + 12$

Page 311   Graphing Calculator Applications

15. C   **17.** F   **19.** D

Page 314   Lesson 9-6

1. c; yes   **3.** a, b, c; no   **5.** b, c; yes   **7.** b; no

Page 317   Lesson 9-7

1. yes   **3.** no   **5.** 13, 15, 17   **7.** $-13, -16$ **9.** 16, 21, 31
11. $\frac{1}{9}, \frac{1}{27}, \frac{1}{81}$ **13.** 1, 0, 1   **15.** $-5, -7, -9$
17. 5, 5.5, 6   **19.** $-5.2, -4.7, -3.7$ **21.** $y = 4x$; 20, 24
23. $n = 2m + 1$; 3, 5, 7   **25.** $b = 4a + 3$; 19, 27, 35
27. $y = 15 - x$; 11, 10, 9, 8
29. $b = 3 - 5a$; $-2, -7, -17, -32$
31. $s = \frac{1}{2}r + 1$; 2, 3, 4, 5   **33.** $y = x^2$ **35.** $y = 16 - x^2$
37. $xy = 24$

Page 321   Chapter Review

1. IV   **3.** I
9. $D = \{3, 2, 5\}$ $R = \{5, 6, 7\}$ $\{(5, 3), (6, 2), (7, 5)\}$
11. $D = \{-3, 4\}$ $R = \{5, 6\}$ $\{(5, -3), (6, -3), (5, 4), (6, 4)\}$
13. $\{(-1, 4), (3, 4), (4, 6), (0, -4)\}$
$D = \{-1, 3, 4, 0\}$ $R = \{4, 6, -4\}$
15. $\{(3, 6), (3, -3), (4, 8), (5, 8), (5, 11), (5, 3)\}$
$D = \{3, 4, 5\}$ $R = \{6, -3, 8, 11, 3\}$
17. $\{(-2, -1), (-1, -1), (0, 1), (1, 2), (3, 2)\}$
$D = \{-2, -1, 0, 1, 3\}$ $R = \{-1, 1, 2\}$ **19.** $y = 7 - 3x$
21. c   **23.** $\{(-4, -11), (-2, -3), (0, 5), (2, 13), (4, 21)\}$
25. $\{(-4, \frac{21}{2}), (-2, \frac{15}{2}), (0, \frac{9}{2}), (2, \frac{3}{2}), (4, -\frac{3}{2})\}$ **27.** no
29. yes   **35.** no   **37.** yes   **39.** 3   **41.** 1   **43.** 3
49. $y = 3x + 5$

CHAPTER 10   LINES AND SLOPES

Page 327   Lesson 10-1

1. 3   **3.** -1 **5.** $\frac{2}{1} = 2$ **7.** $\frac{-1}{1} = -1$ **9.** $\frac{-4}{3} = -\frac{4}{3}$
17. -1 **19.** 0   **21.** $-\frac{1}{5}$ **23.** -2 **25.** lines go from
upper left to lower right   **27.** undefined   **29.** 2
31. $-\frac{1}{2}$ **33.** $\frac{7}{4}$ **35.** $-\frac{5}{3}$ **37.** -2 **39.** -2 **41.** -1
43. undefined   **45.** $-\frac{3}{4}$ **47.** $\frac{25}{3}$ **49.** 7   **51.** 7
53. 2, 3   **55.** $-6, 4$

Page 331   Lesson 10-2

1. 3; (5, 2)   **3.** -2; $(-1, -5)$ **5.** $-\frac{3}{2}$; $(-5, -6)$

Selected Answers

7. 2; $(-2, -4)$ **9.** $-\frac{3}{5}$; $\left(-6, -\frac{3}{8}\right)$ **11.** 0; $(0, -2)$
13. $2x - y = -6$ **15.** $2x - 3y = -1$
17. $5x - 3y = -37$ **19.** $2x + 3y = 22$
21. $2x - 3y = 15$ **23.** $2x - y = 11$
25. $3x - 2y = -20$ **27.** $y = 7$ **29.** $x = 1$
31. $x + y = 9$ **33.** $5x + y = 31$ **35.** $x + 4y = -4$
37. $x + 2y = -4$ **39.** $4x + 7y = -18$ **41.** $x = 5$
43. $4x + 10y = 13$ **45.** $2x + 3y = 6$ **47.** $2y = 1$
49. $x = -2$ **51.** $x = -2$

Page 335   Lesson 10-3

1. 5, 3 **3.** 3, -7 **5.** $\frac{1}{3}$, 0 **7.** $\frac{3}{5}$, $-\frac{1}{4}$ **9.** $\frac{3}{5}$, $\frac{1}{4}$ **11.** $\frac{1}{4}$, $\frac{3}{4}$
13. 6, 5 **15.** $\frac{3}{2}$, $\frac{7}{2}$ **17.** $\frac{1}{2}$, $\frac{7}{4}$
19. $x = c$ (c is any number) **21.** $y = 3x + 1$
23. $y = -3x + 5$ **25.** $y = 4x - 2$ **27.** $y = \frac{1}{2}x + 5$
29. $y = -\frac{5}{4}x + 3$ **31.** $y = -3.1x + 0.6$ **33.** 2, 3
35. 2, 10 **37.** 8, 6 **39.** $-\frac{5}{3}$, $\frac{5}{2}$ **41.** $\frac{1}{2}$, 4
43. x-int., -2; no y-int. **45.** $\frac{8}{3}$, 16 **47.** 8, $-\frac{8}{3}$
49. 0.5, 2.2 **51.** $m = -\frac{2}{5}$, $b = 2$, $y = -\frac{2}{5}x + 2$
53. $m = 5$, $b = -15$, $y = 5x - 15$
55. $m = -\frac{2}{5}$, $b = \frac{8}{5}$, $y = -\frac{2}{5}x + \frac{8}{5}$
57. $m = \frac{5}{4}$, $b = -\frac{11}{4}$, $y = \frac{5}{4}x - \frac{11}{4}$
59. $m = -\frac{4}{3}$, $b = \frac{5}{3}$, $y = -\frac{4}{3}x + \frac{5}{3}$
61. $m = \frac{13}{11}$, $b = -2$, $y = \frac{13}{11}x - 2$
63. $m = -6$, $b = 15$, $y = -6x + 15$
65. $m = -2$, $b = 12$, $y = -2x + 12$
67. $m = \frac{1}{3}$, $b = -\frac{2}{3}$, $y = \frac{1}{3}x - \frac{2}{3}$
69. $m = \frac{3}{2}$, $b = \frac{7}{2}$, $y = \frac{3}{2}x + \frac{7}{2}$
71. $m = \frac{2}{3}$, $b = -\frac{7}{3}$, $y = \frac{2}{3}x - \frac{7}{3}$
73. $m = \frac{11}{2}$, $b = -16$, $y = \frac{11}{2}x - 16$
75. $m = 11$, $b = 14$, $y = 11x + 14$
77. $m = -9$, $b = 2$, $y = -9x + 2$
79. $m = \frac{1}{42}$, $b = \frac{6}{7}$, $y = \frac{1}{42}x + \frac{6}{7}$
81. $m = \frac{44}{31}$, $b = 0$, $y = \frac{44}{31}x$ **83.** $y = \frac{4}{5}x + \frac{3}{2}$
85. $y = \frac{2}{5}x + 12$ **87.** $y = 9$ **89.** $(4, 8)$ **91.** $(7, 2)$

Page 340   Lesson 10-4

1. 2, 8 **3.** 2, $\frac{3}{2}$ **5.** 4, -8 **7.** $\frac{10}{7}$, 5 **9.** $\frac{2}{5}$, -4
11. $\frac{1}{3}$, -1 **13.** $\frac{5}{2}$, $\frac{5}{3}$ **15.** 0, 0 **17.** $(1, 3)$, $(2, 5)$
19. $(1, -3)$, $(2, -10)$ **21.** $(7, 3)$, $(9, 4)$
23. $(-2, 14)$, $(-4, -2)$ **49.** they are equal; $(0, 3)$
51. they are equal; they are parallel

Page 344   Lesson 10-5

1. -5 **3.** 9 **5.** -5 **7.** $-\frac{1}{3}$ **9.** $\frac{1}{3}$

11. 2, -2, $y = 2x - 2$ **13.** -4, 4, $y = -4x + 4$
15. $-\frac{2}{3}$, 2, $y = -\frac{2}{3}x + 2$ **17.** $y = 2x - 10$
19. $y = \frac{1}{2}x + \frac{1}{2}$ **21.** $y = -3x + 16$ **23.** $y = -\frac{5}{3}x - 10$
25. $y = -x + 6$ **27.** $y = -\frac{2}{3}x + 4$ **29.** $y = -1$
31. $y = -\frac{1}{8}x + \frac{15}{4}$ **33.** $y = -\frac{5}{6}x + \frac{20}{3}$
35. $y = -\frac{7}{9}x - \frac{8}{3}$ **37.** $y = 1.524x - 1.881$
39. $y = 1.610x + 1.986$ **41.** $y = -0.686x + 4.301$

Page 348   Lesson 10-6

1. 5, $-\frac{1}{5}$ **3.** 2, $-\frac{1}{2}$ **5.** $\frac{2}{3}$, $-\frac{3}{2}$ **7.** undefined, 0 **9.** $\frac{5}{3}$, $-\frac{3}{5}$
11. 4, $-\frac{1}{4}$ **13.** $\frac{2}{3}$, $-\frac{3}{2}$ **15.** -1, 1 **17.** $y = -6x + 9$
19. $y = -\frac{2}{3}x + \frac{14}{3}$ **21.** $y = \frac{4}{3}x - \frac{16}{3}$ **23.** $y = -\frac{1}{3}x - \frac{13}{3}$
25. $y = -\frac{3}{5}x + \frac{14}{5}$ **27.** $y = -\frac{1}{3}x + 1$ **29.** $y = -\frac{1}{5}x - 1$
31. $y = \frac{7}{3}x - \frac{16}{3}$ **33.** $y = -\frac{1}{2}x - 4$

Page 351   Lesson 10-7

1. 6 **3.** 14 **5.** 5 **7.** $\frac{3}{2}$ **9.** $\frac{5}{2}$ **11.** $(5.5, 1)$ **13.** $(-3, 6)$
15. $(-3, -4)$ **17.** $(1, -5.5)$ **19.** $(13, 4)$ **21.** $(14, 3)$
23. $(15, 1)$ **25.** $(1, -3)$ **27.** $\left(6, \frac{1}{2}\right)$ **29.** $(1, 5)$
31. $\left(6, \frac{7}{2}\right)$ **33.** $\left(\frac{1}{2}, 7\right)$ **35.** $(4x, 2y)$ **37.** $\left(\frac{1}{2}, \frac{1}{3}\right)$
39. $\left(3\frac{9}{10}, 2\frac{5}{6}\right)$ **41.** $B(7, -19)$ **43.** $A(-5, 20)$
45. $P\left(8, -\frac{15}{2}\right)$ **47.** $A(10, 17)$ **49.** $B(11, -10)$
51. $A(5, 5)$ **53.** $(-2, -7)$ **55.** $(3, -2)$ **57.** $(6, 2)$
59. $\left(\frac{7}{2}, -\frac{7}{2}\right)$ **61.** $(5, 4)$ **63.** $(-1, 7)$ **65.** $\left(\frac{15}{4}, -\frac{13}{4}\right)$
67. yes **69.** yes

Page 355   Chapter Review

1. 0 **3.** undefined
5. $y - 1 = -\frac{4}{11}(x - 8)$; $4x + 11y = 43$
7. $y - 5 = \frac{5}{2}x$; $5x - 2y = -10$ **9.** $\frac{3}{2}$, $-\frac{7}{2}$ **11.** -8, 4
13. $(5, 0)$, $(0, 2)$ **19.** $y = 4x - 26$ **21.** $y = -\frac{5}{2}x + 7$
23. $y = 4x - 9$ **25.** $y = -\frac{7}{2}x - 14$ **27.** $(6, 1)$

CHAPTER 11   SYSTEMS OF OPEN SENTENCES

Page 360   Lesson 11-1

1. $(6, 3)$ **3.** $(1, 3)$ **5.** $(0, 0)$ **7.** $(6, 3)$ **9.** $(3, -3)$
11. a, c, d **13.** a, d **15.** $(3, 1)$ **17.** $(5, 3)$ **19.** $(2, 2)$
21. $(2, 3)$ **23.** $(0, 0)$ **25.** $(-1, 2)$ **27.** $(-1, -5)$
29. $(-3, 1)$ **31.** $(3, 4)$ **33.** $(-2, -3)$ **35.** $(3, 2)$
37. $(4, -7)$

Page 363   Lesson 11-2

1. $m = -1$, $b = 4$; $m = -\frac{2}{3}$, $b = 3$; one

3. $m = -1, b = 6$; $m = -1, b = 1$; no solution
5. $m = -\frac{1}{2}, b = \frac{5}{2}$; $m = -\frac{1}{2}, b = \frac{5}{2}$; infinitely many
7. $m = -3, b = 0$; $m = \frac{1}{6}, b = -\frac{19}{3}$; one
9. $m = \frac{3}{7}, b = \frac{6}{7}$; $m = -\frac{1}{2}, b = \frac{11}{2}$; one
11. $m = \frac{3}{8}, b = -\frac{1}{2}$; $m = \frac{3}{8}, b = -\frac{21}{8}$; no solution
13. one, (3, 3) **15.** one, (9, 1) **17.** one, (6, 2)
19. one, (2, 0) **21.** no solution **23.** one, (2, 3)
25. no solution **27.** infinitely many **29.** one, (7, −3)
31. no solution **33.** infinitely many **35.** one, (0, −4)
37. $\left(-\frac{1}{2}, 2\right)$ **39.** $\left(4\frac{1}{2}, -1\frac{1}{2}\right)$

Page 367   Lesson 11-3

1. $x = y + 1$; $y = x - 1$
3. $x = \frac{1}{2}(6 - 3y)$; $y = \frac{1}{3}(6 - 2x)$
5. $x = 2(3y - 7)$; $y = \frac{1}{3}\left(7 + \frac{1}{2}x\right)$
7. $x = -5 - y$; $y = -5 - x$
9. $x = \frac{3}{2}\left(10 + \frac{1}{2}y\right)$; $y = 2\left(\frac{2}{3}x - 10\right)$
11. $x = -\frac{1}{0.75}(0.8y + 6)$; $y = -\frac{1}{0.8}(0.75x + 6)$
13. $x + (3 + 2x) = 7$; $x = \frac{4}{3}$
15. $\left(-6 + \frac{2}{3}y\right) + 3y = 4$; $y = \frac{30}{11}$ **17.** $\frac{2}{3}x = 4x$; $x = 0$
19. one, $\left(\frac{8}{5}, \frac{16}{5}\right)$ **21.** one, $\left(3, \frac{3}{2}\right)$ **23.** infinitely many
25. no solution **27.** one, (0, 0) **29.** one, (−9, −7)
31. one, (4, 2) **33.** one, (4, 3) **35.** one, (1, 1)
37. width, 2 m; length, 10 m **39.** 23, 24 **41.** 21 males,
16 females **43.** (36, −6, −84) **45.** (−14, 27, −6)

Page 370   Lesson 11-4

1. yes **3.** no **5.** yes **7.** no **9.** yes **11.** (10, 15)
13. (3, 0) **15.** $\left(\frac{14}{3}, \frac{10}{3}\right)$ **17.** $\left(2, \frac{1}{2}\right)$ **19.** (0, 0)
21. $\left(\frac{4}{3}, -2\right)$ **23.** (24, 4) **25.** 19, 26 **27.** 11, 53
29. 42 ft, 30 ft **31.** 3, 12 **33.** 179 acres, 97 acres

Page 374   Lesson 11-5

1. x: multiply first equation by −3, add; y: multiply
second equation by −2, add **3.** x: multiply first equation
by −4, add; y: multiply second equation by 4, add
5. x: multiply second equation by −4, add; y: multiply
first equation by 2, add **7.** x: multiply second equation
by −8, add; y: multiply second equation by −3, add
9. x: multiply second equation by −2, add; y: multiply
first or second equation by −1, add **11.** x: multiply
second equation by −1.2, add; y: multiply first
equation by 2, add **13.** $\left(\frac{11}{2}, -\frac{1}{2}\right)$ **15.** $\left(\frac{5}{2}, 5\right)$
17. (1, 1) **19.** $\left(-1, \frac{9}{2}\right)$ **21.** $\left(\frac{20}{9}, \frac{5}{3}\right)$ **23.** (5, 1)
25. (8, −2) **27.** (−9, −2) **29.** (6, 2) **31.** (3, 3)
33. (10, 25) **35.** (4, 16) **37.** (0, 0) **39.** (0, −2)
41. 14, 34 **43.** 4, 48 **45.** 6, 2 **47.** 28 cm, 112 cm
49. Layla, 30 years; Diana, 10 years **51.** 13 cm, 7 cm

53. 19, 5 **55.** Kari, 16 years; Kari's mother, 36 years
57. $\left(\frac{1}{5}, \frac{1}{2}\right)$ **59.** The system is inconsistent if a false
statement, such as 2 = 4, occurs.

Page 378   Lesson 11-6

1. yes **3.** yes **5.** B **7.** C **33.** $x \le 2$; $y \ge 2$
35. $x \le -1$; $y \ge x$ **37.** $y \ge x - 3$; $y \le -x - 3$
39. $y < 3 - x$; $y < 3 + x$ **41.** $y > \frac{1}{5}x$; $y < 2x + 2$
43. yes **45.** no

Page 383   Lesson 11-7

1. 4 mph, 1 mph **3.** 7 lb, 3 lb **5.** 20, 10 **7.** 550 mph,
50 mph **9.** 300 gal of 25%; 200 gal of 50% **11.** $1000
at 10%; $3000 at 12% **13.** 280 mi **15.** 4 mph
17. 1 mph

Page 386   Lesson 11-8

1. 24 **3.** 82 **5.** 16 **7.** 48 **9.** 51 **11.** 35 **13.** 94
15. 72 **17.** 603 **19.** 370 **21.** 39
23. 40, 51, 62, 73, 84, 95 **25.** $\frac{21}{12}, \frac{42}{24}, \frac{63}{36}, \frac{84}{48}$

Page 391   Chapter Review

1. (1, 1) **3.** (6, 5) **5.** no solution **7.** one **9.** (3, −5)
11. (4, −2) **13.** $\left(\frac{1}{2}, \frac{1}{2}\right)$ **15.** (2, 0) **17.** (2, −1)
19. (5, 1) **21.** (−9, −7) **23.** (2, −1) **25.** (−4, 6)
27. 16 cm, 9 cm **31.** 5 mph **33.** 35

CHAPTER 12   RADICAL EXPRESSIONS

Page 397   Lesson 12-1

1. 100 **3.** 49 **5.** 0.09 **7.** $\frac{1}{4}$ **9.** $\frac{49}{64}$ **11.** 11 **13.** −9
15. $\frac{2}{3}$ **17.** $\pm\frac{7}{11}$ **19.** 0.04 **21.** 7 **23.** 4 **25.** $\frac{5}{6}$
27. $\frac{6}{14}$ or $\frac{3}{7}$ **29.** 0.03 **31.** 0.06 **33.** 6 **35.** −10
37. ± 5 **39.** 0.6 **41.** 23 **43.** 26 **45.** −21 **47.** ± 32
49. 27 **51.** 42 **53.** $-\frac{17}{10}$ **55.** 3 **57.** $\frac{16}{19}$ **59.** $\pm\frac{31}{27}$
61. $\frac{8}{9}$ **63.** $\frac{3}{4}$

Page 400   Lesson 12-2

1. $\mathscr{L}, \mathscr{Q}$ **3.** $\mathscr{N}, \mathscr{W}, \mathscr{L}, \mathscr{Q}$ **5.** $\mathscr{Q}$ **7.** $\mathscr{N}, \mathscr{W}, \mathscr{L}, \mathscr{Q}$ **9.** $\mathscr{I}$
11. $\mathscr{Q}$ **13.** 1.414 **15.** 4.472 **17.** 256 **19.** 5.568
21. 9.434 **23.** 4356 **31.** $\mathscr{I}$ **33.** $\mathscr{Q}$ **35.** $\mathscr{Q}$ **37.** $\mathscr{I}$
39. 9.17 **41.** 8.12 **43.** −4.5 **45.** −7.1 **47.** 6.63
49. 3.61 cm **51.** $\mathscr{Q}$; 82 **53.** $\mathscr{I}$; 61, 62 **55.** $\mathscr{I}$; 29, 30
57. $\mathscr{Q}$; 95 **59.** $\mathscr{I}$; 84, 85 **61.** $\mathscr{Q}$; 38 **63.** 7.9 ft × 23.7 ft

Page 402   Lesson 12-3

1. Inverse operations result in the original number.
3. 9.22 **5.** 24 **7.** 7.77 **9.** −12.21 **11.** ± 13.89
13. 11.463 **15.** ± 68 **17.** 13.251 **19.** −0.781
21. 0.068

1. $2\sqrt{2}$ **3.** $2\sqrt{5}$ **5.** $2\sqrt{6}$ **7.** $4\sqrt{3}$ **9.** $|m|$ **11.** $x^2\sqrt{x}$
13. $2a\sqrt{2a}$ **15.** $ab\sqrt{ab}$ **17.** 6   **19.** $5\sqrt{2}$ **21.** 11
23. $3+\sqrt{6}$ **25.** $\sqrt{35}-7$ **27.** $3\sqrt{3}$ **29.** $3\sqrt{5}$
31. $6\sqrt{2}$ **33.** $3\sqrt{10}$ **35.** $8\sqrt{2}$ **37.** $10\sqrt{5}$ **39.** $12\sqrt{5}$
41. 72   **43.** $4|x|\sqrt{2}$ **45.** $2b^2\sqrt{10}$ **47.** $6|ab|$
49. $2|a|\sqrt{30ab}$ **51.** $2|x|y^2\sqrt{15}$ **53.** $2|m|n^3\sqrt{5n}$
55. $8m^2|n^3|\sqrt{5}$ **57.** $|x|\sqrt{21y}$ **59.** $4\sqrt{3}$ **61.** 10
63. $14\sqrt{15}$ **65.** 150   **67.** $60\sqrt{3}$ **69.** $3+3\sqrt{2}$
71. $\sqrt{42}-3\sqrt{2}$ **73.** $5\sqrt{2}-\sqrt{10}$ **75.** $\sqrt{15}+9$
77. $12+4\sqrt{21}$ **79.** 49 cm^2

1. 2   **3.** 8   **5.** $\sqrt{3}-4$; -13 **7.** $6-\sqrt{8}$; 28
9. $\sqrt{2}-\sqrt{5}$; -3 **11.** $2\sqrt{5}+\sqrt{6}$; 14   **13.** $\frac{\sqrt{5}}{5}$ **15.** $\frac{\sqrt{6}}{6}$
17. $\frac{\sqrt{5}}{5}$ **19.** $\frac{3-\sqrt{7}}{3-\sqrt{7}}$ **21.** $\sqrt{7}$ **23.** $\frac{\sqrt{70}}{7}$ **25.** $\frac{\sqrt{3}}{3}$
27. $\frac{\sqrt{21}}{3}$ **29.** $\frac{\sqrt{35}}{10}$ **31.** $\frac{\sqrt{15}}{3}$ **33.** $\frac{\sqrt{21}}{7}$ **35.** $\frac{\sqrt{3a}}{3}$
37. $\frac{|a|\sqrt{5}}{5}$ **39.** $\frac{3\sqrt{3}}{|b|}$ **41.** $\frac{n^2\sqrt{5mn}}{2|m^3|}$ **43.** $\frac{|a|\sqrt{6}}{4|b|}$
45. $\frac{7+\sqrt{3}}{46}$ **47.** $\frac{11\sqrt{2}-55}{-23}$ **49.** $6\sqrt{3}-6\sqrt{2}$
51. $\frac{20a+10a\sqrt{a}}{4-a}$ **53.** $\frac{-6\sqrt{5}-2\sqrt{30}}{3}$ **55.** $\frac{9\sqrt{2}+9}{2}$
57. $\frac{3\sqrt{35}-5\sqrt{21}}{-15}$ **59.** $\frac{12\sqrt{10}+3\sqrt{35}}{25}$
61. $\frac{30+6\sqrt{6}-2\sqrt{21}-5\sqrt{14}}{-38}$ **63.** $\frac{a-2\sqrt{ab}+b}{a-b}$

1. $5\sqrt{3}$, $3\sqrt{3}$ **3.** $3\sqrt{12}$, $5\sqrt{12}$ **5.** $-3\sqrt{3}$, $12\sqrt{3}$
7. $2\sqrt{10}$, $-6\sqrt{10}$, $7\sqrt{10}$ **9.** $2\sqrt{5}$, $3\sqrt{5}$, $-5\sqrt{5}$
11. $-3\sqrt{3}$, -5.20 **13.** $11\sqrt{6}$, 26.94   **15.** $11\sqrt{y}$
17. $3\sqrt{15}-2\sqrt{5}$, 7.15   **19.** $26\sqrt{13}$, 93.74
21. $40\sqrt{19}$, 174.36   **23.** $-5\sqrt{3a}$, $-8.66\sqrt{a}$
25. $11\sqrt{11}$, 36.48   **27.** $-4\sqrt{13}$, -14.42
29. $17\sqrt{7}$, 44.98   **31.** $9\sqrt{3}$, 15.59   **33.** $\sqrt{7}$, 2.65
35. $-10\sqrt{5}$, -22.36 **37.** $13\sqrt{3}+\sqrt{2}$, 23.93
39. $4\sqrt{6}+5\sqrt{7}-6\sqrt{2}$, 14.54   **41.** $\sqrt{3}+\sqrt{5}$, 3.97
43. $4\sqrt{3}$, 6.93   **45.** $-\sqrt{7}$, -2.65 **47.** $35\sqrt{5}$, 78.26
49. $-18\sqrt{7}$, -47.62 **51.** $8\sqrt{2}+6\sqrt{3}$, 21.71
53. $48\sqrt{2}+\sqrt{5}$, 70.12   **55.** $20\sqrt{3}$, 34.64
57. $\frac{8\sqrt{7}}{7}$, 3.02   **59.** $\frac{2\sqrt{3}}{3}$, 1.15   **61.** $4\sqrt{3}-3\sqrt{5}$, 0.22
63. $\frac{85\sqrt{7}}{14}$, 16.06

1. 9   **3.** none   **5.** 4   **7.** none   **9.** 25   **11.** 3   **13.** none
15. 25   **17.** none   **19.** $\frac{9}{2}$ **21.** 2   **23.** 3   **25.** 144
27. 3   **29.** 80   **31.** 45   **33.** $\pm\sqrt{22}$ **35.** 11   **37.** 7
39. 6   **41.** 169   **43.** 175   **45.** 78.4 m   **47.** 16
49. (9, 0)

1. true   **3.** false   **5.** 5   **7.** 13   **9.** 8   **11.** 8   **13.** 13
15. $3\sqrt{3}$ or 5.196   **17.** 9   **19.** 16   **21.** 21   **23.** 4

25. no   **27.** yes   **29.** $\sqrt{89}$ or 9.434 m   **31.** $4\sqrt{6}$ or
9.798 m

1. 3, 6, 4, 8   **3.** -4, 4, 2, 17   **5.** -3, 2, 5, 7
7. -8, -3, -4, 8   **9.** 3, -2, 7, -5 **11.** 2, 5, 2, -1
13. 17   **15.** $\sqrt{29}$ or 5.385   **17.** 13   **19.** $3\sqrt{5}$ or 6.708
21. $\sqrt{185}$ or 13.601   **23.** $\frac{5}{2}$ **25.** $\frac{\sqrt{85}}{3}$ **27.** $\frac{13}{10}$
29. 13, -9 **31.** 17, -13 **33.** 10, 0   **35.** -7, -3
37. -9, 5

1. 11   **3.** $\pm\frac{2}{9}$ **9.** rational   **11.** rational
13. 22 and 23   **15.** -20 and -19 **17.** ±48
19. $3\sqrt{2}$ **21.** $12\sqrt{5}$ **23.** $2|b|$ **25.** $2|a|\sqrt{11b}$ **27.** 15
29. $5\sqrt{2}-\sqrt{15}$ **31.** $\frac{2\sqrt{35}}{7}$ **33.** $\frac{27-9\sqrt{2}}{-33}$
35. $\frac{12\sqrt{42}+147}{-247}$ **37.** $5\sqrt{15}$, 19.365
39. $36\sqrt{3}$, 62.354   **41.** $\frac{2\sqrt{3}}{3}$, 1.155   **43.** $\frac{11}{3}$ **45.** 150
47. 3   **49.** $5\sqrt{5}$ or 11.180   **51.** 17 m   **53.** 17

CHAPTER 13   QUADRATICS

1. up   **3.** down   **5.** down   **7.** up   **9.** down   **11.** $x=2$
13. $x=-3$ **15.** $x=-2$ **17.** $x=-\frac{7}{10}$
19. $x=\frac{5}{2}$; $\left(\frac{5}{2}, \frac{49}{4}\right)$ **21.** $x=2$; (2, 9)
23. $x=\frac{3}{2}$; $\left(\frac{3}{2}, \frac{137}{4}\right)$ **25.** $x=-1$; $(-1, -20)$
27. $x=0$; (0, -9)   **29.** $x=-1$; (-1, 1)
31. $x=1$; (1, 22)   **33.** $y=0$; (0, 4)   **35.** $x=1$; (1, 42)
37. $x=2$; (2, -9)   **39.** $x=3$; (3, 14)
41. $x=0$; (0, -3)   **43.** $x=0$; (0, 3)
45. $x=\frac{1}{2}$; $\left(\frac{1}{2}, -12\frac{1}{4}\right)$ **47.** $x=-3$; (-3, 0)

1. -1, 1   **3.** no real roots   **5.** 0, 2   **15.** -3, 4
17. -1, 6   **19.** -2, 2   **21.** 3, 7   **23.** no real roots
25. $0<x<1$; $3<x<4$ **27.** 4
29. $-2<x<-1$; $0<x<1$ **31.** no real roots
33. $-1<x<0$; $2<x<3$ **35.** no real roots
37. $0<x<1$; $2<x<3$ **39.** $1<y<2$; $2<y<3$
41. no y-intercepts   **43.** 0, 6

1. no   **3.** yes   **5.** no   **7.** 16   **9.** 25   **11.** 1   **13.** $\frac{49}{4}$
15. $\frac{169}{4}$ **17.** -1, -7 **19.** $2\pm\sqrt{6}$ **21.** $4\pm2\sqrt{5}$
23. -3, -4 **25.** 4, 1   **27.** $2\pm\sqrt{3}$ **29.** $3\pm\sqrt{5}$
31. $\frac{5}{2}$ **33.** $\frac{5}{2}$, -4 **35.** $\frac{3}{2}$, -1 **37.** $\frac{2}{3}$, -1 **39.** 4, $-\frac{3}{2}$
41. $-0.125\pm\sqrt{0.516}$ **43.** $3\pm\sqrt{3}$ **45.** $4\pm\sqrt{29}$

47. $\frac{1 \pm \sqrt{5}}{4}$ **49.** $-2 \pm \sqrt{4 - c}$ **51.** $\frac{-b \pm \sqrt{b^2 - 4ac}}{2a}$
53. $y, -2y$

Page 447   Lesson 13-4

1. 1, 7, 6 **3.** 1, 4, 3 **5.** 4, 8, 3 **7.** 1, 0, -25 **9.** 2, 8, 0
11. 3, 5, 2 **13.** 49 **15.** 81 **17.** 36 **19.** 49
21. 2500 **23.** 256 **25.** $-1, -6$
27. $-2 \pm \sqrt{2}$; $-0.59, -3.41$ **29.** $\frac{5}{2}, -3$
31. $\frac{1 \pm \sqrt{33}}{4}$; 1.69, -1.19 **33.** $-\frac{1}{2}, -3$ **35.** ± 7
37. $\frac{5 \pm \sqrt{13}}{6}$; 1.43, 0.23 **39.** $\frac{1}{3}, -4$ **41.** $-1, -9$
43. $\frac{13 \pm \sqrt{41}}{2}$; 9.70, 3.30 **45.** 7, -5 **47.** 5, 0
49. 1, $-\frac{1}{3}$ **51.** $-\frac{2}{3}, -7$ **53.** $\frac{5}{6}, -\frac{3}{4}$ **55.** $\frac{3}{5}, -\frac{2}{7}$
57. 1.5, -0.4 **59.** $\frac{1 \pm \sqrt{17}}{4}$; 1.28, -0.78
61. $3 \pm 2\sqrt{2}$; 5.83, 0.17 **63.** $\frac{-3 \pm \sqrt{5}}{2}$; $-0.38, -2.62$
65. $\frac{-2 \pm \sqrt{5}}{2}$; 0.12, -2.12 **67.** $\frac{-2 \pm \sqrt{34}}{2}$; 1.92, -3.92
69. $-1 \pm \sqrt{3}$; 0.73, -2.73

Page 452   Lesson 13-5

1. 25; 2 real **3.** 49; 2 real **5.** 1; 2 real **7.** 1.44; 2 real
9. 48; 2 real **11.** 73; 2 real **13.** 0; 1 real **15.** 0;
1 real **17.** $2 \pm \sqrt{3}$ **19.** $-4 \pm \sqrt{21}$ **21.** 6, 1
23. $-4 \pm 2\sqrt{7}$ **25.** $-\frac{1}{2}, -1$ **27.** $\frac{-1 \pm \sqrt{13}}{6}$
29. $\frac{1 \pm \sqrt{61}}{6}$ **31.** 8 **33.** $\frac{1 \pm \sqrt{133}}{22}$ **35.** $\frac{1}{3}$
37. 1.28, -0.78 **39.** 7, 2 **41.** $-\frac{2}{3}, -2$
43. $-0.61, -2.72$ **45.** 0.65, -0.22 **47.** 1, $\frac{2}{3}$
49. 3.22, 0.78 **51.** 4.74, -0.74 **53.** none

Page 455   Lesson 13-6

1. factoring **3.** factoring **5.** formula **7.** formula
9. formula or completing the square **11.** 6 **13.** 10
15. 100 **17.** $-5 \pm 3\sqrt{3}$; 0.20, -10.20
19. $\frac{7 \pm \sqrt{85}}{6}$; 2.70, -0.37 **21.** $-6, -7$ **23.** $-\frac{2}{3}, 3$
25. $\frac{-1 \pm \sqrt{41}}{4}$; 1.35, -1.85 **27.** $\frac{5 \pm \sqrt{53}}{2}$; 6.14, -1.14
29. no real roots **31.** $\frac{5 \pm \sqrt{13}}{6}$; 1.43, 0.23
33. $\frac{15 \pm \sqrt{17}}{2}$; 9.56, 5.44 **35.** $2 \pm \sqrt{6}$; 4.45, -0.45
37. $\frac{3}{4}, \frac{1}{4}$ **39.** $-\frac{1}{3}, -\frac{2}{3}$ **41.** $\frac{1}{5}, \frac{2}{5}$ **43.** $\frac{3}{4}, -\frac{2}{5}$
45. $2 \pm \sqrt{14}$; 5.74, -1.74

Page 457   Lesson 13-7

1. 7 m, 8 m **3.** 10 yd, 13 yd **5.** 4 m, 14 m
7. 10 in., 20 in. **9.** 5 in., 9 in. **11.** 10 m, 16 m
13. 6, 8 **15.** -15, 9 or 15, 9 **17.** 12 cm, 36 cm
19. 3 cm, 6 cm **21.** 1.6 m, 2.4 m **23.** 5.92 m, 9.47 m
25. 1 m

Page 460   Lesson 13-8

1. 5, 6 **3.** $-12, -28$ **5.** $-5, -4$ **7.** $-1, -\frac{35}{4}$
9. $\frac{5}{6}, -\frac{7}{2}$ **11.** yes **13.** no **15.** no **17.** yes **19.** no
21. $x^2 - 7x + 10 = 0$ **23.** $x^2 - 11x + 24 = 0$
25. $x^2 + 5x - 6 = 0$ **27.** $x^2 - 14x + 44 = 0$
29. $4x^2 + 4x - 3 = 0$ **31.** $x^2 + 0.3x - 0.18 = 0$
33. 5, -24 **35.** $-12, -28$ **37.** $-\frac{2}{21}, -\frac{8}{21}$ **39.** $-1, \frac{1}{4}$
41. 10, 23 **43.** $-\frac{13}{2}, -\frac{9}{4}$ **45.** $\frac{1}{5}, -\frac{1}{300}$ **47.** $\frac{7}{24}, \frac{1}{48}$
49. $-\frac{\sqrt{2}}{2}, -3$ **51.** $x^2 - 11x + 28 = 0$
53. $x^2 + x - 56 = 0$ **55.** $x^2 - 5x - 24 = 0$
57. $x^2 + 19x + 34 = 0$ **59.** $2x^2 - 9x + 10 = 0$
61. $4x^2 - 29x - 24 = 0$ **63.** $x^2 - 2 = 0$
65. $x^2 - 4x + 1 = 0$ **67.** $x^2 - 8x + 6 = 0$
69. $2x^2 - 2x - 3 = 0$ **71.** $x^2 - (q + r)x + qr = 0$

Page 464   Chapter Review

1. $x = \frac{3}{2}; \left(\frac{3}{2}, -\frac{25}{4}\right)$ **3.** $x = -\frac{9}{4}; \left(\frac{9}{4}, -\frac{9}{8}\right)$ **5.** -3 **7.** 36
9. $\frac{25}{4}$ **11.** $-3 \pm \sqrt{5}$; $-0.76, -5.24$ **13.** $-\frac{3}{2}, -\frac{5}{2}$
15. $\frac{3}{2}, -5$ **17.** $\frac{2 \pm \sqrt{5}}{3}$; 1.41, -0.08
19. $\frac{-9 \pm \sqrt{21}}{10}$; $-0.44, -1.36$ **21.** no real roots
23. 2 real roots **25.** formula; $-5 \pm 4\sqrt{2}$
27. formula; $\frac{11}{9}, -1$ **29.** formula; no real roots
31. factoring; 2, 0.3 **33.** 6, 15 **35.** $\frac{3}{2}, \frac{11}{4}$
37. $2x^2 + 5x - 12 = 0$

CHAPTER 14   RATIONAL EXPRESSIONS AND APPLICATIONS

Page 469   Lesson 14-1

1. $\frac{1}{y - 4}$; $y \neq -4, 4$ **3.** r; $r \neq -3$
5. $\frac{x}{x^2 - 4}$; $x \neq 0, 2, -2$ **7.** $\frac{a + 4}{a - 2}$; $a \neq 2, 4$
9. $(t - 2)(t + 3)$; $t \neq -2, 3$ **11.** $\frac{-1}{w + 4}$; $w \neq \frac{2}{3}, -4$
13. $\frac{1}{y + 3}$; $y \neq -3, 3$ **15.** $\frac{a - 5}{a - 2}$; $a \neq -5, 2$
17. r^2; $r \neq 1$ **19.** $\frac{n^2 - 2}{n - 1}$; $n \neq 1$ **21.** $\frac{m^2}{2m - 1}$; $m \neq 0, \frac{1}{2}$
23. $\frac{1}{x + 3}$; $x \neq -3$ **25.** $\frac{g + 2}{g - 2}$; $g \neq 1, 2$
27. $\frac{m - 6}{m - 1}$; $m \neq -6, 1$ **29.** $\frac{2}{y + 5}$; $y \neq -5, 2$
31. $\frac{k - 1}{k + 1}$; $k \neq -1$ **33.** $\frac{a + 3}{a + 2}$; $a \neq -2, 3$
35. -1; $x \neq 3$ **37.** $\frac{4y - 1}{8y - 1}$; $y \neq -2, \frac{1}{8}$
39. $\frac{3}{4}$; $m \neq -2, -1$ **41.** $\frac{t + 3}{t - 4}$; $t \neq 4, \frac{7}{2}$
43. $\frac{1}{(b + 2)(b - 4)}$; $b \neq \pm 4, \pm 2$ **45.** $\frac{x + 4}{2x - 1}$; $x \neq \frac{1}{2}$
47. $\frac{x^2 + 1}{(x - 2)(x + 2)}$; $x \neq \pm 2, \pm 1$ **49.** $\frac{2s + 7}{s - 6}$; $s \neq \frac{2}{3}, 6$

51. $\frac{c-5}{c(c+6)}$; $c \neq 0, -4, -6$

53. $(a-1)(a+2)$; $a \neq -1, 2$ **55.** $\frac{2x+3}{2(x+2)}$; $x \neq 0, \frac{1}{2}, -2$

Page 473 Lesson 14-2

1. $\frac{5}{24}$ **3.** $-\frac{35}{48}$ **5.** $\frac{35}{72}$ **7.** $\frac{a^2}{15}$ **9.** $\frac{1}{3}$ **11.** $-\frac{1}{6}$ **13.** $-\frac{1}{6}$

15. $\frac{11}{28}$ **17.** $\frac{4}{27}$ **19.** 20 **21.** $\frac{4}{9}$ **23.** $\frac{512}{125}$ **25.** $\frac{a^2}{bd}$

27. $\frac{4n^2}{5x}$ **29.** $\frac{10ac}{3b^2}$ **31.** $\frac{4y^2z}{3x}$ **33.** 15 **35.** 6 **37.** $\frac{3}{2}$

39. $4a-4b$ **41.** $\frac{3m+9}{4}$ **43.** 7 **45.** $\frac{y-2}{y-1}$

47. $\frac{r^2+s^2}{r^2+2rs+s^2}$ **49.** $\frac{3k}{k-3}$ **51.** $\frac{-xy-x^2}{y}$ **53.** $\frac{9m^2}{n}$

55. $\frac{2}{x^2+3x}$ **57.** $\frac{1}{3}$ **59.** $\frac{z-3}{z-4}$ **61.** $\frac{1}{z+6}$ **63.** t

65. $\frac{2m^2+11m-21}{3m^2-11m+6}$ **67.** $\frac{4x^2+12x+9}{x^4}$

69. $\frac{6x^2+27x+27}{2x-3}$ **71.** $\frac{9x^2-42x+49}{16x^2}$

73. $\frac{-2(a+b)(y+1)}{y}$ **75.** $\frac{x^2+x-2}{x^2+3x}$

Page 476 Lesson 14-3

1. $\frac{4}{3}$ **3.** $-\frac{3}{8}$ **5.** $\frac{2}{m}$ **7.** $\frac{2p}{5}$ **9.** $\frac{1}{6}$ **11.** $\frac{1}{a^2}$ **13.** $\frac{1}{2bc}$

15. $\frac{x-y}{x+y}$ **17.** $\frac{5}{3a^2b^2}$ **19.** $\frac{3}{8} \cdot \frac{4}{1}$ **21.** $\frac{1}{3} \cdot \left(-\frac{1}{6}\right)$

23. $\frac{2x}{4-2a} \cdot \frac{b-2}{a^4}$ **25.** $\frac{5}{4}$ **27.** $-\frac{5}{2}$ **29.** $\frac{a^4}{b^4}$ **31.** $\frac{y^2}{a^2}$

33. a **35.** $\frac{3m}{m^2-m-2}$ **37.** $\frac{b+3}{4b}$ **39.** y **41.** $\frac{-p}{y+2}$

43. $-\frac{x}{7}$ **45.** $\frac{m^2-1}{2}$ **47.** $\frac{2m+2n}{3m^2-3mn}$ **49.** $\frac{y-4}{y+8}$

51. $\frac{3t+12}{2w-6}$ **53.** $\frac{x+3}{x}$ **55.** $\frac{x^2+x-12}{x^2-3x-18}$ **57.** -1

59. $\frac{x+5}{x+7}$ **61.** $\frac{3x^2+13x+4}{3x^2-13x+4}$ **63.** $\frac{4x^2+10x-24}{2x^2+11x-6}$

Page 480 Lesson 14-4

1. a **3.** $4m^2$ **5.** x^2 **7.** $x+4$ **9.** $a+5$

11. $c+3$ R9 **13.** $r-5$ **15.** $2x+3$

17. x^2+2x-3 **19.** t^2+4t-1 **21.** $3c-2+\frac{4}{9c-2}$

23. $3n^2-2n+3+\frac{3}{2n+3}$ **25.** $3s^2+2s-3+\frac{1}{s+2}$

27. $5t^2-3t-2$ **29.** $8x^2-9$ **31.** -20

Page 483 Lesson 14-5

1. $\frac{7}{8}$ **3.** $\frac{7}{a}$ **5.** $\frac{b+2}{x}$ **7.** $\frac{2k}{t}$ **9.** $\frac{1}{11}$ **11.** $\frac{a-b}{5}$ **13.** $-\frac{2}{x}$

15. $\frac{k}{m}$ **17.** y **19.** $\frac{x}{12}$ **21.** $\frac{9}{x}$ **23.** $-\frac{4}{x}$ **25.** $\frac{9a}{7}$

27. $y-3$ **29.** $-\frac{1}{6}$ **31.** $\frac{x+1}{x-1}$ **33.** $\frac{2}{y-2}$ **35.** 1 **37.** 0

39. 0 **41.** $\frac{r^2+s^2}{r-s}$ **43.** $m+n$ **45.** 4 **47.** 4

49. $a+b$ **51.** $r+3$ **53.** $\frac{x+1}{x-1}$ **55.** 0 **57.** $\frac{1}{x+1}$

59. $\frac{3}{m+1}$ **61.** $-\frac{1}{t+1}$ **63.** $p=2x-4$

Page 488 Lesson 14-6

1. 24 **3.** 55a **5.** ab **7.** ab^3 **9.** $120a^2b^2$ **11.** $120a^4n^2$

13. $(a+5)(a-3)$ **15.** $3(x-2)$ **17.** $\frac{13}{14}$ **19.** $-\frac{5}{24}$

21. $\frac{11}{20}$ **23.** $\frac{1}{36}$ **25.** $\frac{13t}{21}$ **27.** $\frac{8n-15m}{20}$ **29.** $\frac{6b+a}{5b}$

31. $\frac{13-x}{2a}$ **33.** $\frac{5z+6x}{xyz}$ **35.** $\frac{6x-5}{x^2}$ **37.** $\frac{7b-ax+a}{ab}$

39. $\frac{m^2-5m+5n}{m(m-n)}$ **41.** $\frac{14}{3x-9}$ **43.** $\frac{-3a-6}{a^2-5a}$

45. $\frac{y^2+12y+25}{y^2-25}$ **47.** $\frac{3a-1}{3a-6}$ **49.** $\frac{2x^3+5x^2-3x}{(2x-3)(2x+3)^2}$

51. $\frac{1}{y^2-2y+1}$ **53.** $\frac{3a^2+3ab-b^2}{(a-b)(2b+3a)}$ **55.** $\frac{2m^2+7m-1}{(m+1)(2m+5)}$

57. $\frac{3a+20}{a^2-25}$ **59.** $\frac{-x^3+x^2+3xy-3y}{9xy^2}$ or $\frac{(3y-x^2)(x-1)}{9xy^2}$

61. $\frac{x^2+6x-11}{x^2-2x-3}$ **63.** $\frac{7}{5-p}$ **65.** $\frac{a^3+7a^2+8a+12}{(a+2)^2(a-2)}$

67. $\frac{10a^2+11a+1}{(3a-2)(2a+1)}$ **69.** $\frac{7a^2+3a-12}{(3a+2)(a-4)^2}$

Page 492 Lesson 14-7

1. $\frac{4x+2}{x}$ **3.** $\frac{xy+x}{y}$ **5.** $\frac{2m^2+m+4}{m}$ **7.** $\frac{b^3-2b^2+2}{b-2}$

9. $\frac{14}{19}$ **11.** $\frac{y^3(x+4)}{x^2(y-2)}$ **13.** $\frac{a-b}{x-y}$ **15.** $\frac{x+y}{x-y}$ **17.** $\frac{x+1}{x-2}$

19. $\frac{1}{y+4}$ **21.** $\frac{a+2}{a+3}$ **23.** $\frac{(x+3)(x-1)}{(x-2)(x+4)}$ **25.** $\frac{8x^2-27y^2}{x^2-4y^2}$

27. $\frac{x+1}{x+5}$

Page 495 Lesson 14-8

1. 6 **3.** 8 **5.** r^2-1 **7.** $4x(x-1)$ **9.** $(k+5)(k+3)$

11. $(x+1)(x-1)$ **13.** -3 **15.** $\frac{5}{4}$ **17.** $\frac{5}{4}$ **19.** $\frac{41}{10}$

21. $\frac{1}{4}$ **23.** $-\frac{1}{4}$ **25.** $-\frac{3}{2}$ **27.** $\frac{1}{2}$ **29.** 6, -1 **31.** $-5, 3$

33. 20, 10 **35.** -3 **37.** 1 **39.** 5, 10 **41.** 7 **43.** 5

45. 3, 1 **47.** 6

Page 498 Lesson 14-9

1. $\frac{1}{8}, \frac{3}{8}, \frac{x}{8}$ **3.** $\frac{1}{8}, \frac{x}{8}, \frac{1}{10}, \frac{x}{10}, \frac{9}{40}, \frac{9x}{40}$

5. a. $120 + s$; $\frac{420}{120+s}$; 420; $120 - s$; $\frac{300}{120-s}$; 300

b. $\frac{420}{120+s} = \frac{300}{120-s}$ **7.** $3\frac{3}{7}$ hours **9.** 6 hours

11. 30 hours **13.** 9 hours **15.** 15 mph **17.** 25 mph

Page 502 Lesson 14-10

1. Multiply by 2, divide by n, then subtract t.

3. Multiply by abf, subtract af, then divide by $b - f$.

5. $t = \frac{v}{a}$ **7.** $v = \frac{2s-at^2}{2t}$ **9.** $M = \frac{Fd^2}{Gm}$ **11.** $V = \pm\sqrt{\frac{fgR}{W}}$

13. $P = \frac{A}{1+rt}$ **15.** $P = \frac{36,000}{IR+365}$ **17.** $y = \frac{r}{2a+0.5}$

19. $R = \frac{H}{0.24I^2t}$ **21.** $R_1 = \frac{R_T R_2}{R_2 - R_T}$ **23.** $n = \frac{IR}{E-Ir}$

25. $m = \frac{y-b}{x}$ **27.** $y_2 = mx_2 - mx_1 + y_1$

29. $R = P - DQ$ **31.** $n = 2b$ **33.** $n = \frac{acd-ab}{c}$

35. $n = \frac{ac+bc}{a+b+1}$ **37.** $n = \frac{b^2-2c}{c-b}$ **39.** $n = \pm\sqrt{ab}$

41. $n = \frac{rk-k^2}{2}$

Page 506 Lesson 14-11

1. $3\frac{3}{7}$ ohms **3.** 4 ohms **5.** 8 ohms, 4 ohms

7. 18 ohms **9.** $1\frac{1}{3}$ ohms **11.** $7\frac{1}{17}$ ohms

13. $2\frac{11}{12}$ ohms **15.** $7\frac{2}{3}$ ohms

1. $\frac{x}{4y^2z}$; $x \neq 0$, $y \neq 0$, $z \neq 0$ **3.** $\frac{x+3}{x(x-6)}$; $x \neq 0$, 6, -7

5. $\frac{3axy}{10}$ **7.** $\frac{7ab(x+9)}{3(x-5)}$ **9.** $x^2 + 4x - 2$ **11.** $\frac{1}{x-1}$ **13.** 2

15. $\frac{10axy - 3}{6x^2y}$ **17.** $\frac{8x-9}{x^2-4}$ **19.** $\frac{x^2-3x}{(x+5)^2}$

21. $\frac{(x-5)(x+13)}{(x+2)(x+6)}$ or $\frac{x^2+8x-65}{x^2+8x+12}$ **23.** $\frac{25}{7}$ **25.** $-\frac{23}{2}$

27. $21\frac{9}{11}$ hours   **29.** $a = \frac{bf}{b-f}$ **31.** 10 ohms, $2\frac{2}{5}$ ohms

CHAPTER 15   STATISTICS AND PROBABILITY

Page 514   Lesson 15-1

1. Burger King Whopper   **3.** Possible answers include Burger Chef Hamburger, Arthur Treacher's Fish Sandwich, McDonald's Fillet-O-Fish, or Long John Silver's Fish.   **5.** There appears to be a high correlation between calories and protein.

Page 516   Lesson 15-2

1. From 4 to 10 – intervals of 1   **3.** From 10 to 50 – intervals of 10   **7.** Blue catfish   **9.** Yes. Bluegill and White Perch; Channel Catfish and Carp   **13.** 91   **15.** 121

Page 518   Lesson 15-3

1. 45   **3.** 12   **5.** 1, 2, 3, 4, 5   **7.** 2, 3, 4, 5, 6, 7   **9.** 100   **11.** 2   **13.** 39   **15.** 34   **17.** teens   **19.** $190 – 199   **21.** 3   **23.** truncated

Page 522   Lesson 15-4

1. mean, 7.2; median, 6; no mode   **3.** mean, 9; median, 9; no mode   **5.** mean, 96.8; median, 50; no mode   **7.** 94; 82   **9.** 116.5; 119   **11.** $339   **13.** 26.385 s; 26.585 s   **15.** $138.8\overline{3}$; 137

Page 526   Lesson 15-5

1. median, 17; UQ, 20.5; LQ, 14   **3.** median, 40; UQ, 45; LQ, 34   **5.** median, 65; UQ, 85; LQ, 45   **7.** median, 87; UQ, 100.5; LQ, 79   **9.** median, 73; UQ, 77; LQ, 62   **11.** interquartile range, 25; outliers, 241 and 142   **13.** range, 296; interquartile range, 138.5   **15.** range, 918; interquartile range, 313

Page 529   Lesson 15-6

1. 25%   **3.** 50%   **5.** 95   **7.** 25%   **9.** 50   **11.** X   **13.** Y

Page 531   Lesson 15-7

1. Answers will vary.   **3.** negative   **5.** none   **7.** positive   **9.** positive   **11.** Yes. Some players have more opportunity to tackle because of the position played and mobility on the field.   **15.** Portland, ME

Page 536   Lesson 15-8

1. Answers will vary.   **3.** 1   **5.** Answers will vary.   **7.** $\frac{1}{6}$

9. 0   **11.** $\frac{5}{6}$ **13.** 1:1   **15.** 1:2   **17.** 4:3   **19.** $\frac{14}{23}$

21. $\frac{26}{207}$ **23.** female   **25.** $\frac{1}{13}$

Page 538   Lesson 15-9

1. yes   **3.** no   **5.** yes   **7.** $\frac{1}{36} = 0.028$ **9.** $\frac{2}{36} = 0.056$

11. $\frac{6}{36} = 0.167$ **13.** $\frac{1}{8} = 0.125$ **15.** $\frac{3}{8} = 0.375$

17. $\frac{4}{8} = 0.5$ **19.** Answers will vary.   **21.** Answers will vary.

Page 540   Lesson 15-10

1. $\frac{1}{2} = 0.5$ **3.** $\frac{1}{4} = 0.25$ **5.** $\frac{1}{4} = 0.25$ **7.** $\frac{3}{8} = 0.375$

9. $\frac{2}{7} = 0.286$ **11.** 0   **13.** $\frac{1}{14} = 0.071$ **15.** $\frac{1}{3} = 0.\overline{3}$

17. $\frac{1}{24} = 0.042$ **19.** $\frac{1}{8} = 0.125$

Page 544   Chapter Review

1. Driver's License Fee   **3.** Gasoline   **5.** mean, 145; median, 144.5; no mode   **7.** median, 257; UQ, 244.5; LQ, 261.5; interquartile range, 17   **9.** 65   **11.** 50% **13.** $\frac{9}{13} = 0.692$ **15.** $\frac{1}{4} = 0.25$

CHAPTER 16   TRIGONOMETRY

Page 549   Lesson 16-1

1. 5°   **3.** 77°   **5.** 35°   **7.** 79°   **9.** $(90 - x)°$ **11.** $(50 - 2x)°$ **13.** 50°   **15.** 53°   **17.** 90°   **19.** 72°   **21.** $(180 - y)°$ **23.** $(175 - 3x)°$ **25.** 48°; 138°   **27.** none; 55°   **29.** 0°; 90°   **31.** 69°; 159°   **33.** none; 81°   **35.** $(90 - 3y)°$; $(180 - 3y)°$ **37.** $(128 - x)°$; $(218 - x)°$ **39.** $x°$; $(90 + x)°$ **41.** 122°   **43.** 85°   **45.** 1°   **47.** 19°   **49.** $(180 - x - y)°$ **51.** $(190 - 2y)°$ **53.** 37°; 106°   **55.** 26°; 64°   **57.** 25°; 35°; 120°   **59.** 50°   **61.** 30°

Page 552   Lesson 16-2

1. 4 m   **3.** 6.5 mm   **5.** $2\frac{1}{4}$ in.   **7.** 8.18 m   **9.** 14 m   **11.** 8.70 mm   **13.** $12\frac{3}{4}$ in.   **15.** 7.72 cm   **17.** 4 ft, 8 ft   **19.** 8 m, 16 m   **21.** 7 yd, 14 yd   **23.** $4\sqrt{3}$ or 6.928 cm   **25.** 7 units   **27.** $\frac{1}{2}$ m   **29.** 3 m, $3\sqrt{3}$ or 5.196 m   **31.** $1\frac{3}{4}$ in., $\frac{7}{4}\sqrt{3}$ or 3.031 in.   **33.** 13 m, $6.5\sqrt{3}$ or 11.258 m   **35.** 4 m, 2 m   **37.** $\sqrt{37}$ m   **39.** $\frac{60 + 20\sqrt{3}}{3}$ yd; 31.547 yd

1. $\angle A$ and $\angle D$; $\angle B$ and $\angle E$; $\angle C$ and $\angle F$; AC and DF; AB and DE; BC and EF
3. $\angle Q$ and $\angle Q$; $\angle S$ and $\angle P$; $\angle T$ and $\angle R$; QS and QP; QT and QR; ST and PR **5.** $b = \frac{25}{7}$, $c = \frac{30}{7}$
7. $a \approx 2.78$, $c \approx 4.23$ **9.** $d = 10.2$, $e = 9$
11. $c = \frac{7}{2}$, $d = \frac{17}{8}$ **13.** $\frac{125}{9}$ ft; 13 ft 11 in.   **15.** 11 cm

1. $\frac{8}{17}$ **3.** $\frac{8}{15}$ **5.** $\frac{15}{17}$ **7.** $\frac{9}{41}$ **9.** $\frac{9}{40}$ **11.** $\frac{40}{41}$ **13.** $\frac{4}{5}$ **15.** $\frac{4}{3}$
17. $\frac{3}{5}$ **19.** yes; yes   **21.** They are equal.
23. $\sin D \approx 0.969$; $\sin E \approx 0.246$; $\cos D \approx 0.246$; $\cos E \approx 0.969$; $\tan D \approx 3.938$; $\tan E \approx 0.254$
25. $\sin P \approx 0.882$; $\sin Q \approx 0.471$; $\cos P \approx 0.471$; $\cos Q \approx 0.882$; $\tan P \approx 1.875$; $\tan Q \approx 0.533$
27. $\sin A \approx 0.753$; $\sin B \approx 0.658$; $\cos A \approx 0.658$; $\cos B \approx 0.753$; $\tan A \approx 1.146$; $\tan B \approx 0.873$
29. $\sin X \approx 0.709$; $\sin Z \approx 0.709$; $\cos X \approx 0.709$; $\cos Z \approx 0.709$; $\tan X = 1.000$; $\tan Z = 1.000$

1. $90°$; $0°$ **3.** $45°$ **5.** no maximum or minimum values
7. 0.6009   **9.** 0.2493   **11.** 0.3746   **13.** 0.4452
15. 0.1584   **17.** 0.9511   **19.** 1.1918   **21.** 0.8660
23. 0.5774   **25.** 0.500   **27.** $61°$ **29.** $12°$ **31.** $74°$
33. $60°$ **35.** $68°$ **37.** $80°$ **39.** $3°$ **41.** $3°$ **43.** $30°$
45. 1   **47.** 1   **49.** 1   **51.** 0

1. $\sin 46° = \frac{b}{15}$; $\cos 46° = \frac{a}{15}$
3. $\sin A = \frac{5}{13}$; $\cos B = \frac{5}{13}$
5. $\angle B = 69°$; $AB = 13.9$ in.; $BC = 5$ in.
7. $\angle B = 20°$; $AB = 9.6$ cm; $AC = 3.3$ cm
9. $\angle A = 38°$; $AC = 18.1$ cm; $BC = 14.2$ cm
11. $\angle B = 50°$; $AC = 12.3$ m; $BC = 10.3$ m
13. $\angle B = 59°$; $AC = 10$ m; $AB = 11.6$ m
15. $\angle A = 48°$; $BC = 7.4$ in.; $AC = 6.7$ in.
17. $\angle A = 38°$; $\angle B = 52°$; $BC = 8.7$ ft
19. $\angle A = 50°$; $BC = 7.2$ cm; $AB = 9.3$ cm
21. $\angle B = 45°$; $AC = 1$ ft; $BC = 1$ ft

1. $\angle CAB$; $\angle DCA$ **3.** $\angle JKL$; $\angle MJK$ **5.** $4°$ **7.** 65 ft
9. 300 ft   **11.** 65 ft   **13.** 1716 cm   **15.** 6143 ft; 5808 ft
17. $3.4°$ **19.** $\angle B = 105°$, $BC = 254.6$ m, $AC = 491.8$ m

1. $24°$; $114°$ **3.** no complement; $32°$ **5.** $92°$ **7.** $53°$
9. 4 cm; $4\sqrt{3}$ or 6.928 cm   **11.** 7 in.; $3.5\sqrt{3}$ or 6.062 in.
13. $b = \frac{70}{11}$; $c = \frac{30}{11}$ **15.** $d = \frac{48}{5}$; $e = \frac{36}{5}$ **17.** $\frac{28}{53}$ **19.** $\frac{45}{53}$
21. $\frac{28}{45}$ **23.** 0.4848   **25.** 0.2309
27. $\angle B = 28°$; $AC \approx 3.7$ m; $AB \approx 7.9$ m
29. $\angle A = 54°$; $BC \approx 6.5$ yd; $AC \approx 4.7$ yd
31. 12.26 cm; 10.28 cm

Index

A

Abscissas, 289
ABS function, 589
Absolute value, 46, 73, 152, 310-311, 378
  BASIC, 589
Addition, 4, 36
  identity, 12, 20, 37
  on number lines, 44, 49, 73
  with integers, 44-47, 73, 284
  with polynomials, 194, 224
  with radical expressions, 411-412, 426
  with rational numbers, 49-50, 53, 73, 482, 486, 507
  with square roots, 411-412, 426
Algebraic fractions, 467-468, 471-477, 482-483, 486-493, 508
  complex, 490-492, 508
  excluded values, 467-468
  in equations, 494-495, 508
  mixed expressions, 490
  simplifying, 282, 284, 467-468, 508
  undefined, 467
  with polynomials, 468
Angles, 547-548
  complementary, 197, 547, 572
  corresponding, 554, 572
  of depression, 567-568
  of elevation, 567, 570
  of triangles, 197, 548, 572
  radians, 571
  supplementary, 197, 547, 572
Application
  acceleration, 92
  airplane flying time, 345
  angle of elevation, 570
  area, 234, 309
  break-even analysis, 361
  checking accounts, 219
  compound interest, 269
  consumer price index, 168
  escape velocity, 416
  F-stops, 500
  hexadecimals, 29
  maximum profit, 435
  tolerance, 148
  weighted mean, 523
  wind chill factor, 66
Area, 234, 309
  of circles, 24
  of polygons, 212-213
  of rectangles, 23, 278-279
  of trapezoids, 24
  of triangles, 23
  surface, 24
Axis of symmetry, 431-437, 463
  equations of, 432-437, 463

B

Bases, 4, 36, 110-111, 120, 157, 198
Base two, 258
BASIC, 576-590
  ABS, 589
  assignment statements, 578-579
  conditional branches, 151
  COS, 571
  DATA statements, 578-580
  DIMENSION statements, 588
  E notation, 184, 588
  END statements, 577
  equals sign, 579
  FOR-NEXT statements, 319, 585-587
  GO TO statements, 586
  IF-THEN statements, 151, 582-583, 586
  independent loops, 586
  input statements, 577
  INT, 589-590
  LET statements, 579-580
  line numbers, 577
  loops, 319, 586
  nested loops, 586
  order of operations, 576
  OUT OF DATA, 580
  output, 577
  parentheses, 576
  powers, 184
  PRINT statements, 577
  READ statements, 578-580
  RUN, 578
  SIN, 571
  SQR, 424, 589
  symbols in, 576, 582
  TAN, 571
Binary system, 258
Binomials, 191, 224
  conjugates, 410, 427
  factoring, 232, 235-236, 246, 256, 259-260
  multiplying, 201-202, 224
  squares of, 205-206, 224
Bits, 258
Boundaries, 312, 321
Boyle's law, 181
Bytes, 258

C

Calculators
  change-sign key, 64-65
  checking solutions, 64-65, 255, 368
  comparing numbers, 144
  decimal coefficients, 461
  evaluating sums and differences, 489
  factoring, 268
  finding factors, 481
  irrational roots, 443-444, 446
  order of operations, 7
  percent, 110, 112
  powers of numbers, 5
  proportions, 106
  recall keys, 208
  reciprocal key, 85
  scientific notation, 166
  simplifying expressions, 17
  solving equations, 64-65
  square root keys, 402
  store keys, 208
  trigonometric functions, 562
  verifying answers, 412
  writing equations, 343
Capacity, 109
Celsius degrees, 72, 332
Chapter reviews, 38, 74-75, 120-121, 153, 188, 225, 260-261, 285, 321-322, 355, 391-392, 426-427, 464, 508-509, 544-545, 573
Chapter summaries, 36-37, 73-74, 119-120, 152, 187, 224, 259-260, 284, 320-321, 354, 391, 425-426, 463, 508, 543-544, 572
Chapter tests, 39, 76, 121, 154, 189, 226, 261, 286, 323, 356, 393, 428, 465, 510, 545, 574
Circles
  area of, 24
  center, 423
  equations of, 423

Circuits
  parallel, 504-505
  series, 504-505
Closure, 41
Coefficients, 16, 37
Collinear points, 353
Completing the square, 442-444,
  454, 463
Complex fractions, 490-492, 508
  simplifying, 491-492, 508
Compound interest, 223, 269
  program for, 223
Compound sentences, 140-142,
  145-146, 152
  absolute values, 145-146, 152
Compound statements, 127
  conjunction, 127
  disjunction, 127
Computers
  adding fractions, 507
  assignment statements, 578-579
  bar graphs, 542
  BASIC, 576-590
  bits, 258
  bytes, 258
  conditional branches, 151
  DATA statements, 578-580
  DIMENSION statements, 588
  END statements, 577
  E notation, 186, 588
  formulas, 35
  FOR-NEXT statements, 319,
  585-587
  GO TO statements, 586
  hexadecimal numbers, 29
  IF-THEN statements, 151,
  582-583, 586
  independent loops, 586
  input, 577
  LET statements, 579-580
  line numbers, 577
  loops, 319, 586
  nested loops, 586
  order of operations, 576
  output, 577
  powers, 184
  programs, 577-590
  READ statements, 578-580
  square roots, 424
  successive discounts, 118-119
  sums of integers, 284
  symbols, 576, 582
  temperature, 72
  trigonometric functions, 571
Conjugates, 408, 425
Conjunctions, 127

Consistent equations, 362
Constants, 157, 185
  degree of, 191, 224
  of variations, 176
Coordinate planes, 289-290
Coordinates, 42
Cosine, 557-558, 571-572
  in BASIC, 571
Cramer's rule, 388-389

D

DATA statements, 578-580
Decimals, 398
Degree, 191-192
  of constants, 191, 224
  of monomials, 191, 224
  of polynomials, 192, 224
Delta, 563
Denominators
  least common, 486-488, 491,
  494-495, 508
  like, 482-483, 508
  unlike, 486-488, 508
  zero in, 467, 508
Dependent equations, 362
Depression, 567-568
Determinants, 388-390
Diagrams, 104
Difference of squares, 207, 224
  factoring, 235-236, 256, 259-260
DIMENSION statements, 588
Direct variations, 178-179, 187
  constant of, 178
  linear equations, 178-179, 187
  proportions, 179
Discounts, 114-116
  successive, 118
Discriminant, 450-452, 463
Disjunctions, 127
Distance formula, 421-422, 426
Distributive properties, 15, 20, 37
  in factoring, 232, 259
Divide-and-average method, 403
Dividends, 3, 36
Divisibility rules, 233
Division, 3-4
  property of equality, 87, 119
  property for inequalities, 132, 152
  quotients, 3, 36
  rule, 84, 119
  with monomials, 162-163

with polynomials, 479-480
with powers, 162-163, 187
with radical expressions, 396,
  407-408, 425
with rational expressions, 475-476,
  508
with rational numbers, 83-84, 119,
  475, 508
with square roots, 396, 407-408,
  425
Divisors, 3, 36
Domains, 292-293, 296-298, 320

E

Electricity, 504-505
Elements, 11
Elevation, 567, 570
Elimination, 369, 372-373
Empty set, 11
END statements, 577
E notation, 184, 588
Equalities, 12-13, 37
Equal sign, 9, 23, 37
Equations, 9, 37
  in BASIC, 579
  in problem solving, 100-101
  in two variables, 296-298, 301-302,
  320
  roots, 438, 451, 453, 463
  solving, 60-65, 74, 86-98, 210,
  263-264, 270-274, 284, 359-370,
  372-376, 388-391, 415-416, 426,
  436-444, 454-455, 462-463,
  494-495, 508
  solving using factoring, 270-274,
  284
  using calculators, 64-65, 343
  writing, 30
Equivalent equations, 60
Eratosthenes, 231
Excursions
  capacity in the metric system, 109
  checking solutions, 255
  comparing numbers, 144
  Cramer's rule, 388-389
  distributive property, 99
  divide-and-average, 403
  divisibility rules, 233
  does 2 = 1?, 478
  equations of circles, 423
  factoring, 268
  factoring differences of cubes, 246
  formulas, 503
  grouping three terms, 249

history, 34, 200
imaginary roots, 453
inequalities, 281
length, 14
magic squares, 71
mass in the metric system, 238
misleading graphs, 315
parabolic surfaces, 441
radical sign, 397
reciprocals, 85
rise over run, 337
sets, 11
sieve of Eratosthenes, 231
slopes and grades, 349
squaring numbers that end in five,
218
systems with three variables, 376
Exponents, 4, 36, 157, 198
negative, 163, 185
zero, 162-163, 185
Expressions, 3-7
algebraic, 3-7, 36
binomial, 408
equivalent, 406
evaluating, 6-7
mixed, 490
numerical, 3, 6
radical, 395-396, 404-409, 411-412
rational, 467-468, 471-484,
486-496
simplest form, 16
verbal, 3-5
Extremes, 105, 120

F

Factoring, 229-237, 239-240,
243-249, 251-253, 256-257,
259-260, 266
binomials, 232, 235-236, 246, 256,
259-260, 266
by grouping, 247-248, 260, 266
differences of cubes, 246
differences of squares, 235-236,
259-260
greatest common factor, 230-232,
259-260, 266
grouping three terms, 249
in simplifying algebraic fractions,
282, 284, 467-468, 508
in solving equations, 270-274
integers, 229-230
monomials, 230
perfect squares, 239-240, 259-260,
266
prime factorization, 229-230, 259
prime polynomials, 245, 259

trinomials, 232, 239-240, 243-245,
251-253, 256, 259-260
using the distributive property,
232, 259, 266
Factors, 3, 36, 229
finding, 481
greatest common, 230, 259
prime, 229, 259
program for, 590
Fahrenheit degrees, 72, 332
Focal length, 501
Focus, 441
FOIL method, 201-202, 224
Formulas, 23-24, 37
acceleration, 92
area, 309
area of circles, 24
area of rectangles, 23, 278-279
area of trapezoids, 24
area of triangles, 23
Boyle's law, 181
compound interest, 269
distance, 421, 426
escape velocity, 414
focal length, 501
for falling objects, 394
F-stops, 500
midpoints, 350-351, 354
Pythagorean Theorem, 418, 426
quadratic, 445, 463
simple interest, 216
slope, 326, 354
surface area of rectangular solids,
24
uniform motion, 220, 224, 497
using, 501
using computers, 35
velocity, 278
FOR-NEXT statements, 319,
585-587
Fractional equations, 494-495
Fraction bars, 3, 7
Fractions
adding, 482, 486-488, 507-508
algebraic, see algebraic fractions
dividing, 475-476, 508
improper, 490
least common denominator,
486-488, 508
mixed numerals, 490
multiplying, 471-472, 508
subtracting, 482-483, 487, 508
with like denominators, 482, 508
with unlike denominators,
486-488, 508
Fulcrums, 184

Functional values, 307
Functions, 305-308, 320-321
absolute value, 46, 73, 152,
310-311, 378
functional values, 307
greatest integer, 589-590
quadratic, 433-439, 463
trigonometric, 557-558, 561-562,
564-572

G

GO TO statements, 586
Grams, 109, 238
Graphing calculators
absolute value function, 310-311
break-even analysis, 361
regression lines, 533
Graphs
area, 309
axis of symmetry, 431-437, 463
bar, 542
boundaries, 312, 321
box plots, 528-529
coordinate planes, 289-290
estimation, 332
for solving quadratic equations,
438-439, 454, 463
half-planes, 312-313, 321
inequalities in two variables,
312-313, 321
misleading, 315
of inequalities, 125, 141-142, 146
of linear equations, 301-302,
338-339
of ordered pairs, 290
of sine function, 571
of systems of equations, 359-362
of number lines, 41-42, 73,
124-125, 141-142
parabolas, 431-437, 463
quadrants, 290, 320
reading, 300
regression lines, 533
relations, 292-293, 301-302
scatter plot, 530-531
systems of inequalities, 377-378
Greatest common factors of integers,
230, 259
of monomials, 230
Greek letters, 563
Grouping symbols, 6-7, 37
brackets, 7, 37
fraction bars, 3, 7

in inequalities, 134-135
in solving equations, 94
parentheses, 7, 22, 37

H

Half-plane, 312-313, 321
  boundaries, 312, 321
  closed, 312, 321
  open, 312, 321
Hamilton, Sir William Rowan, 34
Hexadecimal numbers, 29
History
  Hamilton, 34
  Noether, 200
  Pythagoras, 418
  radical sign, 397
  Rudolff, 397
Hypotenuse, 418

I

Identities, 94, 119
  additive, 12, 20, 37
  multiplicative, 12, 20, 37
IF-THEN statements, 151,
  582-583, 586
Imaginary unit, 453
Inconsistent equations, 362
Independent equations, 362
Inequalities, 123-152
  absolute values, 145-146, 152
  addition property for, 128, 152
  comparing rational numbers,
  136-137, 152
  compound sentences, 140-143,
  145-146, 152
  division property for, 132, 152
  graphs of, 125, 141-142, 146,
  312-313, 321, 377-378
  in two variables, 312-313, 321
  multiplication property for, 131,
  152
  problem solving, 149
  solving, 128-135, 281
  solving using factoring, 281
  subtraction property for, 128, 152
  symbols of, 123, 152
  systems of, 377-378, 391
  tolerance interval, 148
  with grouping symbols, 134-135
Input, 577
Integers, 41-42, 73, 398
  adding, 44-47, 73, 284
  consecutive even, 101

consecutive odd, 101
even, 284
factoring, 229-230
greatest, 589-590
greatest common factors, 230, 259
problems, 275-276
Interest
  compound, 223, 269
  principal, 216
  rate, 216
  simple, 216
INT function, 589-590
Inverses
  additive, 55, 73, 195
  multiplicative, 84, 119, 475
Inverse variations, 181-182, 187
  product rule for, 182
  proportions, 182
Irrational numbers, 398-400, 425
Isosceles right triangles, 548

K

Kilograms, 109, 238

L

Least common denominators,
  486-488, 508
Least common multiples, 486-487
Legs, 420
Length, 14
LET statements, 579-580
Lever problems, 184-185, 187
Like terms, 15-16, 37
Linear equations, 292-293, 338-339
  direct variations, 178-179
  graphs of, 301-302, 338-339
  point-slope form, 329, 354
  slope-intercept form, 333-334, 339,
  354
  standard form, 330, 339, 354
  writing, 342-343
Lines
  horizontal, 326, 330, 338
  parallel, 346, 354, 362, 366
  perpendicular, 347-348, 354
  point-slope form, 329, 354
  slope-intercept form, 333-334, 339,
  354
  slope of, 325-331, 333-340,
  342-343, 346-349, 354
  standard form, 330, 339, 354
  vertical, 326, 330, 338

Line segments, 350-351, 354
Liters, 109
Loops, 319
  independent, 586
  nested, 586

M

Magic squares, 71
Mappings, 292-293
Mass, 109, 238
Maximum points, 432, 434-435, 463
Mean, 105, 120
  statistical, 520
  weighted, 523
Means-extremes property of
  proportions, 105-106, 120
Meters, 14
Metric system
  capacity in, 109
  centimeters, 14
  force, 414
  grams, 109, 238
  kilograms, 109, 238
  length, 14
  liters, 109
  mass, 109, 238
  meters, 14
  metric tons, 238
  milliliters, 109
  milligrams, 238
  Newtons, 414
  prefixes of, 14
Metric tons, 238
Midpoints, 350-351, 354
Milligrams, 238
Milliliters, 109
Minimum points, 431-435, 463
Mini-reviews, 13, 21, 34, 48, 58, 71,
  82, 89, 99, 108, 130, 144, 167, 183,
  193, 215, 237, 254, 268, 280, 299,
  315, 337, 349, 364, 380, 401, 417,
  441, 458, 478, 493, 503, 519, 532,
  541, 553, 566
Mixture problems, 171-176, 381, 383
Monomials, 157, 160-162, 187, 191,
  224
  constant, 157, 187
  degree of, 191, 224
  dividing, 162-163
  factoring, 230
  greatest common factors, 230
  multiplying, 157-158, 160-161
  multiplying by polynomials, 199

powers of, 160-161
Motion
  rate, 220
  uniform, 220-221, 497-498
Multiples, 486-487
Multiplication, 3-5, 36
  associative property for, 19-20, 37
  by zero, 12, 20, 37
  commutative property for, 19-20, 37
  factors, 3, 36
  FOIL method, 201-202, 224
  identity, 12, 20, 37
  of polynomials by monomials, 199
  products, 3, 36
  property of equality, 86-87, 119
  property for inequalities, 131, 152
  with binomials, 201-202, 224
  with monomials, 157-158, 160-161
  with polynomials, 199-207, 224
  with powers, 157-158, 160-161, 187
  with rational expressions, 396, 425, 471-472, 508
  with rational numbers, 79-80, 119, 471, 508
  with square roots, 396, 425
Multiplicative inverses, 84, 119, 475
Multiplicative property of negative one, 80, 119
Multiplicative property of zero, 12, 20, 37

N

Natural numbers, 398
Negative exponents, 163, 185
Negative numbers, 41
  negative one, 80, 119
Negative square roots, 395
Nested loops, 586
Nested polynomials, 481
Newtons, 414
Noether, Amalie Emmy, 200
Notations
  E, 184, 588
  scientific, 165-166, 187
Null set, 11
Number lines, 124-125
  adding on, 44, 49, 73
  comparing numbers on, 125, 152
  completeness property, 399
  graphs on, 41-42, 73, 124-125, 141-142

  midpoints, 350-351, 354
  real numbers, 399, 425
Numbers
  composite, 229, 259
  comparing on number lines, 125, 152
  consecutive, 101
  counting, 398
  even, 284
  hexadecimal, 29
  integers, 41-42, 73, 398
  irrational, 398-400, 425
  line, 577
  natural, 398
  negative, 41
  prime, 229, 259
  rational, 49-51, 53-57, 73, 79-84, 119, 124-127, 398, 471, 475, 482, 486, 490, 507-508
  real, 399, 425
  whole, 41, 398
Numerals
  decimals, 398
  mixed, 490
Numberical coefficients, 16, 37
Numerical expressions, 3, 6

O

Odds, 535
Ohms, 504
Open half-planes, 312, 321
Open sentences, 9, 37
  solving, 9-10, 37
Operations
  with integers, 44-46, 73
  with monomials, 157-158, 160-163, 185
  with polynomials, 194-196, 199-207, 224, 479-480
  with powers, 157-158, 160-163, 185
  with radical expressions, 396, 407-408, 411-412, 425-426
  with rational expressions, 471-476, 482-483, 486-488, 507-508
  with rational numbers, 49-50, 53-57, 73-74, 79-84, 119, 471, 475, 482, 486, 507-508
  with square roots, 396, 407-408, 411-412, 425-426
Opposites, 55, 73
Ordered pairs, 289-298, 320
  abscissas, 289

  as solutions to equations, 296-298
  domain, 292-293, 296-298, 320
  graphs of, 290
  ordinates, 289
  relations, 292-298, 320
  ranges, 292-293, 296, 320
  x-coordinates, 289, 320
  y-coordinates, 289, 320
Order of operations, 6, 36
  grouping symbols, 6-7, 37
  in BASIC, 576
  using calculators, 7
Ordinates, 289
Origins, 289
OUT OF DATA, 580
Output, 577

P

Parabolas, 433-439, 463
  axis of symmetry, 433-439, 463
  focus, 441
  maximum points, 432, 434-435, 463
  minimum points, 433, 463
Parabolic surfaces, 441
Paraboloid of revolution, 441
Parallel circuits, 504-505
Parentheses, 7, 22, 37, 406
  in BASIC, 576
Percentages, 110-111, 120
Percent keys, 112
Percents, 110-120
  as proportions, 110-111, 120
  base, 110-111, 120
  consumer price index, 168
  discounts, 114-116, 118
  of decrease, 114
  of increase, 114
  percentages, 110-111, 120
  rates, 110-111, 120
  sales taxes, 115
  simple interest, 216
  successive discounts, 118
Perfect squares, 239
Perfect square trinomials, 239-240, 259
Perimeters, 212
Pi, 563
Planes
  completeness property for points in, 290, 320
  coordinate, 289-290

half, 312-313, 321

Points
  collinear, 353
  maximum, 432, 434-435, 463
  minimum, 433, 463

Polygons
  area of, 23-24, 212-213, 278-279
  perimeters of, 212

Polynomials, 191-208, 224
  adding, 194, 224
  ascending order, 192, 224
  binomials, 191, 201-207, 224, 232, 235-236, 246, 256, 259-260, 408, 425
  degrees of, 191-192, 224
  descending order, 192, 224
  dividing, 479-480
  FOIL method, 201-202, 224
  finding factors, 481
  in rational expressions, 468
  monomials, 157-158, 160-163, 187, 191, 199, 224, 230
  multiplying, 199-207, 224
  multiplying by monomials, 157-158, 160-161
  nested, 481
  prime, 245, 259
  squares of, 205-206, 224
  subtracting, 194-196, 224
  trinomials, 191, 224, 232, 239-240, 243-245, 251-253, 256

Powers, 4-5, 157-166, 185
  bases, 157, 198
  dividing, 162-163, 185
  E notation, 184, 588
  exponents, 157,198
  in BASIC, 184
  multiplying, 157-158, 160-161, 187
  reading, 198
  scientific notation, 165-166, 185
  using calculators, 5
  with negative exponents, 163, 185
  with zero exponents, 162-163, 185

Prime factorization, 229, 259, 486

Probability, 534-535
  compound event, 539-540
  empirical, 537
  equally likely events, 537
  experimental, 537
  simple event, 539

Problem solving, 26, 66-67, 212-213
  age problems, 169-170
  area problems, 212-213, 279
  consecutive number problems, 101
  current problems, 382, 496
  digit problems, 385-386
  inequalities, 149
  integer problems, 275-276
  lever problems, 184-185
  mixture problems, 171-176, 381, 383
  percent problems, 114-116
  perimeter problems, 212
  quadratic equations, 456
  resistance problems, 504-505
  simple interest problems, 216
  uniform motion problems, 220-221, 497-498
  using equations, 100-101
  using systems of equations, 382-383
  using trigonometry, 567-568
  velocity problems, 278
  work problems, 497
  writing equations, 30-31

Problem solving strategies
  diagrams, 104
  graphs and estimation, 332
  guess and check, 52
  identifying subgoals, 470
  list possibilities, 449
  look for a pattern, 159
  using charts, 383
  using tables, 265

Product rule for inverse variations, 182

Products, 3, 36
  of a sum and a difference, 207, 224

Product property of square roots, 396-425

Programs, 577-590
  for absolute value, 589
  for adding fractions, 507
  for comparisons, 151
  for compound interest, 223
  for determining positive and negative numbers, 583
  for factors, 590
  for finding square roots, 424
  for finding zeros, 462
  for graphing quadratic functions, 462
  for ordered pairs, 586-587
  for printing integers, 585
  for sine function, 571
  for solving systems of equations, 390
  for squares and square roots, 426, 589
  for sums of even integers, 284
  for sums of two numbers, 577
  for testing collinear points, 353
  loops, 319, 586

Properties
  additive identity, 12, 20, 37
  additive inverse, 55, 73, 195
  associative, 19-20, 37
  commutative, 19-20, 37
  comparison, 123, 136, 152
  completeness, 290, 320, 398
  density, 137
  distributive, 15, 20, 37, 232, 259
  for inequalities, 128, 131-132, 152
  means-extremes, 105-106, 120
  multiplication by zero, 12, 20, 37
  multiplicative identity, 12, 20, 37
  multiplicative inverse, 84, 119
  of equality, 12-13, 37, 60-63, 73-74, 86-87, 119
  of levers, 184-185, 187
  of negative one, 80, 119
  of square roots, 396, 425
  reflective, 12, 37, 123-124
  substitution, 13, 20, 37
  symmetric, 12, 37, 123-124
  transitive, 12, 37, 124, 152
  zero product, 263, 284

Proportions, 105-107, 120
  direct variations, 178-179, 187
  extremes, 105, 120
  inverse variations, 181-182, 187
  means, 105, 120
  percents, 110-111, 120
  similar triangles, 554-555
  using calculators, 106

Protractors, 547

Pythagoras, 418

Pythagorean Theorem, 418-420, 426
  distance formula, 421-422, 426

Quadrants, 290, 320

Quadratic equations, 438-440, 442-446, 450-463
  completing the square, 442-444, 454, 463
  complex roots, 453
  finding, 459-460, 463
  in problem solving, 456
  nature of roots, 451-452, 463
  quadratic formula, 445, 463
  roots, 438, 451-453, 459-463
  solving, 438-439, 442-446, 453-455, 463
  solving by graphing, 438-439, 454, 463

Quadratic formula, 445-446, 450-455
  discriminant, 450-452, 463

Quotient property of square roots, 396, 425

Quotients, 3, 36

R

Radians, 571

Radical equations, 415-416, 426

Radical expressions, 395-396, 404-409, 411-412
  conjugates, 408, 425
  negative square roots, 395
  principal square roots, 395
  product property of square roots, 396, 425
  quotient property of square roots, 396, 425
  radical sign, 395, 425
  radicand, 395, 425
  simplifying, 396, 404-405, 407-408, 426-427
  rationalizing the denominator, 407-408, 425
  verifying answers, 412

Radical signs, 395, 425
  history of, 397

Radicands, 395, 425

Range, 292-293, 296, 320
  interquartile, 524-525
  lower quartile, 524-525
  upper quartile, 524-525
  outlier, 525

Rates, 110-111, 120

Rational expressions, 467-468, 471-484, 486-496
  adding, 482-483, 486-488, 508
  complex, 490-492, 508
  dividing, 475-476, 508
  excluded values, 467-468
  in equations, 494-495
  least common denominator, 486-488, 508
  mixed expressions, 490
  multiplying, 471-472, 508
  simplifying, 282, 284, 467-468, 508
  subtracting, 482-483, 486-488, 508
  undefined, 467
  with like denominators, 482-483, 508
  with polynomials, 468
  with unlike denominators, 486-488, 508

Rationalizing the denominator, 407-408, 425

Rational numbers, 49-50, 53-56, 73, 398
  adding, 49-50, 53, 73, 482, 486, 507
  comparing, 136-137, 152
  comparison property for, 123, 136, 152
  dividing, 83-84, 119, 475, 508
  least common denominators, 486-488, 508
  mixed numerals, 490
  multiplying, 79-80, 119, 471, 508
  repeating decimals, 398
  subtracting, 55-56, 73, 482, 508
  terminating decimals, 398
  with unlike denominators, 486-488, 508
  with like denominators, 482-483, 508

Ratios, 105-107

Reading Algebra
  compound statements, 127
  equivalent expressions, 406
  graphs, 300
  Greek letters, 563
  parentheses, 22
  powers, 198
  reading speed, 250

READ statements, 578-580

Real numbers, 399, 425

Real roots, 450-452

Reciprocals, 84, 119, 475

Rectangles, 23, 278-279

Rectangular solids, 24

Reflective property, 12, 37, 123-124

Relations, 292-303, 305-306, 316-317, 320
  finding equations, 316-317
  functions, 305-307, 320-321
  equations as, 296-298
  graphs of, 292-293, 301-302
  vertical line test, 306-307, 321

Replacement sets, 11

Resistances, 504-505

Right triangles, 418-420, 426
  hypotenuse, 418
  isosceles, 548
  legs, 418
  solving, 564-565
  30°-60°, 550-551, 572
  trigonometric ratio, 557-558, 561-562, 564-570, 572

Rise, 325, 337

Roots, 438, 463
  complex, 453

irrational, 443-444
  nature of, 451-452, 463
  real, 450-452
  sum and product of, 459-460, 463
  using calculators, 443-444, 446

Rudolff, 397

Run, 325, 337

RUN, 578

S

Sales taxes, 115

Scientific notation, 165-166, 186
  E notation, 186, 588

Series circuits, 504-505

Set-builder notation, 129

Sets, 11
  notation, 129
  set-builder notation, 129
  solution, 125
  subsets, 11

Sieve of Eratosthenes, 231

Sigma, 563

Similar triangles, 554-558, 572
  corresponding angles, 554, 572
  corresponding sides, 554, 572
  trigonometric ratios, 557-558

Simple interest, 216
  principal, 216
  rate, 216

Simplest form, 16
  for radical expressions, 408, 425
  using calculators, 17

Sine, 557-558, 571-572
  graph of, 571
  in BASIC, 571
  program for, 571

Slope-intercept form, 333-334, 339, 354

Slopes, 325-331, 333-340, 342-343, 346-349, 354
  negative, 325-326
  of horizontal lines, 325, 331
  of parallel lines, 346, 354
  of perpendicular lines, 347-348, 354
  of vertical lines, 326, 331
  positive, 325-326
  rise, 325, 337
  run, 325, 337

Solution sets, 125

Solutions, 9, 37

of equations in two variables, 296-298, 301-302, 320
of systems of equations, 359-369, 372-373
of systems of inequalities, 128-135, 281
SQR function, 424, 589
Square roots, 395-405, 407-408, 411-426
adding, 411-412, 426
BASIC, 424, 589
conjugates, 408, 425
divide-and-average method, 403
dividing, 396, 407-408, 426
keys, 402
multiplying, 396, 426
negative, 395
principal, 395
product property of, 396, 425
program for, 424, 589
quotient property of, 396, 425
rationalizing the denominator, 407-408, 425
simplifying, 396, 407-408, 411-412, 426
subtracting, 411-412, 426
tables of, 399-400, 592
using computers, 424
verifying answers, 412
Squares
difference of, 207, 224
factoring, 235-236, 256, 259-260
magic, 71
of binomials, 205-206, 224
of differences, 207, 224
of sums, 205-206, 224
perfect, 239
program for, 589
tables of, 399-400, 592
trinomials, 239-240, 259
Squaring, 395
Standard form, 330, 339, 354
Statements
compound, 127
conjunction, 127
disjunction, 127
Statistics
box plots, 528-529
data, 513
line plot, 515
mean, 520
median, 520-521
mode, 521-522
range, 524-525
regression lines, 533
scatter plots, 530-531
stem and leaf plot, 517-518

Stem and leaf plot, 517-518
back-to-back, 517
Store keys, 208
Subsets, 11
Substitution, 365-366
Substitution property, 13, 20, 37
Subtraction, 4
property for inequalities, 128, 152
property of equality, 61, 74
with polynomials, 194-196, 224
with radical expressions, 411-412, 426
with rational expressions, 482-483, 486-488, 508
with rational numbers, 55-56, 73, 482, 508
with square roots, 411-412, 426
Successive discounts, 118
Sum and product of roots, 459-460, 463
Surface area, 24
Symmetric property, 12, 37, 123-124
Systems of equations, 359-369, 372-376
algebraic methods, 365-369, 372-373
consistent, 362
Cramer's rule, 388-389
dependent, 362
elimination, 369-373
graphing, 359-362
inconsistent, 362
independent, 362
in problem solving, 382-383
parallel lines, 362, 366, 390
program for, 390
substitution, 365-366
solving, 359-362, 365-369, 372-373
using computers, 390
with three variables, 376
Systems of inequalities, 377-378, 391
with no solutions, 378

T

Tables
relations, 292
square root, 399-400, 592
squares, 399-400, 592
trigonometric, 561, 592
using, 265

Tangent, 557-558, 571-572
in BASIC, 571
Terms, 15
like, 15-16, 37
Theta, 563
Tolerance interval, 148
Transitive property, 12, 37, 124, 152
of order, 124, 152
Trapezoids, 24
Triangles
area of, 23
corresponding angles, 554, 572
corresponding sides, 554, 572
equilateral, 548
isosceles right, 548
Pythagorean Theorem, 418-420, 426
right, 418-420, 426, 548, 550-551, 557-558, 561-562, 564, 570, 572
similar, 554-558, 572
sum of angles, 197, 548, 572
trigonometric ratios, 508-509, 512-522
Trigonometric ratios, 557-558, 561-562, 564-572
cosine, 557-558, 571-572
in problem solving, 567-568
sine, 557-558, 571-572
solving right triangles, 564-565
tables of, 561, 593
tangent, 557-558, 571-572
using calculators, 562
using computers, 571
Trinomials, 191,224
factoring, 232, 239-240, 243-245, 251-253, 256, 259-260
perfect squares, 239-240, 259

U

Uniform motion problems, 220-221, 497-498

V

Variables, 3, 36
Variations, 178-182
Velocity, 278
Verbal problems, 26
Vertical line test, 306-307, 321

W

Whole numbers, 41, 402

X

x-axis, 289, 320

x-coordinates, 289, 320
x-intercept, 333-335, 354

Y

y-axis, 289, 320
y-coordinates, 289, 320
y-intercept, 333-335, 354

Z

Zero
   additive identity, 12, 20, 37
   exponent, 162-163, 185
   in denominator, 467, 508
   multiplication by, 12, 20, 37
   product property, 263, 284
   reciprocal of, 79